Plant Growth Regulators in Tropical and Sub-tropical Fruit Crops

Plant growth regulators or plant bio-regulators have emerged as a powerful tool for improving the performance of horticultural crops in general and fruit crops in particular. This book provided recent information on role of plant hormones, how their concentrations are regulated, and how they modulate the various plant processes. 'Plant Growth Regulators in Tropical, Sub-tropical Fruit Crops' is a comprehensive book covering function of plant growth regulators in propagation including micro-propagation, growth, flowering and fruiting behaviour, yield, quality, shelf life and stress management etc. This book has 26 chapters covering most of the tropical and sub-tropical fruit crops like aonla, avocado, banana, ber, citrus, custard apple, date palm, fig, grape, guava, jamun, kokam, litchi, mango, mulberry, papaya, passion fruit, sapota, phalsa, pomegranate and strawberry.

Dr. S.N. Ghosh (Dr. Satyanarayan Ghosh) started his career as Junior Horticulturist (Assistant Professor) in All India Co-coordinated Research Project in Cashew, where he worked for a period of almost more than 7-years after which he joined as Associate Professor and later as Professor of Horticulture on 27.07.1998.

Dr. Ranjan Kumar Tarai was born on 30th May, 1977 and obtained his M.Sc (Hort.) in Fruits and Orchard Management during 2002 and Ph.D. (Hort.) in Fruits and Orchard Management during 2006 from Bidhan Chandra Krishi Viswa Vidyalaya, Mohanpur, Nadia, West Bengal.

Dr. T. R. Ahlawat is presently serving as an Associate Professor in the Department of Fruit Science, ASPEE College of Horticulture and Forestry with the additional charge of Associate Director of Research in Navsari Agricultural University, Navsari, Gujarat.

Plant Growth Regulators in Tropical and Sub-tropical Fruit Crops

Prof. (Dr.) S. N. Ghosh

Faculty of Horticulture
Bidhan Chandra Krishi Viswavidyalaya
Mohanpur, Nadia- 741252, West Bengal, India

Dr. R. K. Tarai

College of Horticulture, Chiplima
Orissa University of Agriculture and Technology
Sambalpur-768025, Odisha, India

Dr. T. R. Ahlawat

Navsari Agricultural University
Navsari-396 450, Gujarat, India

NARENDRA PUBLISHING HOUSE
DELHI (INDIA)

First published 2022
by CRC Press
4 Park Square, Milton Park, Abingdon, Oxon, OX14 4RN

and by CRC Press
6000 Broken Sound Parkway NW, Suite 300, Boca Raton, FL 33487-2742

CRC Press is an imprint of Informa UK Limited

British Library Cataloguing-in-Publication Data
A catalogue record for this book is available from the British Library

Library of Congress Cataloging-in-Publication Data
A catalog record has been requested

ISBN: 978-1-032-29172-7 (hbk)
ISBN: 978-1-003-30034-2 (ebk)

DOI: 10.1201/9781003300342

List of Contributors

A K Srivastava
Banda University of Agriculture and Technology
Banda – 210001, Uttar Pradesh, India
Email: srivastavahort@yahoo.com

A. K. Godara
Department of Horticulture, College of Agriculture
Choudhury Charan Singh Hayana Agricultural University
Hisar, Haryana, India
Email: hod_hort@hau.ernet.in

A. K. Panda
Department of Fruit Science, Post Graduate Department of Horticulture
Dr. Rajendra Prasad Central Agricultural University
Pusa, Samastipur, Bihar, India
Email: asish.panda014@gmail.com

A. K. Sahoo
College of Horticulture
Orissa University of Agriculture and Technology, Chiplima
College of Agriculture Chiplima, Odisha
Email: sahooajit.md@gmail.com

A. K. Singh
Central Horticultural Experiment Station, (ICAR)
Vejalpur (Godhra), Panchmahals, Gujarat
Email: aksbicar@gmail.com

A. Nagaraja
Division of Fruits & Horticultural Technology
Indian Council of Agricultural Research-Indian Agriculture Research Institute
New Delhi-110012, India
Email: anrciah@gmail.com

A.K. Pandey
Department of Fruit Science
ASPEE College of Horticulture and Forestry
Navsari Agricultural University, Navsari, Gujarat

Ankit Kumar Singh
Department of Horticulture
Udai Pratap College, Varanasi, Uttar Pradesh, India
Email: ankitp130n@gmail.com

B T. Patil
Agriculture Technical School, K. Digraj
Dist. Sangli 416305, Maharashtra, India
Email: btpatilveg@rediffmail.com

B. K. Sethy
College of Agriculture, Chiplima
P.O. College of Agriculture Chiplima Sub- Post Office
Dist- Sambalpur, Odisha
Email: bijayasethy2010@gmail.com

Binayak Chakraborty
Assistant Professor (Horticulture)
College of Agriculture, Navsari Agricultural University
Waghai-394 730 (The Dangs), Gujarat, India

C. Kavitha
Department of Fruit Crops, Faculty of Horticulture
Tamil Nadu Agricultural University, Coimbatore, Tamil Nadu, India
Email: ck77@tnau.ac.in

C.M. Muralidharan
Date Palm Research Station
Sardarkrushinagar Dantiwada Agricultural University
Mundra-Kachchh, Gujarat- 370421, India
Email: muralidharancm@yahoo.com

C.Y. Dhumal
Zonal Agricultural Research Station (Plain Zone)
Ganeshkhind, Pune-411067, Maharashtra, India

D.S. Mishra
Central Horticultural Experiment Station
Vejalpur-389340, Godhra, Gujarat
Email: dsmhort@gmail.com

Dharmishtha Patel
Department of Fruit Science, ASPEE College of Horticulture and Forestry
Navsari Agricultural University
Navsari 396451, Gujarat
Email: dmpatel134@gmail.com

H S Singh
ICAR-NCIPM, New Delhi
Email: singhhs21@rediffmail.com

I. Panigrahi
Division of Vegetable Science
ICAR-IARI, New Delhi
Email: ipsitapanigrahi.hort@gmail.com

Kapil Mohan Sharma
Date palm Research Station,
Sardarkrushinagar Dantiwada Agricultural University
Mundra-Kachchh, Gujarat- 370421, India
Email: k.m.sharma456@sdau.edu.in

Kundan Kishore
Central Horticultural Experiment Station (ICAR-IIHR)
Aiginia, Bhubaneswar – 751019, Odisha
Email: kkhort12@gmail.com

M.B. Shete
Zonal Agricultural Research Station (Mahatma Phule Krishi Vidyapeeth, Rahuri)
Ganeshkhind, Pune-411067, Maharashtra, India
Email: mohanshete64@gmail.com

M.M. Baghel
Division of Fruits & Horticultural Technology
Indian Council of Agricultural Research-Indian Agriculture Research Institute
New Delhi-110 012, India
Email: murli.manohar.baghel@gmail.com

Nidhi Sharma
Department of Horticulture, College of Agriculture
Choudhury Charan Singh Haryana Agricultural University
Hisar-125004, India
Email: nidhi.nandani@gmail.com

Nidhika Thakur
Department of Fruit Science
Banda University of Agriculture and Technology
Banda – 210001, Uttar Pradesh, India

Om Prakash
Department of Fruit Science
Banda University of Agriculture and Technology
Banda – 210001, Uttar Pradesh, India

P. K. Nimbolkar
IIHR, Hessaraghatta Lake post
Bengaluru-560 089
Email: prashantnimbolkar111@gmail.com

Parmpal Singh Gill
Department Fruits Science
Punjab Agricultural University
Ludhiana-141001, Punjab

Pragnyashree Mishra
College of Horticulture, Chiplima
Orissa University of Agriculture and Technology
Sambalpur, Odisha-768025, India
Email: pragnyashree.mishra@gmail.com

Priyam Chattopadhyay
Department of Horticulture and Post Harvest Technology
Palli Shiksha Bhavana, Visva-Bharati, Sriniketan-731236, West Bengal
Email: priyamchattopadhyay4@gmail.com

R M Kurian
Division of Fruit Crops, ICAR - Indian Institute of Horticultural Research
Hassarghatta Lake Post, Bengaluru
Email: rkurian@iihr.res.in

R. Sri Hari Babu
Acharya N.G. Ranga Agricultural University
Hyderabad-500 030, Andhra Pradesh, India

R. K. Tarai
College of Horticulture, Chiplima
Orissa University of Agriculture and Technology, Sambalpur
Odisha-768025 India
Email: ranjanouat@gmail.com

R. R. Sharma
Division of Food Science and Postharvest Technology
Indian Council of Agricultural Research-Indian Agriculture Research Institute
New Delhi-110012, India
Email: rrs_fht@rediffmail.com

Sayan Sau
Department of Fruits and Orchard Management
Faculty of Horticulture
Bidhan Chandra Krishi Viswavidyalaya
Mohanpur, Nadia, West Bengal, India

S . A . Ranpise
Department of Horticulture, Post Graduate Institute
Mahatma Phule Krishi Vidyapeeth
Rahuri – 413722, Dist. Ahemadnagar, Maharashtra, India
Email: shrimantranpise12@gmail.com

S. Pradhan
Division of Fruits & Horticultural Technology
ICAR-IARI, New Delhi-110012
Email: satyapdhn@gmail.com

Sanjay Singh
Central Horticultural Experiment Station, (ICAR)
Vejalpur (Godhra), Panchmahals, Gujarat
Email: sanjaysinghicar@gmail.com

Shivraj Kumar Verma
Department of Horticulture
Udai Pratap College, Varanasi, Uttar Pradesh, India
Email: shivraj.vns@gmail.com

Soubhagya Kumar Sahu
Department of Horticulture
Institute of Agricultural Sciences, Banaras Hindu University
Varanasi - 221005, Uttar Pradesh, India
Email: soubhagyahorti202@gmail.com

Suchismita Jena
Department of Fruit Science, ASPEE College of Horticulture and Forestry
Navsari Agricultural University
Navsari, Gujarat- 396451
Email: suchismitajena2594@gmail.com

Sudip Kumar Dutta
ICAR-Research Complex for North Eastern Hill Region
Mizoram Centre, Kolasib, Mizoram-796081, India
Email: sudipiari@rediffmail.com

Sujata Tetali
Cytogeneticist, Scientist 'C' of Genetics Group
Agharkar Research Institute
GG Agarkar Road, Pune 411004, Maharashtra, India
Email: sujatatetali@gmail.com

T. R. Ahlawat
Navsari Agricultural University
Navsari-396 450, Gujarat, India
Email: tahlawat4@gmail.com

Tridip Kumar Hazarika
Department of Horticulture, Aromatic and Medicinal Plants
School of Earth Sciences and Natural Resources Management
Mizoram University, Tanhril, Aizawl, Mizoram
Email: tridip28@gmail.com

Trina Adhikary
Department Fruits Science
Punjab Agricultural University, Ludhiana-141001, Punjab
Email: trinaadhikaryhort@gmail.com

U. S. Nayak
Regional Research and Technology Transfer Station
Ranital, Orissa University of Agriculture and Technology
Bhadrak - 756111, India
Email: usnayak74@gmail.com

V. K. Sharma
Department of Horticulture
Choudhury Charan Singh Hayana Agricultural University, Hisar
Email: vksharma087@gmail.com

V. S. Supe
Zonal Agricultural Research Station (Plain Zone)
Ganeshkhind, Pune-411067, Maharashtra, India
Email: vssupe@gmail.com

Veena Joshi
College of Horticulture
Sri. Konda Laxman Telangana University of Horticulture
Rajendranagar, Hyderabad – 500030, Telangana, India
Email: drveenahorti@gmail.com

V. Srilatha
Department of Horticulture
S.V. Agriculture College, ANGRAU
Tirupati – 517 002, Andhra Pradesh, India
Email: latha_scientist@yahoo.com

Contents

Preface

Plant growth regulators or plant bio-regulators have emerged as a powerful tool for improving the performance of horticultural crops in general and fruit crops in particular. Diverse group of plant growth regulators like auxins, gibberellins, cytokinins, abscisic acid, ethylene and growth retardants, polyamines, etc. at various concentrations are reported to influence flowering, fruit set, retention, plant growth and development, ripening, shelf life and fruit quality characters in tropical and subtropical fruit crops. Agricultural universities and institutes across the country have scientifically demonstrated the potential of plant growth regulators in increasing the profitability of fruit growers. Many scientists have worked on the nature and action of plant growth regulators on different fruit crops for the last few decades and as a result remarkable recommendations have been made for increasing production and productivity of many tropical and subtropical fruit crops. This necessitated the need to present this data in a clear, concise and comprehensive manner for the benefit of the students, scientists and farmers.

In this book we have tried to focus the available literature in a comprehensive manner. Wherever possible, we have highlighted useful reviews which provided background information, along with recent publications that have contributed significantly to the literature. This book provided recent information on role of plant hormones, how their concentrations are regulated, and how they modulate the various plant processes. 'Plant Growth Regulators in Tropical, Sub-tropical Fruit Crops' is a comprehensive book covering function of plant growth regulators in propagation including micro-propagation, growth, flowering and fruiting behaviour, yield, quality, shelf life and stress management etc. This book has 26-chapters covering most of the tropical and sub-tropical fruit crops like aonla, avocado, banana, ber, citrus, custard apple, date palm, fig, grape, guava, jamun, kokam, litchi, mango, mulberry, papaya, passion fruit, sapota, phalsa, pomegranate and strawberry. This book is the result of painstaking effort and keen abiding interest to improve the use of plant growth regulators in tropical and subtropical fruit crops so as to improve their quality and productivity.

Prof. (Dr.) S.N. Ghosh
Dr. T. R. Ahlawat
Dr. R. K. Tarai

1

Plant Growth Regulators

R. Sri Hari Babu, V. Srilatha and Veena Joshi

1. Introduction

Plant growth regulators (PGRs) / phytohormones (endogenously produced growth substances-called hormones or phytohormones) are organic compounds other than nutrients that modify plant biological processes. They are signal molecules and act as messengers. Phytohormones are produced within the plant in extremely low concentrations. Generally their site of biosynthesis and site of action are different. They are vital for growth and development of plants and are normally active at very low concentrations. They act inside the plant cells to stimulate or inhibit specific enzymes or enzymic synthesis and help to regulate plant metabolism. Most of the PGRs exhibit a broad spectrum influence and thus a single PGR may influence several entirely different plant processes. PGRs cause changes in physiological and biochemical characters and thereby improve crop productivity.

PGRs (synthetic hormones) are useful because they can modify plant processes. This may be interfering with the biosynthesis, metabolism or translocation of phytohormones or the PGRs may replace or supplement the plant hormones, when their endogenous levels are below the threshold levels, to change the course of plant development.

The importance of PGRs was first recognised in the 1930's. Since that time, natural and synthetic compounds, that alter / regulate functions in plants have been discovered and tested from time to time. Today specific PGRs are identified and used to modify crop growth rate and growth and development pattern during the various stages of development from seed germination, through harvest and even post-harvest preservation of different commodities.

2. Plant Growth Regulators and their Role

2.1. Auxins

Auxin is a generic term used to denote a group of compounds, characterized by their activity to induce cell division and cell elongation in shoot cells. They were the first plant hormones to be discovered about a century ago. Auxins include both natural (endogenous) and synthetic ones. Indole-3-acetic acid (IAA) was the first auxin discovered in plants. There after several auxins were synthesized and tested for their utility in agricultural and horticultural practices. Synthetic auxins exhibit similar functions or effects of natural auxin on growth and development in plants. Synthetic auxins include: Indole butyric acid (IBA), Naphthalene acetic acid (NAA), 2, 4-Dichlorophenoxy acetic acid (2, 4-D), 2, 4, 5-Trichlorophenoxy acetic acid (2, 4, 5-T), 4-chlorophenoxy acetic acid (4-CPA) etc. The general functions / roles of auxins are as follows:

1. Stimulate cell division and cell elongation.
2. Initiate root formation.
3. Simulate cambial activity. Hartman and Kester (1972) stated that auxins are known for their involvement in several physiological activities in plants including activation of cambial cells.
4. Enhance apical dominance
5. Stimulate differentiation of cells, xylem and phloem.
6. Play specific roles in seed germination, parthenocarpy, tissue culture etc.
7. Increase nucleic acid activities
8. Induce flowering, improve fruit set and development and increase in fruit size (weight and volume)
9. Hasten or delay fruit maturity.
10. Influence abscission of plant appendages, particularly leaves and fruits.

2.2. Gibberellins

A gibberellin may bc defined as a compound that has a gibbane skeleton that stimulates cell division or cell elongation or both. Gibberellins are named after the fungus, *Gibberella fujikuroi* that caused rice plants to grow abnormally tall (lankey). They are synthesized in apical portions of stems including leaves and roots. They are second most important plant growth hormones. So far over 125 types of gibberellins are known and named serially as GA_1, GA_2, GA_3 of which about 100 types are found in higher plants, and remaining ones are found in fungi and bacteria (Crozier *et al.*, 2000). Gibberellins are also synthesized and made available. The well known functions / effects of gibberellins are enumerated below:

1. Break the dormancy of seeds and buds.
2. Hasten the seed germination.
3. Stimulate the vegetative growth, particularly in the internodes.
4. Enhance apical dominance.
5. Stimulate the growth of dwarf plants.
6. Inhibit flowering in woody plants.
7. Have a role in sex determination.
8. Increase the size of the fruits and growth of seedless fruits.
9. Stimulate RNA and protein synthesis.
10. Can over come the effect of viruses.
11. Regulate the action of auxin-exhibit antagonism to auxins in affecting the rooting of cuttings (Haissig, 1972).

2.3. Cytokinins

Cytokinins (CKs) are plant growth substances that cause cell division, but have no role in extension of growth. The credit of control of cell division by CKs is difficult to define, because auxins and gibberellins also induce cell division. It was felt that cell division is controlled not by one growth substance but by more than one and auxins and CKs share the role in cell division. CKs are adenine derivatives represented by Kinetin, zeatin and 6-benzyl amine purine. CKs are found not only in plants but also in fungi (Crozier *et al.*, 2000., Heldt *et al.*, 2011). CKs are synthesized in root tips, cambium and developing seeds (Davies, 2004) and move into the stem through xylem. CKs were isolated from young developing seeds and germinating seeds. The biological roles / functions of CKs include the following:

1. Induce cell division, cell enlargement and cell differentiation.
2. Increase the leaf size as a result of cell enlargement.
3. Cause elongation in etiolated stems.
4. Influence the seed germination in dark.
5. Overcome apical dominance.
6. Have role in morphogenesis and organogenesis.
7. Promote bud growth by terminating the rest in buds.
8. Play an important role in inducing fruit set and fruit growth by inducing cell division (Letham, 1967).
9. Prolong shelf life by slowing down the break down of proteins and other plant constituents.
10. Delay senescence in plant tissues by maintaining the synthesis of proteins and nucleic acids (DNA and RNA).
11. Antagonise the action of ethylene.
12. Cause mobilization of assimilates, besides auxins and gibberellins.

2.4. Polyamines

Of late 'Polyamines' have been implicated in wide range of plant growth and developmental processes, by which they gained importance in crops. Polyamines (PAs) are low cationic compounds implicated in an overwhelming array of plant growth and developmental processes. PAs are important group of naturally occurring low molecular weight polycations, diphatic nitrogenous compounds present in cells. They are found in plant cells at all levels significantly higher than those of other plant hormones. Naturally occurring PAs are diamines (1, 3 diamine propanes, putrescine, spermine, spermidine etc.). The ubiquity and abundance of PAs in plant tissues indicate their growth regulator role

(Galston, 1983). PAs are organic polycations with two or more primary amino (NH_2) groups and are ubiquitous in bacteria, plants and animals. PAs occur in free form conjugated with phenolic compounds or bound to other low molecular weight compounds or macro-molecules (Galston and Kaur – Sawhney, 1995). PAs are synthesized endogenously in plants from basic amino acids : arginine, ornithine and lysine, which form basic carbon skeleton while methiomine contributes propyl amino group to putrascine to form spermidine and spermidine to form spermine (Bagni, 1986). PAs are positively changed diphatic auxins. They are transported throughout the plant in the phloem and xylem sap (Antognom *et al.*, 1998).

The importance of PAs in plants is due to their participation in a multitude plant metabolic functions including photosynthesis, enzyme activities and maintenance of cell proliferation, division and differentiation, morphogenesis, embryogenesis as well as organogenesis (Evans and Malamberg, 1989; Galston *et al.*, 1987, Kakkar *et al.*, 2000). In addition, PAs are involved in the correct conformation of nucleic acid, gene expression and translation, mediating hormone action, modulation of cell signalling, membrane stabilization and ion channel regulation as well as heat shock protein and macro – molecular synthesis. Besides all the above functions, PAs also act as messengers in leaf senescence and more importantly for biotic and abiotic stresses (Kumar *et al.*, 1997; Waldon *et al.*, 1997; Malanberg *et al.*, 1998; Bouchereau *et al.*, 1999; Liu *et al.*, 2000; Konigbiofer and Lechner, 2002; Ioannidis and Kotzalasis, 2007; Kusano *et al.*, 2008).

PAs decrease the aging of plant tissues. A general decrease in PA levels occurs during ripening of avocado fruits (Cutting *et al.*, 1990). PA levels are increasingly correlated with senescence and was further claimed, that large endogenous PAs may have effect similar to that of exogenous application of PAs (Dibble *et al.*, 1988).

Further, ripening or senescence activity involves PAs and some of their physiological effects are reported to be opposite to the action of ethylene, which may be due to competition between the two for a common precursor, S-adenyl methionine (SAM) (Roberts *et al.*, 1984). Ethylene is a senescence inducer, while PA is a senescence inhibitor.

PAs are involved in a number of physiological and biological processes in plants including plant growth and development, flowering, fruit growth and development, fruit ripening, stress responses and senescence (Kakkar and Rai, 1983; Malik and Singh, 2003a). PAs have a role in flower differentiation and initiation, pollen viability, root growth etc. (Bouchereau *et al.*, 1999).

2.5. Growth inhibitors

The growth inhibitors are a group of plant growth substances that inhibit / retard physiological or a biochemical processes in plants. This inhibitory action helps to keep plant from becoming too large. They affect dormancy, growth, reproduction etc.,. Growth inhibitors occur in all plants, in flowers, seeds etc. Roots or leaves of some plants excrete growth inhibitors. Most of the naturally occurring inhibitors are aromatic organic substances. Many of them are phenol compounds including benzoic acid. Abscisic acid (ABA) is one of the growth inhibitors which is found in almost all the plant parts and is also synthesized. Maliec Hydrazide (MH) is another growth inhibitor, a synthetic one, which is widely used to prevent sprouting of potatoes and onions.

ABA is synthesized endogenously in plastids and roots. It is also produced in response to stresses. It is commonly found in fleshy fruits, where it prevents seeds from germinating when the fruits are still on the plant ABA is the most widely used growth inhibiter in plant kingdom, because of the following roles / functions:

1. Inhibits the growth of plants and plant parts like buds.
2. Prolongs the rest of seeds and buds and delays seed germination.
3. Terminates the rest of buds.
4. Induces flowering.
5. Regulates stomatal opening and closure.
6. Acts as a general inducer of abscission and senescence.
7. Inhibits the stimulatory effects of other hormones to inhibit cell growth.

2.6. Growth retardants

These are organic chemical compounds or synthetic ones that retard cell division and cell elongation in meristem of the shoot apices and thus regulate plant height physiologically without causing malformation of leaves and stems. These compounds slow the rate of cell division and cell elongation in shoots and control the height of the plants without causing stem bending and leaf malformation. Growth of treated plants, however, is not completely suppressed unlike in case of growth inhibitors and neither their rate of organ development is affected. They have little effect on the production of leaves or on root. Growth retardants usually inhibit the action of gibberellins by inhibiting the synthesis of GA through interfering in the pathway of gibberellin synthesis. AMO 1618, Phosphdon-D, SADH, Cycocel and Paclobutrazol are some of the growth retardants which inhibit gibberellin synthesis in higher plants. Synthetic plant growth retardants fall into 3 classes (i) the gibberellin inhibitors [Daminozide (N-diemethyl amino-succenic acid)] also known as Alar or B9. ii). The Orium type-chloremequat chloride (2-chloroethyl trimethyl ammonium chloride) also known as Cycocel or CCC), iii) The steroid synthesis inhibiting triazoles : Paclobutrazol also known as Cultar, PP333, Sadabahar, Parly, clipper etc. and Uniconazol also known as Suneaze and XE-1019 (Rademacher, 1991). The latter two classes of compounds inhibit Kaurune synthesis, an enzyme in gibberellin synthesis pathway (Nickel, 1983; Dalzial and Lawrence, 1984; Radamacher, 1991).

Paclobutrazol (PBZ) and other triazole analogues are probably the most effective growth retardant compounds to date for controlling shoot elongation in woody fruit trees. Biological functions of growth retardants are as follows :

1. Retardation of stem elongation, through prevention of cell division in the sub-apical meristem usually without affecting the apical meristem, which is just opposite to that of gibberellins.
2. Induce and hasten flowering in woody plants.
3. ncrease the number of flowers, fruits and yield.

Paclo. utrazol (PBZ) is one of the triazole groups of growth retardants. It is a broad spectrum pla t growth retardant. It is an inhibiter of gibberellin synthesis and thus it is often referred to as ' nti-gibberellin'. It is well known as an effective growth inhibiter and promoter of flowering in voody species. Its impact on vegetative and reproductive growth, adaption to abiotic stresse (drought, chilling and heat shock), aging and pathogenic infections is well known and docu ented (Wang and Lin, 1992). It has been successfully employed around the world to enhan yield potential of field and fruit crops. On application on to plant PBZ performs the follow g functions:

1. Inhibits cell di sion and cell elongation by interfering biosynthesis of gibberellins.
2. Inhibits germir tion of seeds and pollen grain by blocking or counteracting gibberellins.
3. Induces, increases, lvances and extends flowering by checking vegetative growth, by checking the gibb rellin synthesis.

4. Reduces fruit size due to decreased cell division and increased fruit number.

2.7. Ethylene

Ethylene is an unsaturated hydrocarbon and is a volatile compound. It is a colourless gas at physiological temperature under which a plant can live and survive. It diffuses in the plant rapidly. It is produced by all living tissues, particularly in the nodes of stems, about-to-abscise plant organs, ripening fruits and dying organs. Ethylene is acclaimed as a "ripening hormone" since ethylene is invariably produced by the ripening fruit and is also essentially needed for ripening. Auxins increase ethylene levels in plants. Hence, actions of auxins are often attributed to ethylene, such as increase in percentage of female flowers, apical dominance and leaf epinasty. Besides being a ripening hormone, however, ethylene has other functions in the plants.

1. Pre-treatment of ethylene inhibits sprouting of bulbs, tubers etc and seed germination, but when sprouted bulbs and tubes or germinating seeds are treated with ethylene, it inhibits their further growth, stem elongation and bud development.
2. Induces flowering in some plant species.
3. Regulates sex – determination.
4. Induces and hastens ripening of harvested fruits.
5. Induces degreeing of fruits.
6. Induces premature abscission of plant organs.
7. Encourages senescence of plant organs, weakens cell membranes, and thereby softens cell walls.
8. Causes epinasty of leaves and thickening of shoots.

3. Plant Growth Regulators in Propagation

3.1. Physiological basis of root initiation in cuttings and layers

Certain natural (endogenous) plant growth substances (phytohormones) should be present in the plant parts for rooting of cuttings and layers. In the absence of these growth substances, cuttings and layers do not root readily. Not only their presence, but they should be in optimum quantities and proper balance for induction of rooting. Of the essential phyto-hormones like auxins, gibberellins, cytokinins etc., auxins, as root promoters are more important for rooting.

3.1.1. Auxins

Auxins promote callus and root formation, which can improve establishment from cuttings and layers. Naturally occurring auxins like indole-3-acetic acid (IAA) are involved in the rooting process. Synthetic compounds (auxins) like indole butyric acid (IBA), naphthalene acetic acid (NAA), 2, 4-Dichloro phenoxy acetic acid (2, 4-D) etc. also induce rooting in cuttings and layers following their application. Of these, IBA is more effective and important for rooting, hence, it is often referred to as 'Rooting hormone'.

3.1.2. Gibberellins

Gibberellins are not generally required for the initiation of rooting. They inhibit root formation (antagonistic) by stimulating vegetative growth through resultant competition for assimilates. They are generally used to stimulate vegetative growth after rooting for better establishment of rooted cuttings and layers. Gibberellins probably prevent the cell division in the meristamatic tissues. Poor rooting was observed in cuttings taken from 10-year-old 'langra' trees sprayed with gibberellic acid (GA_3) 200 ppm (Sandhu, 1992). Gibberelins exhibit strong inhibitory effect on rooting, which could not root even after application of auxin (Sandhu, 1992).

3.1.3. Cytokinins

Both natural (endogenous) and synthetic Ck (Adenine, Kinetin, 6-Benzyladenine) are employed. Auxins cause root initials and their further growth is favoured by CKs. However, CKs at higher concentration (above 15 ppm) inhibit root formation. Therefore, CKs should be used at less than 15 ppm. Further, there should be favourable balance between auxins and CKs for rooting. When the ratio between auxins and CKs is high rooting occurs, when the ratio is low, shoots develop and at intermediate ration callus (undifferentiated cells) forms. Thus, the physiological basis of root formation in cuttings and layers is under the control of balance between auxins and cytokinins.

Auxins and CKs are not the only factors which stimulate rooting in cuttings and layers, there are other factors (Co-factors) which in combination with auxins and CKs enable the cuttings and layers to root. The source of these co-factors is 'leaves' that's why the loss / absence of leaves on cuttings and layers greatly reduces rooting. The sugars and nitrogenous materials produced in the leaves probably are the rooting co-factors. Besides leaves, the vegetative buds on cuttings / layers also aid in influencing rooting. These buds stimulate rooting by virtue of being meristematic, synthesize and supply growth substances required for root initiation.

3.1.4. Endogenous inhibitors

The above presentation supports the theory that in difficult-to-root species, like mango, cuttings / layers fail to root because of lack of the necessary co-factors, besides the required phytohormones, particularly auxins and CKs and their optimum balance. Another possibility is that such cuttings / layers may contain some kind of inhibitory substances in amounts high enough to mask the effects of endogenous root promoting substances. One alternate to induce rooting in such species is to allow leaching of such inhibitors by some means.

3.2. Use of plant growth regulators to stimulate rooting

Several types of plant growth regulators are used for stimulating rooting in cuttings and layers. Commercial use of rooting with auxins is very common. Auxins promote callus and root formation, which can improve establishment from cuttings and layers. The most commonly used auxin to stimulate rooting is the 'Indole butyric acid' (IBA), belonging to indole group of auxins. It has a weak activity and is destroyed relatively slowly by auxin destroying enzyme. IBA translates poorly in the cuttings / layers, hence, it is retained at the site of application for long. If an applied growth regulator moves readily away from the site of application in the cuttings and layers rooting will be either poor or nil.

Another excellent auxin frequently employed for root promotion is "naphthalene acetic acid" (NAA). Although NAA has higher potentiality for rooting, it is not used singly, as it is toxic to the plant tissues at higher concentrations. According to Sulladmath and Kololgi (1969) NAA has synergestic effect on the rooting when mixed with IBA. Another auxin, which is also frequently used in rooting is "Indole-3-acetic acid" (IAA). However, it is less effective than IBA and NAA. Further, it is very unstable in plants and decomposes rapidly in plants.

The amides of IBA and NAA are also very effective rooting agents. The amide form of NAA *viz.*, napthalene acetamide (NAD) is less toxic than NAA and thus safer to use.

Many phenoxy auxin compounds promote root formation at low concentrations. At high concentrations, they produce thick and stunted roots. Though 2, 4-Dichloro phenoxy acetic acid (2, 4-D) promotes rooting in certain species, being readily translocated, it usually tends to inhibit shoot development and causes injury to the shoots. Several other phenoxy compounds *viz.*, 2, 4, 5-Trichlorophenoxy acetic acid (2, 4, 5-T), 2, 4, 5-Trichlorophenoxy propionic acid (2, 4, 5-TP), 2, 4, 5 – Trichlorophenoxy buteric acid (2, 4-5-TB) etc. have proved effective and produced good rooting at low concentration, without being toxic. Unfortunately, the safe (effective) and toxic range of concentrations of these compounds being quite narrow, they are not frequently used for rooting purposes.

References

- Antognom, F.S., Fomale, C., Grimmer, E., Komer, E. and Bagui, N. 1998. Long distance translocation of polyamines in phloem and xylem of *Recinus communis* L. plants. *Planta*, **204** : 520-527.
- Bagni, N. 1986. The function and metabolism of polyamines in plants. *Acta Horticultura*, **179**: 95-103.
- Bouchereau, A.A., Aziz, A., Lather, F. and Martin-Tanguy, J. 1999. Review : Polyamines and environmental changes. Recent Developments. *Plant Science*, **140**: 103-125.
- Crozier, A., Kamiya, Y., Bishop, G. and Yokota, T. 2000. Biosynthesis of hormones and eliciter molecules. p. 850-929. *In. B.B. Buchannar W. Gruissem and R.L. Jones (Eds). Biochemical and Molecular Biology of Plants American Society of Plant Physiology*. Rockville, Maryland, USA.
- Cutting, J.G.M., Bower, J.R., Wolstenhome, B.N. and Hoffman, P.J. 1990. Changes in ABA polyphenol oxidase, phenolic compounds and polyamines and their relationship with mesacarp discolouration in ripening avocado (Persea Americana). *Journal of Horticultural Science*, **65**(4) : 465-471.
- Dalzial, J. and Lawrence, D.K. 1984. Biochemical and biological effects of Kauren oxidase inhibition, such as Paclobutrazol. *British Plant Grown Regulator Group Monograph*. **4**: 1-14.
- Davies, P.J. 2004. The plant hormones : their nature, occurrence and functions. *Plant Hormones*. Springer, Netherlands. pp. 1-15.
- Dibble, A.R.C., Davies, P.J. and Mustachler, M.A. 1988. Polyamine content of long keeping Alcobaca tomato fruit. *Plant Physiology*, **83** : 338-340.
- Evans, P.T. and Malamberg, R.L. 1989. Do polyamines have role in plant development. *Annual Review of Physiology and Plant Molecular Biology*, **40**: 234-269.

- Galston, A.N. 1983. Polyamines as modulators of plant development. *Bio Science*, **33**: 382-388.
- Galston, A.W. and Kaur-Sawhney, R. 1995. Polyamines and senescence in plants. pp. 167-181. *In: Plant senescence: Its biochemistry and physiology* (eds) WW.Thomson, E.A. Nothnagal and R.C. (Haffakin). American Society of Plant Physiology. Rodeville, M.D.
- Galston, A.W., Kaur-shawney, R., Altabella, T. and Tiburio, A.F. 1987. Plant polyamines in reproductive activity and response to abiotic stress. *Botanic Acta*,**110**: 197-207.
- Haissig, B.E. 1972. Meristamatic activity during adventitious entitious root primordia development. Influence of endogenous auxin and applied gibberellic acid. *Plant Physiology*, **49**: 886-892.
- Heldt, H.W, Piechulla, H. and Heldt, F. 2011. *Plant biochemistry*. Academic Press Burlington. Massachusetts, U.S.A.
- Ioannoidis, N.E. and Kotzalasis, K. 2007. Effects of polyamines on the functionality of photosynthetic membrane *in vivo* and *in vitro*. *Biochim. Biophys. Acta*, **1767** : 1232-1382.
- Kakkar, R.K. and Rai, V.K. 1983. Plant polyamines in flowering and fruit ripening. *Phytochemistry*, **33**: 1281-1288.
- Kakkar, R.K., Nagar, P.K., Ahuja, P.S. and Rai, V.K. 2000. Polyamines and plant morphogenesis. *Biol. Plant*, **43**: 1-4.
- Konigshofer, H. and Lechner, S. 2002. Are polyamines involved in the synthesis of heat shock proteins in cell suspension culture of tobacco and alfadfa in response to high temperature stress ?. *Plant Physiology and Biochemistry*, **40**: 51-59.
- Kumar, A., Taylor, M., Altabeta, T. and Tiburcio, A.F. 1997. Recent advances in polyamine research. *Trends in Plant Science*, **2**: 124-130.
- Kusano, T., Berberich, T., Tateola, C. and Takahashi, Y. 2008. Polyamines essential factor for growth and survival. *Planta*, **228**: 367-384.
- Letham, D.S. 1967. Chemistry and physiology of Kinetin-like compounds. *Annual Review of Plant Physiology*, **13**: 349-364.
- Liu, K., Fu, H., Bei, Q. and Luan, S. 2000. Inward potassium channel in guard cells as a target of polyamine regulation of stomatal movements. *Plant Physiology*, **124**: 1315-1325.

- Malamberg, R.L., Watson, M.B., Galloway, G.L., Yu, W. and Yu, W. 1998. Molecular genetic analysis of plant hormones. Critical Reviews. *Plant Sciences,* **17**: 199-224.
- Malik, A.U. and Singh, Z. 2003a. Abscission of mango fruitlets as influenced by biosynthesis of polyamines. *Journal of Horticultural Science and Biotechnology,* **78**: 721-727.
- Nickell, L.G. 1983. Plant growth regulating chemicals. CRC Press, Boca Raton.
- Rademacher, W. 1991. Biochemical effects of plant growth retardants. *In : Gausman, H.W. (ed.). Plant Growth Regulators.* Marcel Dekker, Newyork. pp. 169-200.
- Roberts, D.R., Walker, M.A., Thompson, J.E. and Dumbroft, E.N. 1984. The effects of inhibitors of polyamine and ethylene biosynthesis on sensescnce ethylene production and polyamine levels in cut Carnation flowers. *Plant Cell Physiology,* **25**: 315-322.
- Sandhu, M.K. 1992. Pre-treatments of stock plants with different growth regulating chemicals to induce rooting of mango cuttings. Acta Horticulturae. 321 : 356-365.
- Sulladmath, U.V. and Kologi, S.D. 1969. Synergistic influence of 3-IBA and NAA on the rooting of air layers of chicke var. Kalipetti. South Indian Horticulutre. 17 : 9-17.
- Waldon, R.A., Cordeiro, A. and Tiburcio, A.F. 1997. Polyamines, small moleules triggering pathways in plant growth and development. *Plant Physiology,* **113**: 1009-1013.
- Wang, L.H. and Lin, C.H. 1992. Effect of paclobutrazol on physiological and biochemical changes in the primary roots of pea. *Journal of Experimental Botany,* **43**: 1367-1372.

2

Brassinosteroids and their Use in Fruit Crops

R. R. Sharma, M. M. Baghel and A. Nagaraja

1. Introduction

The plant growth and development processes are regulated by complex and coordinated actions of the endogenous hormones which have the potential of increasing plant productivity and quality by influencing various metabolic processes. Plant hormones are a group of some naturally occurring organic substances that influence several physiological processes at a low concentration (Davies, 1987). Plant hormones play several important roles in controlling the growth and development of plants. While metabolism is responsible for the power and building blocks of a plant, on the other hand, hormones regulate the growth of the individual plant part and their integration in the form of a plant which became evident after the studies of Julius von Sachs and Charles Darwin who demonstrated that various plant growth processes are regulated by 'substances' which were later called as hormones. The major phytohomones identified were: auxins, cytokinins, gibberrelic acid (GAs), brassinosteroids (BRs), abscisic acid (ABA), and ethylene (Gray, 2004). Brassinosteroids (BRs) is a group of plant steroid hormones, which regulate many processes in plant growth and development, including cell elongation, cell division, vascular differentiation, reproduction, rhizogenesis, flowering, fruit ripening, tolerance response to various biotic and abiotic stresses, and senescence (Clouse and Sasse 1998; Divi and Krishna 2009; Unterholzner *et al.*, 2015). The biosynthetic and signaling pathways of BRs offer exciting potential for enhancing the productivity and quality of fruits and other crops (Rao *et al.*, 2002).

Fruits play a crucial role in human diets and as a commercial commodity in the trade. Like major fruits (mango, apple, banana, pear, citrus, guava, etc.), minor fruits (aonla, mangosteen, karonda, bael etc.) are also becoming popular because of medicinal and nutritional properties (Parthsarathy and Nandakishore, 2014). India having varied agro-climatic conditions produces almost all types of fruits in one or other part of the country. At present, India is the second largest producer of fruits in the world after China which has become possible after standardization of several fruit production techniques and their adoption by the Indian farmers. Although, modern production techniques (fertilizers, insecticides, fungicides etc.) have paved the way for higher production but their indiscriminate and unscientific use is affecting the quality production of fruits (Sharma and Krishna, 2014). Furthermore, some of them are unsafe to the environment and the ecosystem. Among the eco-friendly means, the use of the plant growth regulators is common and BRs in particular can prove beneficial for plant productivity especially under stress conditions (Ali, 2017).

In this chapter, an attempt has been made to provide information on understanding the biosynthesis, metabolism, pathways and functions of BRs and their possible role in fruit crops. Discussion will also be made on the role of BRs in plant development and architecture, photosynthesis, transition to flowering and fruit set in fruit crops. In addition, role of BRs in improving plant resistance against biotic and abiotic stresses caused by drought, extreme temperatures, salinity, oxidative stress, heavy metals and pathogens will also be discussed.

2. Brassinosteroids (BRs) and their Occurrence in Plants

Mitchell and co-workers identified a steroidal substance in the year 1970 and called it as 'brassins'. Later, it was chemically extracted by Grove *et al.*, (1979) from bee-collected pollen of rapeseed plant (*Brassica napus* L.) and termed it as '*brassinolide*' (Figure 1). They could get only 10 mg of a solid crystalline form of brassinolide (BL) from 230 kg of pollen grains. Later in 1982, another growth substance similar to brassinolide was discovered from the insect galls of chestnut (*Castenea crenata*) which was termed as '*castasterone*'. Both these inventions led to the further investigations towards such compounds in other plant species. All these steroidal substances having growth promoting activity were cumulatively referred as '*brassinosteroids*' (BRs). BRs is a group of naturally occurring polyhydroxy steroids (Rao *et al.*, 2002).

BRs are considered as the 6th important group of phytohormones in plants, which are considered essential for normal growth and several plant development processes such as cell division and elongation, vascular differentiation, pollen tube growth, seed germination, rhizogenesis, flowering, senescence, abscission, maturation regulation of gene expression, enzyme activation and photosynthesis (Hayat and Ahmad, 2011). These steroidal compounds are also involved in inducing defence mechanism against various biotic and abiotic stresses, such as water, temperature, oxidative stresses and high salinity. BRs are natural, non-toxic, non-genotoxic, bio-safe, and eco-friendly phytohormone, which can be used in agri-horti plants to improve the growth, yield, quality etc (Coll *et al.*, 2015).

BRs act together with other phytohormones such as auxins, cytokinins, ethylene, gibberellin, jasmonic acid, abscisic acid, salicylic acid and polyamines for regulating various physiological and developmental activities in plants (Saini *et al.*, 2015). As of now, more than 70 BRs have been isolated from different plant species, out of which, only three i.e, brassinolide, 24-epibrassinolide and 28-homobrassinolide are the biologically most active forms and are extensively used in physiological studies.

Several BRs have been extracted from a large number of plant species, as these steroidal chemicals are very ubiquitous in distribution throughout the plant taxa including, fruit crop species (Table 1). These growth-inducing steroidal hormones were isolated from different plant parts including, reproductive organs such as pollen grains, anthers, vegetative parts like stems, leaves, roots, flowers, and seeds. In addition, BRs have also been extracted from crown galls of chestnut plant (*Castanea crenata*). The highest BR-concentrations were measured in pollen and immature seeds (1–100 g/ kg) fresh mass) (Bajguz and Tretyn, 2003).

3. Chemical Structure of BRs

Naturally occuring BRs are polyhydroxylated steroidal hormones, which has a common 5α-cholestane skeleton. These steroidal components have been classified as C-27, C-28 or C-29 BRs based on the alkyl-substitutions in the side chain of brassinosteroids. Oxygen at C-6 and hydroxyls on the side chain of C-22 and C-23 are essential for activity. Among them, the C_{28}-BRs that harbour a C-24 methyl group are major BRs in the plant kingdom. Castasterone (CS) and brassinolide (BL) belonging to the C28-BRs and are the most active BR. Brassinosteroids (BRs) consist of a precise class of low abundance plant steroids which bear an oxygen moiety at C-3 and additional ones at one or more of the C-2, C-6, C-22 and C-23 carbon atoms (Bishop and Yokota 2001; Sasse, 2003; Bajguz and Hayat, 2009; Hayat and Ahmad, 2011). The chemical structures of some important brassinosteroids are depicted in Figure 2.

4. Biosynthesis and Metabolic Pathway Regulation of BRs

The main BR biosynthetic pathway has not yet been studied (Divi and Krishna, 2009). However, at first, the BR biosynthesis was reported to be based on the following two parallel pathways: the early C-6 and late C-6 oxidation pathways (Fujioka *et al.*, 1998; Li *et al.*, 2010). Later, the comprehensive study revealed that the pathways can diverge at campesterol to establish an early C-22 hydroxylation pathway (Choe *et al.*, 2001; Fujioka and Yokota, 2003). In general, the BL biosynthetic pathways are triterpenoid pathways, where squalene is cyclized into cycloartenol *via* a plant-specific enzyme system (Hartmann, 1998). Most popular pathways of BRs biosynthesis have been depicted in the figure 3.

The various approaches have been tested to elucidate the biosynthetic pathways of brassinosteroids. The BR-biosynthetic and signaling pathways have been well studied in *Arabidopsis thaliana*, rice and tomato (Divi and Krishna, 2009). However, the BR biosynthetic pathways were primarily elucidated by *in vivo* feeding tests of labeled precursors of brassinolide in cell lines of periwinkle (*Catharanthus roseus*) (Fujioka and Sakurai, 1997). The pathways were later confirmed by investigating the endogenous levels of BRs in BR-deficient mutants in the late 1990s (Fujioka *et al.*, 1997). These combined approaches led to the identification of the complete metabolic pathways underlying the biosynthesis and inactivation of BRs. Several enzymes that catalyze brassinosteroid biosynthesis were revealed based on BR-deficient mutant analysis in *Arabidopsis*. DE-ETIOLATED 2 (DET2), a 5*a*-reductase, catalyses the conversion of 4-en-3-one to 3-one and 22-OH- 4-en-3-one to 22-OH-3-one (Fujioka *et al.*, 1997).

5. Mode of Action of BRs

BRs participate in diverse activities of plant growth and development. Plants which lack in brassinosteroid biosynthesis and/or malfunctioning in the BR signal transduction, many times may lack in sex expression or may develop into unusual phenotypes. It demonstrates the significance of BR biosynthesis and their signal transduction pathway in controlling biological processes in plants (Yang *et al.*, 2011). By means of various approaches such as genetics, molecular and genomic study, several components involved in BRs biosynthetic and signaling pathway have been identified. A number of physiological phenomenon in plants including cell elongation, root, anther and pollen development and formation, stem elongation, vascular differentiation etc., are regulated by these steroidal compounds. Novel techniques have been developed in many laboratories to detect the additional components in BRs signaling pathway. Among these, proteomics (Tang *et al.*, 2008) and mass spectrometry (Hink *et al.*, 2008), bioinformatics and screening tools (Nam and Li, 2004) have provided essential data in identifying new components. It includes BRI-1 interacting proteins such as BRI-1-associated receptor kinase 1 (BAK-1), BRI-1 kinase inhibitor 1 (BKI-1), transthyretin-like (TTL) and BR signaling kinases (BSKs) (Yang, *et al.,* 2011).

6. Role of BRs in Fruit Crops

BRs is a new group of phytohormones, which occur naturally, and plays significant and diverse biological role with growth promoting properties in plant organisms (Clouse and Sasse, 1998; Deng *et al.*, 2016). The role of BRs in plant function has been intensively studied in the last few years. Owing to the multiple effect of BRs on different aspects of physiological growth and development of any plant, right from seed germination to senescence, it is considered as "pleiotropic hormone" (Clouse, 2011). In addition, BRs also act to combine stimuli perceived from the atmosphere into endogenous developmental programs and thereby confer plants an adaptive potential to various environmental factors and changes. In this way, it improves biotic and abiotic stresses. Unlike other plant hormones, application of Brassinosteroids in agriculture and horticulture, especially in fruit crops would not cause any residual effects on food, as these are reported to be nontoxic, non-genotoxic, biosafe, eco-friendly as well as natural synthesized in various plant parts (Ali, 2017). The growth inducing and physiological roles of BRs in fruit crops are depicted in figure 4 and discussed below:

6.1. Micropropagation of hard-to-root fruit crops

During last three decades, plant micropropagation industry has emerged as one of the most significant contributions of biotechnology to the horticulture sector. The possibility of mass propagation of elite, endangered, or difficult-to-propagate genotypes and/or varieties, in relatively small areas and time period, is considered as the most important and useful characteristic of micropropagation technique. In most of the woody fruit species, micropropagation techniques have failed to stimulate shoot proliferation which is a major hindrance to use this technique commercially in woody perennials. It has been reported that some BRs like 28-homocastasterone, a fluoro derivative (5F-HCTS) stimulates branch elongation in *in-vitro*-cultured shoots of marubakaido apple rootstock. It has also been suggested that this hormone has a potential use in efficient tissue culture technique for woody species and can also be useful for stimulating branch elongation (Pereira-Netto *et al.*, 2009). Explants from coconut (*Cocos nucifera* L.) also respond well to the brassinosteroid, which helps the growth in the initial callus, embryogenic callus and somatic embryos (Azpeitia *et al.*, 2003).

6.2. Root formation in propagules

Adventitious rooting is an indispensable step in the vegetative propagation of economically important woody horticultural species. Clonal propagation is of practical significance to horticulture and forestry crops, because genetic improvement in perennial woody species with lengthy generation cycles is often limiting (Legue *et al.*, 2014). BRs regulate root meristem size and lateral root development in a concentration-dependent manner. Biochemical and physiological analyses proved that application of low concentration of BRs can promote root growth, while high amount of BRs can slow it down (Gonzalez-Garcia *et al.*, 2011; Gupta *et al.*, 2015; Lee *et al.*, 2015). Exogenous application of BRs is reported to increase root formation on stem cuttings (Swamy and Rao, 2006).

6.3. Promotion of vegetative growth

Extensive studies have indicated that cell division and cell elongation are significantly influenced by BRs alone or in combination with other phytohormones viz. auxins, and gibberellins (Matusmoto *et al.*, 2016). The significant increase in the growth, and promotion of cell elongation, could be associated with BRs involvement in increasing the carbohydrate supply. It has been reported that BRs up-regulate the activity of an extra cellular invertase which is responsible for increasing carbohydrate supply (Nakajima and Toyama, 1999; Goetz *et al.*, 2000). In strawberry, one spray of brassinosteroids (0.01 ppm) improves the number of leaves, petiole length, total leaf area and number of crowns by more than hundred times (Pipattanawong *et al.*, 1996).

6.4. Induction of flowering

Previous reports have proved that hormones like GAs play significant role in transition of vegetative phase of a plant to reproductive phase, whereas ABA delays flowering. Now, it has been established that BRs also influence flowering in some plants including fruit crops (Clouse and Sasse 1998; Clouse, 2008). For example, exogenous application of brassinosteroids significantly increased the number of flowers in strawberry (Pipattanawong *et al.*, 1996), and in grapes (Yoshiok *et al.*, 1990). Brassinosteroids increased the number of flowers in autumn but reduced in late winter crop. BRs also help in determining branching and flower formation processes in plants through modulating metabolic pathways and nutrient allocation and/or by interacting with other signaling pathways (Mussig, 2005). Exogenous application of EBR (epi-brassinosteroids) increased precocity in bearing as well as increased the production of female flowers in cucumber (Papadopoulou and Grumet, 2005).

6.5. Fruit setting

It has been established that novel hormones like BRs are very effective even at very low concentration in doing so. For example, spraying of brassinolide @ 0.01ppm considerably increased fruit set in Navel orange (*Citrus sinensis* L.) cv. 'Morita' (Sugiyama and Kuraishi, 1989). Similarly, fruit set of Japanese persimmon was increased by exogenous application of BRs. BRS @ 0.01ppm improved fruit set in Japanese persimmon (var. Fuyu) by 76.2% and 70.60% when applied at 7 days before blooming and at blooming time, respectively (Watanabe *et al.*, 1997).

6.6. Fruit growth and development

Several reports have indicated BRs are involved in the growth and development of fleshy fruits (Rao *et al.*, 2002; Fu *et al.*, 2008). Both brassinosteroid and GA_3 increased cell elongation and division in grape when applied at 3-4 mm berry growth stage. Similarly, exogenous application of CPPU and 0.4 mg/1 brassinosteroid considerably enhanced berry size, cluster weights and berry numbers (Bhat *et al.*, 2011). Further, exogenous application of BRs as aqueous solution considerably enhanced grape cluster weight, berry weight, length, and diameter. Chai *et al.* (2013) investigated that brassinosteroids might be involved in initial fruit development by stimulating cell division of strawberry fruits.

6.7. Fruit yield

Several reports have suggested that BRs are responsible for escalating crop yields either through altering the plant metabolic activity or providing extra tolerance to plants from any environmental stresses. For instance, genetic manipulation of BRs activity may lead to increase in crop yield by 20–60% (Divi and Krishna 2009). BRs stimulate photosynthetic process by accelerating fixation of CO_2, increasing biosynthesis of protein in case of wheat and mustard. In grapevines, the application of 24-epibrassinolide at a concentration of 0.01 and 0.1 ppm improved the number of grapes/bunch and total yield by 66.7 and 29.9% over untreated vines (Pozo *et al.*, 1994). BB-16 a *brassinosteroid* analogue, also stimulated greater CO_2 assimilation. The application of brassinosteroid analogue, BR-3 @ 5 ppm caused 65% increased predictable yield of the yellow passion fruit plants (12 t/ha) over control and BB-16 (4 ppm) when sprayed during a period of reproductive development in passion fruit, increased the number of fruits per plant drastically (Gomes *et al.*, 2006). Similarly, increased yield was also observed in Navel orange and sweet cherry (Sugiyama and Kuraishi, 1989; Roghabadi and Pakkish, 2014). Therefore, commercial application of some BRs like Epin and Tianfengsu, have been recommended for potato, cucumber, pepper, tomato and barley to improve their production and quality in countries like Belarus, Japan, Russia and China (Moiseev, 1998).

6.8. Fruit ripening

Fruit ripening is an unique plant developmental process that has direct implications not only on our food supply, but also on nutrition, and our health. The ripening process is positively correlated with the respiration rate and ethylene evolution rate of a fruit which are usually higher during the ripening process of a fruit. Stimulation of fruit ripening by inducing ethylene biosynthesis is considered as one of the important physiological response of BRs in fruit crops. Only exogenous application of these compounds improves fruit ripening (Table 2). But, any alteration in the endogenous level of BRs does not have any prominent impact on ripening of climacteric fruit such as mango (Zaharah *et al.*, 2012). BRs also play a role in fruit softening during the ripening process by stimulating biosynthesis of ethylene between SAM (S-adenosyl methionine) and ACC (1-amino-cyclopropane 1- carboxylic acid) pathways as BRs sprays to grape berries were very effective in promoting fruit ripening. Furthermore, Zaharah and Singh (2010) suggested that the postharvest application of epibrassinolide (Epi-BL) not only promoted fruit softening and colour development in mango but also hastened fruit ripening process. However, in a study, it has been shown that Chinese ber (*Ziziphus jujuba* cv. Huping), fruits treated with BR showed a considerably lower ethylene level during storage as compared to the untreated fruits (Zhu *et al.*, 2010).

6.9. Fruit quality

Exogenous application of BRs as aqueous solution considerably enhanced grape cluster weight and diameter, berry weight, length, and diameter. Even under cold storage, BRs helped in maintaining skin colour, reducing the rate of soluble solids and titratable acidity degradation (Champa *et al.*, 2015). Foliar applications of BRs also increased skin colour of fruits by increasing anthocyanin content, organic acids, ascorbic acid, and phenol content (Roghabadi and Pakkish, 2014). BRs at the concentration of 0.4 mg/L enhanced total anthocyanin content in grapes over control fruits. Similarly, the exogenous application of HBR, a type of BRs led to early maturation of sweet cherry cv. 'Tulare' and 'Bing' with increased fruit firmness, skin colour development and also increased the force required to remove from stem (Mandava and Wang, 2016) (Table 3).

6.10. Amelioration of abiotic stresses

To cope up with various forms of abiotic factors (drought, cold, heat, high salinity) as well as biotic stresses, plants have developed their own complex defense mechanisms to boost their stress tolerance (Leavitt, 1980). Besides five classical phytohormones (auxins, gibberrelic acid, abscisic acid, cytokinins, ethylene), BRs possess extensive variety of adaptive responses in plant adaptation to several biotic and abiotic stresses (Kagale *et al.*, 2007). It is evident from previous reports that BRs have a shielding effect on enhancing tolerance or resistances against several extreme conditions, such as, temperature water stress/drought, salt stress, toxicity of heavy metals (Cd, Cu, Al, and Ni) and disease causing microorganisms (Ali *et al.*, 2008; Gomes *et al.*, 2013). All these stresses affect plant growth and development. BRs are reported to control stress response by activating or holding back the key enzymatic reactions of different biochemical pathways, induction of protein biosynthesis, and the production of a wide range of chemical defence compounds.

7. Conclusions

After three decades of the discovery of brassinosteroids (BRs), there is a rapid increase in research and development of this steroidal hormone in different horticultural crops. Altering the endogenous level of BRs within the plant system by exogenous application have significant influence on growth and development by cell division, cell elongation, roles in reproductive development, vascular differentiation, stress tolerance and pathogen resistance, fruit set, yield and quality etc. Based on the evidences of different research findings, it could be concluded that besides the classical growth hormones, brassinosteroids and their analogues are the sixth group of phytohormones, which proved their role in modern fruit production. Exogenous applications of BRs have specific physiological influences on different aspects of fruit crops such as micropropagation, growth and development of plants including fruit quality and yield. BRs also influence the defence mechanism of the fruit trees and provides tolerance/resistance to various biotic and abiotic stresses, and thus can play a significant role in the fruit industry.

References

- Ali, B. 2017. Practical applications of brassinosteroids in horticulture—Some field perspectives. *Scientia Hortic.*, **225**: 15-21.
- Ali, B., Hasan, S. A., Hayat, S., Hayat, Q., Yadav, S., Fariduddin, Q., and Ahmad, A., 2008. A role for brassinosteroids in the amelioration of aluminium stress through antioxidant system in mung bean (*Vigna radiata* L. Wilczek). *Env. Exp. Bot.*, **62**: 153-159.
- Azpeitia, A., Chan, J. L., Saenz, L., and Oropeza, C. 2003. Effect of 22(S), 23(S)

homobrassinolide on somatic embryogenesis in plumule explants of *Cocos nucifera* (L.) cultured in vitro. *J. Hort. Sci. Biotechnol.*, **78**: 591–596.

- Bajguz, A., and Hayat, S. 2009. Effects of brassinosteroids on the plant responses to environmental stresses. *Plant Physiol. Bioch.*, **47**: 1-8.
- Bajguz, A., and Tretyn, A. 2003. The chemical characteristic and distribution of brassinosteroids in plants. *Phytochemistry,* **62**: 1027-1046.
- Bhat, Z.A., Reddy, Y.N., Srihari, D., Bhat, J.A., Rashid, R., and Rather, J.A. 2011. New generation growth regulators- brassinosteroids and CPPU improve bunch and berry characteristics in 'Tas-A-Ganesh' grape. *Intl. J. Fruit Sci.,* **11**: 309-315.
- Bishop, G.J., and Yokota, T. 2001. Plants steroid hormones, brassinosteroids: Current highlights of molecular aspects on their synthesis/metabolism, transport, perception and response. *Plant Cell Physiol.*, **42**: 114-120.
- Chai, Y. M., Zhang, Q., Tian, L., Li, C. L., Xing, Y., Qin, L., and Shen, Y. Y. 2013. Brassinosteroid is involved in strawberry fruit ripening. *Plant Growth Regul.*, **69**: 63-69.
- Champa, W. H., Gill, M. I. S., Mahajan, B. V. C., Aror, N. K., and Bedi, S. 2015. Brassinosteroids improve quality of table grapes (*Vitis vinifera L.) cv.* Flame Seedless. *Trop. Agri. Res.,* **26**: 368-379.
- Choe, S., Dilkes, B. P., Fujioka, S., Takatsuto, S., Sakurai, A., and Feldmann, K.A. 1998.. The DWF4 gene of Arabidopsis encodes a cytochrome P450 that mediates multiple 22α-hydroxylation steps in brassinosteroid biosynthesis. *The Plant Cell,* **10**: 231-243.
- Choe, S., Fujioka, S., Noguchi, T., Takatsuto, S., Yoshida, S., and Feldmann, K. A. 2001. Overexpression of DWARF4 in the brassinosteroid biosynthetic pathway results in increased vegetative growth and seed yield in Arabidopsis. *The Plant J.*, **26**: 573-582.
- Clouse, S. D. 2008. The molecular intersection of brassinosteroid- regulated growth and flowering in *Arabidopsis. Proc. Natl. Acad. Sci.* **105**: 7345-7346.
- Clouse, S. D. 2011. Brassinosteroids. *The Arabidopsis Book,* **9**: e0151.
- Clouse, S. D., and Sasse, J. M. 1998. Brassinosteroids: essential regulators of plant growth and development. *Annu. Rev. Plant Physiol. Plant Mol. Biol.*, **49**: 427-451.
- Coll, Y., Coll, F., Amorós, A., and Pujol, M. 2015. Brassinosteroids roles and applications: an up-date. *Biologia,* **70:** 726–732.

- Davies, P. J. 1987. Plant hormones and their role in plant growth and development. Martinus Nijhoff Publishers, Dordrecht.
- Deng, X. G., Zhu, T., Peng, X. J., Xi, D. H., Guo, H., Yin, Y., Zhang, D. W., and Lin, H. H. 2016. Role of brassinosteroid signaling in modulating tobacco mosaic virus resistance in *Nicotiana benthamiana. Sci Rep.*, **6**: 20579.
- Divi, U. K., and Krishna, P. 2009. Brassinosteroid: a biotechnological target for enhancing crop yield and stress tolerance. *New Biotechnol.*, **26**:131-135.
- Fu, F. Q., Mao, W. H., Shi, K., Zhou, Y. H., Asami, T., and Yu, J. Q. 2008. A role of brassinosteroids in early fruit development in cucumber. *J. Exp. Bot.*, **59**: 2299-2308.
- Fujioka, S., and Sakurai, A. 1997. Biosynthesis and metabolism of brassinosteroids. *Physiol. Plant.*,**100:** 710-715.
- Fujioka, S., and Yokota, T. 2003. Biosynthesis and metabolism of brassinosteroids. *Ann. Rev. Plant Biol.,* **54**: 137-164.
- Fujioka, S., Li, J., Choi, Y. H., Seto, H., Takatsuto, S., Noguchi, T., Watanabe, T., Kuriyama, H., Yokota, T., Chory, J., and Sakurai, A. 1997. The Arabidopsis deetiolated2 mutant is blocked early in brassinosteroid biosynthesis. *Plant Cell,* **9**: 1951–1962.
- Fujioka, S., Noguchi, T., Yokota, T., Takatsuto, S., and Yoshida, S. 1998. Brassinosteroids in Arabidopsis thaliana. *Phytochemistry,* **48**: 595–599.
- Goetz, M., Godt, D., and Roitsch, T. 2000. Tissue-specific induction of the mRNA for an extracellular invertase isoenzyme of tomato by brassinosteroids suggests a role for steroid hormones in assimilate partitioning. *Plant J.*, **22**: 515-522.
- Gomes, M. D. M. A., Campostrini, E., Leal, N. R., Viana, A. P., Ferraz, T. M., do Nascimento Siqueira, L., Rosa, R. C. C., Netto, A. T., Nunez-Vázquez, M., and Zullo, M. A. T. 2006. Brassinosteroid analogue effects on the yield of yellow passion fruit plants (*Passiflora edulis* f. flavicarpa). *Scientia Hortic.*, **110**: 235-240.
- Gomes, M. D. M. D. A., Torres Netto, A., Campostrini, E., Bressan-Smith, R., Zullo, M. A. T., Ferraz, T. M., Siqueira, L. D. N., Leal, N. R., and Núñez-Vázquez, M. 2013. Brassinosteroid analogue affects the senescence in two papaya genotypes submitted to drought stress. *Theoret. Exp. Plant Physiol.*, **25**: 186-195.
- González-García, M. P., Vilarrasa-Blasi, J., Zhiponova, M., Divol, F., Mora-García, S., Russinova, E., and Caño-Delgado, A. I. 2011. Brassinosteroids control meristem size

by promoting cell cycle progression in *Arabidopsis* roots. *Development,* **138**: 849-859.

- Gray, W. M. 2004. Hormonal regulation of plant growth and development. *PLoS Biol.,* **2**: E311.
- Grove, M. D., Spencer, G. F., Rohwedder, W. K., Mandava, N., Worley, J. F., Warthen, J. D., Steffens, G. L., Flippen-Anderson, J. L., and Cook, J. C. 1979. Brassinolide, a plant growth promoting steroid isolated from *Brassica napus* pollen. *Nature,* **281**: 216-217.
- Gupta, A., Singh, M., and Laxmi, A. 2015. Interaction between glucose and brassinosteroid during the regulation of lateral root development in *Arabidopsis. Plant Physiol.*, **168**: 307–320.
- Hartmann, M. A. 1998. Plant sterols and the membrane environment. *Trends Plant Sci.*, **3**:170-175.
- Hayat, S., and Ahmad, A. 2011. Brassinosteroids: a class of plant hormone. Springer Science & Business Media.
- Hink, M. A., Shah, K., Russinova, E., De Vries, S. C., and Visser, A. J. 2008. Fluorescence fluctuation analysis of Arabidopsis thaliana somatic embryogenesis receptor-like kinaseandbrassinosteroid insensitive 1 receptor oligomerization. *Biophys. J.,* **94**: 1052-1062.
- Kagale, S., Divi, U. K., Krochko, J. E., Keller, W. A., and Krishna, P. 2007. Brassinosteroid confers tolerance in Arabidopsis thaliana and *Brassica napus* to a range of abiotic stresses. *Planta,* **225**: 353-364.
- Leavitt, J. 1980. Responses of plants to environmental stresses, Vol II. Academic, New York.
- Lee, H. S., Kim, Y., Pham, G., Kim, J. W., Song, J. H., Lee, Y., Hwang, Y. S., Roux, S. J., and Kim, S. H. 2015. Brassinazole resistant 1 (BZR1)-dependent brassinosteroidsignalling pathway leads to ectopic activation of quiescent cell division and suppresses columella stem cell differentiation. *J. Exp. Bot.*, **66**: 4835-4849.
- Legue, V., Rigal, A., and Bhalerao, R. P. 2014. Adventitious root formation in tree species: involvement of transcription factors. *Physiol. Plant.*, **151**:192-198.
- Li, J., Li, Y., Chen, S., and An, L. 2010. Involvement of brassinosteroid signals in the floral-induction network of *Arabidopsis. J. Exp. Bot.,* **61**: 4221-4230.
- Mandava, B., and Wang, Y. 2016. Effect of brassinosteroids on cherry maturation firmness and fruit quality. *Acta Hortic.*, **1139**: 451-458.

- Matusmoto, T., Yamada, K., Yoshizawa, Y., and Oh, K. 2016. Comparison of effect of brassinosteroid and gibberellin biosynthesis inhibitors on growth of rice seedlings. *Rice Sci.,* **23**:51-55.
- Moiseev, A. 1998. Preparation epin: Water of life. *Eureka*, **68**.
- Mussig, C. 2005. Brassinosteroid-promoted growth. *Plant Biol.*, **7**: 110-117.
- Nakajima, N., and Toyama, S. 1999. Effects of epibrassinolide on sugar transport and allocation to the epicotyl in cucumber seedlings. *Plant Prod. Sci.*, **2**: 165-171.
- Nam, K. H., and Li, J. 2004. The *Arabidopsis* transthyretin-like protein is a potential substrateofbrassinosteroid-insensitive 1. *Plant Cell* **16**: 2406-2417.
- Papadopoulou, E., and Grumet, R. 2005. Brassinosteriod-induced femaleness in cucumber and relationship to ethylene production. *Hort Sci.* **40**: 1763-1767.
- Parthsarathy, U., and Nandakishore, O.P. 2014. A Study on nutrient and medicinal compositions of selected Indian *Garcinia* species. *Curr. Bioact. Compounds,* **10**: 55–61.
- Pereira-Netto, A. B., Roessner, U., Fujioka, S., Bacic, A., Asami, T., Yoshida, S., and Clouse, S.D. 2009. Shooting control by brassinosteroids: metabolomic analysis and effect of brassinazole on *Malus prunifolia*, the Marubakaido apple rootstock. *Tree Physiol.*, **29**: 607-620.
- Pipattanawong, N., Fujishige, N., Yamane, K., and Ogata, R. 1996. Effects of brassinosteroid on vegetative and reproductive growth in two day-neutral strawberries. *J. Japan Soc. Hort. Sci.*, **65**: 651-654.
- Pozo, L., Rivera, T., Noriega, C., Iglesias, M., Coll, F., Robaina, C., Velázquez, B., Rodríguez, O. L., and Rodríguez, M. E. 1994. Algunos resultadosen el cultivo de losfrutalesmediante la utilizacio´n de brasinoesteeroides o compuestosana´logos. *Cult. Trop.*, **15**: 79-92.
- Rao, S. S. R., Vardhini, B. V., Sujatha, E., and Anuradha, S. 2002. Brassinosteroids – a new class of phytohormones. *Curr. Sci.*, **82**: 1239-1245.
- Roghabadi, M. A., and Pakkish, Z. 2014. Role of brassinosteroid on yield, fruit quality and postharvest storage of 'TakDanehe Mashhad' sweet cherry (*Prunus avium* L.). *Agric. Commun.*, **2**: 49-56.

- Saini, S., Sharma, I., and Patil, P. K. 2015. Versatile roles of brassinosteroid in plants in the context of its homoeostasis, signaling and cross talks. *Front. Plant Sci.*, **6**: 1-17.
- Sasse, J. M. 2003. Physiological actions of brassinosteroids: an update. *J. Plant Growth Regul.*, **22**: 276-288.
- Sharma, R.R. and Krishna, H. (2014). *Fruit Production: Major Fruits.* Daya Publishing House, Delhi, India.
- Sugiyama, K., and Kuraishi, S. 1989. Stimulation of fruit set of' 'Morita' navel orange with brassinolide. *Acta Hortic.*, **239**: 345-348.
- Swamy, K. N., and Rao, S. S. R. 2006. Influence of brassinosteroids on rooting and growth of geranium (*Pelargonium* sp.) stem cuttings. *Asian J. Plant Sci.* **5**: 619-622.
- Tang, W., Kim, T. W., Oses-Prieto, J. A., Sun, Y., Deng, Z., Zhu, S., Wang, R., Burlingame, A. L., and Wang, Z. Y. 2008.. BSKs mediate signal transduction from the receptor kinaseBRI1 in *Arabidopsis. Science,* **321**: 557-560.
- Unterholzner, S. J., Rozhon, W., Papacek, M., Ciomas, J., Lange, T., Kugler, K. G., Mayer, K. F., Sieberer, T., and Poppenberger, B. 2015. Brassinosteroids are master regulators of gibberellin biosynthesis in arabidopsis. *The Plant Cell,* **27**: 2261-2272.
- Watanabe, T., Noguchi, T., Kuriyama, H., Kadota, M., Takatsuto, S., and Kamuro, Y. 1997. Effects of brassinosteroid compound [TS303] on fruit-setting, fruit-growth taking roots and cold-resistance. *Acta Hortic.*, **436**: 267-270.
- Yang, C. J., Zhang, C., Lu, Y. N., Jin, J. Q., and Wang, X. L. 2011. The mechanisms of brassinosteroids' action: from signal transduction to plant development. *Mol. Plant.*, **4**: 588-600.
- Yoshiok, T., Nesumi, H., and Ito, Y. 1990. *Ibid,* **59**: 44-45.
- Zaharah, S. S., and Singh, Z. 2010. Role of brassinosteroids in mango fruit ripening. *Acta Hort.* **934**: 929-935.
- Zaharah, S. S., Singh, Z., Symons, G. M., and Reid, J. B. 2012. Role of brassinosteroids, ethylene, abscisic acid, and indole-3- acetic acid in mango fruit ripening. *J. Plant Growth Regul.,* **31**: 363-372.
- Zhu, Z., Zhang, Z., Qin, G., and Tian, S. 2010. Effects of brassinosteroids on postharvest disease and senescence of jujube fruit in storage. *Postharvest Biol. Technol.,* **56**: 50-55.

Table 1: The occurrence of brassinosteroids in fruit tree species

S.No.	Family/species	Plant part	Type of BR
1	Loquat (*Eriobotrya japonica*)	Flower buds	CS
2	Satsuma mandarin (*Citrus unshiu*)	Pollen	BL
3	Sweet orange (*Citrus sinensis*)	Pollen	BS, CS
4	Bael (*Aegle marmelos*)	Leaves	24-epiBL
5	Chestnut (*Castanea crenata*)	Galls	CS, BL, 6-deoxoCS
6	Date palm (*Phoenix dactylifera*)	Pollen	24-epiCS

[**Abbreviations**- CS (castasterone), BL (brassinolide), 24-epiBL (24-epibrassinolide)]

Table 2: Effect of BRs on ripening of different fruit crops

S.No.	Fruit crops	Type of BRs	Effects
	Non-climacteric fruit crops		
1	**Grape (*Vitis vinifera* L.)**	**BR**	Significantly promoted ripening
2	**Jujube (*Zizyphus jujube*)**	Brassinolide	**Delayed ripening**
	Climacteric fruit crops		
3	**Strawberry (*Fragaria x ananassa* Duch.)**	**BR**	**Role in ripening**
4	**Mango (*Mangifera indica* L.)**	epibrassinolide (Epi-BL)	Hastened fruit ripening

Table 3: Effect of BRs on post harvest quality of different fruit crops

S.No.	Fruit crop	Quality attribute
1	Orange (*Citrus sinensis*)	Increased sugar content
2	Litchi (*Litchi chinensis*)	Increased fruit firmness
3	Yellow passion fruit (*Passiflora edulis*)	Enhanced TSS
4	Chinese jujube (*Zizyphus jujube*)	Maintained higher TSS and TA & increased fruit firmness
5	Sweet cherry (*Prunus avium*)	Increased fruit firmness; maintained higher TSS and TA; improved peel colour, anthocyanins, extended postharvest life
		Enhanced peel colour development; increased stem removal force
6	Grape (*Vitis vinifera*)	Increase in cluster weight, berry weight, reduced rate of berry softening; maintained external colour and stabilized anthocyanins; increased total phenolics; reduced rate of TSS and TA degradation during low temperature.

Figure 1. BRs compounds "B**rassinolide (BL)**" was first extracted from pollen of *Brassica napus* L. (a) Inflorescence of *Brassica napus* L., (b) Single flower of *B. napus* L., (c) Anthers of *B. napus* L.

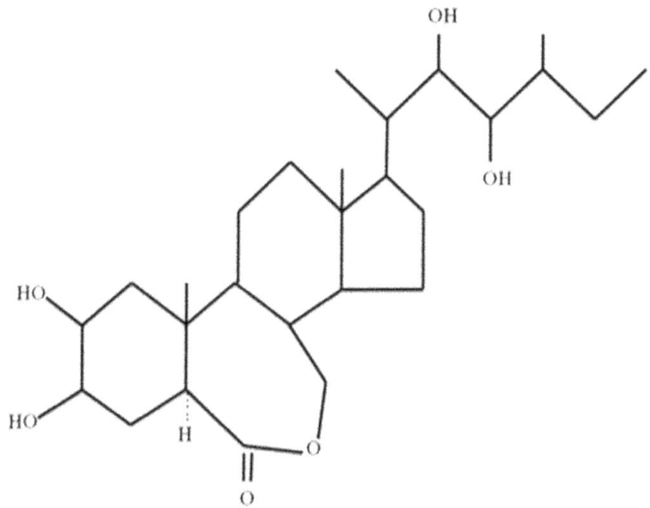

Brassinolide (BL)

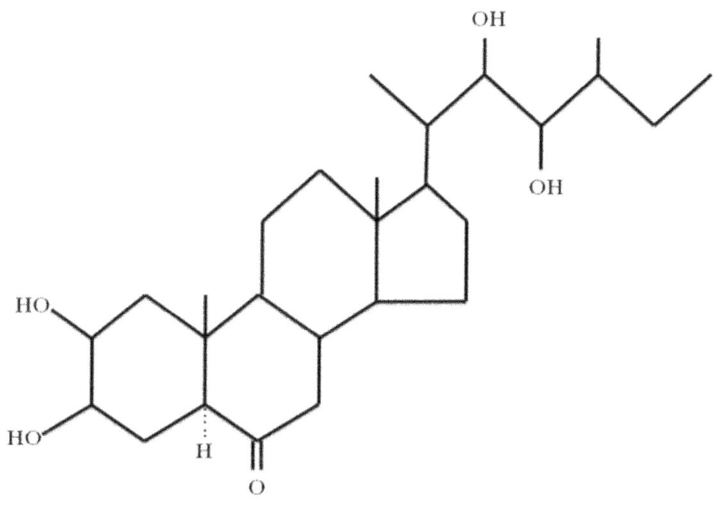

Castasterone

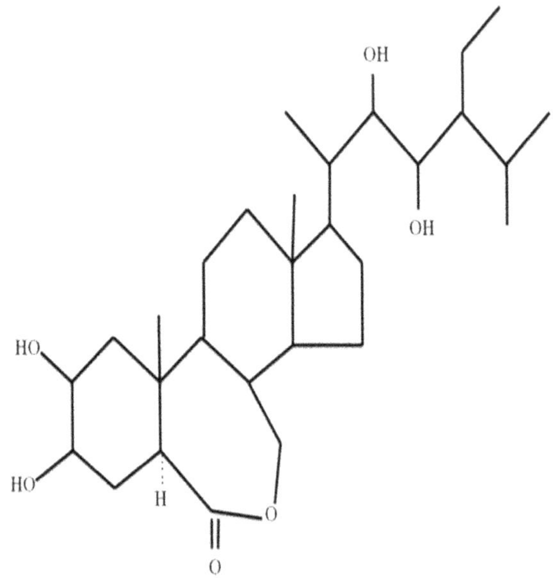

28-homobrassinolide (HBL)

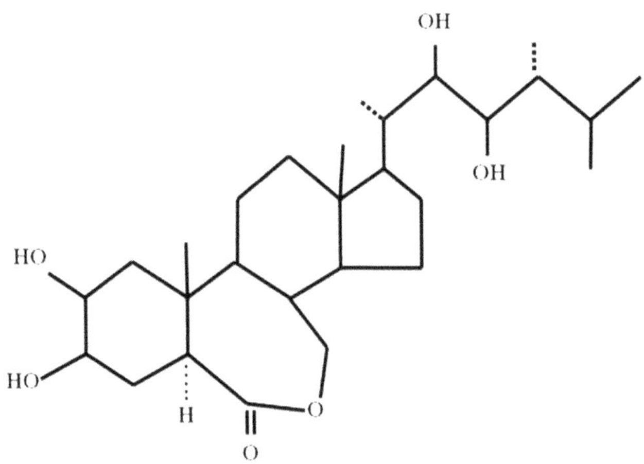

24-epibrassinolide (EBL)

Figure 2. Chemical structure of some important BRs

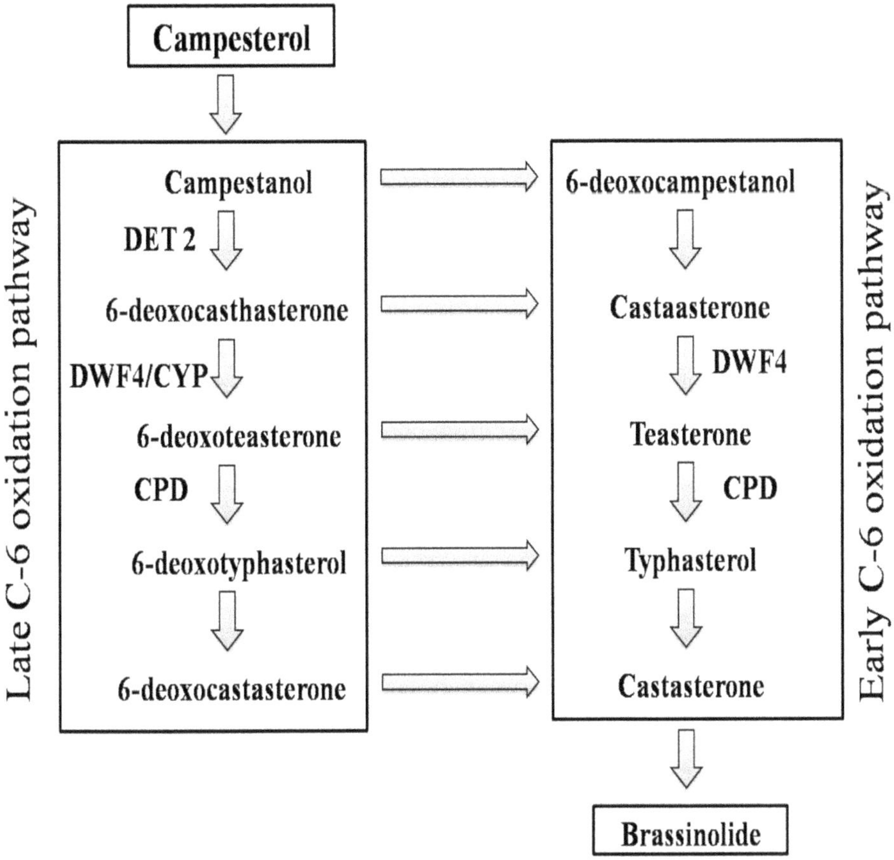

Figure 3. A simplified diagram of the BR biosynthesis pathway (adapted from Divi and Krishna, 2009). (Abbreviations: DET2 – De-etiolated-2, which is a 5α-reductase; DWF4 (CYP90B1) - steroid 22-α-hydroxylase; CPD – Constitutive photomorphogenesis and dwarfism; CYP92A6 - typhasterol/6-deoxotyphasterol 2α-hydroxylase)

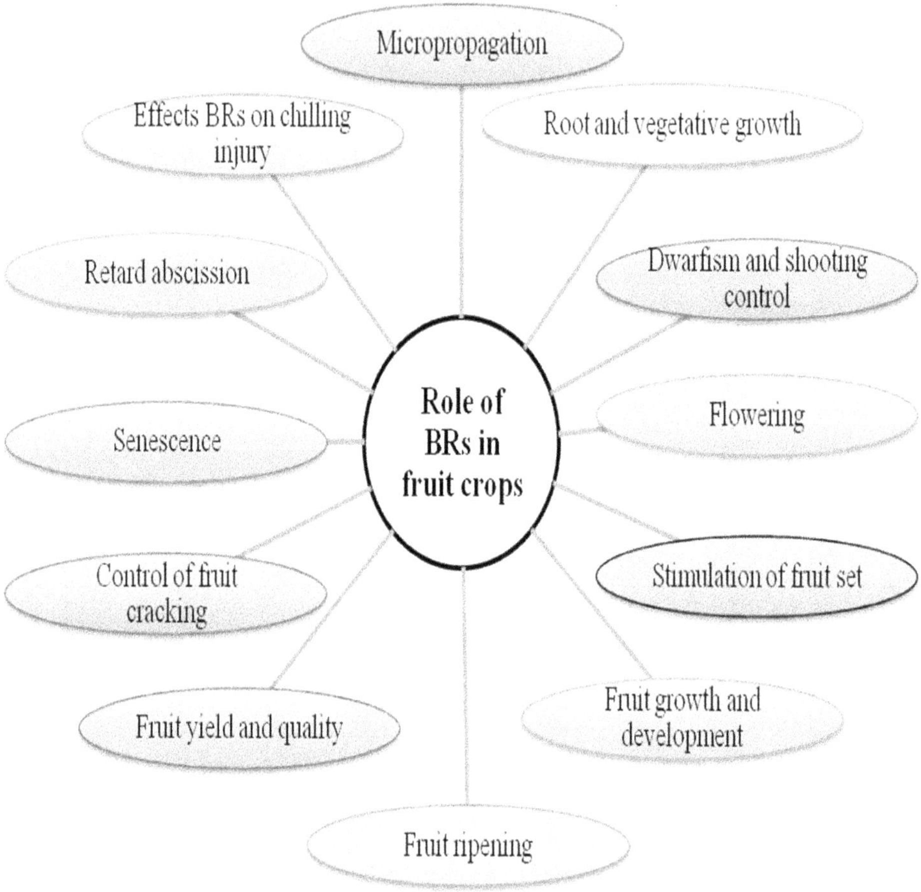

Figure 4. Schematic representation of pleiotropic influences of BRs on fruit crops

3

Role of Paclobutrazol in Fruit Crops and its Residual Fate

Kundan Kishore, H S Singh and R M Kurian

1. Introduction

Growth retardants have become an integral part of crop production system in many countries. They are primarily used to reduce the shoot growth by lowering down the rate of cell elongation and cell division without being phytotoxic. Paclobutrazol is an N-containing heterocycle triazole which inhibits gibberellin (GA) biosynthesis and in turn retards vegetative growth (Rademacher, 2000). It blocks the cytochrome P_{450}-dependent monooxygenases, thereby inhibiting oxidation of *ent*-kaurene into *ent*-kaurenoic acid of GA pathways (Rademacher, 2000). Among various triazoles (uniconazole-P, ancymidol, flurprimidol, paclobutrazol) developed as plant growth regulators, paclobutrazol ($C_{15}H_{20}ClN_3O$) [2*RS*,3 *RS*,1-(4-chlorophenyl)-4,4-dimethyl-2-(1H-1,2,4-triazol-1-yl pentan-3-ol)] is the most widely used. Paclobutazol holds considerable promise in manipulation of flowering, yield and vigour in fruit crops. However its high potency, hazardous nature and slow mobility raise genuine concerns over its long term use (USEPA, 2007).

Fruit crops like mango, citrus, litchi, apple, pear, nuts, etc. suffer from the intricate problem of irregular bearing or cropping periodicity as well as staggered or erratic flowering behaviour, leading to considerable loss of their production potential. Among fruit crops, mango has been the worst sufferers of cropping periodicity. The erratic rhythm of flowering could be due to environmental factors, crop genetics, orchard management practices or hormonal imbalances, either alone or in combination. In addition to irregular bearing, management of tree vigour and canopy, especially under tropical climate has become a challenge (Kurian *et al.*, 2013). Several strategies suggested overcoming the problems of flowering periodicity and tree vigour in sub-tropical and tropical regions, which are: use of dwarfing rootstock, shoot pruning and growth regulators. The later approach *i.e.*, use of growth regulators is most promising in managing canopy and ensuring regularity in flowering and enhancing fruit yield under commercial cultivation (Allan *et al.*, 1993).

A fundamental understanding of flowering in the tropics and sub-tropic is essential to bring sustainability in the production system. The nature of flower triggering viz. initiation, induction and elongation, involves several factors and among them, the balance between growth regulators and retardants is vital (Davenport, 2007). It has been reported that enhanced synthesis of gibberellin like substance in the plant system promotes vegetative growth and restricts the development of generative shoots. In such circumstances, gibberellin inhibitors like triazoles are used to initiate flowering by converting vegetative phase into reproductive one (Nartvaranant *et al.*, 2000). Paclobutrazol, a gibberellins inhibitor, has been effectively used in reducing canopy volume and increasing flower intensity in peach, plum, almond grapes and most importantly, in mango. Paclobutrazol is effective not only in flower induction but also in early and off season flower induction in different mango cultivars (Upreti *et al.*, 2013; Burondkar *et al.*, 2013). However the action of plant growth regulators (PGR) is highly specific to plant species, cultivar, organ and stage of development, and strongly dependent on its rate of application and environmental conditions (Hoffmann, 1992).

Paclobutrazol (PBZ), a non-polar broad spectrum growth regulator, has been characterized as an environmentally stable compound in soil and water environments (Table 1) with a long half-life under both aerobic and anaerobic conditions. Moreover PBZ is unlikely to volatilize to any significant extent owing to a low estimated vapour pressure (1.9 × 10-6 Pa). Paclobutrazol is translocated acropetally via xylem in plants (Hamid and Williams, 1997), although phloem translocation has also been reported (Witchard, 1997).

Table 1 Environmental fate properties for mobility and persistence of paclobutrazol

Parameter	Value	Source
Hydrolysis	Stable: <6% degradation after 30 d at pH 4,7, and 9	USEPA, 2007
Photolysis in water	Stable: < 5% degradation after 10 d at pH 7	USEPA, 2007
Aerobic soil metabolism (half-life)	> 1 yr	USEPA, 2007
Anaerobic soil metabolism (half-life)	> 1 yr	USEPA, 2007
Field dissipation (half-life)	450-950 days in orchard US soils 175 – 252 days in agricultural US soils	USEPA, 2007 EFSA, 2006
Aquatic metabolism (half-life)	164 days	EFSA, 2007
Soil Adsorption Coefficient (K_D) mL/g	1.3 – 23.0 0.8 – 21.3 (mean of 4.3)	USEPA, 2007 EFSA, 2006

Paclobutrazol has been registered in 1985, however it has now been permitted (cultar, ICI Americas, Goldsboro, N. C) for use on food crops in Australia, New Zealand, South Africa, India, Philippines, Vietnam, Canada, USA (California), Finland, Hungry, Greece, Cyprus, Denmark and Netherlands (Davis and Curry, 1991). In India PBZ, has been registered as a plant growth regulator under the section 9(3) of Insecticides Act, 1968 in November, 2009 by Central Insecticides Board & Registration Committee and is available in the market with various trade names.

2. Roles of Paclobutrazol

2.1. Regulation of flowering, yield and fruit quality

Paclobutrazol has demonstrated its usefulness by regulating traits of agronomic interest in various crops including cereal, vegetables, fruits and ornamentals. It has been effectively used for flower regulation in mango, peach, plum almond, apple, avocado and grapes. But its application has become common in mango to bring regularity in flowering (Reddy and Kurian, 2008).

The mechanism of flowering is a complex phenomenon depending on the interaction of internal and external factors regulating vegetative and reproductive shoot initiation and induction (Davenport, 2000). Low temperature and shoot age are considered as the crucial factor for trigging flowering in mango in tropical and subtropical conditions, respectively. But flowering does not occur even with the prevalence of low temperature and attainment of optimum shoot age, which gives an indication of hormonal regulation of flowering. Floral induction is considered to be the result of elevated levels of up-regulated florigenic promoter (FP) and down-regulated vegetative promoter (VP), whereas the reverse condition promotes vegetative growth. The production of vegetative shoots in place of reproductive shoots is due to the elevated level of gibberellin which is considered as a vegetative promoter. Paclobutrazol, a gibberellin inhibitor, reduces VP level and thereby increases FP/VP ratio which stimulates flowering shoots in weakly inductive shoots of mango (Adil *et al.*, 2011). Moreover PBZ modifies the source-sink relationships in mango to support fruit growth with lesser leaf area requirement and alters shoot anatomy towards higher phloem to xylem proportion.

The efficacy of paclobutrazol (PBZ) in mango flower manipulation was first assessed in Baiganapalli and Deshehari. The application of paclobutrazol in July- August has become a standardized technology and recommended practice for induction of regular flowering in mango. The optimized PBZ concentration has been 0.50 - 1.0 g a.i./ meter canopy diameter for inducing flowering without accompanying reduction in shoot length but higher concentration promoted the same degree of flowering with reduced shoot length. Singh (2000) reported that the soil application of cultar (20–40 g/tree) in October resulted in reduced shoot length, tree height and spread and increased the percentage of hermaphrodite flowers, flowering intensity, along with cauliflory and axillary flowering, fruit set and yield in Deshehari. However the treated plants had reduced panicle size. The application of cultar (2 ml/tree) in Dashehari during off-year enhanced the fruit yield significantly under high density planting. PBZ advanced flowering, time of fruit harvest and increased flower and fruit densities, fruit set, crop density, yield and yield efficiency in 'J.H. Hale' and 'Red Skin' peach. However, fruit size was smaller than control (Arzani *et al.*, 2009). PBZ in combination with nitrogen significantly shortened the fruit maturity period and increased the yield efficiency of peach, whereas, paclobutrazol alone was effective in reducing vegetative growth (George and Nissen, 1992). Paclobutrazol effectively controlled the vegetative growth and improved the fruiting efficiency of grapevines. However the effectiveness was dependant on stage of the vine development as the application of paclobutrazol at bud bursting and two weeks before anthesis of grape increased the yield significantly (Christov *et al.*, 1995).

The efficacy of PBZ varied with the intensity of bloom. Paclobutrazol reduced flushing and increased flowering in moderately bloomed trees of *Litchi chinensis*. However the panicle development, fruit set and fruit quality of litchi were less influenced by PBZ (Faizan *et al.*, 2000). Paclobutrazol also initiated flower buds in satsuma mandarin and increased flower cluster and fruit numbers in young apple plants (Yamshita *el al.*, 1997). Paclobutrazol was also effective in enhancing fruit retention and yield in guava (Brar and Bal, 2011).

Enhanced dose of PBZ caused unacceptable panicle compaction in 4-5 year-old trees of mango cv. Kensington Pride whereas optimal dose significantly increased the flowering and yield (Winston, 1992). Application of PBZ @ 5 g a.i. around the main stem in 3 year old mango plants spaced at 4 x 4m accelerate the flower induction, panicle density and fruit number and weight, but the length and width of inflorescences was shorter than the control (Husen *et al.*, 2012). Application of PBZ in September at 1.0 g a.i/tree/meter canopy diameter of ten-year old Dashehari trees was highly effective in flowering, reducing vegetative growth, increasing fruit set and yield as compared to August and October applications (Karki and Dakal, 2003). PBZ usually takes at least 90 days to express its physiological impact in plants. Hence its application is recommended three about months before anticipated flowering.

Physico-chemical properties of fruits were by and large unaffected by the lower doses. Under tropical climate, application of paclobutrazol as soil drench for three consecutive years and then its discontinuation for the subsequent three years appears to be appropriate for mango (Reddy and Kurian, 2008). Soil application was more efficacious than foliar application in inducing flowering and fruiting in mango. The application of paclobutrazol as a soil drench around the tree trunk (collar drench) is the most effective method, as it ensures proper uptake. Moreover trees should be allowed to develop a good canopy before PBZ treatment. In high density planting, paclobutrazol should be applied early when trees are about three years old. However when trees are spaced farther apart, early application with paclobutrazol will reduce canopy size and the fruit bearing area. In such a situation, treatment can commence when trees develop good canopy. It has also been observed that large trees respond more slowly than young (Kulkarni *et al.*, 2006). The response to PBZ varied with cultivar and crop load. The effectiveness of PBZ in promoting flowering in *Citrus sp.* depends on the crop load as the heavy fruit load trees scarcely flowered. In medium to low fruit load trees PBZ significantly increased the percentage of sprouted buds and floral shoots and reduced the number of vegetative shoots. Variation in the advancement in flowering and harvesting with the time of application of PBZ is also common. PBZ had a significant effect on early flowering when combined with KNO_3 however efficacy varies with the growing conditions. There is a conflicting report on the impact of PBZ on fruit quality. Paclobutrazol had shown no improvement in fruit quality in grapes, strawberries, apple and peach, whereas the juice content, TSS and ascorbic acid of lemon were increased with the paclobutrazol application (Jain *et al.*, 2002).

2.2. Tree architecture

Canopy size is the most critical aspect of paclobutrazol application for yield sustainability. It is evident that apart from flower induction, paclobutrazol also restricts tree vigour hence trees should be allowed to develop a good canopy before treatment commences. The efficacy of paclobutrazol in regulating canopy size of mango has been reported by many workers. PBZ effectively regulated tree canopy and induced flowering in six year old pruned mango trees planted under high density system (2.5m x 2.5m). The canopy regulation was due to the shortening of third and fourth flushes and internodal lengths of the treated trees. The treated plants not only flowered more profusely but also were considerably earlier (Charnvichit *et al.*, 1994). However the treatment effect lasted for only one year. Pacloburazol was found effective in reducing tree vigour and in promoting flowering, fruit set and yield in mango. Paclobutrazol has also been reported to be effective in regulating vegetative growth of peach, apple, citrus and guava. However the efficacy of paclobutrazol in terms of shoot growth and production efficiency depends on the time of pruning.

2.3. Root activity and nutrient level

There is a significant increase in the root activity toward the trunk and close to soil surface and sparser root activity in the subsoil zone and in drip line area in paclobutrazol treated plants of mango (Kotur, 2006). Rieger and Scalabrelli (1990) reported that PBZ decreased N, P, K, Fe and Mo and increased the levels of Ca, Mg, B and Mn in peach. On the other hand, Werner (1993) observed an increase of N, Ca, Mn, Zn and B contents and decrease of P, K, and Cu contents in PBZ treated mango trees. On the contrary, Wieland and Wample (1985) observed no influence of PBZ on foliar content of N, P, K and Mg in apple trees. Soil application of paclobutrazol for two consecutive years in mango increased the levels of phosphorus, potassium and calcium at lower doses but decreased at higher dose. A similar trend was recorded in soil microbial level. The findings indicate inhibitory effect of paclobutrazol at higher concentration on soil nutrient status and microbial population (Singh *et al.*, 2005). PBZ also promotes the avoidance of salt stress in mango by increasing the levels of photosynthetic pigments, water content, K^+ uptake and accumulation, and uptake of harmful Na^+ and Cl^- ions (Kishor *et al.*, 2009). The influence of paclobutrazol on leaf nutrient content lacks consistency as it showed variation with the crop species and soil conditions. In addition to leaf nutrient content paclobutrazol also influences soil properties.

2.4. Metabolites and plant hormones

Paclobutrazol is basically a gibberellin inhibitor. A significant decline in the GA content was observed in the shoots of PBZ-treated mango plants. However, there was no difference in the level of GA between control and treated plants after one year (Protacio et al., 2000). There was an increase in cytokinin levels and total phenolic compounds in leaves of paclobutrazol treated mango trees. Starch and sucrose levels were increased by the paclobutrazol treatment. Paclobutrazol besides affecting gibberellins, also increases ABA and cytokinin contents concomitant with C: N ratio and leaf water potential in mango buds to elicit flowering responses. Singh and Sharma (2008) also recorded increase in C: N ratio, leaf *water potential, chlorophyll* content, total sugar, total protein, nitrate reductase activity, ABA *and cytokinins* in mango. Paclobutrazol retards cell elongation and internode extension by inhibiting gibberellins biosynthesis whereas the synthesis of abscisic acid, phytol and chlorophyll components increases with its application. PBZ also induces morphological modifications such as enhanced leaf specific weight, stomatal density, leaf thickness, root-to-shoot ratio and root density that strengthen stress tolerance capacity in plants.

3. Residual Effect and Risk Assessment

3.1. Residual effect on plant growth

Although paclobutrazol has been found efficacious in flower induction and canopy management, its residual aspect needs due attention. Concerns have been expressed over the use of paclobutrazol as it inhibits gibberellin bio-synthesis which is responsible for cell elongation and internode extension. The treated tree produces compressed panicle which does not dry out well and can develop powdery mildew or anthracnose even with the slight increase in the humidity. Furthermore, application of PBZ may cause severe stunting of plants in pruned plants. The stunting effect may sustain for years as paclobutrazol persists in the soil. The residual effect of PBZ was also observed in *Citrus sp.* on the subsequent growth of the scions with the application of PBZ (Hadlow and Allan, 1989), however lower dose had no residual effect on the growing scions. No residual or cumulative effect of PBZ was detected in avocado with regards to tree vigour, and yield (Adato, 1990). PBZ reduced the root hydraulic conductivity and altered the nutrient uptake in peach which appeared to be side-effects of PBZ treatment. Paclobutrazol causes morphological alteration in young roots of citrus and peach by increasing thickness and decreasing length. Such morphological alteration may influence water and nutrient uptake as these processes occur most actively in young roots Furthermore, PBZ treatment may reduce drought avoidance of under rainfed condition by decreasing the volume of rhizosphere.

3.2. Residual effect on fruits

Paclobutrazol has prolonged persistence in plant system due to its slow rate of metabolism. However it has differential translocation rate among different plant parts. PBZ does not translocate into the mango fruits, as assumed from its mode of translocation (Costa *et al.*, 2012). On the other hand, Srivastava and Ram (1999) confirmed the translocation of PBZ into fruits and seeds of mango though the concentration was below permissible limits. Similar findings were also obtained by Sharma and Awasthi (2005) who found residue of PBZ in unripe mango with the application of PBZ @ 5 to 10 g a.i. per tree for three consecutive years. Moreover, the level further decreased to 0.001 mg kg^{-1} at fruit maturity.

3.3. Residual effect in soil

Paclobutrazol is characterized by moderate potential of mobility in soil and water environments which enables it is applied in soil unlike other growth regulators, however its mobility varied with the soil type. Studies conducted in USA indicate that half-lives of paclobutrazol residues ranged from 450-950 days for orchard soils and 175-252 days in agricultural soils (Table 1) which indicates poor degradation rate of PBZ. Paclobutrazol showed low soil adsorption coefficient, however adsorption appeared to increase with soil organic matter and a decrease in soil pH. Studies conducted in USA revealed that less than 10 per cent of total PBZ applied were detected in soils between the depths of 60-120 cm, whereas the PBZ ketone metabolite was predominately detected in the subsurface soil layers though at insignificant levels. Sharma and Awasthe (2005) detected residues of paclobutrazol in the tree basin soil (0-15 cm) at the end of each season followed by a slight increase in the amount of residues with the year of applications. Reddy and Kurian (2008) also observed residual influence of PBZ in soil if applied continuously for three consecutive years and suggested discontinuation of application or to taper down its dose. Sharma *et al.*, (2008) could not detect paclobutrazol residues above quantifiable levels (0.01 ppm) either in tree basin surface soils or in the fruits even after more than five years continuous application. However, they further reported that the residues increased with the increase of the application rate Jaradrattanapaiboon *et al.* (2008) reported spatial difference of paclobutrazol residue in soils as they observed high concentration of PBZ residue in upper soil layer (0-5 cm) and low residue level in lower soil layer (10-20 cm). Ochoa *et al.*,(2009) expressed the possibility of environmental contamination with the regular application of paclobutrazol. The adsorption and leaching of the residues is dependent upon the soil physical and chemical characteristics as well as environmental factors such as rainfall.

3.4. Effect on microorganisms and earthworms

Paclobutrazol may remain active for many years in soils if applied directly, and can severely affect the growth and development of subsequent crops or even interact, in a harmful way, with soil microorganisms. Soil application of paclobutrazol in mango orchards, affected soil microbial community by reducing total number of bacteria, fungi and actinomycetes and additionally affected the dehydrogenase activity. However lower dose has neither affected soil microbial community nor dehydrogenase activity (Silva *et al.*, 2003). Studies on the impact of PBZ on the activity of earthworm (*Eisenia foetida*) showed that there was no death and behavioural abnormality in the earthworm with PBZ, however reduction in body weight was observed (MDAR, 2012). The residue of paclobutrazol influences the activity of microbes and earthworms which in turn affects soil fertility.

3.5. Residual effect on ground water

Much work has not been carried out on the role of PBZ in ground water contamination. Modelling study on ground water exposure to PBZ showed that the concentration of the degradate did not exceed 0.1 μg/L except in one of the six scenarios. The study concluded that the potential for the degradate hydroxyl triazole to reach groundwater at high concentrations is low (EFSA, 2010). Long term studies on the potential of paclobutrazol to impact groundwater from its use indicated low rate of ground water contamination. The maximum allowable concentration of PBZ in drinking water is 66.0 μg/L.

3.6. Residual effect on aquatic life

Paclobutrazol has been found hazardous to aquatic life. Significant increase in the enzymatic activity of lactate dehydrogenase (LDH) and glutathione S-transferase (GST) was observed in the liver of Brazilain fish (*Metynnis argenteus*) when fish was exposed to paclobutrazol contaminated water for 28 days. Moreover PBZ residues in fish increased at the end of the exposure period and suspected significant risk to humans consuming contaminated fish (Jonsson, 2002). He suggested that the measure of LDH and GST activities could be used as biomarkers of paclobutrazol exposure. Paclobutrazol has high octanol-water partitioning coefficient which indicates its potential to bioaccumulate in fish (EFSA, 2010). Therefore much caution has to be exercised to prevent contamination of the water bodies with paclobutrazol while using it for certain advantages in mango cultivation. To avoid soil and water contamination, trunk injection and bark application have been attempted). Bark application of paclobutrazol has been effective in sweet cherry to avoid contamination of soil (Jacyna and Dodds, 1999). Bark application of growth retardants was simple, efficient and environmentally safe, but not many tree species responded to this. Exploring the possibility of alternate methods of application of PBZ (trunk injection and bark application) could be a feasible approach in minimising the soil and water contamination.

3.7. Risk to human health

According to the "Globally Harmonized System of Classification and Labeling of Chemicals" (GHS), paclobutrazol has been classified under category 4 and considered moderately hazardous for human being (WHO, 2010). It can cause harmful effect if enters either through oral route (LD_{50} = 300 - 2000 mg/kg bw) or dermal route (LD_{50} = 1000 - 2000 mg/kg bw). Skin irritation studies indicated that PBZ is mildly irritating to skin and eye and is not a skin sensitizer. On the other hand, paclobutrazol was neither found genotoxic nor carcinogenic and developmental toxicant up to maternally toxic dose levels (USEPA, 2007). Based on *in vitro* and *in vivo* mutagenicity tests, it was concluded that PBZ is not genotoxic. There was no evidence of bioaccumulation and bio-retention of PBZ. It also has a low potential for volatilization with an estimated atmospheric half-life of less than 2 days. Therefore, long-range transport through the atmosphere is not expected (EPSA, 2010). PBZ is unlikely to volatilize to any significant extent owing to a low estimated vapour pressure. The possible route of ecological contamination of paclobutrazol has been schematically described (Figure 1). After entering soil system, paclobutrazol can have multi-directional movement; a) absorption by plants, b) soil adsorption, c) movement towards groundwater through percolation and leaching, d) movement towards water bodies through run off and sub-surface flow and e) vapourization. However, slow mobility and low vapour pressure limit the movement of residue of PBZ in groundwater and water bodies, and in the air. The residue of paclobutrazol may reach to the human being by following ways; a) through direct contact, b) through contaminated groundwater, c) through contaminated water bodies, d) by consuming contaminated fish, e) by consuming fruits containing residue and f) through inhalation and may affect human health. Findings clearly indicate that application of optimal dose of PBZ significantly minimises the chance of contamination but the residue may impose a risk to human health if applied in high dose. Moreover the chance of contamination of groundwater and water bodies can be aggrevated with the increase in the slope of the land and intensity of rainfall and irrigation (MDAR, 2012).

Fig. 1 Schematic presentation on the movement of paclobutrazol residue in the environment

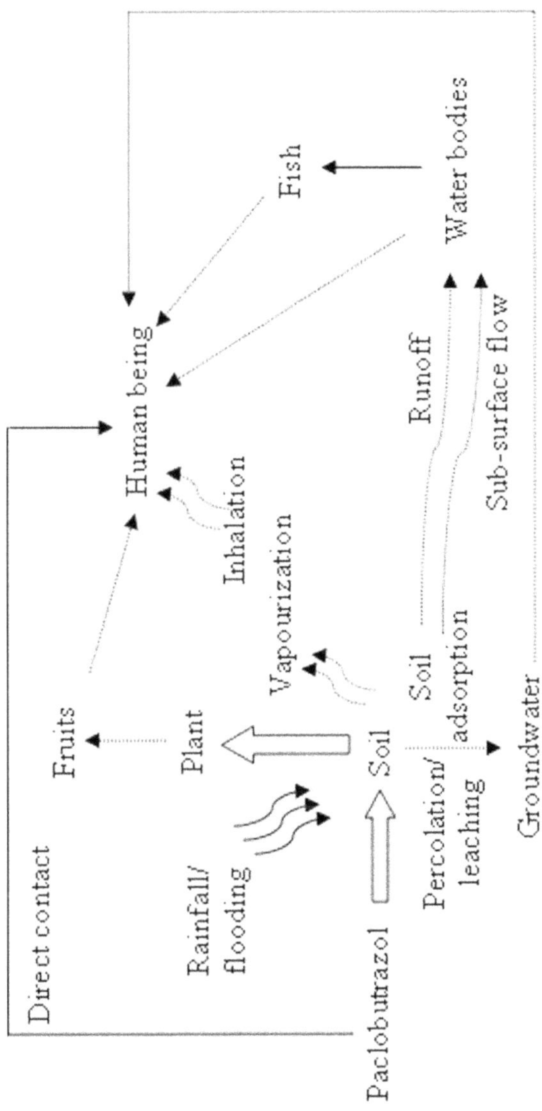

4. Conclusion

Paclobutrazol has been effective in terms of regularity and synchronization in flowering, yield enhancement and regulation of tree vigour. On the other hand, if used excessively, it causes stunting effect to plants and affects the root and microbial activities in soil. In spite of low mobility, paclobutrazol residue has been detected in soil and fruits, though below quantifiable levels in stray cases. It has low potential to contaminate, surface water and groundwater; however the risk to aquatic life is high and consequently significant risk is suspected if human consumes contaminated fish though paclobutrazol has been considered moderately hazardous for human being with remote chance of being genotoxic and carcinogenic. The important question is whether the continuous use of paclobutrazol is safe. Though paclobutrazol has remote chance of ecological contamination due to its low mobility and high persistence, the risk cannot be completely ruled out. In order to minimise residue threats substantially, it is advisable to use the optimal dose at the right time. Due care should be taken to avoid indiscriminate soil application especially in sloppy areas as treated soil is likely to be washed away by rainfall or irrigation runoff. All care should be taken in such areas to avoid erosion of contaminated soil to water bodies. Moreover terracing and bunds could be the effective approach in such locations. The chance of environmental contamination may also be minimized if soil properties, varietal response and climatic conditions are taken into consideration before extensive commercial adoption of PBZ. Alternate methods of paclobutrazol application such as bark application or trunk injection may also be attempted for their effectiveness.

References

- Adato I. 1990. Effects of paclobutrazol on avocado (*Persea americana* Mill.) cv. Fuerte. *Scientia Horticulturae,* **45**: 105-15.
- Adil O S, Rahim A, Elamin O M and Bangerth F K. 2011. Effects of paclobutrazol (PBZ) on floral induction and associated hormonal andmetabolic changes of beinnially bearing mango (*Mangifera indica* L.) cultizars during off year. *ARPN Journal of Agricultural and Biological Science,* **6**:55-67.
- Allan P, George A P, Nissen R J, Rasmussen T S and Morley-Bunker M J. 1993. Effects of paclobutrazol on phenological cycling of low-chill 'Flordaprince' peach in subtropical Australia. *Scientia Horticulturae,* **53**: 73-84.
- Arzani K, Bahadori F and Piri S. 2009. Paclobutrazol reduces vegetative growth and enhances flowering and fruiting of mature 'J.H. Hale' and 'Red Skin' peach trees. *Horticulture, Environment and Biotechnology,* **50**: 84-93.

- Brar J S and Bal J S. 2011. Effect of spacing and growth retardants on guava (*Psidium guajava* L.) plant vigour. *Journal of Research*, **48**: 34–41.
- Brar J S. 2010. Influence of paclobutrazol and ethephon on vegetative growth of guava (*Psidium guajava* L.) plants at different spacing. *Notulae Scientia Biologicae*, **2**: 110-3.
- Burondkar M M, Rajan S, Upreti K K, Reddy Y T N, Singh V K, Sabale S N, Naik M M, Ngade P M and Saxena P. 2013. Advancing Alphonso mango harvest season in lateritic rockysoils of Konkan region through manipulation in time of paclobutrazol application. *Journal of Applied Horticulture*, **15**: 178-82.
- Burondkar M N, Gunjate R T. 1993. Control of vegetative growth and induction of regular and early cropping in 'Alphonso' mango with paclobutrazol. *Acta Horticulturae*, **341**: 206-15.
- Charnvichit S, Tongumpai P, Saguansupyakorn C, Phavaphutanon L, Subhardrabandhu S. 1994. Effect of paclobutrazol on canopy size control and flowering of mango, cv. nam dok mai twai no. 4, after hard pruning. *Acta Horticulturae*, **291**: 60-6.
- Christov C, Tsvetkov I and Kovachev V. 1995. Use of paclobutrazol to control vegetative growth and improve fruiting efficiency of grapevines (*Vitis vinifera* L.). *Bulgarian Journal of Plant Pysiology*, **21**: 64–71.
- Costa M A, Torres N H, Vilca F Z, Nazato C and Tornisielo V L. 2012. Residue of ^{14}C paclobutrazol in mango. *IOSR Journal of Engineering*, **2**: 1165-7.
- Davenport T L. 2000. Processes influencing floral initiation and bloom: The role of phytohormones in a conceptual flowering model. *Hort. Technology*, **10**: 733–9.
- Davenport T L. 2007. Reproductive physiology of mango. *Brazilian Journal of Plant Physiology*, **19**: 363-6.
- Davis T and Curry E. 1991. Chemical regulation of vegetative growth. *Critical Review in Plant Science*, **10** : 204–16
- EFSA. 2006. Paclobutrazol Draft Assessment Report (DAR). Report and proposed decision of the United Kingdom made to the European Commission.Extension document FNR-252-W, European Food Safety Authority.
- EFSA. 2010. Conclusion on the peer review of the pesticide risk assessment of the active substance paclobutrazol. European Food Safety Authority (EFSA), Parma, Italy.

- Faizan A, Mohammad A and Ganesh K. 2000.Effect of paclobutrazol on growth, yield and quality of litchi (*Litchi chinensis*). *Indian Journal of Horticulture,* **57**: 291–4.
- George A P and Nissen R J. 1992. Effects of water stress, nitrogen and paclobutrazol on flowering, yield and fruit quality of the low-chill peach cultivar, 'Flordaprince'. *Scientia Horticulturae*, **49**: 197-209.
- Hadlow A.P. and Allan P. 1989. Effect of paclobutrazol on vegeiative growth in citrus nursery trees. *South African Journal of Plant and Soil,* **6**: 50-2.
- Hamid M M and Williams R R. 1997. Translocation of paclobutrazol and giberellic acid in Sturt's desert pea (*Swainsonia formosa*). *Plant Growth Regulation,* **23**:167-71.
- Hoffmann G. 1992. Use of plant growth regulators in arable crops: Survey and outlook, (*In*) *Proceedings 14th International Conference on Plant Growth Substances*, pp. 798-808. Karssen C M, Van L C and Vreugdenhil D. (Eds.). Kluwer Acad. Publ., London, UK
- Husen S, Kuswanto S A and Basuki N. 2012. Induction of flowering and yield of mango hybrids using paclobutrazol. *Journal of Agriculture and Food Technology,* **2**:153-158.
- Jackson M J, Line M A and, Hasan O. 1996. Microbial degradation of a recalcitrant plant growth retardant-paclobutrazol (PP333). *Soil Biology and Biochemistry,* **28**: 1265–7.
- Jacyna T and Dodds K G. 1999. Effect of method of application of paclobutrazol in high density cherry orchards on tree performance and apparent soil residue. *Jounal of Horticultural Science and Biotechnology,* **74**: 213-8.
- Jain S K, Singh R and Mishra K K. 2002. Effect of paclobutrazol on growth, yield and fruit quality of lemon *(Citrus limon). Indian Journal of Agricultural Sciences,* **72**: 488-9.
- Jaradrattanapaiboon A, Sruamsiri P, Kathrin R, Martin L, Jons B, Neidhart S and Carle R. 2008. Method Validation and Investigation of Paclobutrazol in Soil Using SPME-GC-MS Technique. *CMU Journal of Natural Science,* **7**: 279-93.
- Jonsson C M, Feracini V L, Paraiba L C, Rangel M and Aguir S R. 2002. Biochemical changes and accumulation in pacu fish (M*etynnis argenteus*) exposed to paclobutrazol. *Scientia Agricola,* **59**: 441-446.
- Karki K B and Dhakal D D. 2003. Effect of paclobutrazol on off-year induction of flowers in mango. *Journal of the Institute of Agriculture and Animal Science,* **24**: 51–7.
- Kishor A, Srivastav M, Dubey A K, Singh A K, Sairam R K., Pandey R N. Dahuja

A and Sharma R R. 2009. Paclobutrazol minimises the effects of salt stress in mango (*Mangifera indica* L.). *Journal of Horticultural Science & Biotechnology*, **84**: 459–65.

- Kotur S C. 2006. Effect of paclobutrazol on root activity of mango (*Mangifera indica*). *Indian Journal of Agricultural Sciences*, **76**:143-4.
- Kulkarni V J, Hamilthon D and Mc Mahon G. 2006. Flowering and fruiting in mangoes in top end with paclobutrazol. Crops, Forestry and Horticulture, Darwin, Northern Territory Govt. p. 1-3.
- Kurian R M, Singh H S and Kishore K. 2013. Canopy management in mango is a challenge in some parts of Odisha-A preliminary survey report. (*In*) *Canopy management and high density planting in subtropical fruit crops*. pp 148-153. Singh V K and Ravishankar H (Eds). CISH, Lucknow.
- MDAR. 2012. Paclobutrazol. Massachusetts Department of Agricultural Protection Boston.
- Nartvaranant P, Subhadrabandhu S and Tongumpai P. 2000. Practical aspects in producing off-season mango in Thailand. *Acta Horticulturae*, **509**: 661-668.
- Ochoa J, Franco A, Banon S and Fernández J A. 2009. Distribution in plant, substrate and leachate of paclobutrazol following application to containerized *Nerium oleander* L. Seedlings. *Spanish Journal of Agricultural Research*, **7**: 621-8.
- Protacio C M, Bugante R D, Quinto J, Molinyawe G and Paelmo G. 2000. Regulation of flowering in Carabao mango tree by paclobutrazol. *Philippine Journal of Crop Science*, **25**: 27-33.
- Rademacher W. 2000. Growth Retardants: Effects on gibberellin biosynthesis and other metabolic pathways. *Annual Review of Plant Physiology and Plant Molecular Biology*, **51**: 501–31.
- Reddy Y T N and Kurian R M. 2008. Cumulative and residual effects of paclobutrazol on growth, yield and fruit quality of 'Alphonso' mango. *Journal of Horticultural Science*, **3**: 119-22.
- Rieger M and Scalabrelli G. 1990. Paclobutrazol, root growth, hydraulic conductivity, and nutrient uptake of'Nemaguard' peach. *Hort. Science*, **25**: 95-8.
- Sharma D and Awasthi M D. 2005. Uptake of soil paclobutrazol in mango (*Mangifera indica* L.) and its persistence in fruit and soil. *Chemosphere*, **60**: 164-9.

- Sharma D, Shalini L and Raja M E. 2008. Uptake and persistence of soil applied paclobutrazol in 'Alphonso' mango and soil in the Konkan region of Maharashtra, India. *Toxicology and Environmental Chemistry,* **90**: 577-83.
- Silva C M M S, Vieira R F and Nicolella G. 2003. Paclobutrazol effects on soil microorganisms. *Applied Soil Ecology*, **22**: 79-86.
- Singh D K. 2000. Effect of paclobutrazol on yield and quality of different cultivars of ber *(Zizyphus mauritiana). Indian Journal of Agricultural Sciences,* **70**: 20-2.
- Singh V K and Sharma K. 2008. Physiological and biochemical changes during flowering of mango (*Mangifera indica*). *International Journal of Plant Developmental Biology,* **2**: 100-5.
- Singh V K, Garg N and Bhirigwanshi S R. 2005. Effect of paclobutrazol doses on nutritional and microbiologieal properties of mango *(Mangifera indica)* orchard soils. *Indian Journal of Agricultural Sciences,* **75**: 738-9.
- Singh Z. 2000. Effect of (2RS, 3RS) paclobutrazol on tree vigour, flowering, fruit set and yield in mango. *Acta Horticulturae,* **525**: 459-62.
- Srivastava M. and Ram S. 1999. Paclobutrazol residues in the fruits of mango cultivars. *Journal of Applied Horticulture,* **1:** *27-28.*
- Upreti K K, Reddy, Y T N, Shivu Prasad S R, Bindu G V, Jayaram H L and Rajan S. 2013. Hormonal changes in responseto paclobutrazol induced early flowering in mango cv. Totapuri. *Scientia Horticulturae,* **150**: 414-418.
- USEPA. 2007. *Paclobutrazol summary document: Registration review.* Docket number EPA-HQ-EPA-2006-0109. United States Environmental Protection Agency, Washington DC.
- Werner H. 1993. Influence of Paclobutrazol on Growth and Leaf Nutrient Content of Mango (cv. Blanco). *Acta Horticulturae,* **341**: 225-231.
- WHO. 2010. The WHO recommended classification of pesticides by hazard and guidelines to classification: 2009. Geneva
- Wichard M. 1997. Paclobutrazol Is Phloem Mobile in Castor Oil Plant (*Ricinus communis* L.) *Journal of Plant Growth Regulation*, **16**: 215-17.
- Wieland W F and Wample R L. 1985. Root growth, water relations and mineral uptake of young 'Delicious' apples. *Scientia Horticulturae,* **26**: 139-47.

- Winston E C. 1992. Evaluation of paclobutrazol on growth, flowering and yield of mango cv. Kensington Pride. *Australian Journal of Experimental Agriculture*, **32**: 97–104.
- Yamashita, K.; Kitazo, K. and Iwasaki, S. 1997. Flower bud differentiation of Satsuma mandarin as promoted by soil drenching treatment with IAA, BA and paclobutrazol solution. *Journal of the Japanese Society for Horticultural Science*, **66**: 67-76
- Yeshitela T, Robbertse P J and Stassen P J C. 2004. Effects of various inductive periods and chemicals on flowering and vegetative growth of 'Tommy Atkins' and 'Keitt' mango (*Mangifera indica*) cultivars. *New Zealand Journal of Crop and Horticultural Science*, **32**: 209–15.

4

Plant Growth Regulators in Aonla

**Sanjay Singh, D.S. Mishra, A. K. Singh,
Ankit Kumar Singh, Shivraj Verma and Pragnyashree Mishra**

1. Introduction

Aonla (*Emblica officinalis* Gaertn), the Indian gooseberry has emerged as one of the important minor fruits of India owing to its nutritive, medicinal values and suitability for processing of diversified value added products (Pathak, 2003). It is the rich source of vitamin 'C' (500-600 mg/ 100 gm) among fruits and ranks second after Barbados cherry (*Malpighia glabra* L.) (Asengo, 1953). It has oxidation of which is retarded by the presence of leucoanthocanins in the fruit itself. Fruits are rich source of minerals like iron, calcium and phosphorous and dietary fiber. The demand for aonla fruits has been progressively increased due to extensively utilized by processing industry for making trifla and chyavanprash. Other preserve products are, pickle, candy, jelly, jam and squash. Fruits can also be dried and powered to be used subsequently in the preparation of inks, hair dye and hair oils (Tiwari *et al.*, 2008). It posses diverse medicinal (useful in haemorrhages, diarrhoea, dysentery, anaemia, jaundice, dyspepsia and cough) and industrial (Trifla and chavanprash) uses. Besides fruits, leaves, bark and even seeds are being used for various purposes.

The use of PGRs in modern fruit culture is well established, more particularly in aonla industrially where its application has been used for breaking seed dormancy to rooting of cuttings and micro-propagation. In addition to this, PGRs have been advantageously used for improving flowering, fruit setting, shelf-life and reducing fruit drop in aonla.

The major areas where PGRs are used in aonla industry are:

- Plant propagation especially breaking seed dormancy and enhanced seedling growth
- Micro-propagation and embryogenesis
- Improvement in flowering and fruit set
- Fruit drop control
- Improvement of fruit yield and quality
- Extending the shelf life of fruits

Plant growth regulators in very minute quantities can induce above mentioned responses dramatically. The response to PGR's however varies with the cultivar, age of the plant, light, temperature, availability of mineral nutrients, vigour of the plant and its endogenous hormonal content. Use of PGRs on above mentioned responses in aonla production is mentioned below:

2. Use of PGRs in Propagation

2.1. Propagation by seed and use of PGRs

In recent years asexual methods of propagation such as wedge grafting and patch budding have been found to be quite successful and their use has been advocated. However, even for grafting and budding seedlings have to be raised. Seedling trees bear fruits of variable size and quality but such trees are generally long-lived. For raising rootstocks, seeds are used to raise seedlings on which the desired variety or scion are grafted. Most of the fruit crop seeds as is the case of aonla also, germinate poorly and unevenly and require more time for seedling emergence. The dormancy in seeds might be due to hard seed coat, impermeability to water and gases, physiological immaturity of embryo, deficiency of some endogenous growth promoters or excess of endogenous growth inhibitors. Freshly harvested seeds of aonla do not germinate even if exposed to favourable conditions of germination owing to seed dormancy (Srimathi *et al.*, 2000).

The fruit of aonla is fleshy and drupaceous and the seeds are found within the hardened endocarp of the fruit known as stone. The seeds do not germinate easily owing to seed coat related dormancy (Mawalagedera *et al.*, 2014) and the germination rate is also very poor, necessitating pretreatments before sowing, however, the seeds can remain viable for a longer period under natural conditions (Pushpakumara *et al.*, 2007). It has been observed that the concentration of auxin increased in the fruit with the onset of dormancy and decreased to a low level prior to dormancy break and exposure to GA3 500 ppm for 24hrs could have counter downed the auxin concentration. Besides, the impermeable layer in seed coat allowed water and oxygen to enter the seed and permitted the embryo to overcome the mechanical restriction of surrounding tissue by providing uniform germination when immersed in tap water for 24hrs.

Different methods like water soaking, scarification and PGR treatments are used for breaking dormancy in such seeds to improve germination (Kumar *et al.*, 2012). The some plant growth regulators have been helpful in germination of aonla seeds by increasing water uptake and exerting an effect on membrane permeability. These results indicate that use of plant growth regulators might have helped to break the embryo dormancy and induction of synthesis of alpha amylase and other hydrolytic enzymes (Looney, 1983). The exogenous application of Gibberlic acid antagonizes the ill effect of inhibitors and increases endogenous gibberellin like substances. GA_3 helps in the synthesis of enzymes and one of them is α-amylase which converts the starch into simple sugars during the process of germination. These sugars provide energy that is required for various metabolic and physiological processes associated with germination.

Pre-seed germination treatments to overcome dormancy, different treatments viz., tap water for 24 hours, thiourea (2%) for 24 hours, GA3 500 ppm for 12 hours, GA3 500 ppm for 24 hours, stratification at 5°C for 10 days, acid (conc. HCl) scarification for 30 seconds were imposed on aonla seeds. Bhujbal (1975) reported highest germination per cent (92.50%) with minimum period when dried stones of aonla were treated with 500 ppm GA_3

Dhankhar and Singh (1996) studied the percentage of germination and subsequent growth parameters of aonla. Days to germination were least with 250 ppm GA3, 250 or 500 ppm thiourea, and distilled water. GA3 at 250 ppm gave the highest percentage germination (75.98 and 64.14% in the laboratory and pot study, respectively). Significant effects over other treatments on plumule and radicle length at 35 days after sowing (laboratory study), and on seedling height (28.84 cm), seedling girth (0.90 cm), and seedling FW and DW (1.205 and 0.260 g) at 75 days after sowing in the pot with GA3 at 250 ppm. Maximum number of roots at 35 days after sowing was attained with thiourea at 750 ppm.

Pawshe *et al.*, (1997) observed the effect of pre germination seed treatments on the germination and vigour of aonla seeds. Treatments included gibberellic acid (GA3 at 50 and 100 ppm, soaking in water for 24 hr and hot water soaking at 60 degrees C for 5 minutes. GA3 at 50 and 100 ppm increased the percentage seed germination. The tallest plants were obtained following seed treatment with 100 ppm GA3 and soaking for 24 hr.

Wagh *et al.,* (1998) observed the germination of seeds of *phyllanthus emblica* soaked for 12 hr in 100-400 ppm gibberellic acid (GA3) or water, or not soaked (controls). Treatment with 400 ppm GA3 resulted in the highest percentage of germination (87.25%). Seedling development (plant height, number of leaves/plant and root development) was also best with seed treatment with 400 ppm GA3.

Gholap *et al.,* (2000) also observed better germination and seedling growth with GA3 200 ppm in aonla. Such response is obvious as GA enhances cell elongation, so the radical can push through the endosperm and seed coat that restrict its growth (Hartman and Kester, 1979).

Rajamanickam and Anbu (2001) evaluated the effect of biofertilizers, chemicals and growth regulators on germination and seedling growth of aonla. Treatment of fresh seeds of aonla with Azospirillum + Phosphobacteria+0.5% KNO3 for 8 hr recorded the highest germination percentage (52.08%), and one-year-old seeds treated with Azospirillum+Phosphobacteria+200 ppm GA3 for 8 hr showed higher germination (49.17%), which was significantly superior than all other treatments. The same treatment was found to induce higher shoot length, root length, dry matter production and vigour index. In the case of time taken to reach buddable thickness, fresh seeds treated with Azospirillum+Phosphobacteria+0.5% KNO3 for 8 hr took lesser time compared to other treatments. One–year-old seeds treated with Azospirillum+Phosphobacteria + 200ppm GA3 for 8 hr took lower days compared to other treatments.

Rajamanickam *et al.,*(2002) studied the seed germination of aonla cv. BSR-1. Earliness in seed germination and germination percentage of Aonla recorded in fresh (6.30 days and 69.33%) and 1-year-old seeds (6.64 days and 46.00%) soaked in 0.5% KNO3 for 8 hours and in 200 ppm GA3 for 8 hours, respectively.

Chemical treatment of seed exposed to $GA_3$500 ppm for 24hous resulted better germination than other treatments. Imbibitions percent increased in treated seeds upto 90 % in contrast to 70% in non-treated seeds (control). Treatment of seeds with GA_3 500 ppm and thiourea were effective in breaking seed coat dormancy (Laishram and Sahoo, 2015).

Chiranjeevi *et al.,* (2017) evaluated the influence of growth regulators on germination, seedling growth and vigour attributes of aonla. The growth regulators like GA3 and NAA were used as treatments at different concentrations for 12 hours. The seeds pre-soaked with GA3 200 ppm solution recorded the earliest germination (8.33 days), highest germination percentage (88.88 %), faster rate of germination (1.07), maximum seedling height (28.47 cm), seedling stem girth (1.26 cm), seedling biomass (2.28 g) and highest vigour index of seedlings (202.6) compare to other treatments.

2.2. Use of PGRs in Micropropagation

Micro-propagation is the technique of *in vitro* multiplication of large number of plants from its any part (leaves, seeds, nodes etc). The concept of totipotency is the basis for micropropagation. Micro-propagation is used for the production and multiplication of large number of novel plants, which are genetically similar and disease free. Usually, *in vitro* propagation is carried out in two ways: direct and indirect. Callus production from explant followed by induction of shoots and roots is one method, while direct shooting on the axillary explant like nodes, shoot tips followed by rooting is another method. Micro-propagation is the application of tissue culture for efficient clonal plant production has been used commercially since the 1960s and is possibly the oldest example of commercial Biotechnology (Rai, 2010). The technique of *in vitro* propagation is applied with the objective of enhancing the rate of multiplication. Through tissue culture over a million plants can be grown from small piece of plant tissue. Plants can be multiplied throughout the year irrespective of the season. Using this method stock of germplasm can be maintained for many years. It is a matter of the fact that through *in vitro* methods pathogen free plants can be raised and maintained economically. However, there are few reports on the micropropagation of aonla (Mishra *et al.*, 1999, Verma and Kant, 1996 and Goyal and Bhadauria, 2008). In micro-propagation studies, PGRs are used in addition to the supply of other salts, mineral, vitamins and sugars to get the desirable response to achieve the set of objectives. Auxins are a class of PGRs which cause cell elongation, apical dominance and root initiation. The most frequently used auxins are 2, 4-D, NAA, IAA and IBA of which IAA occurs naturally in plants. NAA and 2, 4-D are the most effective auxins to initiate callus. Cytokinins are phytohormones which promote cell division, proliferation of tissues. The most widely used cytokinins are kinetin, BAP and 2-IP. Auxins or cytokinins alone cannot show organogenesis through cell division or callus *in vitro*. The relative concentration of auxin and cytokinin decides the plant morphogenesis *in vitro* (Skoog and Miller, 1957).

Verma and Kant (1996) cultured nodal segment explants of aonla on modified MS medium supplemented with BA (3-5 mg/l) and NAA (0.5 mg/l) during February-April and August-October. The bud break observed in only 8-10 per cent of explants. Leaching of phenolic compounds resulted in the loss of almost 90 per cent in cultures. The regenerated shoots were elongated on hormone free medium and subsequently rooted on half strength MS medium supplemented with IBA (3.0 mg/l) and sucrose (1.5%). Again Kant *et al.* ,

(1999) reported 3-4 multiple shoots from nodal explants from mature tree on MS medium supplemented with BAP (5 mg/l) and IAA (0.5 mg/l) along with antioxidants. These shoots were elongated on hormone free MS medium. Rooting of the shoots was achieved on half strength MS medium fortified with IBA (2 mg/l) and sucrose (1.5%).

Micropropagation studies in aonla were carried out from nodal explants by Mishra *et al*, (1999) to develop the protocol for mass multiplication of true-to-type plants. The better shoot proliferation (33-37.5%) on modified MS medium supplemented with kinetin 0.4 mg/l + GA 1.0 mg/l than in WPM. Higher GA3 concentration (3 mg/l) caused complete defoliation and dropping of determinate shoots. Regenerated shoots from 3 week old cultures in MS media supplemented with growth regulators and antioxidants failed to produce roots.

Kadam *et al.,* (2006) undertook a study to come out with optimal culture conditions for high frequency plant regeneration from nodal segment along with axillary bud and shoot tip explants of the aonla genotype Krishna. Seeds of the genotype Krishna were raised *in vivo* in sterilized soil rite and supplemented with MS basal liquid medium. Shoot tip and Nodal segment explants were collected from these seedlings and cultured aseptically in MS as well as WPM medium fortified with growth regulators viz., BAP, NAA, GA3 and TDZ. The nodal segment explants showed better response than shoot tip for micropropagation of aonla. The maximum number of multiple shoots per culture (13.67) were recorded by nodal segments cultured in MS + BAP 0.25 mg/l +NAA 0.1 mg/l + GA3 2 mg/l. However, normal rhizogenesis could not be achieved with supplement of IBA and NAA to half MS media and abnormal root induction was noticed which failed to establish during subsequent hardening.

3. Use of PGRs on Fruit Retention and Fruit Yield

In recent past, a set of factors including feasibility of commercial aonla cultivation in marginal lands, availability of improved varieties and huge possibilities for the value-addition of fruits have enabled rapid coverage of vast area under aonla cultivation in many parts of India (Pathak, 2003). However, fruit bearing in aonla trees often suffer from heavy fruit drop which significantly lowers down the yield. Fruit drop often causes poor fruit set in majority of the cultivars and may result in substantial crop loss in susceptible ones. Among the major causes noted for fruit drop are self-incompatibility, inadequate pollination, nutritional deficiency, water stress, insect-pest and disease infestations and hormonal imbalances (Allemullah and Ram, 1990 and Singh *et al.*, 2008). Amongst the reasons, suboptimal biosynthesis and poor translocation of phytohormones may be a major cause of fruit drop (Ram and Rao, 1981).

Good fruit set in aonla is prevented by adverse weather which hinders pollen production, pollination and fertilization and also due to low level of auxin. The auxin from the pollen grain and pollen tube might be responsible for the early stage of fruit growth. However, small amount of pollen in the grain may not carry enough auxin to account for early fruit development. The growing pollen tube may secrete auxin which helps in fruit growth (Muir, 1942). In case of aonl, sufficient auxin may not be available during fruit growth which resulted in fruit drop and low yield. To check the pre-mature fruit drop in aonla and to improve the yield and fruit quality, various investigations were carried out with the application of plant growth regulators.

Results of two years of investigation undertaken by Ghosh *et al.* (2009) revealed that spray of NAA at 10 ppm was the best to increase fruit retention, followed by NAA 20 ppm, Vermiwash and Borax, which consequently resulted in the highest fruit yield of 54.9, 52.0, 46.8 and 36.2 kg/plant, respectively, against 13.8 kg in the control. Beneficial effect of NAA application in reducing fruit drop may be explained from the fact that it maintains the on-going physiological and biochemical process of inhibition of abscission (Tomaszewska and Tomaszewska, 1970). Fruit weight was the maximum with 0.5% ZnSO4 spray, followed by NAA 10 ppm. Fruit quality with regard to TSS, total sugar and ascorbic acid content was better in all the treated fruits compared to control.

Singh and Singh (2015) also reported that application of PGRs on bearing trees of aonla cultivar Narendra Aonla-6 significantly increased fruit retention. The minimum fruit drop was recorded with the use of NAA (15 ppm) + Thiourea (0.1%) followed by 2,4-D (10 ppm) + Thiourea (0.1%) and GA3 (50 ppm) + Thiourea (0.1%). The maximum increase in the size, weight and volume of fruits and the maximum fruit yield were noted on trees sprayed with NAA (15 ppm) + Thiourea (0.1%). The improvement in fruit quality parameters (TSS, ascorbic acid, sugars and acidity) was highest with the application of GA3 (50 ppm) + Thiourea (0.1%) which showed non-significant difference with NAA (15 ppm) + Thiourea (0.1%). Results revealed that foliar application of NAA (15 ppm) + Thiourea (0.1%) or GA3 (50 ppm) + Thiourea (0.1%), twice during mid-May and mid-July, may effectively overcome the problem of fruit drop leading to higher yield of quality fruits.

4. Role of PGRs in Improving Fruit Quality

Patel (2017) observed the total soluble solids was found significantly higher in treatment with NAA 40 mg/l + $GA_3$50 mg/l. Significantly the minimum total soluble solids was found in treatment Control (Water spray). However, The total soluble sugars was found significantly higher in treatment NAA 40 mg/l + $GA_3$50 mg/l, which was followed by NAA 40 mg/l, NAA 20 mg/l + $GA_3$25 mg/l, and NAA 20 mg.

Patel (2017) concluded from the experiment that the plant growth regulators are known to increase in total soluble solid in treated fruits which might be due to rapid transformation of complex carbohydrates into soluble sugars and also fast mobilization of metabolites from source to sink. The growth substances like GA_3 which promote quick metabolic transformation of starch and pectin into soluble sugars and rapid mobilization of photosynthetic metabolites and minerals from other parts of the plant. Similarly NAA increase the TSS which might be due to synthesis of auxin in plant, it increase the physiological activities leading to increased TSS in fruits, while total sugars increase might be due to activation of enzymes which affect the physiological processes, which in turn hydrolyzed the starch and helps in metabolic activity during the change in available starch into sugar and soluble solid content. NAA had shown significant increase in the total sugar of aonla fruits and this might be due to synthesis of auxin in plant, it increased the physiological activities. Content of acidity can also be increased by applying plant growth regulator that might be due to catalytic influence of growth regulators on its bio-synthesis from its precursor glucose-6-phosphates throughout the development of fruits which is thought to be precursor of vitamin- C. The application of gibberellic acid, NAA may have favorably influenced the metabolic activities possibly due to their increased endogenous level which increased the ascorbic acid of aonla fruit. The shelf life of aonla fruits can be increased that might due to antagonistic effect of GA3 which inhibit ethylene production and delayed the conversion of starch to sugar (Patel, 2017).

5. Role of PGRs on Shelf Life of Aonla

Due to perishable nature of aonla fruits, it becomes difficult to store or transport aonla over a long distance. Shelf life of aonla can be extended by checking respiration, transpiration and microbial infection. The different plant growth regulators exhibited significant influence on physiological loss in weight of aonla fruits. Among the different treatments, *i.e.* NAA 40 mg/l+ GA3 50 mg/l at pin head and pea stage was found significantly superior over the rest of the treatments and lowest physiological loss in weight after 12th and 24th day storage period for anola fruit. Similar trend was also recorded after 24th day storage period. The reduction of weight loss in the fruits treated with GA3 might be due to its anti-senescence action. The plant growth regulators and boron treatment resulted in to decrease in the tissue permeability and thereby reduces the rate of water loss. The reduced PLW in fruits during storage is mainly due to reduced rate of transpiration and respiration.

An experiment was conducted during 2001-2002 to see the effect of different post-harvest treatments on storability of Indian gooseberry (*Emblica officinallis* Gaertn) during storage at ambient temperature. Increase in physiological loss in weight, spoilage percentage, total soluble solids, total sugar and reducing sugar and decrease in acidity, ascorbic acid with advancement of storage period were general phenomena in all the treatments. Fruits treated with calcium nitrate 1.5 % + perforated polyethylene bag and GA_3 100 ppm + perforated polyethylene bag recorded the least physiological loss in weight (2.12-16.00 and 2.15-16.34%) and spoilage loss (2.40-15.00% and 2.50-15.60 %) and exhibited 11 days of storage life. The same treatments also showed lowest respiratory activity (72.10-82.00 mg CO 2/kg/hr and 72.00-82.10 mg CO 2/ kg/hr), on the last day of storage (day 13) (Singh *et al*, 2005). Fruits were dipped in GA for 10 minutes.

References

- Allemullah, M and Ram, S. 1990. Causes of low fruit set and heavy fruit drop in Indian gooseberry (*Emblica officinalis* Gaertn). *Indian Journal of Horticulture,* **47**: 270-77.
- Anonymous, 2013. Indian Horticulture Database, National Horticulture Board, Ministry of Agriculture, Government of India.
- Asengo, C. F. 1953. The west Indian Cherry, Bull. Del collegio, de quimicosde, Puret Rico. 10: 8-11.
- Bhujbal, B.G., 1975. Improvement of seed propagation of aonla. *Journal of Mahatma Phule Agricultural University,* **6**: 73-75.
- Chiranjeevi, M.R., Muralidhara,B.M., Sneha, M.K., Shivan and Hongal 2017. Effect of growth regulators and biofertilizers on germination and seedling growth of aonla (*Emblica officinalis* Gaertn). *Int. J. Curr. Microbiol. App. Sci.,* **6**(12): 1320-1326
- Dhankhar, G.S. and Singh, M., 1996. Seed germination and seedling growth in aonla (Phyllanthus emblica L.) as influenced by GA_3 and Thiourea. *Crop Research,* **12**(3): 363-366.
- Gholap, S.V., Dod, V.N., Bhuyar, S.A. and Bharad, S.G. 2000. Effect of plant growth regulators on seed germination and seedling growth in aonla (*Phyllanthus emblica* L.) under climatic condition of Akola. *Crop Research,* **20**(3): 546-548.
- Ghosh, S.N., Bera, B., Roy, S., Kundu, A.and Roy, S.K.D. 2009. Effect of nutrients and plant growth regulators on fruit retention, yield and physicochemical characteristics in aonla cv. NA-10. *J. Hortl. Sci.,* **4** (2): 164-166.

- Goyal, D. and Bhadauria, S. 2008. *In vitro* shoot proliferation in *Emblica officinalis* var. Balwant from nodal explants. *Indian J. Biotechnol.*, **7**:394-397.
- Hartmann, H.T. and Kester, D.E. 1979. Plant Propagation: Principles and Practices. Fourth Edition, Prentice Hall of India, Ltd., p.407.
- Kadam, J.H., Karale, A.R. and Patil, S.C. 2006. In Vitro micropropagation of *Emblica officinalis* Gaertn. cv. Krishna. *Internat. J. Agric. Sci.*, **2**(2): 610-612.
- Kant, U., Vemra, B. and Kishor, P.B.K., 1999. Propagation of *Emblica officinalis* Gaertn. through tissues culture. *Proceedings of Symposium on Plant Tissue Culture and Biotechnology*: Emerging Trends, held at Hyderabad 29-31st January 1999, pp.106-111.
- Kumar, R., Misra, K.K., Mishra, D.S. and Brijwal, M. 2012. Seed germination of fruit crops: a review. *Hort. Flora Research Spectrum,* **1**(3): 199-207.
- Laishram Lilabati and U.K.Sahoo 2015. Effect of pre-treatments on seed germination and seedling vigor of *emblica officinalis* gaertn. *Global Journal of Advance Research*, **2**(10): 1570-1576.
- Looney, N.E. 1983. Plant Growth Regulators Chemicals, CRC Press Inc. Boco.Raton., USA. 1-26.
- Mawalagedera, S.M.U.P., Perera, G.A.D. and Sooriyapathirana, S.D.S.S. 2014. Prolonged seed dormancy in *Phyllanthus emblica* L. can be overturned by seed scarification and gibberellins pretreatment. *Open Journal of Forestry,* **4**(1): 38-41.
- Mishra, M., Saxena, R.P., Pathak, R.K. and Srivastava, A.K. 1998. Studies on micropropagation of Aonla (Emblica officinalis Gaertn). *Prog. Hort.,* **31**(3-4):116-122.
- Mishra, M., Saxena, R.P., Srivastava, A.K. and Pathak, R.K., 1999a, Micropropagational studies in aonla (*Emblica officinalis* Gaertn.). *Annual Plant Physiology*, **13** (2): 11-117.
- Muir, R. M. 1942. Growth hormone as related to the setting and development of fruit in Nicotiana tabacum. *Am. J. Bot.*, **29**: 716-720.
- Patel, J. S. 2017. Influence of plant growth regulators and boron on nutritional quality and shelf life of aonla fruit. *Int.J.Curr.Microbiol.App.Sci.,* **6**(4): 2533-2540.
- Pathak, R.K. 2003. Status Report on Genetic Resources of Indian Gooseberry-Aonla (Emblica officinalis Gaertn.) in South and Southeast Asia. IPGRI Office for South Asia National Agriculture Science Centre (NASC) DPS Marg, Pusa Campus, New Delhi 110 012, India, pp. 1-96.

- Pawshe, Y.H, Patil, B. N. and Patil, N.P 1997. Effect of pre germination seed treatments on germination and vigour of seedlings in Aonla (*Emblica officinalis* Gaertn.). *PKV Research Journal*, **21**(2): 152-154.

- Pushpakumara, D. K. N. G., and Heenkenda, H. M. S. 2007. Nelli (Amla) Phyllanthus emblica L. In: D. K. N. G. Pushpakumara, H. P. M. Gunasena, and V. P. Singh (Eds.), Underutilized fruit trees in Sri Lanka. World Agroforestry Centre, pp: 180-221. New Delhi: South Asia Office.

- Rai, K.M. 2010. Review: Biotechnological strategies for conservation of rare and endangered medicinal plants. *Biodiversitas,* **11**(3): 157-166.

- Rajamanickam, C , Anbu, S. and Balakrishnan, K. 2002. Effect of chemicals and growth regulators on seed germination in aonla (*Emblica officinalis* G.). *South Indian Horticulture,* **50**(1/3): 211-214.

- Rajamanickam, C.and Anbu, S. 2001. Effect of bio-fertilizers and growth regulators on seed germination and seedling vigor in amla. *Madras Agricultural Journal*, **88**(4/6): 295-297.

- Ram, S. and Raja Rao, T. 1981.Naturally occurring Auxins and inhibitors and their role in fruit growth of aonla (*Emblica officinalis* Gaertn.). *New Phytol,***88**(1):53–66.

- Singh Sanjay, Singh AK and Joshi HK 2005. Prolong storability of Indian gooseberry (*Emblica officinalis* Gaertn) under semi- arid ecosystem of Gujarat. *Indian Journal of Agricultural Sciences,* **75** (10): 647-50.

- Singh, A. and Singh, H.K. 2015. Application of plant growth regulators to improve fruit yield and quality in Indian gooseberry (*Emblica officinalis* Gaertn.). *Journal of Agri. Search,* **2**(1): 20-23.

- Singh, I. S. 2012. Aonla-Production, Handling and Processing, Westville Publishing House, New Delhi.

- Singh, J.K., Prasad, J., Singh, H.K. and Singh, A. 2008. Effect of micronutrients and plant growth regulators on plant growth and fruit drop in aonla (*Emblica officinalis* Gaertn.) fruits cv.'Narendra Aonla-10. *Plant Archives*, **8**: 911-13.

- Skoog, F. and Miller, C.O. 1957. Chemical regulation of growth and organ formation in plant tissues cultured in vitro. *Symposium Society of Experimental Biology*, **11**: 118-131.

- Tiwari J.P., Mishra, D.S., Chand S, and Kumar, R. 2008. Aonla: A Potential Fruit Crop of India. Indian *Farmer's Digest,* **41**(2): 12-14.

- Tomaszewska, E. and Tomaszewska, M. 1970. Endogenous growth regulators in fruit and leaf abscission. *Zeszyty Nauk Biol.,* Copernicus Univ. Torun Pol. **23**: 45-53.
- Verma, B. and Kant, U. 1996. Micropropagation of *Emblica officinalis* through mature nodal explants. *J. Phytol. Res.*, **9:**107-109.
- Wagh, A.P., Choudhary, M.H., Kulwal., L.V., Jadhav., B.J. and Joshi, P.S. 1998. Effect of seed treatment on germination of seed and initial growth of aonla seedling in polybags. *PKV Research Journal,* **22**(2): 176-178.

5

Plant Growth Regulators in Avocado

A.K.Panda, C.Kavita, S.Pradhan and R.K.Tarai

1. Introduction

Avocado (*Persea americana* Mill.) is a nutritious and healthy fruit containing all food elements (carbohydrates, proteins and fats), wide spectrum of vitamins (A, B, C, D, E and K) and minerals. It has gained an immense popularity over the last few decades as a luxurious fruit and has become a very important tropical horticultural crop in the modern world (Silva *et al.*, 2014). As a result, research on use of plant growth regulators in avocado is less advanced. Plant growth regulators (PGRs) are the most powerful tools available for manipulating tree growth and yield to solve production problems in an established orchard. However, a number of factors have further delayed progress towards commercialization of plant growth regulator strategies for avocado production. In avocado production, the use of plants growth regulators remains under developed despite the enormous potential that plant growth regulators offer for maximizing yield, optimizing fruit size and quality, and increasing net return to the grower. In contrast, for a wide variety of tree crops, there are many examples of the successful use of plant growth regulators to solve production problems. Plant growth regulators have been used successfully as foliar sprays to increase flowering, synchronize bloom, or change the time of flowering to avoid adverse climatic conditions or to shift harvest to a time when the market is more economically favourable (Lovatt, 2005). Due to poor uptake by avocado leaves, higher plant growth regulator concentrations than those used with other crops are required to draw out a response.

2. Use of Plant Growth Regulators in Propagation

Avocados are commonly propagated from seeds but the seeds do not breed true to type and this should be taken into consideration while going for planting of avocado. The avocado can be propagated by budding and grafting. Vegetative propagation of avocado cuttings is of primary importance if uniformity is desired with respect to genetic characteristics like tolerance of rootstocks to a saline condition, highly calcareous soil, and resistance to root rot diseases etc. In many of the experiments, the plant hormones used did not affect the rate of rooting but they improve the development of the root system.

Gustafson and Kadman (1969) conducted an experiment on the effect of some plant hormones in the rooting capacity of avocado cuttings (in two varieties Fuchs 20 and Gvar-Am-13) in a glasshouse, under intermittent mist spray during daytime, controlled automatically by an electric timer and valve, which supplies a relative humidity of nearly 100%. Constant bottom heat of 25°C was maintained by means of an electric cable with concrete heating plates buried in the rooting medium (vermiculite). The cuttings were selected and cut in to a length of about 30 cm and the two basal pairs of leaves removed. The cuttings were treated with the various hormones. Three different plant hormones of the auxin groups were used: 1) K.I.B.A. (Potassium salt of I.B.A. (indole butyric acid)) 2) N.A.A. (Naphthalene acetic acid) and 3) I.A.A.(indole acetic acid). Different concentrations (10,100 & 500 ppm) of each were prepared in aqueous solution. In addition, two controls, water and a dry, were used as checks for comparison with the treated cuttings. Altogether there were 26 treatments in groups of twenty cuttings each, totaling 560 cuttings. Results of these preliminary tests indicate that Gvar-Am 13, a Mexican type rootstock, generally roots better than the West Indian type, Fuchs 20. There were individual differences between variety and treatment. For instance, using potassium-indole-acetic acid (K.I.B.A.) at a concentration of 500 ppm on Gvar-Am 13, a 70% rooting occurred. With the Fuchs 20 variety, a 60% rooting took place with K.I.B.A. at 10 ppm concentration. The 500 ppm concentration of K.I.B.A. on Fuchs 20 resulted in only 5% rooting, while 10 ppm K.I.B.A. Gvar-Am 13 gave a 20% rooting. Gvar-Am 13 variety rooted with similar percentage when treated with Indole-acetic acid (IAA), naphthalene acetic acid (NAA) and potassium Índolebutyric acid (KIBA). The use of potassium-indole butyric acid resulted in a higher percentage of rooting with Gvar-Am 13, even though the rooting with IAA and NAA was generally good, compared with the same treatments on the Fuchs 20 variety.

Salazar-Gracia *et al.*, (2004) conducted an experiment to determine the rooting time and success rate of 38 Avocado and two Chinini (*Persea schiedeana*) selections using stem bark ringing and basal air layering. The evaluation was carried out in a 50% shade nursery in Xalisco, Nayarit, Mexico. Selections were veneer grafted onto 7-month-old seedling rootstocks. When scions were 15 to 20 cm in length a 2-cm wide bark ring was made at the base of the stem and a solution of indole-3-butyric acid (2000 mg/ L) plus alpha-naphthalene acetic acid (1000 mg/ L) was applied above the ringed area. Root production varied among selections with 10 to 100% of the selections successfully rooted, although overall rooting averaged 66%. The rooting response of drought- and salinity-tolerant selections was similar (69%), whereas for *Persea cinnamomi* tolerant selections was 57%.

3. Role of Plant Growth Regulators in Micro propagation of Avocado

Being planted as a grafted tree, propagation of avocado refers to propagation of rootstock cultivar, then graft it with scion from a mature desired commercial cultivar. Elite cultivar propagation is critical to maintain the quality of fruit and farm management practices. Rooting of cuttings is only possible through a complex, lengthy and expensive process called "Frolich and Platt method". These create hindrance on rapid industry expansion due to scarcity and high price of quality planting materials in many countries. Alternative propagation methods are sought over 5 decades. Commercial application of micro propagation for avocado will undoubtedly boost the industry around the globe (Hiti-Bandaralage *et al.,* 2017).

Different growth regulators with varying concentration plays major role in avocado micro propagation. Effects of different growth regulator concentrations and their combinations for shoot regeneration and root induction from nodal explants of mature avocado is given in Table 1 and Table 2 (Hiti-Bandaralage *et al.,* 2017). Auxins, when added in appropriate concentrations, may regulate cell elongation, cell division, tissue swelling, formation of adventitious roots, inhibition of adventitious and axillary shoot formation, callus initiation and growth, and induction of embryogenesis. The synthetic auxin 2, 4-D is very often used to support callus formation and maintenance. However, Ahmed *et al.* (2001) have shown that callus induction from avocado embryonic tissue depends on the hormone regime; the minimum level being 2.0 mgL^{-1} BAP or 2.0 mgL^{-1} NAA.

Cytokinins promote shoot formation, but need auxins as well to promote mitosis. At varying concentrations, they may regulate cell division, stimulate axillary and adventitious shoot proliferation, regulate differentiation, inhibit root formation, activate RNA synthesis and stimulate protein and enzyme activity. Benzyladenine (BA) is the most commonly used cytokinin,either alone, at concentrations in the range 0.1 - 2 mgL^{-1} (Cooper, 1987; Schall, 1987; Barcelo-Munoz *et al.,* 1990), or combined with gibberellic acid (GA3), (Capote *et al.,*

2000). The cytokinin kinetin, combined with indole-3-butyric acid (IBA) has also been used in the initiation phase (Gonzalez Rosas and Salazar, 1984). Barcelo-Munoz *et al.* (1999) developed a procedure for successful micropropagation of the avocado rootstock (IV-8) using different basal medium having varying concentration of various growth regulators. Basal shoots obtained after pruning were used as cultures. Buds sprouted in MS solid medium with macro-elements at half strength and a 1.3 μM benzyladenine supplement. For shoot proliferation, shoots were cultured for 2 weeks in liquid medium in a rollordrum with the Gamborg salt formulation and 1.3 μM benzyladenine, followed by 6 weeks in double phase conditions (solid medium with a layer of liquid medium on the top) using the same salt formulation and two different benzyladenine supplements, 2.8 μM in the solid phase and 0.4 μM in the liquid phase. Ninety per cent of the shoots rooted after a 3-day culture in liquid medium in a rollordrum with MS macro-elements at 0.3× and 4.9 μM indolebutyric acid, followed by transfer to solid medium in the absence of auxin. Seventy per cent survival was obtained during the acclimatization in the greenhouse.

Although auxins and cytokinins will continue to be the major plant hormones used in tissue culture, it is clear that there are many other growth active substances that can be used to enhance growth and differentiation *in vitro.*

Table 1. Effects of different growth regulator concentrations and their combinations for shoot regeneration from nodal explants of mature avocado.

Plant Growth Regulators	Shoot Regeneration Results	References
Meta-topolin (0.1 mgL^{-1}) + GA3 (0.1 mgL^{-1})	100% shoot regeneration	Bandaralage *et al.* (2015)
BAP (0.5 mgL^{-1}) + IBA (0.1 mgL^{-1})	3 to 5 shoots per explants and 71.3% shoot regeneration	Martinez-Pacheco *et al.* (2010)
No hormone + 2% (W:V)) peptone	44% shoot regeneration	Nhut *et al.* (2008)
BAP (2 mgL^{-1})	0% shoot induction	Barrera-Guerra *et al.* (1998)
Alternate culture in liquid and double phase medium, BAP (1.3 μM) in liquid medium and BAP (0.4 μM) in liquid phase and BAP (2.8 μM) in solid phase	3.2 shoots per explants	Barcelo-Munoz *et al.* (1999)
BAP (0.65 mgL^{-1})	70% shoot regeneration	Castro *et al.* (1995)
BAP (1 mgL^{-1})	0% shoot regeneration	Cooper (1987)

Table 2. Effects of different growth regulator concentrations and their combinations for root induction for nodal explants of mature avocado.

Plant Growth Regulators	Root Induction (%) Results	References
IBA-0.5 mgL^{-1} MS macro nutrients with and Gamborg vitamins Activated charcoal- 1 gL^{-1} Ascorbic acid-0.4 gL^{-1}	57.5	Martinez-Pacheco *et al.*,(2010)
IBA-1 mgL^{-1} MS macro nutrients 0.3 times	90	Barcelo-Munoz *et al.*, (1999)
Shoots grafted to juvenile rootstock IBA-1 mgL^{-1} MS macro nutrients 0.3 times Activated charcoal-1 gL^{-1}	50	Pliego-Alfaro and Murashige (1987)

4. Use of Plant Growth Regulators in Growth and Development

The discovery of compounds such as gibberellins inhibitors which act in preventing the synthesis of high levels of gibberellins has been reported to limit the vigorous vegetative growth phase in avocado (Woistenholme *et al.*, 1990). Paclobutrazole was reported to have growth retarding effects in avocado. But Uniconazole was found to have more retarding effect than of paclobutrazol (Kohne and Kohne, 1989). According to Adato (1990) the spray of Cultar (a commercial formulation containing 25% Paclobutrazole) did not alter the numbers of vegetative flushes but they did affect the shoot elongation causing a reduction of 30 % in shoot length. Lovatt (2006) reported increased shoot elongation with GA_3 (25 ppm) at 30 days after treatment.

5. Use of Plant Growth Regulators in Flowering, Fruit Set, Fruit Growth and Development, Control of Fruit Drop

Effects of canopy sprays of gibberellic acid (GA_3) or prohexadione calcium (a gibberellic acid biosynthesis inhibitor) applied at different stages of tree phenology on inflorescence development, time of anthesis, date of legal maturity for harvest of 'Hass' avocado fruit, yield, and fruit size were quantified. But no plant growth regulator treatment influenced the time of anthesis. Mature avocado trees have been reported to produce up to 1.6 million flowers with a fruit set of 0.001 to 0.23% (Cameron *et al*, 1952). The majority of the flowers are shed within a month of anthesis, suggesting inadequate pollination and fertilization (Adato and Gazit, 1977). Floral initiation occurs in autumn with flowering in late winter or early spring. To complete its cycle, the avocado flower opens on two consecutive days. The first day of opening, the flower is functionally female and the second day, functionally male. Sedgley (1980) reported that 90% of the flowers and fruit lets which abscised during the first week following the end of flowering were unfertilized. Inadequate pollination can be improved by the use of beehives in the groves.

Three major areas of opportunities for control of development and growth regulation by plant growth hormones in avocado are vegetative growth control, floral induction, and fruit set and retention (Davenport, 1987). The vegetative flushes are an extremely strong sink for minerals, water and photo assimilates. As a result, developing fruits are deprived of the nutrients required for their growth leading to production of poor quality fruits of smaller size. Poor yield, small fruit size and substandard quality in avocado are related to the strong flush which arises as a result of competition between vegetative and reproductive growth (De Villiers and Joubert, 2011). They also reported that this competition can lead to poor fruit set and retention. By reducing vegetative vigour, growth inhibitors can eliminate a lot of this competition, resulting in better fruit set and retention as well as an improvement in fruit size and possibly quality (De Villiers and Joubert, 2011).

Auxins, Gibberellins and Cytokinins are required for most of the normal course of fruit development but their function during ovary development prior to fertilization is poorly understood (Gillaspy *et al.*, 1993). Poor pollination leads to incomplete fruit set which in turn results in undersize fruits or ovary abortion (Nitsch, 1970). Wolstenholme *et al.* (1985) showed that mesocarp auxin concentrations of 'Fuerte' avocado fruit were highest early in the developmental programme and then steadily decreased towards maturity. Blumenfeld and Gazit (1972) noted high levels of GA activity in seed and seed coat of developing avocados, the latter decreasing with fruit growth, but no measureable GA-like activity in fruit flesh. Salazar-Garcia and Lovatt (2000) conducted an experiment on the role of GA_3 to manipulate flowering in Avocado cv. Hass. The experiment was initiated approximately seven months before an anticipated "off" bloom in an attempt to increase flowering intensity and yield. GA_3 (25 or 100 mg/l) was applied to separate sets of trees in September (early stage of inflorescence initiation), November (early stage of inflorescence development), January (initial development of the perianth of terminal flowers), March (cauliflower stage of inflorescence development; only 25 mg/l), or monthly from September through January (only 25 mg/1). Control trees did not receive any treatment. GA_3 (100 mg/l) applied in September reduced inflorescence number in both years, but not yield. GA_3 (25 or 100 mg•L-1) applied in November before the "on" bloom reduced inflorescence number with a concomitant increase in vegetative shoot number and 47% yield reduction compared to control trees. This treatment might provide avocado growers with a tool to break the alternate bearing cycle by reducing yield in an expected "on" crop year to achieve a higher yield the following year. GA_3 (25 mg/l) applied in November or January stimulated early development of the vegetative shoot of indeterminate inflorescences. January and March applications did not affect the number of flowering or vegetative shoots produced either year. GA_3 (25 mg/l) applied in March at the start of an "off" bloom increased 2-fold the production of commercially valuable fruit (213 to 269 g per fruit) compared to the control.

Adato (1990) reported that a spray of 3 % Paclobutrazole increased yield by 90% although other researchers have found no significant yield increase with application to branches (Kohne and Kohne, 1989) or to whole trees.

Mid-anthesis foliar sprays of paclobutrazol (PB) at 2.5, 1.25 and 0.62 g a.i. I^{-1} reduced spring growth and altered dry-matter distribution in fruiting spring shoots of 'Hass' avocado. *(Persea americana* Mill.) trees. There was a decrease in dry matter allocation to vegetative components and an increase in allocation to fruit. The starch concentration in the wood of spring shoots on trees treated with 2.5 g a.i. I^{-1} foliar PB was increased by ca. 70% compared with spring shoots from trees of other treatments. The mid-anthesis foliar treatments also reduced summer growth by ca. 20%. Trunk injection of PB at 0.2 g a.i. m^{-2} of canopy silhouette area when the spring growth matured, in addition to a mid- anthesis foliar spray of 1.25 g a.i. I^{-1}, reduced the length of summer shoots by 36% when compared with untreated trees. The spray treatments of 2.5 and 1.25 g a.i. T^{1} PB at mid- anthesis increased the mean fruit size at harvest by 16 and 11% respectively. Fruit yield was not significantly affected by PB applications on an annual basis. However, the PB spray treatments of 1.25 and 0.62 g a.i. I^{-1} significantly increased the two year cumulative yield by ca. 63%. There was no significant effect from the trunk injection treatment on fruit yield in either the season of treatment or the following season (Whiley *et al.*, 1991).

The positive aspects of application of Paclobutrazole in these studies were nearly negated by a heavy summer fruit drop (Wolstenholme *et al.*, 1988), although application tended to increase yield in the 'off' year in these trees (Adato, 1990).

6. Use of Plant Growth Regulators in Fruit Maturity and Ripening and Fruit Yield

According to Salazar-Gracia et *al.*, (2007) a single or double foliar application of GA_3 (50 mg/l) at 4 months (July) before the expected date of main harvest (November) resulted in 'Hass' avocado fruit reaching legal maturity (mesocarp dry matter 21.5% or greater) 24.8 to 28.2 days earlier than those of untreated control trees with no negative effect on yield or fruit size. The economic yield of avocado depends on the balance of vegetative growth flushes, flowering, pollination, and fruit retention. Another potential effect of GA_3 treatments is on fruit size. An increase on size of 'Hass' avocado was obtained in Israel with a canopy spray of 200 mg/l of GA_3 at the time of fruit set (May) (Zilkah *et al.*, 1987). In California, a lower dosage of GA_3 (25mg/l) applied when inflorescences of the 'Hass' avocado were at the cauliflower stage (March) at the start of an "off" bloom year increased 2-fold the production of commercially valuable large size fruit (213–269 g/ fruit compared with the control) (Salazar-Garcıa and Lovatt, 2000). A delay in color break and blackening of the avocado fruit was observed for late harvested fruit of GA_3 treated trees. In another study in California (Lovatt, 2004), GA_3 (25 mg/l) applied when 'Hass' avocado were at the cauliflower stage had no significant effect on the number of days for fruit to ripen after harvest. Lovatt and Salazar-Gracia (2006) reported that application of GA_3 (25 mg/L) at the cauliflower stage of inflorescence development during March significantly increased the 2-year cumulative total yield with commercially valuable large size fruit (178 g to 325 g/fruit).

Avocado is an example of a climacteric fruit. The onset of ripening corresponds to a marked rise in respiration. Auxins have been reported to be the agent to be responsible for the inability of avocado fruit to ripen on the tree (Twinga and Young, 1975). The endogenous levels of fruit auxins represent a resistance factor in ripening which must be deactivated before ripening can occur (Frenkel and Dyck, 1973). If exogenous auxins are applied to the surface of the fruit, ethylene levels are stimulated and fruit ripening is accelerated; but if auxins are infused into the fruit tissue by the vascular system, ethylene levels are limited and ripening is inhibited. Abscisic acid (ABA) levels were nearly constant during avocado fruit maturation but rose dramatically during ripening, apparently as a result of synthesis rather than release from the bound form (Cutting *et al.*, 1985).

Biennial bearing has been reported in avocado. Paclobutrazol is recommended to overcome the concerned problem of biennial bearing in many perennial tree fruits. It was tried in avocado by Adato (1990). He reported an increase in the yield during the off year with reference to the on year. The additional yield of about 379 and 546% were obtained in two different groves, when cultar (25% PBZ) was sprayed at the stages of inflorescence elongation and incipient anthesis. It generally retards the excessive vegetative growth in order to enhance the fruit yield. Salaza–Garcia *et al.* (2007), attempted to manipulate the vegetative and reproductive growth of avocado with foliar application of plant bioregulators to shift the date of flowering and harvesting period. Though the time of anthesis could not be altered, single or double application of GA_3 on foliage @50mg/ l four months before main harvest resulted in early fruit maturity without impairing the yield or fruit set.

References

- Adato, I. 1990. Effects of paclobutrazol on avocado (Persea americana Mill.) cv. Fuerte. *Scientia Horticulturae,* **45**: 105-110.
- Adato, I., and S. Gaz'rt. 1977. Role of ethylene in avocado fruit development and ripening. 1. Fruit drop. *J. Expt. Bot.,* **26**:636-643.
- Ahmed, M. F., Kantharajah, S. A. and Holford, P. 2001. Factors affecting adventitious shoot and root formation from the Avocado cultivar 'Hass' (*Persea Americana* Mill.) *in vitro. Gartenbauwissenschaft*, **66**, 293-297.
- Bandaralage, J. C. A. H., Hayward, A., O'Brien, C., and Mitter, N. 2015. Gibberellin and cytokinin in synergy for a rapid nodal multiplication system of avocado. ***In***: *Proceedings of the World Avocado Congress VIII, Lima*, 13-18.
- Barcelo-Munoz A, Pliego-Alfaro F., Barea J.M. 1990. Micropropagacion de aguacate (*Persea americana* Mill.) en fase juvenil. *Actas de Horticultura*, **1**, 503-506.
- Barcelo-Munoz, A., Encina, C. L., Simon-Perez, E., and Pliego-Alfaro, F. 1999. Micropropagation of adult avocado. *Plant Cell, Tissue and Organ Culture*, **58**(1), 11-17.
- Barrera-Guerra, J. L., Ramirez-Malagon, R., and Martinez-Jaime, O. A. 1998. *In vitro* propagation of avocado (*Persea drymifolia* Ness.). ***In***: Proceedings of the Induced Mutations in Connection with Biotechnology for Crop Improvement of the FAO/IAEA, Division of Nuclear Techniques in Food and Agriculture, Peru, 63-69.
- Blumenfeld, A. and S. Gazit. 1972. Gibberellin-like activity in the developing avocado fruit. *Physiol. Plant*, 27:77-82.
- Cameron, S. H., R. T. Mueller, and A. Wallace. 1952. Nutrient composition and seasonal losses of avocado trees. Calif. Avocado Soc. Yearbook 37:201-209.

- Capote M., Rodriguez N. N., Blanco M. 2000. *In vitro* propagation of avocado. *Tropical Fruits Newsletter,* **36/37**: 3-7.
- Castro, M., Oyanedel, E., and Cautin, R. 1995. *In vitro* shoot proliferation in avocado (*Persea americana* Mill.) induced by CPPU. ***In***: *Proceedings of the World Avocado Congress,* **3**, 223-226.
- Cooper, P. A 1987. Advances in the micropropagation of avocado *(Persea americana* Mill.). *Acta Horticulturae,* **212**: 571-575.
- Cutting, J. G. M., Wolstenhulme, B. N. , Hofman, P. J. and Lishman, A. W.1985. Theoretical and practical implication of plant growth substance trends in developing 'Fuerte' avocado fruits. *S. Afr. Avocado Grow. Assn. Yearbook,* **8**:257-289.
- Davenport, T. L. 1987. Opportunities for tropical fruit crop growth regulation. *Plant Growth Regulator Society of America Meeting.* pp:406-412.
- De Villiers, E.A. and Joubert, P.H. 2011. The Cultivation of Avocado. 2nd Edition. ARC-LNR, Nelspruit.,132-133 pp.
- Frenkel, C, and R. Dyck. 1973. Auxin inhibition of ripening in Bartlett pears. *Plant Physiol.,* **51**:6-9.
- Gillaspy, G., H. Ben-David and W. Gruissem. 1993. Fruits: A developmental perspective. *Plant Cell,* **5**:1439-1451.
- Gonzalez-Rosas H., Salazar S. G. 1984. Root induction and vegetative development from avocado plantules *(Persea americana* Mill.). *California Avocado Society Yearbook* **46**:167-171.
- Gustafson, C.D. and Kadman, A. 1969. Effect of some plant hormones on the rooting capacity of avocado cuttings. *California Avocado Society Yearbook,* **53**: 97-100
- Hiti-Bandaralage, J. C., Hayward, A., and Mitter, N. 2017. Micropropagation of avocado (*Persea americana* Mill.). *American Journal of Plant Sciences,* **8**(11), 2898.
- Kohne,J.S.and Kohne, S.K. 1989. Comparison of growth regulators paclobutrazol and uniconazole on avocado. *South African Avocado Growers' Association Yearbook,* **12**:38-39.
- Lovatt, J.C. and Salaz-Gracia, S.2006. Plant growth regulators for avocado production. *In : Proceedings of 33rd PGRSA Annual Meeting.*pp.98-107.
- Lovatt, C.J. 2004. Use of plant growth regulators to increase fruit set, fruit size and

yield and to manipulate vegetative and floral shoot growth, California Avocado Research Symposium, Riverside, Calif. (30 Oct. University of California, Riverside and California Avocado Commission. *pp* 96–107.

- Martinez-Pacheco, M. M., Suarez-Rodriguez, L. M., Hernandez-García, A., Salgado-Garciglia, R., Fernandez, I.V., Palomares, M.E.A., Cortes-Rodriguez, M.A. and Lopez-Gomez, R. 2010. *In vitro* propagation of mexican race Avocado (*Persea americana* Mill. var Drymifolia). *Acta Horticulturae*, **923**: 47-52.
- Nhut, D. T., Thi, N. N., Khiet, B. L. T., and Luan, V. Q. 2008. Peptone stimulates *in vitro* shoot and root regeneration of avocado (*Persea americana* Mill.). *Scientia Horticulturae*, **115**(2), 124-128.
- Nitsch, J.P. 1970. Hormonal factors in growth and development. In: A.C. Hulme (Ed.). The Biochemistry of Fruits and their Products, pp. 427-472. Academic Press, Amsterdam.
- Pliego-Alfaro, F. and Murashige, T. 1987. Possible rejuvenation of adult Avocado by graftage onto juvenile rootstocks *in vitro*. *Hort. Science*, **22**: 1321-1324.
- Rostron. P. J.H. 1988. Paclobutrazol trials in avocado orchards: initial results from Queensland and Natal. *South African Growers Association Year book*, **11**:57-59.
- Salazar-Garcia,S.and Lovatt,J.C. 2000. Use of GA3 to Manipulate Flowering and Yield of `Hass' Avocado. *J. Amer. Soc. Hort. Sci.*,**125** : 25-30.
- Salazar-Gracia, S., Cossio-Vargas,L.E., Gonzalez-Duran, I.J.L.and Lovatt,J.C. 2007. Foliar-applied GA3 Advances Fruit Maturity and Allows Off-season Harvest of 'Hass' Avocado. *Hort Sci.*, **42**(2):257–261.
- Salazar-Gracia, S., Velasco-Cardenas J, de J. Medina-Torres, R. Gomez-Aguilar, J. R. 2004. Avocado selections with potential use as rootstocks. II. Rooting response to air layering. *Revista Fitotecnia Mexicana*, **27**(2): 183-190.
- Schall, S. 1987. La multiplication de l'avocatier (*Persea americana* Mill. cv. Fuerte) par microbouturage *in vitro*. *Fruits* **42**(3), 171-176.
- Sedgley, M. 1980. Anatomical investigation of abscissed avocado flowers and fruitlets. *Ann. Bot.* **46**:771-777.
- Tingwa, P. O., and Young. R. E. 1975. The effect of indole-3-acetic acid and other growth regulators on the ripening of avocado fruits. *Plant Physiol.*, **55**:937-940.

- Whiley, A.W.; Saranah, J.B.; Nigel Wolstenholme, B.; Rasmussen, T.S. 1991. Use of paclobutrazol sprays at mid-*anthesis for increasing fruit size and yield of avocado (Persea americana Mill. cv. Hass). Journal of Horticultural Sciences*, 66(5):593-600.
- Wolstenholme, B. N., A. W. Whiley, and J. B. Saranah. 1990. Manipulating vegetative reproductive growth in avocado (*Persea americana* Mill.) with paclobutrazol foliar sprays. *Scientia Horticulturae*, **41**:315-327.
- Wolstenholme, B.N., P.J. Hofman, J.G. Cutting and A.W. Lishman. 1985. Theoretical and practical implications of plant growth substance trends in developing 'Fuerte' avocado fruits. *South African Avocado Growers Association Year Book*, **8**:92-96.
- Wolstenholme, B.N.; Whiley, A.W.; Saranah, J.B.; Symons, P.R.R.; Hofman, P.J. and Rostron, H.J. 1988. Paclobutrazol trials in avocado orchards: initial results from Queensland and Natal. South African Avocado Growers' Association Yearbook, **11**:57-59.
- Zilkah, S., I. Klein, S. Feigenbaum, and S.A. Weinbaum. 1987. Translocation of foliar applied urea 15N to reproductive and vegetative sinks of avocado and its effects on initial fruit set. *J.Amer. Soc. Hort. Sci.*, **112**:1061–1065.

6

Plant Growth Regulators in Banana

Tridip Kumar Hazarika

1. Introduction

Banana (*Musa × paradisiaca* L.) is an important fruit crop of tropical and sub-tropical regions of the world. It is dessert fruit for millions, is used in different regions as staple food owing to its rich and easily digestible carbohydrates. It is rich source of vitamins, minerals and has several medicinal properties. Among all the fruits, banana is plentiful, most nourishing and relatively a cheaper source of nearly all essential nutrients including vitamins and minerals and rich source of energy. It provides a more balanced diet containing sufficient amount of carbohydrates and essential nutrients including minerals, vitamins (Hazarika 2011).

The plant hormones are identified as promoters (auxins, gibberellin and cytokinin), inhibitors (abscissic acid and ethylene) and other hypothetical growth substance (florigen, flowering hormone, *etc.*). The term growth retarding or growth retardant is that the chemical slows cell division and cell elongation of shoot tissue and regulate plant height physiologically without formative effects. Eg. AMO 1618, phosphon-D, CCC, MC and Alar. These do not occur naturally in plants and acts in retardation of stem elongation, preventing cell division. Inhibitors suppress the growth of plants. There are phenolic inhibitors and synthetic inhibitors and Abscissic acid (ABA).

2. Role of PGRs on Propagation

Propagation of banana through *in vitro* techniques has been reported by several workers using different explants sources and methods. In tissue culture, plant growth regulators (PGRs) are critical media components in determining the developmental pathway of the plant cells. Cytokinins such as benzyl amino purine (BAP) and Kinetin are generally known to reduce the apical meristem dominance and induce both axillary and adventitious shoots formation from meristematic explants in banana. The most established banana shoot-tip culture system was achieved by using BAP as a supplement to basal media. The effectiveness of BAP over other cytokinins in inducing multiplication of shoot tip cultures has been reported in different cultivars of bananas. BAP has a marked effect in stimulating the growth of axillary and adventitious buds and foliar development of shoot tip cultures. Meanwhile, combinations of BAP with auxins such as indole acetic acid (IAA) or indole-3-butyric acid (IBA) were also used for *in vitro* multiplication of bananas. Although BAP stimulates shoot proliferation in bananas, it is also known to have mutagenic effects at high concentration producing off type plantlets. Appearance of the off type plantlets from *in vitro* multiplication process is considered a great disadvantage.

Different *in vitro* micropropagation protocols, using different adenine-based cytokinins, have been used in several Musa spp. of divergent genomic constitution and ploidy. The cytokinins, that have been commonly used, include benzyl aminopurine (BAP); isopentenyl adenine (2iP); zeatin (ZN) and kinetin (KN). However, only BAP has been tested for AAA-EA bananas. The proportion of auxin±cytokinin is a determinant for meristem formation and the hormone balance that becomes established between growth regulators determines the type of buds induced. A BAP concentration range of 8.9±22.2 mM is recommended for Musa *in vitro* propagation. The use of rates beyond this range induces increased rates of somaclonal variants. Diphenyl urea derivatives (DPU) appear not to have been used in Musa spp.

Tissue culture propagation of banana has gained attention due to its potential to provide genetically uniform, pest and disease free planting material. In spite of its massive use, the use of shoot tips for in vitro culture of banana cultivars often results in severe genetic defects. These genetic defects result in the production of large numbers of off-types or variants referred to as soma clonal variants. Understanding the causes of soma clonal variation and eliminating off-types before field establishment, preferably at the in vitro stage, has great significance for in vitro banana propagation. Although there are different factors (culture medium composition, rate of multiplication, primary explants origin, formation of adventitious shoots, increased duration in culture, and certain genotypes) reported to cause soma clonal variation in banana, the reports on effects of plant growth regulators seem to be inconsistent. Some authors report that these factors have no effect, while others report that they affect the rate of soma clonal variation both directly and indirectly.

Murali and Duncan (1995) grown micro propagated Banana shoots from floral apices in Stage III culture media with 0, 1, 2 or 4 mg 1^{-1} of triazoles (triadimefon or uniconazole) for 1 month. The resulting plantlets were then transferred to a peat moss and sand (1:1, v/v) potting mix and post-transplant performance was evaluated. In vitro triadimefon treatment (2 mg1^{-1}) acted as a conditioning agent and obviated the need for hardening or weaning the plantlets under a polyethylene tent. Triadimefon treated plants were turgid and healthy compared with control plants. Control plants which were not acclimatized wilted, and the leaves dried up, whereas plants treated with triadimef on, particularly with 2 mg1^{-1} remained turgid. All plants treated with uniconaxole dried up upon transplanting.

In another study, Arinaitwe *et al.*, (2000) cultured shoot-tips (3 mm) of banana (Musa spp.) cultivars Kibuzi (AAA-EA), Bwara (AAA-EA) and Ndiziwemiti (ABB) on modified Murashige and Skoog nutrient salts. The modifed medium was supplemented with various equimolar concentrations (16.8, 20.8, 24.8 and 28.8 mM) of BAP, TDZ, ZN, 2-iP and KN to determine suitable concentration ranges of the cytokinins for micropropagation of banana cultivars. Three sub-culture cycles were used and after each subculture, the shoots per explant were counted. To select a suitable and cost effective cytokinin, the optimum concentration, its cost in US$ and its corresponding proliferation rate were determined. The results showed that shoot proliferation was significantly dependent on cytokinin type, its concentration and the banana cultivar. The responses of cultivars to BAP were significantly better than other adenine-based cytokinins (ZN, KN and 2-iP).

The effect of liquid pulse treatment of growth regulators on *in vitro* propagation of banana (*Musa* spp. AAA) was studied by Madhulatha *et al.* (2004).).In liquid pulse treatment, the surface sterilized seedling tops were placed in sterile 50-ml plastic centrifuge tubes containing filter sterilized liquid PGRs at various concentrations in sterile, deionized water, at pH 5.6) and agitated on a shaker at 150 rpm for 2 h at room temperature. After incubation, the seedling tops were removed and rinsed by dipping in sterile water. Cotyledons were excised with a scalpel and placed in 100- x 15-mm plastic petri plates on solid medium consisting of SH mineral salts and vitamins, plus 30 g/L sucrose, 0.1 g/L myo-inositol, 7 g/L BRL Ultrapure agarose (except for the gelling-agent experiment), and no plant growth regulators. The pH was adjusted to 5.6 with NaOH before adding the gelling agent, and the medium was autoclaved for 20-25 min at 121°C. All subsequent transfers were made to this same medium. Optimal shoot proliferation rates were achieved due to the pulse treatment of 6-benzyl aminopurine (BA) and kinetin combination (1:1) at the concentration of 50 mg l^{-1} for 60 min. Similarly high frequency of root induction was obtained due to pulse treatment with a NAA and IBA combination (1:1) at the concentration of 100 mg l^{-1} for 60 min.

Bairu *et al.,* (2006) studied the effect of the type and concentration of plant growth regulators and sub-culturing on somaclonal variation in Cavendish banana cv. 'Zelig'. Auxins (IAA, IBA and NAA) and cytokinins (BA and TDZ) were used to multiply shoots for 10 generations. Bands generated through RAPD-PCR were scored according to whether they were present (1) or absent (0) to determine the extent of somaclonal variation. Results indicated that treatments with higher multiplication rates produced more variants; sometimes as high as 72%. Dwarf off-types accounted for 88% of the variation.

The influence of growth regulators on the root induction of the Musa genus plants cultivated within in vitro condition in 'Cavendish' cultivar by using different concentrations of growth regulators (NAA, IAA, 6-BAP, 2, 4-D) were analysed by Viehmannová *et al.* (2007). The induced roots were evaluated in conditions in vitro and ex vitro. The amount and length of the roots were evaluated, as well as the capacity of absorption of the roots by conductivity was determined. The experiments have proven that the most roots are created by using naphthalene acetic acid (5.4 µM), but the longest roots provide the control variant (MS medium). After 7 weeks of the transfer to ex vitro conditions the plants that were growing on medium with addition indole-3-acetic acid have the best vitality and root absorption.

Jafari *et al.,* (2011) studied micropropagation in banana by using MS media supplemented with benzyl aminopurine (BAP) and reported that the number of bud formation in shoot cultures of *Musa acuminate* cv. Berangan during the initiation stage increased proportionately with the concentrations used (11, 22 and 33 µM). However, the highest concentration of BAP (33 µM) simultaneously increased the formation of abnormal shoots. After the first apical bud appeared, explants were transferred to MS medium with lower concentrations of BAP either with or without indole acetic acid (IAA). Proliferation media supplemented with IAA showed enhanced shoot multiplication and elongation but did not help to reduce the abnormality index that occurred.

The effect of different concentrations of BAP on shooting in different species of Banana viz. Ardhapuri, Basrai, Shrimanti revealed that sword suckers with medium size were inoculated on MS medium supplemented with different concentrations of BAP (3mg/l,

5mg/l, 7mg/l, 9mg/l) cultures were incubated at 25+1°C with a 16 hr photoperiod (2000 lux) provided by cool white florescent tubes. The pH of medium was adjusted to 5.8 prior to autoclaving. The materials were sub-cultured at 30 day interval in same medium to produce multiple shoots. The effect of different concentration of regimes of BAP on bud initiation and shoot multiplication were investigated. After the initial bud was sub-cultured, multiple adventitious buds were produced from the base of the explants after 30 days. The number of multiple shoots developed at different levels of BAP in variety Basrai. The frequency of bud formation doubled and the fresh weight increased about four times higher in media with BAP at 7mg/l when compared to media supplemented with 3mg/l. The frequency of bud formation increased in media with 7 mg/l BAP as compared to BAP 3 mg/l. The mean number of multiple shoots developed at various levels of BAP revealed that 7 mg/l BAP shows increased average number of shoots (Bhosale *et al.*, 2011).

Devendra Kumar *et al.,* (2013) studied the effect of different concentrations of BAP on shoot induction and shoot proliferation in Banana cultivar Pisang Jayee. Explants obtained from sword suckers were inoculated on MS medium supplemented with different concentrations of BAP (5,10, 20 and 30 mg/L) and the cultures were incubated at 25 ± 1°C with a 16 hr photoperiod (2000 lux) provided by cool white florescent tubes. The best result was obtained at 20 µMBAP concentration.

Similarly, Anbazhagan *et al.* (2014) conducted an experiment for the standardization of *in vitro* culture technique for the mass propagation of *Musa* sp. by using shoot tips were used as explants and obtained the best results from MS medium supplemented with BAP+IAA at the concentration of 3.0 mg/l and 0.5mg/l respectively. Best root formation of *in vitro* developed shoots could be achieved on half strength MS medium supplemented with IBA at concentration 1.0 mg/l.

3. Role of PGRs on Growth and Development

A common problem of banana plants cultivated under protected environment is the excessive growth of their pseudostem, which can reach more than six meters for some cultivars, requiring more planting spacing and causing, in some cases, the rupture of the plastic l m surrounding the roof. Thus, the use of growth regulators to reduce plant height may be a solution for plants cultivated under protected environment (El Otmani *et al.* 1992, Jeyakumar *et al.* 2003). Maia *et al.* (2009) evaluated the action of paclobutrazol on the growth of the pseudostem of banana plants and other characteristics of their development by using five paclobutrazol doses (0; 0.5; 1.0; 1.5; and 2.0 g of a.i. plant^{-1}) with two banana plant cultivars (`Prata Anã' and `FHIA-01'). Paclobutrazol did not affect the number of days from planting to flowering, total leaf area and leaf emission rate. However, it increased leaf permanence on the plants and sucker number, reduced the foliar area of the leaves emitted after its application and provided an average reduction in pseudostem height of 26%.

Yadlod and Kadam (2010) studied the effect of plant growth regulators on growth and development of banana (*Musa* spp.) cv. Basrai by using different concentration of IAA and GA. The highest height of plant (179.35 cm) and maximum number of leaves (14.35) were recorded due to application of IAA 80 ppm, maximum girth of pseudostem was measured in GA3 80 ppm (66.53 cm). Significantly early maturity was observed due to application of IAA 80 ppm (121.50 days).

4. Role of PGRs on Fruit Set and Fruit Growth and Development

Fruit set represents the very first step of fruit development. In angiosperms, it depends on the successful completion of pollination and a unique double fertilization event where one of the pollen nuclei fertilizes the egg cell, whereas the other fuses with two haploid polar nuclei in the central cell. These events set the formation of the seed which eventually controls the cell division and fruit growth in a synchronized manner. Current evidence supports that combined action of three hormones, auxin, gibberellins (GAs), and cytokinin, plays a major role in the regulation of fruit set. Individually, any of these hormones can only initiate the fruit development to a certain extent; however, their combined application has been found to induce normal fruit growth even in the absence of fertilization in both dry and fleshy fruits.

Biswas and Lemtur (2014) studied the effect of growth regulators and organic sprays on bunch characters of banana variety Robusta. The growth regulators and organic sprays were applied one month after emergence of the bunch. The bunch weight, bunch length, weight of hands, weight of fingers, pulp weight increased while bunch maturity period decreased by the application of the treatment consist of removing one apical hand and spraying 50 ppm GA, one month after bunch emergence. The results also showed that bunch shape index, finger girth increased whereas when the treatment consist of removal of two apical hands and spraying of 50 ppm GA one month after bunch emergence was imposed. The finger length and number of fingers falling in the acceptable range were also improved by application of GA 50 ppm clubbed with one or two apical hands.

Mulagund *et al.*, (2015a) studied the effect of post-shoot application of SOP (2%), 50 ppm GA3, 25 ppm 2, 4-D, CPPU and Brassinostroid @ 2 ppm in Banana cv. Nendran with aiming to improve the bunch and fruit yield characters. The combined foliar sprays of

2 per cent SOP and 2 ppm Brassinosteroid significantly increased the bunch characters viz., bunch weight (11.35 kg), finger weight (215.40 g), finger length (29.10 cm), pulp weight (180.22 g), pulp to peel ratio (5.13) and total bunch yield (29.38 tonnes/ ha) with relatively higher benefit: cost ratio (2.87).

5. Role of PGRs on Fruit Yield and Quality

Yadlod and Kadam (2010) studied the effect of plant growth regulators on quality of banana (*Musa* spp.) cv. Basrai by using different concentration of IAA and GA, The PGRs were sprayed on the plant just before flowering by using a hand sprayer. The showed that the maximum weight of mature finger (150.33 g) and ripe finger (143.44 g) with the application of GA3 80 ppm. Minimum per cent loss in weight during ripening (3.84%) was observed in waxol 6% while maximum in control (7.92%). Maximum pulp to peel ratio (2.83) was recorded in IAA 80 ppm. Maximum total sugar (15.90 %) and reducing sugar (12.09%) was observed in GA3 80 ppm. Highest TSS (20.71%) were recorded in IAA 80 ppm. Highest non reducing sugar (3.92%) and maximum Vit. C (0.96) was recorded in GA3 80 ppm.

In another study, Bauri *et al.* (2014) studied the effect of micronutrients and growth regulators for yield and quality in banana cv. Martaman (Musa AAB, Silk) during 2011-2012 by applying two sprays of PGRs at 5 and 20 days after last hand opening. They obtained maximum vitamin C content (8.60 mg/100g pulp), shelf life (6.40 days) and benefit: cost (1.48) by the treatment of 2, 4-D (30 ppm).The same treatment was also found to be next most influencing treatment in increasing bunch weight, yield, hands number and finger length. It may be concluded that the application of 2,4-D @ 30 ppm was effective in modifying the finger 2 4-characters thereby improving the yield, fruit quality and B: C ratio of banana cv. Martaman.

6. Role of PGRs on Shelf-Life of Fruits

Banana fruit ripens rapidly in accompany with significant changes of physico-chemical and biochemical attributes such as transformation of starch to sucrose, mineral reduction and flavor formation once harvested, which causes green banana fruit into notably perishable. As a typical climacteric fruit, the quality attributes of banana fruit after harvest such as colour, texture and flavour, deteriorate easily, leading to a short shelf life (about 6–8 days under ambient conditions). Currently, the postharvest handling, transportation and marketing of banana fruit need sophisticated techniques and facilities. Therefore, great efforts to develop new and simple post-harvest technology to be required to maintain quality and extend shelf life of banana fruit during storage and transportation.

Huang and Ziang (2012) treated freshly harvested banana fruits with CPPU and GA3 in different concentrations, and then stored at 25 ± 2 ◦C, 75–95% RH after air dry. The freshly harvested banana bunches were immersed in GA3 and CPPU for 10 minutes. The results showed that, as compared with the control fruits, 10 mg l^{-1} CPPU with different concentrations of GA3 had good effect on the fruit and vegetable. Especially in the group treated with 10 mg l^{-1} CPPU + 50 mg l^{-1} GA_3, surface color change was obviously retarded, fruit respiration rate was inhibited, and the natural browning was significantly inhibited as well. In addition, the accumulation of soluble sugars in the fruit was reduced, and the water loss rate (4.55 ± 0.04%) was also inhibited. These treatments not only delay maturing and softening of banana, but also can maintain the colour and resist the fungal infection to extend the shelf life.

Huang *et al.*, (2014) treated banana fruits with 10 mg L^{-1} CPPU in combination with GA3 at 25, 50 or 100 mg L^{-1} and then stored under ambient conditions (23±2 ◦C and 75–90% relative humidity). The results exhibited that the combined treatments of CPPU and GA_3 significantly suppressed fruit softening in association with the delayed peaks of respiration and ethylene production rates and retarded decreases of hue value and the maximal chlorophyll fluorescence of banana fruit during storage. Furthermore, the accumulation of soluble reducing sugars and losses in contents of ascorbic acid and total phenols were delayed. The most beneficial effect of quality maintenance and shelf life extension of postharvest banana fruit was obtained by application of 10 mg L^{-1} CPPU and 50 mg L^{-1} GA_3. These results clearly indicated that the combined treatment of CPPU and GA_3 could contribute to ripening inhibition, shelf life extension and quality maintenance of banana fruit during storage.

Duguma *et al.*, (2014) studied the efficiency of postharvest treatments of GA_3 to improve the storage of banana fruits by dipping in gibberellic acid (GA3) for 15 minutes. All fruits treated with GA3 delayed in ripening in concentration dependent manner when compared to control fruits. The ripening of fruits treated with 300ppm GA3 delayed more than other treatments. The treatment with higher concentration of gibberellic acid delayed peel color changes, weight loss, ethylene, CO2 production and total sugar content as compared with control groups. In the meanwhile, the pulp to peel ration and total soluble solids were increased with increasing the concentration of gibberellic acid treatments.

Mulagund *et al.*, (2015) studied the effect of different chemicals and growth regulators in combination with Thiabendazole (Fungicide) on post-harvest quality and shelf life of banana cv. Nendran (French Plantain Musa AAB). Various concentration of, GA3, BA, CaCl2 and Hot water treatment either alone or in combination with Thiabendazole was applied to fruits after harvest to study their impact on post-harvest quality and shelf life. Exogenous application of chemicals, growth regulators in combination with Thiabendazole significantly decreased post-harvest disease incidence leading to increase in post- harvest fruit quality and shelf life. Among the treatments, post-harvest dipping of fruits in 150 ppm GA3 + 200 ppm Thiabendazole significantly increased firmness (3.70 kg/cm2) titrable acidity (0.09%) and storage life of fruits (18.00 days). While the parameters like PLW (13.53%), TSS (25.45 oB), total sugars (18.88%), reducing sugars (17.11%) and percent disease incidence (5.55) was observed minimum in the same treatment after 12th day of storage.

7. Role of PGRs on Embryogenesis

Most of the edible bananas are highly sterile, triploid and seedless. These features are barriers for the implementation of breeding strategies. Nevertheless, a very efficient approach is the synthesis of triploid varieties from diploid clones. Briefly, the basic scheme consists of selecting good diploid clones to be chromosome-doubled by colchicine treatment to form tetraploid plants. These tetraploid clones are then crossed with other diploids to deliver large, segregating triploid progenies (Bakry *et al.*, 1997). Unfortunately, some of the more interesting diploid clones and their derived tetraploids are fairly sterile, thereby impeding access to new interesting parental combinations. In this context, protoplast fusion can accelerate or facilitate attempts to obtain tetraploid parents.

Very few reports exist on banana protoplast cultures. Bakry (1984) reported the first successful isolation of viable banana protoplasts from inflorescence callus of *Musa* cv. Cavendish (AAA). Matsumoto *et al.* (1988) were able to induce cell divisions and the formation of microcalli from protoplasts of bract tissues. The successful establishment of cell suspensions and their successful regeneration (Novak *et al.*, 1989; Dhed'a *et al.*, 1991; Marroquin *et al.*,1993) have led to them being used as a source for protoplast isolation (Megia *et al.*, 1992). The donor material for embryogenic cell suspensions can be pseudo-stemand rhizome tissues (Novak *et al.* 1989), meristematic shoot tips (Dhed'a *et al.* 1991), immature zygotic embryos (Marroquin *et al.*, 1993), immature male flowers(Ma, 1991; Côte *et al.*, 1996; Grapin *et al.* 1996) and, recently, immature female flowers (Grapin *et al.* 2000).Using this kind of suspension as a source, Megia *et al.*(1992) obtained callus formation from protoplasts of the wild diploid Long tavoy (AA). Whole cv. Bluggoe plants (ABB) (e.g. cooking bananas for local consumption) have been regenerated from protoplasts isolated from embryogenic cell suspensions and cultured on feeder layers (Megia *et al.*, 1993; Panis *et al.*, 1993).

Assani *et al.*, (2001) achieved protoplast culture and plant regeneration of the dessert banana cultivar Grande Naine (*Musa* spp., through somatic embryogenesis. Protoplasts were isolated from cell suspensions at a yield of 3×107 protoplasts/ml packed cell volume (0.5 g). For the induction of cell divisions, two banana cell suspensions, SF265 (AA) and IRFA903 (AA), were used as feeder layers. SF265 (AA) was found to be more efficient for inducing cell divisions than IRFA903 (AA). The first embryogenic cell suspensions were established from protoplast-derived microcalli. The transfer of microcalli and protoplast-derived cell suspensions onto regeneration medium containing plant growth regulators slightly increased the number of embryos relative to those maintained on a feeder layer with growth regulators. Plant regeneration was achieved in the same regeneration medium.

Dayarani *et al.*, (2014) studied the influence of maturity and hormonal factors on germination and regeneration of *Musa ornata* seeds through embryo culture and embryo rescue. Embryos extracted from seeds harvested at various maturity stages were cultured in MS media with different concentrations of plant growth regulators. Medium rich in auxins led to callus formation at all maturity levels, leading to indirect regeneration. Good direct regeneration was observed from 100% mature embryos, in media supplemented with 6-benzylaminopurine. Study revealed that zygotic embryos of *M. ornata* could be rescued and regenerated through callus when harvested at 80% maturity and media augmented with Kinetin gave the best regeneration.

Tran *et al.*, (2016) studied the role of auxin and cytokinin on somatic embryogenesis from cell suspension culture of banana cultivar 'Cau Man' (AA genome, Sucrier subgroup). A highly proliferating meristem culture (CHP) of banana cultivar 'Cau Man' was initiated on CHP medium containing 0.17 mg L-1 indol acetic acid (IAA), 2.5 mg L-1 N6-benzyladenine (BA) and 1 mg L^{-1} zeatin. Somatic embryos were obtained on Ma3 medium containing 0.1 mg L^{-1} kinetin, 0.2 mg L-1 6-dimethylallylamino purine (2-iP) and 0.05 mg L^{-1} zeatin, at initial cell density of 15 μL mL^{-1} SCV (settled cell volume). Cytokinin treatment increased the number of mitochondria observed under transmission electron microscope (TEM). IAA and zeatin levels in somatic embryos in different stages were determined by high performance liquid chromatography (HPLC). IAA level increased at the globular embryo stage. Use of TIBA (2,3,5- triiodobenzoic acid), an inhibitor of auxin polar transport, showed that auxin polar transport played an important role in the vertical axis formation of embryo organs as well as their development. The combination of auxin (1.5 mg L-1 IAA) and cytokinin (0.227 mg L-1 BA or 1 mg L-1 zeatin) at the volcano stage was necessary to complete the development and germination of somatic embryos.

8. Role of PGRs on Seed Germination

In the genus *Musa,* germination is highly variable and relatively difficult. Even more difficulties are faced when producing hybrids. The seed yield of hybrids in breeding programs is usually low and to ensure the viability and survival of seeds it is often necessary to attempt to germinate a large excess of these seeds. Several reports have indicated the limited and variable seed germination exhibited by *Musa* (Simmonds 1952, Pancholi *et al.* 1995). Therefore, *in vitro* embryo culture represents a potential tool for improved recovery of hybrid germplasm in a short time.

Banana breeders are under increasing pressure to produce disease-resistant cultivated varieties. This and the use of seedling banana plants in breeding programs necessitate elucidation of factors affecting germination of these seeds. The seed yield of hybrids is usually low, and it is imperative that a maximum number of these seeds be germinated. Several current research programs require a continuous supply of small and uniform plants, which can be obtained most readily from seeds. However, since classical breeding programs are complicated by the sterile nature of cultivated *Musa* and the low seed yield of hybrid plants, moreover germination is highly variable and relatively difficult, new strategies are preferred. The majority of these are based on *in vitro* tissue culture methods associated with traditional techniques.

Ahmed *et al.*, (2006) studied the *in vivo* and *in vitro* germination of *M. balbisiana* seeds or mature embryos under various conditions and their study revealed that, *in vitro* darkness, 26°C temperature and 2.2 µM BAP were more suitable than any other culture medium and regime tested for the seed germination of banana.

9. Role of PGRs on Stress Management

Maintaining crop yields under adverse environmental stress conditions is probably a major challenge facing modern agriculture. Further, in their natural environment, plants encounter a vast array of pathogenic microorganisms such as fungi, bacteria, viruses and nematodes. These diverse pathogens deliver effecter molecules (also called virulence factors) into the plant cell to promote virulence and cause disease. To survive under such conditions, plants have evolved intricate mechanisms to perceive external signals, allowing optimal response to environmental conditions. Phyto hormones, such as salicylic acid (SA), jasmonic acid (JA), ethylene (ET) and abscisic acid (ABA) are endogenous, low molecular weight molecules that primarily regulate the protective responses of plants against both biotic and abiotic stresses *via* synergistic and antagonistic actions, which are referred to as signaling crosstalk

(Tuteja and Sopory 2008). Most of the abiotic stress is responded by variation in the levels of abscissic acid whereas, biotic stress enhances ethylene and jasmonic acid levels. It is now believed that other hormones such as gibberellins (GAs), auxin, cytokinin (CYT) and recently brassino-steroids (BRs) do also participate in various types of signal transduction during stress in plants.

Environmental stress severely restricts the distribution and productivity of plants; particularly drought is a major abiotic factor that limits crop productivity. Drought stress remains an ever growing environmental problem that severely limits crop production worldwide and causes important agricultural losses particularly in arid and semi-arid areas. Water stress enforces a serious threat to banana productivity. The consequences of water deficit include its adverse effects on plant phenology, development, assimilate partitioning, carbon assimilation, growth, and plant reproduction processes. Drought induced osmotic stress triggers a wide range of perturbations ranging from growth and water status disruption to the modification of ion transport and uptake systems. Upon water deficit exposure, plants exhibit physiological, biochemical and molecular responses at both cellular and organismal.

Utilization of tissue culture techniques for quantifying stress tolerance of various crops has been increasing rapidly. Tissue culturing systems are useful for the evaluation of tolerance to environmental stresses because stress conditions can be easily controlled in vitro. Moreover, in vitro cultures provide a uniform population of synchronously developing plant cells without involving regulatory mechanisms that naturally are repaired at the organismal level. Also, In vitro culture techniques minimize environmental variations due to defined nutrient media, controlled conditions and homogeneity of stress application. In addition, the simplicity of such manipulations enables the study of large plant population and stress treatments in a limited space and a short period of time. Simulation of drought stress under in vitro conditions during the regeneration process constitutes a convenient way to study the effects of drought on the morphogenic responses. Tissue culturing is also an efficient procedure to study the effect of abiotic stress on the cell metabolism.

Said *et al.*, (2015) investigated the adverse effects of drought stress on growth, yield and endogenous phyto hormones of the banana (*Musa acuminata*); and focused on the progress towards the development of stress tolerant lines through tissue culturing based on in vitro selection. Drought stress was induced by applying Polyethylene Glycol (PEG) solutions with elevated strengths of (1, 2 & 3%). Tolerance effect was acquired by pre-treating the plantlets with trehalose using different concentrations (20, 60 & 100 mM), to test which concentration provides the most eminent outcome. Results exposed that trehalose has positive significant effects combating drought stress.

References

- Ahmed, K. Z., Remy, S., Sági, L. and Swennen, R. 2006. Germination of *musa balbisiana* seeds and embryos. Proceedings of XVII International Meeting of the Association for the Cooperation in Investigations of Banana in Tropical America and the Caribbean, October 15 to 20, 2006, Joinville, Brazil.
- Anbazhagan, M., Balachandran, B.and Arumugam, K. 2014. *In vitro* propagation of *Musa* sp (Banana). *Int. J. Curr. Microbiol. App. Sci.,* **3**: 399-404.
- Arinaitwe, G., Rubaihayo P.R., Magambo M.J.S. 2000. Proliferation rate effects of cytokinins on banana (Musa spp.) cultivars. *Scientia Horticulturae,* **86**:13-21
- Assani A., Haicour R., Wenzel, G., Côte, F. , Bakry, F., Foroughi-Wehr, B., Ducreux, G. , Aguillar M.E. Grapin A. 2001. Plant regeneration from protoplasts of dessert banana cv. Grande Naine (Musa spp., Cavendish sub-group AAA) via somatic embryogenesis. *Plant Cell Rep.,* **20**:482–488
- Bairu, M. W., Fennell Catherine, W., Staden Johannes van. 2006. The effect of plant growth regulators on somaclonal variation in Cavendish banana (Musa AAA cv. 'Zelig') *Scientia Horticulturae,* **108**:347–351.
- Bakry, F. 1984. Choix du matériel à utiliser pour l'isolement de protoplastes de bananier (*Musa* sp.). *Fruits,* **39**:449–452
- Bakry, F., Carreel, F., Caruana. M.L., Côte. F.X., Jenny, C., Tezenas. du Montcel H. 1997. Les bananiers. In: Charrier A, Jacquot M, Hamon S, Nicola D (eds) L'amélioration des plantes tropicales. Repères, CIRAD Montpellier, France pp 109–139.
- Bauri, F. K., De, A., Misra, D. K., Bandyopadhyay, B, Debnath, S., Sarkar S. K. and Avani. P. 2014. Improving yield and quality of banana cv. Martaman (*Musa* AAB, Silk) through micronutrient and growth regulator application. *Journal of Crop and Weed,* **10**:316-319.
- Bhosale, U. P., Dubhashi, S. V., Mali N. S. and Rathod H. P. 2011. *In vitro* shoot multiplication in different species of Banana. *Asian Journal of Plant Science and Research,* **1**:23-27.
- Biswas, P. J. and Lemtur, K. 2014. Effect of growth regulators and certain organic sprays on bunch characters in banana cv. Robusta. *The Asian Journal of Horticulture,* **9**:269-271.
- Côte, F.X., Domergue, R., Monmarson, S., Schwendiman, J., Teisson, C., Escalant, J.V. 1996. Embryogenic cell suspensions from the male flower of *Musa* (AAA) cv. Grande Naine. *Physiol Plant,* **97**:285–290.

- Dayarani M., Dhanarajan M.S., Arun K., Uma S. and Padma Narayani. 2014. Embryo culture and embryo rescue studies in wild Musa spp. (*Musa ornata*) *Journal of Applied Horticulture*, **16**: 126-130.

- Devendrakumar, D., Anbazhagan, M. and Rajendran, R.. 2013. Effect of Benzyl Amino Purine (BAP) concentration on *in vitro* shoot proliferation of Banana (*Musa* spp.). *International Journal of Research in Biotechnology and Biochemistry*. **3**: 31-32.

- Dhed'a, D., Dumortier, F., Panis, B., Vuylsteke, D., de Langhe E. 1991. Plant regeneration in cell suspension cultures of cooking banana cv. "Bluggoe" (*Musa* spp. ABB group). *Fruits*, **46**:125–135.

- Duguma, T., Egigu Meseret C., Muthuswamy, M. 2014. The effects of gibberellic acid on quality and shelf life of banana (musa spp.) *Int J Cur Res Rev.*, **6**: 63-69.

- El Otmani, M, Jabri, K and Sedki, M. 1992. Paclobutrazol effect on development of greenhouse-growth banana: 2- year assessments. *Acta Hort*, **296**: 89–96.

- Grapin, A., Ortiz, JL., Lescot, T., Ferriere, N., Côte, F.X. 2000. Recovery and regeneration of embryogenic cultures from female flower of False Horn Plantain (*Musa* AAB). *Plant Cell Tissue Organ Cult.*, **61**:237–244.

- Grapin, A., Schwendiman, J and Teisson C. 1996. Somatic embryogenesis in plantain banana in vitro. *Cell Dev Biol.*, **32**:66–71.

- Hazarika, T K. 2011. Evaluation of Integrated Nutrient Management Packages for growth, yield and quality of banana cv. Grand Naine in Mizoram. Ph. D. Thesis, Mizoram University, Aizawl, India.

- Huanga, H. and Jiang Yueming. 2012. Effect of plant growth regulators on banana fruit and broccoli during storage. *Scientia Horticulturae*, **145**: 62–67.

- Huanga, H., Jing G., Wanga H., Duana X., Qua H., Jianga Y., 2014. The combined effects of phenylurea and gibberellins on quality maintenance and shelf life extension of banana fruit during storage *Scientia Horticulturae*, **167**: 36–42.

- Jafari, N., Yasmin O. R. and Khalid N. 2011. Effect of benzylaminopurine (BAP) pulsing on *in vitro* shoot multiplication of *Musa acuminata* (banana) cv. Berangan. *African Journal of Biotechnology*. **10**: 2446-2450.

- Jeyakumar, P., Kumar, N. and Kavino, M. 2003. Physiological response of banana cv. 'Robusta' (AAA) to foliar applied plant growth regulators on productivity. *Madras Agric J.*, **90:** 702–706.

- Karthiayani, A, Mathod, P.F, Varadhraju, N and Kennedy, Z.J. 2006. Studies on preservation of banana stored under MAC condition using diffusion channel. *J. Food Sci. Technol.*, **43**: 314-317.
- Ma, S.S. 1991. Somatic embryogenesis and plant regeneration from cell suspension culture of banana. In: Department of Agriculture, National Taiwan University (eds) *Proc Symp Tissue Cult Hortic Crops.* National Taiwan University, Taipei, Taiwan, pp 181–188.
- Madhulatha, P., Anbalagan, M., Jayachandran, S. & Sakthivel, N. 2004. Influence of liquid pulse treatment with growth regulators on *in vitro* propagation of banana (*Musa* spp. AAA). *Plant Cell, Tissue and Organ Culture,* **76**: 189–191
- Maia, E., Siqueira D.L., Salomão Luiz C.C., Peternelli Luiz A., Ventrella Marília C. and Cavatte Rithiely P.Q. 2009. Development of the banana plants `Prata Anã' and `FHIA-01' under the effect of paclobutrazol applied on the soil *Annals of the Brazilian Academy of Sciences*, **81**: 257-263
- Marroquin CG, Paduscheck C, Escalant JV, Teisson C. 1993. Somatic embryogenesis and plant regeneration through cell suspensions in *Musa acuminata. In Vitro Cell Dev. Bio.* **29**:43–46.
- Matsumoto, K, Crepy, L, Teixeira, J.B., Ferreira, F.R. 1988. Isolation, culture and fusion of banana bract protoplast. In: International Rice Research Institute (IRRI) and Academia Sinica (eds) Genetic manipulation in crops. Cassell Tycooly, Philadelphia, pp 414–415.
- Megia, R., Haicour, R., Tizroutine, S., Bui Trang, V., Rossignol, L., Sihachakr, D. and Schwendiman, J. (1993) Plant regeneration from cultured protoplasts of the cooking banana cv. Bluggoe (*Musa* spp., ABB group). *Plant Cell Rep.*, **13**:41–44.
- Megia, R., Haicour, R., Rossignol, L. and Sihachakr, D. 1992. Callus formation from cultured protoplasts of banana (*Musa* spp.). *Plant Sci.*, **85**:91–98.
- Mulagund, J., Porika, H., Soorianathasundaram, K. and Deepika, C. 2015. Influence of Growth Regulators Combined with Chemicals to Improve Postharvest Fruit Quality in Banana cv. Nendran (Musa AAB) *J Food Process Technol.*, **6**:3.
- Mulagund, J., Kumar, S., Soorianathasundaram, K. and Porika, Harikanth. 2015a. Influence Of Post-shooting sprays of sulphate of potash and certain Growth Regulators on Bunch characters and fruit yield of Banana cv. Nendran (French Plantain Musa Aab). *The Bioscan.* **10**: 153-159.

- Murali, T.P. and Duncan, E.J. 1995. The effects of in vitro hardening using triazoles on growth and acclimatization of banana. *Scientia Horticulturae,* **64**: 243-251.
- Novak, F.J., Afza, R., van Duren, M., Perea-Dallos, M., Conger, B.V. and Xiaolang, T.1989. Somatic embryogenesis and plant regeneration in suspension cultures of dessert (AA and AAA) and cooking (ABB) banana (*Musa* spp.). *Biotechnology,* **7**:154–159.
- Pancholi, N., Wetten A. and Caligari 1995. Germination of *Musa velutina* seeds: comparison of *in vivo* and *in vitro* systems. *In Vitro Cell Dev. Biol. Plant,* **31**: 127-130.
- Panis, B., Wauwe A.W.and Swennen R. 1993. Plant regeneration through direct somatic embryogenesis from protoplasts of banana (*Musa* spp.). *Plant Cell Rep.,* **12**:403–407.
- Said EM, Mahmoud RA, Al-Akshar R and Safwat G. 2015. Drought Stress Tolerance and Enhancement of Banana Plantlets *In Vitro. Austin J Biotechnol Bioeng.,* ;**2**(2): 1040.
- Simmonds N.W. 1952. The germination of banana seeds. *Trop Agric.,* **29**: 35-49.
- Tran, T.H., Bui, T.V. and Feng, T.Y. 2016. Role of plant growth regulators on shoot development of shoot apical meristems of banana genotypes. *Acta Hortic.* **1114,** 211-218.
- Tuteja Narendra and Sopory Sudhir K. 2008. Plant signaling in stress Gprotein coupled receptors, heterotrimeric Gproteins and signal coupling via phospholipases. *Plant Signaling & Behavior,* **3**:2, 79-86.
- Viehmannová, I., Fernández C.E., Hnilička F. and Robles, C.D. 2007. The influenceof growthregulatorsonroot induction i*n vitro* of the *Musa* genus. *Agricultura tropicaet subtropica.* **40** :115-119.
- Yadlod, S.S. and Kadam, B.A.. 2010. Effect of plant growth regulators and micronutrients on quality of banana cv. BASRAI. *The Asian Journal of Horticulture,* **5**: 19-21.

7

Plant Growth Regulators in Ber

Ajit Kumar Sahoo, Ranjan Kumar Tarai,
A.K. Srivastava, Nidhika Thakur, Om Praksh and Sayan Sau

1. Introduction

Ber is a quick growing and early bearing fruit which yields a heavy crop every year and perhaps requires the least inputs and care. In spite of that heavy fruit drop after setting, small fruit size and poor fruit quality in many commercial cultivars is a common phenomenon in many ber growing areas in the country. In order to minimize these problems, comprehensive researches are being conducted and observed that the plant growth regulators have immense potential to increase the yield and also improve the fruit quality.

Growth and development of the plant is controlled by two sets of internal factors, namely, nutritional and hormonal. Nutritional factors constitute the raw materials required for growth. The utilization of these substances for proper development of the plant is controlled by certain chemical messengers called plant growth regulators, which in minute amounts increase or decrease or modifies the physiological processes in plants. In *ber*, the PGRs are widely used for increasing fruit set, controlling fruit drop, enhancing quality and uniform maturity.

2. Use of Plant Growth Regulators in Propagation

Earlier ber was propagated by seeds during earlier period. But the main disadvantage of t his method is of heterozygosity and variability in seedling progeny. Presently, superior varieties of ber is propagated by budding on seedling rootstock. Sometimes, seed germination is not uniform or may be poor mainly due to its hard seed coat and internal factors. Plant growth regulators played a vital role in improving seed germination, seedling growth and also enhancing seed viability in storage in ber.

Murthy and Reddy (1989) carried out laboratory and field experiments using kernels of *jujube* cv. Umran. In the experiment kernels were treated with GA_3 (100 or 200 ppm), benzyladenine (50 or 100 ppm), thio-urea (0.5 or 1%) or KNO_3 (0.5 or 1%). Germination was recorded daily and they found that thiourea treatment resulted in significantly higher percentage of germination compared with the other treatments.

A study was conducted to evaluate the effect of different types and concentrations of priming solution on seed germination and seedling performance of *Ziziphus spina-christi*. *Spina-christi* is one of the wild species of *Ziziphus* that grows in the form of shrub or little bushes that strongly resists heat and drought. It develops very deep taproot and has an extraordinary regenerative power. It has edible fruit and in addition different parts of plant such as leaf, bark and root have been used in folk medicine. The seedlings are used as a rootstock for cultivars with high fruit quality. The seeds of Z. *spina-christi* possess hard woody seed-coat which hampers germination. Moreover, slow growth of the seedlings in the first year of sowing is one of the problems that put a limit on the growing of this species. Different techniques have been used to improve seed germination and seedling growth. Having seedling with enhanced and identical growth is the main step in producing grafted sapling in nursery (Takhti and Shekafandeh, 2012). One seed treatment method that has been proved to be successful in improving seed germination and emergence is priming. Seed priming is a controlled hydration treatment at low water potential that allows pre-germinative metabolism to proceed, but prevents radicle emergence. This metabolic advancement at low water potential is expressed as a more rapid radicle emergence rates when seeds are transferred to higher water potential. Different substances are used as osmotica which lower the water potential such as PEG, NaCl, KNO3, ZnSO4 and CaCl2, etc. Seed priming has been successfully demonstrated to improve germination and emergence in seeds of many crops under normal as well as stress condition (Takhti and Shekafandeh, 2012).

The seeds were primed in solutions of sodium chloride of 0, 4, 8, 12 and 16 dS/m, zinc sulfate of 0, 4, 8, 12 and 16 dS/m (osmopriming), distilled water (D.W.) and distilled water at 40C° (D.W. 40°C) (hydro-priming). The seeds were soaked in above solutions for 12 hr and a lot of seeds were used as control (without priming). All priming treatments1 showed a germination rate higher than control ($P < 0.05$). The highest germination rate (18.11%), stem length, stem fresh and dry weight and root fresh weight occurred in ZnSO4 solution with electrical conductivity (EC) of 8dS/m. Also, the results showed that seedlings originated from the seeds primed at different solutions of ZnSO4 had better performance and produced more leaves and spins than those originated from the seeds primed with different solutions of NaCl. The maximum leaves and spins produced in 12 dS/m of ZnSO4 (Takhti and Shekafandeh, 2012).

For raising rootstocks, seeds of *Zizyphus rotundifolia* is used. Germination of ber seed is quite low due to presence of stony shell (endocarp) around the seed. Seeds collected from dropped fruits contain 50-70% non-viable seeds. Seeds should be dipped into a salt solution of 17-18% concentration for 24 hour before sowing and the seeds which are floating should be discarded as they are dead seeds. The viable seeds then may be treated with different plant growth regulators to enhance the germination process. If the seeds are treated with GA_3@500 ppm gives about 90% germination followed by 94% and 92% after treatment with GA_3@200 ppm + 5% sucrose or thio-urea at 500 ppm, respectively. The germination of freshly harvested seeds can also be improved by soaking in concentrated sulphuric acid for 6 minutes (Singh and Choudhary, 2012).

3. Effect of Plants Growth Regulators on Micro-Propagation

The micropropagation of ber (*Ziziphus mauritiana* Lam.) was studied through tissue culture by **Xu** *et al.* **(2003.)** The results show that it is a perfect time for cutting explants from healthy growing plants in March and April. Because in this period it can improve the rate of germination for stem explants culture. The optimum media were screened out through studies on the condition of initiation, proliferation and rooting culture. Main achievements were shown as follows: MS medium supplemented with BA 0 8 mg/L and IBA 0 4 mg/L was suitable for ber initation. MS medium supplemented with BA 1 2 mg/L and IBA 0 5 mg/L was the best for shoots muliplication. But the medium favourable for rooting was 1/2MS basal medium supplemented with IBA 3.0 mg/L. Furthermore, the paper illustrates the relation between generation and ratio of proliferation or rate of rooting. The results show that eight generations are suitable for shoot multiplication. After eight generations rooting culture should be carried out.

The experiment was conducted during 2005-2006 on "Micropropagation studies in ber (Ziziphus mauritiana Lamk.)" in the laboratory of Centre for Research and Application in Plant Tissue Culture CRAPTC CCS HAU, Hisar Haryana. The efforts were made to optimize conditions for mass-multiplication of ber cultivars Gola and Umran. The explants comprised of shoot tip and nodal segment with different combination of growth regulators and additives (BAP, NAA, IBA, Kinetin, KNO3, NH4NO3) supplementing the MS basal medium. Sterilization treatment of 0.1% $HgCl_2$ supplemented with ascorbic acid and citric acid for 3 minutes duration was observed best for cvs. Gola and Umran. The minimum number of days taken for shoot initiation was observed in MS+BAP 0.5 mg/l in both cultivars. The maximum shoot regeneration, shoot length and average number of shoots per plant was observed on MS+ KNO3 3800 mg/l + NH4NO3 2475 mg/l + BAP 2.5 mg/l + IAA 0.5 mg/l media in cvs. Gola and Umran. The shoots were transferred for rooting on different rooting media. Rooting was not observed in both cultivars on any of the rooting media tried (Godara, 2007).

The tissue culture of Chinese jujube (*Ziziphus jujuba* Mill.) was studied by Dai *et al.* (2009) with different explants including stem tip, stem segment, leaves and embryo. The results showed that: (1) MS medium could be used as basal medium for all above mentioned explants in Chinese jujube; (2) The optimum kinds and concentrations of growth regulators in medium were quite different among explants. High concentration of 6- BA (about 0.5-3.0 mg/L) plus low concentration of IBA (about 0.1-0.5 mg/L) were suitable for initiation and proliferation culture of stem tip and stem segment. TDZ of 1.0 mg/L was best for shoot regeneration from leaves. No or low concentration of growth regulator (0.1 mg/L 6-BA+0.5 mg/L IBA+0.1 mg/L NAA) was fitted for 20-70 days old embryo culture; and (3) 6-BA of low concentration (0-0.1 mg/L) and IBA of high concentration (about 0.5-1.5 mg/L) or IAA of 1.0 mg/L were suitable for rooting culture of plantlet originated from different kinds of explants.

Plant growth regulators are indispensable in culture media and are important to the plant tissue induction, organ differentiation and growth. Therefore, the effects of seven plant growth regulators such as BA, TDZ, KT, NAA, IBA, IAA and GA_3 with different concentrations on the leaf regeneration, subculture of shoots and rooting were studied in *Ziziphus jujube* cv. 'Dongzao'. The results indicated that the efficiency of TDZ (thidiazuron) was significantly higher than that of BA in the induction of adventitious bud from leaf. Leaves should be first induced on MS medium supplemented with TDZ (1.0 mg/L) and IBA (0.1 mg/L) for 28 days, and then transferred to medium MS+IBA 0.1 mg/L+GA_3 0.05 mg/L. In this way, the regeneration rate reached 92.45%. MS+BA 1.0 mg/L+KT 0.5 mg/L+IBA 0.1 mg/L was suitable for subculture of shoots, with the multiplication coefficient of 3.64. Supplements of GA_3 at 0.5 mg/L in the medium could significantly increase the elongation of shoots. The efficiency of IAA was best among that of IAA, IBA and NAA in the induction of rooting.

The regenerated plantlets rooted well in 1/2 MS medium plus IAA (1.5 mg/L), with rooting percentage of 95.3% (Zhou and Liu, 2009).

Chen *et al.,* (2009) used the young tender leaves from Chinese date Mu-Zao in North of Shaanxi province as explants for tissue culture, the calli were induced from leaves by inoculated in MS or 3/4 MS medium with the affixture 6-BA 1.0 mg.L-1, IBA 0.2 $mg.L^{-1}$, TDZ 0.015 $mg.L^{-1}$, agar 0.55% and sugar 3%. Then, the induced calli were transferred into CY_I basic medium with the affixture 6-BA 1.0 $mg.L^{-1}$, IBA 0.2 $mg.L^{-1}$, TDZ 0.015 $mg.L^{-1}$, agar 0.55% and sugar 3% for the differentiating culture. After cultured continuously for 40 days, the somatic embryos were developed. The histological observations showed that Mu-Zao leaf somatic embryoids could be derived from a single somatic embryonic cell, the embryo sac mother cell or a cluster of embryonic cells. Moreover, the development of Mu-Zao leaf somatic embryoids underwent different stages including a single embryonic cell, and then came through two cell embryo, quadrant embryo, many cell embryo, globular-shaped, heart-shaped, torpedo-shaped, mature embryo and budded stages. During the development process, the conducting tissue could be observed evidently in the area concentrating with embryonic cells, and new-formed meristem in calli was also observed. The active regions of occurrence of somatic embryogenesis and meristem were at surface layer of calli. Therefore, cultured with the selective medium, the calli and meristems could be induced from the tender leaves of Chinese date Mu-Zao in North of Shaanxi province. The development process of the cultured Mu-Zao somatic embryo was similar to that of zygotic embryo of dicotyledonary plants.

A study was carried out by Bekir San *et al.* (2014) to establish protocols for the *invitro*-propagation of Jujube (*Ziziphus* spp) and contribute to the conservation of plant genetic resources. In the study, shoot tips of two selected jujube genotypes (20-C-10 and 20-C-22) were used as a material. MS (Murashige and Skoog) medium supplemented with both TDZ (thidiazuron) and BAP (benzylaminopurine) as a cytokinin was used for micropropagation of jujube genotypes. The highest percentage of explants forming shoots (93.3 %) and the highest number of shoot per explants (5.7) was obtained on the MS medium containing 0.1 mg/l TDZ+0.5 mg/l BAP+0.1 mg/l IBA (indolebutyric acid)+0.3 mg/l GA3(gibberellic acid) in 20-C-10 jujube genotype. The highest rooting percentages of 20-C-10 and 20-C-22 jujube genotypes (83.3 % and 80.0 %, respectively) were obtained on half-strength MS medium supplemented with 2.0 mg/l IBA.

3.1. Role of PGRs on embryogenesis

In order to overcome embryo abortion Wang *et al.* (2009) used male sterile germplasm as female parent in cross breeding of Chinese jujube. Embryo rescue system was explored and established using germplasm 'male sterile No.1 (JMS 1), male sterile No.3 (JMS 3)' and Daxuezao'. The results showed that plant regeneration from immature embryo was obviously influenced by genotype, embryo age and hormones. Among them, embryo age was the key factor. Plants were directly regenerated from embryos at late cotyledonal stage. In contrast, plants regenerated from invisible or barely visible young embryos through two steps, i.e., cotyledonal embryo formation and plant regeneration. A successful rescue for very young embryo took about 60 days. Firstly, the young embryos were cultured in medium MS+ IAA 0~2.0 mg/L+ BA 0~0.5 mg/L + sugar 50 g/L+ agar 3.3 g/L +LH 800 mg/L+ activated carbon 1 g/L until it developed into cotyledonal stage about 30 days later. Secondly, the cotyledonal embryos were transferred into medium MS+sugar 50 g/L+agar 3.3 g/L+ LH 500 mg/L+IBA 0.1 mg/L +BA 0.5 mg/L+ NAA 0.1 mg/L, and regenerated plants were obtained about 30 days after. The rate of cotyledonal embryo formation in JMS 3 and 'Daxuezao' reached as high as 61.0 and 75.6%. Plant regeneration rate of JMS 3 was only 6.4%. In addition, cotyledonal embryo formation in JMS 3 and 'Daxuezao' was not significantly accelerated by 10 days dark treatment. To date, there was few data about the calli development of Chinese date (Chinese jujube).

4. Role of PGRs on Growth and Development (Promotion vs. Dwarfness)

Bhosale (2012) observed that an application GA_3 at 20 ppm, significantly increased the plant height at harvest (3.64m) and plant spread at harvest (3.59 and 3.94m) minimum plant height (2.43) and plant spread (2.68 and 2.86) found in treatment 2, 4-D 20ppm, while treatment 2, 4-D 20ppm recorded the maximum number of main and subsidiary branches.

5. Role of PGRs on Flowering

The inflorescence of ber is an axillary cyme. Each inflorescence consists of 15-30 numbers of flowers. The inflorescence is born on both mature as well as on current season's growth. The process of anthesis begins from 6 AM and may continue upto 4 PM. The dehiscence starts just after anthesis and is completed within 4-5 hours. Generally, ber tree produces two types of flowers, male and hermaphrodite flower. But the percentage of the hermaphrodite flower is few in the early part of the flowering season, however its percentage increases with the advancement of season. Flowering occurs during the month of September and continues upto November in North Indian condition. In Western Indian, flowering starts in August. Flowering period varies from 50-65 days depending upon cultivar and climate conditions of the locality. Cross-incompatibility is also not uncommon. There are some insect pollinators in ber like Honeybee, House fly and yellow wasp. In early varieties, fruits are ready for harvesting in 120-130 days whereas in late varieties 160-180 days.

The production of vegetative shoots in place of reproductive shoots is due to the elevated level of gibberellin which is considered as a vegetative promoter. Paclobutrazol, a gibberellin inhibitor, reduces VP (vegetative promoter) level and thereby increases FP/VP ratio which stimulates flowering shoots in weakly inductive shoots of fruit crops. Foliar application of paclobutrazol (200 ppm) was effective in increasing yield and minimizing fruit drop and fruit cracking in ber (Singh 2000).

6. Use of Plant Growth Regulators in Fruit Set

Ber trees can withstand extremely high summer temperatures and are found to grow well in regions having maximum temperatures of 39 to 42° C and can tolerate temperatures as high as 49-50° C. However, fruit set is adversely affected at temperatures above 35° C. The trees shed leaves and enter dormancy during the extremely hot summers in the subtropics of northern India. This appears to be an adaptive mechanism to escape damage through desiccation during hot weather. Various growth regulators like IAA, IBA, IPA, NAA, 2, 4-D, 2, 4, 5-T and GA have been found to improve fruit set in many crops. Among these chemicals 2, 4-D and NAA (Planofix) have been found to be most effective. In ber, NAA was found most effective and its dose of effectiveness varied in different location and cultivars as reported by many investigators. Singh *et al.* (2001) sprayed varying concentrations of NAA on ber cv. Umran and revealed that NAA @ 10 ppm significantly increased the fruit set. The effects of foliar sprays of NAA were also evaluated by Bhati and Yadav (2004) also. The highest fruit retention (66.44 %) was observed with NAA @ 20 ppm. The application of NAA increased the fruit set significantly than control with all doses of 10, 20 and 30 ppm. Angamuthu *et al.* (2004) reported that in ber cv. Tikdi, fruit set increased with the application of IAA (Indole Acetic Acid) @ 100 ppm but decreased with IAA @ 200 ppm.

Bhosle (2009) has reported that the maximum fruit set percentage (26.67%), fruit retention percentage (48.00%) and decreased the fruit drop percentage (52.00%) and fruit drop numbers per plant (180) found in treatment with GA3 20 ppm.

Application of different plant growth regulators like 2,4-D or 2,4,5-T and GA_3 at

10-15 ppm increased fruit set. These growth regulators may be sprayed twice, once at full bloom stage and second spraying is done one month later. Both the hormone Auxin and GA_3 induces pollen germination and pollen tube growth (Singh and Choudhary, 2012).

Arora and Singh (2014) conducted one experiment on ber, and they have taken eight years old healthy and uniform trees of ber cultivar Umran and seven growth regulator treatments used were NAA 10, 20 and 30 ppm, GA3 10, 20 and 30 ppm and control (water spray), applied in 3rd week of November at active growth phase with four replications each. They found NAA@30 ppm gave maximum fruit set and reduced fruit drop. Gill and Bal (2009) also noted that spraying of NAA@ 30 ppm once in October and second in November, drastically reduces the fruit drop and improves the fruit set percentage

7. Use of Plant Growth Regulators in Controlling Fruit Drop

Fruit drop refers to natural tendency of trees to shed some of their immature fruits. Fruit trees often set more flowers than they need for a full crop, to offset sudden loses from weather or other factors. Fruit trees set fruit so that they can produce seed. A large crop strains the tree's resources and result in smaller fruits, possibly of lesser quality. So the tree protects itself and its seed by thinning the crop, once it senses weather and growing conditions are stable. The immature fruits compete for same food and water and only the strongest retained. Fruit trees actually start this thinning process earlier in the season by shedding some flowers that are not pollinated. At this stage, it is not noticed because the flowers are expected to drop. But when there is drop of fruits, it causes a serious problem. Fruit drop is a major and serious problem in ber production. A heavy fruit drop is a major obstacle for expansion of ber in certain region. The bulk of fruit drop occurs at early stage of fruit development *i.e.* during second fortnight of December. Many reasons such as hormonal imbalance, abortion of embryo and inclement weather have been ascribed to drop immature fruits in *ber* (Singh *et al.*, 1991).

Kumar *et al.*, (2017) concluded that fruit drop is a serious problem associated with ber. It fruits in the months of September-October and fruits ripe during February-March under North Indian conditions. During their long period of stay on tree, fruits face the vagaries of climatic conditions changing from humid rains to chilling winter and ultimately spring season. These changes may be attributed as a major cause of fruit drop, however attack of powdery mildew, forced scarcity of nutrients and plant hormones in developing fruits may further add to this quantum of premature fruit loss.

Generally the number of fruit set is very high, but the extent of fruit retention varies according to the cultivar type and on the level of production of endogenous plant hormones. Several studies have been done on fruit drop and level of fruit retention and the past records have shown that PGRs can reduce the fruit drop by overcoming the two major reasons of fruit drop i.e. embryo abortion and premature abscission. After successful pollination followed by fertilization, the ovary of the flower develops into fruit and the ovules into seed. In the absence of pollination, the ovary abscises and falls. Therefore, pollination supplies the necessary stimulus for the development of ovary. The stimulus is in the form of auxin. At the outset, pollen grains themselves constitute a source of auxin (Fitting, 1909). The amount of auxin supplied by pollination, however, is quite small and it is barely sufficient for future fruit development. At this stage if auxins are supplied exogenously they help in normal fruit setting and development. The amount of auxins is higher in developing fruits but as the fruits mature, the auxin level decreases and if it falls below a critical point called as 'auxin gradient' which resulted in fruits tends to fall (Addicott and Lynch, 1951). The foliar application of auxins prevent formation of premature abscission layer. Before abscission the activity of cellulase and pectinase enzymes increases which cause formation of abscission layer. The exogenous supply of auxin across the abscission zone prevents cellulase formation which delays abscission and thus reduces fruit drop.

Foliar application of KNO3 (1.5 %) alone or in combination with urea at pea size stage increased fruit size and quality in cv. Umran (Yadav, 2001). It is established that plant growth regulators can significantly change the hormonal status of the plant resulting good fruit retention and thereby improve the fruit yield and quality.

Karole (2014) carried out an experiment to find the effect of plant growth regulators on ber cultivar Gola under Malwa plateau conditions and observed that the maximum fruit set and minimum fruit drop can be obtained with the application of NAA @ 30 ppm at fruit set stage and superimposed after one month of fruit setting. All the treatments of varying doses of NAA and GA_3 improved the fruit setting significantly than control, which implied that, the application of growth regulators have immense scope in improving the fruit setting of ber. The significant reduction in fruit drop by foliar application of GA_3 is also observed. This is due to the role of exogenous GA_3 in inhibition of ethylene production (Cooper and Henry, 1973) and higher endogenous level of gibberellin probably counteract the effect of endogenous ABA which has been described to cause fruit drop in several fruits. Singh *et al.* (1987) conducted an experiment in which ber cv. Banarasi Karaka was sprayed with GA_3 at 20,40,60 and 80 ppm at flowering time and found maximum fruit set (20.07%), least fruit drop (56.10%) and higher retention of fruit (4.70%) with 80 ppm GA_3. Ram *et al.* (2005) studied the effects of the plant growth regulators NAA (10, 15 and 20 ppm) and gibberellic acid (15, 25 and 35 ppm GA_3) on the fruit retention of *ber.* The spray of NAA and GA_3 at different concentrations was done and the most suitable concentrations were 15 ppm NAA and 25 ppm GA_3 which reduced the fruit drop and led to fruit retention. Singh *et al.* (2001) also reported the effect of plant growth regulators on fruit drop, yield, and quality of ber cultivar Umran and reported that GA_3 at 60 ppm recorded the lowest fruit drop (75%) and highest fruit set (16.6%).

8. Role of PGRs on Fruit Growth and Development

Auxins are known to play fundamental role in determining fruit growth. Two sprays of GA_3 (40 ppm), first on October and second on November significantly increased weight of fruit and pulp weight (Singh *et al.*, 2001).

9. Role of PGRs on Parthenocarpy

Seedless fruit production is yet to achieve in ber, however seed size can be reduced by applying different plant growth regulators. Spraying of GA_3 (30 ppm) during fruit set reduces the stone weight and stone weight percentage of fruit (Banker and Prasad, 1993).

Two spraying of NAA at 10 ppm, once at full bloom stage and second spraying 15 days later reduces the stone weight of fruit, which ultimately increased the pulp percentage (Kale *et al.*, 2000). Maximum decrease of 21.45 per cent in stone percentage over control obtained with the application of NAA 30 ppm followed by 18.59 per cent decrease by GA3 50 ppm. (Arora and Singh, 2014).

10. Use of Plant Growth Regulators in Fruit Size and Quality Improvement

The larger fruit size can attract immediate attention in the market but to convert on looker to a buyer, the fruits must be having better organoleptic properties which include fruit palatability, fruit colour, surface texture and uniformity. Thus in improvement of any fruit, taste and organoleptic qualities cannot be compromise. Ber fruit size should be such that the weight of individual fruit is more than 20 g, fruit colour is bright golden, fruit surface is firm, glossy and smooth and fruit specific gravity is less than one. The better sized and tasty fruits cannot provide full value to consumer money if after consuming the fruit; they encounter bigger stone within the fruit. Thus an ideal ber fruit must have smaller sized stone. It is a common misconception that bigger sized ber fruits have bigger stone. The application of exogenous plant growth regulators can bring changes in chemical properties of fruits which can deteriorate or improve the flavour. The PGR effect the total soluble solids content of ber. The exogenous application of growth regulators had direct effect on assimilation of metabolic compounds within the fruits which improved the TSS content significantly which in turn enhance fruit flavour. The improvement in TSS is very important from processing point of view, as products like ber candy, jelly, dried ber etc. have direct association with TSS percentage of the fruit.

Ber is one of the richest sources of vitamin C and is valued for in nutritional properties. Ber the poor man's fruit is one of the cheapest sources of vitamin C and likewise other biochemical constituents vitamin C content is also directly influenced by application of PGRs. NAA causes improvement in ascorbic acid content in ber due to the higher synthesis of some metabolic intermediary substances which promote the greater synthesis of glucose-6-phospahte, which is a precursor of ascorbic acid (Orzorek and Angell, 1974).

To improve fruit quality with plant growth regulators, ber plant of different cultivars were sprayed with GA_3 and NAA (10 and 20 ppm) at full bloom and 15 days after flowering. The average fruit weight and size were increased appreciably with 20 ppm GA_3. The maximum decrease in stone weight was recorded with 10 ppm NAA. Fruit quality was improved with 20 ppm GA_3. This concentration induced higher increases in total soluble solids and reducing sugars and a higher decrease in acid content compared with other treatments (Kale *et al.*, 2000).

Total Soluble Solids, reducing sugar and total sugar significantly increased by spraying of ethephon @400 ppm. Spraying of NAA@30 ppm during November improved fruit set percentage appreciably. Fruit weight, pulp: stone ratio, TSS and shelf-life increased at lower dose of paclobutrazol in ber (Singh 2000).

Ram *et al.*, (2005) also found the effects of the plant growth regulators NAA (10, 15 and 20 ppm) and gibberellic acid (15, 25 and 35 ppm GA_3) on the physicochemical parameters of ber. The spray of NAA and GA_3 at different concentrations improved the length and width of fruits, pulp weight, total soluble solids and total sugar in the ber fruit. The most suitable concentrations was 15 ppm NAA and 25 ppm GA_3. This increase in fruit length can be attributed to the involvement of PGRs in cell division, cell expansion and increased volume of inter-cellular spaces in the mesocarpic cells. The application of NAA has a role in increasing the auxin level of fruits which, in turn help in the development of fruit components as there is direct correlation between auxin content and fruit growth in several plants. The fruit weight was found to be significantly improved by higher concentration of GA_3 and

NAA. However, the combination of GA_3 and NAA was not much effective while 20 ppm GA_3 was most effective in inducing highest increase in total soluble solids as compared with other treatments. Rachna (2009) observed maximum increase in fruit size (length and breadth) with the application of NAA 30 ppm.

GA_3 (20 ppm) improves TSS and total sugar content of the fruit. Fruit volume can be improved by spraying of GA_3 20 ppm (Bhosale, 2012). Significant increase in fruit size traits i.e. fruit length, breadth, weight and volume was recorded with application of NAA 30 ppm (Arora and Singh, 2014). Growth regulator such as 2, 4, 5-T has immense potential to improve the quality of ber fruit. Two time spraying of 2, 4, 5-T first on October and second one month later on November enhance the reducing sugar, vit-C content in the pulp and improves Brix-Acid ratio and reduces the pulp acidity content along with also improve the fruit size and increase the fruit pulp weight (Suresh *et al.*,2017).

11. Use of Plant Growth Regulators on Fruit Yield

The yield of ber depends upon soil, nutrition, irrigation and precautions against serious pests and diseases mainly fruit fly and powdery mildew. Plant growth regulators play a significant role in many physiological phenomena which in turn improved the fruit yield. Various types of plant growth regulators like IAA, IBA, NAA, 2,4-D, 2,4,5-T, GA and TIBA are used for improving the flowering, fruit set, size and quality of fruit which ultimately affect the fruit yield. One cannot compromise yield for higher fruit set or better sized tasty fruits. There should be a balance between yield and fruit quality.

Several investigations implicating yield increase as a result of growth regulators have been reported. Pandey (1999) reported that GA_3 @ 15 ppm followed by NAA @ 20 ppm resulted in highest fruit yield in ber cv. Banarasi Karaka. Banarasi Karka cultivar of ber is the choicest variety in the state of West Bengal. The cultivar produces good crop and quality fruits during January-February. One of the main constraints in fruit production in this cultivar is occurrence of high natural fruit drop (about 70%). It is established that plant growth regulators can significantly change the hormonal status of the plant resulting good fruit retention and thereby improve the fruit yield. Bhati and Yadav (2004) confirmed these findings by applying NAA @ 20 ppm at pea stage followed by second spray one month after in ber cv. Gola.

A study was taken up on a 6-year-old Banarasi Karka cultivar of ber by Ghosh *et al.* (2009). They raised the plants through *in situ* budding on 2 year old rootstock with the spacing of 3.5

m (row to row) and 4.5 m (plant to plant). The auxin and giberellins were thoroughly sprayed three times just after fruit set at 21 days interval. Results of two years of investigation revealed that application of NAA at 25 mg/l gave significantly highest fruit retention which resulted in highest fruit yield of 120.5 quintals as against 64.7 quintals/ha in control. However, Ghosh *et al.* (2008) signified increase in yield of Banarasi Karaka cultivar of ber with application of NAA at 25 ppm but there was no beneficial effect of GA_3 on improving fruit yield. The improvement in yield with application of PGR can be contributed to increased fruit set and decreased fruit drop with PGR application. Moreover, the yield is incremented by use of PGR which enhanced fruit length, fruit breadth, fruit weight and fruit volume. Two sprays of GA_3 (40 ppm), first on October and second on November also decreased the fruit drop (74.25%) and enhances fruit set percentage (Tripathy *et al.*,2009). Bhosale (2012) observed maximum yield kg per plant (23.04 kg/plant), kg per hectare (6382.08 kg/ha) and ton per hectare (6.38 ton/ha) with the application of GA_3 at 20 ppm.

Kumar *et al.*, (2017) conducted a study with different growth regulators to record the physical attributes and yield of ber (*Zizyphus mauritiana* Lamk) fruit of cv. Banarasi Karaka. Solutions of NAA, GA3, and 2,4,5-T mixed with sticker such as teepol and sprayed the trees at fruit set stage during after-noon of clean day with the help of hand automizer in month of September. Results of two years showed maximum fruit yield with 2,4,5-T at 25 and 20 ppm followed by NAA 30 and 20 ppm. However, minimum yield was recorded under GA 25 ppm.

Thirty years of thirty three uniform ber trees of *cv.* Banarasi Karaka giving maximum yield with the spraying of 2,4,5-T at 35 ppm (Suresh *et al.*, 2017).

12. Role of PGRs on Fruit Maturity

Several physiological and biochemical changes in maturing ber have been assessed as possible maturity indices. Such changes include an increase in total soluble solids (TSS), loss of chlorophyll, decrease in titratable acidity, accumulation of carotenoids and increase in ascorbic acid content. At maturity, ber fruits contain sucrose, glucose and fructose in the ratio of 3:1:1. On maturity, the TSS, acidity and ascorbic acid content varies between 12–21%, 0.13–1.42% and 39–166 mg/100g respectively in different cultivars and regions. There is huge potential to improve market acceptability of ber with application of growth regulators.

The specific gravity when approached less than one, the fruits were in harvest maturity and attained golden yellow to deep golden yellow colour. Bal *et al.*, (1992) observed maximum percent maturation and colour development with higher concentration of ethephon 400 and 500 ppm; SADH 2000 and 3000 ppm and morphactin and TIBA at 50 ppm.

Spraying of 100 ppm CCC in mid July induce early flowering by 10 days in ber *Cv.* "Katha" but its CCC spray results in reduced size of the ber fruits. Application of combined spray with 100 ppm GA_3 and 10 ppm NAA counteract it, but this application delay the maturity of the fruits. However, if the above spray is followed by two applications of ethrel@500 ppm + 10 ppm NAA at 7 days interval, the fruit maturity could be obtained in time (Singh and Choudhary, 2012).

13. Use of Plant Growth Regulators in Fruit Ripening

Ripening is the composite of the processes that occur from the later stages of growth and development through the early stages of senescence, and which result in characteristic aesthetic or food quality, as evidenced by changes in composition, colour, texture or other sensory attributes. It is possible that the high rate of respiration and ethylene production during ber fruit ripening are responsible for the short storage life under room temperature conditions (Abbas *et al.*,1997).

The plant growth regulators can be employed to hasten or delay fruit ripening. Application of ascorbic acid (300 ppm) and KMnO4 (0.05% and 0.1%) increased the ethylene evolution and enhances ripening process. Application of KNO3 (1.0 and 1.5%) and ZnSO4 (0.5%) resulted attractive yellowish green colour of ber fruit (Siddiqui and Gupta, 1995).

Sharma *et al.*, (2000) studied pigmentation changes during ripening of ber fruits of cv. Umran harvested at 5 different stages: green (stage I, mature), yellowish green (stage II, colour turning), yellowish (stage III, half life), brownish yellow (stage IV, full ripe) and dark brown (stage V, overripe). They reported that there was a progressive decrease in total chlorophyll and chlorophyll a and b contents in the peel with increase in the ripening of fruit. However, there was no significant change in the content of carotenoids during ripening. An investigation was conducted by Bal *et al.*, (1992) on the effect of ethephon, SADH, morphactin and TIBA at colour break stage to enhance maturity and uniform ripening of ber cv. Umran. The maximum percent maturation and colour development was observed with higher concentration of ethephon 400 and 500 ppm, SADH 2000 and 3000 ppm and morphactin and TIBA 50 ppm.

14. Role of PGR's on Shelf-Life of Fruits

The storage life of ber fruits is extremely short and the rapid perishability of the fruits is a problem. At ambient temperature, shelf-life of 2–4 days is common. Due to surplus of fruits in the local markets during peak season, a substantial quantity goes to waste, resulting in heavy postharvest losses.

Maturity determination at harvest is one of the most important factors that determine storage life and fruit quality. Immature fruits are more subject to shrivelling and mechanical damage, and are of inferior quality when ripe. Overripe fruits are likely to become soft and mealy with insipid flavour soon after harvest. Fruits harvested either too early to too late are more susceptible to physiological disorders and have a shorter storage life than fruits picked at the proper maturity. Ber fruits that are allowed to ripen on the tree normally have a shorter shelf-life, and the best results were obtained when fruits were picked before the onset of ripening. The development of fruit skin colour is one of the most reliable maturity indices. Fruits are harvested at the mature golden yellow stage for Umran and Kaithli cultivars, and at the mature green stage for Mallaey and Bam- badawi cultivars. In general ber fruits required about 150–154 days from fruit set to maturity.

The colour of fruit was slightly improved during cold storage period. The change of fruit colour is due to chlorophyll degradation (Sankhla *et al.* 2006). Similarly, Tembo *et al* (2008) reported that ber fruits stored at 5°C maintained high colour chroma than at ambient storage.

Use of certain chemicals as pre-harvest applications has been reported to affect the weight loss of fruits during storage. Efforts have been made to increase the shelf-life and quality of fruits using calcium compounds. Calcium is involved in a number of physiological processes of plants including cell walls, membranes, chromosome structure and enzyme activities. Ber contain calcium as pectate, carbonate, oxalate and phosphate. Spraying ber fruits 10 days before harvest with $CaCl_2$ (1.7 g/L) with 1% Teepol as a surfactant, reduced PLW (Physiological loss in weight), delayed colour development and maintained good quality ber fruits during the storage period. Reduction in respiration rate was associated with increased calcium con- tent due to exogenous application of calcium nitrate. Spraying of 1% CaNO3 at the colour turning stage improved the shelf-life of fruits at room temperature (Gupta *et al.*, 1987).

Singh *et al.*, (1989) observed that a pre-harvest spray of 0.03% boron and 0.05% zinc along with 50 ppm NAA on Pewandi ber improved fruit quality. Ripening in harvested Umran fruits was delayed by treatments with $CaCl_2$ (4.5 or 9 g Ca/L) solution. Dipping

Gola fruits in 500 ppm thiabendazole, Captan and Dithane M-45 improved the shelf-life by reducing rate of respiration. Dipping of Umran fruits in 0.05 and 0.1% $KMnO_4$ also reduced decay loss during storage. Postharvest dipping of cultivar Gola in 1,000 ppm of $KMnO_4$ at colour turning stage gave the best result, prolonging the shelf-life of fruits to 14 days at room temperature. Fruits of Umran cultivar were dipped for 15 min in a solution of KMnO4 (0.05, 0.1%) using Teepol as surfactant, resulting in reduced PLW, over ripening and decay loss compared with controls at the 6th day of storage (Siddiqui and Gupta, 1995).

The fruits after harvest at peak maturity period along with pedicel and dipped in ascorbic acid, boric acid and sodium carbonate (1, 2, 3 %) each for 5 min. Treated fruits are packed in Corrugated Fibre Board boxes of 2 Kg capacity and kept in cold chamber at 7°C temperature and 90-95 per cent RH. The lowest physiological loss in weight was found in ber fruits treated with ascorbic acid 2 per cent and sodium carbonate 1 per cent. So, Umran ber kept with pedicel could be stored satisfactorily for 21 days in cold chamber (7°C) without any major effect on their quality (Kumar, 2011).

Treatment of ber fruits with cycocel, ascorbic acid and $KMnO_4$ before storage increased shelf life by delaying the losses in weight, vitamin C content and acidity (Siddiqui and Gupta 1995). Fruit quality parameters like palatability, fruit weight, fruit colour, TSS, acidity, sugars, ascorbic acid and total phenolics are highly affected during storage.

Umran fruits in 0.17 per cent Ca salts [CaC12 and Ca (NO3)2] solution extended the shelf life of ber fruits by three days and maintain market acceptability at room temperature (Panpatil *et al.*, 2000).The pre-harvest spray of Bavistin 0.1 per cent extend the storage life of Umran ber up to six days at room temperature by maintaining good quality and highest palatability rating as compared to control (Mukherjee *et al.*, 1999). The general conditions of ber fruits *cv.* Umran was 'excellent' up to 20 days and 'good' up to 30 days in CFB cartons, when stored at 0°C-3.3°C and 85-90% RH (Bal *et al.*, 2002). The rate of ethylene evolution by the fruits of cultivar Umran was reduced by the application of ascorbic acid (150 ppm) and cycocel (500 and 1,000 ppm) throughout the storage period.

Treatments with 500 ppm cycocel for 15 min;1,000–2,000 ppm cycocel for 10 min and cycocel at 200 ppm have been reported to result in lower weight loss and improvement in quality. Cycocel was found to be the most suitable chemical, compared with ascorbic acid and $KMnO_4$. The maximum spoilage observed in untreated fruits kept without pedicel. Bal *et. al.*, (2002) observed little spoilage in ber fruits *cv.* Umran up to 20 days of storage at 0 to 3.3°C and 80-90% RH.

15. Role of PGRs on Stress Management

The plants under stress show inductions in the activities of enzymes associated with the ABA biosynthesis and relative induction in mRNA leading to ABA accumulations. Besides the inhibition of ABA catabolism under stress is also a factor for induced ABA accumulation. Most of the abiotic stress is responded by variation in the levels of abscissic acid whereas, biotic stress enhances ethylene and jasmonic acid levels. It is now believed that other hormones such as gibberellins (GAs), auxin, cytokinin (CYT) and recently brassinosteroids (BRs) do also participate in various types of signal transduction during stress in plants. Recent reviews by Bari and Jones (2009) and López *et al.* (2008) clearly indicate that all phytohormones get activated during any kind of stress and either up regulate or down regulate several genes to let plant exert its effect.

References

- Abbas, M.F. Jujube, I.N. and Mitra (Ed.) S.K. 1997.Postharvest Physiology and Storage of Tropical and Sub tropical Fruits: CAB International, London. 405–415.
- Addicott, F.T. and Lynch, R. S. 1951. Acceleration and retardation of abscission by IAA. *Hort. Science,* **144:** 688-689.
- Angamuthu, M., Sharma, V. P., Singh, S. K., Jindal, P. C., Saxena, S. K. and Nath, V. 2004. Effect of IAA on compatibility and fruit development in ber (*Zizyphus mauritiana* Lamk.). *Indian Journal of Horticulture,* **61:** 359-361.
- Arora, A. And Singh, S. 2014. Effect of Plant Growth Regulators and Nutrients on Fruit Drop and Quality in Ber (Zizyphus mauritiana Lamk.) cv. Umran. *Agric. Sci. Digest.,* **34** (2): 102 – 106.
- Bal J S. 2002. Effect of pre-harvest spray of growth regulators on post-harvest physiology of ber in cold storage. *Proc. National Workshop Post harvest Management of Horticultural Produce.* NPC, Chandigarh pp 161-67.
- Bal, J.S., Kahlon, P.S., Jawanda, J.S. and Sandhu, S.S. 1992. Effect of pre harvest spray of growth regulators at turning stage on the maturity of ber fruits (*Zizyphus mauritiana* Lamk.) *Acta Horticulturae,* **321**: 318-325.
- Banker, G. and Prasad, R.N.1993. Effect of Gibberellic acid and NAA on fruit set and quality of fruits in ber cv. Gola. *Progress in Horticulture,* **22** (1-14), 60-62

- Bari, R. and Jones, J.D.G.2009. Role of plant hormones in plant defence response. *Plant Molecular Biology*, **69**, 473-488.
- Bekir San, Adnan Nurhan Yıldırım, Fatma Yıldırım, Fevzi Mustafa Ecevıt. 2014. A Study on Micropropagation as a Tool for Sustainable Utilization of Jujube (*Zizyphus jujuba* Mill.) Genotypes. *Journal of Science and Technology*, 2(1): 85-94.
- Bhati, B.S. and Yadav, P.K. 2004. Effect of foliar application of urea and NAA on quality of ber cv. Gola. *Haryana J. Hort. Sci.,* **32** (1-2):32-33.
- Bhosale, G. H. 2009. Effect of plant growth regulators on growth, yield and quality of ber (*Zizyphus mauritiana* Lamk.) *Cv.* Mehrun under saurashtra region. JAU (http://krishikosh.egranth.ac.in/handle/1/5810023628).
- Bhosale, G. H. 2012. Effect of plant growth regulators on growth, yield and quality of ber (*Zizyphus mauritiana* Lamk.) *Cv.* Mehrun under saurashtra region. JAU (http://krishikosh.egranth.ac.in/handle/1/5810023628).
- Chen, Z.L., Wang, X.J., Liu, S.P., Cao, W., Ning, X.J., Qi, X.Y. and Zhang, X.Q. 2009. Development of somatic embryo and adventitious buds from the differentiating cultured calli of chinese jujube Mu-Zao. *Acta Hortic.* **840**: 273-282.
- Cooper, W.C. and Henry, 1973. Chemical control of fruit abscission. In: *Shedding of plant parts.* T.T. Kozlowski (ed.) Academic Press. pp. 476-524.
- Dai, L., J. Zhao and M.J. Liua .2009. Tissue culture of Chinese Jujube using different explants. *Acta Hort.,* 840: 293-296.
- Fitting, H. 1909. Die beeinflussung der orchideenbluten durch die diebeestandung und durch anders umstande) *Zeitschrift für Botanik,* **1:** 1-86.
- Ghosh, S. N., Bera, B., Kundu, A. and Roy, S. 2008. Effect of plant growth regulators on fruit retention, yield and physico-chemical characteristics of fruits in ber cv. Banarasi Karka grown in close spacing. *Proceedings of 1st International Jujube Symposium.* pp 18. Agricultural University of Hebei, Baoding, China.
- Ghosh, S.N., Bera, B., Kundu, A. and Roy, S. 2009. Effect of plant growth regulators on fruit retention, yield and physic-chemical characteristics of fruits in ber Banarasi Karka grown in close spacing. *Acta Horticulturae,* **840**: 357-362.
- Gill, P.P.S. and Bal, J.S. 2009. Effect of growth regulator and nutrients spray on control of fruit drop, fruit size and quality of ber under sub-montane zone of Punjab. *J. Hort. Sci.,* **4** (2): 161-163.

- Godara, Jitender Kumar. 2007. Micropropagation studies in ber (*Ziziphus mauritiana* Lamk.) M. Sc. Dissertations. CCS Hayana Agricultural University.
- Gupta, O.P.; Siddiqui, S. and Chauhan K.S. 1987. Evaluation of various calcium compounds for increasing the shelf life of ber *(Ziziphus mauritiana* Lamk.) fruit. *Indian Journal of Agricultural Research,* **21**:65–70.
- Jawandha, S. K.; Randhawa, J. S.; Gill, P. P. S. and Singh, J.2008. Effect of post-harvest treatment on storage quality in 'Umran' ber fruit. *J. Hortl. Sci.* Vol. **3** (1): 48-52.
- Kale, V.S., Dod, V.N., Adpawar, R.N. and Bharad, S.G. 2000. Effect of PGR on fruit characters and quality of ber. *Crop Research*, **20** (2): 327-333.
- Karole Baliram. 2014. Effect of preharvest spray of plant growth regulators and urea on growth, yield and quality of ber (*Zizyphus mauritiana* Lamk.) cv.Gola under Malwa Plateau. M.Sc. Thesis, Department of Fruit Science, Rajmata Vijayaraje Scindia Krishi Vishwa Vidyalaya, Mandsaur.
- Kumar Suresh, Kumar Upesh and Naresh Prem. 2017. Effect of plant growth regulators on yield of indian ber *(Zizyphus mauritiana* L.) fruit. *Int. J. Pure App. Biosci.*, **5**(1): 966-969.
- Kumar Suresh, Upesh Kumar and Prem Naresh. 2017. Effect of Plant Growth Regulators on Yield of Indian Ber (*Zizyphus mauritiana* L.) Fruit. *Int. J. Pure App. Biosci.*, **5** (1): 966-969.
- Kumar, S. 2011. Effect of chemicals on storage life and quality of ber *cv.* Umran.Sc. Thesis, Department of Horticulture, PAU Ludhiana.
- Lopez, M.A; Bannenberg, G, and Castresana, **C.**2008. Controlling hormone signalling is a plant and pathogen challenge for growth and survival. *Current Opinion in Plant Biology*, **11**, 420-427.
- Mukherjee S, Paliwal R and Gupta N K. 1999. Effect of pre-harvest application of some fungicides on the shelf life of ber (*Zizyphus mauritiana* Lamk) fruits cv. Umran. *South Indian Hort.*, **47** (1-6): 172-173.
- Murthy, B.N.S. and Reddy, Y.N. 1989. Improvement of seed germination in jujube (*Z. mauritiana* L.) with growth regulators and nitrogenous compounds. *Indian Hotic. J.*, **15** (1/2): 3-8.
- Orzorek, M. J. and Angell, F. F. 1974. Effects of ethephon on ascorbic acid and soluble solids in processing tomato cultivars. *Hort Science,* **9:** 302-306.

- Pandey, V. 1999. Effect of NAA and GA_3 spray on fruit retention, growth, yield and quality of ber *(Zizyphus mauritiana* Lamk.) cv. Banarasi Karaka. *Orissa J. Hort.*, **27**(10): 69-73.
- Panpatil N V, Munjal S V, Patil S R and Damame. 2000. Effect of post-harvest application of calcium salts on physico-chemical parameters and shelf life of ber (*Zizyphus mauritiana* Lamk) fruits at ambient storage. *Orissa J. Hort.*, **28**(1): 25-30.
- Rachna. 2009. Effect of Plant Growth Regulators and Nutrients on Fruit Drop and Quality in Ber (*Zizyphus mauritiana* Lamk.) cv. Umran. Guru Nanak Dev University, Amritsar-143005, Punjab (India). Ph.D. thesis.
- Ram, R. B., Pandey, S. and Kumar, A. 2005. Effect of plant growth regulators on fruit retention, physicochemical parameters and yield of ber (*Zizyphus mauritiana* Lamk.) cultivar Banarasi Karaka. *Biochemical and Cellular Archives,* **5:** 229-232.
- S. Takhti and A. Shekafandeh. 2012. Effect Of Different Seed Priming On Germination Rate And Seedling Growth of *Ziziphus Spina-Christi* . *Advances in Environmental Biology*, **6**(1): 159-164.
- Sankhla N, Gehlot P, Agarwal P, Choudary R and Joshi S. 2006. Post –harvest ripening of Zizyphus mauritiana fruits: Effect of 1- methylcyclopropene. *Proc 33rd PGRSA Annual meeting,* Canada.
- Sharma, R. K., Gupta, O. P. and Siddiqui, S. 2000. Pigmentation changes during ripening of ber (*Zizyphus mauritiana* Lamk.) fruits on tree and in storage. *Haryana Journal of Horticultural Sciences,* **29:** 201-202.
- Siddiqui S and Gupta O P 1995. Effect of post-harvest aaplications of some chemicals on shelf life of ber(*Zizyphus mauritiana* Lamk) fruits. *Haryana J. Hort. Sci.,* **24**(4): 19-23.
- Silva, P. O., Medina, E. F., Barros, R. S., and Ribeiro, D. M. 2014. Germina- tion of salt-stressed seeds as related to ethylene biosynthesis ability in three Stylosanthes species. *J. Plant Physiol.*, **171:** 14–22. doi: 10.1016/j.jplph.2013. 09.004.
- Singh D K. 2000. Effect of paclobutrazol on yield and quality of different cultivars of ber (Zizyphus mauritiana). Indian Journal of Agricultural Sciences 70: 20–2.
- Singh, A. R, Shukla PK and Singh K. 1989. Effect of boron, zinc and NAA on the chemical composition and metabolites of ber (*Ziziphus mauritiana* Lamk.) fruit. *Haryana Journal of Horticultural Sciences*, **18**:23–28.

- Singh, R., Godara, N.R. and Dahiya, S.S. 2001. Responses of foliar application of growth regulators and nutrients in ber cv. Umran. *Haryana J. Hort. Sci.*, **30** (3-4):161-164.
- Singh, S. P. and Choudhary, M. R. 2011.Production technology of fruit crops in wasteland. Scientific publication (India). Pp 54-55.
- Singh, S. P. and Choudhary, M. R. 2012.Production technology of fruit crops in wasteland. Scientific publication (India). Pp 54-58.Spoel, S.H; Dong, X.2008. Making sense of hormone crosstalk during plant immune response. *Cell Host and Microbe*, **3**, 348-351.
- Singh, S.P., Singh, I.B. and Singh, V.N. 1987. Effect of GA_3 on fruit setting, fruit retention and quality of jujube cv. Banarsi Karaka. *Prog. Hort.*, **19** (1-2): 61-64.
- Singh, Z., Dhillon, B.S. and Sandhu, A.S. 1991. Relationship of embryo degeneration with fruit drop and its pattern in different cultivars of ber. *Ind. J. Hort.*, **48**:277-281.
- Suresh, K., Upesh, K. and Prem N. 2017. Effect of Plant Growth Regulators on Yield of Indian Ber (*Zizyphus mauritiana* L.) Fruit. *Int. J. Pure App. Biosci.*, **5** (1): 966-969.
- Tembo L, Chiteka Z A, Kadzere I, Akinnifesi F K and Tagwira F. 2008. Storage temperature affects fruit quality attributes of ber (*Zizyphus mauritiana* Lamk) in Zimbabwe. *African J. Biotech.*, **7**(8): 3092-99.
- Tripathy, D., Pandey, A. K., Pal, A.K., Yadab, M. P. 2009. Studies on effect of plant growth regulators on fruit drop, development, quality and yield of ber (*Zizyphus mauritiana*) cv. Banarasi Karaka. *Progressive Horticulture*, **41**: (2):184-186.
- Wang, J.R., Zhang, M. and Liu*, M.J. (2009). Plant regeneration from immature embryo in chinese jujube (*Ziziphus jujuba* mill.). *Acta Hortic.*, **840:** 262-272.
- Xu Jiuru; Li Chunli; Sun Jianshe. 2003. Micropropagation of ber (*Ziziphus mauritiana* Lam.). *Journal of Beijing Forestry University*, 2003-03. School of Biological Science and Biotechnology, Beijing For. Univ., 100083,P.R.China.
- Yadav, P.K. 2001. Note on foliar application of urea and potassium sulphate on yield and quality of ber. *Curr. Agri.*, **25**:129-130.
- Zhou, R.J. and Liu, M.J. 2009. Effect of plant growth regulators on tissue culture in chinese jujube. *Acta Hortic.*, **840:** 309-314.

8

Plant Growth Regulators in Citrus

Trina Adhikary, Priyam Chattopadhyay and Parmpal Singh Gill

1. Introduction

Citrus fruits have special importance due to their distinct flavours and therapeutic value. Citrus fruits are an excellent source of vitamin C with fair amounts of vitamins A and B and minerals such as calcium, phosphorus and iron (Nawaz *et al.*, 2008). Like all other fruit crops, the growth and development in citrus is governed by both the intrinsic as well as extrinsic factors. The intrinsic factors comprise of the genotype and the endogenous hormone balance, whereas, the extrinsic factors are the environmental conditions. There are also some other factors, viz., planting density, irrigation and fertilization, which may affect the growth and development of the plant. Since the discovery of the plant growth regulators, they have been used to manipulate plant growth and development for the improvement of quality and quantity of the produce in order to enable the fruit growers to meet to pressure of increasing demand for food of high quality. In citrus, a large number of plant hormones have been evaluated for their commercial application (Rodriguez *et al.*, 2005, Bons *et al.*, 2015). A plant hormone is a chemical signal which is produced in one part of the plant and then transported through vascular bundles to another part where it triggers a response. Hormones regulate plant responses to various biotic and abiotic stimuli.

PGRs are synthetic analogues of naturally occurring plant hormones (Table 1). Foliar applied PGRs are routinely used in various fruit crops for flower and fruit thinning, improving fruit set, growth, and development, controlling vegetative growth, and reducing fruit drop. Plant growth regulators (PGRs) are a tool used to manipulate vegetative and reproductive growth, flowering, and fruit growth and development. Citrus is no exception to the use of PGRs, which can provide significant economic advantages to citrus growers when used appropriately. For example, gibberellins' influence on citrus flowering, fruit set, seedlessness, colour development, and pre-harvest fruit drop varies with those factors listed above (Rokaya *et al.* 2016, Singh *et al.*, 2003, Thapa *et al.*, 2002, Ladaniya *et al.*, 2001).

In order to effective, plant growth regulators must be absorbed by plant tissue. Good spray coverage is essential and climatic conditions that favour absorption (warm and humid conditions) are therefore desirable. Some factors are considered like tree size, canopy density, location of fruit, and type of spray equipment when deciding how much spray material will be required to achieve good coverage. All spray materials should be apply uniformly to the fruiting canopy. Both 2,4-D and GA_3 seem to be compatible with urea, potassium foliar sprays, zinc and manganese micronutrient sprays, and neutral copper sprays, but the timing of growth regulator applications may not coincide with the best time for nutrient sprays.

Table 1. Major plant growth regulator classes, associated function(s), and practical uses

Class	Associated Function(s)	Practical Uses
Auxins	Shoot elongation	Fruitlet thinning; increase rooting and flower formation; sprout inhibitor
Gibberellins	Stimulate cell division and elongation	Increase shoot length, fruit size, and fruit set
Cytokinins	Stimulate ell division	Prolong storage life of flowers and vegetables and stimulate bud initiation and root growth
Ethylene	Ripening, abscission and senescence	Induces ripening and loosens fruit
Abscisic acid	Seed maturation, dormancy	Regulates plant stress
Jasmonates	Plant defence	Wound response
Salicylic acid	Systemic acquired resistance (SAR)	Defence against pathogenic invaders
Brassinosteroids	Developmental processes	Regulate germination and other developmental processes
Strigolactones	Suppress branching and promote rhizosphere interaction	Suppress branching, promote secondary growth, and promote root hair growth

2. General Consideration for Use of PGRs

Since PGRs function directly influencing plant metabolism, plant response can vary considerably depending on the variety and plant stress level. Therefore, it is highly recommended that growers become familiar with PGRs and their effects before application, and should test them on small plots before treating significant acreage. Most PGRs work best when used with an adjuvant (surfactant, sticker, or spreader). PGRs are regulated as pesticides; therefore, the label should be followed. Label is the law regarding use of different PGRs. Table 2 summarizes some of the PGRs that are known to be effective in Florida citrus production.

Table 2. Plant growth regulator sprays-Florida citrus. (The Florida Legislature. 2017)

Variety	Response	Time of Application	Growth Regulator and Formulation
Orange, Temple, and Grapefruit	Pre-harvest fruit drop	November–December. Do not apply during periods of leaf flush.	Citrus Fix 2, 4-D Dichlorophenoxyacetic acid (Isopropyl ester 3.38 lbs/gal)
Navel	Reduction of summer-fall drop.	6–8 weeks after bloom or August– September for fall drop. Do not make late application when fruit is to be harvested early. Do not apply during periods of leaf flush.	Citrus Fix 2, 4-D Dichlorophenoxyacetic acid (Isopropyl ester 3.38 lbs/gal)
Tangerine and Murcott	Fruit thinning; activity is temperature-dependent. Severe over-thinning may result from applications made to trees of low vigor and/or under stress conditions	Mid-May.	Naphthaleneacetic acid, NAA (Fruit Fix 800 g/gal liquid amine formulation
Seedless Grapefruit	Delay of rind aging process and peel color development at maturity; combine with 2, 4-D for fruit drop control	November–January.	Gibberellic acid, GA3 (Pro-Gibb 4% liquid concentrate)2
Tangelos	Improvement of fruit set; can result in small size and leaf drop	Full bloom.	Gibberellic acid, GA3 (Pro-Gibb 4% liquid concentrate)2
Minneola Tangelo	Delay of stem-end rind deterioration	Early December.	Gibberellic acid, GA3 (Pro-Gibb 4% liquid concentrate)2

3. Caution

Growth regulators may cause serious problems if misused. Excessive rates, improper timing, and fluctuating environmental conditions can result in phytotoxicity, crop loss, and/or erratic results. Under certain environmental conditions, 2, 4-D may drift onto susceptible crops in surrounding areas. Observe wind speed restrictions and follow all label directions and precautions.

4. Endogenous Growth Substances of Citrus Tissues

In citrus there is a positive correlation between the level of endogenous growth substances and the degree of growth and development in the developing tissues. Presence of high concentrations of a growth substance in a developing tissue is often taken as evidence that the growth promoting substances play an important regulatory role in that tissue, but the concentrations can still be associated with high rates of both synthesis and consumption. The application of exogenous hormones depends on the presence of endogenous level of growth hormones which ensures the probable positive impact of plant growth and development.

4.1. Endogenous growth substances in leaf

Endogenous level of auxin plays a dominant role in leaf abscission. That is, a high level of auxin in leaf blades retards abscission, while its decrease induces abscission.

4.2. Endogenous growth substances in flowers

Citrus trees, once pass the juvenile phase, bloom every year. Annual flowering of adult trees is affected by several exogenous and endogenous factors. Apart from the exogenous factor, plant hormones or PGRs have an influential role in floral induction of citrus (Nishikawa *et al.*, 2013). Large amounts of auxin were also found in flower organs. The role of gibberellins as native inhibitors of flowering in citrus has been demonstrated through careful treatments with growth retardants and gibberellins. Such a comprehensive view has been achieved for the development of vegetative and generative shoots whose gibberellin-like substances and auxin contents correlate with their elongation. Gibberellin treatment widely used to inhibit flowering in citrus trees. A study revealed that during bud development in citrus, the application of GA_3 has shown to inhibit flower production, leading to greater ratio of terminal flowers in the leafy shoots thus higher development of fruits (Iglesias *et al.*, 2007). These results were also shown by use of ethychlozate and GA_3 for flower induction in citrus fruits but GA3 caused inhibitory effect (Takahara *et al.*, 2001).

Cytokinins were found in all flower organs, but petals were found to be particularly rich in cytokinins, although the expansion of petals is generally ascribed to cell enlargement rather than to cell division.

4.3. Endogenous Growth Substances in Fruits

The situation in fruits is apparently far more complex and might be related to the complex structure of citrus fruit. High auxin activity can be found in ovaries and fruitlets during anthesis and fruit set, ovaries of seedless cultivars containing more auxin than seeded ones. Recent studies based on bioassay of purified extracts from *Citrus unshiu* showed a sharp peak in auxin activity 10 days after full bloom followed by a rapid decrease approaching zero levels at 40 days after full bloom. In developing fruits the GA levels are low just before and after anthesis and approximately double at anthesis. This transitory rise in GA levels can be detected in seeded genotypes as well as in seedless cultivars possessing high or normal ability for setting (Talon *et al.,* 1990). In seeded cultivars, GA increases at anthesis are therefore induced by pollination whereas in parthenocarpic species, the rise is developmentally regulated. Determinations of cytokinins in rough and smooth skinned fruit showed trends similar to those found with gibberellins, rough skinned fruit containing more cytokinin level than smooth skinned fruit. Peel and flavedo has more level of endogenous cytokinins than albedo. Again, fruits approaching maturation had relatively low activity of cytokinin. Natural as well as ethylene-induced maturation involve an upsurge in ABA concomitant with a decrease in gibberellin-like substances. The young fruitlet is an extremely active meristem containing large amounts of auxins, gibberellins, cytokinins, ABA and other growth inhibitors all of which act in concert to control fruit development. Young fruitlets show a peak in ABA concentration during the first few days after anthesis. Fruits seem to maintain a stable, relatively low level of ABA subsequently but levels rise slowly when the fruit approach maturation. A decrease in ABA-like growth inhibitors has been found in regreening Valencia fruits. Benzyladenine which delays peel colour change also reduces the accumulation of ABA. ABA concentrations in mature fruit peel are among the highest reported in the literature. Citrus tissues appear to be unique in the high proportion of "bound ABA," presumably the ABA-glucose ester, which attains 4-10 fold higher levels than free ABA in senescent fruit peel. Increase in ABA-like growth inhibitors and decrease in gibberellin-like substances during ripening and senescence of citrus fruits. The kinetics of ABA accumulation in ethylene-treated fruit and in water-stressed leaves indicates that free ABA starts increasing up to a certain level, beyond which it is transformed into the apparently inactive bound ABA. It has also been observed that pigmented plant tissues usually contain more growth inhibitors than pigmentless tissues. Findings represents that the pigmented flavedo has more inhibitors than the albedo and that the pigmentless petals have low inhibitor contents. Talon *et al* (1992) studied two species of seedless mandarins differing in their tendency to develop parthenocarpic fruits. Satsuma is a male-sterile cultivar that shows a high degree of natural parthenocarpy and a high fruit set. Seedless Clementine varieties are self-incompatible, and in the absence of cross-pollination show a very low ability to set fruit. Satsuma mandarin have high levels of endogenous GAs which are associated with elevated parthenocarpic ability, and in Clementine mandarin, low levels of GAs are associated with reduced parthenocarpic ability.

4.4. Endogenous growth substances in stem/bark

In figure 1 it is mentioned that the total auxin content was remarkably high (800 µg/kg), showing the maximum peak six days after full bloom. The change of acidic auxin content was similar to that of the total auxin content, while that of the neutral auxin showed two peaks (70 µg/kg and 20 µg/kg) six and thirty-two days after full bloom, respectively. The content of ABA reached maximum seven and thirty five days after full bloom and continued to gradually increase after this period.

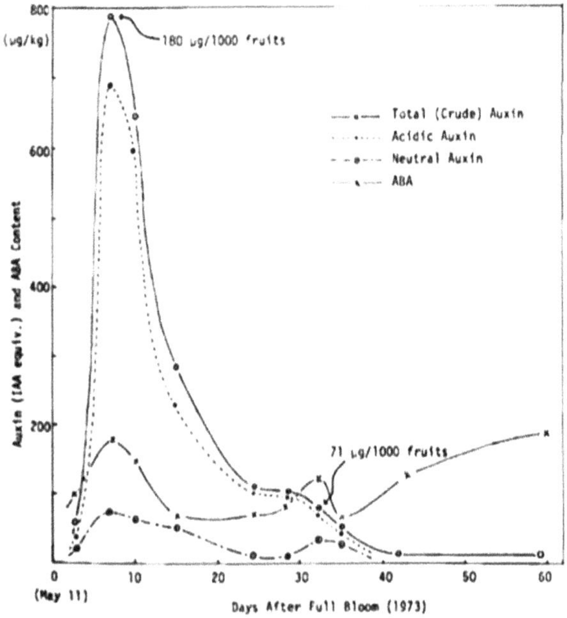

Fig.1. Kinetics of acidic, neutral and total auxin activities and ABA contents in developing *Citrus unshiu* fruits (Yamaguchi *et al.*, 1975).

5. Role of PGRs on Seed Propagation

Several plant growth regulators have been tested as aids in citrus propagation and seedling growth. Objectives are shortening the time of seed germination, increasing seed germination percentage, increase growth rate of seedling. Best results were obtained by soaking seeds for 12 h in 40 mg/l solutions of GA or NAA. 2, 4- D, 2,4- T IAA also showed significant effect.

6. Role of PGRs in Rooting of Cutting

IBA and NAA are still the most widely used auxins for rooting in stem cuttings. It has been repeatedly confirmed by several workers that auxin is required for initiation of adventitious roots on stems and indeed it has been shown that divisions of the first root initiation cells are dependent upon either applied or endogenous auxins (Tanabe, 1982). Anatomical studies of rooting in lemon cuttings revealed that roots can originate from secondary phloem, pericycle, cambium and medullary rays (Sadhu, 1997). Thangji and Wunnachit (1994) observed that the maximum rooting (50%), average root length (3.09 cm) and shooting (33.33%) in pummelo was recorded under 6000ppm concentration IBA. Sandhu and Singh (1986) showed that the percentage of callus formation in stem cuttings of sweet lime (*Citrus limettioides* Tanaka) was highest in IBA 200 ppm and the amount of callus per cutting was maximum in NAA 100 ppm treatment. The proportion of rooted cuttings was higher with IBA 100 ppm. However, number of roots per cutting and root length was more with 200ppm NAA. They also found that there was no correlation between callusing and rooting. Bhusa *et al.,* (2001) showed that the rooting abilities varied among species and varieties (0-100%). Rough lemon and lemon had 100% rooting while Tengu and Kuno Satsuma mandarin had 0%. In common trifoliate orange, the maximum rooting (100%) was found in June cutting followed by July cutting (90%) and the minimum rooting (30%) was recorded in September. There is a Synergistic improvement in rooting of cuttings is due to treatment with IAA or IBA + CCC and IAA + ascorbic acid as well in lemon. It is observed that IBA 1000-2000mg/1 with 2% ascorbic acid was the best treatment for increasing rooting cuttings of sweet lime. Rooting procedures also includes, quick dipping of cuttings, rooting induced chemical powder or solutions of IBA or NAA. 2, 4- D, 2, 4- T, IAA, etc. (Aboutalebi Abd and Tafazoli, 2006-2007).

Leaf blade, petiole and a short piece of stem bearing axillary buds), taken from the apical shoots of lemon (*C. limon*) cv. Assam, were treated with IBA (1000, 2000, 3000 and 4000 ppm) and planted in two rooting media (sand and soil media). Rooting and root characteristics were recorded 70 days after planting, while survivability of rooted cuttings was studied one month after replanting in nursery bed. Results showed that the highest percentage of rooting (98.5%) and survival (100%) were recorded for cuttings treated with 3000 IBA ppm under sand medium, while the lowest rooting and survival were obtained in untreated cuttings. Cuttings grown in sand with 3000 ppm IBA also proved to be significantly superior to the rest of the treatments with regard to number of primary roots, root length dry weight of roots, number of leaves and shoot length per cutting. IBA at higher concentration (above 3000 ppm) showed an inhibitory effect on rooting and root characters (Nath, 2000).

Table 3. Use of stem cuttings in different citrus species

Citrus spp. (Cultivar)	Type of stem cutting	Method of treatment	Optimum concentration (mg/l or ppm)	References
Citrus medica L. (Corsian)	Semi hard wood	Quick dip	IBA 500 and 1000 ppm	Kako Al-Zebari *et al.*, (2015)
Citrus aurantifolia L (Kagzi lime)	Hardwood	Quick dip	IBA 500 ppm	Bhatt and Tomar (2011)
Citrus limon B (Pant lemon -1)	Hardwood	Quick dip	IBA 2000 ppm	Singh *et al.*, (2013)
Citrus aurantifolia L (Mexican lime)	Hardwood	Quick dip	IBA 6000 ppm	Elsheikh, 1999
Citrus latifolia Tan. Lima acida. (Tahiti)	Semi hard wood	Quick dip	IBA 2000 ppm	Prati *et al.*, 1999
Citrus limon B (Baramasi)	Semi hardwood	Quick dip	IBA 2000 ppm and pHBA 2000ppm	Rakesh-Kumar *et al.*, (1995)
Citrus limon B (Baramasi)	Hardwood	Quick dip	IBA 2000 ppm	Kumar *et al.*, (1995)

7. Role of PGRs on Layering

PGRS improved root production in air layers of seedless lemon, using containing 1 g/l of either IBA or IAA.

Dutta (2000) studied the effect of auxin and non-auxinic compounds on root initiation in air-layers of sweet oranges. Application of 5000 ppm IBA + 200 ppm PHB (P-hydroxybenzoic acid) promoted 100% rooting 45 days after layering and 98.00% plant survival 6 months after planting of rooted layers. Dubey and Yadav (2003) carried out an experiment on sweet orange in mid-altitude of Arunachal Pradesh, India. The rooting ability in terms of number of primary roots, length and diameter of primary roots were carefully studied.

Application of NAA was significantly affected in successful production of air layers. The highest success of 31.20% was recorded at 400 ppm NAA followed by 200 ppm (22.63%) and the least was in the highest concentration of NAA (20.45%). The final survival of air layers in the nursery after 90 days of severing was also recorded maximum at 400 ppm NAA (78.20%), followed by application of 600 ppm NAA (73.77%) whereas the lowest was recorded at 200 ppm NAA (71.10%). It was concluded that rooting could be enhanced in layers by the exogenous application of NAA and the 400 ppm NAA may be the best concentration for getting high success and survival of sweet orange air layering.

Lime plants of uniform vigour and size were selected for air layering that were treated with IBA (500, 750, 1000 ppm) IAA (500, 750, 1000 ppm), NAA (500, 750, 1000 ppm), and combination of IBA 500ppm+ IAA 500ppm, IBA 750ppm+ IAA 750ppm, IBA 1000ppm+ IAA 1000ppm, IBA 500ppm+ NAA 500ppm, IBA 750ppm+ NAA 750ppm, IBA 1000ppm+ NAA 1000ppm, IAA 500ppm+ NAA 500ppm, IAA 750ppm+ NAA 750ppm, IAA 1000ppm+ IAA 1000ppm, in rainy season. IBA at 1000 ppm was found to be most suitable for increasing the callusing, rooting and survival of air layers of kagzi lime as single application of growth regulator. Application of IBA 1000 ppm + NAA 1000 ppm was most effective amongst the treatments of combined as well as alone application of growth regulators to obtain more callusing, rooting and survival of air layers of kagzi lime (Dhakar 2014).

8. Role of PGRs on Budding

Buds are often excised in orchard trees to induce vegetative growth responses to various growth hormones. The effect of exogenous IAA, GA and cytokinins on the time required for sprouting varied at different periods along the season, defined as concept of "critical levels" in the endogenous balance of hormones. In summer buds, IAA delayed sprouting, while G A enhanced it and caused shoot elongation. Cytokinins specifically induced the formation of numerous adventitious buds, whereas ABA completely inhibited sprouting; this inhibition, however, was reversible. A marked decrease in total protein and in their synthesis was evident during the first 20 days of sprouting induction and early bud growth.

Success of bud grafting increased bud growth, and shortening the time period between bud grafting and outgrowth of bud using PGRs also reported. IBA, IAA improve bud grafting success whereas 6-benzyladenine (BA) and 6-(benzylamine)-9-(2-tetrahydropyranyl-) 9H-purine increased bud growth.

9. Role of PGRs in Micro-Propragation

Among the different citrus species, Mandarin is difficult to root, IBA is commonly used for rooting. Other rooting aids includes NAA, IAA. Growth regulating chemicals are used extensively as a supplement to invitro media for propagation and regeneration of citrus. In the laboratory, 2, 4-D (2mg/l) and kinetin (0.5mg/l) proved successful in callus production for regeneration of triploids (seedlessplants) from citrus endosperm (Jaskani *et al.,* 1996). An experiment conducted by Tallón *et al.* (2012) on 'Cleopatra' and 'Alemow' mandarin at Murcia, Spain. The influence of various basal medium and plant growth regulators on the efficient micropropagation of nodal explants from mature trees of alemow, and 'Cleopatra' mandarin citrus rootstocks was studied. Rootstocks shoot cultures showed a preference for high-salt media, like Murashige and Skoog or Driver and Kuniyuki Walnut medium. Several combinations of N6 -benzyladenine (BA) and adenine (AD), kinetin (KIN) or gibberellic acid (GA) were tested to optimize the shoot proliferation phase. BA/GA combinations improved the proliferation of all the rootstocks studied, especially Alemow. The addition of BA and AD to the culture medium improved shoot proliferation in sour orange and 'Cleopatra' mandarin in the same way as BA and GA. The addition of different combinations of BA/KIN did not result in further improvement of any of the studied variables. The transfer of in vitro shoots to rooting media, containing different concentrations of indolebutyric acid (IBA) and indoleacetic acid (IAA), resulted in regeneration of complete plantlets. Alemow and 'Cleopatra' mandarin shoots rooted well using these plant growth regulators. To improve rooting 'Cleopatra' mandarin, different combinations of naphthaleneacetic acid (NAA) and IBA were tested. All NAA/IBA combinations produced higher rooting percentages than did the IBA/IAA combinations, and in sour orange nearly 100% of explants developed roots.

Citrus jambhiri is an important rootstock for a number of Citrus fruit crops including lemons, oranges, grape fruits, kinnows and mandarins. Because of its high vigour, well adaptation to warm-humid areas with deep sandy soils and resistance to citrus tristeza virus. It grows easily in old citrus soil. However, it is very cold sensitive and susceptible to phytophthora. The conventional breeding approaches have their own limitations to solve this problem of citrus industry because of perennial nature of the crop, long juvenility, nucellar polyembryony, heterozygocity etc. Recently, various biotechnological approaches have been developed for improvement by creating the variations and selecting the variants for potential needs and it has been established that the plants multiplied through in vitro cultures can serve as sound plantation stock. For establishment of micropropagation of *C. jambhiri* via callus induction and regeneration, explants like leaf segments, nodal segments and root segments are used. Different concentrations and combinations of plant growth regulators i.e. BA, KN, ME, NAA and IAA are used for regeneration. Maximum callus induction (98.66%) was observed from leaf segments on MS medium supplemented with 2, 4-D (4 mg/L). For nodal segments, maximum callus induction (96%) was observed with 2, 4-D (1 mg/L) and from root segments, it was 48.66% on MS medium supplemented with 2, 4-D (2 mg/L). Callus raised from leaf segments showed maximum regeneration (57%) with NAA (0.5 mg/L) + BA (1 mg/L) whereas nodal segments showed better regeneration (71.89%) with NAA (0.5 mg/L) + BA (3 mg/L). However, callus from root segments did not regenerate. Regenerated shoots were rooted on MS medium supplemented with different plant growth regulators and best response (71%) was observed with NAA (0.5 mg/L) reported by Savita *et al.* (2010).

10. Role of PGRs on Flowering

Mehouachi *et al.*, (2000) found that GA levels did not differ between treatments from anthesis until 84 days later when fruitlet abscission was induced in Clausellina Satsuma mandarin by removing either old or young leaves. Okada (2000) found increased levels of ABA in leaves of defruited Aoshima satsuma mandarin trees, a heavy alternate bearer, during the flower induction period when compared to control trees. The control trees produced almost no flowers the next spring and Okuda (2000) suggested that a minimum ABA threshold level is necessary during this period for adequate flowering in the next spring. Since "on" trees usually have higher ABA levels during the flower induction period, ABA cannot be the deciding factor in flowering when considering the evidence to the contrary.

11. Role of PGRs on Fruit Set

Thirugnanavel *et al.*, (2007) reported higher fruit set by application of GA_3 which might have influenced pollen germination and pollen tube growth. PGR ProGibb (active ingredient: Gibberellic Acid 4%) used in citrus, by spraying thorough fruit wetting. In most of the cases, higher water volumes causes some drop of older mature leaves. However, application in low vigour or under stress condition (pest, nutritional, or water etc.) has sometimes caused severe leaf and or fruit drop. Application of copper fungicides and or oils within 3 weeks before or after the ProGibb 4% application has known to result in significant leaf and fruit drop control. Higher percentage of fruit set was found with the application of growth regulators such as 2, 4-D (10, 20 and 30 mg/l), GA_3 (5, 10 and 20 mg/land) NAA (5, 10 and 20 mg/l) during the 2nd week of March at full bloom (3.5 l/tree) on 10-year-old 'Kinnow' mandarin plants grafted onto 'Rough lemon' (*Citrus jambheri* Lush) rootstock. Exogenous application of all the PGRs significantly increased fruit set percentage (from 43.89% with 10 mg/l 2, 4-D to 82.77% with 10 mg/l NAA) leading to an increase in the total number of fruits/plant and fruit weight/plant (yield). Juice percentage, total soluble solids, acidity and vitamin C concentrations in fruit juice were also improved following the application of PGRs. PGRs at lower concentrations enhanced fruit set more efficiently than higher concentration as reported by Azher Nawaz *et al.* (2011). El-Sese (2005) confirmed that the number of fruits per tree and the total yield increased with GA_3 as compared to control. The fruit set was significantly affected by the GA3 treatment individually and also in combination with auxin and gave the maximum fruit set of 32.3 per cent (Saleem *et al.*, 2005). Talon *et al.* (1990) showed that initial fruit set of sterile satsuma mandarins fruit is triggers by GA levels at anthesis that is absent in self-incompatible clementine mandarins. They also found that satsuma mandarin trees set 23% of the initial flowers formed compared to the 3% in clementine trees. Therefore, it would seem that Clementine trees do not reach a certain GA threshold at anthesis that is conducive to sufficient fruit set. To illustrate this, exogenous application of GA_3 at, or shortly after, anthesis increases set in Clementine trees (Guardiola, 2000).

An experiment conducted by Ranganna *et al.* (2017) at AICRP on Citrus, Citrus Research Station, Tirupati, Andhra Pradesh during the year 2015 under Dr. YSR Horticultural University, to find out Effect of plant growth regulators on fruit set and yield parameters of summer crop in acid lime (*Citrus aurantifolia* Swingle) *cv.* Balaji. Plants were sprayed with treatments viz. T1- Spraying GA3 @ 50 ppm during June + CCC @1000 ppm during September + KNO3 2% during October, T2- Spraying CCC @ 200 ppm (Chloromequat chloride) 1st in August, 2nd in September followed by light pruning in September and heavy irrigation in mid-October, T3- Application of Paclobutrazol @ 5 ml per meter canopy 4 times at 45 days interval starting from July to December, T4- Spraying NAA @ 200 ppm during December, T5- Spraying 2, 4-D @ 40 ppm during November + 20 ppm during February, T6- Imposition of 30 days of soil moisture stress during October followed by 50 ppm Ascorbic acid spray at release of stress, T7- Spraying ethephon @ 200 ppm during the month of October, T8- With holding irrigation (Bahar) for 30 days in September followed by application of recommended dose of fertilizers and irrigation in the month of November and along with T9- control (Not spray). It was recorded that higher fruit set was obtained from the treatment application of 2, 4-D @ 40 ppm during November + 20 ppm during February (T5). Highest number of fruits per shoot was recorded with application of Paclobutrazol @ 5 ml per meter for four times starting from July to December (T3), number of fruits per tree was recorded with the treatment application of Paclobutrazol @ 5 ml four times starting from July to December (T3) and Spraying GA3 @ 50 ppm during June + CCC @ 1000 ppm during September + KNO3 @ 2 % during October (T1) has recorded significantly highest fruit yield per tree.

12. Role of PGRs on Fruit Growth and Development

In case of Citrus, the rate of application of the plant growth regulator varies with cultivar, stage of fruit development and climatic region. It is important to ensure optimum tree nutrition and irrigation as well as to provide full tree coverage during application. The mode of action of these plant growth regulators has already been summarized, but it is important to note that they increase vesicle size, not number and that their effect on vesicle size is through an enhancement of cell enlargement, not cell division (Saleem *et al.*, 2005, Jain *et al.*, 2014). Developing fruits are important sinks for water and photo assimilates; their sink strength is increased as a result of auxin treatment, with all fruit tissues increasing in proportion. Auxin application increases fruit peduncle size through an increase in phloem cell size consistent with increased sink strength and carbohydrate movement into the fruit. This indicates that GA is important for fruit development during the initial cell division period, possibly by increased mobilization of metabolites to fruitlets containing high GA levels.

13. Role of PGRs in Control of Fruit Drop

Growth regulators are known to control fruit drop by balancing the internal status of auxin responsible for inhibiting the formation of abscission layer in citrus fruits and improve the productivity as well as quality of acid lime (Jain, 2010). An experiment conducted by Patel *et al.*, (2013) at Jawaharlal Nehru Krishi Vishwa Vidyalaya, Jabalpur on Fifteen years old Kagzi lime trees which indicated that foliar application of urea, NAA, GA_3, 2,4-D, ZnSO4 sprayed at full bloom and pea stage of fruit, significantly increased the fruit retention at various stages of fruit growth and development over unsprayed trees. Fruit retention was maximized (63.89) with the application of NAA 20 ppm followed by urea 2 percent (53.10). On the other hand fruit drop was reduced and noted minimum fruit drop with 20 ppm NAA followed by urea 2 percent in all three waves of fruit drop. The GA_3 and 2, 4-D are used in abscission to control the overloading of fruit for better quality of fruits (Iglesias *et al.* 2007). Reduced fruit growth rate during the early stages of fruit development was positively correlated with fruit abscission, particularly when fruit were treated during the June drop period. The application of 2, 4-D after fruit set results in acceleration of abscission (Lee 2003). The role of 2, 4-D depends on external and internal factors. According to Iglesias *et al.* (2007), decrease in GA and increase the ABA level shows that both these plant growth substances control the initial fruit set and June drop along with other essential component. GA application during flowering increases the initial growth rate of ovaries and decreases the extent of abscission shortly after anthesis; however, GA application at 4 to 6 weeks later has little or no effect on physiological fruit drop (Guardiola, 2000). It has also been observed that the fruit drop percentage was reduced with spray 2, 4-D @ 5 – 20 ppm and

GA3 @ 20 – 40 ppm in Pant lemon-1 and in Navel orange 2, 4-D @ 10-20 ppm and GA3 @ 20 ppm. It has been observed that decrease in fruit drop with spraying 2, 4-D @ 15-20 ppm in mandarins. The reason for reduction of fruit drop due to 2, 4-D application as it is reputed for its inhibition of fruit drop in various fruit crops, by delaying the development of abscission layer through prevention of loss of pectin material in the middle lamella, so that stem still alive at longer time. In the studies of Ranganna *et al.(*2017) (Treatments detail mentioned earlier) reported lowest fruit drop percent with spraying of 2, 4-D @ 40 ppm during November + 20 ppm during February (T5).

14. Role of PGRs on Parthenocarpy/ Parthenopic Fruit Development

This shows that GA could be responsible for the difference in set between cultivars with strong and poor parthenocarpic ability. The application of 20 mg/l GA3 in Kinnow plant during the 2nd week of March at full bloom (3.5 l/tree) significantly reduced the number of seeds/fruit (15.82) compared to the control (19.63) as reported by Azher Nawaz *et al.,* (2011).

15. Role of PGRs on Fruit Maturity and ripening, Fruit Yield and Quality

Fruit quality is a concept that varies according to the final use of the fruit and at what point from orchard to consumer the fruit is evaluated. Ethephon, an ethylene-releasing agent, is the only commercially available chemical that causes no damage to the fruit peel that has shown promising results as a fruit-loosening agent. This compound was registered in Florida on lemons, tangerines, and tangerine hybrids at concentrations of 200 to 250 mg/l. Besides its fruit-loosening effect, ethephon enhances fruit colour. 'Valencia' orange becomes unresponsive to treatment during regreening. Ethephon is used to facilitate both manual and mechanical harvesting. For manual harvesting, ethephon is used on 'Eureka' lemon to reduce the force necessary to separate the fruit from its stem and increase picker efficiency (i.e., number of trees harvested per day), and on 'Dancy' mandarin and 'Orlando' tangelo to facilitate separation of the fruit from the stem, which results in less plugging damage to the fruit (Perez *et al.,* 1992).

According to Kaur *et al.*, (2000) fruit weight increased with increase in amount of 2,4-D in trees of Kinnow mandarin. Eman *et al.*, (2007) carried out an experiment with ten years old Washington navel orange trees (*Citrus sinensis* L. Osbeck) budded on Volkamer lemon (*C. volkamerina* L.) rootstock, grown in sandy soil at Sadat, Egypt. GA at 10 or 20 ppm was sprayed twice (beginning of April and beginning of June) also, chelated zinc at 0.4% was sprayed three times (mid February, beginning of April and beginning of June). GA sprays at 20 ppm were more effective than that at 10 ppm in term of fruit set, fruit retention and yield as fruit number or weight (Kg) per tree. Spraying 0.4% chelated zinc alone or with GA especially at 20 ppm significantly increased fruit set, fruit retention, decreased fruit drop and subsequently improved the yield as well as the physical and chemical fruit characteristics. So, spraying 0.4% 3 chelated zinc at mid-February followed by spraying 0.4% chelated zinc + 20 ppm GA twice at (beginning of April and beginning of June) seems to be the promising treatment for increasing productivity of low yield Washington navel orange trees grown under sandy soil condition. The fruit weight and peel thickness also increased with the application of Zn alone or in combination with GA_3.

Yıldırım *et al.*, (2012) conducted an experiment on Valencia oranges, 29-year-old trees budded on sour orange rootstock in Adana, Turkey. Three concentrations (50, 100 and 150 ppm) of 2,4-DP (dichlorophenoxypropionic acid) were applied during the June drop, when the mean fruit diameter was 13-15 mm. The results indicated that the application of 2,4-DP increased the fruit size of Valencia oranges without reducing yield. The application had a significant effect on fruit yield. Compared with the control trees, applications of 2,4-DP increased the number of large, commercially valuable Valencia oranges (75-80 mm). Moreover, an increase in the concentration of 2,4-DP increased the yield and weight of the fruits. Except for the titratable acid content (TA), which increased with an increase in the 2,4-DP concentration, the internal characteristics of the fruit were not affected by treatment with 2,4-DP. Thus, the results of the present study revealed that 2,4-DP can be used to improve the fruit size of Valencia fruitlets during the June drop.

One study was undertaken by Nawaz *et al.*, (2011) to assess the ability of plant growth regulators (PGRs) to improve fruit set percentage and quality, and to reduce the number of seeds in 'Kinnow' mandarin (*Citrus reticulata* Blanco). Various concentrations of 2,4-dichlorophenoxy acetic acid (2,4-D) (10, 20 and 30 mg/l), gibberellic acid (GA 3) (5, 10 and 20 mg/l) and 1-naphthalene acetic acid (NAA) (5, 10 and 20 mg/l), were applied at the full bloom stage on thirty 10-year-old 'Kinnow' mandarin plants grafted onto 'Rough lemon' (*Citrus jambheri* Lush) rootstock. Exogenous application of PGRs significantly increased fruit set percentage (from 43.89% with 10 mg/l 2,4-D to 82.77% with 10 mg/l NAA) that leading to an increase in the total number of fruits/plant and fruit weight/plant (yield). Juice percentage, total soluble solids, acidity and vitamin C concentrations in fruit juice were also improved following the application of PGRs. The application of 20 mg/l GA_3 significantly reduced the number of seeds/fruit (15.82) compared to the control (19.63).

An experiment conducted by Patel *et al.*, 2013 at Jawaharlal Nehru Krishi Vishwa Vidyalaya, Jabalpur on Fifteen years old Kagzi lime trees which indicated that foliar application of urea, NAA, GA3, 2,4-D, ZnSO4 sprayed at full bloom and pea stage of fruit. Highest (30.38Kg) fruit yield per tree was recorded with the application of 20 ppm NAA which was found to be significantly superior over control and all the other treatments.

In the studies of Ranganna *et al.* (2015) (Tratments details mentioned earlier), highest number of fruits per shoot was recorded with application of Paclobutrazol @ 5 ml per meter for four times starting from July to December (T3), number of fruits per tree was recorded with the treatment application of Paclobutrazol @ 5 ml four times starting from July to December (T3) and Spraying GA3 @ 50 ppm during June + CCC @ 1000 ppm during September + KNO3 @ 2 % during October (T1) has recorded significantly highest fruit yield per tree. According to Sourour (2000) both Zn and EDTA increased the number and weight of fruits per tree. The application of Zn and K increased fruit weight as compared with control.

Trees treated with gibberillic acid yielded more than 10% juice than the control ones (Davies *et al.*, 2001). Such increase was economically beneficial to Florida citrus growers because processed fruit value increased with juice yield.

Granulation of fruit is a physiological condition in which the juice sacs become gelled with little extractable juice. It develops pre-harvest as well as post-harvest. A reduction of the disorder by GA_3 sprays has been reported, but this effect has not been reproduced consistently. Fruit weight increased with increase in amount of 2,4-D in trees of Kinnow mandarin. The fruit weight and peel thickness has been increased with the application of Zn alone or in combination with GA3 in 'Washington Navel' orange. According to Monga and Josan (2000) the highest TSS was obtained from tree sprayed with Zn (0.3%) alone. Treatment with 2, 4-D at 20μg/ml and gibberllic acid at 20 μg/ml increased the TSS in grapefruit juice.

16. Role of PGRs on Shelf-Life of Fruits

Prolonging citrus fruit shelf-life is the goal of researchers working in the area of postharvest and this subject is approached from different angles. Plant growth regulators improve citrus fruit quality reported by EI-Otmani and Ait-Oubahou (1999b). GA3 and 2, 4-D are used commercially to maintain rind firmness and avoid loss of the calyx and thus reduce fruit decay. Conversely, ethylene has a deleterious effect on the storage life of fruit. In their study initiated in 1992 in California, to determine the possible effects of two growth regulators, 2,4-dichlo- rophenoxy acetic acid (2,4-D) and gibberellic acid (GA3) on the quality of ontree-stored and cold- stored fruits of cultivar 'Clementine' mandarin (*Citrus reticulata* Blanco) and cultivar 'Washington' navel orange (*Citrus sinensis* (L.) Osbeck). Treatments were applied at fruit colour break as whole tree sprays, and fruits were either analysed immediately after harvest or stored at 4°C and 85% relative humidity and analysed at 4 and 8 weeks of storage. Regardless of the cultivar, GA_3 alone or in combination with 2,4-D significantly slowed the rate of rind colour change and the rate of rind softening, both on the tree and post-harvest. However, the rate of change in colour and firmness of the rind on the tree and post-harvest was variety dependent. When used alone, 2,4-D had no effect on these factors. No growth regulator effects were noted on weight loss, extractable juice, juice titratable acidity (TA) and total soluble solids (TSS), or the ratio TSS/TA.

An experiment reveled that Sathgudi fruits were treated with the 2,4-D 500ppm + wax 6%, GA3 500ppm + wax 6%, Benzyl adenine 50ppm + wax 6% and wax 6%. Various physico-chemical parameters like physiological loss in weight (PLW), juice content, peel content, firmness, spoilage, colour index, shelf life, total soluble solids, acidity, sugars and ascorbic acid were analyzed by Hemalatha *et al.*, (2015) at an interval of 5 days. It was observed that the fruits treated with BA 50ppm + wax 6% recorded the lowest PLW, spoilage, colour index thereby improved the shelf life compared to control fruits. The highest juice content, peel content and firmness was recorded in BA 50ppm + wax 6% compared to other treated fruits and lowest in control fruits. The lowest TSS, total sugars and highest ascorbic acid content was observed in BA 50ppm + wax 6% treated fruits. Thus, the analysis of various physico-chemical parameters indicated that BA 50ppm + wax 6% preserved the quality of sweet orange compared to other treated fruits. From the present study, it can be concluded that the fruits treated with BA 50ppm + wax 6% has improved the shelf life (18.48 days) by 45.86 percent over control (12.67 days). There was an increase in the spoilage from 5th day (11.41) to 25th day (76.63). Spoilage increased with increase in duration of storage period due to Shrivelling, Alternaria rot, Blue mold and Black mold similar to observed by Ladaniya (2008) in citrus fruits.

17. Role of PGRs on Crop Regulation, Fruit Thinning

Flowering, like all other physiological processes in plants, is under the control of cultivar genes, the environment, and cultural practices. Most citrus cultivars, under optimal conditions flower profusely, particularly in the spring of the "on-crop year" in alternate- or biennially bearing cultivars, and it is common to find more than 100,000 flowers per tree.

In an experiment by Tripathi and Dhakal *et al.*, (2005) suggested that the paclobutrazol applied on 17 July was the most effective in inducing early flowering at fourth week of December, which was 70 days ahead of normal flowering date. The subsequent applications on September, October and December also advanced flowering time by 59, 41 and 32 days, respectively. The earlier (July) application of paclobutrazol was superior among the treatments under Chitwan condition to induce and advance early flowering for offseason market. Thirugnanavel *et al.*, (2007) revealed that application of GA_3 50 ppm in June + cycocel 1000 ppm in September + KNO two percent in October showed better performance in delaying of flowering, number of flower shoot , initial fruit -1 set, fruit retention, number of fruits and yield in acid lime. Martinez-Fuentes *et al.* (2013) observed that the floral bud inductive period, PBZ promotes flowering in Citrus. PBZ significantly increased the percentage of sprouted buds and leafless floral shoots (both single-flowered shoots and inflorescences) and reduced the number of vegetative shoots. By contrast, heavy fruit load trees receiving the same amount of PBZ in the same season or at floral bud differentiation period scarcely flowered. Fruit nullified the effect of PBZ irrespective to treatment date (inductive period or bud burst) as well as the dose applied (1, 10 or 15 g tree−1) or the treatment method (soil application or canopy spraying). In conclusion, the effectiveness of PBZ in promoting flowering in Citrus depends on the fruit load since the tree showed a cultivar-dependant threshold value above which PBZ is unable to promote flowering. Depending on the cultivar, 90 to 99% of the flowers abscise and final set is frequently less than 1% (El-Otmani *et al.*, 1999). During the "off-crop year" flowers are rather scarce. Plant growth regulators have been tested for the purpose of regulating tree production either by reducing vegetative growth and enhancing flowering at "off-year. Small fruit size can be brought about by many factors, including competition, which is particularly important for cultivars that set a large number of fruit such as 'Satsuma' mandarin, 'Wilking' mandarin and 'Murcott' tangor. These cultivars, as well as others such as 'Valencia' and 'Shamouti' oranges, grown under less than optimal conditions may have a considerable degree of alternate bearing. In the "off-crop year" there may be a too large, course rough-peeled fruit, whereas in the "on- crop year" thousands of very small fruit per tree are produced. After extremely heavy crops, alternate bearing can result in tree decline and even collapse in such extreme cases as 'Kinnow' mandarin and 'Murcott' tangerine trees.

There is a competition for carbohydrate reserve among fruits in a tree. Competition between fruits affects final fruit size in citrus. There is an inverse relationship between fruit size and flower number and fruit number per tree. Consequently, flower and fruit thinning, both manually and chemically, have been used to improve fruit size. Hand thinning is costly and time consuming. Thus, chemical thinning has been the desired strategy for cultivars that set excessively and for minimizing the effects of alternate bearing. A significant increase in fruit size occurs only if fruit thinning is considerable and performed sufficiently early in fruit development. Depending on the cultivar, thinning can lead to a serious reduction in yield (kg fruit/ tree).

The most widely tested growth regulators used for thinning include ethephon, an ethylene releasing agent, NAA and ethyl 5-chloro-3 (IH)-indazolylacetate (ethychlozate; Figaron). Auxins such as 2,4- D, 2,4,5-T,2,4-dichlorophenoxypropionic acid (2,4-DP) were also tested but thinning was either highly variable or the compounds caused no thinning at all. NAA, ethephon, and ethychlozate were more consistent as thinning agents, but their efficacy was affected by factors such as concentration, cultivar, timing, tree factors (nutrition and water-deficit stress) and environmental conditions (temperature and relative humidity). According to published literature, NAA is used commercially on citrus, particularly on mandarins, in Israel and in Florida in the U. S. In Japan NAA was used to thin fruit until the discovery of ethychlozate as a fruit thinner and its secondary benefit as a maturation enhancer. For the success with ethychlozate, it is important to ensure that target trees are well watered and are not under any kind of stress .To avoid over thinning, treatments should be applied early in the morning or late in the afternoon and only during a period in which excessive temperatures (>30°C) are not common. These precautions must also be observed with use of ethephon 200 mg/l to thinning as is done in Agentina. Fruit size is an important factor in timing applications. If small fruit have little value, a large reduction in fruit number can be afforded as long as price in the market place is based on size of individual fruit. In addition to an increase in fruit size, advantages of thinning also include prevention of tree collapse, which ensures tree survival. In this case, even a loss in immediate profit by production of low yield over survival of plant can be justified. A compound that was tested recently as a growth regulator is the auxin 3, 5, 6-trichloro-2-pyridyloxyacetic acid (3, 5,6-TPA). It was reported to have a great thinning effect, particularly when used at concentrations >20 mg^{-1}. Intensity of thinning was dependent on the time of application, greater thinning is achieved when the treatment was applied during June drop compared with later applications.

18. Role of PGRs on Growth and Development

In the review of Bons *et al.,* (2015, they mentioned that the fruit size is a desirable character for the marketability of citrus. The variation in fruit size results from difference in cell number or cell size or a combination of both. The potential for increasing fruit size by enhancing cell division early is a relatively undeveloped approach. An ability to increase fruit size is desirable, because for some cultivars small size fruit are of little value in the fresh market. Small fruit size is very common in mandarins. Fruit of unmarketable size are eliminated at the packinghouse causing serious losses in grower profits. Fruit size is mainly determined by the genetic make-up of the cultivar but can be affected by cultural practices, factors related to the tree condition, and climatic factors. Differences in fruit sizes result from differences in cell number or cell size or a combination of both. The relative contribution of each component to final fruit size is unknown. At first cell number is increased secondly cell size increases. Treatment with plant growth regulators such as gibberellins and cytokinins shortly after flower opening enhances early fruit growth but does not always result in an increase in fruit weight at harvest. This early increase in fruit size is due mainly to a transient increase in cell division in the ovary wall. Early application of auxins increases final fruit size more consistently (Agusti *et al.*, 1996, 1998). The auxins 3, 5, 6-TPA and phenothiol are also being used with success on sweet oranges. The rates vary with cultivar, plant growth regulator, stage of fruit development, and climatic region. The mode of action of these plant growth regulators has been summarized elsewhere, but it is important to note that they increase vesicle size, not number, and that their effect on vesicle size is through an enhancement of cell enlargement, not cell division. Developing fruits are important sinks for water and photoassimilates; their sink strength is increased as a result of auxin treatment, with all fruit tissues increasing in proportion. Auxin application increases fruit peduncle size through an increase in phloem cell size. Efficacy of these growth substances is greatest towards the end of the June drop period. After this period, fruit sensitivity to treatment is reduced and higher concentrations may be necessary to obtain an increase in fruit size. It must be noted that early application of these compounds reduces fruitlet growth. Reduced fruit growth rate during the early stages of fruit development was positively correlated with fruit abscission, particularly when fruit were treated during the June drop period. When fruit abscission is increased, competition among the remaining fruit is reduced and fruit size is increased, but yield may be significantly diminished. Auxin application increases fruit peduncle size through an increase in phloem cell size (Agusti *et al.*, 1996) consistent with increased sink strength and carbohydrate movement into the fruit. Reduced fruit growth rate during the early stages of fruit development was positively correlated with fruit abscission, particularly when fruit were treated during the June drop period (Agusti *et al.*, 1995).

19. Role of PGRs on Embryogenesis

In spite of the importance attained by somatic embryogenesis and of the many studies that have been conducted on this developmental process, there are still many aspects that are not fully understood. Among those features, the involvement of plant hormones and plant growth regulators on determining the conversion of somatic onto embryogenic tissues, and on allowing progression and maturation of somatic embryos, are far away from being completely comprehended. Part of these difficulties relies on the frequent appearance of contradictory results when studying the effect of a particular stimulus over a specific stage in somatic embryogenesis. Recent progress achieved on understanding the interaction between exogenously added plant growth regulators over the concentration of endogenous hormones, together with the involvement of sensitivity of the tissues to particular hormone groups, might help clarifying the occurrence of divergent patterns in somatic embryogenesis, and in tissue culture in general. The aspects described above, emphasizing on the effect of the concentration of plant hormones and of the addition of plant growth regulators during the different phases of somatic embryogenesis. An important plant growth regulator for the regeneration of citrus somatic embryos is the cytokinin BA. The use of 13.3 μM BA stimulated the embryogenic response of stigmas and styles while levels of BA higher than 0.44 μM repressed embryogenesis in ovules. In some cases the presence of BA in the culture medium is not necessary to induce the formation of somatic embryos in citrus cultured on Murashige and Skoog medium (MS).

An experiment was conducted by Husain *et al.*, (2015) in Indonesia to develop somatic embryogenesis in citrus (*Citrus nobilis).* 6-benzylaminopurine (BAP) 3 mg/L supplemented to Murashige and Skoog (MS) modified medium was induced direct somatic embryos (SE) 4 weeks after treatments. MS modified medium with single BAP 3 mg/L has resulted the average of eight SE. Nevertheless, average number of SE which resulted from combination BAP 3 mg/L with some level of concentration 2.4-D (0.1, 0.3, and 0.5 mg/L) decreases with increasing the concentration of 2.4-D. Abscisic acid (ABA) is added at concentration 0.5-3.5 mg/L cannot increase the number of SE, globular early-stage, at 4 and 8 weeks. Gibberellic acid (GA3) added at concentration of 2.5 mg/L can increase the growth in the number of normal plantlets in citrus at 8 weeks. Media of enlargement of Tangerine SE plantlets used half-strength MS modified medium concentration. This medium can increase the average in height of plantlets at 4 weeks and the average in a number of leaves at 8 weeks. The regenerated plants were acclimatized under the 60% sunlight block and to keep the humidity remains high the plants were lid with UV plastic. The results indicated that the plant regeneration via somatic embryogenesis from mature seed of *C. nobilis* an efficient and effective technique for producing more SE and plant production in large scale. The next sub-cultured in medium containing GA3 and then the half-strength concentration of medium can increase the number of normal and enlargement plantlets, respectively.

20. Role of PGRs on Seed Dormancy, Germination and Seed Viability

Plant growth regulators at low concentrations (1-10 mg / ml) in promoting, inhibiting or modifying growth and development. The growth regulators were also found to be effective in changing the biochemical properties including protein and amino acids. The growth regulators both from organic (Panchgavya, lemon) and inorganic sources (GA_3, IAA) along with some traditional growth regulators were tried to increase the germination and vigour of citrus seeds.

An experiment conducted by Gurjar (2015) on "Effect of growth regulators and fungicide on seed germination and seedling vigour of Kagzi lime (*Citrus aurantifolia* Swingle)" was conducted during the year 2014-15 at Rewa (Madhya Pradesh, India). The combined application of GA_3+IAA+Ridomil Gold resulted in earliest seed germination in 15 days and its completion in 20 days as compared to the remaining treatments. This treatment has given highest seed germination upto 84.33% and the lowest seed mortality only upto 14.00%. On the other hand, the untreated check took maximum period of 20 days to start germination, 27 days taken to complete germination, lowest (59%) seed germination and highest (37.67%) seedling mortality. All the applied treatments were significantly superior to untreated check.

Shinde *et al.,* (2008) found that the seeds of Rangpur lime were treated with distilled water (control); gibberellic acid (GA3) at 40, 60 and 80 ppm; NAA at 40, 60 and 80 ppm; thiourea at 1.0, 1.5 and 2.0%; KNO3 at 1.0, 1.5 and 2.0%; cow urine; and cow dung paste. The earliest seed germination rate was observed in 10 seeds on the 13th day after sowing due to GA3 at 80 ppm, followed by 9 and 8 seeds due to NAA at 80 ppm and cow dung paste treatments, respectively. The highest rate of seed germination was recorded in 12 seeds on the 18th day after sowing in the case of the 3 treatments. GA at 80 ppm showed 93.33% seed germination, followed by 90.00 and 86.66% seed germination due to NAA at 80 ppm and cow dung paste treatments, respectively. The 3 treatments were at par with each another and statistically significant over the remaining treatments.

Dhaka and Pal (2009) conducted an experiment during 2005-06 in Kanpur (Uttar Pradesh) where three concentrations of GA_3 (450, 12 500 and 550 ppm) were applied by long dip method for 30, 35 and 40 hours to lime seeds. After long dipping, seeds were sown in post in the month of July. Data were recorded for germination of seeds, length and diameter of seedling, number of leaves per seedling, length and width of leaves, length and diameter of root and survival of sapling. GA3 at 500 ppm for 40 hours resulted in better germination, growth and survival compared with the other treatments.

The effect of pre-sowing seed treatment of Gibberellic acid (GA)) solutions on rate of germination and final percentage germination of grapefruit (*C. paradisi* Macf.), kinnow mandarin (*C. reticulata* Blanco) and rough lemon (*C limon* L.) seeds was investigated by Khan *et al.* (2002) Faisalabad, Pakistan. The seed size, coat thickness, thousand seed weight and polyembryony was found to be higher in case of grapefruit followed by kinnow mandarin and rough lemon while moisture content was higher in Kinnow mandarin followed by grape fruit. No significant effect of GA treatment on final percentage germination was observed and only had it a marginal effect on the rate of germination of both the citrus species.

21. Role of PGRs on Stress Management

Endogenous plant growth regulators play an important role in regulating plant responses to stress by sensitizing growth and developmental processes. While the physiological and molecular mechanisms linked to the role of ABA and cytokinins in stress tolerance are well explained, there is growing interest to elucidate the associations of auxins, ethylene, gibberellins, brassinosteroids, and polyamines in stress tolerance mechanism and also on possible cross talk mechanism among different growth regulators during stress tolerance acquisition. Identification and characterization of the gene regulating synthesis of different endogenous growth regulators and recent progresses on hormonal signaling, mutant research, and physiological actions have provided scope for manipulating their biosynthetic pathways for developing transgenic crop plants with enhanced abiotic stress tolerance. Researchers have also provided some leads in exploiting the potential of growth regulators in enhancing the resistance to abiotic stresses of crops.

22. Conclusion

Plant growth regulators have been in use in citriculture for more than half a century. Their use is dictated by the immediate objectives and needs of the citrus industries of the various countries. Some production problems (and thus objectives) are common to many countries and the compounds used to solve these problems are often similar, but concentrations used and amounts applied vary. A few of the compounds have been questioned with regard to their effects on human health and the environment. Most plant growth regulators are used at very low concentrations, which result in low residues and low to no hazards to human health or the environment. Laws regulating the use of plant growth regulators in the different citrus-producing countries are extremely varied and beyond the scope of discussion. An important goal of researchers has been to find ways to increase or enhance the efficacy of plant growth regulators in order to reduce the quantity applied.

Reference

- Aboutalebi Abd, A. and Tafazoli, E. A. 2006-2007. Effects of Cutting Time and Auxin on Rooting of Sweet Lime (*Citrus Limetta* L.) *J. Agric. Sci. Nat. Resources,* **13(5)**:29-37.
- Agusti, M., Almela, V., Zaragoza, S., Juan, M, Trenor, I., Alonso, E., Primo-Millo, E. 1998. Tecnicas para mejorar el tamano del fruto de naranjas y mandarinas. Cuadernos TEchologia Agraria. Serie Citricultura. No. 3, Generalitat Valenciana. *Conselleria de Agricultura, Pesca y Alimentacion,* 15 pp.
- Agusti, M., Almela, V., Zaragoza, S., Primo-Millo, E., El-Otmani, M. 1996. Recent findings on the mechanism of action of the synthetic auxins used to improve fruit size of citrus. *Proc. of Inter. Soc. of Citriculture,* **2**:922-28.
- Agusti, M., El-Otmani, M., Aznar, M., Juan, M., Almela, V. 1995. Effect of 3, 5, 6 trichloro-2-pyridyloxyacetic acid on Clementine early fruitlet development and on fruit size at maturity. *J. Hort. Sci.,* **70:**955-62.
- Azher Nawaz, M., Afzal, M., Waqar, A., Ashraf, M., Jaime, A., Silva, T., Akthar, N., Shahzad, S. M., Hameed, U. and Zahoor, H. 2011. Exogenous application of 2, 4-D, GA3 and NAA at flowering improves yield and quality of Kinnow mandarin (*Citrus reticulate* Blanco). *Asian Australian J. Plant Sci. Bio tech.,* **5(1):** 17-21.
- Bhatt, B. B., Tomar, Y. K. 2011. Effect of IBA and growing conditions on vegetative performance of *Citrus aurantifolia* (Swingle) cuttings. *J. of Hill Agric.,* **2(1):**98-101.
- Bhusa R. C., Moon F.D. G., Rutto K. L. 2001. Propagation of citrus by stem cutting and seasonal variation in rooting capacity. *Pakistan Journal of Biological Science,* **4(11):** 1294-98.
- Bons, K. H., Kaur, N., Rattanpaul, H. S., 2015. Quality and Quantity Improvement of Citrus: Role of Plant Growth Regulators, *Intl. J. of Agr, Env, and Biotech.,* **8(2)**: 433-47.
- Davies, F. S., Campbell, C. A., Fidelibus, M. W. 2001. Gibberellic acid tank mix and adjuvant effects on peel quality and juice yield of Hamlin oranges. *Hort. Tech.,* **11**:171-74.
- Dhaka, S. S. and Pal, S. L. 2009. A study on lime (*Citrus aurantifolia*) seed germination as affected by gibberelic acid. *Annals Hort.,* **2 (2):** 228-229.
- Dhakar, S. K. 2014. Effect of different concentrations of plant growth regulators on growth and survival of air layering of lime (*Citrus auriantifolia*).M.Sc. Thesis, Department of Horticulture, Rajmata Vijayaraje Scindia Krishi Vishwa Vidyalaya, Gwalior, Madha Pradesh.

- Dubey, A.K. and D.S. Yadav 2003. Effect of NAA on rooting of air layers in different varieties of sweet orange. *Orissa J. Hort.*, **31(1):** 51-53.
- Dutta, P. 2000. Effect of growth regulators on air-layering of sweet orange Citrus sinensis (L.) Osbeck. *Environ. Ecol.*, **18(4):** 899-01.
- El-Otmani, M. and Ait-Oubahou, A. 1999a. Plant growth regulators and citrus fruit quality. In: 1st Int/. Symp. Fruit Crops: Production and Quality of Citrus Fruits, Botucatu, Sao-Paulo, Brazil. Pp 221- 49.
- El-Ottnani, M. and Ait-Oubahou, A. 1999b. Improving and or maintaining quality of citrus fruit by pre- and postharvest application on plant growth regulators. In: Schirra, M. (ed.), Recent Advances in Postharvest Diseases rind Disorder Control of Citrus Fruit, Research Signpost, Trivandrum, India. pp 35-55.
- EL-Sese, A. M. A., 2005. Effect of gibberellic acid (GA3) on yield and fruit characteristics of Balady mandarin. *The J. of Agri. Sci.*, **36**: 23-35.
- Elsheikh, S. E. M. 1999. Propagation of lime (*Citrus aurantifolia* L.) by Stem cuttings technique. M.Sc. Thesis, Department of Horticulture, University of Sudan.
- Eman, A. A., Abd, E. M., Abid, E. M. O. and Ismail, M. M., 2007. GA3 and zinc sprays for improving yield and fruit quality of Washington navel orange trees grown under sandy soil conditions. *Res. J. of Agri. and Biol. Sci.* **3**(5): 498-03.
- Guardiola, J. L. 2000. Regulation of flowering and fruit development: Endogenous factors and exogenous manipulation. *Proc. Intl. Soc. Citricult.*, **9:** 342-46.
- Gurjar, B. 2015. Effect of growth regulators and fungicides on seed germination and seedling vigour of kagzi lime (*Citrus aurantifolia* Swingle) M. Sc. Thesis Submitted to the Jawaharlal Nehru Krishi Vishwa Vidyalaya, Jabalpur, M. P. India.
- Hemalatha, V., Babu, J. D. and Sankar, A. S., 2015. Influence of Growth Regulators on the Shelf Life of Sweet Orange Cv. Sathgudi, *Plant Archives.* Vol. 15 No. 2, 2015 pp. 1101-06.
- Hume, H. H. 1957. The citrus fruits. The MC-Millan Co., New York.
- Husain, I., Utami, F. N., Purwito A., Husni A., Mutaqin, K. H. and Susanto, S. 2015. Plantlet regeneration via somatic embryogenesis from mature seed of Citrus nobilis variety SoE from SoE NTT. *Hort. Biotech. Res.*, **1:** 23-29.

- Iglesias, D.J., Cercos, M., Colmenero-Flores, J.M., Naranjo, G., Rios, E., Carrera, Ruiz-Rivero O Leiso, I., Morillon, R., Tadeo, F.R. and Talon, M., 2007. Physiology of citrus fruiting. *Brazil J. of Plant Phys.,* **19**:333-62.
- Jain, M.C., Choudhary, H.D., Sharma, M.K. and Singh, B., 2014. Yield and quality attributes of Nagpur mandarin as affected by use of different plant growth regulators. *Env. Ecol.,* **32**:1141-45.
- Jain, V. K. 2010. Fundamentals of Plant Physiology S Chand & company LTD, New Delhi.
- Jaskani, M. J., Khan, L. A., and Yakub, M. S. 1996. Regeneration of triploids from citrus endosperm. *Proc. Intl. Soc. Citriculture,* **1:**128-29.
- Kako Al-Zebari, S. M., Ali Al-Brifkany, M. A. A. 2015. Effect of Cutting Type and IBA on Rooting and Growth of Citron (*Citrus medica* L). *American J. Exp. Agric.,* **5(2):**134-138.
- Kaur, N., Monga, P. K., Thind, S.K., Thatai, S. K. and Vij, V.K., 2000. The effect of growth regulators on periodical fruit drop in Kinnow mandarin. *Haryana J. of Hort. Sci.,* **29**: 39-41.
- Khan, M. M., Usman M., Waseem R. and Ali M. A. 2002. Role of gibberellic acid (GA_3) on citrus seed germination and study of some morphological characteristics. *Pak 1. Agri. Sci.,* **39:** 2.
- Kumar, R., Gill, D.S., Kausik, R. A. 1995. Effect of indole butyric acid, p-hydroxy benzoic acid and season on the propagation of lemon *cv.*Baramasi from cuttings. *Haryana Hort. Sci.,* **24(1):**13-18.
- Ladaniya, M. S. 2008. *Citrus fruit - Biology, Technology and Evaluation.* In: Preparation for fresh market. Acadamicpress - Elsevier Publications. pp 250 - 51.
- Ladaniya, M. S., 2001. Post-Harvest Technology of Fresh Citrus Fruits. In: Singh, S. and Naqvi, S.A.M.H., Eds., *Citrus, International Book Distributing Co.*, Lakhnow, 459-73.
- Lee, I. J., 2003. Practical application of plant growth regulators on horticulture crops. *J. of Hort. Sci.,* **10**:179- 87.
- Martinez-Fuentes, Mesejo, C., Mu˜noz-Fambuena, N., Reig, C., Gonzalez-Mas, M. C., Iglesias, D. J., Primo-Millo, E., and Agusti, M. 2013. Fruit load restricts the flowering promotion effect of paclobutrazol in alternate bearing Citrus spp. *Sci. Hort.,* **151**:122–27.
- Mehouachi, J., D. J., Iglesias, F. R., Tadeo, M., Agusti, E., Primo, M., and. Talon, M.

2000. The role of leaves in citrus fruitlet abscission: Effect on endogenous gibberellin levels and carbohydrate content. *J. Hort. Sci. Biotech.*, **75**:79-85.

- Monga, P. K., Josan, J. S. 2000. Effect of micronutrients and leaf composition, fruit yield and quality of Kinnow mandarin. *J. Appl. Hort.*, **2**(2):132-33.
- Nath, J. C. 2000. Effect of rooting media and IBA on rooting of leaf-bud cuttings of Assam lemon (*Citrus limon* Burm.). *Horticultural Journal*, 13(2):83-86.
- Nawaz, M. A. Waqar Ahmad, Saeed Ahmad and M. Mumtaz Khan. 2008. Role of growth regulators on preharvest fruit drop, yield and quality in Kinnow mandarin. *Pakistan J. Bot.*,**40** (5): 1971-81.
- Nawaz, M. A., M. M. 2011. Exogenous Application of 2,4-D, GA 3 and NAA at Flowering Improves Yield and Quality of Kinnow Mandarin (*Citrus reticulata* Blanco). Conference: *The Asian and Australasian Journal of Plant Science and Biotechnology*, Japan 5(1):
- Nishikawa, F., 2013. Regulation of floral induction in citrus. *J. of Japan Soc. Hort. Sci.* **82**:283-92.
- Okuda, H. 2000. A Comparison of IAA and ABA levels in leaves and roots of two citrus cultivars with different degrees of alternate bearing. *J. Hort. Sci. Biotech*,**75:** 355-59.
- Patel, N., Pandey, S. K. and. Jain, P. K. 2013. Effect of growth promoting substances and PGRs on fruit retention, drop and yield of acid lime (*Citrus aurantifolia* Swingle) cv. Kagzi. *JNKVV Res. J.*, **47(1):** 70-75.
- Perez, M., Pozo, L., Afion, R., Oliva, H., Noriega, C. 1992. Growth regulators on citrus cropunder Cuban tropical conditions. *Proc. Inti. Soc. Citriculture*, **1:** 503-08.
- Prasanna Lakshmi, N. 2005. Studies on the effect of post-harvest application of chemicals and growth regulators in extending shelf life of mango. *M.Sc. Thesis*, Acharya N.G. Ranga Agricultural University, Hyderabad, India.
- Prati, P., Mourão Filho, F. A. A., Dias C. T. S. 1999. Scarpare Filho JA. Estaquia semi-lenhosa: um método rápido e alternativo para a produção de mudas de lima acida 'Tahiti'. Scientia Agricola. **56(1):**85-190.
- Rakesh-Kumar, R., Gill, D. S., Kaushik, R. A. 1995. Effect of indolebutyric acid, p-hydroxybenzoic acid and season on the propagation of lemon cv. Baramasi from cuttings. *Haryana J. Hort. Sci.*, **24:**13- 18.

- Ranganna, G., Venkataramana, K. T., Mukundalakshmi, L., Swarajyalakshmi, K. and Sudhakar, P. 2017. Effect of Plant Growth Regulators on Fruit Set and Yield of Summer Crop in Acid Lime (*Citrus aurantifolia* Swingle) cv. Balaji. *Intl. J. Current Microbio. App. Sci.*, **6:** 2208-14.
- Rodriguez, V. A., Mazza, S. M., Martinez, G. C. and Ferrero, A. R., 2005. Zn and K influence on fruit sizes of Valencia orange. *Revista Brasileira de Fruitcultura,* **27**: 132-35.
- Rokaya, P. R., Baral, D. R., Goutam, D. R., Sreshtha, A. K. and Paudayal, K.P., 2016. Effect of Pre-Harvest Application of Gibberellic Acid on Fruit Quality and Shelf Life of Mandarin (*Citrus reticulata* Blanco). *American J. Plant Sci.*,**7:** 1033-39.
- Sadhu MK. 1997. Citrus. In: Propagation of Tropical and Subtropical Horticultural crops. Eds. T.K. Bose, S.K. Mitra, M.K. Sadhu and P. Das. Naya Prokash, Calcutta, pp166.
- Saleem, B.A, Ziaf, K., Farooq, M. and Ahmed, W., 2005. Fruit set and drop patterns as affected by type and dose of fertilizer application in mandarin cultivars (*Citrus reticulate* Blanco). *Inter. J. of Agri. and Biology,* **7**: 962-65.
- Sandhu AS, Singh Zora. 1986. Effect of Auxins on the Rooting and Sprouting Behaviour of Stem Cuttings of Sweet Lime (*Citrus Limettioides* Tanaka). *Indian J. of Hort.*,**43**(3&4): 224-26.
- Savita, V., Virk, G.S. and Nagpal, A. 2010. Effect of Explant Type and Different Plant Growth Regulators on Callus Induction and Plantlet Regeneration in *Citrus jambhiri* Lush. *Environ. We Int. J. Sci. Tech.*, **5**: 97-06.
- Shinde, B. N., Kalalbandi, B. M. and Gaikwad, A. R. 2008. Effect of pre- sowing seed treatment on seed germination, rate and percentage of Rangpur lime. *Intl. J. Plant Sci.*, **3(1):** 321-22.
- Singh, A. K., Singh, R. and Mann, S. S., 2003. Effect of Plant Regulators and Nutrients on Fruit Set, Yield and Quality of Pear CV. Le-Conte. *Indian J. of Hort.*, **60**, 34-39.
- Singh, V. P., Mishra, D. S., Mishra, N. K., Rai, R. 2013. Effect of growing season, PGRs and rooting media on survival of hard wood stem cuttings of lemon (*Citrus limon* Burm.) cv. Pant Lemon-1. *Hort Flora Res. Spectrum,* **4(4):**347-50.
- Sourour, M. M. 2000. Effect of some micronutrients forms on growth, yield, fruit quality and leaf mineral composition of valencia orange tree grown in North sinai. *Alexandria J. of Agri. Res.* **45**:1269-85.

- Takahara, T., Ogata, T., Koshita Y., Fujisawa, H. 2001. Effect of ethychlozate, paclobutrazole and other plant growth regulators on the promotion of flowering in Citrus. *Bull Natl First Tree Fruit Sci.,* 35.
- Tallón, C. I., Porras, I. and Pérez-Tornero, O. 2012. Efficient propagation and rooting of three citrus rootstocks using different plant growth regulators. *In Vitro Cell. Dev. Biol.—Plant,* **48:** 488–99.
- Talon M., Hedden P, Primo-Millo E 1990a. Gibberellins in Citrus sinensis: a comparison between seeded and seedless varieties. *J. Plant Growth Regul.,***9:** 201-206.
- Talon, M., Zacarias, L and Primo-Millo. E. 1990. Hormonal changes associated with fruit set and development in mandarins differing in their parthenocarpic ability. *PlantPhysiol.,***79:** 400- 06.
- Talon, M., Zacarias, L. and Primo-Millo. E. 1992. Gibberellins and Parthenocarpic Ability in Developing Ovaries of Seedless Mandarins. *PlantPhysiol.,***99:** 1575-81.
- Tanabe MJ.1982. Single node stem propagation of Alyxia olivarformis. *Hort Sci.,***17:**50.
- Thangjit, S., Wunnachit, W. 1994. Effect of growth regulators on cutting of pummel cultivar Hom Hat Yai. Songklanakarin. *J. Sci. Tech.,* **16**(1):25-30.
- Thapa, B. B., and Gautam, D. M. 2002. Pre-Harvest Application of Gibberellic Acid and Maturity Stage on Fruit Quality and Shelf Life of Mandarin. *J. of the Inst. of Agri. and Anim. Sci.,***23:** 41-47.
- The Florida Legislature. 2017. "§487.021." The 2017 Florida Statutes. Accessed on January 3, 2018.http://leg.state.fl.us/STATUTES/index.cfm?App_mode=Display_Statute&Search_String=&URL=0400-0499/0487/Sections/0487.021.html
- Thirugnanavel, A., Amutha, R., BabyRani, W., Indira, K., Mareeswari, P., Muthulaksmi, S., and Parthiban S., 2007. Studies on regulation of flowering in Acid Lime (*Citrus aurantifolia* Swingle.) *Res. J. Agr.Biol. Sci.,***3**(4): 239-41.
- Tripathi, K. M. and Dhakal, D. D. 2005. Effect of paclobutrazol on off-season flower induction in acid lime (*Citrus aurantifolia* Swingle) landraces under chitwan condition. *J. Inst. Agric. Anim. Sci.,***26:**87-92.
- Yamaguchi, T., Kono, K., Igoshi, K. H., and Sazuki K. 1975. Characterization of plant growth substances in *Citrus unshiu* and their change in fruit development. *Plant Cell. Physiol.,***16:** 1101-11.

- Yıldırıma, B., Yes¸ilog˘lua, T., Incesua M., Kamilog˘lub,M. U.,Cimen, B., and Tamerc, S. 2012. Effects of 2,4-DP (2,4-dichlorophenoxypropionic acid) plant growth regulator on fruit size and yield of Valencia oranges (*Citrus sinensis* Osb.). *New Zealand J. Crop and Hort. Sci.*,**40(1):** 55-64.

9

Plant Growth Regulators in Custard Apple

Vikas Kumar Sharma, Anil Kumar Godara, A.K. Panda and R.K. Tarai

1. Introduction

Custard apple (*Annona squamosa* L.) is a delicious and important minor fruit crop cultivated in tropical and sub-tropical climate. It comes under the family "Annonaceae" and its native is West Indies. The genus *Annona* contains more than 50 species; among them few produce edible fruits of some importance. In India it is cultivated especially in Andhra Pradesh, Bihar, Madhya Pradesh, Maharashtra, Rajasthan, Tamil Nadu, and Orissa states.

The successful commercial cultivation of custard apple depends on many factors such as variety, climate, soil, irrigation, pollination and fertilization, plant density per unit area, age of plant etc. However, poor fruit set which leads to low productivity, is a major constraint in the expansion of its cultivation. Therefore, improvement in fruit set and fruit retention and thus increasing yield is one of the most important practical aspects of custard apple cultivation. Among the various scientific tools applied for the betterment of the crops, plant growth regulators have been widely used. The plant hormones or regulators are the organic chemical compounds, which modify or regulate physiological processes in an appreciable measure in the plants when used in small concentrations. They are readily absorbed and move rapidly through the tissues, when applied to different plant parts. These substances when applied as spray have been found to regulate, modify or accelerate various growth processes of the plants. In custard apple, promising results due to the exogenous application of growth regulators have been reported. Growth regulators have been used to increase flowering, fruit set and fruit retention, yield and improve the quality of custard apple (Chaudhari *et al.*, 2016). The work carried out in India as well as abroad on custard apple is very scanty. Generally, GA_3 and NAA are more popular and more effective and thereby frequently used in custard apple. GA_3 play significant role in cell elongation, breaking seed dormancy, induce flowering, while NAA is effective for cell division and apical dominance in plant. This chapter describes the interference on plant growth regulators on custard apple with following heads.

2. Role of PGRs on Micro-Propagation

Propagation of *Annona* species is usually done by grafting or budding with selected scions on to seedling rootstocks. Seedling rootstocks are highly variable in vigor and disease resistance and consequently scion growth and productivity are also variable (George and Nissen, 1987). However, variability in seedling rootstock performance is a major cause of scion yield and fruit quality reduction (George, 1984). In Asia, soursop is usually propagated from seed (Campbell and Phillips, 1983; Bejoy and Hariharan, 1992). With the exception of a few cultivars, clonal propagation of sugar apple (*A. squamosa*), cherimoya (*A. cherimola*) and *Annona* hybrids by cuttings or marcottage has not been successful (George and Nissen, 1987). A reliable *in vitro* method for the propagation of the *atemoya* hybrid would have considerable benefits for the horticultural industry by allowing the rapid clonal propagation of superior yielding, disease resistant varieties, as well as enabling clonal multiplication of superior rootstock material in areas where grafted custard apple plants are required. There have been a few published accounts of the tissue culture of custard apple (Nair *et al.*, 1984a; Jordan, 1988; Jordan *et al.*, 1991) but rapid multiplication with sustained proliferation has not been obtained. Another problem with the use of micro-propagation for this crop is that *in vitro* rooting has always proved to be difficult. The use of kinetin and BAP during *in vitro* culture of *A. glabra* favoured the development and survival of plants during acclimatization. TDZ had toxic effects on the chloroplasts of this species, decreasing their chlorophyll a content and the photosynthetic potential of their leaves. The use of a sucrose-free rooting medium did not beneficially affect acclimatization of plants upon transfer to the *ex vitro* condition (Oliveira *et al.*, 2008).

BAP and KN act synergistically to produce healthy and vigorous shoots from leaves of *Annona squamosa* in culture. Best response was observed when KN and BAP were added in combination each at 0.5 mg/l. Callus formation was observed with 2,4-D at 0.5 and 1.0 mg/l and with NAA at 1.0 mg/l with simultaneous reduction in the number of shoot buds (Nair *et al.*, 1984b). Haploid plants were induced from anther callus of *Annona squamosa* L.

(Custard apple) on a Nitsch basal medium supplemented with 6-benzylaminopurine and naphthalene acetic acid. When naphthalene acetic acid was replaced by indole-3-acetic acid only multiple shoots were obtained (Nair *et al.*, 1983). Double hormone treatments (Kinetin and BAP) in equimolar concentration had a profound influence on the initial leaf pulvinus abscission and the time taken for shoot bud induction. Kinetin 0.6 mg/l + BAP 0.6 mg/l, whereas, with Kinetin 0.8 mg/l + BAP 0.8 mg/l slight callusing was observed with decrease in number of buds (Nair *et al.*, 1984b).

Table 1: Effect of growth regulators on micro-propagation in *Annona spp*

S. No.	PBR	Concentration	Response/Effect	Reference
1.	IBA	39 μM	Rooting was obtained in sugar apple when sub-cultured shoots were preconditioned for 2 weeks in medium with 10 g activated charcoal before treatment	Lemos and Blake, 1996
2.	BA + NAA	0.2 + 0.1 mg/l	Highest degree of axillary shoot proliferation from nodal explants 80% with 6.0 ± 0.42 shoots per culture	Muhammad *et al.*, 2002
3.	IBA	0.1 mg/l	Highest percentage of root formation with 2.00 ± 0.21 roots per cutting	Muhammad *et al.*, 2002
4.	BAP Kinetin IAA	Alone or in combination	Large scale *in vitro* soot formation has been successfully achieved in *Annona squamosa* L in 7-14 days	Farooq *et al.*, 2002
5.	Zeatin	2.28 μM	Axillary shoot proliferation was obtained after culturing nodal sections from *Annona cherimola* cv. 'Fino de Jete', on MS medium	Padilla and Encina, 2004
6.	BA	0.1 μM	For axillary shoot proli-feration, of *in vitro* nodal segments of 0.5 cm medium is recommended for Multi-plication of *Annona emarginata*	Rodrigo *et al.*, 2016

3. Role of PGRs on Dormancy Removal and Seed Germination

A. squamosa is usually propagated sexually. Its asexual propagation is by grafting used to multiply the most productive clones (Almeida *et al.*, 2010), but this form of propagation is not yet the most widespread, although it is most correct for quality seedling production. Seeds are used to form the root stock. Seed germination of custard apple is uneven and irregular making sexual propagation difficult. Much experimental evidences support the concepts that specific endogenous growth promoting and inhibiting compounds are involved directly in the control of seed development, dormancy and germination (Black, 1980). Custard apple seeds requires 35-50 days for potential germination (Hernandez, 1983). Irregular germination, in custard apple seeds may be due to the morphological dormancy or due to hard seed coat (Adeniji *et al.*, 2014; Baskin and Baskin, 2014). Morphological dormancy occurs when a rudimentary embryo occupies less than 30% of the seed volume at the time of seed dispersion, with an endosperm (endospermic seeds) that fills most of the seed and a differentiated embryo that needs specific conditions to complete its growth before germination (Baskin and Baskin, 2004). According to Pinto (2005), seed dormancy in Annonaceae is due to the degree of seed maturity and other factors like climatic conditions during seed maturation on the mother plant and genetics of the mother plant. Therefore, difficulties in sexual propagation of the species of this family could be partially explained by seed dormancy. The dormancy state in seeds is manifested by delay in germination for longer or shorter periods of time even if the environmental conditions (light, temperature, water, oxygen) are favourable for germination (Ferreira *et al.*, 2009). Weaver (1987) reported that dormancy may be the result of the hormone balance among growth promoters and inhibitors. Nevertheless, dormancy can be overcome and germination started using plant regulators together with imbibitions. Hormones play a fundamental role in the seed germinating process, especially gibberellins, cytokinins and ethylene (Taiz and Zeiger, 2009).

The erratic and slow germination of custard apple seeds has been reported by several authors, who proved that this can take around 30 days (Morton, 1987; Cruz, 2002), 50 d (Hernandez, 1983) or 90 d (Hayat, 1963). Additionally, Stenzel *et al.* (2003) indicated that annona seeds have slow germination, with values that, after 2 months of germination, reach 1 to 3.8%. Such seeds may require special treatments like scarification, soaking in water, growth regulators etc. to overcome dormancy. Thus, the use of plant growth regulators may improve the germination of these seeds. To break dormancy and/or increase the germination process, the hormonal balance between germination inhibitors, such as abscisic acid (ABA), and promoters, such as GA_3, must be changed (Weaver, 1987; Kigel and Galili, 1995). The highest sugar apple seedling emergence speed and growth rate were obtained when the seeds were subjected to the method of scarification and soaking in GA_3 for 12-24 h. Scarification and soaking in water for 24 h resulted in greater economy, practicality and financial savings (Chagas *et al.*, 2013). Improvement in seed germination and seedling growth due to activation of dormant embryo of seeds with gibberellic acid and also gibberellic acid treatment helps to increase cell division, cell elongation and cell multiplication which might have reflected into maximum seedling growth. These results are in accordance with results obtained by and Gawade (2008).

Soursop seeds lose viability easily and do not store for a very long time and therefore, best sown without delay (Ken and Robert, 2011). Soursop has thick black seed coat that reduces water inhibition during the first stage of germination and therefore requires some pre-sowing treatments to enhance germination and seedling emergence (Okoli *et al.*, 2016). *Annona macroprophyllata* and *Annona purpurea* for both species, the dormancy period varies between six and eight months (Gonzalez- Esquinca *et al.*, 2015; Gonzalez- Esquinca *et al.*, 2016). Additionally, Gomez-Castaneda *et al.* (2003) reported that *A. purpurea* seeds have a dormancy period of more than six months of storage; after this period, only seeds treated with the gibberellin GA_3 are able to germinate.

Maximum germination percentage was recorded when seeds soaked in GA_3 might be due to fact that GA_3 involved in the activation of cytological enzymes like amylase which convert in soluble starch into soluble sugars and it also initiates the radical growth by removing some metabolic blocks. This might be due to fact that GA_3 play an important role in leaching out of the inhibitors into the soaking medium, might be due to GA_3 is a known growth regulator for breaking seed dormancy. However, these results disagreed with those obtained by Ferreira *et al.* (2002b), who did not detect significant effects on abnormal seedling percentage for sweetsop (*Annona squamosa* L.) subjected to 250 mg/l GA_3, and Oliveira (2004), who did not observe an increase in the percentage of *atemoya* cv. Gefner abnormal seedlings under application of GA_3 and ethephon. Various workers attempted to increase seed viability and their germination percentage of custard apple by using different growth regulators and their results have been presented in the Table 2.

Table 2: Effect of growth regulators on dormancy removal and seed germination in *Annona spp*

PBR	Concentration	Response/Effect	Reference
GA_3	1000 ppm	Minimized the days (10.27) taken for germination and improved the germination percentage (83.33%) and survival percentage (72.81%)	Ferreira *et al.*, 2002a
NAA	20 ppm	Most effective in retaining seed viability up to and beyond 90 days	Ghosh *et al.*, 2003
GA_3	50 – 100 ppm	Resulted in the highest percentage germination in cultivars Gefner (55-67%)	Stenzel *et al.*, 2003
GA_3	400 ppm	Treatment for 12 hrs gave the highest seed germination percentage (69.00%) and earliest days for initiation of germination (16.00 days)	Ratan and Reddy, 2004
GA_{4+7}	143 mg/l	Broke dormancy in *A. crassiflra* (43% germination)	Da Silva *et al.*, 2007
GA_3	50 – 750 mg/l	germination percentage ranged from 83 to 98% in *A. squamosa* seeds	Sousa *et al.*, 2008
GA_{4+7}	329 mg/l	95% germination in atemoya,cv Gefner	Braga *et al.*, 2010
GA_3	778 mg/l	Increased the germination of atemoya seeds by 85.5%	Oliveira *et al.*, 2010
GA_3 + Ethephon	778 mg/l + 75-100 mg/l	Increases the germination speed index and the percentage of normal seedlings	Marcos *et al.*, 2010
GA_3	520 mg/l	Led to 89.44% germination in the germinative process of *atemoya* cv. Gefner seeds	Braga *et al.*, 2010
GA_3	400 ppm	For 12 hrs treatment gave maximum germination percentage	Gharge *et al.*, 2011
GA_3	100 – 250 mg/l	100 mg/l for 24 h immersion and 250 mg/l for 5 h resulted in higher germination percentages and better seedling development	Menegazzo *et al.*, 2012
GA_3	50 ppm	For 48 hrs was helpful to get higher germination and seedling growth	Jadhav *et al.*, 2015
GA_3	600 ppm	Increased germination percentage (73.89%)	Jain, 2015
GA_3	500 ppm	Best as compare to other GA_3 concentration but increase the concentration of GA_3 resulting decreases survival percentageof seedling	Rahangdale, 2015

PBR	Concentration	Response/Effect	Reference
GA_4 + GA_7+BA	200 mg/l	Promoted the highest germinations (77%) in *A. Macroprophyllata*	Gisela *et al.*, 2016
GA_3	150 mg/l	Minimized the days (18.00) taken for germination, improved the germination percentage (84.33%) and increased survival percentage (75.97%)	Patel *et al.*, 2016
GA_3	200 mg/l	Soaking seed for 12 hours recorded the minimum days taken to germinate (24.00 days) and maximum germination percentage (63.99%)	Parmar *et al.*, 2016
GA_3	500 ppm	For 24 hours achieved maximum value for germination percentage and vigour index	Singh and Maheshwari, 2017
GA_3	500 ppm	Improved germination	Yadav, 2017

4. Role of PGRs on Growth Parameters

Plant growth and development involves the integration of many environmental and endogenous signals that, together with the intrinsic genetic program, determine plant form. Plant biomass lost to herbivores and has been associated with increased photosynthetic rates and mobilization of stored resources from source organs to sinks during active vegetative growth period. Plant hormones play a crucial role in controlling the way in which plants growth and development in custard apple and regulate the speed of growth of the individual parts and integrate these parts.

GA_3 treatment boost the growth of seedling by increasing cell multiplication and cell enlargement which ultimately result into shoot length in custard apple (Gharge *et al.*, 2011). It is natural that when treatment has boosted the growth of seedling above ground, it might have given the same effect of below ground. GA_3 increase in fresh weight of seedling by mobilizing the water and nutrient transport at higher rate from rooting medium during the period of cell elongation which might have promoted more production of photosynthetic products and translocated them to various plant parts which might have resulted in better growth. Application of GA_3 @ 150 mg/litre improved growth *viz.*, girth of plant, number of leaves, leaf area, fresh and dry weight of plants (Patel *et al.*, 2016). The number of leaves increased in custard apple with the application of GA_3 1000 ppm at 60, 120, 180 DAS. The probable reason for increasing number of leaves might be due to activity of GA_3 at the apical meristem, resulting in more synthesis of nucleoprotein responsible for increasing leaf initiation (Jadhav *et al.*, 2015). Higher stem girth in case of seedlings pre-soaked in GA_3 solution might be due to the fact that it plays a vital role in stimulation of cambium and its immediate cell progeny resulted in cell elongation and cell division of stem portion there by increased girth of seedling. However, contrasting results observed by Prajapati *et al.* (2016), they found growth parameters like tree spread (E-W and N-S), number of primary branches per tree, number of secondary branches per tree and girth of primary branch (cm) as affected by plant growth regulators were found non-significant.

5. Role of PGRs on Flowering, Pollination and Fruit Set

Custard apple fruit production and quality may be affected by several factors. However, the factors involving the pollination process are undoubtedly the most important. The difficulty involving natural pollination is due to the protogynous dichogamy phenomenon. Although custard apple trees have hermaphrodite flowers, self-pollination is nearly impossible because the stigma becomes receptive or viable long before the pollen is released (Campbell and Phillips,1994). This botanical phenomenon is a factor associated with a low degree of pollination and fertilization, affecting not only the quantity but also the quality of the fruits produced (Pereira *et al.*, 2011). There is a problem with fruit set in custard apple because of incomplete pollination, in which plant growth regulators may be useful to increase the fruit set. Plant growth regulators sprayed near the terminal buds of trees increased the rate of growth by stimulating enzymatic activities during its growth. Therefore, Plant growth regulators are used mainly to delay and reduce unwanted fruit abscission (fruit drop), to delay the senescence and to inhibit the growth of suckers on the trunk. Adequate supply of auxins in the plant body plays a vital role in the flowering and fruiting of plants.

Merely increasing fruit set is not important, but their retention up to maturity is also important. Savedra (1979) observed that the application of GA_3 at anthesis in custard apple increased fruit set at all concentrations, but fruits dropped at variable periods. However, longer fruit retention was observed by increasing the number of applications and concentrations of GA_3. The fruit retention was generally more with 5 applications of GA_3.

It has been reported that auxin level in the flowering plants has to fall before flower initiation but relatively higher level is required for the flower development. GA produced more flowers and had better fruit retention but increment in yield by GA non-significant and also not influence flowering as well as harvesting period. NAA increased flower number, fruit set, fruit retention and the yield. The flowering and harvesting duration and physico-chemical characters of fruit were not influenced by NAA. 2,4-D increased number of flowers, fruit set, fruit retention and yield (Kulkarni, 1993). Stigma receptivity enhanced with the application of auxin, which to be one of the causes for low fruit set in custard apple (Kumar *et. al.*, 1977). A number of works clearly indicated that plant bio-regulators (PBR) have beneficial effect on flowering, pollination and fruit set in *Annona spp*. (Table 3).

Table 3: Effect of growth regulators on flowering, pollination and fruit set in *Annona spp*

PBR	Concentration	Response/Effect	Reference
NAA	5-25 mg/l	Dipping the flower soon after anthesis, for about a minute increased fruit set in custard apple cv. 'Mammoth'	Sundararajan *et al.*, 1968
GA + NAA	----	Increased fruit set in *A.cherimoya* when apply three times at weekly interval	Durate *et. al.*, 1976
GA_3	20-100 ppm	*A. cherimoya* once at anthesis and 5 times on the fruits, starting from anthesis at three weeks interval increased fruit set	Savedra, 1979
GA_4 + GA_7	100 ppm	Higher fruit set than natural and hand pollination in atemoya cv. Genifer and African Pride when used just prior to the opening of flowers	Campbell, 1986
NAA	20 mg/l	Applied four times at eight days interval during flowering, gave the highest fruit set in custard apple	Keskar *et al.*, 1986
2,4-D	30 ppm	Fruit set percentage in first fortnight of June was the highest (2.95%)	Kulkarni *et al.*,1995
NAA	20 mg/l	Enhanced fruit set and fruit retention as well as increased flowering duration in custard apple (*Annona squamosa* Linn.) cv. 'Sindhan'	Patel *et al.*, 2010
CPPU	2 ppm	Increase number of flowers per shoot followed by NAA @ 15 ppm	Memane, 2013
NAA	200 ppm	Maximum number of flower per shoot	Chaudhari *et al.*, 2016
GA	50 ppm	Highest number of fruit set per shoot, number of fruit retention per shoot and lowest number of fruit drop per shoot were noted	Chaudhari *et al.*, 2016

6. Role of PGRs on Yield and Yield Attributes

The successful commercial cultivation of custard apple depends on many factors such as variety, climate, soil, irrigation, pollination and fertilization, plant density per unit area, age of plant etc. However, use of plant growth regulators may also be helpful to increase the yield and improve the quality of custard apple. Adequate supply of auxins in the plant body plays a vital role in the flowering and fruiting of plants. It has been reported that auxin level in the flowering plants has to fall before flower initiation but relatively higher level is required for the flower development. Growth regulators as a means for enhancing flowering, fruit set, control of fruit drop, fruit size, yield and fruit quality have become important in agriculture. There are also encouraging results due the use of plant growth regulators, such as a means of NAA 2,4-D, 2,4,5-T and GA_3 in custard apple with respect to improvement in fruit retention, fruit size, yield etc. (Table 4). However, research work on this aspect in custard apple is very scanty and sporadic. Sundararajan *et al.* (1968) reported that dipping the flower soon after anthesis, for about a minute, in different concentration of GA_3 (10, 25 and 50 mg/l) increased fruit set, size and weight of fruits in custard apple cv. 'Mammoth'. Among these, treatment of GA_3 @ 50 mg/l was most effective. On economic basis, 50 ppm GA followed by 20 ppm NAA gave highest net income in custard apple (Patel *et al.*, 2010).

Table 4: Effect of growth regulators on yield and yield attributes in *Annona spp.*

PBR	Concentration	Response/Effect	Reference
NAA	10 mg/l	Resulted in higher number of fruits, fruit size and weight of fruits	Keskar *et al.*, 1986
NAA	30 mg/l	Gave significantly maximum average fruit weight and number of fruits per tree	Kulkarni *et al.*, 1995
2,4-D	30 ppm	Gave highest fruit weight (148 g)	Kulkarni *et al.*,1996
GA_3	400 – 1600 ppm	Increase in weight due to application of GA_3 at higher concentrations. However, side effects like small, cracked and deformed fruits found higher concentration	Ezzahouan *et al.*, 1998
GA_3	50 ppm	Increased 21% fruit yield of custard apple cv. Sindhan	Patel *et al.*, 2010
GA_3 + NAA	50 + 20 ppm	Increased number of fruits per tree (67.30), fruit yield per tree (14.83 kg) and fruit yield per hectare (9.27 t) in custard apple cv. 'Balanagar'	Karbhari, 2011
NAA	15 ppm	Recorded highest number of fruits/ plant and yield and found to be economically beneficial	Memane, 2013

PBR	Concentration	Response/Effect	Reference
GA_3 + NAA	20 + 15 mg/l	Found beneficial for getting number of fruits set/ branch (5.20), number of fruits/ tree (118.33), average fruit weight (227.67 g) and fruit yield/ tree (26.94 kg)	Prajapati *et al.*, 2016
GA	50 ppm	Increased number of fruit, fruit yield per tree and fruit yield per hectare	Chaudhari *et al.*, 2016
GA	50 mg/l	Increased fruit weight (159.60 g), volume (96.85 cc) and pulp weight per fruit (79.54 g) at harvest	Bagul and Masu, 2017

7. Role of PGRs on Production of Seedless Fruits (Parthenocarpy)

Custard apple fruits develop normally with the uniformity desired by the consumer market when they are pollinated artificially. However, these fruits have a large number of seeds and high costs associated with labour (Janick and Paul, 2006). The biological function of fruiting is to produce and disseminate seeds. However, seedless fruits have some advantages, including the ease of consumption and the increased added value of the final product and fruit quality. Fruits produced by hand pollination alone or combined with GA_3 had a lower percentage of fruit peel. The use of 1,500 ppm of gibberellic acid is efficient for producing seedless fruits in custard apple with high physical and chemical parameters (Santos *et al.*, 2016). Savedra (1979) published one of the first reports on the use of gibberellic acid (GA_3) in Annonaceae and observed plant-growth regulator promoted the production of seedless fruits and increased fruit length and diameter in the Concha Lisa cherimoya variety. Pino (2008) showed that the combination of gibberellins and cytokines was also efficient in the generation of seedless fruits in *cherimoya*. Yonemoto *et al.* (2004), working with 'Big Sister' and 'Suiho' *cherimoya* cultivars, succeeded in growing seedless fruits with the same quality as the pollinated fruits. Recent studies also showed that applying 1000 ppm GA_3 without hand pollination enabled the production of parthenocarpic custard apple fruits (Pereira *et al.*, 2014). Sundarajan *et. al.* (1968) observed that the dipping of custard flowers in different concentrations of GA, NAA, 2,4-D soon after anthesis for about a minute increased fruit set in custard apple. GA at 50 ppm was found to be most effective treatment for low seed content in custard apple.

8. Role of PGRs on Fruit Quality Parameters

Spraying of plant growth regulators like GA_3 50 ppm and NAA 200 ppm had significant effect on TSS, acidity, reducing sugar, non-reducing sugar, total sugar and ascorbic acid on custard apple. TSS, reducing sugar and total sugar increased with the application of 50ppm GA_3 while, acidity, non-reducing sugar and ascorbic acid were noted the lowest with the same application in custard apple. This might be due to its action on converting complex substances into simple ones. This enhances the metabolic activities in fruits and ultimately results in increased TSS, total sugar and reducing sugar and ascorbic acid content in custard apple (Patel *et al.*, 2010). Sundararajan *et al.* ,(1968) reported that spraying of GA_3 @ 50 mg/l in rainy season increased the total soluble solids in custard apple cv. 'Mammoth'. The physico-chemical characteristics of custard apple cv. 'Mammoth' significantly improved with NAA @ 30 mg/l (Kulkarni *et al.*, 1995), whereas, acidity not influenced by these treatment (Kulkarni *et al.*, 1996). A trial conducted to study the influence of different plant growth regulators on quality of custard apple. Maximum TSS, reducing sugar and total sugar were found significantly highest and acidity, non-reducing sugar and ascorbic acid were noted lowest with application of 50 ppm GA_3 (Chaudhari *et al.*, 2017). Four applications of 1500 ppm GA_3 in flowers and fruits of 'Gefner' custard apple are efficient in producing seedless fruits with high physical and chemical parameters (Santos *et al.*, 2016). Combine application of GA_3 30mg + NAA 30 mg/l was recorded maximum ascorbic acid content. However, reducing sugar, non reducing sugar and total sugars were recorded highest in combination spray of GA_3 40mg + NAA 45 mg/l. For economic point of view, GA_3 20mg + NAA 15 mg/l followed by NAA 90 mg/l treatment gave highest net income (Prajapati *et al.*, 2016).

9. Role of PGRs on Shelf Life Improvement with Pre and Post-Harvest Treatment

The fruits of custard apple are very delicate and highly perishable. Being climacteric in nature, the biochemical changes in the fruit after harvest occur at a faster rate and the fruits become unfit for consumption within short period of transportation to distant market. Due to its delicate nature spoilage is more pronounced at each and every stage after harvest. Ripe fruits can be stored only 1-2 days without decay (Kumar *et al.*, 2016). Attack of fungus on the fruit after harvest is common, which is main cause of its deterioration and spoilage. These funguses not only cause rapid fruit spoilage but also lower down its market value due to poor appearance. Of the several economically cheaper techniques, the use of plant growth regulators for short term storage of fresh fruits at room temperature has become quite popular. Among various plant growth regulators the use of auxin and inhibitors/retardants (MH, B-nine) are quite common for improving quality and enhance the shelf life.

MH was found to be significantly effective in improving the shelf-life and maintaining the post-harvest quality of custard apple fruits during storage up to 9th day for some characteristics

viz., rotting, ascorbic acid and non-reducing sugar content of fruits. The combined effect of UV radiation (5 minutes) + 200 ppm MH + 1 per cent $KMnO_4$ soaked blotting paper (U1M2K2) was found to be most effective treatment to enhance the post-harvest shelf-life of custard apple fruits up to 9th day of storage and superior to their individual application. (Tanwar, 2004). Bhadra and Sen (1997) reported most effective treatment for prolonging the shelf-life of custard apple fruits by dipping in gibberellic acid (GA). Bhadra and Sen (1999) carried studies to increase the post-harvest shelf-life of custard apple fruits by NAA (100-200 ppm) and gibberellic acid (50-100ppm), when stored at 10°C and 85-90 per cent RH. Pre-harvest application of 2,4-D (20 ppm) and packaging in corrugated boxes recorded higher shelf life, firmness, less decay and physiological loss in weight with better quality of fruit with respect to maximum TSS, sugars, ascorbic acid content and organoleptic score with low acidity upto the 8th day of storage (Kumar *et al.*, 2016). NAA 100 ppm as well as GA_3 50 ppm were found most effective in extending the shelf-life, respectively over no growth regulator (1.75 and 1.42 days), while they helped in maintaining the marketability and tended to reduce the fruit weight loss (Patel *et al.*, 2011).

References

- Adeniji, I.T., Adio, A.F., Iroko, O.A., Kareem, A.A., Jegede, O.C., Kazeem-Ibrahim, F., Adewole, T.O. and Adeosun. A.O. 2014. Pre-treatment of seeds of *Annona squamosa* (sugar apple) a non timber forest product. *Res. Plant Sci.*, **2**: 50-52.

- Almeida, L.F.P., Alencar, C.M., Yamanishi, O.K. and Kiyoshi, O. 2010. Propagation by grafting of 'Thompson' atemoia on Rollinia species. *Revista Brasileira de Fruticultura*, **32**(2): 653-656.

- Bagul, A.A. and Masu, M.M. 2017. Effect of preharvest application of chemicals and plant growth regulators on physical parameters and shelf-life of custard apple (*Annona squamosa* L.). *International Journal of Agricultural Sciences*, **13**(2): 371-377.

- Baskin, C.C. and Baskin, J.M. 2014. Seeds: ecology, biogeography, and evolution of dormancy and germination. 2nd ed. Academic Press, San Diego, CA.

- Baskin, J.M. and Baskin, C.C. 2004. A classification system for seed dormancy. *Seed Sci. Res.*, **14**: 1-16.

- Bejoy, M. and Hariharan, M. 1992. *In vitro* plantlet differentiation in *Annona muricata*. *Plant Cell, Tissue and Organ Culture*, **31**: 245-247.

- Betancur-Galvis, L., Saez, J., Granados, H., Salazar, A. and Ossa J. 1999. Anti-tumor and anti-viral activity of Colombian medicinal plant extracts. *Mem Inst Oswaldo Cruz.*, **94**(4): 531-535.

- Bhadra, S. and Sen, S.K. 1997. Post-harvest storage of custard apple (*Annona squamosa* L.) fruit var. 'Local Green' under various chemical and wrapping treatments. *Journal of Inter academicia,* **1**(4): 322-328.
- Bhadra, S. and Sen, S.K. 1999. Post-harvest storage of custard apple (*Annona squamosa* L.) fruit var. Local Green under various chemical and wrapping treatments. *Environment and Ecology,* **17**(3): 710-713.
- Black, M. 1980. Role of indogenous harmones in germination and dormancy of fruit crops. *Isreal J. Bot.*, **29**: 181-192.
- Braga, J.F., Ferreira, G., Pinho, S.Z., Braga, L.F. and Sousa, M.P. 2010. Germination of atemoya (*Annona cherimola* Mill. × *A. squamosa* L.) cv. Gefner seeds subjected to treatments with plant growth regulators. *International Journal of Science and Nature,* **1**(2): 120-126.
- Campbell, C.W. 1986. Effect of gibberellin treatment and hand Pollination on fruit set of atemoya (*Annona* hybrid). *Proceedings of the Tropical Region, American society for Horticultural science*, University of Florida, **23**: 122-124.
- Campbell, C.W. and Phillips, R.L. 1983. The sugar apple. Fruit Crops Fact Sheet. Florida Cooperative Extension Service, University of Florida, pp. 1-3.
- Campbell, C.W. and Phillips, R.L. 1994. The atemoya. Gainesville, Florida: Horticultural Sciences Department, Florida Cooperative Extension Service, Institute of Food and Agricultural Sciences, University of Florida, p3.
- Chagas, P.C., Sobral, S.T.M., Oliveira, R.R., Chagas, E.A., Pio, R. and Santos, V.A. 2013. Physical and chemical methods to breach seed dormancy of sugar apple. *Rev. Cienc. Agrar.*, **56**: 101-106.
- Chaudhari, J.C., Patel, K.D., Yadav, L., Patel, U.I. and Varu, D.K. 2016. Effect of plant growth regulators on flowering, fruit set and yield of custard apple (*Annona squamosa* L.) cv. Sindhan. *Advances in Life Sciences,* **5**(4): 1202-1204.
- Chaudhari, J.C., Yadav, L. and Varu, D.K. 2017. Effect of plant growth regulators on quality of custard apple (*Annona squamosa* L.) cv. Sindhan. *International Journal of Chemical Studies,* **5**(5): 1036-1037.

- Cruz, P.E. 2002. Anona cultivation. Technical Bulletin No. 7. National Center for Agricultural and Forestry Technology (Centa), San Salvador, El Salvador.
- Da-Silva, E.A., De-Melo, D.L., Davide, A.C., De-Bode, N., Abreu, G.B., Faria, J.M. and Hilhorst, H.W. 2007. Germination ecophysiology of *Annona crassiflora* seeds. *Annals of Botany,* **99**(5): 823-830.
- Durate, O., Ramirez, A. and Franciosi R. 1976. Improving cherimoya fruit set by use of plant growth regulators. 459 Univ. Nacional Agraria, La Mo Lina, Lima, Peru (Plant Growth Regulators Abstr. 2: 1617).
- Ezzahouan, A., Amazzal, G., Debbagh, S. and Mahhou A. 1998. Effects of gibberellic acid (GA_3) on fruit production of cherimoya (*Annona cherimola* Mill). *Actes Inst. Agron. Vet. (Marco),* **18**(2): 117-120.
- Farooq, S.A., Farooq, T.T. and Rao, T.V. 2002. Micropropagation of *Annona squamosa* L. using nodal explants. *Pakistan Journal of Biological Sciences,* **5**: 43-46.
- Ferreira, G., Erig, P.R. and Moro, E. 2002 (a). Use of gibberellic acid in seeds of *Annona squamosa* fruit for the production of seedlings in different packages. *Revista Brasileira de Fruticultura,* **24**(1): 178-182.
- Ferreira, G., Erig, P.R. and Moro, E. 2002 (b). Rootstock production (*Annona squamosa* L.) with the use of plant regulators. *Revista Brasileira de Fruticultura,* **24**(3): 637-640.
- Ferreira, M.D.G.R., Santos, M.R.A.D., Silva, E.D.O., Gonçalves, E.P., Alves, E.U. and Bruno, R.D.L.A. 2009. Overcoming dormancy in biriba seeds (*Rollinia mucosa* (Jacq.) Baill) *anonna. Revista Brasileira de Sementes,* **31**(4): 95-99.
- Gavamukulya, Y., Abou-Elella, F., Wamunyokoli, F. and AEl- Shemy, H. 2014. Phytochemical screening, anti-oxidant activity and *in vitro* anticancer potential of ethanolic and water leaves extracts of *Annona muricata* (Graviola). *Asian Pac. J. Trop. Med.,* **7**(3): 55-63.
- Gawade, U.S. 2008. Seed viability, germination and seedling growth studies in custard apple. *M.Sc. (Ag.) Thesis,* Dr. Panjabrao Deshmukh Krishi Vidyapeeth, Akola, M.S. (India).
- George, A.P. 1984. Custard apple. InroicTarlee Fruits for Australia'. (Ed. P.E. Page.) pp. 35-41. (Queensland Department of Primary Industries: Brisbane.)
- George, A.P. and Nissen, R.J. 1987. Propagation of *Annona species*: a review. *Sci. Hortic.,* **33**: 75-85.

- Gharge, V.R., Kadam, A.S., Patil, V.K., Lakade, S.K. and Dhomane, P.A. 2011. Effect of various concentrations of GA_3 and soaking periods on seed germination of custard apple (*Annona squamosa* L.). *Green Farming*, **2**(5): 550-551.

- Ghosh, S.N., Manna, S. and Mathew, B. 2003. Effect of pre-sowing seed treatment on germination, seedling growth and viability of custard apple (*Annona squamosa* L.) seeds. *Environment and Ecology*, **21**(1): 4-7.

- Gisela, F., Ivan, C. and Rosa, G.E. 2016. Overcoming seed dormancy in *Annona macroprophyllata* and *Annona purpurea* using plant growth regulators. *Rev. Bras. Frutic., Jaboticabal*, **38**(3): 1-10.

- Gomez-Castaneda, J.A., Ramirez, H., Benavides-Mendoza, A. and Encinarodriguez, I. 2003. Germination and seedling development of soncoya (*Annona purpurea*) in relation to gibberelins and abscisic levels. *Revista Chapingo Serie Horticultura*, **9**(2): 243-253.

- Gonzalez-Esquinca, A.R., De-Lacruz C.I. and Dominguez-Gutu, L.M. 2015. Dormancy and germination of *Annona macroprophyllata* Donn. Sm (Annonaceae): The importance of micropylar plug and seed position. *Botanical Sciences, Ciudad de Mexico*, **93**(3): 1-7.

- González-Esquinca, A.R., De-la-Cruz, C.I., Castro-Moreno, M. and Riley-Saldana C.A. 2016. Phenological strategies of *Annona* from deciduous forest of Chiapas, Mexico. *Botanical Sciences. Ciudad de Mexico*, **94**(3): 531-541.

- Hayat, M.A. 1963. Morphology of seed germination and seedling in *Annona squamosa*. *Botan. Gazette*, **124**: 360-362.

- Hernandez, L.V. 1983. The reproduction sexually multiplication & vegetative growth of the Anonaceas (Spanish) veracruzana University, vera cruz, Mexico. 102-122.

- Jadhav, A.C., Bhagure, Y.L. and Raundal, R.M. 2015. Effect of PGR, chemicals and plant extracts on seed germination and seedling growth of custard apple (*Annona squamosa*). *Asian J. Hort.*, **10**(1): 184-186.

- Jain, S. 2015. Studies on effect of GA_3 and growing media on seed germinability, growth and survivability of custard apple (*Annona squamosa* L.). *M.Sc. (Ag.) Thesis*, Jawaharlal Nehru Krishi Vishwa Vidyalaya, Jabalpur.

- Janick, J. and Paul, R.E. 2006. Encyclopedia of fruit and nuts. Cabi, Wallingford, United Kingdom, p 56.

- Jaramillo, M.C., Arango, G.J., Gonzalez, M.C., Robledo S.M. and Velez, I.D. 2000. Cytotoxicity and anti-leishmanial activity of *Annona muricata* pericarp. *Fitoterapia.*, **71**(2): 183-186.
- Jordan, M. 1988. Multiple shoot formation and rhizogenesis from cherimola (*Annona cherimola* L.) hypocotyls and petiole explants. *Gartenbauwissenschaft,* **53**(5): 234-237.
- Jordan, M., Iturriaga, L., Rovemro, C. and Goreux, A. 1991. Promotion of cherimola *in vitro* shoot morphogenesis as influenced by antioxidants. *Gartenbauwissenschaft,* **56**(5): 224-227.
- Karbhari, T.U. 2011. Influence of growth regulators on yield and quality of custard apple (*annona squmosa* L.) cv. Balanagar. *M.Sc. (Ag.) Thesis*, Vasantrao Naik Marathwada Krishi Vidyapeeth, Parbhani.
- Katyal, S.L., Krishnamurti S. and Singh, S. 1963. Chemical properties of custard apple tree. *Fruit Culture in India,* pp. 239-246.
- Ken, L. and Robert, E.P. 2011. Soursop. *Fruits and Nuts,* **22**: 1- 6.
- Keskar, B.G., Bhore, D.P. and Masalkar, S.D. 1986. Effect of NAA and mulch on yield of custard apple. *J. Maharashtra Agric. Univ.*, **11**(2): 235-236.
- Kigel, J. and Galili, G. 1995. Seed development and germination. Marcel Dekker, New York, NY.
- Kim, G.S., Zeng, L., Alali, F., Rogers, L.L., Wu, F.E., Sastrodihardjo, S. and McLaughlin, J.L. 1998. Muricoreacin and murihexocin C, mono-tetrahydrofuran acetogenins, from the leaves of *Annona muricata*. *Phytochemistry*, **49**(2): 565-571.
- Kulkarni, S.S. 1993. Effect of cultural and chemical treatments on fruit set, yield and quality of custard apple (*Annona squamosa* L.). *M.Sc. (Ag.) Thesis*, Mahatma Phule Krishi Vidyapeeth, Rahuri-413722 , Dist : Ahmednagar.
- Kulkarni, S.S., Desai, U.T. and Masalkar, S.D. 1995. Fruit set studies in custard apple under semi arid conditions of western Maharashtra. *J. Maharashtra Agric. Univ.,* **20**(2): 317-318.
- Kulkarni, S.S., Desai, U.T., Masalkar, S.D. and Desale, S.B. 1996. Effect of cultural and chemical treatments on yield and physicochemical characteristics of custard apple. *J. Maharashtra Agric. Univ.,* **21**(3): 474-475.

- Kumar, R., Hoda, M.N. and Singh, D.K. 1977. Studies on floral biology of custard apple (*Annona squamosa* Linn.). *Indian J. Hort.*, **34**(3): 252-256.
- Kumar, S., Kaushik, R.A., Rathore, R.S. and Sarolia, D.K. 2016. Effect of pre-harvest application of plant growth regulators and packaging materials on shelf-life of custard apple (*Annona squamosa* L.). *Green Farming*, **7**(6): 1460-1464.
- Lemos, E.E.P. and Blake, J. 1996. Micro-propagation of juvenile and adult *Annona squamosa*. *Plant Cell, Tissue and Organ Culture*. **46**: 77-79.
- Marcos, C.O., Gisela, F., Vandeir, F.G. and Glaucia, B.D. 2010. Germination of atemoya seeds (*Annona cherimola* mill. × *A. squamosa* L.) cv Gefner subjected to treatments with gibberellic acid (GA_3) and ethephon. *Rev. Bras. Frutic., Jaboticabal*, **32**(2): 544-554.
- Memane, Y.S. 2013. Effect of different plant growth regulators on fruit set, yield and quality of custard apple (*Annona squamosa* L) under ecological conditions of north Maharashtra. *M.Sc. Thesis*, Mahatma Phule Krishi Vidyapeeth, Rahuri- 413722, Ahmednagar Maharashtra, India.
- Menegazzo, M.L., Oliveira, A.C., Kulczynski, S.M. and Silva, E.A. 2012. Effects of methods of overcoming dormancy in custard apple seeds (*Annona squamosa* L.). *Revista Agrarian*, **5**(15): 29-35.
- Morton, J. 1987. Sugar apple (*Annona squamosa*). pp. 69-72. In: Morton, J. (ed.). Fruits of warm climates. Julia F. Morton Miami, FL
- Muhammad, N.A., Kamrun, N., Firoj, A. and Ahmad, S. 2002. Micropropagation of *Annona Squamosa* Linn. using explants (shoot tip and node) of field grown mature plants. *Pakistan Journal of Biological Sciences*, **5**(4): 394-397.
- Nair, S., Gupta, P.K. and Mascamnhas, A.F. 1984(a). *In vitro* propagation of Annona hybrid (*Annona squamosa* L. × *Annona cherimola* L.). *Indian J. Hortic.*, **41**(3/4): 160-165.
- Nair, S., Gupta, P.K. and Mascarenhas, A.F., 1983. Haploid plants from *in vitro* anther culture of *Annona squamosa* Linn. *Plant Cell Rep.*, **2**: 198-200.
- Nair, S., Gupta, P.K., Shirgurkar, M.V. and Mascarenhss, A.F. 1984(b). *In vitro* organogenesis from leaf explants of *Annona squamosa* Linn. *Plant Cell Tissue Organ Cult.*, **3**: 29-40.
- Okoli, N.A., Obiefuna, J.C., Alagba, R.A. and Emma-Okafor L.C. 2016. Improvement of germination and seedling vigour of soursop (*Annona muricata* L.) using hydro-priming and dehydration treatments. *Futo J. Series*, **2**(1): 57-61.

- Oliveira, L.M., Renato, P., Jose, R.F.S., Eduardo, A., Rairys, C.N. and Flavia, D.P. 2008. Effect of cytokinins on *in vitro* development of autotrophism and acclimatization of *Annona glabra* L. *In Vitro Cell.Dev.Biol.,* 123-129.

- Oliveira, M.C. 2004. Germination of atemoya seeds (*Annona Cherimola* Mill. × *A. squamosa* L.) cv 'Gefner' submitted to treatments with Giberelic acid (GA_3) and ethephon. Dissertation (Master in Agrarian Sciences) Center of Agrarian Sciences, UNIOESTE, Marechal Cândido Rondon.

- Oliveira, M.C., Ferreira, G., Guimaraes, V.F and Dias, G.C.B. 2010. Germination of atemoya seeds (*Annona Cherimola* Mill. × *A. squamosa* L.) cv 'Gefner' submitted to treatments with Giberelic acid (GA_3) and ethephon. *Revista Brasileira de Fruticultura,* **32**(2):5 44-554.

- Padilla, I.M.G. and Encina, C.L. 2004. Micropropagation of Adult Cherimoya (*Annona cherimola* Mill.) cv. Fino de Jete. *In Vitro Cellular & Developmental Biology* (Plant), **40**(2): 210-214.

- Parmar, R.K., Patel, M. J., Thakkar, R.M. and Tsomu, T. 2016. Influence of seed priming treatments on germination and seedling vigour of custard apple (*Annona squamosa* L.) cv. Local. *The Bioscan,* **11**(1): 389-393.

- Patel, D.D., Gaikwad, S.S. and Patel, K.D. 2016. Effect of seed priming treatments on germination and seedling vigour of custard apple (*Annona squamosa*). *Current Horticulture,* **4**(2): 21–24.

- Patel, N., Naik, A.G. and Arbat, S.S. 2011. Response of post-harvest chemical treatments on shelf-life and quality of custard apple cv. Balanagar. *Indian Journal of Horticulture,* 68(4): 93-98.

- Patel, N.M., Patel, D.K., Verma, L.R. and Patel, M.M. 2010. Effect of cultural and chemical treatments on fruit set and fruit yield of custard apple (*Annona sqamosa* Linn.) cv. Sindhan. *Asian Journal of Horticulture,* **5**(2): 498-502.

- Pereira, M.C.T., Crane, J.H., Nietsche, S., Montas, W. and Santos, M.A. 2014. Growth regulators in the effective fruiting and quality of parthenocarpic fruits of atemoya 'Gefner'. *Pesq Agropec Bras,* **49**: 281-289.

- Pereira, M.C.T., Nietsche, S., Costa, M.R., Crane, J.H., Corsato, C.D.A. and Mizobutsi, E.H. 2011. Anonaceas: pine, atemóia and graviola. *Informe Agropec.,* **32**: 26–34.

- Pino, G. 2008. Obtaining parthenocarpic fruits of cherimoya (*Annona cherimola* Mill.) Through the use of crescent regulators. *Thesis (MSC. in Agronomy)*, Pontificia Universidad Católica de Valparaíso, Chile. pp. 51.
- Pinto, A.C.Q. 2005. Origin and distribution. In: Williams, J.T., R.W. Smith, A. Hughes, N. Haq, and C.R. Clements. (eds.). Annona species. International Centre for Underutilised Crops, University of Southampton, Southampton, UK. pp. 17-20.
- Pinto, A.C.Q. and Silva, E. M. 1994. Graviola for export, aspects Production technicians. Embrapa-SPI, Brasília.
- Prajapati, R.D., Laua, H.N., Solanki, P.D. and Parekh, N.S. 2016. Effect of plant growth regulators on flowering, fruiting, yield and quality parameters of custard apple (*Annona squamosa Linn.*) cv. Local. *Ecology, Environment and Conservation,* **22**: 177-179.
- Rahangdale, P. 2015. Effect of GA_3 and date of sowing on seed germination, growth and survival of Custard apple (*Annona squamosa* L.) Seedlings. *Thesis (M.Sc. Ag.),* Submitted to the Jawaharlal Nehru Krishi Vishwa Vidyalaya, Jabalpur.
- Ratan, P.B. and Reddy, Y.N. 2004. Influence of gibberellic acid on custard apple (*Annona squamosa* L.) seed germination and subsequent seedling growth. *J. Res. ANGRAU*, **32**(2): 93-95.
- Rodrigo, T.F., Renato, P., Nadia, A.C., Luciano, C.S., Rony, S. and Bart, P. 2016. *In vitro* culture of *Annona emarginata*: A rootstock for commercial annonaceae species. *Plant Cell Cult. Micropropag., Lavras*, **12**(1): 1-6.
- Santos, R.C., Marlon, C.T.P., Debora, S.M., Raquel, R.S.S., Silvia, N., Gisele, P.M., Bruna, H.C.S. 2016. Gibberellic acid induces parthenocarpy and increases fruit size in the 'Gefner' custard apple (*Annona cherimola* × *Annona squamosa*). *Australian Journal of Crop Science*, **10**(3):314-321.
- Savedra, E. 1979. Fruit set and growth of *Annona cherimola* Mill. Fruit obtained by hand pollination and chemical treatment. *J. Amer. Soc. Hort. Sci.,* **104**(5): 668-673.
- Singh, S.J. and Maheswari, T.U. 2017. Influence of pre-sowing seed treatments on the performance of soursop (*Annona muricata* L.) Seedlings. *Plant Archives,* **17**(2): 1215-1218.
- Singh, S.P. 1992. Fruit crop for wasteland. *Published by Scientific Publishers*, Jodhpur. pp. 49-64.
- Sousa, S.A., Dantas, A.C.V.L., Pelacani, C.R., Vieira, E.L. and Ledo, C.A.S. 2008. Overcoming dormancy in custard apple seeds. *Revista Caatinga*, **21**(4): 118-121.

- Stenzel, N.M.C., Murata, I.M. and Neves, C.S.V.J. 2003. Overcoming of dormancy in atemoia seeds and fruit of the earl. *Revista Brasileira de Fruticultura*, **25**(2): 305-308.
- Sundarajan, S., Shanmugavelu K.G. and Muthuswami, S. 1968. Effect of plant growth regulators on custard apple (*Annona Squamosa* L.). *South Indian Hort.* **16**(3&4): 63-64.
- Taiz, L. And Zeiger, E. 2009. Plant physiology. 4. ed. Porto Alegre: Artmed, p. 819.
- Tanwar, R.K. 2004. Effect of UV radiation, maleic hydrazide (MH), potassium permanganate ($KMnO_4$) and their combination on physio-chemical changes and shelf life of custard apple (*Annona squamosa* L.) during storage at ambient condition. *Thesis (M.Sc. Ag.)*, Maharana Pratap University of Agriculture and Technology Rajasthan College of Agriculture, Udaipur.
- Wang, L.Q., Min, B.S., Li, Y., Nakamura, N., Qin, G.W. Li, C.J. and Hattori, M. 2002. Annonaceous acetogenins from the leaves of *Annona montana*. *Bioorg Med Chem.*, **10**(3): 561-565.
- Weaver, R.J. 1987. Regulators of the growth of plants in agriculture. 5thed. Mexico: Trillas, p. 622.
- Yadav, R.S. 2017. Effect of gibberellins (GA_3) and cow urine on seed germinability, growth and survivability of custard apple (*Annona squamosa* L.). *M.Sc. (Ag.) Thesis*, Jawaharlal Nehru Krishi Vishwa Vidyalaya, Jabalpur.
- Yonemoto, Y., Koura, S., Hasegawa, K. and Yamamoto, Y. 2004. Fruit set and fruit growth of seedless cherimoya (*Annona cherimola* Mill.) induced by GA_3 under greenhouse cultivation in Japan. *Acta Hortic.*, **653**: 63– 66.

10

Plant Growth Regulators in Date Palm

Kapil Mohan Sharma, T. R. Ahlawat, C.M. Muralidharan, Suchismita Jena and Dharmishtha Patel

1. Introduction

Date palm (*Phoenix dactylifera* L.) is one of the oldest cultivated fruit crops in the world and is believed to have originated around 6000 BC (Johnson *et al.*, 2013) from current Iraq. Present day date palms are a result of centuries of selection. It is deeply rooted in the traditions, history and culture of the Arabian Peninsula. It thrives best in arid areas characterized by hot climate and low humidity. It can tolerate high temperature, drought and high salinity levels and can grow in a wide range of soils but prefer alkaline-aerated soils. Although date palm can survive prolonged dry spells, it does need water for commercial production. Date palm can be cultivated in all five continents of the world, however its cultivation is concentrated in the Middle East and North Africa which account for 89% of the global production.

Date fruits are relished for their sweet, succulent and exotic flavor. Date fruits can thus be regarded as an important source of energy and nutrients. It can be a good source of antioxidants and polyphenolics. Processed products from date primarily include dry dates, date bars, date syrup, date juice concentrate, date jam, date butter, date candy and date wine.

In the last few years, India has witnessed an increase in area and production of date palm mainly due the availability of tissue cultured plants and adoption of scientific package of practices. In Gujarat, the total date palm population is nearly twenty lakhs, most of which are of seedling origin. This includes one lakh tissue cultured plantlets of cultivar 'Barhee' which has shown good demand in Indian markets. 'Deglet Nour' and 'Medjool' varieties are highly valued in international markets. Date palm can be an ideal crop for arid regions in the states of Gujarat, Rajasthan, Haryana and Punjab. Since 1990s, horticulturists in the Middle East and North Africa have focused their efforts at increasing production, improving quality and manipulating ripening in date palm using plant growth regulators so as to make its farming more profitable. Work done across the globe in this regard is reviewed and discussed in the following section.

2. Use of PGRs on Propagation

Date palm can be propagated by seeds, offshoots and tissue culture techniques. Seed propagation is not suitable because of heterozygosity and dioecious nature of the date palm which may result in an off type plant. Due to hard stone, germination in date palm is also a long drawn process. Further, germination is inconsistent and plant growth is slow in the initial phase. Singh and Bhargava (2009) at Central Institute of Arid Horticulture, Bikaner evaluated the effect of seed treatments on growth and germination of date palm under hot arid conditions. Seeds were treated with cow dung for 24 hours, cow urine for 24 hours, GA_3 at 100 ppm for 12 hours, PEG at 100 ppm for 12 hours and soaking in water. Seeds treated with GA_3 at 100 ppm exhibited better seedling growth with higher germination percentage, number of leaves per plant, number of roots per plant and root length.

Another mode of propagation is through offshoots, which are axillary buds growing out of the main stem. Offshoots are basically of two types, those growing near the ground area called as "ground offshoots" and others which grow on aerial parts called as "aerial offshoots". Survival of aerial offshoots is comparatively lower than that of ground offshoots, which makes it important to improve rooting before planting. Al-Mana *et al.* (1996) reported that NAA improved rooting in ground offshoots and is also essential for good root formation in aerial offshoots.

Afzal *et al.*, (2011) tried to induce rooting particularly in aerial offshoots of cv. 'Hillawi' using IAA, IBA, NAA, 2,4-D and GA_3 each at 1000, 2000 and 3000 mg L^{-1} along with their various combinations at University of Agriculture, Faisalabad, Pakistan. Treatments were applied both through quick dip method as well as though injection and offshoots were planted in two different environmental conditions, *viz.* open field and plastic tunnel. All the growth hormones induced rooting except GA_3 which induced rooting only at 2000 ppm under plastic tunnel conditions. Among all the combinations, IBA at 3000 ppm resulted in more number of roots and root hairs in both the methods of application. Whereas, the root thickness was higher in IAA at all levels. El-Hodairi *et al.,* (1992a) observed a reduction in root diameter on the application of IBA through injection whilst, NAA increases it significantly. Nasir (1996) also reported a decrease in thickness of date palm roots with an increase in IBA concentration. IBA at higher levels induced early rooting and also increased the percentage of rooting in treated offshoots as compared with untreated ones.

Normally young date palm offshoots have to withstand severe drought conditions until the root system is sufficiently developed for sustainable water absorption. The use of rooting hormones to enhance root formation, considerably shorten the period during which the offshoots may be subjected to adverse drought conditions. A similar research was also done by Al-Khateeb, *et al.* (2015) using Isoprothiolane (IPT) at 0, 25, 50, 75, 100, 200 and 500 g/ offshoot hole to induce rooting in 'Khalas' cv. of date palm at Alhassa farm, Al-shemia area, Saudi Arabia. IPT was injected in the offshoot hole using a disposable syringe. The best rooting percentage, length of root, root fresh and dry weights were recorded under IPT at 75 and 100 g/ offshoot. Four applications of Paclobutrazol and Uniconazole @ 1000 ppm, at an interval of 30 days significantly increased offshoot production in 'Tafsirt' date palm at Sebha, Libya (El-Hodairi *et al.*, 2001).

Zirari and Ichir (2010) studied the effect of exogenous application of IBA on rooting and leaf growth of small date palm offshoots derived from adult *vitro* plants of 'Najda' cultivar at Morocco. Offshoots from adult 'Najda' *vitro* plants grown in the Zagora Date Palm Experimental Station were separated in 4 weight groups (0-2, 2-4, 4-6 and 6-8 kg). Prior to planting in pots, offshoots were dipped for 15 min in 4 IBA solutions (2, 5, 10 and 15 mg L^{-1}). They reported that more than 96% of offshoots rooted, if treated with 10 or 15 mg L^{-1} IBA solution. During this 8-months experiment, offshoots treated with 10 or 15 mg L^{-1} IBA solutions produced an average of 4 new leaves and an increment of 9 cm in the central leaf length.

Darwesh *et al.,* (2013) studied the impact of IBA and Paclobutrazol on rooting of date palm off-shoots cv. 'Zaghloul' at Central Laboratory for Research and Development of Date palm, ARC, Egypt. Three categories of off-shoots weights (4, 8 and 12 kg) were subjected to different growth regulators *i.e.* IBA (1000, 2000 and 4000 ppm) singly and in combination with 0.4 mg/l Paclobutrazol. They observed that treatment with IBA at 4000 ppm/L singly and IBA at 4000 ppm/L in combination with 0.4 mg/L pbz for off-shoots weighing 8-12kg/ offshoot resulted in the highest rooting percentage of transplanting off-shoots in date palm nursery. Abate (2009) studied the effect of offshoot size and IBA on rooting of Date palm.

Different sizes of offshoots (10 ± 2.5, 15 ± 2,5 and 20 ± 2.5 cm) were treated with rooting hormone IBA (0, 1000, 2000 and 3000 ppm). They observed that application of IBA to the basal portion of offshoots with 20 ± 2.5 cm diameter was effective in promoting rooting and subsequent survival of date palm offshoots.

Offshoot propagation is a slow method and mortality is also high (Alkhateeb *et al.*, 2006). The survival rate of offshoots in the field is low and chances of disease transmission are also high (Abahmane, 2011). Besides, offshoot production is limited to a certain period in the palm's life span. Under such circumstances, plant tissue culture technique can emerge as an efficient alternative for mass production of date palm plants. Normally, date palm is propagated *in vitro* by two methods: the first method is by embryogenesis in which vegetative embryos can be raised from embryogenic callus (McCubbin *et al.*, 2000). The second technique is organogenesis which provides date palm buds that eventually gives plantlets without passing through the callus stage. However, since plantlets are produced directly from tissues of mother plant without passing through callus stage, they are typically identical to mother plant (Beauchesne, 1983; Beauchesne *et al.*, 1986; Aaouine, 2000). The utilization of bioreactors is also making a significant contribution toward micropropagation of different plant species (Othmani *et al.*, 2011).

Somatic embryogenesis is considered as the most potent tool to achieve large-scale production. Recently, Mazri *et al.* (2017) reported somatic embryogenesis from the adventitious bud of date cv. 'Najda', where the MS medium was supplemented with 2, 4-D, kinetin (KN) or 6-dimethylallylamino purine (2iP). To produce elite cultivars from rapid clonal propagation, direct organogenesis techniques have been widely adopted (Khierallah and Bader, 2007). To induce embryogenic callus in most of the cases for either shoot tip or inflorescence, a high concentration of auxins are used. Most researchers suggested 2,4-dichlorophenoxyacetic acid (2,4-D) as the most efficient auxin to induce embryogenic callus at 100 mg L^{-1} concentration (Al-Khayri, 2005; Al-Khayri, 2010; Eshraghi *et al.*, 2005; Naik and Al-Khayri, 2016).

Khierallah and Bader (2007) mentioned that date palm cv. 'Maktoom' showed higher shoot-bud multiplication in MS medium with a hormone combination of 1 mg L^{-1} NAA, 1 mg L^{-1} NOA, 4 mg L^{-1} 2iP and 2 mg L^{-1} BAP. Mazri and Meziani (2013) found that half-strength MS medium augmented with 0.5 mg L^{-1} NOA and 0.5 mg L^{-1} Kinetin produced 23.5 shoot buds per explant after 3 months of multiplication in cv. 'Najda'. Al-Mayahi (2014) reported production of an average of 18.2 buds per culture in cv. 'Hillawi', in the MS medium containing 1 mg L^{-1} BAP and 0.5 mg L^{-1} TDZ.

Al-Khateeb (2006) studied the effect of reducing the hormone concentration on bud and shoot multiplication of date palm (*Phoenix dactylifera* L.) cv. 'Sukry' at Date palm Research Center, King Faisal University, Kingdom of Saudi Arabia. Media was supplemented with different concentrations of Auxin (IAA, NOA, NAA) and Cytokinin (Kinetin, 2ip, BAP). The results indicated that low level of hormones promoted the formation of buds and shoots while hormones at higher levels inhibited the formation of shoots and buds. Medium supplemented with 0.20 kinetin + 0.10 2ip + 0.10 BAP + 0.10 IAA + 0.10 NOA + 0.10 mg L^{-1}NAA produced the highest fresh and dry weight.

Sometimes micropropagated plants carry advantageous characters *i.e.* somaclonal variation, which acts as a raw material for the mass propagation, cryopreservation and synthetic seed production. Date palm micropropagation and somatic variation introduces new genotypes with stress tolerance, disease resistance and with high quality fruits (El Hadrami and El Hadrami, 2009; Jain, 2001). The use of high concentration of plant growth regulators induced somaclonal variation in date palm (Al-Mazroui *et al.*, 2007). Fki *et al.* (2011b) reported that high concentrations of 2, 4-D causes somaclonal variations. Many researchers applied low concentration of 2, 4-D such as 1.5, 5 and 10 mg L^{-1} to induce somatic embryos (El Hadrami *et al.*, 1995). High concentration of 2, 4-D induced 25% somaclonal variation, changes in the leaf morphology, poor flower pollination ultimately leading to low quality date fruit (Fki *et al.*, 2011b). To reduce the risk of somaclonal variation researchers suggested using juvenile explants, low auxin concentration, especially 2, 4-D and minimum numbers of subcultures (El Hadrami *et al.*, 2011; Fki *et al.*, 2011a, 2011b; Khan and Bi, 2012).

Accumulation of water content in the cultured tissue (Hyperhydricity) is common physiological disorder in somatic embryogenesis and organogenesis of date palm (Mazri, 2015; Mazri and Meziani, 2013; McCubbin and Zaid, 2007). Kriaa *et al.*, (2012) reported when hydrated calli were cultured on the low concentration of 2, 4-D for a long period, calli will induce somatic embryos or shoots. Factors such as plant hormones, liquid media and concentration of ammonia used are responsible for hyperhydricity (Al-Khateeb, 2008; Mazri and Meziani, 2015). The use of high agar concentration in the semisolid media and low concentration of plant hormones and ammonium concentration will help in the reduction of formation of hyperhydricity in the cultured tissue (Al-Khateeb, 2008).

Doaigey *et al.*, (2013) studied the role of PGRs in leaf anatomy of date palm cv. 'Khedri' employing GA_3 and 2,4-D, both at a concentration of 10^{-5}M and 10^{-3}M as foliar application at King Saud University, Riyadh, Saudi Arabia. They found that on application of GA_3, the size of midrib increased along with number of vascular bundles. Although, 2,

4-D inhibited the formation of midrib, it increased the number of hypodermal cells.

3. Role of PGRs on Water Management

Research on water management in date palm has revealed that irrigation using saline water retarded plant growth and productivity (Aljuburi and Al-Masry, 2000; Hassan and El-Samnoudi, 1993; Sane *et al.*, 2005). Aljuburi *et al.*, (2010) investigated the role of NAA in regulation of mineral concentration of date palm cv. 'Shabeeby' seedlings under salt stress conditions at Experimental Station of Qatar University, Rawdat Al-Faras region in Qatar State. Plants were treated with NAA alone (0, 150, 300 mg/L) and in combination with NaCl (0, 6000, 12000, 18000 mg/ L) and observed after 115 days. They found that concentration of P, Cu, Mn, Zn and Fe decreased in leaves, but concentration of K and Na underwent an increase, when plants were irrigated with high saline water. However, application of NAA at high concentration along with high saline water prevented a reduction in mineral nutrients. Furthermore, it increased the concentration of Na, Cl, N, P, Ca and Mg in roots. Moreover, seedlings treated with NAA in saline affected irrigation water did not show any symptom of NaCl stress.

4. Role of PGRs on Pollen and Pollination

As date palm is a dioecious crop, it is very much essential to manage pollens and use them efficiently on female inflorescences. Although date palm is naturally wind pollinated, but human have assisted in this pollen transfer since antiquity by limiting quantity of pollen grains as basis to justify the use of mechanical pollination using duster and sprayers. Therefore artificial pollination is imperative to ensure consistent and high yield. Additionally, xenia and metaxenia have been reported in date palm, thus making it important to select the right male for better yield and quality.

El-Hamady *et al.*, (2010) in Balteem district, Kafer El-Sheikh Governorate, Egypt conducted an experiment, where the female inflorescence was pollinated with two different males (M_1 & M_2). A higher presence of IAA, GA_3 and ABA was observed during the initial growth phase 'Hayany' date palm when pollinated with M_1 as compared to M_2. Inflorescence pollinated with pollen (M_1) having higher level of IAA and GA_3 in initial stage exhibited better growth and fruit size with larger fruit weight at later stages, which represents the influence of metaxenia.

An experiment was conducted by Dietz (1998) using NAA, GA_3, BA, HFCA, TIBA and Daminozide along with pollen application. He found that GA_3 induced higher pollen germination and number of fertilized ovaries ultimately improving fruit set. Pollination efficiency was improved on mixing pollen with natural growth activator Milagro (extracted from pollen flowers of cabbage and reflects the effect of Auxin, Cytokinin, Gibberellins, Ethylene, Hydrogen Cyanamide and Humic acid and has a broad effect on different crops) in Zagloul and Samany date palm cultivars at Abo-Rawash region, Giza Governorate, Egypt (Hafez *et al.*, 2014). These two cultivars were pollinated during the 1st week of April as soon as they opened by dusting the female inflorescence with pollen and growth activators. Pollination efficiency was highest in the treatment involving pollen grains at 1 g + Milagro 1 g. It also resulted in the highest fruit set and fruit retention. Physical and chemical characteristics of the fruits also showed a marked improvement.

5. Role of PGRs on Fruit Growth, Development and Ripening

Kader and Hosain (2009) specified five distinct stages in the development of date fruits from fruit set to maturity. These stage are *Hababouk, Kimri, Khalal, Rutab* and *Tamar*. The *Hababouk* stage is characterized by slow growth rate and lasts for 4-5 weeks after fruit set. Fruits are round in shape and colour is green. Moisture content is around 85%. The *Kimri* stage lasts for about 9-14 weeks in which the fruits gain weight and elongate in size. Fruit colour is still green and the tannin concentration is high. The *Khalal* stage is also known as *Bisir* stage in which the fruit colour changes from green to yellow or red depending on the variety and they accumulate sugars. This stage lasts for 3-5 weeks during which fruits attain maximum weight and size. Moisture content is between 45 to 65%. In the *Rutab* stage, tip of the fruit starts ripening and it takes 2-4 weeks. Fruit colour turns dark. Dates at this stage lose water but continue to accumulate sugars. Moisture content ranges between 30-45%. In the *Tamar* stage, dates have fully ripened and completely softened. The fruit colour further darkens, sugar concentration is maximum and the moisture content is less than 30%. This stage lasts for 2-4 weeks and the dates are fit for long term storage.

El-Hodairi, *et al.*, (1992b) at Sebha Agricultural Research Centre, Libya assessed the effect of growth regulators on fruit set of date palm cv. 'Sembelbel, Tallis and Adwi'. Freshly opened clusters were sprayed with Indole acetic acid (10 ppm), Gibberellic acid (10 ppm), 6-Furfurylaminopurine plus gibberellic acid (10 ppm). Results indicated that growth regulators applied significantly increased the fruit weight, fruit length and formation of seedless fruits. Fruit set also increased significantly. However, varieties responded differently to different growth regulators.

In a study at Sebha, Libya, El-Hodiari *et al.* (2001) stated that four applications of paclobutrazol and uniconazole at 1000 mg L^{-1} at an interval of 30 days significantly increased the number of fruit clusters, fruit weight per cluster and the total fruit weight per tree in 'Tafsirt' date palm.

A study was conducted by Aljuburi *et al.* (2001a) at Kuawaytate Experiment and Research Station, Department of Agriculture and Live Stock, Al Ain, UAE to assess the relative effectiveness of GA_3 (150 gm L^{-1}), NAA (100 gm L^{-1}), ethephon (1000 gm L^{-1}) and a mixture of growth regulators on some fruit characteristics and productivity of 'Barhee' date palm cultivar. Flower clusters were sprayed 20 days after pollination. They concluded that application of NAA alone or in combination with GA_3 and ethephon reduced fruit dry matter percentage and ripening. However, it increased the fruit flesh percentage and production per bunch and per tree.

The relative effectiveness of GA_3 (150 gm L^{-1}), NAA (100 gm L^{-1}), ethephon (1000 gm L^{-1}) and a combination of these growth regulators was assessed by Alujuburi *et al.* (2002) on some fruit characteristics and fruit yield of date palm cultivar 'Fardh' at Al Ain, UAE. They deduced that spraying 'Fardh' date palm flower clusters with NAA or mixture of growth regulators, 20 days after pollination was more effective in reducing total soluble solids, dry matter percentage, fruit ripening and increasing the fruit yield (production per bunch and per tree).

Different GA_3 concentrations (0, 50, 100 and 150 ppm) were sprayed 50 days after full bloom on 'Sakkoty' cultivar of date palm during 2004-05 to study their effect on yield and fruit characteristics at Kom-Ombo, Aswan Governorate, Egypt. Sakkoty is the most important dry-date cultivars in Upper Egypt. Bunch weight increased with the application of 100 ppm GA_3 in both the seasons. GA_3 application at higher concentrations increased average fruit weigh, flesh weight, fruit length, fruit diameter and fruit moisture content percentage in both seasons. It was concluded that apraying GA_3 at 100 and 150 ppm resulted in the highest yield and best quality (Soliman and Ali , 2009).

Ahmed *et al.,* (2010) conducted a study at the Experimental Research Station, Faculty Agriculture Giza, Egypt to study the effect of plant growth regulators on yield and fruit characteristics of date palm cv. 'Samani'. It is an important commercial cultivar of soft date in Egypt. Different levels of GA3 and BA (0, 50, 100 and 150 ppm) and CPPU (0, 5, 10, and 20 ppm) were sprayed two weeks after pollination in 'Samani' date palm. GA3 and BA applications at higher concentrations increased average bunch weigh, fruit weigh, flesh weight, fruit length, fruit diameter, total soluble solids, total sugars and decreased total acidity. They concluded that GA3 and BA at a concentration of 150 ppm gave the highest yield with the best quality and fruit characteristics.

Kaseem *et al.*, (2010) conducted a trial to extend the harvest season and to maintain fruit quality for better marketability of 'Barhee' date palms growing in Riyadh, Saudi Arabia. Palms were subjected to preharvest foliar sprays at the *Hababouk* stage and at the beginning of fruit color break with 10 ppm from a new cytokinin related substance (CPPU or N-(2-chloro-4-pyridinyl)-N'-phenyl urea), known as cytofex, 8 mM putrescine (Put) and 50 ppm each of GA_3, NAA, benzyl adenine "BA" and salicylic acid "SA". The harvest date was delayed by one month with NAA and GA_3, three weeks with Put and SA and by two weeks with CPPU and BA from the commercial harvest date. All treatments decreased skin color intensity and carotenoids content and increased acidity as compared to control. Fruits turning to postharvest *Rutab* stage during storage at 0°C and 85-90% RH were reduced by all treatments. GA_3, NAA and Put treatments had a significant impact on extending the shelf life and decreasing the weight loss.

Mohammadi and Khankahdani (2010) investigated the effect of plant growth regulators on quantitative and qualitative traits of 'Shahani' date. Date palm trees were treated with different concentrations of BA (40 and 80 mg/L), GA_3 (25 and 50 mg/L), NAA (30 and 60 mg/L) and 2,4-D (50 and 100 mg/L), alone or in combination and water as control, in two stages: *Hababouk* and *Kimri* (6 and 10 weeks after pollination, respectively). BA treatments with low concentrations of auxin and GA_3 were suitable for improvement in quantity and quality traits of 'Shahani' date fruits.

Hassaballa *et al.* (1983) earlier reported that non-pollinated clusters in 'Zaghloul' date cultivar treated with BA (20 mg/L) in opening spath time and 2,4-D (100 mg/L) in early June and early July, have higher weight than pollinated clusters. They also reported that sugar content and TSS in treated fruits was lower than in control fruits.

In order to induce the increase fruit size and yield of Zahedi date palm in the Farrashband region in Fars province an experiment was conducted by Abdolhossein and Abdolhossein (2015) to evaluate the effects of 2,4-D, NAA and potassium nitrate. 36 Zahedi date palm trees were selected. 2,4-D (20, 30 and 40 mg/L), NAA (30, 60 and 90 mg/L) and potassium nitrate (1% and 2%) were applied on fruit brunches at the beginning of Kimiri stage (6 weeks after pollination) followed by another application 4 weeks later. Four trees were untreated and left as controls. The results showed that 2,4-D (40 mg/L) and NAA (30 and 60 mg/L) at stages significantly increased fruit length, diameter and weight compared with untreated ones. These treatments resulted fruit ripening delayed and significantly decreased TSS as compared with control. Generally, NAA (30 and 60 mg/L) increased fruit size and delayed fruit ripening for 4 to 5 weeks.

El-Hodairi *et al.*, (2001) reported that application of 10 mg/L 6-phurphuril amino purin + GA_3 on pollinated clusters of three date cultivars ('Simbebel', 'Talis' and 'Edvi') significantly increased fruit weight and length and produced seedless fruits.

Al-Qurashi and Awad (2011a) at King Abdul-Aiz University, Jeddah, Saudi Arabia studied the effect of NAA (40, 60, 80 and 100 ppm) on bunch weight and fruit quality in 'Barhee' cv. of date palm at late *Kimri* stage. Significantly higher bunch weight, fruit weight, flesh weight and TSS were recorded by the application of NAA at 60, 80 and 100 ppm. Application of NAA significantly decreased fruit ripening as measure by *Rutab* percentage and significantly increased fruit length and diameter. They concluded that NAA application at 60 ppm showed promotive effects on yield and quality of 'Barhee' date under hot arid climate.

Awad and Al-Qurashi (2012) assessed the effect of Gibberellic acid (50, 100 and 150 ppm) and bagging (0.14 mm thickness) on bunch weight and fruit quality of 'Barhee' date palm cultivar under hot arid conditions at Hada Al-Sham, Saudi Arabia. Bagging was done about 4 weeks from pollination and gibberellic acid was sprayed at late *Kimri* stage, about 15-16 weeks after pollination. Gibberellic acid at all selected levels increased bunch weight of 'Barhee' date palm cultivar compared to control. The highest bunch weight was observed in GA_3 at 150 ppm. Bunch bagging also increased bunch weight compared to control but this was lower than GA_3 treatments. Bunch bagging and GA_3 at 50 ppm synergistically increased bunch weight and vitamin C concentration. GA_3 sprays delayed fruit ripening, as reflected by lower *Rutab* percentage and higher acidity, phenols and tannins concentration. Fruit length was significantly increased by GA_3 spray at 100 and 150 ppm and by bunch bagging alone or with GA_3 at 50 ppm. GA_3 spray and bunch bagging treatments increased total soluble solid (TSS) and acidity concentration. It was concluded that both GA_3 sprays and bunch bagging have promotive effects on bunch and fruit weight and improve quality of 'Barhee' dates under hot arid conditions.

A major problem in Hada Al-Shame valley of Saudi Arabia is the low fruit set and/ or abnormal flowering accompanied by subsequent high fruit drop percentage in date palm cultivars (Al-Qurashi and Awad, 2011a). Date palms, as most other fruit trees, experiences two waves of fruit drop. The first occurs few weeks following pollination. This drop is due to lack of or incomplete pollination or fertilization. The second drop is more dramatic and occurs 5 to 6 weeks later, at the end of the *Kimri* and beginning of *Bisir* stage, around mid May (Ben, 2001). Under Hada Al-Shame valley conditions both 'Ghur' and 'Rothana' date palm cultivars show normal flowering, fertilization and fruit setting. However, close to the maturation stage (*Bisir* stage, color break around mid May) a severe fruit drop (50 to 80%) occurs leading to heavy economic losses. Al-Qurashi *et al.* (2012) studied the effect of plant growth regulators 2,4-D (50 and 100 ppm), NAA (100 and 150 ppm), GA_3 (100 and 150 ppm) and BA (100 and 150 ppm) at 40 and 70 days after pollination, on pre-harvest fruit drop and quality of 'Rothana' and 'Ghur' dates at King Abdul-Aiz University, Jeddah, Saudi Arabia. In both cultivars, application of growth regulators significantly decreased fruit drop. In this regard, 2,4-D and GA_3 were the most effective treatments followed by BA, while NAA was the least effective. Reduction in fruit drop resulted in a higher bunch weight in the treated fruit than the control. They concluded that under hot arid conditions, application of growth regulators especially 2,4-D (50 ppm) and GA_3 (150 ppm) at 40 and 70 days from pollination is effective in reducing pre-harvest drop and improving fruit quality in 'Rothana' and 'Ghur' cultivars of date palm.

An experiment was carried out at Date Palm Research Station, R.A.U., Bikaner to evaluate the response of ethephon spray on yield and yield attributing characters in date palm cultivars 'Halawy, Khadrawy, Zahidi and Shamran' under hot arid region. Ethephon at 500, 1000 and 1500 ppm was sprayed at colour break stage in each variety and fruits were harvested at at the doka (*Khalal*) stage. Spraying ethephon had a remarkable influence on average bunch weight and yield per tree in variety 'Halawy'. Application of ethephon at different levels significantly increased the fruit length, breadth and volume; however maximum values were recorded under 1000 ppm. It also curtailed the number of days taken to colour development and fruit maturity. Spray of 1000 ppm ethephon at colour break stage was thus found beneficial for uniform maturity, ripening and production of better quality fruits in date palm (Meena *et al.*, 2013).

It is important to prolong the *Khalal* stage in date palm to extend their marketing period. El-Shiekh (2014) in an experiment at Dibba Experiment Station, Eastern Region, UAE, used GA_3 + Glutamic Acid (Glu) in various combinations after fruits set to improve quality in date palm cv. 'Khalas'. Bunches were sprayed thrice at an interval of three weeks. Gibberellic acid delayed fruit ripening by about 4-5 weeks which can help increase duration for marketing. GA3 and GA3 + Glu increased yield, fruit bunch weight, fruit height, fruit thickness, fruit volume and average fruit weight significantly in comparison to other treatments. GA_3 spray on date palm fruits has been reported to improve fruit size and quality (Hussein *et al.*, 1996; Moustafa and Seif, 1996; Moustafa *et al.*, 1996); however, it delayed fruit ripening and decreased TSS and total sugars (Hussein *et al.*, 1996; Moustafa and Seif, 1996). Gibberellins have also been used to increase fruit set, size, retention and yield (Kassem *et al.*, 2010). Earlier reports indicate that Gibberellic acid (GA_3) increased fruit weight and delayed fruit ripening (Moustafa and Seif, 1996).

El-Salhy *et al.*, (2016) studied the response of 'Seewy' date palm to GA_3 (50, 75, 100 and 15 ppm) and CPPU (7.5, 15 and 30 ppm) at Assiut University, Egypt. All treatments were applied two and eight weeks after pollination. They observed that spraying GA_3 at 50 ppm after two weeks of pollination plus either GA_3 at 75 ppm and CPPU at 7.5 ppm or GA_3 at 75 ppm and CPPU at 15 ppm increased yield and improved quality of 'Seewy' dates. GA_3 reduce fruit retention and consequently increased fruit weight and size which ultimately increased the yield. CPPU induced an increase in fruit weight and size and thereby improved yield of 'Seewy' dates. Spraying GA_3 and CPPU significantly improved the chemical properties of fruits in terms of total soluble solids and sugar contents and significantly decreased the total acidity and total soluble tannins percentages compared to unsprayed ones.

In India, the four distinct ripening phases in date palm are known as *Gandora (Kimri)*, *Doka (Khalal)*, *Dang (Rutab)* and *Pind* (*Tamar*), respectively. In India, date fruits are harvested at *Khalal* or *Doka* stage because, if retained on trees for longer duration they get spoiled due to rains and high humidity. Fruit weight and size are critical quality parameters that affect dates marketing because larger size is generally preferred over small ones (Botes and Zaid, 1999; Al-Qurashi and Awad, 2011). Chaudhary *et al.* (2018) studied the effect of GA_3 on fruit yield and quality in date palm at Date palm Research Centre, Swami Keshwanand Rajasthan Agricultural University, Bikaner. The experiment consisted of four date palm cvs. 'Barhee, Halawy, Khadrawy and Zahidi' and five level of GA_3 (control, 50 ppm, 100 ppm, 150 ppm and 200 ppm). GA_3 applications were made on fruits at late Kimri stage. The highest fruit weight, fruit length, fruit diameter, fruit volume and pulp:stone ratio were observed under GA_3 at 200 ppm. However, the high fruit quality characters *viz.* total soluble solid, ascorbic acid, total, reducing and non reducing sugars was recorded by the application of GA_3 at 150 ppm. Thus 150 ppm GA_3 could be used to increase the yield and improve fruit quality of 'Barhee, Halawy, Khadrawy and Zahidi' cultivars.

6. Role of PGRs on Post Harvest Management

Internationally, dates are harvested at the *Khalal*, *Rutab* and *Tamar* stage depending on the variety, agroclimate and market demand. The degree of perishability decreases from *Khalal-Rutab-Tamar*. When dates reach *Khalal* stage, they are regarded as ready for marketing as fresh fruits. In India, dates are harvested at the *Khalal* stage also known as *Doka* stage to avoid damage by rainfall. Keeping in mind the demand by consumers and to avoid losses caused by rainfall, researchers have attempted to hasten or delay ripening in date palm utilizing PGRs.

'Zaghloul' dates are very popular in the Arab market and are mainly consumed at the *Khalal* stage. There is strong consumer preference for fruits with intense color and uniformity. Delayed harvest may result in anthocyanin accumulation in fruits. Farmers are however, forced to harvest their dates before the rains. Rain causes cracking, acceleration of the *Rutab* stage and increasing their susceptibility to infection by pathogens. To resolve this, Farag and Kassem (1998) sprayed modified ethephon formulations two weeks before harvest at Edko, Beheira, UAE in 'Zaghloul' dates. They reported that ethephon had the ability to increase and accelerate anthocyanin content of date palm fruits when applied in a modified formulation. Earlier reports indicate that preharvest application of ethephon enhanced ripening and markedly increased the total soluble solids (Mougheith and Hassaballa, 1979). Ethephon application was also found to improve fruit quality (Maximos *et al.*, 1980).

'Helali' is an important late maturing commercial cultivar of the Gulf region. Owing to the presence of soluble tannins, 'Helali' fruits cannot be consumed at the *Khalal* stage. To reduce losses and facilitate early harvest, it is desirable to promote rutab development in 'Helali' dates. Artificially ripening was tried by Farag and Al-Konaissi (2001) at Al-Kwaitat Experimental Station, Al-Ain, UAE, wherein they subjected fruits of cultivar 'Helali' at early or full *Khalal* stage to ethanol fumes (50 % v/v) in sealed thick plastic bags for 24 hours and then incubated them at $22^0 \pm 2^\circ$ C for 6 days. They concluded that ethanol fumes alone at 50% v/v could be used after harvest to induce *Rutab* development in 'Helali' date fruits at full *Khalal* stage.

'Mishrigi Wad Lagi' and 'Mishrig Wad Khatib' are commercially important semi soft dates in Sudan. In Khartoum region of Sudan, rainfall starts in mid July and lasts till the end of October. Early ripening is a necessary to avoid rotting and fermentation. Musa (2001) applied ethrel through two methods for inducing early ripening in 'Mishrigi Wad Lagi' and 'Mishrigi Wad Khatib' cultivars of date palm in the beginning of June when the fruits begin to change color and two weeks later. In the first method, 2 ml of ethrel was injected by making a small pit on the peduncle and covered by a cellotape. In the second method, 1000 ppm ethrel was sprayed on the fruit bunches. There was significant increase in fruit ripening especially by the first method. In 'Mishrigi Wad Khatib' cultivar, ripening percentage could exceed 90% by beginning of August.

Al-Juburi *et al.*, (2001b) reported that NAA or a combination of growth regulators (NAA, GA_3 and ethephon) when applied 20 days after pollination in 'Khenazi' dates, successfully delayed fruit ripening. Preharvest application of ethrel increased ripening of 'Helali' dates, whereas ABG-3168, an ethylene blocker at 3.33 g L^{-1} inhibited ripening (Awad, 2007).

'Barhee' is a mid-season cultivar grown widely in the Gulf region. 'Barhee' dates are consumed fresh at the *Bisir* stage due to low amount of soluble tannins (Botes and Zaid, 1999). On ripening fruits become soft and sweet, however, their market value decreases. Under agro-climatic conditions prevailing in Kingdom of Saudi Arabia, 'Barhee' dates reach *Bisir* stage in the month of July when the climate is extremely hot. Once harvested, fruits ripen quickly, especially under ambient conditions thereby triggering distress sale. It is imperative to prolong the shelf life so that harvested *Bisir* fruits of cv. 'Barhee' can be marketed at optimal quality with minimal physiological disorders and physical damage. Al-Qurashi and Awad (2011b) at King Abdul-Aziz University, Jeddah, Saudi Arabia evaluated the effect of postharvest dipping in GA_3 (50, 100 and 150 mg L^{-1}), BA (20, 40 and 60 mg L^{-1}) and NAA (50, 100 and 150 mg L^{-1}) on fruit quality of 'Barhee' dates harvested at *Bisir* stage. Treated fruits were kept in perforated (10 perforations) plastic boxes in a refrigerator [(1 ± 1) °C and 90–95 RH]. Fruits treated with BA at 20 mg L^{-1} exhibited the lowest weight loss. Treated fruits had higher vitamin C content as compared to control fruits. After 60 days of storage, GA_3 treated fruits (50 mg L^{-1}) had lower *Rutab* percentage than control fruits. Dipping 'Barhee' dates harvested at the bisir stage in GA_3, BA and NAA solutions at different concentrations showed slight improvement in fruit quality characters and ripening during subsequent cold storage at 1°C. Work done in the past indicates that application of GA_3 in different date cultivars at *Kimri* stage delayed fruit ripening [Abou-Aziz 1983; Hussein *et al.*, 1993]. Further, application of NAA at (40 and 60) mg L^{-1} on 'Zahidi' dates at late *Kimri* stage increased fruit size and delayed ripening by one month [Mohammed and Shabana, 1980]. The application of NAA at (50 and 100) mg L^{-1} on 'Sewy' dates at 10–30 days after fruit set stage increased fruit drop and decreased yield but improved fruit weight and quality compared with control (Moustafa *et al.*, 1993) .

An investigation on effect of GA_3 and storage temperature on shelf life of Date Palm fruits variety 'Khasab' was conducted at Date Palm Research Centre, College of Agriculture, SKRAU, Bikaner. Harvested fruits of date palm variety 'Khasab' were treated with different concentrations of GA_3 (0, 25, 50 and 75 ppm) for 5 minutes and stored at different temperatures (room temperature, 5°C and 0°C and -40°C) for 45 days in polyethylene bags. The minimum physiological loss in weight was observed in fruits treated with 75 ppm GA_3 after 45 days duration of storage. Fruits stored at lower temperatures (0°C and -40°C) secured higher values for organoleptic evaluation and visual appearance with minimum loss in fruit quality characters as compared to fruits stored at room temperature and at 5°C (Meena *et al.*, 2017).

References

- Aaouine, M. 2000. Production of date palm *vitro* plants: The Moroccan experience. *Proceedings of the Date Palm International Symposium*-2000, February 22-25, 2000, Windhoek, Namibia, pp: 46-52.

- Abahmane, L. 2011. Date palm micropropagation *via* organogenesis. **In**: *Date palm Biotechnology* (Eds.) Jain, S. M.; Al-Khayri, J. M. and Johnson, D. V., Dordrecht: Springer, pp: 69-90.

- Abate, T. 2009. Effects of offshoot sizes and Indole-3-Butyric Acid on rooting of date palm (*Phoenix dactylifera* L.). M.Sc. thesis submitted to Haryana Agricultural University, India.

- Abdolhossein Aboutalebi and Abdohossein Mohammadi. 2015. Evaluating the effects of 2,4-D, naphthalene acetic acid and potassium nitrate on yield and fruit quality of date palm (Phoenix dactylifera) cv. Zahedi. *International Journal of Biosciences*, **6** (1): 192-200.

- Abou-Aziz, A. B. 1983. Effect of GA_3 and hand pollination on the yield and quality of 'Sewy' dates. *Proceedings of First Symposium on Date palm*, March 23-25, 1982, Saudi Arabia, pp: 258-267.

- Afzal, M.; Khan, M. A.; Pervez, M. A. and Ahmed, R. 2011. Root induction in the aerial offshoots of date palm (*Phoenix dactylifera* L.) cultivar 'Hillawi'. *Pakistan Journal of Agricultural Sciences*, **48**(1): 11-17.

- Ahmed, M. A.; Hassan, H. S. A. and Soliman, S. S. 2010. Effect of some growth regulators on yield and fruit quality of 'Samani' date palm. *Acta Horticulturae*, **882**: 745-753.

- Aljuburi, H. J.; Al-Masry, H. H. and Al-Banna, M. 2002. Fruit characteristics and productivity of date palm as affected by growth regulators. *Annals of Arid Zone*, **41**(1): 57-63.

- Aljuburi, H. J.; Al-Masry, H. H. and Al-Muhanna, S. A. 2001a. Effect of some growth regulators on some fruit characteristics and productivity of the 'Barhee' date palm tree cultivar (*Phoenix dactylifera* L.). *Fruits*, **58**(5): 325-332.

- Aljuburi, H. J.; Al-Masry, H. H.; Al-Banna, M. and Al-Muhanna, S. A. 2001b. Effect of some growth regulators on some fruit characteristics and productivity of date palm trees (*Phoenix dactylifera* L.) II: 'Khaniezy' cultivar. *Proceedings of the Second International Conference on Date palm*, United Arab Emirates University, Al-Ain, pp: 22-35.

- Aljuburi, H. J.; Maroff, A. M.; Wafi, M. K. and Al-Far, M. 2010. The role of NAA in the regulation of mineral concentration of date palm seedlings ('Shabeeby') under salt stress conditions. *Acta Horticulturae*, **882**: 121-130.
- Aljuburi, H.J. and Al-Masry, H. H. 2000. Effects of salinity and indole acetic acid on growth and mineral content of date palm seedlings. *Fruits*, **55**(5): 315-323
- Al-Khateeb, A. A. 2006. Role of cytokinin and auxin on the multiplication stage of Date palm (*Phoenix dactylifera* L.) cv. 'Sukry'. *Biotechnology*, **5**(3): 349-352.
- Al-Khateeb, A. A. 2008. The problems facing the use of tissue culture technique in date palm (*Phoenix dactylifera* L.). *Scientific Journal of King Faisal University*, **9**: 85-104.
- Al-Khateeb, A. A.; Aljaber, A. M. S. and Aljabr, A. M. H. 2006. *Date palm in Kingdom of Saudi Arabia*. The National Date Palm Research Center, Ministry of Agriculture, Kingdom of Saudi Arabia, pp: 136.
- Al-Khateeb, A. A.; Al-Khateeb, S. A.; Ohmura, T.; Okawara, R. and Ali-dinar, H. 2015. Rooting of date palm (*Phoenix dactylifera* L.) offshoots by Isoprothiolane (IPT). *European Scientific Journal*, **2**: 264-274.
- Al-Khayri, J. M. 2005. Date palm (*Phoenix dactylifera* L.). **In:** *Protocols of somatic embryogenesis in woody plants* (Eds.) Jain, S. M. and Gupta, P. K., Berlin: Springer, pp. 309-319.
- Al-Khayri, J. M. 2010. Somatic embryogenesis of date palm (*Phoenix dactylifera* L.) improved by coconut water. *Biotechnology*, **9**: 477-484.
- Al-Mana, F. A.; Ed-Hamady, M. A.; Bacha, M. A. and Abderahman, A. O. 1996. Improving root development of ground and aerial palm offshoots. *Principes*, **40**(4): 179-181 and 217-219.
- Al-Mayahi, A. M. W. 2014. Thidiazuron-induced *in vitro* bud organogenesis of the date palm (*Phoenix dactylifera* L.) cv. 'Hillawi'. *African Journal of Biotechnology*, **13**: 3581- 3590.
- Al-Mazroui, H. S.; Zaid, A. and Bouhouche, N. 2007. Morphological abnormalities in tissue culture-derived date palm (*Phoenix dactylifera* L.). *Acta Horticulturae*, **736**: 329-335.
- Al-Qurashi, A. D. and Awad, M. A. 2011a. Naphthalene acetic acid increase bunch weight and improve fruit quality of '*Barhee*' date palm cultivar under hot arid climate.

American-Eurasian Journal of Agriculture and Environmental Science, **10**(4): 569-573.

- Al-Qurashi, A. D. and Awad, M. A. 2011b. Quality characteristics of bisir 'Barhee' dates during cold storage as affected by postharvest dipping in gibberellic acid, naphthalene acetic acid and benzyladenine. *Fruits*, **66**: 343-352.
- Al-Qurashi, A. D.; Awad, M. A. and Elsayed, M. I. 2012. Pre-harvest fruit drop, bunch weight and fruit quality of 'Rothana' and 'Ghur' date palm cultivars as affected by some plant growth regulators. *African Journal of Biotechnology*, **11**(81): 14644-14651.
- Awad, M. A. 2007. Increasing the rate of ripening of date palm fruit (*Phoenix dactylifera* L.) cv. 'Helali' by preharvest and postharvest treatments. *Postharvest Biology and Technology*, **43**(1): 121-127.
- Awad, M. A. and Al-Qurashi, A. D. 2012. Gibberellic acid spray and bunch bagging increase bunch weight and improve fruit quality of 'Barhee' Date palm cultivar under hot arid conditions. *Scientia Horticulturae,* **138**: 96-100.
- Beauchesne, G. 1983. Vegetative propagation of date palm (*Phoenix dactylifera* L.) by *in vitro* culture. *Proceedings of the First Symposium on Date Palm*, March 23-25, 1982, Saudi Arabia, pp: 698-699.
- Beauchesne, G.; Zaid, A. and Rhiss, A. 1986. Meristematic potentialies of bottom of young leaves to rapidly propagate date palm. *Proceedings of the Second Symposium on Date Palm*, March 3-6, 1986, King Faisal University, Kingdom of Saudi Arabia, pp: 87-94.
- Ben, S. M. 2001. Study on fruit drop of 'Khenazy' date palm cultivar. *Proceedings of the Second International Conference on Date Palm*, United Arab Emirates University, Al-Ain. p. 161.
- Botes, A. and Zaid, A. 1999. The economic importance of date production and international trade. **In:** *Date Palm Cultivation,* FAO Plant Production and Protection Paper No. 156, (Eds.) Zaid, A. and Arias, E.J. FAO, pp: 45–57.
- Chaudhary, S. K.; Kumar, S.; Meena, R.; Yadav, P. K. and Sudarsan, Y. 2018. Effect of GA_3 on fruit yield and quality of date palm (*Phoenix dactylifera* L.). *International Journal of Current Microbiology and Applied Sciences.*, **7**(2): 3448-3456.
- Darwesh, R. S.; Madbolly, E. A. and Gadalla, E. G. 2013. Impact of Indole Butyric Acid and Paclobutrazol on rooting of Date palm (*Phoenix dactylifera* L.) off-shoots cultivar 'Zaghloul'. *Journal of Horticultural Science and Ornamental Plants*, **5**(3): 145-150.

- Dietz, T. H. 1998. Regulation of Fruit Development by Pollen in the Omani Date palm. Ph.D thesis submitted to University of Nottingham, U.K.
- Doaigey, A. R.; Al-Whaibi, M. H.; Siddiqui, M. H.; Al-Sahli, A. A. and El-Zaidy, M. E. 2013. Effect of GA_3 and 2,4-D foliar application on the anatomy of date palm (*Phoenix dactylifera* L.) seedling leaf. *Saudi Journal of Biological Sciences*, **20**: 141-147.
- El-Hadrami, A.; Daayf, F.; Elshibli, S.; Jain, S. M. and El Haldirami, I. 2011. Somaclonal variation in date palm. **In:** *Date palm Biotechnology,* (Eds.) Jain, S. M.; Al-Khayri, J. M. and Johnson, D. V., Dordrecht: Springer, p. 183-204.
- El-Hadrami, I. and El Hadrami, A. 2009. Breeding date palm. **In:** *Breeding Plantation Tree Crops,* (Eds.) Jain, S. M. and Priyadarshan, P. M., New York: Springer, pp. 191-216.
- El-Hadrami, I.; Cheikh, R. and Baaziz, M. 1995. Somatic embryogenesis and plant regeneration from shoot-tip explants in *Phoenix dactylifera* L. *Biologia Plantarum*, 37: 205-211.
- El-Hamady, M.; Hamdia, M.; Ayaad, M.; Salama, M. E. and Omar, A. K. 2010. Metaxenic effects as related to hormonal changes during date palm (*Phoenix dactylifera* L.) fruit growth and development. *Acta Horticulturae,*, **882**: 155-164.
- El-Hodairi, M. H.; Albahi, A.; Ibrahim, S. B. and Hamza 2001. Interaction effects of the sequential applications of some growth regulators on the growth of date palm trees (*Phoenix dactylifera* L.). *Proceedings of the Second International Conference on Date palm*, United Arab Emirates University, Al-Ain: 78-84.
- El-Hodairi, M. H.; El-Barkouli, A. A. and Bawa, O. 1992b. The effect of some growth regulators on fruit set of date palm (*Phoenix dactylifera* L.) trees. *Acta Horticulturae*, **321**: 334-342.
- El-Hodairi, M. H.; El-Fagih, A. S. and Amer, A. A. 1992a. The effects of Indole Acetic Acid (IAA), Indole Butyric Acid (IBA) and Naphthalene Acetic Acid (NAA) on the growth of 'Taghiat' date palm (*Phoenix dactylifera* L.). *Acta Horticulturae,* **321**: 326-333.
- El-Nabawy, S.M., El-Hammady, A.M., Marie, N.S. and Bondok, A.Z. 1977. Effect of some growth regulators on growth and development of 'Samani' date fruits. *Research Bulletin*, Faculty of Agriculture, Ain-Shams University 729:23, Hort. Abst. **51**(5):4071, 1981.
- El-Salhy, A. M.; Mostafa, R. A. A.; Al-Badawi, A. A. and Abd-Allah, M. G. 2016. Response of 'Seewy' date palm to GA_3 and CPPU spraying. *Assiut Journal of Agricultural Sciences*, **47**(5): 125-137.

- El-Shiekh, A. F. 2014. Effect of gibberellic acid, glutamic acid and pollen grains extract on yield, quality and marketability of 'Khalas' date palm fruits. *Acta Horticulturae*, **1047**: 93-98.

- Eshraghi, P.; Zaghami, R.; Mirabdulbaghi, M. 2005. Somatic embryogenesis in two Iranian date palm cultivars. *African Journal of Biotechnology*, **4**: 1309-1312.

- Farag, K. M. and Al-Konaissi, S. M. 2001. Rutab induction in 'Helali' date fruits by ethanol fumes. *Proceedings of Second International Conference on Date palm*, United Arab Emirates University, Al-Ain, UAE, pp. 733-741.

- Farag, K. M. and Kassem, H. A. 1998. Accelerating and intensifying color formation of 'Zaghloul' date palm fruits using modified ethephon formulations. *Proceedings of the First International Conference on Date palm*, United Arab Emirates University, Al-Ain, UAE, 62-71.

- Fki, L.; Bouaziz, N.; Sahnoun, N.; Swennen, R.; Drir, N. and Panis, B. 2011a. Palm cryo-banking. *Cryo Letters*, **32**(6): 451-462.

- Fki, L.; Masmoudi, R. Kriaa, W.; Mahjoub, A.; Sghaier, B.; Mzid, R.; Mliki, A.; Rival, A. and Drira, N. 2011b. Date palm micropropagation *via* somatic embryogenesis. **In:** *Date palm Biotechnology*, (Eds.) Jain, S. M.; Al-Khayri, J. M. and Johnson, D. V., Dordrecht: Springer, pp: 47-68.

- Gupta, O. P. and N. R. Godara. 1984. Rooting in aerial suckers of date palm. *Haryana Agricultural University Journal of Research*, **14**: 82-84.

- Hafez, O. M.; Saleh, M. A.; Mostafa, E. M. A.; El-Shamma, M. S. and Maksoud, M. A. 2014. Improving pollination efficiency, yield and fruit quality of two date palm cultivars using growth activators. *Journal of Agricultural Research*, **9**(1): 29-37.

- Hassaballa LA., M.M. Ibrahim, M.M. Sharaf, A.Z. Abd-EI-Aziz, and N.A. Hagagy 1983.Fruit physical and chemical characteristics of Zaghloul date cultivar in response to some fruit thinning treatments. Annals of Agric. Sci., Moshtohor. 20: 1-12

- Hassan, M. M. and El-Samnoudi, I. M. 1993. Salt tolerance of date palm trees. *Proceedings of Third Symposium on Date palm* organized at Date Palm Research Center, King Faisal University, Saudi Arabia during January 17-20, 1993, pp: 293-297.

- Hussein, M. A.; El-Agamy, S. Z.; Amin, K. I. A. and Galal, S. 1993. Physiological studies for prolonging harvest date of Samany dates under Assiut Governorate conditions. A- Effect GA_3 and fruit thinning. *Proceedings of Third Symposium on Date palm* organized at

Date Palm Research Center, King Faisal University, Saudi Arabia during January17-20, 1993, pp: 423- 433.

- Hussein, M. A.; Mahmoud, H. M.; Amen, K. I. A. and Mustafa, M. 1996. Changes in the physical and chemical characteristics of Zaghloul dates during development and maturity as affected by GA_3 and CCC under Assiut Governorate conditions. *Proceedings of Third Symposium on Date palm* organized at Date Palm Research Center, King Faisal University, Saudi Arabia during January17-20 1993, pp:389-404.
- IPGRI, INRAA, INRAM, INRAT, FEM, PNUD. 2005. Descripteurs du Palmier dattier (*Phoenix dactylifera* L.). Institut international des ressources phytogénétiques, Rome, Italie; Fonds pour l'Environnement Mondial, Washington, Etats-Unis ; Programme des Nations Unies pour le Développement, New York, Etats-Unis; Institut National de la Recherche Agronomique, d'Algérie, du Maroc et de Tunisie.
- Jain, S. M. 2001. Tissue culture-derived variation in crop improvement. *Euphytica*, **118**: 153-166.
- Johnson, D. V.; Al-Khayri, J. M. and Jain, J. M. 2013. Seedling date palm (*Phoenix dactylifera* L.) as genetic resources. *Emirates Journal of Food and Agriculture*, **25**(11): 809-830.
- Kader, A.A. and Hussein, A. M. 2009. *Harvesting and Postharvest Handling of Dates*. Postharvest Research and Technology Information Centre, UC Davis, California.
- Kassem, H. A.; Al-Obeed, R. S. and Ahmed, M. A. 2010. Extending harvest season, improving fruit quality and shelf life of 'Barhee' Date palm by preharvest sprays. *Acta Horticulturae*, **882**: 147-154.
- Khalifa, A.S.; El-Hammady A. A. M.; El- Hammady, M. M. and Wally, Y. A. 1975. Ripening of date fruits as affected by ethephon. *Egypt Journal of Horticulture*, **2**: 83-102.
- Khan, S. and Bi, T. B. 2012. Direct shoot regeneration system for date palm (*Phoenix dactylifera* L.) cv. 'Dhakki' as a means of micropropagation. *Pakistan Journal of Botany*, **44**(6): 1965-1971.
- Khierallah, H. S. and Bader, M. S. M. 2007. Micropropagation of date palm (*Phoenix dactylifera* L.) var. 'Maktoom' through organogenesis. *Acta Horticulturae*, **736**: 213-223.
- Kriaa, W.; Sghaier-Hammami, B.; Masmoudi-Allouche, F.; Benjemaa-Masmoudi, R. and Drira, N. 2012. The date palm (*Phoenix dactylifera* L.) micropropagation using completely mature female flowers. *Comptes Rendus Biologies*, **335**: 194-204.

- Maximos, S. E., Aziz, A. B. A., Desouky, I. M. and Antoun, N. S. 1980. Effect of GA_3 and ethephon on the yield and quality of 'Seewy' date fruits. Moshtohor. *Annals of Agricultural Science,* **12**: 251-262.
- Mazri, M. A. 2015. Role of cytokinins and physical state of the culture medium to improve *in vitro* shoot multiplication, rooting and acclimatization of date palm (*Phoenix dactylifera* L.) cv. Boufeggous. *Journal of Plant Biochemistry and Biotechnology*, **24**(3): 268-275.
- Mazri, M. A. and Meziani, R. 2013. An improved method for micropropagation and regeneration of date palm (*Phoenix dactylifera* L). *Journal of Plant Biochemistry and Biotechnology*, **22**: 176-184.
- Mazri, M. A. and Meziani, R. 2015. Micropropagation of Date Palm: A Review. *Cell and Developmental Biology*, **4**: 160.
- Mazri, M. A.; Belkoura, I.; Meziani, R.; Mokhless, B. and Nour, S. 2017. Somatic embryogenesis from bud and leaf explants of date palm (*Phoenix dactylifera* L.) cv. 'Najda'. *3 Biotech*, **7**(1): 58.
- McCubbin, M. J. and Zaid, A. 2007. Would a combination of organogenesis and embryogenesis techniques in date palm micropropagation be the answer? *Acta Horticulturae*, **736**: 255-259.
- McCubbin, M. J.; Van, S. J. and Zaid, A. 2000. A Southern African survey conducted for off-types on date palm production using somatic embryogenesis. *Proceedings of the Date Palm International Symposium*, February 22-25, 2000 Windboek, Namibia, pp: 68-72.
- Meena, R. B.; Kumar, S.; Choudhary, S. K. and Yadav, P. K. 2017. Effect of post-harvest treatments of plant growth regulator and temperature on a storage life of date palm. *International Journal of Chemical Studies*, **5**(6): 2348-2350.
- Meena, R.; Yadav, P. K. and Singh, R. S. 2013. Response of date palm cultivars to varying concentration of ethephon on yield and yield attributing characters. *Progressive Agriculture*, **13**(1): 1-3.
- Mohammadi, S. A. and Khankahdani, H. H. 2010. Effects of plant growth regulators on quantitative and qualitative traits of 'Shahani' date (*Phoenix dactylifera*) fruits. *Acta Horticulturae.,* **882**: 717-723.
- Mohammed, S. and Shabana, H. R. 1980. Effects of naphthalene acetic acid on fruit size, quality and ripening of 'Zahdi' date palm. *Hort Science,* **15**(6): 724-725.

- Mougheith, M. G. and Hassaballa, I. A. 1979. Effect of pre-harvest sprays of some plant growth regulating substances on yield and fruit characteristics of 'Hayany' date cultivar. *Research Bulletin*, Faculty of Agriculture, Ain-Shams University, No. 1073: pp. 21.

- Moustafa, A. A. and Seif, S. A. 1996. Effect of ethrel and GA treatments on yield and fruit quality of 'Seewy' date palms, grown in El-Fayoum Governorate. *Proceedings of the Third Symposium on Date Palm* organized at Date palm Research Centre, King Faisal University, Saudi Arabia during January, 17-20, 1993, pp: 379-388.

- Moustafa, A. A.; Seif S. A.; and Abou-El-Azayem, A. I. 1996. Date fruit response to naphthalene acetic acid. *Proceedings of the Third Symposium on Date Palm* organized at Date palm Research Centre, King Faisal University, Saudi Arabia during January, 17-20, 1993, pp: 369-378.

- Moustafa, A. A.; Seif, S. A. and Abou-El-Azayem, A. I. 1993. Date fruit response to naphthalene-acetic acid. *Proceedings of the Third Symposium on Date Palm* organized at Date palm Research Centre, King Faisal University, Saudi Arabia during January, 17-20, 1993, pp: 369-377.

- Musa, S. K. 2001. Early ripening of date using ethrel. *Proceedings of the Second International Conference of Date palm*, United Arab Emirates University, Al-Ain: 36-46.

- Naik, P. M. and Al-Khayri, J. M. 2016. Somatic embryogenesis of date palm (*Phoenix dactylifera* L.) through cell suspension culture. **In:** *Protocols for in vitro cultures and secondary metabolite analysis of aromatic and medicinal plants: Methods in Molecular Biology,* (Ed.) Jain, S. M., New York: Springer, **1391**: 357-366.

- Nasir, M. J. 1996. Effect of IBA on rooted and un-rooted offshoots of date palm *(Phoenix dactylifera* L.) cv. 'Hillawi'. M.Sc. thesis submitted to University of Agriculture, Faisalabad, Pakistan.

- Othmani, A.; Mzid, R.; Bayoudh, C.; Trifi, M. and Drira, N. 2011. Bioreactors and automation in date palm micropropagation. **In:** *Date palm Biotechnology* (Eds.) Jain, S. M.; Al-Khayri, J. M. and Johnson, D. V., Dordrecht: Springer, pp: 119-136.

- Rouhani, I. and Bassiri, A. 1977. Effect of ethephon on ripening and physiology of date fruits at different stages of maturity. *Journal of Horticultural Science*, **52**: 289-197.

- Sane, D.; Ould, K. M.; Diouf, D.; Diouf, D.; Badiane, F. A.; Sagna M. and Borgel, A. 2005. Growth and development of date palm (*Phœnix dactylifera* L.) seedlings under drought and salinity stresses. *African Journal of Biotechnology*, **4**(9): 968-972.

- Sharples, G. C. and Hilgeman, R. H. 1950. Progress report. Effects of hormone spray application upon the development and ripening of date fruit. *Date Grower's Inst. Rep.* 24:14
- Singh, R. S. and Bhargava, R. 2009. Effect of seed treatments on germination and growth behaviour in date palm (*Phoenix species*) under hot arid conditions. *Journal of Tropical Forestry*, **25**(1 & 2): 42-48.
- Soliman, S. S. and Ali, E. A. M. 2009. Effect of GA_3 on yield and fruit characteristics of 'Sakkoty' date palm under Aswan conditions in Egypt. *Green Farming*, **2**(7): 459-462.
- Tisserat. B. 1984. Propagation of date palm by shoot tip culture. *Hortic. Sci.*, **19**: 230-231.
- USDA, National Nutrient Database for Standard Reference, April, 2018. *Basic Report*-09087, Dates, 'Deglet Noor'.
- Zaid, A. and P.F. de Wet 2002. Botanical and Systematic Description of the date palm. **In:** *Date palm Cultivation*, FAO Plant Production and Protection Paper 156A, (Eds.). Zaid and E.J. Arias-Jimenez, FAO, Rome, Italy.
- Zirari, A. and Ichir, L. 2010. Effect of exogenous indole butyric acid (IBA) on rooting and leaf growth of small date palm offshoots (*Phoenix dactylifera* L.) derived from adult *vitro* plants of 'Najda' cultivar. *Acta Horticulturae*, **882**: 839-844.

11

Plant Growth Regulators in Fig

Nidhi Sharma, Vikas Kumar Sharma, Soubhagy Kumar Sahu, A.K. Godara and A K Panda

1. Introduction

Fig (*Ficus carica* L.) is a moderately important fruit crop with an annual estimated global production of one million tons of fruit. Fig is originated from western Asia and spread to Mediterranean region by Greeks and Romans. In India, fig is considered to be a minor fruit crop and the commercial cultivation of common (edible) fig is mostly confined to western parts of Maharashtra, Gujarat, Uttar Pradesh (Lucknow & Saharanpur), Karnataka (Bellary, Chitradurga & Srirangapatna) and Tamil Nadu (Coimbatore). Fig is consumed fresh or in processed form. The dried form being the most popular. The dry fig has a high nutritive value. Fig contains an appreciable amount of vitamin A, Vitamin C and smaller amount of vitamin B and D, but high in sugar content, Fe and Ca. Fig fruits have been prized over countries for its laxative, medicinal and dietary properties due to the presence of many antioxidants. Fig is valued for its laxative properties and is used in the treatment of skin infection. The fruit helps to maintain acid alkali balance of the body. Latex is used to coagulate milk. Vinegar preserved figs used as a medicine in treatment of liver and spleen enlargement in children. *Ficus carica* is highly useful in liver and spleen disorders, gout treatment and to cure piles (Lansky *et al.*, 2008). Some of the most interesting therapeutic effects include anticancer, hepatoprotective, hypoglycemic, hypolipidemic, and antimicrobial activities (Shukranul *et al.,* 2013).

Generally, fig can be cultivated under adverse soil and climatic conditions like alkaline, poorly drained soil, hot and arid climate. Fig trees are drought tolerant and easily manage in the orchard as compared to other tree, so the fig cultivation is gaining importance. Like other fruits, fig gives good response to the application of growth regulators for higher yield and quality fruit production. It easily cultivates in dry lands either as mono crop or in mixed cropping with other fruit trees (Naik *et al.*, 2015). Plant growth regulators (PGRs) are being used to maximize the fruit yield and quality and minimize harvest cost. These are also used to accelerate blooming become more uniform, improve fruit set and size, thinning of fruit or flowers to improve size and quality and fruit maturation at the same time (Bremer *et al.*, 2006). This chapter aimed to point out the effects of plant growth regulators (PGRs) on the various stages of fig under following heads:

2. Effect of Plant growth Regulators on Propagation

2.1. Effect of PGRs on dormancy removal and seed germination

Fig can be propagated by seeds extracted from dried fruits. Seed germination rate in fig about 60 per cent under the greenhouse conditions, zygotic seedlings usually are not true to type and are used in development of new varieties. Plants grown from seed are not guaranteed to have the same traits as either parent. Additionally, only some seeds will produce female trees with edible fruit while the others will produce male fig trees with small inedible fruit. Fertile seeds can be germinated by spreading them on flats of moist soil mix or sphagnum moss (Hartmann, 2011).

For production of the seedling rootstocks and hybrid plants, seeds are necessary. (Hartmann *et al.*, 2002), but the seed germination is affected by a number of external and internal factors as seed coat, undeveloped embryo or chemical inhibitors that cause seed dormancy (Agrawal and Dadlani, 1995). Although it is possible to grow fig from seed it is not recommended unless breeding new cultivars. In fig breeding programs, seeds are used for producing mosaic virus-free seedlings (Caglayan *et al.*, 2010). Under laboratory conditions, it is very difficult to obtain high percentages of *F. carica* seed germination (Ellis *et al.*, 1985). Seeds treated with GA_3 (500-1000 ppm) gives the highest percentages of germination and emergence and also reduce the time to germination and emergence from the seeds in Bursa 'Siyahı' and 'Sarılop' fig cultivars (Caliskan *et al.*, 2012).

2.2. Effect of PGRs on rooting in cuttings and air layering

Fig can be propagated by a number of vegetative propagation methods such as stem cutting, air layering, grafting, budding and by tissue culture method. Asexual propagation is also very useful for replicating clonal planting material of elite plants, germplasm conservation and introduction of fast growing species. Fig is commonly propagated by cuttings to ensure that new trees are female and "true to type", with the same traits as the tree from which cuttings were collected. Dormant cuttings of 2-3 year old wood should be used for propagation. Cuttings should be ½ to ¾ inches in diameter and eight to twelve inches long. For best results cuttings should be prepared in early spring well before bud break. Grow cuttings for one or two seasons (12-15 months) in the nursery before transplanting to a permanent location (Morton, 1987).

Fig is hard- to- root so growth regulators are used for rooting activation and initiation. Plant growth regulators *viz;* auxin used for stimulation and initiation of roots in cuttings, by breaking root apical dominance induced by cytokinin (Cline, 1997). For induction of rooting in stem cuttings, the most important auxin especially 1-Naphthaleneacetic acid (NAA) and Indole-3- butyric acid (IBA), are widely applied either singly or in combination (Andus, 1963). In a large number of plant species, auxin treated cuttings are shown increase the per cent of rooting, hasten initiation and increase the number and quality of roots. Treatment with auxins has been shown to increase rooting percentage and rooting time together with uniformity of rooting (Hartmann *et al.*, 2002). The auxins promoted cell division and their elongation led to differentiation of cambial initials into root primordia and in the mobilization of reserve food material to sites of root initiation there by giving higher number of roots per cutting. According to Topacoglu *et al.* (2016) different hormones and different concentrations have different effects on rooting percentage and morphological characters. Thus, hormones should be selected based on the primary character. For instance, 10 ppm NAA should be preferred when high rooting rate is requested; 100 ppm IAA should be preferred when a big amount of root formation is requested; and 100 ppm GA_3 should be preferred when long root is requested. Growth regulator treatments IBA 2500 ppm + PHBA 2000 ppm recorded highest percentage of rooting, maximum survival percentage more number of roots per cutting and longest root length per cutting in both hardwood and semi hardwood cuttings. However, hardwood cuttings recorded better than semi hardwood cuttings for all the parameters (Reddy *et al.*, 2008). Quick dip treatment of IBA (3000 ppm) increased number of shoots, average shoot diameter, shoot length, number of leaves and leaf area (Kaur *et al.,* 2018). Effect of different plant bio-regulators on rooting of cuttings of fig has been given in Table 1.

Layering should be done in early spring using one-year old branches to ensure sufficient rooting my midsummer (Hartmann, 2011). Application of NAA and IBA found significantly effective in inducing rooting in layers in fig var. Dinakar and Poona fig alone as well as in combination. IBA @ 4000 and 2000 ppm and NAA @ 6000 ppm were found very effective in reducing the time laps of period of rooting in air layering of both the varieties. (Patel, 2009).

Table 1: Effect of growth regulators on rooting of cutting in fig

S. No.	PBR	Concentration	Response/Effect	Reference
1.	IBA	5000 ppm	Highest rooting percentage (76.6%) was obtained in Timla fig stem cuttings	Sundaram and Rangaswam, 1994
2.	IBA	100 ppm	The best root and shoot development was obtained without any stratification	Antunes *et al.,* 1996
3.	IBA	1000-4000 ppm	Green cuttings with 2 to 3 leaves exposed to IBA hormone, rooting of 85% to 100% was obtained in fig cv. Burnswick	Kai *et al.,* 1997
4.	IBA	3000 ppm	Recorded the highest percentage of rooting (64%) and survival percentage	Borah and Das, 2000
5.	IBA	100 mg/l	Cutting treated for 24 hours in IBA solution recorded the highest percentage of rooting (95%)	Norberto *et al.,* 2001
6.	IBA	100-300 mg/l	Treatment for 24 hrs improved rooting process	Pio *et al.,* 2003
7.	IBA or NAA	2500 ppm	Results 100% rooting success in fig air-layers during July and August	Sonpethkar *et al.,* 2004
8.	IBA + NAA	2500 + 2500 ppm	Gave maximum root growth (21.78 cm), percentage of rooted cuttings (52.49%) and survival percentage (82.29%) of in hardwood cuttings	Khapare *et al.,* 2012
9.	IBA	1000 mg/l	Quick dip method for 20 second gave best results with respect to rooting percentage and root volume	Zarei and Aboutalebi, 2013
10.	IBA	2000 mg/l	Most effective with respect to rooting percentage, number of roots per cutting, number of leaves per cutting and per cent survival	Bhuva, 2014

S. No.	PBR	Concentration	Response/Effect	Reference
11.	IBA	3000 mg/l	Results early root initiation (8.73 days), minimum days required for bulk appearance of roots (20.80) and highest number of primary roots (58.87)	Reddy *et al.*, 2014
12.	IBA	6000 ppm	During the month of July performed the best for rooting in wild fig	Mewar and Naithani, 2016
13.	IBA	4000 ppm	Beneficial effect on rooting percentage (73.24%), number of roots (10.87) and root length (10.63cm) at 60 DAP	Patel *et al.*, 2017
14.	IBA + NAA	1000 + 1000 mg/l	Helped in better induction in rooting (58.66%) and exhibited high survival percentage of rooted cuttings (82.50%)	Dahale *et al.*, 2018
15.	IBA	3000 ppm	Quick dip treatment gives the best results in terms of rooting percentage (71.10%), root length (28.12cm) and number of roots per cutting (36.99)	Kaur *et al.*, 2018

2.3. Effect of PGRs on In vitro multiplication

Fig tree (*Ficus carica* L.) has gained more interests in recent years for its economic importance and medicinal virtues. However, several abiotic and biotic constraints limit the development of this fruit crop and contribute to the decrease in revenue and the gradual disappearance of plantations (Saddoud *et al.*, 2011). The lack of selected pollinators according to the needs of female cultivars, and the difficulties of their multiplication in fields, are major constraints (Gaaliche *et al.*, 2013). Virus diseases, especially fig mosaic disease, are actually a threat and spread in all areas of production (Saddoud *et al.*, 2007); which inhibits the development of healthy orchards and limits the production (Ashihara *et al.*, 2004). In case of traditional methods only 20-30% of the cuttings are survive. To increase the cultivated area of fig, there is huge requirement of cuttings for large scale propagation. Though cuttings are best suitable for propagation but it will not be available. Thus tissue culture method has become an important technique for propagation of fig and other woody plants (Mustafa and Taha, 2012). Fig shoot tips and nodes have been successfully micro-propagated using several techniques. Improvement of farming techniques (Fraguas *et al.*, 2004), the sanitation of local varieties (Mars *et al.*, 2008) and the large-scale production of good quality and healthy fig plants and pollinators by meristem or shoot tip culture (Gunver *et al.*, 1998) seem to be basic requirements for successful commercial orchards (Pasqual and Ferreira, 2007). This requires the mastery of rapid micropropagation of local figs.

In general, roots can be propagated from shoot tips by placing them on semi-solid propagation material with IBA. Nodal explants can be propagated by placing them on propagation material with BA (Bapat and Mhatre, 2005). Each fig variety may require different concentrations of growth hormones for successful *in vitro* propagation. Some studies on the propagation of *Ficus carica* via tissue culture conducted by various workers have been presented in Table 2.

Table 2: Effect of growth regulators on *in vitro* propagation in fig

S. No.	PBR	Concentration	Response/Effect	Reference
1.	NAA + IBA	0.5 + 0.5 mg/l	For rooting transfer darkness for 7 days and transfer to the presence of light for 3 weeks	Muriithi *et al.*, 1982
2.	IBA	1 mg/l	Higher rooting responses (80%) were obtained after 28 days in light	Pontikis and Melas, 1986
3.	Kinetin + GA	0.5 + 0.1 mg/l	Shoot multiplication and the rate of propagation of fig was 1: 4 to 1: 6.0	Haelterman and Docampo, 1994
4.	Kinetin	0.5 mg/l	Best for shoot proliferation of *Ficus carica* 'Roxo de Valinhos' plants. The inclusion of BA in the medium induced excessive callus formation and small and vitrified shoots, while GA_3 induced excessive elongation associated with vitrification, chlorosis, and tip-burned shoots.	Fraguas *et al.*, 2004
5.	TDZ + 2iP	2 + 4 mg/l	The highest shoot formation percentage (89%) and highest percentage of shoots forming roots (95%) was obtained with 1.0 mg/l IBA	Soliman *et al.*, 2010
6.	BAP	2.5 mg/l	enhanced shoot multipli-cation and callus formation	Mustafa and Taha, 2012
7.	BA and GA	5 and 1 mg/l	Enhanced *in vitro* shoot multiplication	Darvesh *et al.*, 2014
8.	BAP	0.5 – 3.0 mg/l	Optimal concentration for shoot multiplication from leaf segment, while no beneficial effect was obtained from inclusion of Kinetin in the culture medium	Danial *et al.*, 2014

S. No.	PBR	Concentration	Response/Effect	Reference
9.	BAP + NAA + GA_3	0.2 + 0.1 + 1.0 mg/l	Best initiation of shoot-tips with sizes 0.5, 1 and 1.5 mm	Bayoudh *et al.,* 2015
10.	BAP + NAA	0.5 + 1.0 mg/l	For callus induction, the highest fresh weight in stem explants was 2.28±0.239 obtained	Patah *et al.,* 2018
11.	2,4-D + BAP	0.5 + 0.05 mg/l	100% maximum callus induction from petiole segment in *F. religiosa*	Hesami *et al.,* 2018

3. Effect of PGRs on Vegetative Growth

Naturally growth of aerial parts of the fig trees are abundant, require annual pruning, removal of last year fruited branches drastically. The percentage of pruning is too high, which means wasting of dry matter produced by plants, which could be used for production of better fruit quantity and quality, especially with regard to size (Tadeu Dias *et al.,* 2015). Pfnur (1958) studied that GA can cause stem elongation to a great extent in fig. Crane (1962) opined that use of growth regulator like GA on the plant, it breaks the dormancy of the terminal buds, so that the carbohydrates were utilized in shoot growth instead of in fruit development and the fruits became light in weight and lack of sweetness. Use of the rest breaking agent hydrogen cyanamide (HC) as Dormex (1.5%) enhances the vegetative growth, which influences the yield of fig. It also increased shoot length and advanced bud break by 10 days (Gaaliche *et al.,* 2017).

From the results of Tadeu Dias *et al.* (2015) concluded that the use of ethyl trinexapac, application at 250 mg/l did not affect the plant growth, increase the number of internodes, branch length and height of insertion of the first fruits up to the first harvest of fig fruits (*Ficus carica* L.) cv. Roxo de Valinhos. Fig plants sprayed with Gibberellic acid at 60 ppm with three sprays at an intervals of 15 days starting from bud initiation, resulted in significant increase in shoot length (2.95 cm), number of functional leaves (616.23) in Poona fig (Kurubar *et al.,* 2017).

4. Effect of PGRS on Flowering and fruit Set

Fig trees are much differs from many other plant species in their reproductive growth characteristics. However, the molecular mechanisms underlying this unorthodox physiology of floral induction and fruit setting in fig trees have not been much elucidated. Because generally the floral induction occurs when specific environmental conditions are satisfied, most of the plants bloom and bear during the same season each year. In fig, by contrast, the time interval during which inflorescence (flower bud, fruit) differentiation occurs corresponds to the shoot elongation period.

Most of the fig cultivars have potentially two crops, fruit produced from the first crop are called as brebas and the second or main crop, called as figs which is the commercially important crop for most fig cultivars. Due to less production and lower quality fruit in breba crop, as increase in labor cost it is become uneconomical. Early spring ethephon application at 300 ppm, when breba fruit and leaves are just start to develop and figs are not present, was a safe, less costly and effective way to reduce the breba crop (Crisosto *et al.*, 2010).

5. Effect of PGRs on Yield and Yield Attributes

Fig starts bearing fruits from the second year, economical harvesting thus starts from third year of planting. The yield increases with increase in canopy size of the tree and stabilizes during eighth year. With the application of 2,4,5- trichlorophenoxyacetic acid (20 ppm), it stimulates the production of ethylene by fruit and leaves which responsible for hastened maturation of fruit, epinasty, and senescence of leaves (Maxie and Crane, 19671). GA_3 at 50 ppm and NAA at 25 ppm increased the fruit diameter as compared with BA treatment at 40 ppm and untreated fruits. GA_3 treatment advanced fruit maturity by 3-4 weeks in Abbodi and Abiad Aswan fig fruits (El- Mahdy *et al.*, 2015). Sprayed with Gibberelins (50 and 100 ppm) and chlorophenoxy propion amide (CPA) (250, 500 and 1000 ppm) on figs plants pruned in December, for an out of season production of fruits. CPA at 1000 ppm on treated figs induced on elongation of fruits but did not differ from the without treatment in fruit weight (Sampaio *et al.*, 1983). GA at 15 ppm significantly increased fruit diameter and length but fruits receiving GA at 5 ppm were the heaviest. 2,4,5-T at the concentration of 5 ppm reduced the fruit development, fruit length and at 10 ppm damaged the shoots and fruits, whereas, CCC (500-1000 ppm) decreased fruit diameter and length (Phad *et al.*, 1980). The treatment of Ethrel at concentration of 500 ppm resulted in 82% of the crop being harvested at the first of 3 harvests, compared with 56% from untreated trees. Ethrel may advance the fig harvest so that attack by late-season pests could be avoided. So that the damage caused by insects or fungal infection can be controlled (Ferguson *et al.*, 1990). Effect of plant bio-regulators on fig in respect to yield and yield attributing characters has been presented in Table 3.

Table 3: Effect of growth regulators on yield and yield attributes in fig

S. No.	PBR	Concentration	Response/Effect	Reference
1.	GA	80 ppm	Spray at an interval of one month after bud initiation hastened the maturity of Mission fig by 25 days	Crane and Grossi, 1960
2.	GA	40 ppm	Reduced the maturity period by about 16 days, increase fruit length and decrease fruit diameter and weight in cv. Black Ischia when spray one month after bud initiation	Rane, 1962
3.	Ethephon	500 ppm	Applied near the commercial harvest time, fig ripening was enhanced, while, during the stage I, inhibited fruit growth and triggered fruit abscission within 5 to 6 day	Crane *et al.,* 1970
4.	GA	15 ppm	Reduced the maturity period and increase the fruit length, as increase the conc. of GA in Poona fig decrease fruit diameter and weight, when first spray was given on 14th December and second spray on 21st December	Gaikwad, 1975
5.	GA and IAA	At different concentr-ation	Increased fruit diameter and advanced fruit maturity by 4 week when sprayed on fruits at the end of the first period of fruit rapid growth (6 weeks old fruits)	El-Mahdy *et al.,* 2015
6.	Dormex	2.0%	Spray immediately after pruning in the month of July increase fruit weight and diameter of fig cv. Roxo de Valinhos	Leonel *et al.,* 2015
7.	GA	40 ppm	Plants sprayed at bud initiation stage and 15 days after bud initiation, higher yield (8.9 t/h) was obtained and with increase in its concentration and number of spray, the maturity period was advanced	Kurubar *et al.,* 2017
8.	Dormex	1.5%	Application of dormex at 40 days before bud break increased fruit number per shoot and yield however, fruit diameter remain unaffected	Gaaliche *et al.,* 2017

6. Effect of PGRs on Fruit Quality

In common fig or Adriatic fig, the phenomenon of fruit set depends on Parthenocarpy which can be altered by climatic conditions of a particular location. The parthenocarpic fruit set can be enhanced by spraying 25 ppm of NAA or IBA on the flowers. For Smyrna figs, inter planting of Capri figs should be done for effective fruit set. To promote proper shoot growth and to prevent fruit drop in fig, growth regulators like Gibberellic Acid (GA) @ 30 ml / litre of water is used. Baskaya and Crane (1950) conducted an experiment to compare the PCPA and IBA treated Calimyma Figs with naturally parthenocarpic Kadota and Caprified Kadota fig and it was found that the parthenocarpic figs contained high sugar percentage than caprified one. Ethylene is a growth hormone that initiates a chain of metabolic and physiological events leading to fig fruit ripening during the III stage of fruit growth (Marei and Crane, 1971). Singh and Chayrusia (1991) reported that Ethephon (750 ppm) when applied on mature unripe fruits brought about ripening of fruits 5 times more in 12 days as compared to control. When the fig trees were treated with GA_3 at fruit development stage produced best quality fruits with respect to diameter, length, TSS content of ripe fruits and market appearance as compared to olive oil and ethephon treatments. Quality parameters of fruit such as all sugars and titerable acidity were found to increase with increased concentration of spray at increased intervals (Kurubar *et al.*, 2017).

Crane and Grossi (1960) reported that when Mission figs (a naturally parthenocarpic variety) treated with GA at 80 ppm concentration, the maturity was hastened by 25 days. But at the lower concentration it was hastened up to only 15 days. They noticed that after spraying the fruits became ribbed sweetness would be lowered as compared to control. They also reported that the combination treatment of PCPA (GA antagonist) at 75 ppm with GA at 25 or 50 ppm cause hundred percent parthenocarpic fruit set. At 5 and 15 ppm of 2,4,5-T, the amount of seed content was reduced. Rane (1962) noticed that GA increased the fruit length and reducing sugar content (6.72%) of fruit with increase in its concentration while GA decreased the fruit diameter and weight of the fruit. With increase in concentration of GA, the non-reducing sugar content was decreased in both varieties although in Black Ischia variety of fig, it increased the non-reducing sugars at 40 ppm

7. Effect of PGRs on Post-Harvest and Stress Management

Fig fruit is very perishable in nature so it gets spoiled till it reaches to consumer. The main problem in export is the spoilage of fruits. The total losses due to spoilage in fig fruits may be ranging from 30-40 per cent (Salunkhe and Desai, 1984). Fig has a shelf life of 7-10 days at 0°C at 85-90% relative humidity in cold storage. Under normal room temperature it can be stored only 1 to 2 days. Slightly immature fruits are to be harvested for transporting to distant markets. Ripe fruits are picked either form the tree by twisting the neck at the stem end or by cutting it or gathered after they drop. A combination of pre-harvest and post-harvest treatments was carried out which consists of a spray of 2% calcium chloride 20 days prior to harvest followed by immersion in 50 ppm of kinetin and then in 0.1% carbendazim after storage were found most effective in enhancing the quality. During the storage period, there is slow down increase in TSS, total sugar, reducing sugar and non-reducing sugar content and reduces the decline in fruit acidity. This treatment increases the storage life up to four days at the indicated temperature of 24 to 28°C (Kurubar *et. al,* 2012).

Fig trees often suffer from waterlogging, because they are mainly grown in tropical and subtropical countries where a higher probability of long-term waterlogging and flash floods occur commonly because of extreme rainfall events (Zhou, 2003). ALA (5-aminolevulinic acid) pretreatment improves water logging tolerance of fig plants significantly, and the promoted root respiration, leaf photosynthesis, and antioxidant ability may contribute greatly to improve plant waterlogging tolerance (An *et al.,* 2016).

References

- Agrawal, P.K. and Dadlani, M. 1995. Techniques in seed science and technology. 2nd Ed. India: *South Asian Publishers Limited.*
- An, Y., Qi, L. and Wang, L. 2016. ALA pretreatment improves water-logging tolerance of fig plants. *Plos One,* **11**(1): e0147202. https://doi.org/10.1371/journal.pone.0147202.
- Andus, L.J. 1963. Plant growth substances. Leonard Hill Books Limited, London.
- Antunes, L.E.C., Chalfun, N.N.J., Ramos, J.D., Pasqual, M. and Veiga, R.D. 1996. Influence of different periods of stratification, indole butyric acid concentration and substrate on rooting of fig cuttings. *Ciencia e Agrotecnologia,* **20**(3): 307-314.
- Ashihara, W., Kondo, A., Shibao, M., Tanaka, H., Hiehata, K. and Izumi, K. 2004. ecology and control of eriophyid mites injurious to fruit trees in Japan. *JARQ-Jpn. Agr. Res. Q.,* **38**: 31- 41.
- Bapat, V.A. and Mhatre, M. 2005. *Moraceae.* Biotechnology of fruit and nut crops. *CABI Publishing, Oxford shire,* England.
- Baskaya, M. and Crane, J.C. 1950. Comparative histology of naturally parthenocarpic, hormone- induced parthenocarpic and caprified fig syconia. *Botanical Gazette,* **111**(4): 395-413.

- Bayoudh, C., Labidi, R., Majdoub, A. and Mars, M. 2015. *In vitro* propagation of caprifig and female fig varieties (*Ficus carica* L.) from shoot-tips. *J. Agr. Sci. Tech.*, **17**: 1597-1608.
- Bhuva, S.K. 2014. Effect of plant growth regulators on propagation of fig (*Ficus carica* L.) by hardwood and semi hardwood cuttings. *M.Sc. Ag. Thesis*, Fruit Science Department, ASPEE College of Horticulture and Forestry Navsari Agricultural University, Navsari.
- Borah, A. and Das, R.P. 2000. Efficacy of indole butyric acid (IBA) on rooting of stem cuttings of some minor fruits of Assam. *Haryana Journal of Horticultural Sciences,* **29**(3/4): 196-198.
- Bremer, V., Norton, M., Rivers, D. and Crisosto, C.H. Evaluating Ethephon as a tool to manipulate breba crop load (2006-2007). Department of Plant Sciences, UC Davis, *UCCE Merced.* 1-10.
- Caglayan, K., Medina, V., Gazel, M., Serçe, C.U., Serrano, L., Achon, A., Soylu, S., Caliskan, O. and Gumus, M. 2010. Putative agents of fig mosaic disease in Turkey. Turkish *Journal of Agriculture and Forestry,* **33**(5): 1-8.
- Caliskan, O., Mavi, K. and Polat, A. 2012. Influences of presowing treatments on the germination and emergence of fig seeds (*Ficus carica* L.). *Acta Scientiarum (Agronomy),* **34**(3): 293-297.
- Cline, M.G. 1997. Concepts and Terminology of Apical Dominance. *American Journal of Botany,* **84**: 1064-1069.
- Crane, J.C. 1962. Potential use of gibberellin and parachlorophenoxy acetic acid for settinga Calimyrna figs. *Proc. 16th Annual Res. Conf. Calif. Fig. Inst.,* 11-12.
- Crane, J.C. and Grossi, N. 1960. Fruit and vegetative responses of the Mission fig to gibberellin. *American Society for Horticultural Science,* **76**: 139-145.
- Crane, J.C., Marei, N. and Nelson, M.M. 1970. Ethrel speeds growth and maturity of figs. *California Agriculture,* **24**(3): 8-10.
- Crisosto, C.H., Bremer, V., Norton, M., Ferguson, L. and Einhorn, T. 2010. Pre-harvest ethephon eliminates first crop figs. *Hort. Technology,* **20**(1): 173-178.
- Dahale, M., Ningot, E.P. and Deepa, N.M. 2018. Effect of plant growth regulators on rooting and survival of hard wood cuttings in fig. *Int. J. Curr. Microbiol. App. Sci.,* Special Issue, **6**: 2386-2391.

- Danial, G.H., Ibrahim, D.A., Brkat, S.A. and Khalil, B.M. 2014. Multiple shoots production from shoot tips of fig tree (*Ficus carica* L.) and callus induction from leaf segments. *Int. J. Pure Appl. Sci. Technol.*, **20**(1): 117-124.
- Darvesh, H.Y., Bazaid, S.A. and Samra, B.N.A. 2014. *In vitro* propagation method of *Ficus carica* at Taif governorate using tissue culture technique. *International Journal of Advanced Research*, **2**(6): 756-761.
- Ellis, R.H., Hong, T.D. and Roberts, E.H. 1985. Handbook of seed technology for genebanks. Volume I. Principles and methodology. International Board for Plant Genetic Resources. Hand books for Gene banks.
- El-Mahdy, T.K., El-Akaad, M.M., Gouda, F.M. and Hussein, A.S. 2015. Effect of some growth regulators application on fruit growth and ripening of Abbodi and Abiad Aswan fig cultivars. *Assiut J. Agric. Sci.*, **46**(2): 107-119.
- Ferguson, L., Michailides, T.J. and Shorey, H. (1990). The California fig industry. *Hort. Rev.*, **12**: 409-490.
- Fraguas, C.B., Pasqual, M., Dutra, L.F. and Cazetta, J.O. 2004. Micropropagation of fig (*Ficus carica* L.) 'Roxo De Valinhos' plants. *In Vitro Cellular & Developmental Biology Plant*, **40**(5):471-474.
- Gaaliche, B., Ghrab, M., and Mimoun, M. 2017). Effect of hydrogen cyanamide on vegetative growth, yield, and fruit quality of fig cv. Zidi in a warm production area. *International journal of fruit science*, **17**(1): 63-71.
- Gaaliche, B., Majdoub, A., Trad, M. and Mars, M. 2013. Assessment of pollen viability, germination, and tube growth in eight Tunisian Caprifig (*Ficus carica* L.) cultivars. *ISRN Agronomy*, **1**(4). http://dx.doi.org/10.1155/2013/207434.
- Gaikwad, T.S. 1975. Effect of various growth regulators on growth, maturity period and quality of fig (*Ficus carica* L.) cv. Poona fig. *Doctoral Dissertation*, Mahatma Phule Krishi Vidyapeeth, Rahuri.
- Gunver, G., Ertan, E., Aksoy, U., Ferguson, L. and Hepaksoy, S. 1998. A study on the propagation of figs by the tissue culture techniques. *Acta Hort.*, **480**: 169–172.
- Haelterman, R. and Docampo, D.M. 1994. *In vitro* propagation of mosaic free fig (*Ficus carica* L.) cultivars, using thermotherapy and shoot tip cultures. *Rev. Inv. Agrop.*, **25**: 15-22.
- Hartmann 2011. Hartmann and Kester's Plant Propagation: Principles and Practices. 8th

ed. Chapter 19 Propagation methods and rootstocks for fruit and nut species. p. 738-9. Retrieved Oct. 10, 2011 from: http://aggie-horticulture.tamu.edu/faculty/davies/index.html.

- Hartmann, H.T., Kester, D.E., Davies, F.T. and Geneve, R.L. 2002. Plant Propagation: Principles and Practices (7th edition). *Prentice- Hall,* New Jersey, p 880.
- Hesami, M., Daneshvar, M.H., Mohsen, Y.N. and Milad, A.2018. Effect of plant growth regulators on indirect shoot organogenesis of *Ficus religiosa* through seedling derived petiole segments. *Journal of Genetic Engineering and Biotechnology,* **16**(1): 175-180.
- Kai, M., Zhifen, L., Yan, T. and Wei-Bing, J. 1997. The green cutting propagation techniques for fig trees. *China Fruits,* **3**: 32-38.
- Kaur, A., Kaur, K. and Kaur, A. 2018. Role of IBA and PHB on success of cuttings of fig cv. Brown Turkey. *Asian Journal of Science and Technology,* **9**(5): 8237-8241.
- Khapare, L.S., Dahale, M.H. and Bhusari, R.B. 2012. Effect of plant growth regulators on rooting in cuttings of fig (*Ficus carica* L.) cv. Dinkar. *Asian Science,* **7**(1): 25-27.
- Kumar, V., Radha, A. and Kumar, C.S. 1998. *In vitro* plant regeneration of fig (*Ficus carica* L. cv. Gular) using apical buds from mature trees. *Plant Cell Rep.,* **17**: 717-720.
- Kurubar, A.R., Alloli, T.B., Naik, M.K., and Pujari, B.T. 2012. Effect of pre and post-harvest application of different chemicals on quality traits of' Poona Fig (*Ficus carica* L.). In *VII International Postharvest Symposium,* 1012. pp. 91-95.
- Kurubar, A.R., Allolli, T.B., Naik, M.K. and Angadi, S.G. 2017. Effects of gibberellic acid on growth, yield and fruit quality of fig (*Ficus carica* L.). *Acta Horticulturae,* **1173**: 183-188.
- Lansky, E.P., Paavilainen, H.M., Pawlus, A.D. and Newman, R.A. 2008. *Ficus spp.* (fig): ethnobotany and potential as anticancer and anti-inflammatory agents. *J. Ethnopharmacol.,* **119**(2):195-213.
- Leonel, S., Tecchio, M.A. and Coser, G.M.D.A.G. 2015. Dormancy breaking of the fig tree with hydrogen cyanamide and garlic extrate. *British Journal of Applied Science and Technology,* 1-10.
- Mahmoud, H.M., El-Mahdy T.K. and Gouda, F.M. 2002. Effects of some growth regulators on developmental changes of some fig cultivars grown under Assiut conditions. *Second conference of sustainable agriculture development,* 440-455.

- Marei, N. and Crane, J.C.1971. Growth and respiratory response of fig (*Ficus carica* L. cv. Mission) fruits to ethylene. *Plant Physiol.,* **48**: 249-254.
- Mars, M., Chatti, K., Saddoud, O., Salhi- Hannachi, A., Trifi, M. and Marrakchi, M. 2008. Fig cultivation and genetic resources in Tunisia: An overview. *Acta Hort.,* **798**: 27-32.
- Maxie, E.C. and Crane, J.C. 1967. 2,4,5-Trichlorophenoxyacetic acid: effect on ethylene production by fruits and leaves of fig tree. *Science,* **155**: 1548-1550.
- Mewar, D. and Naithani, D.C. 2016. Effect of different IBA concentrations and planting time on stem cuttings of wild fig (*Ficus palmata* FORSK.). *Plant Archives,* **16**(2): 959-962.
- Morton, J. 1987. Fig. In: Fruits of warm climates. Julia F. Morton, Miami, FL. pp. 47-50. http://www.hort.purdue.edu/newcrop/morton/japanese persimmon.html
- Muriithi, L.M., Rangan, T.S. and Waite, B.H. 1982. *In vitro* propagation of fig through shoot tip culture (*Ficus carica* cultivars, mosaic virus-free plants). *Hort. Science,* **17**: 86–87.
- Mustafa, N.S. and Taha, R.A. 2012. Influence of plant growth regulators and sub culturing *on in vitro* multiplication of some fig (*Ficus carica*) cultivars. *Journal of Applied Sciences Research,* **8**(8): 4038-4044.
- Naik, M.K., Angadi, S.G., Kurubar, A.R. and Allolli, T.B. 2015. Effects of gibberellic acid on growth, yield and fruit quality of fig (*Ficus carica* L.). In *V International Symposium on Fig,* 1173: 183-188.
- Nobre, J. and Romano, A. 1998. *In vitro* cloning of *Ficus carica* L. adult trees. *Acta Hortic.,* **480**: 161-164.
- Norberto, P.M., Chalfon, N.N.J., Pasqual, M., Veiga, R.D., Pereira, G.E. and Mota, J.H. 2001. Effect of grafting time and IBA upon the rooting of fig tree cuttings (*Ficus carica* L.). *Ciecia-e-Agrotechnologia,* **25**(3): 533-541.
- Pasqual, M. and Ferreira, E. A. 2007. Micropropagation of fig tree (*Ficus carica* sp). In: "Protocols for micropropagation of woody trees and fruits", (Eds.): Mohan Jain, S. and Häggman, H.. Springer, PP. 409-416.
- Patah, F.K.A., Hasbullah, N.A., Idris, H., Radzuan, N.S. and Mohammad, M.L. 2018. Micropropagation of Ficus carica L. through tissue culture system. 12th Int'l Conference on Advances in Agricultural, Chemical, Biological & Medical Sciences (AACBMS-18) August 6-8, 2018 Pattaya (Thailand). pp. 19-22.

- Patel, H.R., Patel, M.J. and Singh, S. 2017. Effect of different levels of IBA and NAA on rooting of hardwood and semi hardwood cutting in fig. *International Journal of Agricultural Science and Research,* **7**(4): 519-524.
- Patel, K.K. 2009. Effect of plant growth regulators on rooting in air layering of fig (*Ficus carica* L.). *M.Sc. Thesis* submitted to Department of Horticulture, Anand Agricultural University, Anand (Gujarat).
- Pfnur, E. 1958. Effect of gibberellin in *Ficus carica* stem elongation. *Saatgutwirtach,* **10**: 220-222.
- Phad, V.S., Ballal, A.L., and Patil, V.K. 1980. Development of fruit of fig (*Ficus carica* L.) as influenced by different growth regulators. *Journal of Maharashtra Agricultural Universities,* **5**(2): 142-143.
- Pio, R., Ramos, J.D., Chalfun, N.N.J., Coelho, J.H.C., Gontijo, T.C.A. and Carrijo, E.P. 2003. Rooting of fig tree apical cuttings treated with sucrose and IBA by slow immersion under greenhouse conditions. *Revista Cientifica Rural (Brazil),* **9**(1): 35-38.
- Pontikis, C.A. and Melas, P. 1986. Micropropagation of *Ficus carica* L. *Hort. Science,* **21**(1): 153-154.
- Rane, D.A. 1962. Studies in the fruit differentiation and fruit set in the fig (*Ficus carica* Linn.) varieties Black Ischia and Turkey under Delhi Conditions. *Doctoral Dissertation,* IARI, Division of Horticulture, New Delhi.
- Reddy, K.V., Reddy, C.P., and Goud, P.V. 2008. Role of auxin synergists in the rooting of hardwood and semi hardwood cuttings of fig (*Ficus carica* L.). *Indian Journal of Agricultural Research,* **42**(1): 47-51.
- Reddy, P.P.N., Ray, N.R., Patel, A.D. and Patel, J.S. 2014. Effect of rooting media and IBA (indole butyric acid) levels on rooting and survival of air layering in fig (*Ficus carica* L.) cv. Poona under middle Gujarat agro-climatic conditions. *Asian Journal of Horticulture,* **9**(1): 1-5.
- Saddoud, O., Baraket, G., Chatti, K., Trifi, M., Marrakchi, M., Mars, M. and Salhi-Hannachi, A. 2011. Using morphological characters and simple sequence repeat (SSR) markers to characterize Tunisian fig (*Ficus caraica* L.) cultivars. *Acta Biol. Cracov. Bot.,* **53**: 7-14.
- Saddoud, O., Chatti, K., Salhi-Hannachi, A., Mars, M., Rhouma, A., Marrakchi, M. and

Trifi, M. 2007. Genetic diversity of Tunisian fig (*Ficus caraica* L.) as revealed by nuclear microsatellites. *Hereditas,* **144**: 149-157.

- Salunkhe, D.K. and Desai, B.B. 1984. Postharvest biotechnology of fruits. *CRC Press,* The University of Wisconsin – Madison. pp. 232.
- Sampaio, V.R., Matsue, A.M., Colombo, M.P., Olitta, A.F. and Barbin, D. 1983. Growth regulators application on figs produced out of season. *Anais da Escola Superior de Agricultura Luiz de Queiroz,* **40**(1): 101-108.
- Shukranul, M., Khairana, H. and Ibrahim, J. 2013. *Ficus carica* L. (Moraceae): Phytochemistry, traditional uses and biological activities. *Evidence – based complementary and alternative medicine,* Article ID 974256, 8 pages.
- Singh, S.P. and Chaurusia, S.N. 1991. Induction of early ripening of fig *(Ficus carica* L.) by Ethephone. *J. Hill Res.,* **4**(2): 110-112.
- Soliman, H.I., Gabr, M. and Abdallah, N. 2010. Efficient transformation and regeneration of fig (*Ficus carica* L.) via somatic embryogenesis. *GM Crops,* **1**(1): 47-58.
- Sonpethkar, R.M., Narwadkar, P.R., and Kausadikar, H.K. 2004. Effect of season and growth regulator on air layering of fig (*Ficus carica* L.). *J. Soils and Crops,* **14**(1): 152-155.
- Sundaram, K.S. and Rangaswamy, R. 1994. Studies on rooting of cuttings of Timla fig. *South Indian Horticulture,* **42**(6): 374-375.
- Tadeu Dias, J.P., Suekane, T.W.H., Takahashi, K., Sampaio, A.C. and Orika-Ono E. 2015. Plant growth retardant in fig. *Journal of Agriculture and Life Sciences,* **2**(1): 68-75.
- Topacoglu, O., Sevik, H., Guney, K., Unal, C., Akkuzu, E. and Sivacioglu, A. 2016. Effect of rooting hormones on the rooting capability of *Ficus benjamina* L. cuttings. *Sumarski list,* **1**(2): 39–44.
- Zarei, N. and Aboutalebi, A. 2013. Evaluation the rooting behavior of Caprifig cultivars cuttings in Fars Province. *International Journal of Agriculture Innovations and Research,* **2**(3): 397-400.
- Zhou J. 2003. Introduction and planting trial of *Ficus carica*. *Econ. For. Res.,* **21**: 90–91.

12

Plant Growth Regulators in Grape

Theory and Practical

S. A. Ranpise and B.T. Patil

1. Introduction

Grape cultivation in India is distinctly different from the cultivation elsewhere. Grape being temperate fruit crop, undergoes dormancy during the winter in temperate regions and sub tropical regions where winter is severe and temperature reaches freezing point. In India it is grown under two distinct climatic conditions: (i) sub-tropical climatic conditions of north where winter temperature rarely reaches the freezing point but vines undergo dormancy in winter, and (ii) tropical climatic conditions of peninsular India where the winter is mild and vines do not undergo dormancy and remains evergreen throughout growing period.

Depending upon the temperature and rainy season the grape cultivation is divided into three different regions.

Region- I: In this region mild temperature to subtropical region comprising states like Punjab, Haryana, Utter Pradesh, Rajasthan and Delhi.

Region- II: This region covers the entire Telangana and Rayalseema area of Andra Pradesh, excepting the districts of Chittor and Prakasam, north interior Karnataka and the rain shadow area of Western Ghats in Maharashtra.

Region- III: This region comprises all the grape growing area of Tamil Nadu and districts of Bangaluru, Kolar and Mysore of Karnataka.

Presently, grape cultivation is concentrated in Andra Pradesh, Karnataka, Maharashtra and Tamil Nadu in peninsular India and in Punjab, Haryana and western Utter Pradesh in north.

Plant growth regulators play an important role in grapes. The better quality grapes can be produced through the management of nutrients, water, canopy and more precisely by use of plant growth regulators. To achieve the better results the stage and optimum concentration of bioregulators are very important because the less or over dose causes no effect or adverse effect on vine health and berry quality. Therefore it is always advisable to use the bioregulators judiciously in viticulture.

Plant growth regulators are most commonly employed in grape production since long time on large scale as compared to any other horticulture crops for varied uses such as thinning, hastening or delaying maturity of berries, increasing the quality of berries and inducing the seedlessness in berries. The compound of plant growth regulators like naphthalene compound, phenoxy compounds and gibberellins are commonly used for such various purposes in India and in other countries. In Maharashtra, Thompson Seedless and its mutants like Tas-e-Ganesh, Sonaka, Manik Chaman: Sharad Seedless are the leading varieties. Farmers of Maharashtra are commonly used following plant growth regulators for various purposes with ultimate aim is to increase production of quality berries.

2. Gibberellins

2.1. Role of Gibberellic Acid

1. The amount of GA_3 needed to optimize berry size depends on the cultivar and many other factors, but '10 to 60 g/acre is a typical application rate.
2. The GA_3 may be applied once or several times; on some cultivars, multiple applications in-season applications may increase efficacy.
3. Application timing is critical and also varies by cultivar.
4. In many cultivars, the first application is made at, or when berry diameter is about 3-5 mm.

5. Later applications can delay fruit maturity and inhibit colour development in red- and black-fruited cultivars.
6. Breeding programs have continuously selected for grapes that produce naturally large berries, so many newer cultivars require less GA3 to attain optimal berry size than older cultivars.
7. Seeded grapes are generally less responsive to GA3 treatment than seedless grapes.

2.2. Effect of Excess use of GA_3

1. For export, residual effect should not be beyond 0.015 ppm
2. Excess GA_3 results in unfruitfulness.
3. Peduncle and rachis become tough.
4. Late maturity, reduces shelf life.
5. Blackening of berry during storage.
6. Berry shattering during transport.

The application of GA_3, their stage and mode of application of gibberellic acid is given in following Table.

Table 1: Stages and concentration of GA_3 (under Maharashtra condition)

Sr. No.	Growth stages	Concentration (ppm)	Mode of application
1	Pre-bloom	10	Spraying
2	25 % cap fall	20	Spraying
3	50% cap fall	20	Dipping
4	75 % cap fall	20	Dipping
5	Berry setting	35-40	Dipping

3. Role of Cycocel

It helps in fruit bud differentiation and emergence of fruitful buds and increases photosynthetic rates. The CCC sprays promote the fruitfulness of buds through following ways.

1. By reducing vegetative growth and avoiding shading.
2. By increasing the shoot thickness, reducing shoot length and reducing inter nodal length. These shoots tend to more erect than long shoots as a result the buds will be exposed to light.
3. By reducing the inter-nodal length by counteracting the effect of endogenous gibberellins. The antagonistic effect of CCC on GA is well known.
4. By increasing the endogenous levels of cytokinins.

Plant growth regulators like CCC, 6-BA,, Uracil and hydrogen cynamide is useful for increasing bud break, fruitfulness of buds and reducing shoot growth in grape. Therefore their stage of application and concentration is given in Table 2 and 3.

Table 2: Plant growth regulators after April pruning for increasing bud break and fruitfulness of buds

Sr. No.	Growth stage	PGR	**Concentration**
1	After pruning	Hydrogen cynamide	1.5%
2	5 leaf Stage	CCC	500 ppm
3	10 leaf stage OT 7 + 5 leaf stage	CCC	500 ppm
4	40 days after pruning	6-BA	10 ppm
5	45 days after pruning	Uracil	50 ppm
6	50 days after pruning	6-BA	10 ppm

Table 3: Management of plant growth without using CCC after April pruning

Days after foundation pruning (April)	**Practice to be followed**	**Purpose**
30-40	Nitrogen application to be stopped in case of vigrours vines*	To reduce the shoot growth
40-60	Reduce irrigation level to one-third to that of the vegetative phase**	To reduce the shoot growth

* Petiole-Nutrient guide has to be followed.

** This period generally coincides with the onset of rainy season. Irrigation may or may not be required.

4. Role of 6 BA (Cytokinin)

Main role of BA is cell division. It increases fruit bud differentiation, increase in the berry size and berry weight.

4.1. Time of application of 6 BA

1. After April Pruning: After shoot toping at 14-16 leaf stage, spray 6 BA at 5-10 ppm conc., helps in fruit bud differentiation.
2. After October Pruning: At berry setting stage spraying of 6 BA at 5 ppm with GA3 at 35-40 ppm, helps for cell division and cell elongation.
3. Forchlorfenuron (CPPU), a synthetic cytokinin, also may be applied at fruit set stage to increase berry size.
4. It can be applied by itself, or in combination with GA3, as CPPU and GA3 may have a synergistic effect on berry size.
5. Berries treated with CPPU tend to develop a rounder shape than non-treated berries.
6. This compound is highly potent and normally applied at very low doses (1 to 3 mg/L); excessive doses can delay maturity, retard colouring, and have a negative influence on taste.

4.2. Advantages of CPPU

1. Increases berry size
2. Increase berry weight by 16 % with GA3 application.
3. Spherical berry shape instead of oblong.
4. White-gray colour with shining can be attained which is preferred in European and Gulf countries rather than golden yellow colour.
5. The peduncle becomes strong, berry remains firmly attached to the peduncle.
6. The quality and freshness can be restored after harvest.
7. It resists the anthocyanin development in berry pericarp.

4.3. Disadvantages of CPPU: Delays the maturity by 14-16 days

The GA_3 alone and GA3+CPPU is applied at different stages and various concentrations for berry and rachis elongation and development (Table 4 & 5). The growth regulators and bio-regulators are useful for reducing wet and dry drop as well as increasing cell wall turgidity of grape berries (Table 6).

Table 4: Use of Plant growth regulators for Thompson Seedless varieties

S. No.	Phenological stage of berry	Growth regulator	ppm	Purpose
1	Parrot green colour	GA3	10	Elongation
2	3-4 days after first spray	GA3	15	Rachis Elongation
3	3-4 days after second spray	GA3	20	Rachis Elongation
4	3-4 mm berry size	GA3 + CPPU	40+2	Berry development
5	6-7 mm berry size	GA3 + CPPU	30+1	Berry development

Table 5: Use of Plant growth regulators for Sonaka and Sharad Seedless varieties

S. No.	Stage of bunch development	Conc. of PGR	Purpose
1	10% flowering	10 ppm GA3	Increasing berry length
2	50% flowering	15 ppm GA3	Increasing berry length
3	90% flowering	20 ppm GA3	Increasing berry length
4	3-4 mm berry size	40 ppm GA3 + 2 ppm CPPU Or 50 ppm GA3 + 1 ppm CPPU	Round berry shape Increasing berry length
5	6-7 mm berry size	40 ppm GA3 + 1-2 ppm CPPU	Increasing berry size

Table 6: Use of Bio-regulators to enhance the shelf life in grapes

S. No.	Bio-regulator	Concentration	Stage of application	Purpose
1	NAA (Napthaline acetic acid)	50-100 ppm	10 days prior to harvest	To reduce wet drop
2	GA3 (Gibbrellice acid) + CPPU (Forechlorofenuron) in combination	40-50 ppm GA3 + 1-2 ppm CPPU	3-4 and 6-7 mm berry size	To reduce dry drop
3	Calcium (Nitrate or chloride)	0.5-1.0%	75 or 90 or 105 DAP (Days after pruning)	To increase the cell wall turgidity

5. Optimum Use of plant Growth Regulators

To achieve the better results the stage and optimum concentration of bio-regulators are very important because the less or overdose causes no effect or adverse effects, respectively on the vine health and berry quality. Therefore, it is always advisable to use the bio-regulators judiciously in Viticulture. The bio-regulators are being used for fruit bud differentiation after pruning to control vigour of shoots, for elongation of rachis, berry growth and development etc. after forward pruning (Table 2 and Table 3).

The effect of different plant growth regulators on fruit set, berry size, fruit maturity, bud break and TSS: acidity ratio is discussed hereunder.

6. Effect of plant Growth Regulators (PGRs) in Grape

6.1. Effect of PGRs on fruit set

A number of growth regulators were found to increase berry set in grape (Coombe, 1953; Weaver and Williams, 1950) among them 4-CPA, GA and CCC were found to have marked effect.Among the growth regulators tried napthoxy acetic acid (NOA) was found promising to promote fruit set in all varieties viz., Kishmish (seedless), Pachadraksha, Khandari and Anab-e-Shahi. NOA at 25 ppm recorded higher percentage of fruit set in all the varieties. Gibberellic acid also at lower concentrations (10-25 ppm) increased fruit set in Pachadraksha and Anab-e- Shahi. But naphthalene acetic acid (NAA) did not influence the fruit set in any one of the concentrations tried and it caused only thinning effect as seen in reduction of fruit set (Shanmugavelu *et al.*, 1969).

Weaver and Williams (1950) reported usefulness of 4-CPA in improving fruit set in Black Corienth variety. Later, studies revealed that, under varying environmental conditions, application of 5-40 ppm of 4-CPA before berry shatter stage resulted in excellent fruit set in this variety (Coombe, 1953).

During 1950's, 4-CPA was extensively used as substitute to girdling to improve the berry set in Black Corinth grapes. To obtain good results with this auxin, aqueous solutions containing 2-10 mg per liter are to be sprayed on clusters during the third and sixth day after full bloom. Higher concentrations are injurious to the vines, causing stunted leaves and veins enation (Weaver,1953).

Favorable effects of 4-CPA on berry set under Indian conditions were also observed in Gulabi (Chaparwal, 1965), Bhokri and Italian Eliquina (Sanghavi,1966) and Bengalore Purple (Shrigiriwar,1967). Application of 20 ppm of aqueous solutions within five days after full bloom was found to be most effective. Bajwa *et al.*, (1977) reported that 4-CPA was very beneficial in increasing berry set in Anab-e-Shahi variety of grape.

The effect of gibberellins on fruit set in several grape cultivars has been variable. In general, GA has been found to decrease fruit set (Christodoulou *et al,* 1968). But Bukovac *et al.* (1960) observed a slight increase in fruit set in Concord grapes by the application of GA either during or shortly after anthesis. In India application of GA at full bloom stage was found to increase berry set in Pusa Seedless (Krishnamurthi *et al.* 1959), Bhokri (Sood,1967) and Khalili (Farmahan,1971).The effective concentrations of GA were 10-15 ppm for Pusa Seedless, 50 ppm for Bhokri and 25ppm for Khalili. In Pusa Seedless, GA at lower concentrations of 10 and 25 ppm increased fruit set by 76.5 and 59.1 per cent respectively, but at higher concentrations of 50 ppm, it reduced fruit set by 15.4 per cent.

The mode of action of GA seems to be similar to girdling in increasing the fruit set. Since it attract the carbohydrates to the berries on which it is sprayed (Coombe, 1960), they do not drop due to lack of nourishment even if they have fertilized ovary. In view of this, stage of application is crucial to increase the fruit set by GA. If it is applied before anthesis, it reduces the fruit set, but applied just after anthesis, it can increase fruit set.

Recent work carried out in California indicated that addition of 0.25 to 0.50 Per cent (v/v) Laytron to GA thinning sprays decreased fruit set.

The cytokinins have also marked effect on increasing fruit set in grape varieties. Application of BA and PBA was found to be effective in increasing fruit set in seeded (Muscat of Alexandria, Tokay and Almeria) as well as seedless (Black Corinth and Thompson Seedless) varieties of grapes. Dipping of flower clusters in 100 ppm solution of 6- (benzylamino) -9 (2 tetrahydropyranyl) (BTP) was found to increasing fruit set (Weaver *et al.,* 1966). The more soluble and mobile PBA was always more effective than BA. Cytokinins are believed to increase fruit set by increased mobilization of metabolites to the area of their application (Weaver, 1972). Therefore, treating the clusters at or immediately after full bloom with PBA will be more effective.

Growth retardant play an important role in increasing berry set in grape varieties. The growth retardants like CCC and inhibitors like SADH (Alar or B-9) were also found to increase the fruit set. Pre-bloom application of CCC to the foliage or clusters was found to increase fruit set by more than 20 per cent (Coombe,1965 ; Skene,1969), CCC-4 (750-1250 ppm) increased fruit set in Himrod (Barritt,1970) under conditions of New York. Under north Indian conditions, Bajwa *et al.* (1977) found 500 ppm CCC to be optimum to increase fruit set in same variety. While Farmahan (1971) observed 300ppm CCC to be adequate for increasing fruit set in Thompson Seedless, Dhillion and Sharma (1973) found 1000 ppm of CCC to be optimum for Perlette, while Armugam and Madhavarao (1973) observed 2000 ppm to be optimum for Anab-e-Shahi.

Arumugam (1970) conducted experiment on Anab-e-Shahi variety with different growth retardants and found that two spraying of B-9 (2000 ppm) and CCC (2000 ppm) at 20 days after pruning (20 DAP) and second spray at 50 days after pruning (50 DAP), significantly increased fruit set in both open and self pollinated flowers. Prebloom application of SADH at 2000ppm also increased fruit set in Anab-e-Shahi (Arumugam and Madhavarao, 1973).

The mode of action of growth retardants in increasing fruit set is explained by the inhibition of shoot growth by which the carbohydrates required for shoot growth are diverted to the developing ovaries and drop of set fruits due to lack of nourishment as shatter stage is reduced. This is true because set is improved even when growth retardants are to be sprayed on the foliage without spraying on the clusters. In view of this, growth retardants are to be sprayed before the set to realize their effect on fruit set. Their application after fruit set cannot reduce the drop of fertilized ovaries.

Application of SADH as pre bloom sprays to the seeded cultivar Concord and to the seedless Himrod increased fruit set (Tukey and Fleming, 1967 a, b, 1968). Tafazoli (1977) also reported that treatment with SADH at 7500 ppm increased the number of berries/ clusters in cv. Yaghooti. Girdling of shoot and application of SADH considerably increased fruit set. Brriitt (1970) however did not find any improvement in fruit set in seeded cultivars of *Vitis venifera* by using SADH at pre bloom stage.

6.2. Effect of PGRs on fruit maturity

In general 2, 4 –D at 5-10 ppm and GA 10-100 ppm, hastened maturity of berries of all the varieties tried. It was earlier by 5-12 days. The flower clusters of the variety Kalisahebi in which bunch is large, compact and berries do not ripen simultaneously and uneven in ripening were treated with 2, 4-D and 2,4-5 T at 2 and 5 ppm respectively. These two growth regulators were found to promote uniform ripening of the berries. The extent of berries with uniform ripening was 90 to 92 per cent in 2, 4-D treatment and 83 to 86 per cent in 2, 4, 5-T as against 37 percent in control (Shanmugavelu, and Sundararaj, 1968).

Dokoozlian (1999) in California reported that application of CPPU delayed fruit maturity of Thompson Seedless approximately one week or more when applied at either berry set or during initial stages of berry softening. However, fruit maturation was most delayed when CPPU was applied in combination with GA_3 at fruit set. This indicates that addition of CPPU can delay harvest of Thompson Seedless table grapes two weeks or more depending upon rate.

Ranpise *et al.*, (1999) observed that application of CPPU 3 ppm + GA 25 ppm at 3 mm berry size in Thompson Seedless delayed the harvesting of grapes by 5-6 days as compared to control treatment under Nashik condition of Maharashtra.

Dhatt (1977) observed that Ehtrel 500 ppm advanced the ripening only if applied during 3rd rapid growth stage. In Thompson Seedless and Anab-e-Shahi maturity was advanced by 8 and 10 days, respectively. On the contrary, early application delayed ripening and maximum delay was noted in Aneb-e-Shahi grapes. Ethrel + ringing at one month and two months after fruit set hastened ripening by 7 to 10 days in Bangalore Blue and 10 to 12 days in Muscat (Padmanabhaiah and Madhava Rao, 1977).

GA at 25 to 100 ppm at pre bloom stage and fortnight after fruit set hastened ripening by 15-20 days in Anab-e-Shahi and 14-32 days in Pachadraksha (Manivel and Sundararaj, 1968).

Arora *et al.*, (2012) reported that application of ethephon had significantly higher percentage of uniform coloured bunches of grape variety Flame Seedless as compared to untreated control. They further observed that ehephon @ 400 and 500 ppm on varaison and three days after varaison stage had showed hunter Lab colour values of L=16.25, a=3.74, b=0.50 and L=16.21, a=3.87, b=0.33 respectively as compared to untreated control within recorded values of L=18.26,a= 2.08,b= 0.73 . Likewise application of ethephon was helpful in advancement of ripening by 5-6 days in Flame Seedless variety of grape as compared to control treatment under Punjab condition.

6.3. Effect of PGRs on berry size

Judicious application of growth regulators has been found to supplement and substitute the effect of girdling in increasing the berry weight (size) in grape. Chemicals belonging to auxins, gibberellins, cytokinins, brassinosteriods have been found to increase the berry size.

The beneficial effect of 4-CPA in increasing the berry size in Black Corinth grape was noticed for the first time by Weaver and Williams (1950). Favorable effect of 4-CPA on berry size in this variety was confirmed later by Weaver and Williams (1951), Coombe (1953), Weaver and McCune (1957) and Weaver *et al.* (1957). In Thompson Seedless also, Weaver (1953) reported increase in berry size as a result of post bloom sprays of 4-CPA. Similar, results were also reported by Weaver and Williams (1951), Weaver and Winkler (1952), Weaver (1956), Weaver and MuCune (1957) and Weaver *et al.* (1957). Under Indian conditions, Pareek and Randhawa (1967) and Dass and Randhawa (1968) also observed increase in berry size with post bloom application of 4-CPA in Pusa Seedless. Increased berry weight in Anab-e-Shahi was obtained by Nijjar and Bhatia (1969) with application of 4-CPA at 5-25 ppm at full bloom. The response of different varieties to the application of 4-CPA was different. Seedless varieties responded better than seeded varieties. Within seedless varieties, Black Corinth responded better than Thompson Seedless. Berry size increase with 4-CPA in a cluster was not uniform and not as much as with girdling in Thompson Seedless. Sometimes, it resulted in delay in ripening (Weaver and Wrinkler, 1952). The effective concentration of 4-CPA for Seedless varieties ranged from 5-15 ppm and the stage was immediately after the shatter stage.

Weaver (1957) was first to report the usefulness of gibberellic acid in increasing the berry weight in Thompson Seedless grape. Later, the favorable effect of GA in increasing the berry size in this variety was reported by Stewart *et al.*, (1958), Weaver and McCune (1959), Maxell (1959) and Dahling (1959). Exogenous application of GA was found to increase the berry size in Black Corinth (Weaver,1957; Weaver and McCune,1959; Antcliff, 1967); Concord Seedless (Shaulis,1959); Pusa Seedless (Krishnamurthy *et al.*,1959; Dass and Randhawa,1968; Pareek and Randhawa, 1967; Tripathi,1968); Perlette (Maxwell,1959; Tripathi,1967);Beauty Seedless and Seedless Emperor (Coombe,1959; Rana *et al.*, 1968); Himrod (Iwagaki, 1960); and other Seedless varieties (Chailakhayan and Sarkisova,1963).

Application of GA to seeded varieties of *Vitis vinifera* like Zinfandal, Tokay, Riber and Red Malaga at various development stages did not increase the berry size. On the other hand, when sprayed before anthesis, the proportion of shot berries (seedless berries) increased (Weaver and Mc Cune, 1959). Similar effect of GA was observed by Balha (1963) in White Riesling and Pinot Blank varieties with GA at 100 ppm applied at pre bloom stage. In india Dass and Randhawa (1968 a) observed reduced berry weight in Bhokri and Gros Colman varieties when GA was applied at 50-100 ppm either at pre bloom or near full bloom stages. Reduction in berry weight was due to the formation of increased number of shot berries.

Application of GA was also found to increase the berry weight in certain seeded varieties too. Lavee (1960) observed increased berry size in Queen of the Vineyard variety with application of GA at 20 ppm , 23 days after full bloom stage. The increase in size was only in the berries which had less number of seeds in it. Venkatratnam (1962) reported that, application of GA at 40 ppm at full bloom stage increased length, breadth and weight of Anab-e-Shahi berries marked under Hydrabad conditions. In the same variety, Nijjar and Bhatia (1969) observed increased berry weight by application of GA at 100 ppm at full bloom stage under Punjab condition. Gopalkrishanan and Kerawala (1962) reported marked increase in size and weight of Gulabi berries as result of GA application at 5 ppm after cap fall.

Numerous attempts were made in India during late sixties and the seventies to increase berry size in Seedless grapes by use of GA. Bhalla and Patel (1970), Jawanda *et al.,* (1972), Jindal *et al.,* (1973), Khan *et al.,* (1970) and Sharma *et al.,* (1981) found that application at or after full bloom increased berry size in Thompson Seedless. While, GA application at full bloom increased length of berries, its post bloom application increased girth of berries (Bhalla and Patel,1970). According to the Jindal *et al.* (1973), application at berry set was more effective than the application at early bloom stages in the increasing the berry weight. Application of GA at once at full bloom and followed by second spray at berry shatter stage had additional effect on berry size improvement as compared to single spray at full bloom (Jawanda *et al.,* 1972). The effective concentration ranged from 40-100 ppm in these studies. In Perlette variety, while Nijjar and Gill (1971) found GA at 75 ppm at full bloom stage to increase berry weight, Patil *et al.* (1980) found it to effective at 45 ppm at pin head stage of the berries. In Beauty Seedless, GA at 150 ppm at full bloom was found to be highly effective for increasing berry weight (Dhillion *et al.,* 1973). Berry weight increment was more by berry elongation than by increase of berry girth. According to Nangia *et al.* (1977) combined application of Sevin 1000 ppm + GA 40 ppm repeated again one week proved most effective in improving bunch and berry size. Application of GA 100 ppm at pre bloom stage was effective for increasing the size and weight of berries in Delight variety (Prasad, 1977). Arumugam and Madhava Rao (1977) observed that the application of B-9 (2000 ppm) and Phosphon –D (500 ppm) increased berry size, fruitfulness while MH (500 ppm) and TIBA (500 ppm) reduced the cluster size of Anab-e-Shahi variety under Coimbatore condition. Patil and Gupta (1977) reported girdling + GA (100 ppm) proved to be most effective treatment in increasing bunch weight and bunch length in seeded varieties viz., Bangalore Purple, Gulabi, Phakadi and Anab-e-Shahi under Nagpur condition of Maharashtra. Dass and Randhawa (1972) reported the application of coconut milk to cluster of Thompson Seedless increased berry weight to some extent.

A novel growth regulator CPPU (Forchlorofenuron) for seedless grape production was reported by Retamales *et al.* (1998). CPPU sprayed on Thompson Seedless grapevines at 30% and 80% bloom stage and three times after fruit set in combination with GA resulted in additional increase in berry size.

In Thompson Seedless grape variety El-Shazly (1999) reported that cluster weight, length, diameter and berry weight increased with increased concentration of GA_3.

In California table grape growers commonly attribute the decline in vine fruitfulness over time, particularly in Thompson Seedless due to continue use of GA3. However, when Thompson Seedless canes were exposed to sunlight, even at high rates of GA3 had no significant effect on cluster size in following year.

GA_3 is commonly used to increase the berry size of Red Globe and other seeded cultivars in California. This application is particularly effective on berries that contain only one or two seeds, although most Red globe berries contain three or four seeds.

The greater response of Seedless varieties to exogenous application of GA attributed to the low levels of endogenous gibberellins in Seedless berries. Weaver and Pool (1965) and Iwahori *et al.*, (1968) observed greater quantity of gibberellins in seeded than seedless grapes, which suggest that the seeds are source of gibberellins in grape. That is why normally the berry size in many of seedless varieties is smaller than the berry size of seeded varieties and within a seeded variety, the shot berries are smaller in size than normal seeded ones. Due to lack of adequate levels of endogenous gibberellins in seedless berries, the exogenous application can increase the berry size by about 100-150 per cent. On the other hand, in seeded berries, the increase in berry size due to exogenous application is apparently only in such varieties, where the capacity of the seeds to produce the gibberellins is limited. The number of seeds in berry, size of seed and their inherent capacity to produce adequate levels of GA are the factors that determine the response of variety to GA application. If the seeds are capable producing adequate quantities of gibberellins enough to increase the berry size to the maximum, the exogenous application of GA will have no effect in increasing the berry size, whatever may be the concentration of GA and its stage of application.

Post bloom application of GA between full bloom and berry shatter stage increases the berry size in seedless varieties., mainly by increasing its length, while application just after berry shatter increases the berry size both by increasing its length as well as girth. In varieties like Thompson Seedless, where clusters are compact, application of GA soon after anthesis (full bloom) reduces the berry drop at shatter stage and increases the number of berries in a bunch leading to increased compactness eventually to berry cracking and rotting during ripening. Contrarily, in varieties where berry drop is too high at shatter stage and ultimate berry size is small leading production of straggly bunches, application of GA at full bloom stage followed by the application immediately after berry shatter helps in increasing the filling of clusters by retaining more number of berries in cluster and increasing their size.

It is commercial practice in India to spray GA on Thompson Seedless, Anab-e-Shahi and Perlette grapes to increase the berry size. In Thompson Seedless , GA is sprayed for increasing the berry size 3-4 times at weekly intervals commencing from full bloom until the berry attain 6 mm diameter with increasing concentrations starting from 40-100 ppm. More sprays from full bloom to shatter, particularly in vines where cane size is thin, the shoot elongation was found to cease and resultant berries were thin and slender. On the other hand, if sprays are given after the shatter stage, the shoot growth is adequate to nourish the clusters and bold berries are formed. This has been experience of growers producing Thompson Seedless for table purpose. In vineyard where it is produced for raisin making, it is a practice to spray GA at 10-15 ppm once at pre-bloom stage or some time additional spray at 25 ppm before 50 per cent bloom to produce loose clusters with small and round berries.

In Anab-e-Shahi, it is common practice to spray GA at 40ppm at full bloom stage in Hyderabad and South interior Karnataka to increase berry size.

Perlette grapes are sprayed with GA at 75-100 ppm at full bloom stage to increase the berry size. However, berry thinning by pre bloom application of GA or Sevin application during full bloom is a pre-requisite to get desirable increase in berry size by GA treatment in Perlette variety of grape.

The application of other growth regulators viz., BTP Syn. SD 8339 or PBA (6-benzylamino-9 (2-tetrahydrophyranyl)-9H-purine) was also found to increase the berry size in Black Corinth grapes (Weaver and Van Overbeak, 1963; Weaver *et al.,* 1966). BTP application was better than BA (benzyl adenine) in respect of increased berry size in Black Corinth. In Thomson Seedless, application of BTP at 1000 ppm at full bloom was not associated with any increase in berry size, but application at shatter stage was associated with slight increase in berry weight (Weaver *et al,* 1966). In Pusa Seedless grape, Dass and Randhawa (1968) failed to get any response to kinetin application. BTP at 1000ppm was also found to increase weight of seeded berries of Almeria grape (Weaver *et al.*, 1966).

In recent years, CPPU (N-2-chloro-pyridyl-N-phenyl urea) has been found to be quite effective in increasing berry size in grapes. Application of CPPU at 15 ppm, 10 days after full bloom increased the berry size in stenospermocarpic and seeded grapes Intrieri *et al*, 1992). By post bloom dipping of the clusters in aqueous solution of CPPU at 10 ppm, the berry size was almost doubled in Thompson Seedless, Carignane, Cabernet Sauvignon, Chardonnay and Sauvignan Blanc varieties (Nickell, 1986). Under the peninsular Indian conditions, Shikhamany (1996) observed synergistic effects of CPPU with GA in increasing berry size, particularly its diameter in Thomson Seedless grape.

Ranpise *et al.*, (1999) reported that application of CPPU 3 ppm + GA 25 ppm at 3 mm berry size recorded the highest berry diameter and berry length of Thompson Seedless under Nashik condition of Maharashtra.

Dokoozlian (1999) in California reported effect of CPPU on grape vine involves the regulation of fruit set, berry growth and development when applied after fruit set at 5 to 40 g/ha CPPU stimulate both cell division and cell elongation, resulting in significant increase in berry size.

Applications of auxins or cytokinins immediately after fruit set to increase the cell division followed by GA application to increase the cell size can increase the berry size more than GA application alone after fruit set. Weaver and Mc Cune (1959) obtained more increase in berry size by spraying 4-CPA at 15 ppm at full bloom followed by GA at 50 ppm after 20 days as compared to the mixture applied together or GA application alone. In Pusa Seedless, synergistic effect in increasing the berry size was observed by the combined application of 4-CPA at 10 ppm and GA at 50 ppm (Dass and Randhawa, 1968). Similar synergistic effect was observed by Pareek and Randhawa (1967) in Pusa Seedless by application of mixture of 100 ppm IAA and 50 ppm GA or 10 ppm 4-CPA together with 50 ppm GA. In Thompson Seedless, 25 ppm IAA together with 50 ppm GA applied at bajara grain size of berries was more effective in increasing the ultimate berry size as compared to the application of GA alone. A new class of chemical, Brassinosteriod (BR) when applied at one ppm in combination with 30 ppm GA two weeks after berry set, increased berry size considerably in Dilkhush (a mutant of Anab-e-Shahi) and Anab-e-Shahi. Combination of same chemical at same concentration was also found to be highly effective than single chemical (Shikhamany and Prakash, 1994). Brassinosteriod was found to induce cell proliferation and cell elongation (Michell *et al.*, 1970).

Spraying of mixture of 0.1 % urea phosphate and 15-30 ppm of GA immediately after fruit set produced a greater increase in berry size in Thompson Seedless than spraying GA alone (Shulman and Bazak,1986).

Girdling combined with GA application was also found to increase beneficial effect of GA on berry size. Berry size in Thompson Seedless clusters sprayed with gibberellins and girdled at fruit set are usually results two or three times more in size of berries as compared to not treated with GA nor girdled (Christodoulou *et al.,* 1968). GA sprayed at 100 ppm at full bloom followed by girdling at bajara grain stage was associated with increased effect of GA on berry weight in Thompson Seedless (Bhujbal and Chaudhari,1972). Jawanda and Vij (1973) obtained greater effect of GA (50 ppm) coupled with cane girdling at berry shatter stage in Thompson Seedless. While GA increases the capacity of berries to draw more organic nutrients, girdling provides the same to the berries in increased quantities.

6.4. Effect of PGRs on Fruit Quality (TSS: Acidity ratio)

Exogenous application of growth regulators, namely, auxins, gibberellic acid, ethrel and certain other chemicals during the fruiting season have been found to influence the TSS-acid ratio in number of grape cultivars.

Pre-bloom or bloom time sprays with auxins, namely, 4-CPA (Parachlorophenoxy Acetic Acid), NAA (Naphthalene Acetic Acid), 2-4-D(2,4-Dichlorophenoxy Acetic Acid and 2,4,5-T (2,4,5-Trichlorophenoxy Acetic Acid) were found to increase the TSS content of berries in different varieties. Favorable effect of bloom time sprays of 10 ppm 4-CPA were observed in Anab-e-Shahi (Pareek and Randhawa, 1967), Bengalore Purple (Shringiriwar, 1967), Khilari (Farmahan, 1971) and Thompson Seedless (Jawanda *et al.*, 1974). Bloom time spray of 10 ppm NAA was also found to improve TSS content in Bhokri and Italian Eliquina (Sanghavi, 1966), Perlette (Goyal,1968; Ram *et al.*,1977) and Zinfandel (Weaver,1963), 2,4-D or 2,4,5-T at 2 ppm sprayed during pre-bloom stage increased the TSS content of berries in Kali Sahebi (Shanmugavelu and Sundararaj,1968). As early as in 1959, Weaver and Mc Cune (1960) reported that GA application increased sugar content in Zinfandel, Carigane, Red Malaga and Ribier grapes. In India, variable effects of GA on TSS –acid ratio in grapes have been reported. Increased TSS – acid ratio was observed in Anab-e-Shahi (Manivel and Sundararij,1968; Dhillon,1969; Madhavarao and Palaniswamy,1972), Bhokri (Sanghavi,1966, Manivel and Sundararij,1968, Madhavarao and Palaniswamy,1972) , Delight (Singh and Singh,1980), Gulabi (Moti,1971 a), Himrod (Moti,1971), Italian Eliquina (Sanghavi,1966), Kali Sahebi (Shanmugavelu and Sundararaj,1968), Khalili (Farmahan and Jawanda,1972), Kishmish Chorni (Dhawan *et al,*1981), Motia (Moti,1971; Prasad and Pathak,1975), Perlette (Tripathi,1967; Goyal,1968, Nijjar and Gill.1971; Moti,1971;

Chitkara and Chundawat, 1972), Selection 94 (Kokate,1964) and Thompson Seedless (Bhalla and Patel,1971; Farmahan and Jawanda,1972,1976). Effective stage and concentration of GA application in relation to the increased TSS-acid ratio in different varieties as reported by various workers are given Table 7.

Reduction in TSS-acid ratio as result of GA application was also observed in certain seeded grape varieties by Nijjar and Bhatia (1969) in Anab-e-shai and Jindal *et al.*, (1973) in Thompson Seedless.

Discrepancy in the response to GA with respect to increase in TSS- acid ratio seems to be due to the stage of application. When applied prior to or during bloom, GA normally reduces the number of berries in a bunch without increasing their size considerably and results in increased accumulation of sugars in berries. On the other hand, if applied after berry set , the size of berries increases considerably without reduction in the number of berries in a bunch leading to reduced accumulation of sugars in the berries. Dass and Randhawa (1968 a) observed pre-bloom application of GA to be more effective than post-bloom application in increasing TSS content in Pusa Seedless grapes. Similarly, Jindal *et al.* (1973) have also observed the increased efficacy of post-bloom GA sprays in increasing TSS content of Thompson Seedless as compared to the GA sprays during bloom. Varietal variation was also observed in their response to GA applied at different stages for increasing sugar content in grape berries. While, pre-bloom sprays were more effective than sprays during bloom in Carignane and Zinfandel, bloom time sprays were better in Red Malaga and Ribier varieties (Weaver and McCune, 1959).

Utility of GA sprays after fruit set in increasing TSS content of berries was observed by Dass and Randhawa (1968) in Pusa Seedless, Jadhav (1968) and Jawanda *et al.*, (1972) in Thompson Seedless, and Patil *et al.*, (1980) in Perlette.

Khan *et al.*, (1970) observed the utility of pre-bloom sprays in increasing the TSS content of Thompson Seedless grapes when GA application was repeated after fruit set. On the other hand, post bloom application of GA also found effective in increasing TSS content of berries when it was coupled with cluster clipping in Perlette (Nijjar and Gill, 1971), girdling in Thompson Seedless (Phadnis and Mogal,1972) or ringing in Khalili grapes (Vij *et al.*, 1973).

Table 7. Effective stages and concentration of GA for increasing TSS-acid ratio in grape

Variety	Stage	Concentration	Author (s)
Anab-e-Shahi	Pre-bloom	25 ppm	Manivel and Sundararaj (1968)
	Full bloom	50 ppm	Dhillion (1969)
	Pre/full bloom	20 ppm	Madhavarao and Palaniswamy (1972)
Bhokri	Bloom	20 ppm	Sanghavi (1966)
	Pre-bloom	25 ppm	Manivel and Sundararaj (1968)
	95% capfall	40 ppm	Moti (1971)
	Pre/full bloom	20 ppm	Madhavarao and Palaniswamy (1972)
Delight	Bloom+ berry set	10+40 ppm	Singh and Singh (1980)
Gulabi	95% capfall	40 ppm	Moti (1971)
Himrod	95% capfall	40 ppm	Moti (1971)
Italian Eliquina	Bloom	20 ppm	Sanghavi (1966)
Kali Sahebi	Pre-bloom	50 ppm	Shanmugavelu and Sundararaj (1968)
Khalili	Pre+post bloom	25 + 25 ppm	Farmahan and Jawanda (1972)
Kishmish Chorni	Boolm	40 ppm	Dhawan *et al.,* (1981)
Motia	95% capfall	40 ppm	Moti (1971)
	Pre+post bloom	75+75 ppm	Prasad and Pathak (1975)
Perlette	Pre bloom	50 ppm	Tripathi (1967)
	Pre bloom	100 ppm	Goyal (1968)
	Full+ post bloom	50+ 50 ppm	Nijjar and Gill (1971)
	95% capfall	40 ppm	Moti (1971)
	Pre bloom	40 ppm	Chitkara and Chundawat (1972)
Selection 94	Panicle emergence	50 ppm	Kokate (1964)
Thompson Seedless	Bloom	75 ppm	Bhalla and Patel (1971)
	Pre/post bloom	25 ppm	Farmahan and Jawanda (1976)

Among the all growth substances, ethrel seems to have a convincing effect in increasing TSS content in grape. Spraying the clusters with ethrel at varying concentrations during different stages of berry growth and development was found to increase the TSS content in number of grape varieties, whether they are seeded or seedless, coloured or white. Ethrel at 250 ppm increased sugar / acidity ratio in Bangalore Blue and Muscat (Padmanabhaian and Madhavarao, 1972). The effective stage of application and ethrel concentration for increasing the TSS content in different varieties of grape as observed by different workers is presented in Table 8.

Table 8. Effective stage and ethrel concentration for increasing TSS acid ratio in grape

Variety	Stage	Concentration	Author (s)
Bangalore Blue	2 months after set	250 ppm	Padmanabhaiah and Madhavarao (1977)
Bangalore Purple	Varaison	200 ppm	Chakrawar and Rane (1976)
Beauty Seedless	Varaison	500 ppm	Mehata and Chundawat (1979)
	Varaison	500 ppm	Chauhan and Singh (1981)
Black Corinth	Varaison	500 ppm	Singh *and Khanduja* (1977)
Delight	Varaison	500 ppm	Singh and Chundawat (1978)
Gulabi	Varaison	200 ppm	Chakrawar and Rane (1976)
	Varaison	600 ppm	Dass *et al.* (1976)
	2 months after set	250 ppm	Padmanabhaiah and Madhavarao (1977)
	8 weeks after bloom	250 ppm	Khandiuja and Chaturvedi (1979)
Kishmish Chorni	Varaison	500 ppm	Dhawan *et al.* (1981)
Perlette	2 Months after set	500 ppm	Sidhu and Dhaliwal (1984)
Pusa seedless	Lag phase	250 ppm	Rao *et al.* (1974)
Red Prience	Varaison	250 ppm	Chauhan *et al.*(1981)
Shadipur Local	Pre-harvest	250 ppm	Jawanda *et al.* (1973)

Other chemicals like Sevin (Carbaryl), CCC and TIBA also found to increase TSS-acid ratio in grapes. Nangia and Bakshi (1971), Sarowa and Bakshi (1977), Arora and Singh (1972); Jawanda and Singh (1973) and Ram *et al.* (1977) observed increased TSS and reduced acid content in Perlette grapes as results of spraying 2000 ppm Sevin at full bloom stage. Favorable effects of 2000 ppm Sevin sprays during full bloom on increasing the TSS content was also observed in Beauty Seedless (Singh and Chauhan,1980).

While pre-bloom or bloom time sprays of CCC reduced TSS-acid ratio in Thompson Seedless (Farmahan and Jawanda,1976); Pre-bloon sprays with TIBA at 50 ppm increased the TSS content in Perlette grapes (Goyal,1968). Spraying of homobrassinolide at 1 ppm on Thompson Seedless clusters twice after fruit set was also found to increase the TSS content of berries (Shikhamany and Prakash,1994 a).

6.5. Effect of PGRs on budbreak

Bud break is the basic requirement for the initiation of new growth by vines. The important aspects that are relevant to viticulture are the time, duration and extent of bud break.

Time of bud break refers to the date on which the bud break takes place and also the time lag between pruning and bud break. In north India and also the other temperate grape-growing regions of the world, grapevines are pruned at the flag end of winter and before the onset of the spring depending upon the convenience of the grower, availability of labour and size of vineyard. Bud break was found to start only when the minimum temperature of the atmosphere crosses 10 ^{0}C. Bud break was found to start at 8^0 C also in regions where the grape buds are exposed to more chilling temperatures during their dormancy (Poeneru and Lazarescu, 1959). It indicates that temperatures higher than 10^0 C are required for bud break in regions where the buds are exposed to less chilling during their dormancy and much higher temperatures are required where there is no bud dormancy. Therefore, the threshold level of minimum temperature for bud break in the tropical regions of India and elsewhere could be much more than 10 ^{0} C.

Early termination of dormancy or hastening the bud break in late winter is important to have adequate growth period for ripening. Late bud break in early spring delays the ripening and the crop is lost due to cracking and rotting during berry ripening because of rains.

Termination of bud dormancy could be advanced by 26 days and harvest by 16 days in Pusa Seedless by the application of hydrogen cyanamide at 1.5 per cent on dormant buds in the beginning of January (Pandey, 1989) under Delhi conditions. It helped to harvest grapes by 3rd June before the rains started. Bud break was also advanced in North India in Thompson Seedless, Madeleine Angevine and Pusa Seedless by one to two weeks with the application of 2 per cent thiourea on dormant vines after pruning (Balasubrahmanyam *et al.*, 1975; Pandey *et al.*, 1980). Thiourea (2 or 4 per cent) alone or in combination with potassium nitrate (2.5 per cent) enhanced the bud break without reducing the yield.

Under normal conditions, in the peninsular India, the bud break starts from the 11th day and continues up to 35th day after pruning depending upon the variety and the atmospheric temperatures prevailing after pruning. The time gap in days from the initiation to completion of bud break is the duration of bud break on a vine. The bud breaks is said to be uniform if the gap is narrow and otherwise if it is wide. Uniformity of bud break is important in view of the uniformity in the occurrence of phenological stages of development, and efficacy and ease of cultural practices to be carried out at specific stage, such as GA application, cluster clipping, berry thinning and girdling aimed at increasing the berry and bunch size. Uniform of bud break was obtained by the application of hydrogen cyanamide at 1.5 per cent to the buds of Thompson Seedless soon after pruning under Bangalore (south interior Karnataka) conditions. Heat summation units required for the full bloom of 50 per cent of the panicles from bud break was also significantly less with cyanamide treatment (Shikhamany and Reddy, 1989).

The extent of bud break refers to the percentage of sprouted buds to the total buds on a cane or a vine. While, 12-14 buds sprout on a cane of 16 buds in the temperate conditions irrespective of the number of buds left on a cane at forward (winter) pruning, mostly the terminal bud and sometimes the second bud from the tip of the cane only sprout in peninsular India. Bud break is not a problem after summer pruning in the tropics as the spurs have one bud only and the atmospheric temperatures are high. Failure of bud break is due to poor development of buds on account of inadequate cane maturity. Usually, the failure of bud break is seen in vines in certain pockets of the vineyard. A comparison of the conditions of these vines with normal vines with respect to vigour, crop load, defoliation, foliage density, and soil conditions in respect of drainage will help identify the exact cause for this problem.

7. Reasons for Poor Bud Break in the Tropics are:

1. Cane maturity,
2. Lack of chilling of buds,
3. Reducing temperatures after pruning, and
4. High degree of apical dominance.

While, bud bread takes place from one year old canes under temperate conditions, it takes place from six-month old canes in the tropics. The level of cane maturity and the quantity of carbohydrate reserves are more in one- year old canes as compared to six-month- old canes. Buds are exposed to chilling temperatures prior to their sprouting in the temperate regions. Such exposure does not occur in the tropical conditions. Receding temperatures follow the pruning in peninsular India, while the rising temperatures follow it in North India.

8. Measures for Increasing Bud Break

In view of the sprouting of one or two buds on a cane, the benefits of increased number of fruitful buds on a cane cannot be unless all the fruitful buds are not forced to open. Several means were tried to increase the bud break in grape. They include twisting, ringing and bending of the canes and application of thiourea, hydrogen cyanamide and other chemicals to the buds immediately after pruning.

Cane twisting is done to rupture the phloem tissue without injuring the wood or xylem. This is done to restrict the downward movement of auxins from apical buds to the lower buds which is one of the reasons for apical dominance. Cane twisting is believed to enhance the endogenous levels of ethylene, which is known as wound hormone. Ethylene produced in the cane acts in two ways: (i) it reduces the degree of apical dominance either by its inhibitory effect on besipetal transport of auxin or auxin biosynthesis or both as opined by Philips (1975), and (ii) by increasing the bud break. Weaver *et al.* (1961) observed increased bud break in grape when treated with ethephon. Increased bud break was obtained by cane twisting in Thompson Seedless as observed by by Chittiaichelvan *et al.* (1984) and Shikhamany *et al.* (1989) under Bangalore (south interior Karnataka) conditions. It has been a practice to twist the canes in Anab-e-Shahi and Thompson Seedless in this region. Cane twisting some times leads to injury to the cane due to human error on account of carelessness and fatigue.

Cane bending and ringing were also practiced in the past in Thomson Seedless vines subjected to long cane pruning (12-15 buds) for obtaining increased bud break by inducing bud break at lower nodes. By bending or ringing the canes, the buds at lower nodes or in the middle region of the cane are made apical. When these buds show the symptom of opening, the canes are made straight so that the real apical bud takes the apical position. Since these operations are labour intensive, some sprouted buds get damaged while bringing the canes to original shape.

Chemicals, such as potassium nitrate, thiourea, ethephon and paclobutrazol have been found to increase bud break in grape (Shikhamany *et al.*, 1989). Among these, thiourea was most effective in Thompson Seedless under mild tropical climate conditions of Bangalore (south interior Karnataka). Potassium nitrate at 4% solution was equally effective as 6% solution. Paclobutrazol at 1000 or 1500 ppm was equally effective. Favorable effects of thiourea on budbreak in Thompson Seedless and Madeleine Angevine at 2% (Balasubrahmanyam *et al.*, 1975), Pusa Seedless (Pandey *et al.*, 1980) and Perlette (Sodagar and Chauhan, 1984) were reported under north Indian condition. Under mild tropical conditions of Bangalore (south interior Karnataka), thiourea at 4.5 per cent concentration was more effective than 1.5 or 3.0 per cent (Reddy and Shikhamany, 1992). According to Shivashkandara *et al.*, (1980), thiourea at 2 per cent was not effective in increasing the bud break in Thompson Seedless but was effective in Anab-e-Shahi. Ethrel at 4000 ppm was more effective than thiourea at 2 per cent in Anab-e-Shahi under Bangalore conditions. Ethrel even at 8000 ppm was not effective in Thompson seedless (Shivashkankara *et al.*,1980) while it was found to be effective even at 2000 ppm in the same variety (Shikhamany *et al.*, 1989). This discrepancy could be due to the variation in the atmospheric temperature prevailing after pruning and the positions of buds in which the observations were made. While Shikhamany *et al.*, (1989) considered the bud break in the apical five buds, Shivashankara *et al.* (1980) did it in the buds at fifth and sixth nodal positions from the apex.

In recent years, studies in India and elsewhere have shown the usefulness of hydrogen cyanamide in increasing the bud break in grape (Shulman *et al.*, 1983; Williams, 1987; George *et al.*, 1988; Shikhamany and Reddy, 1989; Pandey, 1989; and Reddy and Shidhamany, 1990). Hydrogen cynamide at 1.5 per cent was more effective than thiourea at 4.5 per cent in Thompson Seedless under mild tropical conditions of Bangalore. Pandey (1989) also reported 1.5 per cent to be optimum concentration of hydrogen cynamide for increasing bud break in north India.

With regard to the time of application of the chemicals for obtaining increased bud break, these have to be applied invariably within 48 hours after pruning under the tropical conditions. In case of temperate conditions, these can be applied on the dormant buds 15-20 days after pruning depending upon the date of pruning but at least 30-35 days prior to the anticipated date of bud break. Care must be taken to avoid their application to the buds in which the sprouting has initiated already, particularly in vines pruned after considerable gap between defoliation in the tropical conditions. In view of this, selective application becomes essential. Smearing of individual buds with cotton/sponge soaked in the aqueous solution of these chemicals is found effective and economical (Reddy and Shikhamany,1990, 1992) . Due to human error, there is a chance of leaving some buds without smearing since it is difficult to identify the treated and untreated buds or the portion of the canes. Therefore, to get a black colour of the thiourea solution, it is mixed with Bordeaux mixture (1-1.5%). This is called Black Bordeaux locally.

The efficacy of thiourea has not been found to reduce with the addition of Bordeaux mixture to it. Dipping of four or five apical buds in the solution contained in a long jar of 10cm diameter and 50 cm length by bending the cane is also being practiced, since these chemicals are corrosive to skin. Use of rubber gloves and protective smearing of vegetable oil, preferably castor oil to exposed skin is a common practice while using these chemicals. Spraying of these chemicals is not only expensive, but also less effective, because more quantity of solution is required to spray per unit area as compared to smearing or dipping. Wetting the required buds effectively is also difficult by spraying due to drift by wind and location of buds alternating on a cane. Yet another disadvantage with spraying is that all the buds, productive and unproductive, burst and many shoots grow from a cane. Removal of the unproductive shoots becomes an additional operation. If the unproductive shoots are not removed, they reduce the efficiency of plant-protection operations, hamper the growth of cluster-bearing shoots, resulting in inadequate foliage for bunch nourishment and protect it from sun burn. Shikhamany and Manjunath (1992) observed increased proportion of water berries as a result of increased bud break and reduced mean shoot length in Thompson Seedless.

Shikhamany and Manjunath (1992) also observed interaction between the cyanamide concentration and atmospheric temperature prevailing after pruning in bud break. At temperatures less than 18^0 C even higher concentrations (3.0%) was not as effective as low concentration (1.0%) of cyanamide at temperatures higher than 18^0C. Therefore, it is necessary to prune Thompson Seedless before the night temperature falls below 18^0 C in the tropics to increase bud break with chemicals.

Ranpise *et al.*, (2001) observed that application of hydrogen cyanamide @ 30 ml/L of water immediately after October pruning found best for increasing bud break 85.76 % within 9.75 days than 52.14 % in Thiourea with 12.50 days in Thomson Seedless grape variety under Nashik condition of Maharashtra.

The plant growth regulators are used by farmers on very large scale. The farmers apply these growth regulators on high concentrations but it affect not only grape berry quality but also rejection of consignment at the time of export. Therefore the various growth regulators, their MRLs and limit of quantification is given in Table 9.

Table 9: MRLs and Grape Test report format for Grape Net

S.No.	Name of Chemicals / Pesticides detected	Residue Content (mg/kg)		Harmonized EU-MRL (mg/kg)	Equipment used for analysis	Limit of Quantification (LOQ) (mg/kg)
		Individual	Sum			
1.	1-Naphthylacetic acid (alphanapthyl acetic acid)	BLQ	BLQ	0.05*	LC-MS/MS	0.02
2.	6-Benzyl adenime	BLQ	BLQ	0.01*	LC-MS/MS	0.01
3.	Chlormequat (CCC)	BLQ	BLQ	0.05*	LC-MS/MS	0.01
4.	Ethephon	BLQ	BLQ	1.0	LC-MS/MS	0.01
5.	Gibberellic acid	BLQ	BLQ	5.00	LC-MS/MS	0.01
6.	Paclobutrazol	BLQ	BLQ	0.05	LC-MS/MS	0.01
7.	Uracil	BLQ	BLQ	1.00†	LC-MS/MS	1.00

(Source: APEDA, Report, 2016)

References

- Annonous, 2014. Indian Horticulture Database, National Horticulture Board, Govt. of India pp. 71-72.
- Antcliff , A.J. 1967. A field trial with Zante Current (Vitis vinifera L.) *Vitis:* 14-20.
- Armugam, R. 1970. Effect of growth retardants and inhibitors on Anab-e-Shahi graps (*Vitis vinifera* L.). M.Sc. thesis, Tamil Nadu Agricultural University, Coimbatore.
- Armugam, R. and Madhavarao, V.N. 1973. Note on the effect of growth retardants on pollen production and fruit set in grapes. *Indian J. Agric. Sci.,* **34** (6): 610-2.
- Armugam,R. and Madhavarao, V.N.1977. Effect of some growth retardants on grapes var. Anab-Shahi (Vitis vinifera L.) Viticulture in Tropics, HSI, Bangalore. pp. 270-77.
- Arora, N.K., Singh, Indrajeet, Gill, M.I.S.,Alam, M.S. and Novjot. 2012. Effect of time of ethephon applications on ripening, colour development and storage life of Flame Seedless grapes in subtropical conditions. 5th Indian Horticulture Congress, held at Punjab Agriculture University during Nov. 6-9, abst. Pp.446.
- Arora, R.K. and Singh, K. 1972. Effect of carbaryl (Sevin) on quality regulation in grapes. *Haryana J. Hort. Sci.,* **1**(1-4): 31-8.
- Bajwa,M.S., Singh, S.N. and Deol, S.S. 1977. Effect of cycocel, gibberellic acid and 4-chlorophynoxy acetic acid on berry setting, bunch and berry size and quality in Himrod grape. Viticulture in tropics : *Proc. Of working group on viticulture,* held at Bengalore 8-14 feb.,pp.258-262.
- Balasubrahmanyam, V.R, Khanduja, S.D . and Abbas, S. 1975. Effect of thiourea on rest period of grapevine buds. *Amer. J. Enol. Vitic.,* **26**: 168-70.
- Barritt, B.H 1970. Fruit set in seedless grapes treated with growth regulators, Alar, CCC and gibberellin. *J. Amer. Soc. Hort. Sci.,* **95**: 58-61.
- Bhalla, P.L and Patel, R.M. 1971. Effect of gibberellic acid on Thompson Seedless grapes (Vitis vinfera L.). *Madras Agric. J.,***58** (7): 673-5.
- Bhalla, P.L. and Patel, R.M. 1970 Effect of gibberellic acid on Thompson Seedless grapes (*Vitis vinfera* L.) *Madras Agric. J.,* **58** (7): 673-5.
- Bhujbal, B.G. and Chaudhari, K.G. 1972. Interaction of gibberellin and girdling on

certain fruit characteristics in Thompson Seedless grape. *Res. J. Mahatma, Phule Agric. Univ.*, **3**(2) : 126-30.

- Blaha, J. 1963. Influence of gibberellic acid on grapevine and its fruit in Czechoslavakia. *Amer. J. Enol. Vitic.*, **14**: 161-3.
- Bukovac, M.J. Larvsen, R.P. and Bell, H.K.1960. Effect of gibberellin on berry set and development of concord grapes. *Quar. Bull. Michi. Agr. Exp. Sta.*, **42**: 503-10.
- Chadha, K.L. and Shikhamany, S.D. 1999. The grape : Improvement, production and post harvest management. Hand book, Malhotra publishing house, New Delhi. Pp. 235-260.
- Chailakhayan, M.K. and Sarkisova, M.M. 1963. The effect of gibberellin on the growth of grapes. *Doi. Akad. Nauk SSSr* **148** : 247-51.
- Chakrawar, V.R., and D.A. Rane, 1976. Effect of growth regulators on uneven ripening of grape (*cv.* Gulabi and Banglore Purple), *Vitis,* 16(2): 97-99
- Chaparwal, B.N. 1965. Effect of GA and 4-CPA on berry set, size, weight and quality of Gulabi grapes M.Sc. (Agric.) thesis, Nagpur University, Nagpur.
- Chauhan, K.S. and Singh I.S. 1981. Effect of vine part treatment with ethaphon on ripening and quality of Beauty Seedless grape. National symposium on tropical and sub- tropical fruit crops, Bangalore (Abstract). pp 109.
- Chauhan, K.S., Jindal, P.C. and Gupta, P.B. 1981. Effect of pve-harvest application of ethapen on ripening and quality of grapes (Vitis vinifera L.) *Haryana J. Hort. Sci.,* **10** (3-4): 190-200.
- Chitkara, S.D. and Chundawat, B.S. 1972. Response of Perlette grapes to combined application of GA and Sevin at various stages of panicle development . *H.A. U.J. Res.,* **2**(4): 233-8.
- Chittiraichelvan, R. Shihamany, S.D. and Chadha, K.L 1984 Effect of cane twisting on apical dominance in Thomson Seedless grape (*Vitis vinifera* L.). *South Indian Hort.,* **32**(4): 234-5.
- Christodoulou, A., Weaver, R.J. and Pool, R.M. 1968. Relation of gibberellin treatment to fruitset, berry development, cluster compactness in Vitis vinifera grapes. *Proc. Amer. Soc. Hort. Sci.*, **92**: 301-10.
- Coombe, B.G. 1953. Setting Currants by spray with PCPA. *South Australian J. Agric.,* **57**:107-10.

- Coombe, B.G. 1959. Fruit set and development in seeded grape varieties as affected by defoliation, topping girdling and other treatments. *Amer. J. Enol. Vitic.*, **10**: 85-100.
- Coombe, B.G. 1960. Relationship of growth and development to changes in sugars, auxins and gibberellin in fruit of seeded and seedless varieties of Vitis vinfera. *Plant Physiol.,* **35**: 241-50.
- Coombe, B.G. 1965. Increase in fruit set of *Vitis vinifera* by treatment with growth retardants. *Nature* (London), **205**: 305-6.
- Coombe, B.G. 1967. Effects of growth retardants on *Vitis vinifera* L. *Vitis,* **6**: 278-87.
- Dahling, R. 1959. Gibberellins versatile in the vineyards. *Fruit Grower,* **13:** 17-8.
- Dass H.C., Randhawa G.S. 1972. Effect of gibberellic acid on berry enlargement, cluster compactness and yield of 'Pusa Seedless' grape (*Vitis vinifera* L.). *Ind J Hort.,* **29**: 158-161.
- Dass, H.C. and Randhawa, G.S. 1968. Effect of gibberellin on seeded *Vitis vinifera* with special reference to induction of seedlessness. *Vitis,* **7**: 10-21.
- Dass, H.C. and Randhawa, G.S. 1968. Response of certain seeded *Vitis vinifera* varieties to gibberellin application at post-bloom stage. *Amer. J. Enol. Vitic.,* **19**: 56-62.
- Dass, H.C. and Randhawa, G.S. 1968. Response of Pusa Seedless grape to 4-CPA, Kinetin, uracil and GA. *Physiologia Plantrrum,* **21**: 298-301.
- Dass, H.C., Randhawa, G.S. and Negi, S.P. 1976. Growth regulators in crop regulation and quality improvement of grapes, Viticulture in Tropics. Proceedings of the working Crops on viticulture in south East Asia. PP 215-18.
- Daulta, B.S., Singh, H.K. and Chohan, K.S. 1981. Effect of Zinc and CCC sprays on flowering, fruiting and Physio-chemical composition on fruit in mango (Mangifera indica L.) cv. Dashehari. *Haryana J. Hort. Sci.,* **10**: 161-165
- Dhatt, A.S. , Dhillon, B.S. and Bajwa, G.S. 1977. Effect of ethrel spray on grape berry development. Viticulture in tropics : Proc. Of working group on viticulture, held at Bengalore 8-14 Feb., Pp-250-253.
- Dhawan, S.S Chauhan, K.S. and Jindal, P.C. 1981. Effetcts of gibberellic acid, ethephon and girding of fruit quality and ripening in Kishmish Chorni grape (vitis vinifera L.). National symp. Sub-tropical and tropical fruit Crops, Bangalore (abst.). 112pp.

- Dhillon, B.S. Gill, A.P.S. and Sandhu, J.S. 1973. Improvement of Beauty Seedless grape with plant regulators. *Indian J. Hort.*, **30** (1-2) : 366-9.
- Dhillon, K.S. 1969. Effect of gibberellic acid on cropping and quality of Anab-e-Shahi grape. M.Sc. thesis, submitted to Punjab Agricultural University, Ludhiana.
- Dhilon, B.S.and Sharma, K.K 1973. Regulation of Perlette grapes with CCC, GA ,2,4,5-T and NAA. *J. Res. PAU,* Ludhiana, **10** (3) : 331-6.
- Dokoozlian, N. 1999. Influence of Gibberellic acid berry sizing sprays on Crimson seedless table grape. University of California Cooperative Extension, Tulare County. http://cetulare.ucdavis.edu/pubgrape/tb897.htm
- El-Shazly, S.M. 1999. Effect of hydrogen cyanamide (Dormex) spray on bud behaviour, growth, yield, fruit quality and leaf mineral composition of Thompson Seedless grapevines. *Alex. J. of Agric. Research*, **44** (2): 221-235.
- Farmahan, H.L. 1971. Effect of para-chlorophenoxy acetic acid, gibberellic acid and 2-chloroethyltrimethyl ammonium choloride on fruitdset and development in vinifefra grapes. M.Sc. thesis, submitted to Punjab Agricutlural University, Ludhiana.
- Farmahan, H.L. and Jawanda, J.S. 1972. Response of seeded grape (Vitis vinfera L.) cv. Khaliil to growth regulators. *J Res. PAU.* **12**(3): 236-40.
- Farmahan, H.L. and Jawanda, J.S. 1976. Effect of 4-CPA, GA and CCC on the fruit set, development and fruit quality in Thompson Seedless grape (Vitis vinfera L.) *J. Res. PAU*, **13**(1): 57-63.
- George, A.P., Nissen , R.J. and Baker, J.A. 1988. Effect of hydrogen cyanamide in manipulating budburst and advancing fruit maturity of table grapes in south-eastern Queensland. *Australian J. Exp. Agric.*, **28**: 533-8.
- Gopalkrishna, N. and Kerawala, D.N. 1962. Pre- and post-blossom sprays of gibberellin and its effect on fruitset, bunch compactness and quality of Gulabi (*Vitis vinifera*) grapes. *Proc. 16th Int. Hort. Cong., Brussels*, **1**: 288.
- Goyal, V.D. 1968. Effect of plant regulators on compactness in perlette grape (*Vitis vinifera* L.).M.Sc. thesis, submitted to Punjab Agricultural University, Ludhiana.
- Intrieri, C., Filippetti, I. and poni, S. 1992. The effect of N- (2-chloro-4-pyridy)-N-phenylurea (CPPU) on berry growth and ripening of stenospermocarpic and seeded table grapes. *Adv. Hort. Sci.*, **6** (3): 137-43.

- Iwagake, H. 1960. The effect of gibberellin on fruit trees. Third Meet. Japan clbberellin Res. Ass., pp. 66-7.
- Iwahori, S., Weaver, R.J. and Pool, R.M. 1968. Gibberelin like activity in berries of seeded and seedless Tokay grapes. *Plant Physiol.,* **43**:333-7.
- Jadhv, J.G. 1968. Studies in the improvement of the quality of some commercial varieties of grapevine (*Vitis vinifera* L.) by chemical and physiological methods. M.Sc. (Agri.) thesis, submitted to Poona University, Pune.
- Jawanda, J.S. and singh, R. 1973. Effect ofgibberellic acid and 1-naphthyl, N-methyl carbamate (Sevin) on cluster loosening and fruit quality in Perlette grapes. *Indian J. Hort.,* **30**(3-4): 481-91.
- Jawanda, J.S. and Vij, V.K. 1973. Effect of gibberellic acid and ringing on fruitset, cluster and berry characters and fruit quality of Thompson Seedless grape. *Indian J. Agric. Sci.,* **43**(4): 346-51.
- Jawanda, J.S. singh, R. and Pal, R.N. 1974. Effect of growth regulators on flower bud drop, fruit characters and quality of Thompson Seedless grape (*Vitis vinifera* L). *Vitis,* **13**(3): 215-21.
- Jawanda, J.S., Pal R.N. and Singh, R. 1973. Effect of pre-harvest application of ethrel and alar on ripening and fruit quality of Shadipur Local grape. *Annual progress Report. Punjab Agric. Unive.,* Ludhiana. 124pp.
- Jawanda, J.S., Singh R. and Chohan, G.S. 1972. The effect of gibbefrellic acid on the grape culticvar, Thompson Seedless (*Vitis vinifera* L.). *J. Res. PAU,* Ludhiana, **9** (4) : 557-63.
- Jawanda, J.S., Singh, R. and Chohan, G.S. 1972. The effect of gibberellic acid on the grape cultivar Thompson Seedless (*Vitis vinifera* L.). *J. Res. PAU,* **9**(4): 557-63.
- Jindal, P.C. Singh, K. And Bakshi, J.C. 1973. Effect of various concentrations of gibberellic acid applied at various stages of panicle development in Thompson Seedless variety of grapes. (*Vitis vinifera* L.) *Haryana J. Hort. Sci.,* **2**(1-2): 30-7.
- Jindal, P.C., Singh, K. and Bakshi, J.C. 1973. Effect of various concentrations of gibbefrellic acid applied at various stages of panicle development in Thompson Seedless variety of grapes (*Vitis vinifera* L.) *Haryana J. Hort. Sci.,* **2** (1-2): 30-7.
- Khan, M.D.Q. Chandola, R.P. and Tyagi, P.C. 1970. Effect of gibberellic acid on bunch and berry size and quality of Sultana grape (*Vitis vinifera* L.). *Punjab Hort. J.,* **10**(1-2): 87-90.

- Khan, MD.Q., Chandola, R.P. and Tyagi, P.C. 1970. Effect of gibberellic acid on bunch and berry size and quality of Sultanina grape (*Vitis vinifera* L.). *Punjab Hort. J.*, **10** (1-2) : 87-90.

- Khanduja, S.D. and K.N Chaturvedi. 1979. Improving fruit quality in grapes. *Indian Horticulture,* **24**:5-6.

- Kokate, A.S. 1964. Study of the efficacy of some growth substances on thinning of clusters of Selectoion-94 (*Vitis vinfera* L.) grape variety. M.Sc. (Agri). Thesis. Poona University, Pune.

- Kokate, A.S. 1979. Studies on the efficacy of cycocel on biochemical and physiological changes during maturity, ripening and cold storage of Perlette grapes (*Vitis vinifera* L.). Ph.D. thesis, Punjab Agricultural University, Ludhiana.

- Krishnamurthi, S., Randhawa, G.S. and Singh, J.P. 1959. Effect of gibberellic acid on fruit set, size and quality in the Pusa Seedless variety of grape. (*Vitis vinifera* L.). *Indian J. Hort.*, **16**: 1-4.

- Lavee, S. 1960. Effect of gibbereliic acid on seeded grapes. *Nature*, **185**: 395.

- Madhava Rao, V.N. and Palaniswamy, K.P. 1972. Studies on the effect of gibberllic acide and 2-3, 5-triiodobenoic acid on Anab-e-Shahi and Pachadraksha grape varieties. Third INtemational Symposium on Sub-tropical and tropical Horticulture, Bangalore (abst.) 195 pp.

- Manivel, L. 1967. Effect of certain fertilizer treatments on the fruitfulness of grape buds in Anab-e-Shahi variety. M.Sc. thesis, Tamil Nadu Agricultural University, Coimbatore.

- Manivel, L. and Sunderraj, J.S. 1968. A note on the effect of gibberllic acid on Anab-e-shahi and Panchadradshka varieties of grapes. *South Indian Hort.,* **16**(3-4): 65-6.

- Maxwell, N.P. 1959. Effect of gibberellin on the size of Thompson Seedless and Black Corinth grapes in South Texas. *J. Rio Grande Valley Hort. Soc.,* **13**: 117-8

- Mehta, P.K. and Chundawat, B.S. 1979. Effect of Ethephon on quality and ripening of Beauty Seedless grape. *Vitis*, **14**:117-121

- Mitchell, J.W., Mandava, N., Worley, J.F., Plimmer, J.R. and Smith M.V. 1970. Brassins - A new family of plant hormones from rape pollen. *Nature (London)*, **225**: 1065-66.

- Moti, 1971. Effect of pruning on yield and quality of grapes (*Vitis vinifera* L.) under Kanpur conditions. *South Indian Hort.*, **19:** 17-23.

- Moti, 1971a. Effect of plant regulators on ripening and chemical composition of grapes. *Punjab Hort. J.*, **11**(3-4): 194-201.
- Nangia, R.P. and Bakshi, J.C. 1971. Fruit crop and quality regulation in Perlette variety of vinifera grapes I. Effects of Sevin (carbaryl) and gibberellic acid. *J. Res. PAU*, **8** (1): 33-46.
- Nangia, R.P., Bhutani, V.P. and Chitkara, S.D. 1977. Fruit quality of Perlette grape as influenced by GA, Sevin alone : in different combinations and coconut milk. Viticulture in tropics : *Proc. of working group on viticulture*, held at Bangalore 8-14 Feb., pp 245-249.
- Nickell, L.G. 1986. The effect of N-(2-chloro-4pyridyl)-N-phenylurea and 3-chloro-benzyl ester of dicamba on the growth and sugar content of grapes. *Acta Horticulture*, 2 **179** : 805-6.
- Nijjar, G.S. and Bhatia, G.G. 1969. Effects of gibberellic acid and parachlorophenoxy-acetic acid on the cropping and quality of Anab-e-Shahi grapes. *J. Hort. Sci.*, **44**: 91-5.
- Nijjar, G.S. and Gill, H.S. 1971. Effect of gibberellic acid on bunch and berry size, quality and berry thinning in Perlette grapes. *Indian J. Hort.*, **28**(3): 198-202.
- Padmanabbhaiah, D.R. and Madhavarao, V.N. 1977. Effect of ringing and ethrel (2-chloro ethyl phosphonic acid) on ripening and quality of grapes. Viticulture in tropics : *Proc. of working group on viticulture*, held at Bangalore 8-14 Feb.,pp. 263-269.
- Pandey, S.N. 1989. Hastening of budbreak and ripening in Pusa Seedless grapes (*Vitis vinifera* L.) with Dormex. *Indian J. Hort.*, **45**: 348-52.
- Pandey, S.N. Pandey, R.M. Yadav, I.S. and Singh, K. 1980. Effect of thiourea on bud-burst and ripening of Pusa Seedless grapes. Annual Seminar, Maharashtra Grape Growers Association, held at Pune, on 28 August 1980.
- Pareek, O.P. and Randhawa, G.S. 1967. Effect of auxins and gibberellin and their combinations on berry development in Pusa Seedless and Anab-e-Shahi grape varieties. proc,. Int. symp. Subtrop. Trop. Hort., HIS, New Delhi, pp. 270-280.
- Patil, A.V. and Gupta, P.K. 1977. Seasonal changes in developing berries on some grape varieties in Viderbha region (Maharashtrea) . Viticulture in tropics : Proc. of working group on viticulture, held at Bengalore 8-14 Feb.,pp.278-288.
- Patil, N.S., Pathak, S.P. and Pundir, J.S. 1980. Effect of gibberellic acid on bunch , berry and juice quality in Perlette grape. *Haryana J. Hort. Sci.*, **9**(1-2): 31-3.

- Phadnis, N.A. and Mogal, D.B. 1972. Improvement of fruit quality and yield in Thompson seedless grape (*Vitis vinifrra* L.). Third International syumposium on Sub-tropical and Tropical Horticulture, Bangalore (ast.). 194 pp.

- Philips, I.D.J 1975. Apical dominance. *Ann. Rev. Plant Physiol.,* **26**: 341-67.

- Poeneru, I. and Lazarescu, V. 1959. The practical importance of an understanding of the leafing out of grape varieties. *Grachina Via Si Livada*, **8** : 23-80.

- Prasad, A. and Pathak, R.A. 1975. Effect of gibberellic acid on thinning, size and quality of grapes. *Agric Agro. Ind. J.,* **8**(1): 25-7.

- Prasad,A. 1977. Studies on the response of gibberellic acid to grapes in wet-sub tropics variety Delight. Viticulture in tropics : Proc. Of working group on viticulture, held at Bengalore 8-14 Feb., Pp. 254-257.

- Ram, B., Prasad, A. and Tripathi, N. 1977. Effecft of Sevin and Planofix on bunch development and quality in grapes (*Vitis vinifera* L.) cv Perlette. *Progr. Hort.,* **8**(4): 43-7.

- Rana, R,S., Bhardwaj, K.N. and Singh, N.P. 1968. Response of Beauty Seedless grapes to gibberllin. *Himachal Hort.,* **9**: 1-8.

- Ranpise, S.A., Birade,R.M. and Patil, B.T. Effect of Kridor (Hydrogen cyanamide) on bud break and yield of grape cv. Thompson Seedless. *Pestology J.,* XXV (11): 22-25.

- Ranpise, S.A., Patil, B.T. and Ghure, T.K. 1999. Effect of CPPU (Forchlorfenuron) and gibberellic acid on physico-chemical properties of Thompson Seedless grapes. *J. Maharashtra Agric. Univ.*, **24** (3): 249-251.

- Rao, M. M., Pandey, R. M. and Singh, R. N. 1974. Effect of ethrel on Pusa seedless grape. *South Indian J. Hort.,* **22**(3-4): 112- 115

- Reddy, N.N and Shikhamany, S.D. 1990. Comparative efficacy of spray and dip treatments with H_2CN_2 on bud break in Thompson Seedless grapevine under tropical Indian conditions. *Gartenbauwissenchaft,* **55**(1):27-30.

- Reddy, N.N. and Shikhamany, S.D. 1992. Effect of thiourea on budbreak in Thompson Seedless grapevines. *J. Maharashtra Agric. Univ.,* **17** (2): 338-40.

- Reddy, N.N. and Shikhamany, S.D. 1990. Effects of thiourea on bud break in Thompson seedless grapevines. *J. Maharashtra Agric. Univ.,* **17** (2): 338-40.

- Retamales, J., Ravis, A., Pinto, M. And Guardicola, J.Z. 1998. *Acta Horticulturae*, **463**: 259-324.
- Sanghavi, K.U. 1966. Effect of certain mechanical and chemical treatments on the improvement in the quality of Bhokri and Italian Eliquina varieties of grapes (*Vitis vinifera* L.) M.Sc. (Agric) thesis, submitted to Poona University, Pune.
- Sarowa, P.S. and Bakshi, J.C. 1972. Fruit crops and quality regulation in Perlette variety of Vinfera grapes. IV. Effect pruning, bunch thinning, girdling, sevin and gibberellic acid treatments. *Punjab Hort. J.*, **12**(2-3): 68-77.
- Shanmugavelu, K.G., and Sundararaj, J.S. 1968. Note on the effect of plant growth regulators on Kali Sahebi variety of grape (*Vitis vinifera* L.) *South Indian Hort.*, **16**(1-2): 24-5.
- Shanmugavelu, K.G., Thamburaj, S., Dhandyutham, A.S. and Sundaraj, J. S. 1969. Studies on the effect of plant growth regulators on certain varieties of graes (*V. vinifera* L.) *Madras Agriculture Journal*, **56:** 691-698.
- Sharma, K.N. Ramarao, B.V., Sampath Kumar, K. and Prasada Rao, K.V.D. 1981. Effect of gibberellic acid in improving yield and quality of Thompson Seedless grape under Hyderabad condititions. National symposium on Tropical and Sub-tropical Fruit Crops, IIHR, Bangalore, 113pp.
- Sharma, S. and Jindal, P.C. 1982. Effect of 4-CPA and benzyl- adenine on colour development in Beauty seedless grapes. *Haryana Agric. Univ. Res. J.*, **12** (4): 679-80.
- Shaulis, N. 1959. Gibberellin trials for New York grapes. *Farm Res.*, **25**:11.
- Shikhamany, S.D. 1996. Comparative effect of different growth regulators on ripening and export quality of Thompson Seedless grape. *Draksha Vritta*, **16**(7):101-4
- Shikhamany, S.D. and Mannjunath, G.O. 1992. Effect of hydrogen cyanamide and date of pruning on bud break and subsequent shoot growth, yield and quality in Thompson Seedless grape. *Proceedings International Symposium on Recent Advances in Viticulture and Oenology*, held at Hyderabad, pp. 181-7.
- Shikhamany, S.D. and Prakash, G.S. 1994. Use of growth regulators to increase berry size in Thompson Seedless. *Drakshavritta*, **14**: 117-20.
- Shikhamany, S.D. and Reddy, N.N 1989. Comparative efficacy of cyanamide and thiourea sprays on bud break in Thompson Seedless grape. *Indian Grape J.*, **(3-5)**: 37-42.

- Shikhamany, S.D. Prakash, G.S. and Reddy, N.N 1989. Effect of certain chemicals and cane twisting on bud break in Thompson Seedless grape. *Crop. Res.*, **2**(1): 1-7.
- Shikhamany, S.D., Saroj, P.L. and Prakash, G.S 1993. Comparison of ripening of Thompson Seedless and Arkavati grapes on different training systems. *Crop Res.,* **6**(2): 276-83.
- Shirigiriwar, D.Y. 1967 Effect of GA and 4-CPA on fruit set, yield and quality of Bangalore Purple grape variety. M.Sc. (Agric.) thesis, Nagpur University, Nagpur.
- Shivashkankara, K.T., Gowda, P.M. and Gowda. J.V.N. 1980. Effect of certain chemicals on bud break of grapes cv. Anab-e-Shahki and Thompsomn Seedless. *South Indian Hort.*, **28** (3): 72-6.
- Shringiriwar, D.Y. 1967. Effecft of GA and 4-CPA on fruit set, yield and quality of Bangalore Purple grape variety. M.Sc (Agri.) thesis, Nagpur University, Nagpur.
- Shulman, Y. and Bazak, H. 1986. Enhanced effect of gibberellin on grape size by addition of urea phosphate. *Aton Hanotea*, **40**(9): 761-6.
- Shulman,Y., Nir, g., Fanberstein, L. and Lavee, S. 1983. The effect of cyanamide on the release from dormancy of grapevine buds. *Sci. Hort.*, **19**:97-104.
- Singh , N.S. and S.D. Khanduja. 1977. Physical and biochemical changes during maturation of grapes (*Vitis vinifera* L.) *Indian J. Hort.* **34**: 354.
- Singh, I.S and singh, H.K. 1980. Timing of application of gibberellic acid on quality of Delight grapes. *Indian J. Agric. Sci.,* **50**(4): 334-47.
- Singh, I.S. and Chundawat, B.S. 1978. Effect of ethaphon on ripening of Delight grapes. *Hort. Sci.*, **13** (3): 251
- Singh. T.S. and Chauhan, K.S. 1980. Quality improvement in grapes. *India Hort.*, **24** (4): 2-4.
- Skene, K.G.M. 1969. A comparison of the effects of cycocel and tipping on fruitset in Vitis vinifera L. *Aust. J. Biol. Sci.,* **22**: 1305-11.
- Sodagar, N.N. and Chauhan, K.S. 1984. Effect of thiourea on rest period of buds and fruiting in grapes. *Punjab Hort. J.*, **24**:121-5.
- Sood, S.K.1967. Effect of gibberellic acid on the fruit set and parthenocarpic development of fruit in Bhokri grapes (*Vitis vinifera* L.) *Lal Bagh,* **12**(2): 25-9.

- Stewart, W.S. Halsey, D.D. and Ching, F.T. 1958. Effect of the potassium salt of gibberellic acid on fruit growth of Thompson Seedless grapes. *Proc. Amer. Soc. Hort. Sci.*, **72**:165-9.
- Tafazoli, E. 1977. *Scientia Hort.*, **6** : 221-224.
- Tripathi, S.N. 1967. Studies on the effect of GA on fruit thinning, bunch size and chemical composition of Perlette grape. *Punjab. Hort. J.*,**7**:60-4.
- Tripathi, S.N. 1968. Effect of GA on bunch, Berry size and quality of Pusa Seedless grape (*Vitis vinifera*). *Punjab Hort. J.*, **8**: 152-5.
- Tukey, L.D. and Flemming, H.K. 1967 a. Pa. *Agric. Exp. Sta.*, **14**: 11.
- Tukey, L.D. and Flemming, H.K. 1967 a. Pa. *Fruit News,* **46**: 12-31.
- Tukey, L.D. and Flemming, H.K. 1968. *Proc. Amer. Soc. Hort. Sci.*, 93: 300-310.
- Venkataratnam, L. 1962. Effect of gibberellic acid on Aneb-e-Shahi grape. *Proc. 16th Int. Hort. Congress, Brussels* **1**: 285.
- Vij, V.K. Jawanda, J.S. and Singh, R. 1973. Use of gibberellic acid and ringing for improving fruit quality in Khalili grape. *Punjab Hort. J.*, **13**(4): 223-6.
- Weaver, R. J., and W. O. Williams. 1950. Response of flowers of Black Corianth and fruit of Thompson Seedless grapes to application of plant growth regulators. *Bot. Gaz.*, **111**: 477-485
- Weaver, R.J 1953. Berry size of seedless grapes. Calif. Agric. 7: 7-15. Weaver, R.J 1956. Plant regulators in grape production. *Calif. Agric. Exp. Sta. bull.*, **752:** 1-26.
- Weaver, R.J 1957. Gibberellimns on grapes. *Blue Anchor,* **34**: 10-11.
- Weaver, R.J 1961. Effect of gibberellin on vine behavior and crop production in seeded and seedless *Vitis vinifera. Hilgardia,* **30**: 425-44.
- Weaver, R.J 1972. Plant growth substances in Agriculture, W.H Freeman & Co. San Francisco. pp 1-562.
- Weaver, R.J and McCune, S.B. 1957. Response of Thompson Seedless grapes to 4 chlorophenoxy acetic acid and benzothizol-2-oxyaceric acid. *Hilgardia,* **27**: 189-200
- Weaver, R.J and Pool, R.M. 1965. Gibberellin like activity in seeded fruit of *vitis vinfera* L. *Naturwiss,* **5L** 111-2.

- Weaver, R.J and Van Overbeeek, J. 1963. Kinins stimulate grape growth. *Calif. Agric.* 17-12.
- Weaver, R.J and Williams, W.O 1951. Response of certain varieties of grapes to plant growth regulators. *Bot. Gaz.,* **113**: 75-85.
- Weaver, R.J and Winkler, A.J. 1952. Increasing the size of Thompson Seedless grapes by means of 4-CPA berry thinning and girdling. *Plant Physiol.,* **27**: 626-30.
- Weaver, R.J Krimbas, B.D. and Davides, U.X. 1957. Response of Black Corinth and Thompson Seedless grapes to plant regulators under Greek conditions. *Proc. Amer. Soc. Hort. Sci.,* **70**: 169-77.
- Weaver, R.J McCune S.B. 1959. Effect of gibberellin on seedless *Vitis vinfera, Hilgardia,* **29**: 247-75.
- Weaver, R.J McCune S.B. 1960. Further studies with gibberellin on vitis vinifera grapes. *Bot. Gaz.,* **121**: 155-62.
- Weaver, R.J Van Overbeek, J. And Pool, R.M. 1966. Effect of kinins on fruit set and development in *Vitis vinifera. Hilgardia,* **37**: 181-201.
- Weaver, R.J. 1963. On thinning grapes with naphthalene acetic acid and dinitro-sec. Butyl phenol. *Vitis,* **4**: 1-10.
- Weaver, R.J. and Mc. Cune, S.B. 1959. Response of certain varieties of *Vitis vinfera* to gibberellin. *Hilgardia,* **28**: 297-350.
- Weaver, R.J., S.B. McCume, and B.G. Coombe. 1961. Effects of various chemicals and treatments on rest period of grape buds. *Amer. J. Enol. Vitic.* 12:131-142.
- Williams, L.E.1987. The effect of cyanamide on budbreak and vine development of Thompson Seedless grapevines in the San Japquin Valley of California. *Vitis,* **236**: 107-13.

13

Role of Plant Growth Regulators in the Cultivation of Grapes

Sujata Tetali

1. Introduction

The plant growth regulator (PGR) is group of chemicals that influence growth and cell division of plants. They are produced naturally by plants or can be artificially synthesized which are needed in small quantities at low concentration. A single PGR influences several physiological processes. In modern agriculture, PGRs are being used extensively in regulating the growth and different physiological processes of plants. Potential use of PGR agriculture was realized since 1930. The grape growers might have been using these PGRs more extensively and systematically than any other fruit crop.

Plant growth regulators are being used in pomology at various stages right from seed germination, juvenile phase, flowering and fruiting stage, etc. Though the duration of vegetative phase is genetically controlled, it can be modified by controlling the environment during this stage, by various cultural practices and by use of growth regulators (Dhatt, 1977). In addition to this, growth regulators may help to overcome problems like emasculation of flowers, incompatibility, making distant crosses, fruit drop, small berry size, uniformity in fruit maturity which ultimately leads to increasing the plant yield and quality.

2. Classes of Growth Regulators used in Grape

Most PGRs whether natural or synthetic, used in grape, fall into one of the following classes:

A. **Auxins** primarily control growth through cell enlargement. They act as both stimulators and inhibitors of growth. Different plant parts respond differently. For example, at low concentrations, the auxins like 2-4-D stimulates cell enlargement, whereas at higher concentrations, it inhibits enlargement or is even toxic to cells, formation of roots on plant cuttings, and the formation of xylem and phloem tissues. Examples of auxins commonly in use are Naphthalene Acetic Acid (NAA), Indole Acetic Acid (IAA), Indole Butyric Acid (IBA), 4-Chlorophenoxy Acetic Acid (4-CPA), etc.

B. **Gibberellins** consisting of about 70 compounds control cell elongation and division in plant shoots. They have been shown to stimulate RNA and protein synthesis in plants (eg. Gibberellic acid - GA_3, GA_4, GA_7, etc).

C. **Cytokinins** help in cell division, cell enlargement, senescence, and transport of amino acids in plants. CPPU (N-2-Chloro-4-Pyridyl-N=Phenyl Urea) is an example of a synthetic cytokinin.

D. **Growth inhibitors/ retardants** which inhibits the catalytic action of a specific enzyme. Example of Growth inhibitors used in viticulture are Chlorocholin Chloride (CCC).

E. **Ethylene** is produced internally by plant. They interact with auxins to regulate many metabolic processes.

A number of growth regulators are found to be useful in Viticulture. Among them, 4-CPA, GA and CCC are found to have clear noticeable effects. Most widely used plant growth regulators at different growth stages of plant and their roles are described here:

3. Role of PGRs on Propagation

Grapes are propagated through seeds, grafts as well as stem cuttings. To ensure true to type plants, cuttings are preferred. Root promoting hormone IBA is commonly used to promote rooting in many horticultural crops. It is an essential plant hormone that influences many aspects of growth and development, including vascular development, lateral root initiation, apical dominance, phototropism, and gravitropism (Davies, 1995).

An experiment conducted on grape hybrid ARI-516 (Catawba x Beauty Seedless) cuttings at Agharkar Research Institute, Pune showed that cuttings from middle portion having size 5±1 mm survived better with growth hormone IBA (200 ppm) treatment (Tetali & Karkamkar, 2014).

4. Role of PGRs on Crop Regulation, Fruit Thinning

Time of application of PGR is very important to get required results. GA when applied at pre bloom stage in grapes helps in thinning of bunches. In grape vineyards, where Thompson Seedless is grown for raisin-making normally GA is sprayed at 10-15 ppm once at pre bloom stage to produce loose cluster with small and round berries.

5. Role of PGRs on Flowering, Fruit Set, Fruit Growth and Development, Control of Fruit Drop, Parthenocarpy

A number of growth regulators were found to increase berry set in grapes, Predominant among them are 4-CPA (4-Chlorophenoxy acetic acid), gibberelic acid and CCC (Chlorocholine chloride). Weaver and Williams (1950) reported usefulness of 4-CPA in improving the fruit set in Black Corinth variety. 4-CPA was tried as a substitute to girdling to improve berry set in Black Corinth grapes (Coombe, 1953). Aqueous solutions (2-10 mg/ lit) when sprayed on clusters during the third and sixth day after full bloom obtained good results. Favourable effect on berry set under Indian conditions were also observed in grape variety Gulabi by Chaparwal in 1965, in Bhokri and Intlian Eliquina by Sanghavi, 1966 and in Bangalore Purple by Shrigiriwar, 1967. Aqueous solution of 20 ppm 4-CPA within five days after full bloom was found to be most effective.

GA has been found to increase as well as decrease the fruit set depending upon the time of application. Bukovac *et al.*, (1960) observed slight increase in fruits set in variety Concord by the application of GA either during or shortly after anthesis. Application of GA at full bloom was found to increase berry set in Pusa Seedless at 10-25 ppm (Krishnamurthi *et al.*, 1959); Bhokri at 50 ppm (Sood , 1967), Khalili at 25 ppm (Farmahan, 1971)

Application of GA to seeded varieties of *Vitis vinifera* like Zinfandel, Ribier and Red Malaga at various developmental stages did not increase the berry size. When GA was sprayed before anthesis, the proportion of shot berries i.e. seedless berries, increased (Weaver and Mc Cune, 1959). Dass and Randhawa (1968) observed reduced berry weight in Bhokri and Gros Colman varieties when GA was applied at 50-100 ppm either at pre-bloom or near full-bloom stages. Reduction in berry weight was due to the formation of increased number of shot berries.

Greater response of seedless varieties to exogenous application of GA is attributed to the low levels of endogenous gibberellins in seedless berries. Weaver and Pool (1965) observed greater quantity of gibberellins in seeded than seedless grapes, which suggests that the seeds are the source of gibberellins in seeded than seedless grapes.

Mode of action of GA seems to be similar to girdling in increasing the fruit set (Chadha and Shikhamany, 1999). Since it attracts the carbohydrates to the berries on which it is sprayed, they do not drop due to lack of nourishment even if they have fertilized ovary. In view of this, the stage of application is crucial to increase the fruit set by GA. If it is applied before anthesis, it reduces the set, but applied just after anthesis, it can increase the set.

Growth retardants like CCC and inhibitors like Succinic acid-2-2-dimethyl dydiazide (SADH or Alar or B-9) were also found to increase the fruit set. More than 20 percent increase in fruit set was observed in pre-bloom application of CCC to foliage or clusters (Coombe, 1965; Skene, 1969, Bajwa *et al.*, 1977). The mode of action of growth retardants in increasing the fruit set can be explained by the inhibition of shoot growth by which the carbohydrates required for shoot growth are diverted to the developing ovaries and the drop of set fruits due to lack of nourishment at shatter stage is reduced.

Chadha and Shikhamany (1999) reviewed the work of various workers on the effect of GA on increasing berry weight in grape. It revealed that seedless varieties responded better than seeded ones, with varied response in seeded varieties. Stage of application, variety and optimal concentration is very much critical with respect to the response of applied GA.

6. Induction of Male Sterility

The induction of male sterility can help to get rid of the tedious practice of emasculations in tiny grape flowers, thereby permitting production of large hybrid seed progenies with little effort. Various growth regulators were tried for the induction of male sterility. Iyer and Randhawa (1965, 1966) induced male sterility in three varieties of vinifera grapes by using MH-500 ppm (Mallic Hydrazide) and TIBA -750 ppm (2,3,5-Triiodobenzoic Acid). Similarly Dhillon (1973) obtained complete male sterility in Anab-e-Shahi grapes with Mallic Hydrazide (750 ppm) sprays.

Weaver and McCune (1960), and Weaver (1961) found reduced pollen germination with the application of potassium gibberellate. Similar effect of GA on pollen germination was reported by Bamzai and Randhawa (1967), and Dass and Randhawa in 1968.

7. Role of PGRs on Fruit Maturity and Ripening; Fruit Yield and Quality

Selection of PGR in relation to their mode of action and stage of application is very important in increasing the berry weight in grapes. Berry size increases mainly by cell division and a little by cell elongation in the pre lag phase, while during the post –lag phase, it is only by cell enlargement by the influx of carbohydrates and water into the cells. Treatment of CPPU has been found to delay the berry ripening. It has been found to be quite effective in increasing the berry size in grapes. Application of CPPU at 15 ppm, 10 days after full bloom increased the berry size in seedless as well as seeded grapes.

Ferrara *et al.*, (2013) observed that application of abscisic acid at 400 mg/L one week and four weeks after veraison positively affected the berry skin color, shifting to a more red-violet in Crimson Seedless in Mediterranean Climate. It also increased the amount of harvestable bunches at the first pick

8. Role of PGRs on Shelf-Life of Fruits

Being a non-climactric fruit, grapes do not develop colour or taste after harvest. There is no much role of PGR on increasing shelf life of grape.

9. Role of PGRs on Embryogenesis

Tang *et al.*, (2009) observed that that exogenous application of all PGRs did not affect the total number of ovules per berries in maturity. CCC increased the ovules weight and proportion of ovules >4 mm in length of three varieties including Centennial Seedless, Thompson Seedless and Crimson Seedless in maturity. Exogenous application of CCC at 100–500 mg l-1 significantly increased percentage of ovules >2 mm in length by 80.0–82.7% in Centennial Seedless, therefore improving embryo formation.

10. Role of PGR's on Seed Dormancy, Germination and Seed Viability

The seeds of some fruits require a specific period of 'after-ripening', i.e. exposure to low temperatures prior to proper germination and growth of seedlings. GA_3 has been found to substitute for this requirement in most of the fruits. The efficiency of the treatment increases, if followed by stratification for few days. Thiourea alone and in combination with GA_3 and partial stratification is also effective in breaking seed dormancy. This helps breeder to sow their seeds soon after the harvesting and rear the seedlings in green house or in open where the temperatures are favourable. 1000 ppm GA after 90 days stratification treatment at 4 ^{0}C and 0.5 M H_2O_2 was found increase the germination of Muscadine grape seeds (Conner, 2008).

Use of different growth regulators, method application and their effects on plant growth followed in viticulture.

Growth Regulator	Concentration (ppm)	Method of application	Growth stage	Benefit
Ethylene	100	Spraying	After April pruning, at 15 days bud sprout	Reduces apical shoot growth
	200	Spraying	At 15-16 leaf stage after April pruning	Increases the cane thickness
	1000-1500	Spraying	3-4 days before October pruning	Induce leaf drop
	250	Spraying	At veraison stage	Increase TSS content
6 BA	10	Spraying	15-16 leaf stage after April pruning	Increase in fruit setting in the buds.
	10	Spraying	At 3-4 mm berry size along with 30-40 ppm GA after October pruning	Increase in berry size.
	10	Spraying	At 6-7 mm berry size along with 30-40 ppm GA after October pruning	Increase in berry size.
CPPU	2	Dipping	At 3-4 mm berry size along with 30-40 ppm GA	Increase the stalk thickness & berry size, promotes round berry shape and maintains the green colour of berry.
	2	Dipping	2nd application at 7-8 mm berry size	
NAA	20	Spraying	Berry formation	Controls flower & berry drop,
	15-20	Dipping bunches	Sugar formation stage of berry	Improves berry luster
	20-25	Spraying	10-15 days before harvest	Prevents berry drop in transport.
IBA	1000-1500	Dipping	Cuttings	Promotes rooting of cuttings
4 CPA	10	Spraying	At 3 mm diameter size of berry	Increases the pedicel thickness
GA	10-15	Spraying	Before flowering	Increases the stalk length of the bunch
	20-25	Spraying	After 25% cap fall	Berry thinning
	20-25	Bunch Dipping	After 50% cap fall	Berry thinning
	20-25	Bunch Dipping	After 75% cap fall	Berry thinning
	35-40	Bunch Dipping	After fruit set	Thinning of bunches
	35-40 GA + cytokinin	Bunch Dipping	At 3 mm berries	Increases bunch size
	35-40 GA + cytokinin	Bunch Dipping	At 6-7 mm berries	Increases bunch size

(Source: NRC-Grapes)

Abbreviated names of growth regulators

BA, Kinetin: 6-Benzyladenine; **4-CPA**: 4-Chlorophenoxy Acetic Acid; **CCC**: 2-chloroethyl trimethyl ammonium chloride. **CPPU :** N-2-chloro-4-pyridyl-N=Phenyl urea; **GA**: Gibberellic Acid; **IAA** : Indole acetic' **IBA** : Indole butyric acid ; **MH**: Mallic Hydrazide; **NAA**: Naphthalene acetic acid. Acid ; **SADH, B9** : Succinic acid-2-2-dimethyl dydiazide; **TIBA:** 2,3,5-Triiodobenzoic Acid; **AAA:** Ascorbic acid; **MCPA:** Methyl chlorophenoxy acetic acid; **Morphactin**: 2-chloro-9-hydroxyfluorene-9-carboxylic acid; **PCPA:** para-chlorophenoxy acetic acid; **2,4-5-T:** 2-4-5-Trichloro phenoxyl acetic acid; **2,4,5-TP:** 2,4,5-Trichloro phenoxy propionic acid; **IAN:** Indole-3-acetonitrite; **NAD**: Naphthalene acetamide; **NOXA:** Naphthoxyacetic acid; **ABA:** abscisic acid **Phosphon-D**: 2-4-Dichlorobenzyl tributylphosphonium chloride. **PP 333, Cultar:** Paclobutrazol:

References

- Bajwa, M.S., Singh, S.N. and Deol, S.S. 1977. Effect of cycocel, gibberellic acid and 4-chrophenoxy acetic acid on the berry setting, bunch and berry size and quality in Himrod grapes. *Viticulure in Tropics*, HIS, Bangalore. pp. 258-62.
- Bamzai, R.D. and Randhawa, G.S. 1967. Effect of certain growth substances and boric acid on germination, tube growth and storage of grape (*Vitis* sp) pollen. *Vitis.*, **6**: 269-77.
- Bukovac, M.J., Larvsen, R. P. And Bell, H.K. 1960. Effect of gibberellin on berry set and evelopment of Concord grapes, *Quar. Bull. Michi. Agr. Exp. Sta.*, **42**:503-10.
- Chadha, K.L. & Shikhamany, S.D. 1999. *The Grape Improvement, Production and Post-harvest Management* Malhotra Publishing house, New Delhi. pp. 213-260.
- Chaparwal, B.N.1965. Effect of GA and 4 CPA on berry set, size, weight and quality of Gulabi grapes. M.Sc.(Agric.) Thesis, Nagpur University, Nagpur.
- Coombe, B.G. 1953. Setting Currants by spray with PCPA. *South Australian J. Agric.*, **57**: 107-10.
- Coombe, B.G. 1965. Increase in fruit set of *Vitis vinifera* by treatment with growth retardants. *Nature,* **205**: 305-306.
- Conner, P.J. 2008. Effects of Stratification, Germination Temperature, and Pretreatment with Gibberellic Acid and Hydrogen Peroxide on Germination of 'Fry' Muscadine (*Vitis rotundifolia*) Seed. *Hort science*, **43**(3):853–856.

- Dass, H.C. and Randhawa, G.S. 1968. Response of certain seeded Vitis vinifera varieties to gibberellin application at post-bloom stage. *Amer. J. Enol. Vitic.*, **19**: 56-62.
- Davies, P. J.1995. The plant hormones: their nature, occurrence, and functions, *In Plant Hormones: Physiology, Biochemistry, and Molecular Biology*, PJ Davies (ed) Kluwer Academic Publishers, Dordrecht, The Netherlands. pp. 1–5
- Dhatt, A.S. 1977: Role of Growth Regulators in Fruit Breeding. In *Fruit Breeding in India.* Nijjar G.S. (ed). Oxford & IBH publishing Co. New Delhi. pp. 9-14.
- Dhillon, B.S. 1973. Plant regulators induced male sterility in seeded grapes. *Act Hort.*, **34**: 279-282.
- Ferrara, G., Mazzeo A., Matarrese A.M.S., Pacucci C., Pacifico A., Gambacorta G., Faccia M., Trani A., Gallo V., Cafagna I and Mastror P.2013 Application of Abscisic Acid (S-ABA) to 'Crimson Seedless' Grape Berries in a Mediterranean Climate: Effects on Color, Chemical Characteristics, Metabolic Profile, and S-ABA Concentration. *J Plant Growth Regul.*, **32**: 491–505.
- Fermahan, H.L. 1971. Effect of para-chlorophenoxy acetic acid, gibberllic acid and 2-chloroethyltrimethyl ammonium chloride on fruit set and development in vinifera grapes M.Sc. Thesis, Punjab Agricultural University, Ludhiana
- Iyer, C.P.A. and Randhawa, G.S.1965. Chemical induction of pollen sterility in grapes. *Curr. Sci.*, **35**: 411-412.
- Iyer, C.P.A. and Randhawa, G.S.1966. Induction of pollen sterility in grapes (*Vitis vinifera*), *Vitis*, **5**: 433-445.
- Krishnamurthi, S., Randhawa, G.S. and Singh, J.P. 1959. Effect of gibberellic acid on fruit set, size and quality in Pusa Seedless variety of grape. (*Vitis vinifera* L.) *Indian J. Hort.*, **16**:1-4.
- Nickell, L.G. 1982. Plant growth regulators, agricultural uses. Springer-Verlag. Berlin, Heidelberg. pp.173.
- NRC-Grapes: http://nhb.gov.in/pdf/fruits/grape/gra005.pdf.
- Sanghavi, K.U. 1966. 'Effects fo certain mechanical and chemical treatments on the improvement in the quality of Bhokri and Italian Elequina varieties fo grape (*Vitis vinifera* L.). M.Sc.(Agric.) Thesis, Poona University, Pune.

- Shrigiriwar, D. Y.1967. Effect of GA and 4-CPA on fruit set, yield and quality of Bangalore Purple grape variety.' MSc.(Agric.)thesis, Nagpur University Nagpur.
- Skene, K.G.M. 1969. A comparison of the effects of cycocel and tipping on fruitset in *Vitis vinifera* L. *Aust. J. Biol. Sci.*, **22**: 1305-1311.
- Sood, S.K. 1967. Effects of gibberellic acid on the fruit set and parthenocarpic development of fruit in Bhokri grapes (*Vitis vinifera* L.) *Lal Bagh*, **12**(2): 25-29
- Tang, D. Wang Y., Cai J. and Zhao R. 2009. Effects of exogenous application of plant growth regulators on the development of ovule and subsequent embryo rescue of stenospermic grape (*Vitis vinifera* L.) *Scientia Horticulturae.*, 120(1):51-57.
- Tetali, S.P. and Karkamkar S.P. 2014. Induction of rooting in grapes (ARI 516), *Bioinfolet*, **11:** 3(b):930-933.
- Weaver, R.J. and Williams, W.O. 1950. Response of flowers of Black Corinth and fruit of Thompson Seedless grape to application of plant growth regulators. *Bot. Gaz.*, 111: 477-85.
- Weaver, R.J. and Mc Cune, S.B. 1959. Further studies with gibberellin on *Vitis vinifera* grapes. *Bot. Gaz.*, **121**:155-62.
- Weaver, R.J. 1961.Effect of gibberellin on vine behaviors and crop production in seeded and seedless *Vitis vinifera. Hilgardia,* **30**: 425-44.
- Weaver, R.J., Pool, R.M.1965. Gibberellin like activity in seeded fruit of *Vitis vinifera* L. *Naturwiss*, **5**: 111-112

14

Plant Growth Regulators in Guava

Sudip Kumar Dutta

1. Introduction

Guava (*Psidium guajava* L.) belongs to the family Myrtaceace, is one of the most exquisite and valuable fruits of the tropics. Guava has earned the popularity as "poor man's apple" because it is available in plenty to every person at very low price during the season. It is at par to apple for its nutritive values. It is also an excellent source of vitamin C and pectin and other nutrients. The fruit is used for the preparation of processed products like jams, jellies and nectar. Guava jelly is very popular for its attractive purplish-red colour, pleasant taste and aroma. The puree can be used in juice, cakes, puddings, sauces, ice-cream, jam and jelly. Fruits can be preserved by canning as halves or quarters, with or without seed core. Good quality salad can be prepared from the shell of ripe fruits. Leaves of guava are used as ethno-medicine for curing diarrhoea and also used for dyeing and tanning.

Guava production in India is insufficient to meet out the present demand of fruits with the increasing population of the country. This condition necessitates increase in production potential of guava under available resources. Among the technologies available for high production, the use of plant growth regulators (PGR's) has been proved as one of the powerful tools to meet the growing demand by influencing fruit production directly or indirectly. PGR's have multifarious role in guava production system like improving seed germination (Kalyani *et al.*, 2014), rooting of cuttings (Tready, 1983), root formation in stooling (Mishra *et al.*, 2007), success in air layers (Roa and Sulikeri, 1997), for good shoot formation (Jain and Dashora, 2007), flowering (Jain and Dashora, 2007), regulation of fruiting (Maji *et al.*, 2015), increased weight of fruits (Lal *et al.*, 2013), increasing vitamin 'C' content in fruits (Garasiya *et al.*, 2013), high TSS (Iqbal *et al.* 2009), shelf life enhancement and reduce spoilage in storage (Katiyar *et al.*, 2008). Therefore, to achieve fullest production potential of guava, use of plant growth regulator is indispensable.

2. Role of PGRs on Propagation

2.1. Role of PGRs on seed propagation

Guava can be propagated by seeds or other vegetative propagation techniques. Seed propagation is important especially in hybridization as well as for raising of rootstocks. The germination of guava seeds is uncertain due to hard seed coat (Singh, 1967), which results in poor germination. Seed germination in guava is inhibited by presence of tannins in seed leading to poor, uneven and delayed germination thus making guava difficult to propagate sexually (Ali *et al.*, 2003). Physical treatment and use of certain chemicals or PGR's have been found to be helpful in enhancing root stimulation in guava. Treatment of guava seeds with 5000 ppm ethephon (Sinha *et al.*, 1973), 1000 and 500 ppm GA_3 (Kalyani *et al.*, 2014) resulted in high seed germination. Early germination with GA_3 might be due to its action on the embryo which causes *denovo* synthesis of hydrolyzing enzymes particularly amylase and protease and this hydrolyzed food is utilized for growth of embryo and thereby enhanced the germination (Paleg, 1965).

2.2. Role of PGRs on propagation by cuttings

Guava can be propagated through hardwood, semi-hard wood and softwood stem cutting under mist conditions with varying degree of success with certain growth regulator treatment. Reddy and Majumder (1970) reported that when cuttings of guava were dipped in IBA at 0-5000 ppm for 15 seconds and then dipped in phenolic compounds at 0-2000 ppm for 15 seconds, rooting was increased from cuttings. Tready (1983) described that rooting percentage could be increased from 0 to 30.5 percent in hardwood cuttings of guava treated with 500 ppm IBA. Khattak *et al.* (1983) tried IBA at different concentration in semi hardwood cuttings of guava and achieved 4% rooting in cuttings of guava with the treatment of 6000 ppm of IBA. Rahman *et al.*, (2003) obtained maximum sprouting per cuttings (71.22%), more number of branches (3.44), maximum root-weight (1.46g) and better survival percentage (57.22%) and more number of branches (3.44) in softwood cutting treated with paclubutrazol at the 100 ppm solution. Abdullah *et al.*, (2006) studied the clonal propagation of guava by stem cutting from mature stock plants. They reported that the highest rooting percentage (60%) was observed in the cuttings treated with 0.4% IBA solution followed by 0.2% IBA and the lowest was in controlled cuttings. Abdul Kareem *et al.*, (2013) experimented on clonal multiplication of guava through softwood cuttings under mists conditions. They reported that maximum rooted cuttings (87.33%) were recorded in concentration 0.4g of IBA/100g talcum powder as compared to NAA (26%). Maximum number of roots per cutting was also noted in softwood cuttings treated with IBA (19.20) as compared to NAA (9.80) and the longest roots were recorded in softwood cuttings that were treated with 0.4g IBA (8.00 cm) as compared to cuttings treated with NAA (6.00 cm with 0.2g). Wahab *et al.* (2001) revealed that, the highest number of roots (23.75) per cuttings in guava was recorded in the cutting treated with IBA at 4000 ppm. Rahman *et al.* (2002) studied on the effect of different growth-regulators and types of cuttings on rooting of guava *(Psidium guajava L).* They observed that the longest roots were observed in softwood cuttings treated with NAA and paclobutrazol. Minimum root length was observed in control. They further reported that, maximum number of roots (59.66) was found in softwood cutting treated with IBA at 1000 ppm while minimum number (15.67) of roots were noticed in control.

2.3. Role of PGRs on propagation by Air-layering

Air-layering is one of the most important methods in practice for propagation of guava. Use of the growth regulators has increased the success rate of air-layering in guava. Auxins like IBA and NAA are frequently used at different concentrations for getting success in air-layering. NAA is not used singly as it is toxic to the tissue, if used in higher concentration. Application of 30 ppm each of IBA and NAA mixture recorded higher success in air layers of the juvenile shoots (Roa and Sulikeri, 1997). Sulladmath and Kololgi, (1969) reported that NAA had synergistic effect on the rooting when mixed with IBA formulation. Sharma *et al.* (1991) found that 10,000 ppm IBA increased success of air-layers and root quality in guava, Singh and Singh (2001) also reported that 20,000 ppm of plant growth regulators (IAA, IBA and NAA) was found optimum for better rooting.

2.4. Role of PGRs on propagation by Stooling

Several workers have reported successful result by the use of IBA in stimulating of root primordial in stool layers of fruit crops but with varying success (Mishra *et al.*, 2007). Stooling is used for quick multiplication of clonal root stock and scion cultivars on their own roots by heaping the normal soil in the juvenile shoots resulting in induction of better root system, which leads to higher degree of establishment under field conditions. Rymbai and Reddy (2010) found that IBA at 7000 ppm proved to be the most effective for rooting, root characters, establishment and growth characters of planlets of stooling in guava. Singh *et al.*, (1996) observed that 5000, 7500 and 10,000 ppm IBA significantly increased the percentage of rooting and survival percentage in the stools of guava. In stooling, growth regulators caused greater mobilization of sugars and nitrogenous substances, which helped in the initiation of root primordia (Detweiler, 1942). Etiolated plants were found to have a higher level of endogenous auxin (IAA) at site of etiolation (Kawase, 1965). The possible explanation for better survival, minimum days taken for sprouting and maximum number of leaves with IBA is might be due to less brittleness, longer root length, more fibrous and more number of primary and secondary roots and higher fresh and dry weight which is sufficient enough to support the plantlets (Tyagi and Patel, 2004).

3. Use of PGRs in Micropropagation

Plant regeneration *via* tissue culture has immense scope, because it is a reliable method for fast multiplication of true to type plantlets. Shoot bud culture has been fairly successful in guava (Amin and Jaiswal, 1988), but its economic feasibility is still in doubt. In guava, organogenesis form shoot tips and nodal segments obtained from mature trees was reported for the first time by Jaiswal and Amin (1987). Regeneration of plants from seedling explants of guava has also been reported by several workers. However, about 67- 95% seeds of guava are contaminated by bacteria and fungi, which impede the use of standard tissue culture methods (Ali *et al.*, 2007).

In micropropagation, auxins are mainly used to facilitate cell division and root differentiation. Commonly used auxins are IBA, NM, NOA, P-CPA, 2, 4-D and 2, 4, 5-T.

The IBA and NAA are widely used for rooting in combination with cytokinins for shoot proliferation. 2, 4-D and 2, 4, 5-T are very effective for the induction and growth of callus (Bhardwaj *et al.*, 2005). In addition to shoot bud culture, somatic embryogenesis also offers the possibility of mass multiplication of elite planting material of guava. Besides being an efficient method, it facilitates scaling up operations, encapsulation, etc. It has additional importance for developing a system for somatic cell genetics, trans genies and *in vitro* selection (Chandra *et al.*, 2004).

Indirect embryogenesis has been used for genetic improvement in many woody crops. Different explant sources have been used for embryogenesis in guava like mesocarp tissue on MS medium fortified with ascorbic acid, glutamine and sucrose and supplemented with 2,4-D (Chandra *et al.*, 2004); floral tissues and zygotic embryos on LPm basal medium supplemented with 2,4-D (Cangahuala-Inocente *et al.*, 2007), and zygotic embryos on MS medium supplemented with 2,4-D (Rai *et al.*, 2007). Among vegetative tissues employed for callus induction, only leaf disc responded to higher callus induction percentage i.e. (90%) on B5 Gamborg's media supplemented with either 2,4 dichlorophenoxy acetic acid (2,4-D) or Napthalene acetic acid (NAA) under long days (LD, 16/8 hr photoperiod) compared with dark conditions (Butt *et al.*, 2013). Rai *et al.* (2008) recently reported the successful storage of encapsulated somatic embryos of guava by ABA application. A micro-propagation method was described by Rai *et al.* (2009) for guava (*Psidium guajava* L.) using nodal explants from somatic embryo-derived young and aseptic plantlets. Multiple shoots were induced from axillary buds on MS medium containing different concentrations of N_6-benzylaminopurine (BAP), either alone or in combination with kinetin (Kn), indole-3-acetic acid (IAA) or α-naphthalene acetic acid (NAA).

4. Role of PGRs on Growth and Development

Productivity of guava is less than productive potential worldwide which is primarily due to traditional system planting and poor canopy management practices. Therefore, management of tree canopy should be done in such a way that leads to accommodation of higher number of plans to get higher production of good quality fruits per unit area.

As guava tree respond well to canopy management with respect to vegetative and reproductive growth (Singh and Chanana, 2005), modification of canopy through pruning and use of certain growth regulators along with increasing the plant density may enhance the production efficiency. A number of effective shoot growth retardants have been found useful in horticulture e.g. paclobutrazol, ethepon, daminozide, triazides, MH, CCC etc. are some of the well known growth retardants. Paclobutrazol and ethephon may be useful in high density planting as paclobutrazol helps in making the plants dwarf by producing a retarding effect on the growth of tree through inhibition of gibberellin biosynthesis, a key plant growth promoter. Similarly, ethephon acts as a ripening hormone and it enhances the ripening process along with its growth retardation effect. Ethephon at higher concentrations (500-3000 ppm) proved to be quite effective in reducing the plant height (Mohammed *et al.*, 1984). Shaban and Haseeb (2009) found that moderate pruning plus spraying ethephon at 600 ppm produced the highest number of new shoots. Dormex at 1.5% with moderate pruning gave the highest average index number of flower bud burst in the first season. Meanwhile, ethephon at 600 ppm with severe pruning produced the highest average index number of flower bud burst in the second season. Urea or NAA or Ethephon has been applied to guava trees in a vigorous vegetative state of growth to change yield patterns (Singh *et al.*, 1991). Jain and Dashora (2007) found that 60 days after treatment the mean maximum increase in shoot length (36.25%) was observed in 100 ppm NAA treatment followed by 200 ppm NAA (36.23%) and maximum increase in shoot diameter (34.95%) was recorded in 500 ppm PBZ followed by 1000 ppm CCC (34.42%) treatments.

5. Role of PGRs on Crop Regulation and Fruit Thinning

There are two major flowering seasons of guava, one during March–May, the fruits of which are harvested in rainy (*kharif*) season and the other in August– September with the fruit harvested during winter (*rabi*) season. In some states of India, like Maharashtra and Tamil Nadu, there is a third crop which flowers during October and is known as *hasta bahar*. Among the three flowering seasons, the rainy season crop is high in fruiting but poor in quality. Fruits of rainy season are rough, insipid, watery and less nutritive, hence the crop regulation is imperative in guava cultivation. Among different crop regulation practices, use of growth regulators (NAA, NAO, 2, 4-D, carbaryl and ethrel) has been found very effective in thinning of flowers and thus manipulating the cropping season. Maji *et al.* (2015) reported that, among different treatments, deblossoming with NAD @60 ppm showed maximum increase in fruit morphological characters *viz.* length, diameter, volume, weight, specific gravity, pulp weight, core weight, pulp percentage and pulp thickness over the control in winter crop of Sardar guava (L-49). Idate *et al.* (2009) reported application of 100 ppm gibberellic acid in the first week of January and June and 1000 ppm cycocel [chlormequat] in the first week of October every year, in addition to 600 ppm ethrel [ethephon] in the first week of February and July in the first year and only in the first week of July in subsequent years, recorded the highest *hasta bahar* yield, monetary returns (Rs 91 438) and benefit:cost ratio (2.54) in guava grown under plain zone conditions of western Maharashtra, India. Bikash *et al.* (2007) found that the net profit was the maximum in case of 2,4-D (40 ppm) followed by 2,4-D (60 ppm), NAA (200 ppm) and NAA (100 ppm) and in case of Lucknow-49, the maximum increase in yield of winter season crop was observed in case of foliar application of NAA (200 ppm). While, Mohammed *et al.* (2006) recorded maximum profit (Rs. 1,42,600/ha) with NAA 600 ppm followed by 15 and 20 per cent urea treatments on 20 years old Sardar guava trees. Abbas *et al.* (2014) observed that if the farmers spray guava plants with Naphthalene Acetic Acid twice at 400 ppm concentration during the flowering season i.e. last week of April and first week of May, maximum flowers of summer crop will drop and the plant will give better yield during the winter season. They also found that, the most suitable concentration of NAA was 400 ppm which produced maximum yield (32.75 kg/plant) during the winter with maximum flower drop percentage (64.92) during the summer.

Ravi Saini *et al.,* (2013) studied crop regulation in guava with pruning and foliar application of paclobutrazol. Thirteen-year old trees of guava hybrid Hisar Surkha planted at a spacing 6 x 6 m were selected. The trees were pruned on 15 March 2010 by removing current season growth up to 15, 30 and 45 cm from the tip. Paclobutrazol was sprayed @ 500, 1000 and 1500 ppm on whole trees after pruning (i. e. first week of April). Minimum number of flowers was recorded with 45 cm pruning level in rainy season, whereas trend was reverse in winter season. Minimum flower drop (11.0 and 12.5%) and maximum fruit set (89.3 and 87.5%) were observed with 45 cm pruning level during rainy and winter seasons, respectively. Fruit drop significantly decreased with increasing severity of pruning in rainy season only. Fruit diameter (5.3 and 7.7 cm) and fruit weight (79.7 and 151.7 g) were found maximum with 45 cm pruning level during rainy and winter seasons, respectively. Severely pruned (45 cm) trees produced minimum yield (60.4 kg/tree) in rainy season and subsequently maximum (153.0 kg/tree) in winter season. Paclobutrazol concentration @ 1500 ppm significantly increased number of flowers and decreased flower drop in both the seasons. Fruit set was significantly increased in winter season with higher concentration of paclobutrazol. Paclobutrazol concentration could not alter fruit drop, fruit diameter and yield in any of the seasons, whereas fruit weight was found significantly higher with higher paclobutrazol concentration in winter season only and it was maximum (145.6 g) with 1500 ppm paclobutrazol concentration. The interaction of pruning and paclobutrazol was found significant only with fruit weight in winter season. Maximum fruit weight (158.2 g) was recorded with 45 cm pruning level and 1500 ppm paclobutrazol concentration.

6. Role of PGRs on Flowering, Fruit Set, Fruit Growth and Development, Control of Fruit Drop

Plant growth regulators can be effectively used for manipulation of flowering, fruit set, fruit growth and development and control of fruit drop. Jain and Dashora (2007) reported that, minimum days taken to initiating of flowering (29.0), maximum number of flowers/shoots (7.77/shoot), maximum fruit set (71.17%), highest fruits retention (73.16%) were recorded in 500 ppm paclobutrazol (PBZ) treatment as compared to control. Similarly, Brahmachari *et al.* (1996) reported that in guava CCC (500 ppm) induced the earliest flowering and increased number of flowers, fruit set retention and yield. The profuse flowering in young guava with ethrel was also observed by Mohammed *et al.* (1984). Kumar *et al.* (2013) reported that foliar spray of ZnSO4 0.8% + borax 0.4% + NAA 50 ppm + GA_3 100 ppm was found best to increase fruit set, fruit retention on six years old guava tree cv. Chittidar. Significant improvement in fruiting of guava as a result of GA_3 spray has been reported by Rajput et *al.* (1977). Spraying of GA_3 15 or 30 ppm in the month of January proved effective in increasing fruit retention and subsequent fruit yield (Rajput *et al.* 1977).

The auxin from the pollen grain and pollen tube are responsible for the early stage of fruit growth. However, small amount of pollen grains necessary to pollinate a flower may not carry enough auxin to account for early fruit development. The growing pollen tube may secrete auxin which helps in fruit growth. Nagar and Rao (1986) studied the early changes in growth regulator content of pollinated guava fruits and observed that endogenous auxins, gibberellins and cytokinins are associated with fruit set of guava. Lal *et al.* (2013) found the maximum fruit length (9.8 cm), fruit girth (10.23 cm), fruit weight (182g) and volume (178.3 cc), minimum fruit drop (38.8%) and highest yield (37.1 kg/plant) under 50 ppm GA3 as foliar spray. Iqbal *et al.* (2009) found that NAA significantly reduced pre-harvest fruit drop of 28 trees of guava (cv. Red Flesh) when sprayed at two fruit development stages (marble and walnut stage). Maximum reduction (8.83%) in fruit drop was observed with 45 ppm spray followed by 30, 60, 75 and 90 ppm. Lal *et al.* (2013) reported that minimum fruit drop (38.8%) of guava was observed with treatment of 50 ppm GA_3.

7. Role of PGRs on fruit Maturity and Ripening; Fruit Yield and Quality

Reyes and Paull (1995) reported that, treating guava fruit with 100 µl l^{-1} ethylene (C_2H_4) at 20 °C for 24 h gave a significant increase in the rate of skin yellowing and softening of immature-green fruit. Lal *et al.*, (2013) studied on the effect of plant growth regulators on flowering and fruit growth of guava (*Psidium guajava* L.) cv. Allahabad Safeda and reported that, highest yield 37.1kg/plant was recorded with treatment of 50ppm GA_3. Katiyar *et al.*, (2009) reported that, highest yield (95.39 kg/tree) was observed with foliar spray of urea 2% + NAA 100 ppm. in guava (*Psidium guavaja*) fruit c.v. Allahabad Safeda. Iqbal *et al.* (2009) also reported that, application of NAA significantly increased yield of guava. Maximum yield (44.80 kg per treatment) was recorded in case of 45 ppm closely followed by 60 ppm (44.60 kg). Abbas *et al.*, (2014) reported that concentration of NAA @ 400 ppm produced maximum yield (32.75 kg/plant) in guava during the winter with maximum flower drop percentage (64.92) during the summer. They also observed maximum TSS (7.13%) in the plants treated with NAA @ 200 ppm and minimum TSS (5.35%) in the plants treated with NAA @ 500 ppm. Garasiya *et al.* (2013) studied on the effect of plant growth regulators on quality of winter season guava (*Psidium guajava* L.) cv. L–49 (SARDAR). They reported that, application of 40 ppm and 20 ppm NAA significantly enhanced the quality parameters *viz.*, TSS (10.95 and 10.87 %), reducing sugar (3.79 and 3.65 %), non-reducing sugar (2.65 and 2.56 %), total sugar (6.44 and 6.21 %), and ascorbic acid content (180.30 and 173.97 mg/100g pulp, respectively). The minimum acidity (0.42 %) was also observed with 40 ppm treatments. Katiyar *et al.*, (2008) reported that, foliar spray of 60 ppm NAA induced maximum

T.S.S. content. Kumar *et al.* (2013) reported that, various quality parameters of guava namely, total sugars, reducing sugars, non reducing sugars, TSS, zinc and iron content in fruit pulp were improved with application of ZnSO4 0.8% + borax 0.4% + NAA 50 ppm + GA_3 100 ppm. Iqbal *et al.* (2009) reported that, NAA application increased quality parameters i.e. pulp/seed ratio, TSS, total sugars, acidity and ascorbic acid contents in fruits of different treatments of guava (cv. Red Flesh). Maximum pulp/seed ratio (11.31), total soluble solids (11%) and total sugars (7.45%) were recorded in 45 ppm treatment.

8. Role of PGRs on Shelf-Life of Fruits

Guava fruits are highly perishable in nature and post harvest biochemical changes soften it leading to spoilage. Therefore, large quantity of fruit is lost after harvest due to inherent biochemical changes. In recent years, plant growth regulators like auxins, gibberellins and growth retardants like cycocel are being used for improving the fruit quality, delaying deterioration in storage and increasing the shelf life Katiyar *et al.*, (2008). Kumar *et al.*, (2011) conducted an experiment to find out the effect of plant growth regulators and nutrients on shelf life and chemical constituent of guava fruit cv. L-49. They reported that, minimum physiological loss in weight, maximum storage stability were recorded with foliar application of GA_3 50 ppm+Borax 0.25 per cent was founds most effective treatment in guava. Katiyar *et al.* (2008) studied the effect of pre-harvest application of plant growth regulators on post-harvest quality of organically grown guava (*Psidium guajava* L.) fruits. They reported that, GA_3 treated fruit stored at ambient temperature expressed relatively slower senescence as compared to control.

Reference

- Abbas, M.M., Ahmad, S. and Javaid, M.A. 2014. Effect of Napthalene Acetic Acid on flower and fruit thinning of summer crop of guava. *J. Agric. Res.*, **52**: (1): 111-116.
- Abdullah, A.T.M., Hossain, M.A. and Bhuiyan, M.K. 2006. Clonal propagation of guava (*Psiduium guajava* Linn.) by stem cutting from mature stock plants. *J. For. Res.* **17**: (4): 301-304.
- Ali N., Mulwa, R.M.S., Norton, M.A. and Skirvin, R.M. 2007. Radical disinfestation protocol eliminates *in vitro* contamination in Guava (*Psidium guajava* L.) seeds. *Plant Cell Tissue Organ Cult.*, **91**: 295-298.
- Ali, N., Mulwa, R.M.S., Morton, M.A. and Skirvin, R.M. 2003. Micropropagation of guava (*Psidium guajava* L.). *Hort. Sci. Biotech.*, **78**: 739-741.

- Amin, M.N. and Jaiswal, V.S. 1988. Effect of phloroglucinol, sucrose, *pH* and temperatures on *in vitro* rooting of guava *(Psidium guajava* L.) microcuttings, *Bangladesh J. Bot.*, **18:** 129-139.
- Bhardwaj, R.L., Meena, R.R. and Mukherjee, S. 2005. Role of plant growth regulator's in guava *(Psidium guajava* l.) - A review. *Agric. Rev.*, **26**: (4): 281 – 287.
- Bikash, D., Vishal, N., Jana, B.R., Kumar, S. and Dey, P. 2007. Evaluation of different methods of crop regulation in guava grown under rainfed plateau conditions of eastern India. *Indian J. Hortic.*, **64**: 3.
- Brahmachari, V. S., Mandal, A. K., Kumar, R., Rani, R. and Kumar, R. . 1996. Effect of growth substances on flowering and fruiting characters of 'Sardar'guava (Psidium guajava L.). *Hort. J.*, 9: 1-7.
- Butt, M., Usman, M. and Fatima, B. 2013. Enhanced Seed Germination and Callogenesis under long days using Leaf Disc as Explant in Guava cultivars. *Biologia (Pakistan)*, **59**: (2): 293-298.
- Cangahuala-Inocente, G.C., Dal Vesco, L.L., Steinmacher, D., Torres, A.C. and Guerra, M.P. 2007. Improvements in somatic embryogenesis protocol in Feijoa (*Acca sellowiana* (Berg) Burret): Induction, conversion and synthetic seeds. *Sci. Hortic.*, **111**: 228-234.
- Chandra, R., Bajpai, A., Gupta, S. and Tiwari, R.K., 2004. Embryogenesis and plant regeneration from mesocarp of *Psidium guajava* L. (guava). *Indian J. Biotechnol.*, **3**: 246–248.
- Detweiler, C. 1942. Effect of hetroauxin upon the formation of growth substances in higher plants. *Planta*, **33**: 258-277.
- Garasiya, V. R., Patel, N. M., Bhadauria, H. S. and Wankhade, V. R. 2013. Effect of plant growth regulators on quality of winter season guava (*Psidium guajava* L.) cv. L–49 (Sardar). *The Asian Journal of Horticulture*, **8**: (1): 347-349.
- Idate, G.M., Desai, U.T., Gaikwad, S. P. and Nimbalkar, C.A. 2009. *Bahar* regulation in guava (*Psidium guajava*) using plant growth regulators. *Indian J. Agr. Sci.*, **79:** (2): 138–140.
- Iqbal, M., Khan, M. Q., Rehman, K. and Munir, M.2009. Effect of foliar application of NAA on fruit drop, yield and physico-chemical characteristics of guava (*Psidium guajava* L.) red flesh cultivar. *Journal of Agricultural Research* (Lahore), **47**: (3): 259-269
- Jain, M.C. and Dashora, L.K. 2007. Growth, flowering, fruiting and yield of guava (*Psidium guajava* L.) cv. Sardar as influenced by various plant growth regulators. *Internat. J. agric. Sci.*, **3**: (1): 4-7.

- Jaiswal, V.S., Amin, M.N. 1987. *In vitro* propagation of guava from shoot cultures of mature trees. *J. Plant Physiol.*, **130**: 7-12.
- Kalyani, M. Bharad, S.G. Parameshwar, P. 2014. Effect of growth regulators on seed germination in guava. *Int. J. Biol. Sci.*, **5**: (2): 81-91.
- Kareem, A., Jaskani, M.J., Fatima, B. and Sadia, B. 2013. Clonal multiplication of guava through softwood cuttings under mists conditions. *Pak. J. Agri. Sci.*, **50**: (1): 23-27.
- Katiyar P.N., Singh J.P., and Singh P.C. 2009. Effect of nutrients and plant growth regulators on physico-chemical parameters and yeild of guava (*Psidium guajava* L.) fruit cv. Allahabad safeda. *Int. J. Agric. Sci.*, **5**: (1):173-174.
- Katiyar, P. N., Singh, J. P., Singh, P. C. and Gangwar, A. P. S. 2008. Effect of pre-harvest application of plant growth regulators on post-harvest quality of organically grown guava (*Psidium guajava* L.) fruits. *The Asian J. Hortic.*, (3): 2: 330-332.
- Kawase, M. 1965. Etiolation and rooting in cuttings. *Phys. Plant.* **18**: 1044-1076.
- Khattak, M.S., Inayatullah, M. and Khan, S. 1983. Propagation of guava from semi hard wood cuttings. *Frontier J. Agri. Res.*, **13**(1): 81-92.
- Kumar, R., Tiwari, R. and Kumawat B.R. 2013. Quantitative and qualitative enhancement in guava (*Psidium guajava* L.) cv. Chittidar through foliar feeding. *Int. J. Agric. Sci.*, **9**: (1): 177-181.
- Kumar, S. Singh, A.K and Singh, A. 2011. Effect of foliar application of various growth regulators nutrients on shelf life and chemical attributes of guava cv. Lucknow-49. *Plant Archives.*, **11**: (1): 107-111.
- Lal, N., Das R.P. and Verma, L.R. 2013. The effect of plant growth regulators on flowering and fruit growth of guava (*Psidium guajava* L.) cv. Allahabad Safeda. *The Asian Journal of Horticulture*, **8**: (1): 54-56.
- Maji, S., Das, B.C., and Sarkar, S.K. 2015. Efficiency of some chemicals on crop regulation of Sardar guava. *Sci. Hortic.*, **188**: 66–70.
- Mishra, D. Lal, B. and Pandey, D. 2007. Clonal multiplication of *Psidium* species with mound layering. *Acta Hort.*, **735**: 339-42.
- Mohammed, S., Wilson, L.A. and Prendergast, N. 1984. Guava meadow orchard: Effect of ultra high density planting and growth regulator on growth, flowering and fruiting. *Trop. Agric.*, **61**: 297-301.

- Mohammed, S., Sharma, J.R., Kumar, R., Pal, R. and Singh, S. 2006. Effect of different chemicals on cropping pattern, and quality of guava cv. Sardar. *Haryana J. Hort. Sci.*, **35**: (3/4): 226-227.

- Nagar, P.K. and Rao, T.R. 1986. Early changes in growth regulator content of pollinated Guava fruits. *Sci. Hortic.*, **29**: (1-2): 139-146.

- Paleg, L. 1965. Physiological effects of gibberellins. *Annu. Rev. Plant Physiol.*, **16**: 291-322.

- Rahman, N. Tehsinullah, Nabi, G. and Jan, T. 2002. Effect of different growth-regulators and types of cuttings on rooting of guava (*Psidium guajava L*). *Quarterly* Science *Vision*, **9**: 1-2.

- Rahman, N., Tehsinullah, Nabi, G. and Jan, T. 2003. Effect of different growth-regulators and types of cuttings on rooting of guava. *Science Vision*, **9**: 1-2.

- Rai, M.K., Akhtar, N. and Jaiswal, V.S. 2007. Somatic embryogenesis and plant regeneration in *Psidium guajava* L. cv. Banarasi local. *Sci. Hortic.*, **113**: 129-133.

- Rai, M.K., Jaiswal, V.S. and Jaiswal, U. 2008. Effect of ABA and sucrose on germination of encapsulated somatic embryos of guava (*Psidium guajava* L.). *Sci. Hortic.*, **117**: 302–305.

- Rai, M.K., Jaiswal, V.S. and Jaiswal, U. 2009. Shoot multiplication and plant regeneration of guava (*Psidium guajava* l.) from nodal explants of *in vitro* raised plantlets. *Journal of Fruit and Ornamental Plant Research*, **17**: (1): 29-38.

- Rajput, C. B. S., Singh, S. N. and Singh, N. P. 1977. Effect of certain plant growth substances in guava. *Haryana J. Hort. Sci.*, **6**: (3-4): 117.

- Ravi Saini, Devi Singh, R. K. Godara and J. K. Sandooja. 2013. Effect of pruning and foliar application of paclobutrazol on crop regulation of guava cv. Hisar Surkha. *Haryana J. Hortic. Sci.*, **42** (1 & 2): 29-32

- Reddy, Y.N. and P.K. Majumder. 1970. Bottom heat treatment a new technique for rooting hard wood cuttings of tropical fruits. *Curr. Sci.*, **44**: (12): 444-445.

- Reyes, M.U. and Paull, R.E. 1995. Effect of storage temperature and ethylene treatment on guava (*Psidium guajava* L.) fruit ripening. *Post harvest biology and technology*, **6**: (3-4): 357-365.

- Roa, M.M. and Sulikeri, G.S. 1997. In: *Fruit Crops in North Karnataka, In. Res. and Dev.*, pp. 43-45.

- Rymbai, H. and Reddy G.S. 2010. Effect of IBA concentrations on guava stooling and plantlets survival under open and polyhouse conditions. *Indian J. Hort.* **67**(Special Issue): 443-446.

- Shaban, A.E.A. and Haseeb, G.M.M. 2009. Effect of Pruning Severity and Spraying Some Chemical Substances on Growth and Fruiting of Guava Trees. *American-Eurasian J. Agric. & Environ. Sci.*, **5**: (6): 825-831.
- Sharma, R.S. Sharma, T.R. and Sharma, R.C. 1991. Influence of growth regulators and line of operation on rooting of air layering in guava (*Psidium guajava* Linn.) cv. 'Allahabad Safeda'. *Orissa J. Hort.*, **19**: (1-2): 41-45.
- Singh, D. K., Maurya, V. N. and Singh, A. R. 1996. *Haryana J. Hort. Sci.*, **25**: 34-36.
- Singh, G. and. Chanana, Y. R. 2005. Influence of pruning intensity and pruning frequency on vegetative and reproductive attributes in guava 'L-49' *Abstract*: 1st International Guava Symposium, CISH, Lucknow.
- Singh, G., Rajan, S., Pandey, D. and Singh, A.K. 1991. Annual report of Central Institute for Subtropical Horticulture, Lucknow, 1991-1992, pp: 13-18.
- Singh, M. and Singh, M. 2001. Efficacy of plant growth regulators, their concentrations and wrappers on rooting success and survival of air-layered guava twigs. *Crop Res.*, **21**: (2): 153-156.
- Singh, R. 1967. Fruits, National Book Trust of India, New Delhi, pp 87-88.
- Sinha, M.M., Verma, J.P. and Koranga, D.S.1973. Studies on the seed germination of guava (Psidium guajava L.). I. Effect of scarification and plant growth regulator treatments. *Prog. Hort.*, **5**: 39-40.
- Sulladmath, V.V. and Kololgi, S.D. 1969. Synergistic influence of 3-indolebutyric acid and alphanaphthalene acetic acid on the rooting of air layers of Chiku (*Achras zapota* L.) var. kalipatti. *Indian Hort.*, **17**: 17-18
- Tready, M.T. 1983. Effect of IBA on kiwi fruit and guava hard wood cuttings. *Plant Propagation*, **28**: (4): 7-10.
- Tyagi, S.K. and Patel, R.M. 2004. Effect of growth regulators on rooting of air layering of guava (*Psidium guajava* L.) cv. Sardar. *The Orissa J. Hort.*, **32**: 58-62.
- Wahab, F., Nabi, G., Ali, N. and Shah, M. 2001. Rooting response of Semi-hardwood cuttings of guava (*Psidium guajava* L.) to various concentrations of Auxins. *Online Journal of Biological Sciences*, **1**: (4): 184-187.

15

Plant Growth Regulators in Jamun (*Syzygium cumini* Skeels)

Sanjay Singh, D.S. Mishra, A.K. Singh, S. Pradhan, R.K. Tarai and I. Panigrahi

1. Introduction

Jamun (*Syzygium cumini* Skeels) belongs to the myrtle family Myrtaceae is highly adapted to diverse environmental conditions, is widely distributed in India up to an altitude of 1600 m. It is well-known for its nutritional and therapeutic attributes (Mishra *et al.*, 2015). Studies have shown that the various extracts of Jamun possess a range of pharmacological properties such as antibacterial, antifungal, antiviral, anti-genotoxic, antiinflammatory, anti-ulcerogenic, cardioprotective, anti-allergic, anticancer, chemopreventive, radioprotective, free radical scavenging, antioxidant, hepatoprotective, anti-diarrheal, hypoglycemic and antidiabetic effects (Kumar *et al.*, 2011, Singh and Singh, 2012). The presences of anthocyanins, fibers and ellagitannins which are present in the pulp are important in reducing the oxidative stress-induced diseases. Jamun berries contain carbohydrates, iron, sugars, minerals, protein and the pharmacologically active hytochemicals like flavonoids, terpenes, and anthocyanins. The powdered seeds have also reputation of being useful in the treatment of diabetes. These seeds are claimed to contain alkaloid, jambosine, and glycoside jambolin or antimellin, which halts the diastatic conversion of starch into sugar and seed extract has lowered blood pressure. The ripe jamun fruits are consumed fresh and can also be processed into many value added products like jam, jellies, squash, cider, nectar, wine, vinegar, RTS etc. Because of its hardy nature and various uses, it has great potential for commercial exploitation in wastelands and dry-land horticulture. Growth regulators commonly employed for rooting, vegetative propagation and overall yield of several plants (Misra and Mishra, 2003). The chapter includes information on various uses of PGRs in Jamun production including plant propagation.

2. Use of PGRs in Propagation

As jamun has longer gestation period, the production of true-to-the-type quality planting material is the most integral part of its cultivation. Thus utmost care is taken during the selection of genuine planting material at the time of planting. Jamun, particularly mono-embryonic one, is heterozygous in nature which does not produce true to the type planting material. However, rootstocks are being raised by the seeds. Due to existence of polyembryony, it comes true to parent through seed. However, seed propagation is not advisable as it results in late bearing. Therefore vegetative methods like softwood grafting, air layering, budding etc. are preferred for commercial orcharding. Since the discovery of root inducing plant growth regulators, many investigators have reported variable success in propagating cuttings of species previously considered difficult to root.

2.1. Use of PGRs in seed propagation

Jamun is commercially propagated through seeds through exploitation of nucellar embryony. Besides soaking in water, several other efforts had been reported for enhancing germination by use of chemicals and growth regulators like gibberllic acid, thiourea, KNO_3 *etc* with varied success (Shanmugavelu, 1970). The mechanism behind better germination per cent by water soaking may be that, it stimulates series of biochemical change in the seed that are essential to initiate the emergence process like break down of dormancy, hydrolysis, metabolism of growth inhibitors, imbibitions, activation of enzymes (Ajouri *et al.*, 2004). But the simplest reason may be the softening of the seed coats, removal of inhibitors which in turn reduces the time required for germination and thus an increased germination percentage was observed (Hartman and Kester, 1997). Similar results was also reported by Swamy *et al.* (1999), who recorded the highest germination (96.6%) when seeds were treated with water for 48 h and GA_3 at 250 ppm while the lowest germination (56.6%) was in control.

Seed propagation is the most common and simplest method of raising the seedling to be used as rootstock for grafting with desirable genotype (Burman *et al.*, 2016). Plant growth regulators are involved in the regulation of various processes of plant growth and development; therefore have profound effect on seed germination and seedling growth. Growth regulators used in pre-sowing seed treatment, play an important role in regulating seed germination and seedling vigour (Khan, 1977). Various seed invigoration treatments have been used to improve seed germination and seedling establishment. These include alternate hydration-dehydration, water soaking and seed priming (Khan, 1992).

Burman *et al.,* (2016) experimented on effect of different doses of gibberellic acid and benzyl adenine on germination and morpho-physiological characters in jamun and found enhanced seed germination, seedling vigour and improved physiological status of the plant with GA_3 and concentration at 100 ppm (soked for 24 hours) was found to be the best. The seedling vigour was enhanced by 79.46% as compared to control which could be attributed to inherent potential of gibberellic acid in promoting cell elongation. Whereas, Hota (2016) under Jabalpur (Madhya Pradesh) condition observed that GA_3 (450 ppm) significantly influenced all germination related parameters *viz.* days taken to start germination, days taken to 50% germination, height of shoots (cm), number of leaves per seedling, Leaf Area Index (LAI) *etc.* The positive effect of GA_3 on improved germination of jamun seeds is attributed to stimulation of hydrolytic enzymes by gibberellins to weaken the barrier tissues such as the endosperm or seed coat, inducing mobilization of seed storage reserves and stimulating expansion of the embryo (Dhoran and Gudadhe, 2012). GA_3 treatment also lead to increased seedling height which may due to the fact that it increased osmotic uptake of nutrients thus, causing cell multiplication and cell elongation in the cambium tissue of the intermodal region which in turns resulted in taller phenotype of the plant (Shanmugavelu, 1966).

Patil *et al.,* (2018) studied the effect of various seed treatment on germination and seedling growth of jamun at Regional Horticultural Research Station, Navsari (Gujarat). Amongst the different treatments (water soaking for 24 h, hot water (100°C) for 5 sec, GA_3 @ 200 mg l^{-1} for 10 min., KNO_3 @ 0.5% for 10 min and thiourea @ 0.5% for 10 min.), jamun seeds soaked in water for 24 h noted to have the lowest number for initiation of germination (7.13) and also the highest germination percentage (95.73%). But seeds soaked in GA_3 @ 200 mg l^{-1} for 10 min. resulted in seedlings with the longest plant height (82.60 cm), number of leaves (42.63) and stem diameter (7.90 mm).

2.2. Use of PGRs in propagation by cutting

The process of root formation at the base of stem cuttings can be divided into three stages: root initiation, elongation of root initials, and root growth and development (Hartmann *et al.,* 2002). Thus three groups of chemicals could be applied to regulate the whole rooting process in following way: i) hormones, which can induce the initial meristematic process (Wiesman and Riov, 1989) and stimulate the elongation and development of roots formed (Hartmann *et al.,* 2002); ii) nutrients which promote growth of the new roots

(Wiesman and Lavee, 1995); iii) Pesticides, which enable the cuttings to resist attack by pathogens during the rooting period.

Indole-3-acetic acid (IAA) is identified as a naturally occurring auxin has got property of promoting roots on stem cutting (Thimann and Koepfli, 1935). However, two synthetic auxins, indole-3-butyric acid (IBA) and a-naphthalene acetic acid (NAA), are found more effective than the naturally occurring or synthetic IAA for rooting, (Zimmerman and Wilcoxon, 1935). IBA and NAA are still the most widely used auxins for rooting stem cuttings and for rooting tissue-culture-produced micro-cuttings (Henrique *et al.*, 2006). It has been variously confirmed that auxin is required for initiation of adventitious roots on stems, and in fact, it has been shown that divisions of the first root initial cells are dependent upon either applied or endogenous auxins. The formation of root primordium cells depends on the endogenous auxins in the cutting and on a synergic compound such as a diphenol (Henrique *et al.*, 2006. These substances lead to the synthesis of ribonucleic acid (RNA), which act upon root primordium initiation (Hartmann *et al.*, 2002). IBA is the best auxin for general use because it is nontoxic to plants over a wide concentration range than NAA, and is effective in promoting rooting of a large number of plant species (Hartmann *et al.*, 2002). Pandey and Pathak (1981) explained that phenols increased the action of auxins on rooting by inhibiting the IAA-oxidase system. The application of exogenous phenolic compounds can present synergism with synthetic auxins, inducing root formation (Henrique *et al.*, 2006).

Under Jabalpur (M.P.) condition, Pandey *et al.* (2011) observed that application of IBA at 3000 ppm was found to be most suitable for rooting and growth of jamum cuttings followed by 2500 ppm concentration. Application of IBA (3000 ppm) on 17th September was found significantly superior to all other interaction (concentration and time of application) combinations.

2.3. Use of PGRs in air-layering

Air layering is one of easiest means employed for plant propagation. Air layers of woody plants initiate callus development followed by root formation at the physiological base of the incision. These steps can be enhanced with the application of auxins (Misra and Mishra, 2003). Shanmugavelu (1967) found that air layered *Eugenia* shoots treated with IBA and NAA produced roots but amongst the two growth regulators used, NAA was found more efficient as it induced 14 roots with an average length of 4.9 cm, whereas IBA treatment induced 3 roots having 2.5 cm average length. Singh (1969) reported 86 per cent rooting when white polythene film and 1 per cent IBA in lanolin paste were used in air layering of Jamun. Saha (1969) recorded 60 per cent rooting when air layers treated with 500 ppm IBA and prepared during spring.

Kumar (2007) studied the influence of IBA on rooting of jamun air layering under Gandhi Krishi Vignana Kendra, University of Agricultural Sciences, Bangalore. He recorded maximum rooting (100 %) in the air layers prepared during June and July months with IBA (7500 ppm) as well as July and August months with IBA (10000 ppm). But the layers treated with IBA (10000 ppm) during September month recorded highest number adventitious roots. However, length of adventitious roots (3.07 cm) and girth (0.94 mm) was maximum when layers treated with IBA (10000 ppm) during the month of July. The number of secondary roots was maximum (5.04) when layers treated with IBA (7500 ppm) during July month. The secondary roots length (2.54 cm) and girth (0.89 mm) was the highest when air layering done during July month with IBA (2500 ppm). Maximum number (5.95) of tertiary roots was found in the month of July layers treated with IBA (10000 ppm). The length and girth of tertiary roots was the highest with IBA (7500 ppm) during the month of July.

Investigations on air layering in jamun revealed that air layers treated with 10000 ppm IBA during September recorded the maximum number of adventitious roots (2.24). However, root length (3.07 cm), root girth (0.94 mm) and number of tertiary roots (5.95) were highest in 10000 ppm IBA treated air layers during the month of July (Gowda *et al.*, 2009).

Vyas *et al.*, (2017) conducted an experiment to study the effect of time of air layering and application of different doses of IBA on red jamun (*Syzygium samarangense* L.) *cv.* Local under Junagadh (Gujarat) condition. Amongst the different IBA concentrations (1000 ppm, 1500 ppm and 2000 ppm), application of 2000 ppm, resulted in minimum days for root initiation (18.1), maximum number of primary (7.88) and secondary roots (20.2) per layer, length of primary (9.88 cm) and secondary roots (7.14 cm), root fresh (0.95 g) and dry weight (0.34 g) per layer, number of shoot (6.0) per layer, shoot length (8.6 cm). The highest rooting (83.2%) was also observed at this IBA concentration. In terms of interaction of IBA with time of layering, the air layer prepared on 1st August along with IBA (2000 ppm) resulted in minimum days for root initiation (16.5), maximum number of primary (10.87) and secondary roots (26.2) per layer, primary (11.6 cm), secondary root length (7.59 cm), root fresh (1.20 g) and dry weight (0.50 g) and the highest rooted layer (92.4%).

2.4. Use of PGRs in micropropagation

Jamun suffers from very low seed viability and poor germination in its natural habitat(Dent, 1948). Thus, propagation through stem cuttings is an alternative, but not feasible for obtaining large quantity of planting materials within a short time interval. In this context it is necessary to standardise a suitable protocol for rapid, clonal propagation of elite genotypes of this species (Remashree *et al.*, 2007). Although tissue culture of herbaceous angiosperms has been extensively studied but it has not received much attention in case of perennial trees like jamun due to inherent problems like phenolic exudation, explant browning etc. Micropropagation method could assist in rapid and mass production of clonal stock of newly released cultivars (Mishra *et al.*, 2007). Micropropagation is an advanced vegetative propagation techniques for producing a large number of genetically uniform and pathogen-free transplants in a limited time and space. In this context it is necessary to standardize a suitable protocol for clonal micropropagation of this species. Woody plant species in general have proved to be difficult for mass propagation by tissue culture due to exudation high level of polyphenols after transferring of explants on culture media, which could be reduced to some extent with use of antioxidants, activated charcoal and initial incubation of cultures in complete dark for 24 hrs (Mishra *et al.*,2009) as well as frequent shifting to fresh media (Mishra *et al.*,2010). Each plant species propagated through *in vitro* techniques has different requirements of plant growth regulators and their concentrations. Depending on the auxin/cytokinin ratios of the induction medium, either shoots or roots can be regenerated (Murasighe and Skoog, 1962). Following may be the major applications of micropropagation in jamun:

i. Large scale propagation of elite plant materials.

ii. Genetic transformation for development of superior cultivars.

iii. Conservation of plant genetic resources.

iv. Production of medicinally important secondary metabolites through cell suspension cultures. In view of above considerations an overview on jamun micro-propagation is presented in Fig.1.

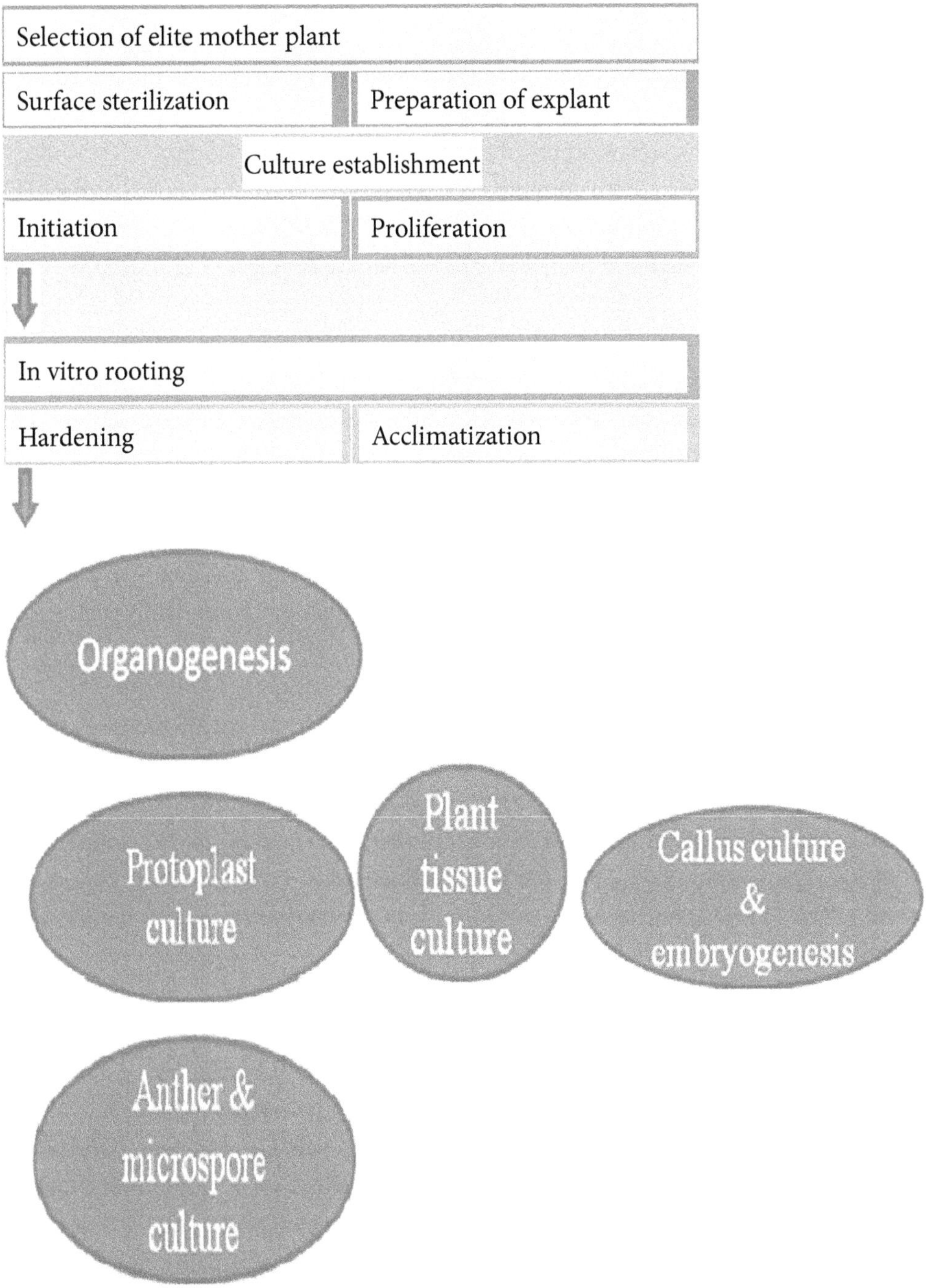

Fig. 1 An Overview of Plant Tissue Culture

2.4.1. Axillary bud proliferation

In vitro propagation of jamun was achieved, using single nodal explants taken from the seedlings. Single nodal explants cultured on half strength MS along with 2mg/l BAP + 3% sucrose + 3% activated charcoal was proved to be better for shoot proliferation. The shootlets were successfully rooted on 1/4th strength MS along with 3% sucrose and 2.5 mg/l IBA. The rooted plantlets were then hardened on vermiculite (32.5 % survival) for six weeks and then planted in the polybags and transferred to the field (Choudhri *et al.,* 2013).

Yadav *et al.*, (1990) standardised a protocol for rapid multiplication of true-to-the-type jamun plantlets form nodal and shoot tip segments. Multiple shoots were obtained from nodal and shoot tip segments of 10-15 day old seedlings of *Syzygium cuminii* L. on modified Murashige & Skoog (MS) medium supplemented with 6-benzyladenine (BA) (0.23-8.90 µM) singly or in combination with α-naphthaleneacetic acid (NAA), indole-3-acetic acid (IAA) or indole-3-butyric acid (IBA). Excised shoots obtained from this medium were placed for root induction on MS medium containing NAA and/or IBA and then transferred to MS basal medium to form complete plantlets. The authors had successfully acclimatized the regenerated plantlets and transferred into the soil.

For cloning of adult tree of jamun, Rathore *et al.*,(2004) took explants from rejuvenated shoots sprouts. Multiple shoots were regenerated by activation of axillary meristems of the explants on MS + 9.0 µM BAP. The shoots were proliferated in culture by repeated transfer of original explants and subculture of shoots on amended medium with combination of BAP (4.5 µM) + kinetin (2.25 µM). The cloned shoots rooted on ½ MS + 0.1% activated charcoal + IBA (10 µM) or NAA (15 µM). The rooted plantlets were then hardened and pot transferred.

Nodal explants, excised from young shoots from mature trees of jamun were cultured on MS medium supplemented with different concentrations of BA or kinetin. Among these, BA (0.5 or 1.0 mg l^{-1}) induced greening and opening of the incipient shoot buds, which however did not elongate. Elongation of the shoot buds was facilitated on MS medium with 1.0 mg l^{-1} BA supplemented with casein hydrolysate (1.5 g l^{-1}) or glutamine (200 mg l^{-1}). Nodal explants (microcuttings), taken from shoots developed *in vitro*, also developed multiple shoots when cultured on MS with1 mg l^{-1} BA. These explants did not require additional supply of reduced nitrogen, for further normal development. Shoots developed from explants from mature trees and microcuttings were rooted by sub-culturing them on Knop's medium supplemented with 2% sucrose and 1 mg l^{-1} IAA. The plants that developed *in vitro* were planted in soil and were transferred to the field after an acclimatization period of 7–8 months. These plants have been thriving well for more than three years and have no apparent phenotypic aberrations (Jain and Babar, 2003).

Jain and Babbar (2003) studied the effect of different carbon source on shoot proliferation potential of epicotyl explants of jamun. Amongst the different source of saccharides studied, sucrose gave the best caulogenic response both in terms of explant response and shoot developing potential. Sucrose at a concentration of 4% was proved to be the best in developing 4.2 shoots per explant.

Remashree *et al.*, (2007) reported a successful protocol for micropropagation and *ex vitro* rooting of *Syzigium cumini* from mature nodal explants through axillary bud proliferation. The authors induced multiple shoots from mature nodal explants using Woody Plant Medium (WPM) supplemented with combinations of BA (8.8 μM), Kinetin (9.3 μM) and NAA (5.37 μM). The culture produced 35-40 shoots by 6-7 weeks from inoculation. These shoots were rooted in WPM supplemented with IBA (4.9 μM). Successful *ex vitro* rooting was also obtained from these shoots. Plantlets established in the field showed identical morphological characteristics as the donor plant with 80% survivability.

Callus development was observed by using a MS basal medium + 6 BAP with additional supplement of IBA and kinetin at concentration 0.60 mg/l – 0.60 mg/l. But MS basal medium + 0.02-0.5 mg/l of IBA & kinetin resulted in shoot development. Although IAA resulted in some growth response in explant but did not induce the callus (Jejurker *et al.*, 2016).

Prabhuling *et al.* (2017) standardized a protocol for *in vitro* regeneration in jamun *cv.* AJG-85 for rapid mass multiplication of healthy and disease free plantlets. Amongst the different explants tried (explants tried includes shoot tip, single nodal segment, double nodal segment, lateral shoot bud, leaf petiole of 1 cm length), shoot tip was found best for establishment of aseptic culture. As compared to ascorbic acid (150 mg/l), activated charcoal (0.50%) was found to better for control for phenolic browning. Amongst the different basal media studied, half strength WPM showed the best response with higher number of shoots per explants, length of shoots and number of leaves per shoot. Amongst the various growth regulators used, BAP (2 mg/l) + NAA (0.1 mg/l) recorded better shoot length, number of leaves and number of adventitious buds per explants. While GA_3 (2 mg/l) showed better performance in terms of shoot length and number of leaves per shoot.

Recently, Naaz *et al.*, (2018) reported successful plant regeneration in *Syzygium cumini* (L.) *via de novo* organogenesis. Best callogenesis (75%) was observed after 4 weeks of inoculation on MS medium supplemented with 2, 4-D (7.5 μM). Best shoot induction (93%) were observed in yellow white friable (YWF) type of callus obtained using cotyledonary node (CN) explants on MS medium enriched with 5.0 μM BA + 0.25 μM NAA + 10 mg/l $AgNO_3$. This combination also resulted in maximum mean shoot number (10.19 ± 0.05) and mean shoot length (4.12 ± 0.32 cm) per clump after 4 weeks of transfer. For *ex vitro* rooting of excised shootlets (4 cm), full or half strength MS medium with varied concentrations of IBA was used. Best rhizogenesis (80%) was achieved with IBA (200 μM) treated shootlets after 4 weeks of transfer to sterilized soilrite. Ninety per cent survivability was observed after establishment of acclimatized regenerated plantlets.

2.4.2. Callus culture

The importance of callus culture approach propagation of woody plants, including fruit plants, for the crop improvement has been discussed by several workers. Significant progress has been made in achieving *in vitro* organogenesis in callus cultures of hardwood trees, however, only a few studies have led to complete plant regeneration (Paranjothy *et al.*, 1990). Microprapogation of the *syzygium cumini* explants (leaves) was tried by using the different kind of MS medium, hormonal changes in the medium show variations in the result (Jejurker *et al.*, 2016). At the appropriate quantity of the hormones, the explant shows a significant amount of development; the basal medium does show the some kind of development into a bud and had no effect on the leaf explant development. The auxin is generally root initiator and cytokinins are the shoot developers. Very low level of PGR concentration does not show the significant result (0.1-mg/l -0.1 mg/l). However, developed callus by using a MS medium of basal medium + 6 BAP with additional supplement of IBA and kinetin at concentration (0.60 mg/l – 0.60 mg/l) and at the (0.02-0.5 mg/l) of IBA & kinetin shoot development observed. IAA only shows some growth in explant but did not develop the callus. However, Yadav *et al.* (2014) induced callus from the seedling explants on MS medium containing 2,4-D (0.05-1.0 mg/l) with BA (0.5-2.0 mg/l). Such calli subsequently, produced shoots when sub cultured on MS medium supplemented with BAP (0.5-2.0 mg/l) singly or in combination with NAA (0.05-0.2 mg/l). Upto 85% of cultures showed shoot formation (25 shoots/culture) from callus cultured on medium containing 1.0 mg/l of BA along with 0.05 mg/l of NAA. Adventitious roots initiated when sub cultured on MS medium containing IBA or NAA (1.0 mg/l). However further growth and development of rooted shoots occurred only when plantlets were transferred to MS basal medium. About 75% of rooted shoots have successfully been established to soil.

2.4.3. Somatic embryogenesis

Somatic embryogenesis was found successful for multiplication of jamun plants (Litz, 1984). Rathore *et al.*, (2004) regenerated multiple shoots by activation of axillary meristems of the explants on MS medium with BAP (9.0 µM) and growth additives. The shoot proliferation was done by repeated transfer of original explants and sub-culturing in media supplemented with BAP (4.50 µM) and kinetin (2.25 µM). Rooting of the cloned shoots was done by either *in vitro* or *ex vitro*. The *ex vitro* was found to be more efficient as 100% rooting was observed in the cloned shoots. A successful micropropagation protocol, of *Syzygium cuminii* has also been standardised at ICAR-CISH, Lucknow using nodal explants from mature trees (Singh *et al.*, 2009).

3. Role of PGRs on Growth and Development

Even though seed propagation is commercial method of propagation of jamun, asexual propagation is commonly recommended to shorten gestation period. The time required to grow jamun seedlings to a suitable size for grafting as a rootstock may be as long as one year. Therefore, shortening this time is considered very important and it can be achieved by enhancing the seedling growth by using various combinations of growth regulator and nutrients. The plant hormones are extremely important agent in the integration of developmental activities (Suman *et al.*, 2017). In this regard, although a very few attempts are available in jamun but it is well studied in other crops like mango (Singh and Singh, 1974), carambola (Marler and Mickelbert, 1992), aonla (Virendra and Shafaat, 1996), khasi mandarin (Govind *et al.*, 2003), valkamer and sour orange (Hoda *et al.*, 2010) *etc.*

Surakshitha *et al.*, (2014) carried out an investigation to reduce the time required to reach graftable size of jamun seedlings using growth regulators and macronutrients at Bengaluru (Karnataka). Amongst the various treatments, foliar application of GA_3 (300 ppm) was observed to be beneficial to increase the seedling height (57.66 cm), diameter (6.06 mm), fresh weight (94.66 g) and dry weight (35.86 g) and to successfully reduced the number of days taken to reach graftable size (132.72 days) as compared to control.

Irregular bearing or cropping periodicity along with erratic flowering behaviour is some of the major constraint of jamun cultivation reducing the production potential. Paclobutrazol (PBZ), a triazole derivative, anti-gibberllin, has been effectively used to induce and manipulate flowering, fruiting and tree vigour in several perennial fruit crops. Soil application of paclobutrazol has been effective in promoting flowering and increasing yield in many fruit crops. A study was conducted by Hegde *et al.* (2018) for exploitation of PBZ to regularize the flowering in jamun *cv.* Chintamani at Regional Horticultural Research and Extension Centre (RHREC), University of Horticultural Sciences Campus, Gandhi Krishi Vignana Kendra, Bengaluru. Amongst the different doses of soil application at pre-flushing stage, PBZ (3.0 g a.i./ plant) resulted less increment in plant height (23.67cm) and canopy spread in N-S (38.50 cm) and E-W (50.83 cm) direction while 2.5 g a.i/ plant resulted in highest number (424) of panicles per plant, number of flowers per panicle (51.33) and length of flowering panicle (16.16 cm). The study revealed that PBZ at 1.5 g a.i/ plant resulted in highest number (292) of new flushes per plant which ultimately led to highest fruit number (4217) and fruit yield per plant (47.13 kg). The application of PBZ significantly influenced the internodal length, number of new flushes and leaf area in jamun *cv.* Chintamani. The reduction in plant height with the use of Paclobutrazole might be due to its inhibitory effect on gibberellins biosynthesis pathway at the sub-apical meristem, which ultimately reduced cell elongation, rate of cell division and decreased the shoot growth. It checks the conversion of ent-kaurene to ent-kaurenoic acid in gibberellins biosynthetic pathway. Similar results were reported in apple (Quinlan and Richardson, 1983), plum and cherries (Quinlan and Webster, 1982), Tommy Atkins mango (Yeshitela *et al.*, 2004), mango (Tandel and Patel, 2011) *etc.*

Patil *et al.*, (2017) studied the effect of seed treatment and foliar spray of chemical substances on seedling growth of jamun (*Syzygium cuminii* L.) at Regional Horticultural Research Station, Navsari Agricultural University, Navsari (Gujarat). Among the seed treatments, GA_3 (200 mg l^{-1}) for 10 min recorded the maximum seedling height, number of leaves, stem diameter, number of shoots/seedling, leaf area, root: shoot ratio, primary root length, number of secondary roots, fresh and dry weight of seedling, fresh and dry weight of root and minimum days to reach graftable size. Amongst the various foliar spray of growth regulators and chemicals, the same GA_3 (200 mg l^{-1}) was found efficient for most of the above growth related parameters. But the seeds treated with GA_3 (200 mg l^{-1}) for 10 min and sprayed with GA_3 (200 mg l^{-1}) was observed most significant for the above growth parameters.

4. Use of PGRs in Improvement in Flowering, Fruit Set, Yield and Quality

PGRs act as messengers and are needed in small quantities at low concentrations. Generally their site of action and biosynthesis are different. Most of the plant growth regulators exhibit a broad spectrum and thus a single PGR may influence various aspects of fruit growth and development. Moreover PGRs enhance the rapid changes in physiological and biochemical characters and improve productivity of fruit crops. Sprays of GA3 have been widely adopted in commercial orchards because they have consistently been shown to increase fruit size and firmness in many fruit crops. In this respect Tuan and Ruey (2013) evaluated the effects of gibberellic acid and 2, 4- dichlorophenoxyacetic acid on flower number, fruit growth and fruit quality of wax apple (*Syzygium samarangenese*). GA_3 and 2,4-D were applied at small bud and petal fall stage. Results indicated that spraying with 10 ppm GA_3 had the best results in number of flower. GA_3 spray at 30 ppm gave the faster rate of fruit growth than the other treatments. Fruit set, fruit size as well as fruit weight markedly improved by spraying 30 ppm GA_3, followed by 10 ppm GA_3 compared to untreated control. Moreover, GA_3 spray at 30 ppm was the most effective and increased total soluble solids, reduced titratable acidity and fruit drop. On the other hand, it was noticed that with 10 ppm 2,4-D application also enhanced the fruit growth rate, improved physiological and biochemical characters of fruit compared to untreated control. It was concluded that both GA3 and 2,4-D spray had positive effects on fruit development, reduced fruit drop, fruit crack and improved fruit quality.

Similar observations were made by Moneruzzaman *et al.* (2011) while investigating the effects of gibberellic acid (GA_3) on the growth and development of the *Syzygium samarangense* fruits grown under field conditions. It was observed that the application of GA_3 at 50 mg/L increased fruit length and diameter. Furthermore, it enhanced faster fruit growth and color development in addition to increased fruit number, weight and yield. It also decreased premature fruit dropping. Same treatment also increased TSS, total sugar, total biomass and total flavonoids content in the fruits. In addition, anthocyanin content, total phenol and antioxidant activity was higher in GA_3 treated fruits. In another study it was observed that application of 50 mg/L GA_3, 10 mg/L NAA, and 5 mg/L 2,4-D once a week from bud development to fruit maturation increased the physiochemical and phytochemical properties of wax apple fruits (Khandaker *et al.*, 2012).

Use of hormones in fruit production became a new paradigm. There is scare information available on the role of plant growth regulators on fruit yield of jamun. Hegde *et al.* (2018) observed significant influence of soil application of PBZ in jamun *cv.* Chintamani at Regional Horticultural Research and Extension Centre (RHREC), Bengaluru. Soil application of PBZ @ 1.5 g a.i. (47.13 kg/plant) at pre-flushing stage resulted in the highest yield per plant while, the lowest yield per plant was recorded in control (22.59 kg/ plant). Per cent increase in yield per plants for PBZ treated plants over control was also maximum in plants treated with 1.5 g a.i./plant (108.63%) followed by 2.5 g a.i./plant (105.27 %), 2.0 g a.i./plant (91.85 %). This may be due to higher flowering intensity which resulted in higher fruit number, more number of fruits/plant which lead to increased fruit yield in PBZ treated plants.

5. Use of PGRs in Shelf Life Improvement

Jamun fruits although non-climacteric but highly perishable, ripen fast during summer and become inconsumable very soon. Due to prevalence of high temperatures (35-42°C) during the time of harvesting, jamun fruits start spoilage rapidly and it is necessary to prolong shelf-life of fruits during storage (Singh *et al.*, 2011). Post-harvest quality deterioration of fruits can be caused due to various factors including high respiration rates, biochemical changes associated with respiratory metabolism, ethylene biosynthesis, compositional changes, anatomical changes associated with growth and development, physical injuries, water loss and pathological breakdown (Kadar, 1966). Post-harvest dipping of fruits in various growth regulators and chemicals with or without fungicides ad bactericides has been used to delay the ripening to reduce post-harvest losses and to improve and maintain the colour and quality by slowing down the metabolic activities of the fruits which thus results in increased shelf life and facilitates longer marketability of the fruits (Rao and Chundawat, 1986; Mulagund *et al.*, 2015).

Gibberallic acid (GA_3) and auxins are effective as a postharvest treatments and used on large scale in a number of fruits. GA_3 acts as anti-senescent agent and thereby enhance the shelf-life of fruits. The Jamun fruits were treated with growth regulators GA3 (50 and 100 ppm), chemical CaC12 (1.0 and 1.5%) and Paraffin wax along with control and kept in with or without perforated polyethylene bag. GA3 (100 ppm with Paraffin wax) effectively reduced the physiological loss in weight as well as spoilage loss and thereby found useful in maintaining good balance between ascorbic acid and sugar content of fruits during storage. The treatment also showed little change in TSS, pH and acidity content and hence, it can be useful in post harvest management of Jamun fruits (Ayar *et al.*, 2011).

Pre-harvest spray of GA_3 restricts the ethylene accumulation in the fruit tissue and has been known to delay the ripening process and to regulate the nucleic acid and protein synthesis. Fruits treated with GA3 (50 ppm) and stored in perforated polyethylene bags, enhanced their shelf-life by restricting the transpiration and respiration at room temperatures (Singh *et al.*, 2011). Singh *et al.* (2011) also noted that fruits treated with GA_3 (50 ppm) and stored in perforated polyethylene bags, enhanced their shelf-life by restricting the transpiration and respiration at room temperatures.

An investigation was carried out at Anand (Gujarat) to study the effect of post harvest treatments (dipping) of GA_3 (50 and 100 ppm) and $CaCl_2$ (1.0 and 1.5%) during storage (Kiran *et al.*, 2011). Amongst the treatments, $CaCl_2$ (1.5%) with perforated polyethylene bag was observed to be the most efficient. This treatment effectively reduced the physiological weight loss as well as spoilage loss and thus maintained a good balance between ascorbic acid and sugar content of fruits during storage. The treatment also showed a little change in TSS, pH and acidity content and hence, it can be useful in post harvest management of *Jamun* fruits.

A study was conducted to evaluate the efficacy of biodegradable coatings of chitosan (CH), alginate (AL), and carboxymethyl cellulose (CMC) for improvement of postharvest shelf life and quality characteristics of Jamun fruit. Jamun samples were dipped for 2 min into the solution of CH (1.5% and 1%), AL (1.5% and 1%), and CMC (1.5% and 1%), while the uncoated fruit served as control. Jamun fruit coated with edible coatings exhibited significant delay in weight loss and positive effects on decay percentage, total soluble solids, pH, titratable acidity, and sugar accumulation as compared to uncoated control fruit. After 16 days of storage, jamun samples treated with CH and CMC coatings maintained good sensory acceptability. The results of the study suggest that a coating of CH 1.5% and CMC 1.5% is useful for extending the shelf life of jamun fruit, which in turn is effective in stabilizing the market demand and proved to be safe and effective at ambient conditions (Gol *et al.*, 2015).

Hiwale (2015) made a very clear observation regarding retardation of ripening of under utilized fruits including jamun by application of various chemicals including GA (200 ppm) followed by packing them in polyethylene bags with $KMnO_4$ -coated silica gel. These treatments also reduced the rate of degradation of various chemicals by retarding the activity of various oxidizing enzymes and ethylene evolution up to 4 days of storage. Thus, it was possible to increase the shelf life of semi-arid fruits up to 4–9 days after harvesting as compared to control.

In a similar study in Junagadh (Gujarat), Dalvadi *et al.* (2018) also reported that $CaCl_2$ (1.5%) significantly enhanced the shelf life of the jamun fruits as compared with the other treatments (GA_3-50 mg/l, GA_3-100 mg/l, $CaCl_2$-1%, $CaCl_2$-1.5%, $Ca(NO_3)_2$-1.5% and control-water). The $CaCl_2$ (1.5%) treated fruits showed significantly reduced physiological weight loss, spoilage loss and took maximum days for shrivelling initiation and maximum firmness and thus, resulted in maximum marketable fruits. GA_3 restricts the ethylene accumulation in the fruit tissue and has been known to delay the ripening process and to regulate the nucleic acid and protein synthesis.

References

- Ajouri, A., Asgedom, H. and Becker, M. 2004. Seed priming enhances germination and seedling growth of barley under conditions of P and Zn deficiency. *Journal of Plant Nutrition and Soil Science*, **167**(5): 630-636.

- Ayar, Kiran, Singh, Sanjay and Chavda, J.C. 2011. Effect of post harvest treatments on quality of Jamun (Syzygium cuminii Skeels) fruits during storage. *Asian J. Hort.*, **6** (3): 297- 299.

- Ayyanar, M. and Subash-Babu, P. 2012. *Syzygium cumini* (L.) Skeels: A review of its phytochemical constituents and traditional uses. *Asian Pacific Journal of Tropical Biomedicine*, **2**(3): 240-246.

- Bose, T. K., Mitra, S. K. and Sanyal, D. 2000. Fruits: Tropical and Subtropical Vol. II, Pub. Naya Udyog, 206, Bidhan Sarani, Calcutta, India, Pp. 643-56.

- Burman, P., Rekha, A. and Paneerselvan, P. 2016. Effect of different doses of gibberellic acid and benzyl adenine on germination and morpho-physiological characters in jamun (*Syzygium cumini* L. Skeels) under different propagation substrates. *J. Farm Sci.*,29(1): 140-142.

- Choudhri, N.A., G.S.K Swamy , R.C. Jagadeesha, Mukesh L Chavan, A.B Mastiholi, G Prabhuling and Basavarajappa, H.R. 2013. Micropropagation Studies in Jamun (Syzygium Cuminii L.). *International Journal of Applied Biotechnology and Biochemistry*, **3**(1): 235-243.

- Dalvadi, A. C., Dodiya, K. D., Bhad, D. M., Chaudhari, T. and Trambadiya, R. 2018. Effect of chemicals on shelf life of (*Syzygium cuminii* Skeels) *cv.* Local. *Agricultural Science Digest*, **38**(1): 68-70.

- Dent, T. V. 1948. *Seed storage with particular reference to the storage of seed of Indian forest plants*. Manager of Publications.

- Dhoran, V. S. and Gudadhe, S. P. 2012. Effect of plant growth regulators on seed germination and seedling vigour in *Asparagus sprengeri* Regelin. *Int. Res. J. Bio. Sci.*,**1** (7): 6-10.

- Gol, N.B., Vyas, P.B. and Rao, T.V.R. 2015. Evaluation of polysaccharide-based sdible coatings for their ability to preserve the postharvest quality of Indian Blackberry (*Syzygium cumini* L.). *International Journal of Fruit Science*, **15**(2): 198-222.

- Govind, S. and Singh, I. P. 2003. Effect of foliar application of NPK on the growth of Khasi mandarin seedlings. *Indian Journal of Horticulture*, **60**: 232-235.
- Gowda, V.N., Kumar, V. and Reddy, P.V.K. 2009. Studies on vegetative propagation in jamun (*Syzygium cumini*). *Acta Hort.*, **890**:107
- Hartmann, H. T. and Kester, E. 1997. Plant Propagation: Principles and Practices. *Prentice Hall of India Private Limited*, New Delhi- 110 001.
- Hartmann, H. T.; Kester, D. E.; Davis Jr, F. T. and Geneve, R. L. 2002. *Plant propagation: principles and practices*. New Jersey: Prentice Hall.
- Hegde, S., Adiga, J. D., Honnabyraiah, M. K., Guruprasad, T. R., Shivanna, M. and Halesh, G. K. 2018. Influence of paclobutrazol on growth and yield of jamun *cv.* Chintamani. *International Journal of Current Microbiology and Applied Sciences*, **7**(1): 1590-1599.
- Henrique, A., Campinhos, E.N., Ono, E.O. and Sheila Zambello de Pinho. 2006. Effect of plant growth regulators in the rooting of *Pinus* cuttings. *Braz. Arch. Biol. Technol.*, **49**(2) http://dx.doi.org/10.1590/S1516-89132006000300002.
- Hiwale, S. 2015. Prolonging Shelf Life of Some Semiarid Fruits. In: *Sustainable Horticulture in Semiarid Dry Lands*. Springer (India) Pvt. Ltd. pp.365-369.
- Hoda, M. M., Abd el-rahman, G. F. and Abd el-rehman, M. E. 2010. Impact of gibberellic acid enhancing treatments on shortening time to budding of citrus nursery stocks. *Journal of American Science*, **6**(12), 410-422.
- Hota, S. N. 2016. Studies on the effect of gibberellic acid and media on germination, growth and survival of jamun [*Syzygium cumini* (L.) Skeels]. M.Sc. Thesis, Department of Horticulture, Jawaharlal Nehru Krishi Vishwa Vidyalaya, Jabalpur, Madhya Pradesh.
- Jain, N. and Babbar, S. B. 2003. Effect of carbon source on the shoot proliferation potential of epicotyl explants of *Syzygium cuminii. Biologia Plantarum*, **47**(1): 133-136.
- Jain, N. and Babbar, S.B. 2003. Regeneration of juvenile plants of black plum, *Syzygium cuminii* Skeels, from nodal explants of mature trees. *Plant Cell, Tissue and Organ Culture*, **73**: 257-263.
- Jejurker, S. H., Tanpure, U. R., Pawar, S. D., Joshi, R. R. and Wadikar, M. S. 2016. Microprapogation of callus from *Syzygium cumini* explant. *International Journal of Pure and Applied Bioscience*, **4**(1): 291-295.

- Jejurker, S.H., Tanpure, U.R., Pawar, S.D., Joshi, R.R. and Wadikar, M.S. 2016. Micropropogation of callus from *Syzygium cumini* explant. *Int. J. Pure App. Biosci.*, **4**(1): 291-295.

- Kadar, A. 1966. Effect of growth regulating substances on ripening and shelf life of mango. *Hort. Science*, **1**: 90-93.

- Khan, A.A. 1977. The Physiology and Biochemistry of Seed Dormancy and Germination, Elsevier Scientific Publications Co. Amsterdam, 477.

- Khan, A.A. 1992. Preplant physiological seed conditioning. *Hort. Rev.*, **14**: 131-181.

- Khandaker Mohammad M., Boyce, A.N., Osman, N. and ABM Sharif Hossain. 2012. Physiochemical and phytochemical properties of Wax Apple (*Syzygium samarangense* [Blume] Merrill & L. M. Perry var. Jambu Madu) as affected by growth regulator application. *The Scientific World Journal*, 2012:1-13, http://dx.doi.org/10.1100/2012/728613.

- Kiran, A., Sanjay, S. and Chavda, J. C. 2011. Effect of post-harvest treatments on quality of Jamun (*Syzygium cuminii* Skeels) fruits during storage. *Asian Journal of Horticulture*, **6**(2): 297-299.

- Kumar, R., Misra, K.K. and Mishra, D.S. 2011. Jamun: A boon for nature. *Indian Farmar's Digest*, **43**(11): 38-39.

- Kumar, V. 2007. Studies on airlayering and softwood grafting in jamun (S*yzygium cumini*). M. Sc. Thesis, Division of Horticulture, Gandhi Krishi Vignana Kendra, University of Agricultural Sciences, Bangalore.

- Litz, R. E. 1984. *In vitro* responses of adventitious embryos of two polyembryonic *Eugenia* species. *Hort. Science*, **19**: 720-722.

- Marler, T. E. and Mickelbart, M. V. 1992. Application of GA_{4+7} to stem enhances carambola seedling growth. *Hort. Science*, **27**(2): 122-123.

- Mishra, D.S. Tiwari, J.P. Kumar, R. Kumar, P.and Mishra, N.K. 2010. *In vitro* clonal propagation studies on the nodal explants of *Psidium cujavillis*. *Annals of Horticulture*, **3** (1), 58-62.

- Mishra, D.S., Singh, A. and Kumar, R., Singh, S., Singh, A.K. and Swamy, G.S.K. 2015. Jamun. In: Tropical and Sub tropical Fruit Crops: Crop Improvement and Varietal Wealth (Ed.) Ghosh, S.N., Jaya Publ. House, Delhi. pp. 375-390.

- Mishra, D.S., Tiwari, J.P. and Kumar, R. 2009. *In vitro* clonal propagation of guava (*Psidium guajava* L.). *Prog. Agric.*, **9**(1):79-84.
- Mishra, D.S., Tiwari, J.P. and Lal, S. 2007. *In vitro* cloning of guava (*Psidium guajava* L.) cv. Pant Prabhat. *Acta Hort.*, **735**:127-132.
- Misra, K.K. and Mishra, D.S. 2003. Role of plant growth substances in forest nursery stock production. In: *Forest Tree Seeds and Nursery Management: Current Trends* (Eds.) Singh, V. and Lavania, S.K. Bishen SMPS Publ&Dis of Scientific Books, Dehra Dun. pp. 193-209.
- Moneruzzaman, K. M., Hossain, A. B. M. S., Normaniza, O. and Boyce, A. N. 2011. Growth, yield and quality responses to gibberellic acid (GA_3) of Wax apple *Syzygium samarangense* var. Jambu air madu fruits grown under field conditions. *African Journal of Biotechnology*, **10**(56):11911-11918.
- Mulagund, J., Porika, H., Soorianathasundaram, K. and Deepika, C. 2015. Influence of growth regulators combined with chemicals to improve post-harvest fruit quality in banana *cv.* Nendran (*Musa* AAB). *Journal of Food Processing and Technology*, **6**(3): 1.
- Murasighe, T. and Skoog, F. 1962. A revised medium for rapidgrowth and bioassays of tobacco tissue cultures. *Physiol Planta*, **15**: 473-497.
- Naaz, A., Hussain, S. A., Naz, R., Anis, M. and Alatar, A. A. 2018. Successful plant regeneration system *via* de novo organogenesis in *Syzygium cumini* (L.) Skeels: an important medicinal tree. *Agroforestry Systems*, 1-11.
- Pandey, D. and Pathak, R. K. 1981. Effect of Rootstocks, IBA and Phenolic Compounds on the Rooting of Apple Cuttings. *Propagation of Horticulture*, **13**: 105-110.
- Pandey, J. 2011. Effect of indole butyric acid and time intervals on rooting and growth of jamun [*Syzygium cuminii* (L.) Skeels] cuttings. M.Sc. Thesis, Department of Horticulture Jawaharlal Nehru Krishi Yashwa Vidyalaya, Jabalpur, Madhya Pradesh.
- Paranjothy, K., Saxena, S., Banerjee,. S., Jaiswal, V.S. and Bhojwani, S.S. 1990. Clonal multiplication of woody perennials. In : S.S. Bhojwani (Editor) Plant Tissue Culture: Applications and limitations, Elsevier Science publisher, Amsterdam, pp. 190-219.
- Patil, H., Tank, R. V., Bennurmath, P. and M. Gotur. 2017. Effect of seed treatment and foliar spray of chemical substances on seedling growth of jamun (*Syzygium cuminii* L.). *International Journal of Chemical Studies*, **5**(5): 1676-1680.

- Patil, H., Tank, R. V., Bennurmath, P. and Patel, M. 2018. Effect of seed treatment on germination and seedling growth of Jamun (*Syzygium cuminii* L.). *International Journal of Current Microbiology and Applied Sciences*, **7**(2): 2654-2659.
- Prabhuling, G., Rashmi, H. and Babu A.G. 2017. *In vitro* regeneration of jamun (*Syzygium cuminii* L.) *cv.* AJG-85. *International Journal of Science and Nature*, **8**(4): 902-907.
- Prince, P. S. M., Kamalakkannan, N. and Menon, V. P. 2003. *Syzigium cumini* seed extracts reduce tissue damage in diabetic rat brain. *Journal of Ethnopharmacology*, **84**(2-3): 205-209.
- Quinlan, J. D. and Richardson, P. J. 1983. Effect of paclobutrazol (PP_{333}) on apple shoot growth. In: *International Workshop on Controlling Vigour in Fruit Trees,* **146,** Pp. 105-112.
- Quinlan, J. D. and Webster, A. D. 1982. Effects of the growth retardant PP_{333} on the growth of plums and cherries. In: *ISHS 21st International Horticultural Congress, Hamburg, Abstract* (Vol. 1071).
- Rao, D. and Chundawat, B. S. 1986. Effect of certain chemical retardants in ripening changes of banana *cv.* Lacatan at ambient temperature. *Progressive Horticulture*, **18**(3-4): 189-195.
- Rathore,V., Shekhawat, N S, Singh, R.P., Rathore, J S and Dagla, H R. 2004. Cloning of adult trrees of Jamun (*Syzygium cuminii*). *Indian J. Biotechnol.*, **3**:241-245.
- Remashree, A.B. Thomas, T.V., Raghu, A.V., Nabeesa, E. and Neelakandan, N. 2007. Micropropagation of *Syzigium cumini* (Linn.) Skeels. A Multipurpose Tree. *Research Journal of Botany,* **2***:* 208-213.
- Saha, A. K. 1969. Some investigations of vegetative propagation of jamun (*Syzygium cumini* Skeels). *Allahabad Farmer,* **18**: 187-189.
- Shanmugavelu, K. G. 1966. Studies on the effect of plant growth regulator on the seedling of some tree plant species. *South Indian Horticulture*, **14**: 24-25.
- Shanmugavelu, K.G. 1967. A note on air layering of Eugenia (*Eugenia jambolana* L.) and (*Azadirachta indica* L.). *South Indian Horticulture,* **15**: 70-71.
- Shnmungavelu, K.G. 1970. Effect of gibberellic acid on seed germination and development of seedling of some tree species. *Madras Agri. J.*, **57**: 311-314.
- Singh, A. K., Bajpai, A., Singh, V. K., Ravishankar, H., and Tandon, D. K. 2009. The Jamun. Central Institute for Subtropical Horticulture (ICAR), Rehmankhera, Lucknow, India. pp. 16-17.

- Singh, A.R. and Singh, J. R. 1974. Effect of foliar application of nitrogen and plant regulators on the growth characters mango. *Hort. Advances.*, **9**: 15-20.
- Singh, R., 1969, Fruits. *National Book Trust,* New Delhi.
- Singh, S., Singh, A. K., Singh, H. P., Bagle, B. C. and More, T. A. 2011. Jamun. Indian Council of Agricultural Research, New Delhi.
- Singh, Sanjay and Singh, A.K. 2012. Enjoy new jamun variety. *Indian Horticulture,* **57** (3): 9-11.
- Singh, Sanjay., Singh, A.K., Singh, H.P., Bagle, B.G. and More, T.A. 2011. Jamun. Indian Council of Agricultural Research, New Delhi.
- Suman, M., Sangma, P. D., Meghawal, D. R. and Sahu, O. P. 2017. Effect of plant growth regulators on fruit crops. *Journal of Pharmacognosy and Phytochemistry,* **6**(2): 331-337.
- Surakshitha, N. C., Mahadevamma, M. and Sharath Kumar, M. 2014. Studies on application of plant growth regulators and macro nutrients on seedling growth in jamun (*Syzygium cuminii* (L.) Skeels). *Indian Journal of Applied Research,* **4**(12): 2-5.
- Swamy, G. S. K., Patil, P. B., Athani, S. I. and Prabhushankar, D. S. 1999. Effect of organic and inorganic substances on germination of Jamun (*Syzygium cumini*) seeds. *Advances in Agricultural Research in India,* **11**: 89-91.
- Tandel, Y. N. and Patel, N. L. 2011. Effect of chemicals on growth, yield and economics of mango (*Mangifera indica* L.). *Karnataka Journal of Agricultural Sciences,* **24**(3):362-365.
- Thimann, K. V. and Koepfli, J. B. 1935. Identity of the growth-promoting and root-forming substances of plants. *Nature,* **135**: 101-102.
- Tuan, N.M. and Ruey, Y.C. 2013. Effect of Gibberellic acid and 2,4- Dichlorophenoxyacetic acid on fruit development and fruit quality of wax apple. *International Journal of Biological, Biomolecular, Agricultural, Food and Biotechnological Engineering,* **7**(5):299-305.
- Virendra, S. and Shafaat, M. 1996. Effect of gibberellic acid and urea on growth and development of seedling rootstock of Aonla (*Emblica officinalis* Gaertn.). *Journal of Applied Horticulture (Navsari),* **2**: 116-121.
- Vyas, S. V., Butani, A. M., Nurbhanej, K. H., Patel, M. S. and Parmar, L. S. 2017. Effect of time of air layering and IBA on Red Jamun (*Syzygium samarangense* L.) *cv.* Local. *International Journal of Pure and Applied Bioscience,* **5**(5): 272-279.

- Warrier, P. K. and Nambiar, V. P. K. 1993. Indian medicinal plants: a compendium of 500 species (Vol. 5). Orient Blackswan.
- Wiesman, Z. and Lavee, S. 1995. Enhancement of IBA stimulatory effect on rooting of olive cultivar stem cutting. *Scientia Horticulturae*, **65**: 189-198.
- Wiesman, Z. and Riov, J. 1989. Paclobutrazol and urea-phosphate increase rooting and survival of peach "Maravilha"softwood cuttings. *Hort. science*, **24**: 908-909.
- Yadav, U., Bisht, S., Siddique, S.S. and Babu, R. 2014. micropropagation of *Syzygium cuminii* L.using *in vitro* technique. *International Journal of Scientific & Engineering Research*, **5**(10): 1517-1528.
- Yadav, U., Lal, M. and Jaiswal, V. S. 1990. *In vitro* micropropagation of the tropical fruit tree *Syzygium cuminii* L. *Plant Cell, Tissue and Organ Culture*, **21**(1): 87-92.
- Yeshitela, T., Robbertse, P. J. and Stassen, P. J. C. 2004. Paclobutrazol suppressed vegetative growth and improved yield as well as fruit quality of 'Tommy Atkins' mango (*Mangifera indica*) in Ethiopia. *New Zealand Journal of Crop and Horticultural Science*, **32**(3): 281-293.
- Zimmerman, P. W. and Wilcoxon, F. 1935. Several chemical growth substanceswhich cause initiation of roots and other responses in plants. *Contrib. Boyce Thompson Inst.*, **7**: 209-228.
- Zimmerman, P. W., Crocker, W. and Hitchcock, A. E. 1933. Initiation and stimulation of roots from exposure of plants to carbon monoxide gas. *Contrib. Boyce Thompson Institute, (United States)*, **5**: 1-17.

16

Plant Growth Regulators in Kokam

T. R. Ahlawat, A.K. Pandey, Dharmistha Patel and Suchismita Jena

1. Introduction

Kokum (*Garcinia indica* Choisy) is an important horticultural crop of South Western India belonging to family Clusiaceae. It is commonly known as kokum butter tree, mangosteen oil tree or brindonia tallow tree and is a native of Southern Maharashtra and Goa. The kokum tree is treated as a "zero attention crop", as it input requirement is minimal. Konkan's kokum is regarded as 'Cool King' of Indian fruits. It is a polygamodioceous, slow growing moderate size tree which reaches a height of about 10 to 15 metres. Kokam is an important fruit tree that has remained under-exploited and neglected. It is generally not cultivated on a commercial scale and is mostly found as a kitchen garden plant or mixed crop in plantations of coconut, arecanut, road side plants or in forest areas. It is mostly found in Konkan region of Maharashtra, Goa, Coastal Karnataka, Kerala, forests of Assam, Khasi Jayanti hills, West Bengal and South Gujarat.

Traditionally, kokum has been used for many years as a medicinal treatment for diarrhea, skin infection and wounds in throughout South Asia. The fruit rind is a rich source of á Hydroxy Citric Acid (HCA) which prevents fat accumulation in body cells and serves as a natural source for production of anti-obesity drugs. (Patil *et al.*, 2005). Fruits also possess anthelmintic, cardiotonic and anti-tumour properties and are useful in the treatment of piles and dysentery (Deodhar Swapna *et al.*, 2014). Kokum fruit contains compounds that have antioxidant, anti-bacterial and antifungal properties. Scientific research indicates activity against several cancer cell lines, including breast cancer, liver cancer and leukemia. In addition, kokum also exhibits antihistamine and anti-inflammatory properties.

There is a dearth of research on the use of plant growth regulators in kokam. Nevertheless, PGRs were evaluated for their effect on germination and micro propagation of kokam where are reviewed under the following headings.

2. Use of PGRs on Growth and Propagation

Kokum is conventionally propagated by seeds. Seeds are recalcitrant in nature due to high sensitivity to desiccation and freezing. Seed germination is not a hurdle in kokum. However, growth regulators like GA and Cycocel can be used to improve seed germination in kokum (Oscar, 1983). Seed treatment with Cycocel at 500 ppm improved germination in kokam (Haldankar *et al.*, 2012). Jadhav (1999) studied the effect of growth regulators on seed germination and seedling growth of kokum at Horticulture College of Agriculture, Dapoli. Foliar application of GA_3 at 200 ppm emerged as the best treatment for improving growth performance of kokum seedling.

Pawar *et al.*, (2018) conducted an experiment at Department of Horticulture, College of Agriculture, Dr. Balasaheb Sawant Konkan Krishi Vidyapeeth, Dapoli, with ten pre-sowing seed treatments. The seed material was collected during the fruiting season. The seeds were extracted from pulp with hand and were washed thoroughly with water to remove all adhering pulp and used for the experiment. The results regarding pre-soaking of seed in bioregulators on growth parameters in kokum showed the significantly varied result at 180 DAS. Soaking of seeds in 50 ppm GA solution for 24 hrs had recorded better result regarding height of seedling (18.94 cm) and girth of seedling (0.37) and number of leaves (17.93). Leaf area (55.29) was significantly highest with the treatment.

3. Use of PGRs in Micropropagation

In most of the tropical tree species *in vitro* establishment and multiplication is difficult. Apomictic nature of seeds and seed segments which can represent the mature trees are ideal explants for *in vitro* establishment and multiplication.

For establishment of sterile cultures, fungal contamination and browning of explants is the major hurdles. Season of plant material collection and culture vessels used for *in vitro* sprouting are also observed to be significant factors in establishment of sterile cultures.

Micropropagation from mature tree explants is often affected by excessive contamination, phenolic exudation, less number of shoot initials, difficulty in elongation and rooting of micro shoots (Purohit *et al.*, 1998). Poor root initiation on response from mature shoots is attributed to lack of juvenile factor in older tissues (Bonga, 1982).

Malik *et al.*, (2005) studied rapid regeneration of plantlets *via* adventitious bud differentiation on mature seeds of *G. indica*. High frequency direct shoot proliferation was induced in seed segments cultured on Murashige and Skoog's medium supplemented with cytokinins (BAP, kinetin and TDZ) alone and in combination with auxin (NAA). Among various combinations, BAP proved to be the most effective. Proliferating shoot cultures were established by repeatedly subculturing the shoot nodes on MS medium supplemented with 5 μM BAP. Multiple shoots were formed within 4-5 weeks of culture. Elongation of induced shoots was achieved on MS basal medium containing 0.2% activated charcoal. Maximum rooting (91.66%) occurred in shoots cultured on half-strength MS medium supplemented with 10 μM IBA. For *in vitro* conservation, shoot cultures were maintained on medium supplemented with 0.5 μM BAP and subculture duration could be enhanced up to maximum of 11 months.

Chabukswar Meera and Deodhar Manjushri (2006) studied the restoration of rooting competence in a mature plant of *Garcinia indica* through serial shoot tip grafting *in vitro* at BSKAU, Dapoli, Maharashtra. Apical and axillary buds from a high yielding, early fruiting elite tree (more than 20 years old) were cultured in woody plant medium supplemented with 0.9 μM BA. Multiple shoots were obtained on WPM basal medium containing 8.9 μM BA and 0.5 μM TDZ. Elongation of axillary shoots was obtained in half-strength WPM medium supplemented with 0.4 μM BA. For restoration of rooting competence, *in vitro* shoots obtained from nodal explants of a 20-year-old elite tree were repeatedly grafted on *in vitro* grown juvenile seedlings. *In vitro* grafts were maintained in half-strength Murashige and Skoog (MS) basal medium. The graft union was seen within 25 days. Six to eight weeks after graft union, regrafting was again done and the procedure was repeated five times. After the fifth regrafting, 75% of scions regained their ability of rooting. The rejuvenated rooted scions were hardened successfully.

Occasionally, mature trees of *G. indica* in nature shows outgrowth of shoots from its roots. Hence, the root tissue has potential for shoot regeneration. Deodhar Swapna *et al.* (2008) studied the *de novo* shoot regeneration from root cultures of *Garcinia indica* Choicy at National Chemical Laboratory, Pune. Half strength MS medium was supplemented with BAP (0.44-2.22 μM) containing 2% of sucrose and 0.8% of agar. Roots of plantlets when cultured for long time on half strength MS medium supplemented with BAP (0.44-2.22 μM) showed production of *de novo* shoots. Shoots were transferred to elongation medium consisting of Woody Plant Medium (WPM) supplemented with BAP (4.44-22.69 μM), IAA (5.71 μM) and kinetin (4.65 μM). It was observed that shoot length increased to 1-2 cm. WPM medium supplemented with NAA (2.69-10.74 μM) and (4.90 μM) induced rooting within 20-25 days. Using this protocol, 20-25 plantlets could be regenerated from single root explants within 3 to 4 months.

Research on direct shoot bud formation from leaf explants of seedling and mature trees have been reported in *G. mangostana*. However, very limited work has been done on regeneration and conservation of *G. indica* using leaf explants. Malik *et al.* (2010) conducted research on regeneration of plantlets *via* adventitious bud differentiation on leaf explants of *Garcinia indica* Choicy at NBPGR, New Delhi. The leaves were cultured on Murashige and Skoog's medium supplemented with cytokinins (BAP, Kinetin and TDZ) alone and in combination with auxin (NAA) for direct shoot proliferation. Maximum number of shoots were formed (63.89%) in explants on MS medium supplemented with 1.0 μM TDZ after 2-4 weeks of culture. Elongation of the induced shoots was achieved on MS basal medium containing 0.2% activated charcoal. Maximum rooting (83.33%) was observed in shoots cultured on half-strength MS medium supplemented with 10 μM IBA.

Baskaran and Krishnan (2011) developed high frequency plant regeneration system for the production of high yielding elite clones of *G. indica via* direct organogenesis. A maximum number of 86.2 shoot buds per explants were induced from the mature seed segments cultured on WPM supplemented with 2.0 mg dm^{-3} BA and 1 mg dm^{-3} IAA. Rooting was achieved on half-strength WPM medium supplemented with 3 mg dm^{-3} IBA.

Kalia *et al.*, (2012) conducted research on *in vitro* morphogenetic studies on three apomictic species of *Garcinia* at Haryna Agricultural University, Hissar. High frequency direct shoot bud differentiation was induced in aseptic seed cultures of *G. indica* and *G. cambogia* on MS medium supplemented with cytokinins (BAP, kinetin or TDZ). Amongst the three cytokinins tested, TDZ (0.1-0.5 μM) was most effective for adventitious bud differentiation in both *G. indica* and *G. cambogia,* however, the proliferating buds failed to elongate. Substantial number of buds induced on BAP supplemented media elongated into shoots after subculture on elongation medium. Addition of NAA along with cytokinins in the induction medium enhanced callusing without improvement in bud induction response. The induced adventitious buds were elongated on MS basal medium containing 0.2% activated charcoal. Direct rooting was achieved in both *G. indica* and *G. cambogia* on auxin supplemented media with best response at 10 μM IBA concentration. The *in vitro* raised plantlets showed 90% survival in the field when transferred after hardening and acclimatization.

Mohan Simi *et al.,* (2012) studied the *in vitro* and *in vivo* adventitious bud differentiation from mature seeds of *G. indica* at IISR, Calicut. MS medium supplemented with various concentrations of BAP (0, 1, 2, 2.5, 3, 4 mg/l). MS medium supplemented with BAP 2.5 mg per litre gave best response and induced multiple shoot initiation. Root initiation took place in MS medium with or without NAA but root elongation was faster in MS medium supplemented with 2 mg per litre of NAA.

Chauhan *et al.,* (2012) developed the high frequency plant regeneration protocol through direct organogenesis for *G. indica* Choicy using leaf explants at National Burro of Plant Genetic Resources, Regional Station Trissure, Kerala. The leaf explants were cultured on MS medium supplemented with different concentrations of BAP and NAA. The highest number of shoots per explants was obtained on the medium supplemented with 22.2 μM BAP. The highest percentage of rooted plants was obtained on MS basal medium (MS + 0.45 μM BAP) supplemented with 0.81-1.07 μM NAA, while maximum root length was recorded on the medium supplemented with 0.05 μM NAA. MS medium supplemented with 0.45 μM BAP and 0.81 μM NAA was recorded optimal for rooting of the shoots.

Vanave *et al.,* (2014) studied *in vitro* propagation in kokam with the help of shoot tip explants in Murashige and Skoogs modified medium. They took various concentration of BA, NAA and GA_3. A combination of 10 ppm BA + 5 ppm NAA + 5 ppm GA_3 was found to be best for establishment of kokam shoot tip cultures. However, various concentration of auxins in MS medium were found ineffective to induce root initiation in shoot cultures.

The use of WPM basal fortified with 2% sucrose and 8.87 μM BAP resulted in maximum sterile and healthy buds. TDZ in the range 0.22-4.54 μM along with BAP induced multiple shoots. Highest 63% rooting was obtained on WPM medium with 4.90 μM IBA whereas, maximum 60 plantlets survived in green house after their transfer to soil (Deodhar Swapna *et al.,* 2014).

4. Use of PGRs in Fruit Ripening

Kokum is a very slow growing tree and flowers in 7 to 8 years after planting. Flowering is observed in November-February and fruits ripen in April-May. Ripe fruits have agreeable flavour and sweetish-acid taste. They contain substantial amount of mallic acid and a little tartaric or citric acid. Kokum fruits are periodically plucked, as all fruits on the tree do not mature at the same time. The number of plucking varies from tree to tree. Generally 6-8 pluckings are required in high yielding plants. It is therefore necessary to reduce the number of plucking so as to reduce the cost and time involved in harvesting. For synchronized harvesting, ethylene is generally used (Singh *et al.*, 1979; Sandhu and Singh, 1983).

Somavanshi *et al.*, (2010) studied the effect of ethrel on yield and quality of kokum at College of Agriculture, Dapoli, Maharashtra. Mature green fruits were sprayed with ethrel at various levels (100, 150, 200, 250 and 300 ppm). They reported that ethrel at 300 ppm reduced the number of plucking, increased fruit yield, TSS, acidity, ascorbic acid, reducing sugar, non reducing sugar and total sugars.

References

- Baskaran, M. and Krishnan, S. 2011. High frequency plant regeneration from the mature seeds of *Garcinia indica*. *Biologia Plantarum*, **55** (3): 554-558.

- Bhuyan, L., Sharma, C.L., Singh, B. and Sharma, M.B. 2017. Vegetative propagation of two important *Garcinia* species of Assam, N.E. India. *J. Medicinal Plants Studies*, **5**(3): 273-277.

- Bonga, J. M. 1982. Vegetative propagation in relation to juvenility, maturity and rejuvenation. **In**: *Tissue Culture in Forestry*, (Eds.) Bonga, J.M.; Durzan, D.Z. Nijhoff, The Hague, pp. 387-412.

- Chabukswar Meera, M. and Deodhar Manjushri, A. 2006. Restoration of rooting competence in a mature plant of *Garcinia indica* through serial shoot tip grafting *in vitro*. *Scientia Hort.*, **108**: 194-199.

- Chauhan, D.K., Thakur, A.K., Dass, A., Lima, J.M. and Malik, S.K. 2012. Direct organogenesis from leaf explants of *Garcinia indica* Choicy: An important medicinal plant. *Indian J. Biotech.*, **11**: 215-219.

- Deodhar Swapna R., Thengane, R.J. and Thengane, S.R. 2008. *De novo* shoot regeneration from root cultures of *Garcinia indica* Choisy. *Indian J. Experimental Bio.*, **46**: 482-486.

- Deodhar Swapna; Pawar, K., Singh Nitasha, Thengane, R.J. and Thengane, S.R. 2014. Clonal propagation of female plants of *Garcinia indica* Choisy: A tree species of high medicinal value. *J. Applied Bio. Biotech.*, **2**(06): 18-25.

- George, S.T., Latha, A.K.B. and Mathew, K.L. 1994. Softwood grafting in kudampuli. *Indian Cocoa, Arecanut Spices J.*, **18**: 51.
- Haldankar, P.M., Pawar, C.D., Kshirsagar, P.J. and Kulkarni, M.M. 2012. Present status and future thrust areas in production technology of Kokum. **In:** *Resource Book on Kokum* (*Garcinia indica* Choisy), (Eds.) Braganza, M., Shirodkar, A., Bhat, J. and Krishnan, S., Western Ghats Kokum Foundation, Panaji – Goa, pp. 46-55.
- Haldankar, P.M., Salvi, M.J., Joshi, G D. and Patil, J.L. 1992. Standardisation of softwood grafting in kokam. *J. Maharashtra Agril. Uni.*, **17**: 227-229.
- Haldankar, P.M., Salvi, M.J., Joshi, G.D. and Patil, J.L. 1993. Factors influencing softwood grafting in kokam (*Garcinia indica* Choisy). *Indian Cocoa, Arecanut & Spices Journal*, **17**(1-2): 15-18.
- Jadhav, V.M. 1999. Effect of growth regulators of growth of kokum (*Garcinia indica* choisy) seedlings. M.Sc. Thesis, Konkan Krishi Vidyapeeth, Dapoli.
- Kalia, R.K.; Malik, S.K. and Chaudhury Rekha 2012. *In vitro* morphogenetic studies on three apomictic species of *Garcinia*. *J. Pl. Biochem. and Biotech.*,, **21**(2): 279-285.
- Malik, S.K., Kalia, R.K. and Chaudhury Rekha 2010. *In vitro* regeneration of *Garcinia indica* using leaf explants. *Indian Journal of Plant Physiology*, **15**(3): 262-266.
- Malik, S.K.; Chaudhury, R. and Kalia, R.K. 2005. Rapid *in vitro* multiplication and conservation of *Garcinia indica*: A tropical medicinal tree species. *Scientia Horticulturae*, **106**(4): 539-553.
- Mohan Simi, Parthasarathy, U. and Babu, K.N. 2012. *In vitro* and *in vivo* adventitious bud differentiation from mature seeds of three *Garcinia* spp. *Indian Journal of Natural Products and Resources*, **3**(1): 65-72.
- Oscar, S.A. 1983. Studies on propagation of kokum (*Garcinia indica* Choisy). M.Sc. thesis submitted to Konkan Krishi Vidyapeeth, Dapoli, Maharashtra, India.
- Patil, B.P., Gawankar, M.S., Sahvekar, V.V. and Jambhale, N.D. 2005. Status of existing kokum plantation in Maharashtra. **In:** *Preview of 2nd National Seminar on Kokum* organized during March 4-5, 2005, Goa University, Goa, India. pp. 17-19.
- Pawar, RM; Golvankar, GM; Agare HR; and Thorat SB. 2018. Effect of soaking of seeds in bioregulators on growth of kokum seedling, *International Journal of Chemical Studies*, **6**(5): 2931-2935

- Purohit, S.D., Kiran, T., Joshi, P. and Samar, D. 1998. Micropropagation of some woody tree species of Aravallis in Rajasthan. **In:** *Plant Tissue Culture and Molecular Biology* Srivastava, P.S. (Ed.) Narosa Publishing House, New Delhi, India, pp. 84-125.

- Rema, J., Krishnamoorthy, B. and Mathew, P.A. 1997. Vegetative propagation of major tree spices – a review. *Journal of Spices and Aromatic Crops*, **6**(2): 87-105.

- Sandhu, A.S. and Singh, Z. 1983. Effect of ethephon on maturation and fruit quality of peach. *Punjab Hort. J.*, **23**(3-4): 172-175.

- Singh, I.S., Singh, H.K. and Chauhan, K.S. 1979. A note on effect of pre-harvest application of ethephonon ripening and quality of Phalsa (*Grewia asiatica*). *Haryana J. Hort. Sci.*, **8**(3-4): 196-197.

- Somavanshi, A.V., Haldankar, P.M., Rangwala, A.D., Joshi, G.D. and Burondkar, M.M. 2010. Effect of foliar application of ethrel on yield and quality of kokum (*Garcinia indica* Choisy). **In:** *Garcinia* Genetic Resources: Linking Diversity, Livelihood and Management (Eds.) Vasudeva, R.; Janagoudar, B.S.; Reddy, B.M.C.; Sthapit, B. and Singh, H. P., pp. 132-134.

- Vanave, P.B., Kunkerkar, R.L. and Khamkar, M.B. 2014. *In vitro* propagation in kokam (*Garcinia indica* Choisy). *Ecology, Environment and Conservation,* **20**(1): 377-379.

17

Plant Growth Regulators in Litchi

D. S. Mishra and P. K. Nimbolkar

1. Introduction

Litchi (*Litchi chinensis* Sonn.) is the most important fruit of Sapindaceae family. It is originated in the Southern region of China and has become widely distributed in the tropic and subtropics. The efficient utilization of plant growth regulators can provide a dramatic change in fruit production. At present, the plant growth regulators commonly used and proved effective on litchi are, paclobutrazol, gibberellin, ethephon, NAA *etc.* These plant growth regulators have the efficacies of growth inhibition, flower induction, fruit setting, ripening and keeping quality of fruits. In India litchi is grown mainly in the states of Bihar, West Bengal and Uttarakhand. It is also grown on limited scale in Tripura, Orissa, Punjab, Himachal Pradesh, Assam and the Nilgiri hills in the south. The problems responsible for low economic potential of litchi cultivation in various litchi growing regions include availability of genuine propagating material, poor fruit set (Sarkar *et al.*, 1984), heavy fruit drop (Singh and Phogat, 1984), fruit cracking (Bhat *et al.*, 1997) and inferior fruit quality (Brahmachari and Rani, 2001). Litchi bears heavy crop in one year and light or no crop in the adjoining year (Pandey and Sharma, 1989). On the basis of above problematic issues of litchi cultivation and production, in this chapter various aspects related use of PGRs in litchi production has been discussed.

2. Role of PGRs on Propagation

As a requirement of faster multiplication rate and availability of planting material, plant propagation methods are of utmost importance in horticultural crops, especially in fruit trees. In some fruit plants, the percentage of success in asexual propagation is extremely low. Litchi can be multiplied sexually but owing to disadvantages of seedling plants, it is chiefly propagated through vegetative means. The absence of easy and reliable clonal propagation method limits large scale cultivation of promising varieties while use of modern techniques like micro propagation has not proved very successful in litchi (Amin *et al.*,1996). Among the various methods of propagation, marcotting or air layering is the most common and convenient method (Bhambota *et al.*, 1968). Nevertheless, the major bottleneck associated with this method of propagation is varying degree of success of air layering, high mortality of detached layers during establishment in nursery on their own root system. (Bhambota *et al.*, 1968, Menzel, 2001; Singh *et al.*, 2009) also suggested that, air layering with use of indole butyric acid (IBA) is most convenient, common and commercial method of propagation for promoting rooting in litchi. It was hypothesized that by applying proper concentration of plant growth regulators and use of suitable growing medium will help in the production of quality planting material with better root system of air-layers.

The use of plant growth regulators especially indole-3-butyric acid (IBA), naphthalene acetic acid (NAA) *etc.* have been advocated for accelerating rooting in litchi layers (Ram and Majumder, 1983). Chand *et al.*(2014) reported that air layering in litchi in the month of August with the use of IBA @600 ppm concentration can be practiced for obtaining higher rooting and better survivability of air layers of litchi cultivars 'Rose Scented' and 'Calcuttia' under *Tarai* conditions of Uttarakhand. Nanda and Kochhar (1985) reported that application of root promoting substances during layering induces profuse rooting within a short period of time and IBA has been found most effective. However, more information regarding use of growth regulators, ways to overcome higher mortality rate and improving the survival rate of litchi air layers needs to be generated.

In litchi cv. 'Purbi' IBA 5000 ppm produced maximum rooting layers (90.00 %) also proved better in survival percentage (86.66 %) and fresh weight of roots (4.37 g) dry weight of roots was found maximum (1.11 g) by IBA 5000 ppm + NAA 5000 ppm treated layers. Finally it was confirmed that the application of IBA 5000 ppm alone or in combination with NAA at higher concentration as IBA5000 ppm plus NAA 5000 ppm were found best for various parameters of root formation, root development, quality and growth of layers in the nursery (Das and Prasad, 2014).

Chawla *et al.*, (2012) found that application of IBA 5000 ppm treated layers took lesser days to root initiation (25.99 days) and also gave better results in relation to per cent rooting (86%), number of first (27.30) and second (41.20) order roots, total length of first order roots (2.14 m), mean root thickness (1.09 mm), fresh (2.89 g) and dry (1.11 g) weight of roots, fresh (46.29 g) and dry (28.00 g) weight of shoot and root: shoot ratio (0.04) of air layer in comparison to all other treatments. They concluded that application of IBA and NAA in combination (IBA 4500 ppm + NAA 200 ppm) had significant influence on rooting of air layers and was closely followed by IBA 5000 ppm.

Litchi also produced roots successfully from softwood cuttings (Ochse, 1952; Hayes, 1957; Ahmed, 1961) and semi-hardwood cuttings (Galloway, 1922; Paxton *et al.*, 1978). Girdling, growth regulators, misting and bottom heat technique improve the success rate for such cuttings. Abutiate and Nakasone (1972) noted that girdling at the base of the shoot 8 weeks before taking the cuttings increased rooting of 'Brewster' in Hawaii compare with control. Several researchers (Bhandary and Shivashankar, 1970; Abutiate and Nakasone, 1972) found that IBA is more effective than IAA, whereas the application of these hormones were not superior to individual applications. Treated shoots with ethephon @ 200 ppm and IBA @ 2000 ppm gave 85 % rooting of cuttings (Mitra and Bose 1991). However, this method of propagating litchi is not become popular amongst nursery growers in comparison to the air-layering method as it requires special structures like mist chambers for easy rooting and high survival percentage. Moreover, rooting is not that much easy as in case of air-layers that too thick and brittle in nature therefore, later on after transplanting in the field survival percentage is less and plants are prone to uprooting during heavy wind blow due to shallow root system.

3. Role of PGRs on Growth and Development

According to Zhang *et al.*, (2004) application of Forchlorfenuron and SS 3307 could inhibit the germination of winter shoot tip of litchi, decrease growth amount of winter shoot tip, reduce nutrient consumption in tree body and increase spike emerging rate and fruit setting rate of litchi, which lays basis for abundant production of litchi, especially for that with vigorously vegetative growth. Moreover, Forchlorfenuron and SS 3307 agent have obviously promoting effect in increasing spike emerging rate, flower spike length, flower spike width, fruit setting number, spike weight, per plant yield. Gao, *et al.*, (2001) studied the internal hormonal relationships between large and small fruits of two litchi cultivars namely Wanxuan No. 5 and Nuomici. The results from their study concluded that the GA+IAA+CTK/ABA hormones responsible for enlarging fruit size by activating the cell division in ovary, pericarp and aril. Li *et al.*(1992) reported that spraying ethephon inhibit shoot tip germination for 22-27 days in litchi varieties of 'Baila', 'Heiye' and 'Guiwei' as well as make the leaves quickly ripened, the photosynthesis intensified, the nutrients accumulated and flower bud differentiation promoted. In conclusion, they suggested that ethephon is involved in the inhibition of germination in winter shoots of litchi.

4. Role of PGRs on Seed Dormancy, Germination and Seed Viability

Prasad *et al.*,(1996) conducted seed germination experiment of litchi cultivars namely 'Deshi', 'Kasba', 'Purbi', and 'Early Bedana' at Sabour (Bhagalpur), India. Seeds of each cultivar were treated with 1, 10 and 100 mM of IBA, GA_3 and ethephon. They found that soaking litchi seeds for 1 h in ethephon improved germinability of 'Deshi' at 10 mM and for 'Early Bedana' at 1 and 100 mM. However, 1, 10, or 100 mM ethephon reduced germination in 'Kasba' and in 'Purbi'. The improving effect of IBA on 'Deshi' seed was similar to that of ethephon. In 'Purbi', 10 or 100 mM IBA improved germination. GA_3 at 100 mM increased the germination of all cultivars and finally suggested that GA_3 plays an important role in litchi germination. Sharma and Dhillon (1986) observed that endogenous levels of gibberellins in litchi (*Litchi chinensis* Sonn) seeds decline at maturity stage. Such a decline may be a limiting factor in maintenance of seed viability and/or germination. Das *et al.* (1999) found 100% germination rate of lychee seeds in *in-vitro* on MS liquid medium supplemented with using 20 mg L^{-1} BAP.

5. Role of PGRs on flowering, Crop Regulation, Fruit Set, Fruit Growth and Development

Kumar *et al.*, (2014) examined that different litchi cultivars show variations in their flowering and bearing habits, may accordingly be classified as regular, irregular, shy bearing *etc.* Litchi bears the flowers mainly staminate, hermaphrodite and pseudo-hermaphrodite. The first flowers to pollen are males, followed by hermaphrodites functioning as females and pseudo-hermaphrodites functioning as males. Kumar *et al.* (2014) mentioned that the two spray of 100 ppm NAA, potassium nitrate (1%) or ethrel (0.05%) during October and November promote flowering and application of paclobutrazol (PP 333) @ 5ml/m canopy diameter at 90 days before the expected date of flowering inhibit the shoot growth and panicle length ultimately increases the fruit set and yield in litchi. Spraying of NAA has been also found to enhance the fruit set, fruit retention and fruit weight. Applications of 100-150 mg/l of the sodium salt of NAA to litchi in Florida in early autumn discouraged vegetative flushes and increased flowering and yields over controls in most years (Ledin, 1953, 1954, 1955; Young 1957). Ethephon also play an important for enhancing the number of flowering shoots in litchi (Sittichaikasem, 1974; Huang and Weng, 1978; Subhadrabandhu, 1986; Subhadrabandu and Koo-Duang, 1987).

Potjanapimon *et al.*, (2000) found that the cytokinin was low in litchi cv. 'Hong Huay' in Chiang Mai 6-8 weeks prior to flowering and increased at 4 weeks to reach a maximum two weeks prior to flowering. Gibberellins were high at four and three weeks before flowering and decreased two weeks prior to flowering to below detectable levels at flower emergence. The flushing habit of litchi varieties was intimately connected with irregular bearing. Problem is generally due to failure of flower initiation which puts forth vegetative growth prior to panicle emergence and flowering, thus eliminating the crop completely. Observations on young as well as old 'Calcuttia' trees showed that vegetative growth after September was at the expense of fruiting in the following year (Mustard and Lynch, 1959).

Several research workers advocated the use of various growth retardants as an alternate approach in litchi to restrict vegetative growth before panicle emergence Chapman *et al.* (1980). The highest amount of flowering was observed with application of ethephon treatment @ 100 ppm in cultivar 'Hong Huay' (Subhadrabandhu and Duang, 1985). However, Negi *et al.* (2012) obtained early panicle emergence, improved fruit retention and yield in 'Rose Scented' litchi when sprayed with TIBA (1 g/l) at monthly interval from September to December. TIBA is considered a polar auxin transport inhibitor and increases the endogenous cytokinin level in the lateral buds. There is may be a positive relationship between cytokinin level and flower formation which may be due to the positive impact of TIBA. Sing *et al.*(2012) reported that soil drench application of PBZ @7.5 ml proved to be the most effective treatment for suppressing shoot growth, panicle size, male flower percentage fruit drop and sex ratio. While same treatment resulted in increased hermaphrodite flower percentage, fruit set and fruit retention in litchi cv. 'Calcuttia'. At the stage of flower bud differentiation, spraying ethephon could remarkably reduce flower number, make flower bud abscission ahead of time, reduce the nutrition consumption of tree and improve flower quality (Li *et al.*,1992). Diao (2006) revealed that spray with mixture of 0.05%-0.06% ethephon and 0.05% PBZ has the inhibitory effect on shoot tip growth and promote flowering.

Li and Zhang (1987) examined that before heading stage, timely spraying with suitable concentration of B9 and ethephon could economize nutrients consumed for flower, spike, and thereby improve flower quality. When buds begin to germinate or sprout for 2-3 cm under low temperature and drought conditions, spraying the mixture of B9 and ethephon has a better effect. Treatment of growth regulators including ethephon and paclobutrazol could remarkably change the contents of various endogenous hormones in 'Jizui' litchi trees.

At the early stage of flower bud differentiation, contents of IAA and GA_3 were remarkably reduced while contents of ABA and ZT were found increased level (Tang, 2006). Girdling of branches 3 to 4 cm in diameter with hardened flush in May or foliar application of 0.5 g paclobutrazol + 0.4 g of ethephon l^{-1} promoted flowering in unproductive litchi trees of cultivar 'Tai So' in Mauritius (Ramburn, 2000). However, related research showed that paclobutrazol treatment was not to a certainly effective in promoting litchi flowering, just effective to the plant with medium flowering proportion (40-60%), to the plant with strong vigour (<30% flowering) this effect was sometimes poor, but to the plant with high flowering proportion (>70 % flowering) it had less effect or even reduced flowering (Menzel and Xiong, 1999).

Chen *et al.*, (1959) reported that gibberellins @ 250 ppm effectively extend the opening of female flowers which helped to get the pollen from male flowers and the same dose encouraged the pollen germination which improved the fruit setting in litchi.

Exogenous CKs could promote flower induction in monocarpic plant as well as polycarpic plant (Bangerth, 2009). In addition, Werner *et al.* (2003) confirmed the essentiality of CKs, using a transgenic CK-deficient Arabidopsis plants showing no flowering. Li and Ji, (1984) studied the effect of cytokinin on 'Nuomici' litchi tree outgrown in 1970-80s, and they found that zeatin and several other endogenous cytokinins played important roles in flower bud differentiation stage, once entered into inflorescence primordium formation stage and rachis formation stage, cytokinins content increased gradually and reached the peak at flower organ formation stage (Wu and Huang, 1986). Chen and Ku (1988) reported that foliar application of kinetin @ 200 ppm to 5 month old 'Heiye' litchi stems in the autumn, about 2 weeks after ethephon @ 200 ppm treatment, stimulated bud break 1 month earlier than in control stems. This treatment resulted in 60 % of the stems flowering, and in combination with ethephon raised the flowering rate to 80 % compared with no flowering in the controls. Chen (1991) found that 100 μg kinetin in a 5μl drop of 1 M acetic acid applied weekly to buds from 6 weeks before initiation of generative shoot development, stimulated bud break 1 week early and increased the proportion of generative shoots over non-treated controls. Thus, manually applying exogenous cytokinins or some growth-inhibitory nutrient substances during flower bud differentiation stage, could promote flower bud differentiation. His experiment from another angle proved the hormone balance theory which holds that flower bud formation promoted by growth inhibitor is probably related with the increase of cytokinin/gibberellin ratio. High cytokinin/gibberellin value promotes the generation of large amount of flower bud, while low cytokinin/ gibberellin value leads to the decrease of flower setting capacity.

6. Role of PGRs on Parthenocarpy

Seedless fruit, botanically termed parthenocarpic, do not have any seeds because the ovary is not fertilized. These types of fruit are rare in commercial varieties. One of the fruit on a twin-fruit stalk is sometimes parthenocarpic, although the companion fruit is usually normal or at least aborted-seeded. It is conceivable that plant growth substances are translocated from the actively growing companion fruit to its neighbour to enable it to grow. Artificial induction of shriveled seeds was achieved in 'Huaizhi' litchi by Liang and Qiu (1998) when maleic hydrazide (MH) was applied @ 1000 mg/l about 2 weeks after bloom, when the liquid endosperm was full. The fruit weighed about the same as controls, but had 10% more aril. Stern *et al.* (1997) reported that shriveled seeds could be induced in Floridian by spraying trichloro-phenoxy propionic acid or 2,4,5-TP at 100 mg/l when the fruit weighed 1 g, whereas the rate of seed abortion induced was less than 30%. Wang *et al.* (1997) found seedless fruit in 'Feizixiao' with 2,4-dichhloro-phenoxy acetic acid or 2,4-D at restricted by the ability of the cultivar to set fruit parthenocarpically.

7. Role of PGRs on Control of Fruit Drop

The phenomenon of heavy fruit drop is typical to litchi (Menzel, 1984; Joubert, 1986; Galan-Sauco and Menini, 1989) and other fruit trees that produce a very large number of pistil-bearing flowers, such as mango (Singh, 1960) and avocado (Papademetriou, 1976). However, the fruit drop in litchi is often excessive. One of the most effective auxin for its control is 2,4,5-trichloro-phenoxypro- pionic acid (2,4,5-TP). Stern *et al.* (1995) found `Mauritius' litchi fruit drop to be reduced and yield significantly increased (42±112%) with 100 ppm 2,4,5-TP applied as Tipimon. Stern and Gazit (1997) also reported that spraying with 50 ppm of 3,5,6-trichloro-2-pyridyl-oxyacetic acid (3,5,6-TPA as Maxim) is as effective (29-550% increase) and unlike Tipimon does not injure the foliage. Stern and Gazit (1999) noted that synthetic auxin 3,5,6-TPA significantly increased fruit retention and yield in `Kaimana' litchi when applied three weeks after full bloom. This appears to be an effective specific to 3,5,6-TPA which is separate from its effect on litchi fruit-drop reduction, as it does not occur after 2,4,5-TP application (Stern *et al.*, 1995, 1997). Kumar *et al.* (2014) also reported that growth substances like NAA, PCPA (parachlorophenoxy acetic acid), GA, BA and CCC (Cycocel) proved beneficial in minimizing the drop and cracking and enhancing quality of litchi fruits. The minimum fruit drop was recorded with 20 ppm PCPA. Fruit cracking was least with 20 ppm NAA. The maximum fruit weight and highest aril percentage were obtained with 100 ppm GA. Fruits sprayed with 500 ppm CCC had the highest total soluble, total sugars and ascorbic acid contents, but had the least acidity. Fruit set, fruit and aril weights, fruit volume and pulp percentage, total soluble solid, total sugar content and fruit production significantly increased with NAA and 2, 4-D.

8. Role of PGRs on Fruit Yield and Quality

Singh *et al.*, (2012) noted that soil drenching of PBZ @ 7.5 ml/tree respond better for increasing number of hermaphrodite flowers, fruit set, fruit retention and sugar content whereas PBZ @ 5.0 ml/tree was responsible of increasing fruit breadth and weight. They also found that the foliar application of CCC @ 2000 ppm in mid-September to mid-November resulted in maximum pulp weight, pulp/stone ratio, total soluble solids and minimum acidity

Ethylene is widely applied in litchi production, which can promote fruit ripening, improves anthocyanin levels in peel, and mean-while decreases the chlorophyll content. Suitable concentration of ethephon treated litchi tree generates fruits with remarkably increased soluble solids, sugar and ascorbic acid contents and meanwhile decreased acidity in their pulps (Sadhume *et al.*, 1992). Dixit *et al*,(2013) reported that GA_3 @ 10 ppm was found effective treatment to increase fruit set, fruit retention and size of fruits. GA_3@ 20 ppm produce maximum number of fruits/tree. Least fruit cracking was also noted with GA_3@ 20 ppm treated fruits. Auxin stimulation due to 2,4-D and GA_3 might be the reason for the accumulation of building block at faster rate and better execution of source-sink relation registering higher fruit setting, retention and less cracking (Kumar *et al.*, 2009).

Concentration of abscisic acid (ABA) were found higher in the pericarp, seed and aril of cracked fruits of litchi than normal fruits while the concentrations of gibberellin (GA_3) were higher in the seeds at the critical period of cracking (Sharma and Dhillon,1986, 1988). Many researchers (Suryanarayana and Das, 1971; Kanwar and Nijjar, 1976; Sharma and Dhillon, 1987; Chandel and Kumar, 1995; Bhat *et al.*, 1997) reported that auxins (2,4-D and NAA) at concentrations (lower than 40 mg/l) reduced cracking. Gibberellin (GA_3) at 10-50 mg/l was partially effective (Suryanarayana and Das, 1971; Sharma and Dhillon, 1987; Sinha *et al.*, 1999). Peng *et al.* (2001) indicated that GA reduced cracking by lowering the activity of cellulose. Shrestha (1981) found that ethephon sprays at 10 mg/l reduced cracking from 12% (control) to 6% in Early Large Red. Stern *et al.*(2001) suggested that synthetic auxins like 2,4,5-TP (as Tipimon) @ 67 ppm and 3,5,6-TPA (as Maxim) @ 50 ppm have very positive effect on yield, fruit size and fruit colour in several litchi cultivars in China.

9. Role of PGRs on Maturity, Ripening and Shelf-Life of Fruits

Even litchi fruit has good demand and vast export potential, yet it is not a major commercial crop, mainly because of its seasonal availability and short shelf-life. Postharvest decay, pericarp browning and desiccation were identified as major problems restricting expansion of the industry in litchi exporting countries (Sivakumar *et al.*, 2007). The removal of ethylene and/or inhibition of ethylene in stored environments are fundamental in maintaining postharvest quality of climacteric produce (Saltveit, 1999). In recent years, however, there has been a paucity of research on developing new and more efficacious ethylene scrubbing materials. Thus, research activity has drifted away from ethylene removal to preventing the actions of ethylene through using 1-MCP or other ethylene absorbers. The coating of chitosan mixture on the surface of 'Huai' litchi fruits showed better fruit coat colour, low rotten rate, less weight loss and low respiratory rate that prolonged the shelf life of litchi fruits (Wu *et al.*, 2001). Coating with chitosan can hind the infiltration of oxygen in atmosphere and emit of CO_2 produced by fruit respiration, but it allows the emit of ethylene that ripens fruit, thus inhibits the fungal propagation and postpones the ripeness of fruit, control the rising of POD activity, thus inhibit the yellowing and browning of fruit, ensuring fruit quality (Lin *et al.*, 2005).

Ozone has strong oxidation and germicidal effect to bacteria and mildew. Findings of Lin (1999) indicated that as ozone could kill pathogens on fruit surface and clear up the effect of metabolites, better controlling the putrefactive degree of litchi fruits. Ozone is effective in degrading ethylene, delaying the aging of fruits and vegetables, also it can break the benzene ring and remove off the pesticide residues like organochlorine and organophosphorus (Zhang *et al.*, 2003) and better ensuring food safety. Li *et al.*, (2004) used different concentration of ozone water to treatment litchi fruits after harvest, and found that ozone water had some effect in maintaining and protecting litchi fruit colour and inhibit browning. Combined with other fresh keeping techniques, pretreatment of 1 mg/L ozone water can prolong preservation duration and enhance fruit quality. Yang *et al.*, (2001) reported that ozone water had effective for precooling, colour protecting and rinsing, and it can kill the pathogens carried by the surface of litchi fruit, which to the largest fruit extent reduces the side effect of mechanical damage. Moreover, ozone could narrow the stoma on litchi fruit surface, hugely reduces weight loss.

Different antioxidants and salicylic acid were tested to overcome pericarp browning and to maintain the postharvest quality of the litchi fruits at ambient storage. It was found that 0.5% salicylic acid, 1% isoascorbic acid and 1% *N*-acetyl cysteine performed better over sulphur dioxide (SO_2) fumigation for most of the parameters under study. Application of 0.5% salicylic acid found superior to reduce the pericarp browning, relative leakage rate, and decay percentage. It was effective in reduction of polyphenol oxidase activity and improvement of anthocyanin pigments of the fruit pericarp over other treatments. Total soluble solid, titratable acidity and ascorbic acid of the litchi fruits were recorded highest with the application of 1% isoascorbic acid followed by 0.5% salicylic acid treatment. Therefore, 0.5% salicylic acid and 1% isoascorbic could be used as an alternative of SO_2 fumigation for quality retention of litchi fruits (Kumar *et al.*, 2013).

Mishra *et al.*, (2012) made an attempt to study the effect of GA_3 and BA on ripening of litchi cultivar 'Rose Scented'. In this attempt, KNO_3 (4%) was sprayed at 1 cm size of panicle in the first week of February. However, other treatments viz. GA_3 (20, 40 ppm) and BA (20, 40 ppm) were applied two weeks before expected date of harvest (on 15th May). They found that KNO3 (4%) advanced the harvesting date only for 2 days in comparison to control. Whereas GA_3@ 20 and 40 ppm delayed the harvest date for 2 and 5 days, respectively and BA @ 20 ppm and 40 ppm delayed the harvest date for 5-6 days. In all the treated trees, fruit weight was found to be more as compared to control. Higher fruit quality attributes were recorded with GA_3 (40 ppm) followed by GA_3 20 ppm over other treatments. They also observed least fruit cracking of GA_3 and BA sprayed trees. Tomer *et al.*(2001) noted that spraying of gibberellic acid (50 ppm) and Magic (Uniconazol 0.5 %) during flowering of litchi cultivar 'Mauritius' which affect fruits with a lower sugar and higher acid contents, which may indicate a delay in ripening. Spraying during fruit development with, gibberellic acid (50 ppm) or KNO3 (4%) had no noticeable or significant effects on fruit ripening and on TSS or organic acids levels.

Litchi is a non-climactic fruit. Yin *et al.*,(2001) studied the changes in ABA and ethylene and their roles in litchi maturation and colouration. Contents of endogenous abscisic acid (ABA) and ethylene in litchi fruit during development were determined and $Ag_2S_2O_3$ (STS) which inhibits synthesis of ethylene was used to study on fruit maturation and coloration. The content of ABA increased rapidly a week before pericarp colouration. With the occurrence of ethylene climacteric, there was a rapid increase in sugar content and invertase activity. STS delayed fruit coloration and maturation. It was suggested that the increase in ABA might have induced the synthesis of ethylene, which consequently promoted fruit maturation and coloration. Results showed that foliage spraying of STS inhibited fruit maturation and coloration, indicating that ethylene is involved in the regulation of the events. Similar observation was also made by Pal and Mishra (2012).

An experiment was conducted to stagger the harvesting period of litchi in cultivar Rose Scented. In this regard, various kinds of treatments were imposed on 25 years old full bearing litchi tress either at flower initiation or few days before harvest of fruits. KNO_3 (4%) was sprayed at 1 cm size of panicle in the first week of February. However, other treatments *viz*. GA (20, 40 ppm) was applied two weeks before expected date of harvest, while shading treatments were given by covering the tree with nylon nets producing 30% and 50% shades respectively, 30 days after fruit set. Shade nets 30% and 50% were most effective in delaying ripening of litchi fruits and delayed the harvest date by 5 and 8-10 days, respectively without deteriorating the fruit quality. GA3 20 and 40 ppm delayed the harvest date for 6-7 and 4 days, respectively while KNO_3 (4%) could not play significant role in advancing/delaying the harvest date of litchi. Higher fruit yield and reduced fruit cracking were obtained with shade net (50%) which was remained at par with shade net (30%). Higher fruit chemical quality attributes were recorded in GA_3 sprays over other treatments (Mishra *et al.*, 2014).

Wang *et al.*, (2001) conducted experiment on ten-year-old trees of 'Fezixiao' litchi in Wantian Orchard, Zengcheng, China. A set of three trees with the same date of full bloom (March 25) was chosen for the determination of fruit colour area (%) and the concentrations of endogenous hormones, anthocyanins, chlorophylls and sugars. Another set of three trees with the same full bloom date (April 7) were chosen to test the effect of 6-BA and ethephon. Ten fruit clusters on each tree were dipped in 0.1g L^{-1} 6-BA or 0.4 g L^{-1}ethephon on 51 DAFB (Days after full bloom). Control fruits were dipped in water. Result showed that 6-BA delayed the degradation of chlorophyll by inhibiting the activity of chlorophyllase and in the meantime anthocyanin formation. These results indicated that delay in peel coloration be linked with the degradation of chlorophylls, which impeded the formation and masked the appearance of anthocyanin.

The concentrations of ABA in the peel and the aril were increased rapidly around 62 DAFB. And this was followed by an accumulation of sugars and anthocyanin. ABA appears to have a major role in maturation. Application of ethylene @ 400 mg/l advanced colouration in Shahi litchi by 8 days compared with controls (Sharma *et al.*, 1986). Ray and Sharma(1986) reported that spraying with GA_3 at 25-50 mg/l, chlormequat @ 2000 mg/l and daminozide @ 100-2000 mg/l delayed ripening in litchi. Dawson *et al.* (1976) showed a strong relation between fruit growth and ABA in cherry. The delay in colouration by 6-BA was in parallel with ABA in the peel with no obvious changes in ACC (1-aminocyclopropane-1-carboxylic acid). Although ethephon increased ACC and slightly accelerated the loss of chlorophyll, ABA appears to be more closely involved in litchi maturation than ethylene. Stern *et al.*(2001) examined the effect of 2,4,5-TP (as Tipimon) @ 67 ppm and 3,5,6-TPA (as Maxim) @ 50 ppm, or their combination (first with 2,4,5-TP and 4 days later with 3,5,6-TPA) on 'FeiZi Xiao', 'Hei Ye' 'Kaimana' and 'Floridian' litchi cultivars. They sprayed the chemicals when fruitlets diameter at 14 mm and found that in cultivars 'FeiZi Xiao' and 'Hei Ye' the red color of the ripe fruit was enhanced by 3,5,6-TPA.

10. Role of PGRs on Embryogenesis

Embryo development in litchi affects fruit set and eating quality, and is presumably associated with changes in endogenous hormones. Chen and Lu (2000) studied the relationship between embryo development and endogenous hormones in litchi, using normal and aborted ovules of 'Lanzhu'. In normal ovules, the concentrations of Indole acetic acid (IAA), gibberellin1+3 (GA_{1+3}) and abscisic acid (ABA) peaked 7 days after anthesis (DAA), and then declined. The concentrations of IAA and GA_{1+3} increased again during globular to torpedo stage, while cytokinin (CTK) peaked before the globular stage. In aborted ovules, ABA remained at a high concentration throughout, while IAA and GA_{1+3} declined to their lowest level, CTK was lower than in normal ovules during the embryo abortion period. The concentrations of zeatinriboside group (ZRs), dehydrozeatinriboside group (DHZRs) and isopentenyladenosine group (iPAs) varied indicating different roles in embryo development.

One of the first works on somatic embryogenesis in litchi was done by Amin and Razzaque (1995), who managed to induce somatic embryogenesis in the cultures of zygotic embryos of lychee using BA (5 mgL^{-1}) and activated charcoal (1 gL^{-1}). Although about 40% of the in vitro formed embryos matured, no plantlets were obtained. Yu and Chen (1998) reported the development and maintenance of highly embryogenic suspensions and protoplast isolation for several lychee cultivars. In another work, Yu *et al.* (2000) managed to successfully culture lychee protoplasts of the cultivar 'Xiafanzhi' from suspensions, only after embedding them in Ca-alginate beads. Although the successful culture of lychee protoplasts could facilitate their use in lychee breeding, the regeneration frequency was low.

As an alternative to air-layering, micro propagation offers an attractive method for vegetative propagation of lychee. It requires only very small amounts of propagating material and has the potential of providing very large numbers of cloned plants. The exogenous application of plant growth regulators also an important aid to stimulate rooting potential during in-vitro explant culture. Poochua (2005) Studied in-vitro plantlet regeneration of litchi (*Litchi chinensis* cv 'Tai So'). Litchi plantlets were successfully regenerated in-vitro using young leaf explants. They examined that 2, 4-D @ 1.5 mg/L) in combination with cytokinin (BAP @ 2.0 mg/L) was more efficient in callus initiation while BAP, in combination with IAA @ 3.0 mg/L was more appropriate for the differentiation of shoots from callus. IBA @ 2.0 mg/L) promoted rooting of the shoots. Although this method of propagation has been improved by the use of younger branches, small earth balls and 1, 4-indole-3-butyric acid (IBA), the process is still slow and inefficient.

Fruits that have aborted seed are termed "chicken tongues" and are preferred by consumers (Lake, 1988), since these fruits have a high flesh to seed ratio. In crosses involving plants which have tendency to produce these fruits as the female parent, many of the most valuable progeny are lost prior to harvest. Production of plants by culturing embryos prior to abortion would be expected to yield progeny with high proportion showing the chicken tongue character (Anon, 1991). This technique has the advantage of providing very clean material as well as producing juvenile tissues which usually respond well in culture (Durzan, 1984). In their work on lychee anther culture, Fu and Tang (1983) used opened staminate and hermaphrodite 'Chenzhi' and GushanJiaohe' litchi flowers containing 3-4 mm long anthers with microspores at the optimum late-uninucleate stage of development were utilized. The anthers were removed and plated on induction medium consisting of MS salts and organic components, 2,4-D @ 5-9 μM, 4.6-9 μM kinetin, 3-5 μM NAA, and 16 g/l sucrose. The cultures were incubated at 25^{0}c with a 10h photoperiod. In order to initiate haploid embryo development and maturation, embryogenic cultures have been subcultured on MS supplemented with 2 μM kinetin, 0.5 μM NAA, 500 mg/l casein hydrolysate, 400 mg/l royal jelly, and 16 g/l sucrose. Early heart-stage embryos were transferred to MS with 2 μM kinetin, 3 μM gibberellic acid (GA_3), 500 mg/l casein hydrolysate, 400 mg/l royal jelly ,1.7 g/l glutamine, and 16 g/l sucrose. The haploid embryos matured and germinated. Higher concentrations of 2, 4-D were avoided because of its possible role in chromosomal aberration and suppression of morphogenesis (Murashige, 1974).

11. Role of PGRs on Stress Management

Under stress conditions plant turns 'on' its ABA synthesis system to synthesize a mass of ABA. Morrillon *et al.,* (2001) mentioned that under stress conditions the ABA responsive genes will activate which later on act as a stress adaptation mechanism by improving water uptake, decreased the leaf extension rate, by closing stomata etc. Zhou *et al.* (2009) performed low temperature stress and ABA treatment on litchi under the conditions of artificially simulated natural low temperature using artificial climate. ABA can combat the stress condition by reducing transpiration rate, cell membrane integrity and by increasing protein levels and enzymatic activities (Ma et al., 1998). Even stressed by 2°C low temperature, ABA treated litchi also had higher flower formation rate consequently, via enhancing endogenous ABA level. ABA treatment could make stomata closed, transpiration weakened, water potential enhanced to some extent alleviating the damage of low temperature. In addition, ABA treatment changes the balance of ABA/IAA, letting ABA/IAA at high level under low temperature. Paclobutrazol is capable of slowing plant growth, promoting plant tillering and enhancing the ability to anti-lodging, effectively controlling the morphogenesis of fruit tree, advancing ripeness stage and increasing stress resistance of crops. At the same time, paclobutrazol is efficient in inhibiting bacteria and fungi, effectively preventing the diseases of various crops, which shows broad application prospect in agriculture and horticulture. In general, the polyamines also play a vital role for combating the adverse effect of stress condition.

Reference

- Abutiate, W. S. and Nakasone, N. Y. 1972. Studies on vegetative propagation of the lychee (*Litchi chinensis*Sonn.) with special references to graftage. *Ghanna Journal of Agricultural Science*, **5**: 201-211.
- Ahmed, S. 1961. The litchi. *Agriculture Pakistan*,**12**: 769-75.
- Amin, M. N. and Razzaque, M. A. 1995..Induction of somatic embryogenesis in the cultures of zygotic embryos of lychee. *Bangladesh J. Bot.*,**24**: 25 – 29.
- Amin, M. N. and Razzaque, M. A. and Anisuzzaman, M. 1996. Regeneration of plants in vitro from seedling explants of lychee (*Litchi chinensis* Sonn.). In: *Plant Tissue Culture*, pp. 21-25.
- Anonymous. 1991.Tissue culture of lychee (*Litchi chinensis* Sonn.).*In*: *CSFRI Information Bulletin* **229**: 1 – 2.
- Bangerth, K.F., 2009. Floral induction in mature, perennial angiosperm fruit trees: Similarities and discrepancies with annual/biennial plants and the involvement of plant hormone. *Sci. Hortic.*,**122**, 53-163.
- Bhambota, J. R., Singh, K. K. and Saini, H. 1968. Propagating litchi plants. *Punjab Horticulture Journal*, **8**: 200-203.
- Bhandary, K. R. and Shivashankar, Y. T. 1970. Propagation of litchi under mist. *South Indian Horticulture*, **18**: 74-76
- Bhat, S.K., Raina, B.L., Chogtu, S.K. and Muthoo, A .K. 1997. Effect of exogenous auxin application on fruit drop and cracking in litchi (*Litchi chinensis*Sonn.) cv. Dehradun. *Advances in Plant Sciences*, **10**: 83-86.
- Brahmachari, V. S. and Rani and Ruby 2001. Effect of growth substances on productivity, cracking, ripening and quality of fruits of litchi. *The Orissa J. Hort.*, **29** (1): 44-45.

- Chandel, S. K. and Kumar, G. 1995. Studies on floral biology, fruit set and its retention and quality of some litchi varieties. *Indian Journal of Horticulture*,**16**: 124-129.
- Chand, S., Sonali, Srivastava, R. and Mishra, D. S. 2014. Response of IBA concentrations and application dates on the performance of air layering in litchi cultivars. *International Journal of Basic and Applied Agricultural Research*, **12**(3): 460-465.
- Chapman, K.R., Paxton, B. and Cull, B. W. 1980.Litchi cultivar evaluation (Project 1) Bienn.Rep. *Maroochy Hortic. Res. Sta.*,**2**: 32-33.
- Chawla, W., Mehta, K. and Chauhan, N. 2012. Influence of plant growth regulators on rooting of litchi (*Litchi chinensis*Sonn.) air layers. *Asian Journal of Horticulture*, **7**(1): 160-164.
- Chen, W. S. 1991. Changes in cytokinins before and during early flower bud differentiation in lychee (*Litchi chinensis*Sonn). *Plant Physiology*, **96**: 1203-1206.
- Chen, W. S. and Ku, M. L. 1988.Ethephon and kinetin reduce shoot length, and increase flower bud formation in lychee. *Hort.Sci.*,**23**: 1078.
- Chen, W. X., Zhung, Y. and Lul, X. 1959.Preliminary report on application of gibberellin to retention of flower and fruit of litchi. *China Fruits*, **5**: 24-25.
- Chen, W. and Lu, L. X. 2000. Endogenous hormones in relation to embryo development in litchi.*Acta Hort.*, **558**:247-250.
- Das, A. K. and Prasad, B. 2014. Effect of plaregulatorntOn rooting survival of air layering in litchi. *Advance Research Journal of Crop Improvement*, **5**(2): 126-130.
- Das, D. K., Shiva P. N. and Bhalla-Sarvin, N. 1999. Multiple shoot induction and plant regeneration in litchi (*Litchi chinensis* Sonn.). *Plant Cell Reports*, **18**: 691 – 695.
- Dawson, R. M., Rudnicki, R. M. and Bukovac, M. J. 1976. Endogenous plant growth substances in developing fruit of *Prunus cerasus* L. V. Changes in inhibitor (ABA) levels in the seed and pericarp.*Journal of the American Society for Horticultural Science*, **101**: 519-23.
- Diao, O. C. 2006. The application of chemistry regulation technology in management of litchi in autumn and winter. *Yunnan Forestry*, **27**(5): 27-28.
- Dixit, A., Shaw, S.S. and Pal, V. 2013. Effect of micronutrients and pant growth regulators on fruiting of litchi. *Hort Flora Research Spectrum*, **2**(10): 77-80.

- Durzan, D. J. 1984. Applications of plant tissue culture to Agriculture and Forestry. Anoverview. Symposium on Plant Cell and Tissue Culture.University of Tennessee, Knoxville, Sept 9 – 14, pp. 232 –256.
- Fu, L. and Tang, D. 1983. Introduction pollen plants of Litchi tree (*Litchi chinensis*Sonn.). *Acta Genetica Sinica,* **10**: 369 – 374.
- Galan-Sauco, V. and Menini, U. G. 1989.Litchi cultivation. FAO, *Plant Production and Protection*, Paper 83, Rome.
- Galloway, B.T. 1922. Improved method of propagating the litchi. *JournaJ of Heredity*, **13**: 201-206.
- Gao, F.F., Chen, D. C., Song, Z.H., Hu, G.B. and Huang, X. M. 2001. A study of some internal factors affecting fruit size in litchi. *Acta Hort.*,**558**: 279-283
- Guang-Wen, Z., Tian, Z. and Shi, Y. 2010. Advances in the use of exogenous Plant Growth Regulators on Litchi. *Hunan-Agricultural Science & Technology*, **11** (6): 65-70.
- Hayes, W. B. 1957. Fruit Growing in India.Kitabistan, Allahabad. pp.18-6.
- Huang, P. and Weng, S. 1978. The effect of delayed shoot treatment of fruit bearing of litchi trees. *Journal of the Chinese Society for Horticultural Science*, **24:** 121-126.
- Joubert, A. J. 1986. Litchi. In: Handbook of fruit set and development. (Monselise, S. P., Ed.). CRC Press, Boca Raton, FL, USA, 233-246.
- Kantharanjah, A. S., Mc-Conchie, C. A. and Dodd, W. A. 1992.In vitro embryo culture and induction of multiple shoots in lychee. *Annals of Bot.*, **70**: 153 – 156.
- Kanwar, J. S. and Nijjar, G. S. 1976.Fruit development in relation to cracking in litchi (*Litchi chinensis*Sonn.) at Gurdaspur (Punjab).*Haryana Journal of Horticultural Sciences*, **5**: 164-168.
- Kumar *et al.*, 2014 www.biotecharticles.com/.../Role-of-Plant-Growth-Regulators-in-Litchi.
- Kumar, D.; Mishra, D.S.; Chakraborty, B. andKumar, P. 2013. Pericarp browning and quality management of litchi fruit by antioxidants and salicylic acid during ambient storage. *Journal of Food Science and Technology*, **50**(4): 797-802.
- Kumar, M., Kumar, R. and Singh, R. P. 2009.Effect of micronutrients and plant growth-regulators on fruiting of litchi. *Intern. J. Agric. Sci.*, **5** (2): 521-524.

- Lake, B. 1988. Lychees growth in queensland industry. *Australian Horticulture*, **863**: 80-81.
- Li, H. Z., Luo, J. S. and Zhang, Y. Y. 1992. Experiment on spraying ethephon to control winter shoots of litchi tree. *China Fruits*, **3**: 26-28.
- Li, J., Zhu, B. Y. and Ding, S. B. 2004. Effects of ozone water on post-harvest physiological and biochemical indexes of litchi. *Subtropical Plant Science*, **33**(4): 15-18
- Li, M. X., Zhang, B. S. 1987. Effect of plant hormone on flowering and fruit setting of litchi and its application in production. *Guangdong Agricultural Sciences*, **1**: 3-6.
- Li, P. W. and Ji, Z.L.. 1984. Cytokinin and differentiation of flower bud in litchi. *Guangdong Agricultural Sciences*, **1**: 5-8.
- Ledin, R. B. 1953. Lychee. Annual Report of the Florida Agricultural Experiment Station, Gainesville, Florida, 314-315.
- Ledin, R. B. 1954. Lychee.Annual Report of the Florida Agricultural Experiment Station, Gainesville, Florida, 285.
- Ledin, R. B. 1955. Lychee. Annual Report of the Florida Agricultural Experiment Station, Gainesville, Florida, 306-307.
- Liang, Y. G. and Qiu, S. d. 1998.A study on the artificially inducing techniques from large seeded fruit to small-seeded and high quality fruit in lychee (*Litchi chinensis* Sonn.).*In*: Proceedings of the XXV international Horticultural Congress, Brussels, Belgium, 355.
- Lin, B. F., Pan, Y.F. and Liang, X. Q. 2005.Effects of chitosan coating on permeability of litchi pericarp and the storage life. *Food and Fermentation Industries*, **31**(9): 103-105.
- Lin, Z. Q. 1999. Application of ozone in cold strorage and quick frozen processing of fruits and vegetables. *Cold storage Technology*, **4**: 27-29.
- Menzel, C. H. and Xiong, W. J. 1999.Effect of paclobutrazol on growth and flowering of lychee (*Litchi chinesis*). *World Tropical Agriculture Information*, **4**: 34-38.
- Menzel, C. M. 1984. The pattern and control of reproductive development in lychee: a review. *Sci. Hortic.*, **22**: 333-45.
- Menzel, C. M. 1983. The control of floral initiation of lychee: A review. *Sci. Hortic.*, **21**: 201 – 215.
- Menzel, C. M. 2001.The Physiology of growth and cropping in lychee. South Agri. Litchi Grow. Assoc. Yearbook, **12**: 9-14.

- Mishra, D.S., Kumar, P. and Kumar, R., 2012. Effect of GA_3 and BA on fruit weight, quality and ripening of 'Rose Scented' litchi. *Hort flora Research Spectrum*,**1**(1): 80-82.
- Mishra D.S., R.L. Lal and Satish Chand. 2014. Delaying the harvesting period of litchi (*Litchi chinensis* Sonn.) cv. Rose Scented. *Prog. Hort.*, **46**(2):273-275.
- Mitra, S. K. and Bose, T. K. 1991. Metabolic changes during adventitious root formation in ethreland IBA treated cuttings of litchi. *Indian Journal of Horticulture*, **48**: 105-107.
- Morillon, R. and Chrispeels, M. J. 2001. The role of ABA and the transpiration stream in the regulation of the osmotic water permeability of leaf cells. *Proc. Natl. Acad. Sci.*, USA, **98** (24): 14138-14143.
- Murashige, T. 1974. Plant propagation through tissue culture. *Annual Review of Plant Physiol.*,**25**: 135-166.
- Mustard, M.J. and Lynch, S.J. (1959).Notes on Lychee panicle development. *Proc. Fla. St. Hort., Soc.*, **72**: 324-27.
- Nanda, K. K. and Kochhar, V. K. 1985.Vegetative propagation of plants. Kalyani Publ., New Delhi (India).
- Negi, U. S., Lal, R. L, Mishra, D. S. and Pal, M. 2012.Effect of chemicals and cincturing on flowering, fruiting and yield of Litchi. *Pantnagar Journal of Research*, **10** (2): 148-152.
- Ochse, J.J. 1952. Observations on the rooting of lychee cuttings. *Proceedings of Florida State Horticulture Society*, **65**: 186-7.
- Pal, M. and Mishra D.S. 2012. Extending harvesting period of litchi (Litchi chinensisSonn.) through chemicals application. *Hort. Flora Research Spectrum*, **1**(3): 235-238.
- Pandey, R.M. and Sharma, H. C. 1989. The Litchi pp. 54-61 ICAR, New Delhi
- Papademetriou, M. K. 1976. Some aspects of the flower behaviour, pollination and fruit set of avocado (*Perseaamericana* Mill.) in Trinidad. California Avocado Society Yearbook, **60**: 106-52.
- Paxton, B.F., Saranah, J. and Chapman, K.R. 1978. The propagation of litchis by cuttings.
- Biennial Report of lI-Maroochy Horticulture Research Station,**1** : 34.
- Peng,J., Xi. J. B., Tnag, X. D., Wang, Y. G., Si, X. M. and Chen, J. S. 2001. Effect of

Ca (No3)2 and GA spray on leaves on the fruit cracking of 'Nuomici' litchi. *Acta Hort. Sinca,* **28**: 348-350.

- Potjanapimon, C., W. Pattanapoo, N. Boonplod, T. Thanyapar, S. Subhadrabandhu, and D. Naphrom. 2000. "Changes in cytokinin and gibberellin-like substances in stem apex of lychee cv. Hong Huay prior to flowering." *Acta Hort.,* **558**:199-203..
- Prasad, J. S., Kumar, R., Mishra, M., Kumar, R., Singh, A. K., & Prasad, U. S. 1996. Characteristics of litchi seed germination. *Hort. Science*, **31**(7): 1187-1189.
- Puchooa, D. 2005. In vitro regeneration of lychee (Litchi chinensisSonn.). *African Journal of Biotechnology*, **3**(11): 576-584.
- Ram, M. and Majumder, P. K. 1983.Effect of indloe butyric acid on stooling in litchi. *Indian J. Hort.*, **40**: 211-212.
- Ramburn, N. 2000. Effects of girdling and growth retardants on flowering and fruiting of litchi in Mauritius.*Acta Hort.,* **558:** 229-232.
- Ray, P. K. and Sharma, S. B. 1986.Delaying litchi harvest by growth regulators or urea spray. *Sci. Hortic.*, **28**: 93-96.
- Sadhume, Su, W. Q. 1992. Effect of a post-harvest fruit dip in ethephon on the ripening of litchi fruits. *Fujian Science &Technology of Tropical Crops*, **60**(2): 70-72.
- Sarkar, S.K., Sinha, M.M. and Misra, R. S. 1984. Effect of NAA on fruit set, fruit drop, cracking, fruit size and quality in litchi cv. Rose Scented. *Prog. Hort.,* **16** (3-4): 301-04.
- Saltveit M E. 1999. Effect of ethylene on quality of fresh fruits and vegetables .*Postharvest Biol. Technol.*, **15**: 279-292.
- Sharma, S. B. Ray, P. K. and Rai, R. 1986. The use of growth regulators for early ripening of litchi (Litchi chinensisSonn.). *Journal of Horticultural Science*, **1**: 1-6.
- Sharma, S. B. and Dhillon, B. S. 1986. Endogenous level of gibberellins in relation to fruit cracking in lithci(*Litchi chinensis* Sonn.). *Journal of Research Punjab Agricultural University*, **23**:432-434.
- Sharma, S. B. and Dhillon, B. S. 1987.Effect of zinc sulphate and growth regulators on fruit cracking in litchi. *Indian Journal of Horticulture*, **44**: 57-59.
- Sharma, S. B. and Dhillon, B. S. 1988.Endogenous level of abscisic acid in relation to fruit cracking in lithci (*Litchi chinensis*Sonn.). *Agricultural Science Digest*, **8**: 55-58.

- Sivakumar, D., Zeeman, K. and Korsten, L. 2007. Effect of a biocontrol agent (*Bacillus subtilis*) and modified atmosphere packaging on postharvest decay control and quality retention of litchi during storage. *Phytoparasitica*, **35**(5), 507-518.
- Shreshtra, G.K. 1981 Effects of ethephon on fruit cracking of lychee (Litchi chinensis Sonn.) Hort Science 16, 498.
- Singh, B., Singh, S. and Sandhu, S. 2012. Effect of growth re tar dants on veg e ta tive growth, flowering and fruit ing of li tchi cv. Calcuttia. *Hort. Flora Res. Spectrum*, **1**(1): 29-33.
- Singh, L. B. 1960. The mango. Leonard Hill, London.
- Singh, O.P. and Phogat, K.P. S. 1984. Effect of growth regulators on fruit drop, size and quality of Litchi cv. Culcuttia. *Punjab Hort. J.*, **24** (1-4): 83-88
- Singh, P. C.,Sukla, H. S. and Katiyar, P. N. 2009. Effect of Indole butyric acid and P-Hydroxybenzoic acid on regeneration of litchi cultivars through air layering. *Annals of Horticulture*, **2**(2): 194-196.
- Sinha, A. K., Singh, C. and Jain, B. P. 1999. Effect of plant growth substances and micronutrients on fruit set, fruit drop, fruit retention and cracking of litchi cv. Purbi. *Indian Journal of Horticulture*, **56**: 309-311.
- Sittichaikasem, S. 1974. The effects of some environmental factors and certain growth regulator treatments on flowering of litchi ChinensisSonn.Var. Hong Huay.M.Sc.thesis, Kasetsart University, Bangkok.
- Stern, R. A., Kigel J., Tomer E. and Gazit S., 1995. 'Mauritius' lychee fruit development and reduced abscission after treatment with the auxin 2,4,5-TP. *J. Hort. Sci. Biot.*,**120**: 65-70.
- Stern, R.A., Nadler M. and Gazit S., 1997. 'Floridian' Litchi yield is increased by 2,4,5-TP spray. *J. Hort. Sci.* **72**: 609-15.
- Stern, R.A., Stern D., Harpaz M. and Gazit S. 2000. Applications of 2,4,5-TP, 3,5,6-TPA and combinations thereof increase lychee fruit size and yield. *Hort. Science,* **35**: 661-664.
- Stern, R, A. and Gazit, S. 1999. The synthetic auxin 3,5,6-TPA reduces fruit drop and increases yield in 'Kaimana' Litchi. *Journal of Horticultural Science and Biotechnology*, **74**(2) 203-205.
- Stern, R. A. and Gazit, S. 1997. Effect of 3,5,6-trichloro-2- pyridyl-oxyacetic acid on fruitlet abscission and yield of `Mauritius' litchi (*Litchi chinensis* Sonn.). *Journal of Horticultural Sciences*, **72**: 659-63.

- Stern, R. A., Kigel, J., Tomer, E. and Gazit, S. 1995. `Mauritius' lychee fruit development and reduced abscission after treat-ment with the auxin 2,4,5-TP. *Journal of the American Society for Horticultural Science*, **120**: 65-70.

- Stern, R. A., Nadler, M. and Gazit, S. 1997. `Floridian' Litchi yield is increased by 2,4,5-TP spray. *Journal of Horticultural Science*, **72**: 609-615.

- Stern, R. A., Stern, D., Miller, H., Xu, H.F. and Gazit, S. 2001. The effect of the synthetic auxins 2,4,5-TP and 3,5,6-TPA on yield and fruit size of young 'feizixiao' and 'heiye' litchi trees in guangxi province, china. *Acta Hortic.*, **558**: 285-288.

- Subhadrabandhu, S. and Duang, A. 1985. Effect of ethephon on flowering of two lychee (*Litchi chinensis* Sonn.) cultivars. *Acta Horticultrae*, **201**: 181-186.

- Subhadrabandhu, S. 1986. Studies of plant growth regulator effects on tropical and subtropical trees fruits of Thailand. *Acta Horticulturae*, **175**: 291-297.

- Subhadrabandhu, S. and Koo-Duang, A. 1987. Effect of ethephon on flowering of two lychee (*Litchi chinensis*Sonn.) cultivars.*Acta Horticulturae*, **201**: 181-186.

- Subhadrabandhu, S. 1990. Lychee and LonganCultivation in Thailand, Faculty of Agriculture, Kasetsart University, Bangkok, 40p.

- Suryanarayana, V. and Das, R. C. 1971. Effect of 2,4-D, NAA, 2,4,5-T on fruit weight and fruit-cracking in litchi (*Litchi chinensis*Sonn.) var. Muzaffarpur. *Plant Science*, **3**: 126-128.

- Suryanarayana, V. and Das, R. C. 1971. Effect of 2,4-D, NAA, 2,4,5-T on fruit weight and fruit-cracking in litchi (*Litchi chinensis*Sonn.) var. Muzaffarpur. *Plant Science*, **3**: 126-128.

- Tang, Z. P. 2006. Effects of ethrel and PP333 one endogenous hormones and flower bud-differentiation of 'Jizui' litchi. *Journal of Hunang Agricultural University*, **32**(2): 135-140.

- Tomer, E., Zipori, I., Goren, M., Shooker, S., Ripa, M. and Foux, Y. 2001. Delaying the ripening of 'mauritius' litchi fruit (preliminary results). *Acta Hortic.*, **558**: 315-317

- Wang, H.C., Huang, X.M. and Huang, H.B. 2001 . Litchi fruit maturation studies: changes in abscisic acid and 1-aminocyclopropane-1-carboxylic acid and the effects of cytokinin and ethylene on coloration in cv. 'Feizixiao'. *Acta Hortic.*, **558**: 267-271.

- Wang, B. Q., Qiu, Y. P., Xiang, X., Yuan, P. Y. and Zhang, Z. W. 1997. Study on the fluctuation of endogenous hormones during litchi (*Litchi chinensis*Sonn.) fruit setting and its induced parthenocarpy. *Acta Horticulturae Sinica*, **24**: 19-24.

- Werner, T., Motyka, V., Laucou, V., Smets, R., Van Onckelen, H., Schmulling, T., 2003.Cytokinin-deficient transgenic Arabidopsis plants show multiple developmental alterations indicating opposite functions of cytokinins in the regulation of shoot and root meristem activity. *Plant Cell*, **15**: 2532-2550.
- Wu, Q., Sun, Y. M. and Xiao, Z. L. 2001. Study on extending the shelf life of litchi with chitosan coating. *Journal of South china Agricultural University*, **22** (2): 83-85.
- Wu, S. F. and Huang, H. B. 1986.Preliminary report on morphological diffentiation of flower bud of litchi. *Guangdong Agricultural Sciences*, **4**: 1-5.
- Yan, Q. Y. 2006. Physiological effects of the new plant growth substance.-L-glutamic acid on litchi. Wuhan: Huazhong Agricultural University,
- Yang, W. L., He, J. F. and Wang, G. Q. 2001. Study on humid cool system in the preservation of litchi. *Science and Technology of Food Industry*, **22**(4): 15-17.
- Yin, J.H., Gao, F.F., Hu, G.B. and Zhu, S.H. (2001). The regulation of litchi maturation and coloration by abscisic acid and ethylene. *Acta Hortic.* **558**, 293-296.
- Young, 1957 Lychee: branch girdling. Annual Report of the Florida Agricultural Experiment Station, Gainesville, Florida, 353-354.
- Yu, C. H. and Chen, Z. G. 1998.Embryogenic suspension culture and protoplast isolation in litchee. *Chinese J. Tropical Crops*, **19**: 16-20.
- Yu, C. H., Chen, Z. G., Lu, L. X and Lin, J. W. 2000. Somatic embryogenesis and plant regeneration from litchi protoplasts isolated from embryogenic suspensions. *Plant Cell, Tissue and Organ Culture*, **61**: 51-58.
- Zhang, B. X. and Zhang Z. H. 2004.Effects of plant growth regulators on flowering and fruit setting in litchi. *Fujian Fruits*, **131**(4): 28-29.
- Zhang, W. H., Chen, D. Y., Yang, H. 2003. Study on degradation of pesticide residues in vegetable by ozone. *Journal of Nanjing Agricultural University*, **26** (3): 123-125
- Zhou LN, Kuang ZS, Ma XJ, Chen JQ, Cao J (1996).The study of factors affecting somatic embryogenesis in young embryo culture of Litchi chinensis. *J. of Agric. Biot.*, **4**: 161-165.
- Zhou, B. Y., Li, Y. B. and Chen, J. Z. 2009.Effects of low temperature stress and ABA on flower formation and endogenous hormone of litchi. *Acta Horticulturae*, **29**(6): 577-578.

18

Plant Growth Regulators in Mango (*Mangifera indica* L.)

R. Sri Hari Babu, V. Srilatha and Veena Joshi

1. Introduction

Mango (*Mangifera indica* L.) is grown in more than 100 countries now. It is one of the most important fruit crops of tropical and sub-tropical regions. India produces about 50 per cent of world mango production with largest area (Galan Sauco, 2013). In India, mango is cultivated in more than 40 per cent of total fruit area. It is grown in 250 million hectares across the country with an annual production of 18 million tonnes (NHB, 2013). However, mango is beset with several problems like erratic flowering, malformation, low production of perfect flowers, poor fruit set, heavy fruit drop, severe incidence of pests, diseases and disorders which directly and indirectly affect the production and cause much economic loss to the grower. Though mineral nutrition contributes much to the production of mango, sometimes even under adequate nutrition and good growing conditions, there will be failure of crop. Under these conditions, it was conceived to take the help of plant hormones to solve the above problems. Plants synthesize their own hormones, however, sometimes their content may be below threshold levels and their balance may be not in favour of physiological and biochemical processes. Hence, it is conceived that by exogenous supply of plant growth regulators and regulating their balances in the plant, erratic flowering etc. may be corrected and production may be increased adequately.

Since the past more than half-a-century several efforts were being made all over the world through the use of plant growth regulators to solve the various problems encountered in mango production. Plant growth regulators are being tested for successful propagation, enhancement and / or regulation of vegetative growth, correction of erratic flowering, enhancement of flowering, perfect flowers and fruit set and reduction of fruit shedding and finally enhancing yield and fruit quality.

2. Role of Plant Growth Regulators (PGRs) on Propagation

2.1. Role of PGRs on seed germination and seedling growth

Mango is amenable for both sexual (seed) and asexual (vegetative) propagation. However, because of heterozygocity, sexual propagation is not adopted for orchard plantations. For commercial orcharding, vegetative propagation is universally adopted in mango. Out of the many vegetative propagation methods, grafting methods are usually and mostly followed for propagation of mango. Grafting requires two plant parts *viz.*, rootstocks and scion shoots. Though raising of rootstocks through cuttings, layers and stooling is possible, because of certain limitations, drawbacks and difficulties these methods of raising rootstocks are not commercially adopted. Hence, the only alternate left, to raise the rootstocks is the usage of seedlings, in spite of heterozygocity and other drawbacks. For raising rootstocks, therefore, seeds are invariably and universally utilised in mango.

Mango seeds (stones) though non-dormant, present some difficulties in germination, like slow germination (taking more time for germination), less percentage of germination, slow growth and less vigorousness of seedlings which take more time to reach the graftable size / stage. Further, the viability of mango stones is comparatively less (12-50 %), because of recalcitrant nature of seed (Chaudhari and Patel, 2012; Vidya *et al.*, 2015). The slow and poor germination of mango stones may be due to hard seed coat (endocarp) as well as presence of high level of inhibitors of germination endogenously. To overcome the above problems of mango seed stones and enhance the percentage of germination, faster growth of seedlings after emergence and improve vigour of the seedlings, certain pre-sowing treatments like dehusking (removal of hard seed coat), soaking in water (Osmopriming) (Venkatarao and Reddy, 2005a, Venkata rao *et al.*, 2006a) or aqueous solutions of chemicals and / or growth regulators are adopted. Of these growth regulators are more commonly used. Plant hormones are more important in propagation, because they act endogenously to regulate plant physiological and biochemical functions and can be applied easily to seeds to induce specific responses such as release of seed dormancy, enhancement of germination and growth and vigour of emerged seedlings. The pre-sowing treatments of seeds may affect and alter enzymatic reactions involved in germination process. Thus, the enhanced enzymatic reaction along with suppression of inhibitors of germination acts in boosting germination of seeds. The relative advantage of pre-sowing treatments of mango seeds exogenously with chemicals or growth regulators is to reduce the period required for germination, to remove the obstructions in the embryo and to boost growth and vigour of seedlings.

Growth regulators like auxins, gibberellins and cytokinins were tested to break the seed dormancy (if any) and to enhance the germination of seeds. Of these GAs were found to be more effective in improving the seed germination and seedling growth. Several investigations in the past have clearly demonstrated that exogenous GA treatments (by soaking) have induced early and faster germination, enhanced the percentage of germination and improved vigour and growth of resultant seedlings (Venkatarao, 2002; Venkatarao and Reddy, 2005b; Yellesh Kumar, 2007; Himabindu and Srihari, 2013; Himabindu *et al.*, 2014; Vidya *et al.*, 2015; Muralidhara *et al.*, 2015).

Mango stones pre-treated with 100 ppm GA3 for 10 minutes increased the percentage of germination from 56.18 in control to 72.15% (Yellesh Kumar *et al* 2007). Germination percentage and number of seedlings per seed increased in some mango cvs (Zebda, Sukkary, Sabre and 13-1) with seed husking and soaking in GA3 at 100 or 200 ppm concentration for 48 hours (Shabini, 2010). Pre-sowing of whole or extracted kernels of Alphonso mango in GA3 at 250 or 500 ppm for 12 hours increased germination percentage from 50.35 in control to 53.55 (Himabindu and Srihari, 2013). Whole and husked kernels of mango cv. Totapuri were soaked for 12 hours. Pre-soaking in GA3 250 or 500 ppm gave quicker germination (8-9 days) against 11 days in control, higher rate of germination (1.16 & 1.24%) against 0.68% in control and higher percentage of germination (62-66%) against 56% in control (Himabindu *et al.,* 2014). Mango stones were pre-soaked for 24 hours in GA3 at 100, 200 and 300 ppm which resulted in maximum rate of germination (Muralidhara *et al.,* 2015). Vidya *et al.,* (2015) achieved significantly maximum germination of 72.25% against 51.67% in control, by pre –treating mango stones with GA3 500 PPM for 10 minutes.

Seed dormancy and germination are complex physiological processes that are controlled by a range of internal and external cues. Endogenously they are regulated by the presence of phytohormones like promoters and inhibitors. Previous studies have shown that gibberellins play an essential role in improving the germination of seeds. Two roles of GA have been proposed in stimulating seed germination (i) GA is necessary to overcome the mechanical barrier conferred by tissues that surrounded the embryo such as the aleurons and the testa *i.e.* GA weakens the tissue that surrounds the embryo to allow the protrusion of the radical

(ii) GA also increases the growth of the embryo as indicated by raised level of GA, in GA deficient embryos (Yellesh Kumar *et al.*, 2007).

The following biological functions of GA in seed germination and seedling growth were also proposed: (a) involvement of gibberellins in the natural process of breaking dormancy and other aspects of germination (b) induction of starch hydrolysis through initiating the synthesis of the enzyme, amylase in the aleuron cells, to help the emerged seedlings in the initial stages, as they have not yet developed their own photosynthesis apparatus and depended on reserved food in the endosperm.

According to Muralidhara *et al.*, (2015) early and enhancement of germination in GA_3 treated mango stones was due to the stimulation effect of GA on the seed as well as enhanced enzymatic processes and suppression of inhibition along with the synthesis of RNA which resulted in the higher germination of seeds, by changing seed coat permeability (Varner and Ho, 1976).

GA, besides improving and enhancing seed germination, influences the growth and vigour of seedlings favourably. Improvement of seed germination and resultant growth of rootstock seedling help in promoting its vigour. The increase in seedling growth and vigour due to pre-sowing treatments of seeds with GA_3 (at different concentrations) have been reported (Padma and Reddy, 1998; Shalini *et al.*, 1999; Venkatarao, 2002; Venkatarao *et al.*, 2006b; Yellesh Kumar, 2007, Shabini, 2010). The higher vegetative growth and vigour of GA treated mango stones was attributed to the effect of pre-sowing treatment of GA, which is known to release the growth stimulating hormones like IAA (Leopold and Kreidemann, 1983). Influence of GA on stem elongation may be due to i) its direct effect on stem elongation by inducing cell wall loosening, by increasing the solute concentration through increasing cell wall extensibility, stimulating the wall synthesis, reducing the rigidity of cell wall by increasing cell division leading to more growth, ii) the indirect effect on stem elongation by increasing the synthesis of IAA (Leopold and Kriedemann, 1983). Increase in seedling height by the exogenous application of GA_3 as pre-sowing seed treatment was reported (Shirol *et al.*, 2005; Venkatarao *et al.*, 2006b; Yellesh Kumar *et al.*, 2007) in mango.

GA_3 increased the number of leaves per seedling (Himabindu *et al.*, 2014). This may be due to vigorous growth induced by GA, which also induced more number of branches, which in turn facilitated better harvest of sunlight by the plant, which resulted in the production of more number of leaves (Khobragate *et al.*, 1999; Venkatarao and Reddy, 2005b, 2006b).

2.2. Role of PGRs on Rootage

Rootage is a method of vegetative propagation in which different plant parts are allowed to strike roots for perpetuation. Detached (cuttage) and undetached (layerage) plant parts are allowed to regenerate roots and the resultant rooted cuttings and rooted layers are utilized for raising new plantation. However, rooting of cuttings and layers under natural conditions is not easy, and sufficient and takes longer time in mango. Hence, it is categorised as a difficult-to-root species. However, rooting in difficult-to-root species can be enhanced with certain pre-treatments (like eliolation or ringing and with plant growth regulators.

The favourable effects of growth regulators on rooting are : i) stimulation of root initiation particularly in shy-to-root and difficult-to-root species, ii) increasing the percentage of rooting and rooted cuttings and layers iii) acceleration of rooting and iv) increase in the survival of rooted cuttings and layers.

The results obtained by various workers indicated that application of plant growth regulators (PGRs) causes greater metabolic activity in cuttings / layers and greater mobilization of sugars and nitrogenous substances from stems and leaves, which help in inducing roots in the cuttings / layers (Cooper, 1936; Pearse, 1938; Dett Weiler, 1942 etc.).

Role of growth regulators (GRS) in inducing or increasing the rooting of treated cuttings/ layers has been attributed to the increased metabolic activity in them after the treatment. Staurt and Marth (1938) found that the stem of the GR treated cuttings maintained a higher percentage of total sugars than the untreated cuttings. Some of the fruit species, which are either difficult-to-root or slow-to-root were made to root easily and rapidly with the use of GR, IBA and NAA were especially found outstanding root promoting GRs.

2.2.1. Effect of PGRs on rooting in cuttage (cuttings)

At Pantnagar, Rajan and Ram (1983a) stimulated rooting, in mango stem cuttings with the application of IBA. At Indian Agricultural Research Institute (IARI), New Delhi, dipping of hard wood cuttings of mango in IBA (2500 and 5000 ppm) in 5 per cent alcohol, increased rooting percentage when applied to both bases and sides. Low concentration (2500 ppm) was more effective than the higher concentration (5000 ppm). IBA was able to mobilize greater amounts of carbohydrates in cuttings from hormone treated shoots of mango. The accumulated food materials appear to have induced better rooting and bud sprouting in mango cuttings (Sinha *et al.*, 1960). At IARI, New Delhi, cuttings of 'Dashehari' seedlings were dipped in IBA (0-5000 ppm) for 15 sec. and then dipped in one of the several phenolic compounds at 0 or 2000 ppm for 15 sec. IBA alone rised the rooting percentage from 31 to 93 per cent. Cent per cent rooting was obtained with IBA followed by dipping in phenotic compounds. However, dipping the cuttings in the phenolic compounds alone had little or no effect on rooting (Reddy and Majumder, 1978). Sandhu (1992) obtained improved rooting in the cuttings from 10-year-old 'Langra' trees sprayed with NAA at 2×10^{-4} M and dipped in IBA at 10^{-2} M.

Wounding with IBA (5000 ppm) gave significantly higher percentage of rooting than wounding without IBA (Basu *et al.*, 1972; 1979; Reddy and Singh, 1987b). Wounding enhances IBA absorption. The beneficial effect of wounding may also be due to destruction of sclerenchyma tissue, which normally depresses the root emergence. Wounding prior to IBA application is beneficial in case of difficult-to-root plant species.

Ringing of shoots resulted in carbohydrates accumulation above the ring. Increased protein synthesis in ringed (shoot) cuttings was associated with increase in DNA and RNA synthesis. Both ringed and IBA treatments resulted in the incorporation of relatively large quantity of free amino acids in bound form. The prevention of such incorporation by means of a protein synthesis inhibitor had an adverse effect on root formation (Basu *et al.*, 1972).

Pre-treatment of stock plants with cycocel or ethrel induced rooting on cuttings and air layers of mango. Cycocel pre-treatment was more effective than ethrel pre-treatment (Sandhu, 1979). The effect of pre-treatment of stock plants with GA_3, Cycocel, CEPA alone or Cycocel + CEPA on rooting of mango cuttings was investigated (Sandhu, 1992). Cuttings taken from GA_3 treated stock plants showed poor rooting even after treatment with auxins. On the other hand, Cycocel, CEPA or Cycocel + CEPA pre-treated cuttings showed significant increase in rooting over control. Rooting was further enhanced when auxins were applied to the bases of pre-treated cuttings in which the effect of IBA and NAA was more pronounced. Application of IBA to the cuttings and air-layers from pre-treated shoots improves rooting. The increased rooting on cuttings taken from pre-treated shoots seems to be associated with higher levels of carbohydrates, phenols and root promoting substances.

Bottom heat technique has been successfully demonstrated for rooting of hardwood cuttings of tropical fruits including mango. The rooting was excellent (85.8%) at 35°C basal temperature, when coupled with wounding and IBA (5000 ppm) treatment, with increased number of roots and their length (Reddy and Singh, 1987a).

2.2.2. Effect of PGRs on layerage (Air-Layering)

Layering is preferred to cuttings in some species because the unit is attached to the mother plant, which allows supply of minerals, hormones and other metabolites and gives higher success in rooting than cuttings. In cuttings, on the other hand, the detached unit has to survive on its own reserves until root system is well established. Earlier studies indicated that mango can be propagated with considerable success by air layering, both for orchard plantation and root stock purposes.

Although 60 per cent of the untreated control air layers rooted but their roots was fewer, smaller and thicker as a result the air layers failed to survive after separation from the mother plant (Srivastava, 1960). Further, as the roots of control air – layers were so brittle that a few air-layers survived on potting (Ledin and Ruehle, 1954).

In view of the above, it becomes necessary to adopt pre-treatment of air-layers either by girdling and eliolating the units. These pre-treatments resulted in increase the concentration of root promoting co-factors in the shoots and thereby help in inducing rooting. In case of difficult-to-root species (mango, ber, avocado, etc), the use of root promoting chemicals becomes necessary to obtain good rooting in air-layers. Combination of pre-treatments with growth regulators was more beneficial than either of them alone.

A few explanations have been advanced for increased rooting in air layers by pre-treatments with or without growth substances. According to Sen and Bose (1966) a steady rise in moisture content is directly related to increase rooting in air layers. Basu *et al.* (1979) reported that the bio-chemical changes that occur during the course of root formation in mango air-layers, caused significant accumulation of carbohydrates in the root forming region of the air layers. Increased protein synthesis due to girdling or treatment with root promoting chemicals was associated with the rise in the levels of the nucleic acids (DNA and RNA) in the regenerating tissues. Girdling of shoots may also have caused greater accumulation of rooting factors in the region above the girdle. In general the content of nitrogen in the bark progressively increases during rooting due to non-utilization of protein – nitrogen during root formation in mango. All these changes might have contributed to better rooting of air layers subjected to pre-treatment of girdling of shoots.

Out of the known plant growth regulators, auxins are more often employed for improving rooting in cuttings / layers. The way in which external application of auxin improves rooting is not clear. The relationship between hormone treatment and increased physiological activities leading to rooting has been studied by many workers. The important role of auxin in the organ formation especially in root formation is well documented. It has been postulated that the processes leading to auxin synthesis in the shoots as well as auxin degradation form a basis for root initiation. More pronounced rooting was obtained when GRs were coupled with certain pre-treatment of cuttings / layers in mango. Ringed cuttings from 20-year-old 'Langra' trees treated with IBA (0.5 %) in lanolin paste showed increased rooting due to increased metabolic activity with rise in reducing and non-reducing sugars, starch and C:N ratio, compared with un-ringed cuttings without IBA treatment. One-year-old shoots of 25-year-old trees of 'Langra' ringed 20 cm below the apex, treated with IBA (20,000 ppm) plus NAA (5000 ppm in lanolin paste) resulted in best rooting (66.6 %) due to synergism (Patel and Singh, 1982).

Ethrel alone, however, was ineffective in cuttings, but promoted rooting in air layers (as effective as did IBA alone). The root growth was promoted when ethrel and IBA were applied jointly. This was attributed to synergism between the two growth regulators and this synergism was more pronounced in the presence of IBA than of IAA (Sandhu and Bose, 1980). They demonstrated significant promotion of rooting by ethrel in mango, guava and wood apple cuttings and air-layers in the presence of IBA.

2.2.2.1. Etiolation and PGRs in air layering

Etiolation is a pre-treatment in which the rooting zone of cuttings / layers is kept away from the exposure to (sun) light. However, etiolation alone is not satisfactory in rooting. Hence, often etiolation is coupled with different GRs for achieving success in rooting. Mukherjee and Bid (1965) reported that air-layering of etiolated shoots in mango even without the application of PGR produced roots, but IBA (10,000 ppm) and NAA (500 ppm) induced 100 per cent rooting in etiolated shoots of 'Langra' mango and their corresponding survival upto one year was 95 per cent and 90 per cent respectively. Bid and Mukherjee (1969) working on the varietal response to etiolation and GR treatment in air layering of mango found that etiolated shoots produced more fibrous roots than the non-etiolated shoots, after treatment with either IBA / NAA or IBA and NAA (1:1) in some cultivars of mango. Sudhir Kumar (1998) observed that etiolated shoots at lower concentration of IBA gave better performance in respect of percentage of rooting and survival, but etiolation alone did not show favourable effect on rooting as well as on survival in mango. Under Bangalore conditions 'Mallika' can be easily propagated by air-layering, when girdled and etiolated prior to treatment with IBA (750 ppm) in lanolin paste. For highest survival of the rooted air layers, they should be conditioned in a mist chamber (Pawan, 1984). Contrary to the above, Kumar and Sharma (2000) reported no effect of etiolation plus IBA treatment on percentage of rooting and survival of air layers of 'Kurukkan' cultivar at IARI, New Delhi. The discrepancy in the results may be attributed to the varieties involved as well as the location of trials.

Higher concentration of NAA (10,000 ppm) was toxic to etiolated shoots, but was conducive to regeneration of roots in non-etiolated shoots of mango (Mukherjee and Bid, 1965). Differences in varietal responses to etiolation plus GR treatments have been documented in mango. In 'Langra' 100 per cent rooting and establishment of air-layers resulted when the etiolated shoots were treated with IBA (10,000 ppm) with IBA + NAA (5000 ppm) (Bid and Mukherjee 1969). In 'Chausa' 98.6 per cent rooting and 90 per cent survival were obtained with etiolated shoots treated with IBA + NAA in combination at 5000 ppm(Bid and Mukherjee 1969). While in "Bombay Green" complete success was achieved when etiolated shoots were treated with 5000 ppm IBA (Bid and Mukherjee 1969).

2.2.2.2. Indole -3- Acetic Acid and rooting in air layering

Indole 3-acetic acid is the natural endogenous auxin discovered in plants and it was expected that exogenous application of IAA would induce rooting in cuttings / layers. With this concept some workers tried IAA for promoting rooting in air layers of mangoes. Talakurh and Dutt (1945) found it possible to successfully raise plants by air-layering of shoots of young plants of 2 to 3 year old age with IAA (1 %) and stated that higher or lower concentration than this would have a depressing effect on rooting. However, IAA at 2 per cent level gave a higher (70 %) rooting in 'Langra' air layers at Kanpur (Singh and Teotia, 1951). Azzouz *et al.* (1969) from Cairo, Egypt in contrast to findings of others stated that IAA was generally superior to IBA and the best root system was found after 90 days in marcots in mango Cv. Arumanis in respect of root number, length and fresh weight. In mango air-layers treated with IAA failed to root in August and October months at Sangareddy (Suryanarayana and Venkateswara Rao, 1984).

2.2.2.3. Naphthalene Acetic Acid (NAA) and rooting in air layering

NAA is a synthetic auxin, which is often employed in plant propagation for its promising effects. At Kanpur, NAA (1 %) induced 100 per cent rooting in air-layered shoots of 'Dashehari' mango (Singh and Teotia, 1951). Sen and Bose (1966) found greater effect of NAA on rooting than IBA in air-layers of mango in West Bengal. NAA induced rooting in rainy season (August, September and October) in mango at Sangareddy, which was attributed to free flow of plant sap and warm temperatures that prevailed during the season (Suryanarayana and Venkateswara Rao, 1984). Contrary to the above results, treatment with 1 per cent μ NAA in lanolin paste did not aid root formation in air-layers of mango in USA (Ledin and Ruchle, 1954).

2.2.2.4. Indole Butyric Acid (IBA) and rooting in air layering

Among the root promoting chemicals, so far tried, IBA has been found to be most effective and is rightly referred to as 'Rooting Hormone' – The beneficial effects of IBA in the rooting of mango air – layers has been reported by several workers. IBA is more stable than other auxins, hence more often used than IAA. Weaver (1972) reported that IBA is destroyed relatively slowly by auxin – destroying enzymes. It is translocated poorly and is retained near the site of application for long. Obviously a chemical that is persistent is very effective as a root generator / promoter. The increase in rooting of mango air-layers treated with IBA was associated with a greater depletion of sugars in the bark and wood of the rooting zone (Sen and Bose, 1966).

IBA was more effective in stimulating rooting in mango air-layers irrespective of concentration. However, lower concentrations upto 10,000 ppm were found to be more beneficial than higher concentrations (Jauhari and Nigam, 1962; Mukherjee and Bid, 1965, Kulkarni and Rameshwar, 1979; Singh *et al.*, 1979; Chatterjee, 1982, Ratnam, 1985 etc.). Higher concentration (above 15,000 ppm) showed depressing effect on rooting (Jauhari and Nigam, 1962). In general 10,000 ppm IBA proved to be the best concentration in respect of all characters studied (Sudhir Kumar, 1998). However, good rooting was obtained in the difficult-to-root 'Langra' cultivar of mango with IBA treatment at 15000 ppm at Pantnagar (Rajan and Ram, 1983a,b). IBA resulted in early rise of rooting co-factors and fall in rooting inhibitors level leading to early rooting of IBA treated mango air-layers (Rajan and Ram, 1988). IBA increases the number of primary and secondary roots with increase in concentration. The increase in the number of roots may be due to hormonal effect leading to accumulation of internal substances and their downward movement as reported earlier (Mukherjee and Majumder, 1963; Srivastava, 1960; Chhonkar and Singh, 1967, Rao, 1983; Chatterjee, 1982; Chandrashekar, 1982). Treatment with IBA (15000 ppm) in the month of October resulted in the highest percentage of successful air layers, maximum percentage of establishment of layers, maximum number of secondary roots and their average length, production of fresh and dry weights and percentage of dry matter while maximum number of primary roots and their lengths in August and September (Ramchand, 1984).

The response of air-layers to IBA differs according to the month of application. Singh *et al.*, (1979) obtained good success of 85-86 per cent in July in air-layers of 2-year-old shoots of mango with 5000 ppm IBA and the success declined from September onwards upto February. However, Rao (1983) obtained maximum percentage of successful layers with IBA in all the months (July, August and September) in mango. IBA induced early rooting in air layers of mango in all the 3 months (August, September and October) at Sangareddy (Suryanaryana and Venkateswara Rao, 1984).

Varietal differences in responses to IBA application were also noted. Rameshwar and Kulkarni (1979) reported that all the 3 dwarf cultivars (Rumani, Mahmooda Vikarabad and Rumani x Manjeera) recorded very high percentage (average 66 %) and profuse rooting as compared to very low (average 12 %) and profuse rooting in vigorous cv. Banganapalli with 5000 ppm IBA in lanolin paste.

The differences in the results with IBA treatments on air-layering of mango between different workers at different places and cultivars may be because of climatic and environmental factors, variety, age of the trees and differences in varietal responses.

2.2.2.5. Combination of PGRs in rooting

In some cases, a combination of different PGRs has been found to give good results than with either of the PGR. Thus application of a mixture of NAA, IAA and IBA gave a high percentage of rooting ability at 2 per cent of each GR (Teotia *et al.*, 1967) and at 0.25 and 0.5 per cent concentration (Rao *et al.*, 1963). Significantly higher percentage of rooting occurred in air-layers of 'Langra' mango with the combination of IBA (20000 ppm) plus NAA (5000 ppm) (Patel and Singh, 1982). Highest percentage rooting was obtained with IBA + NAA at 10000 ppm in girdled and etiolated air layers of 'Mallika' and also highest survival. Whereas, in air layers of 'Alphonso' highest survival was recorded with IBA 10000 ppm followed by IBA + NAA at 7500 ppm and 10000 ppm and NAA at 10 ppm (Al-Khalib, 1986).

2.3. Use of PGRs in Graftage

The major limitations in clonal propagation of mango on their own roots or clonally propagated root stocks are: (a) the difficulty in rooting (b) high mortality of rooted plants on separation from mother and transplanting (c) shallow rooting. In view of the above, graftage has been adopted commercially in mango propagation. Graftage is the art of joining two different plant parts together in such way that they make union and live as one plant in future. Graftage is a method of vegetative propagation in which two parts *viz.*, scion (shoot portion) and stolk (root portion) from two different plants are involved. The success of graftage depends largely on the sound union between the two plants at cambial layers. At times the union does not take place due to one or other reason. Failure of graft union some times is attributed to failure of multiplication of callus tissues. This is here PGR comes to play a role and rescue. Union between stolk and scion is achieved at the cambial region of both stock and scion, through cambial activity. PGRs particularly auxins are known to stimulate cambial activity and aid in the union of stock and scion. There is a definite relationship between callus production, its activity and the level of endogenous growth substances particularly auxins and kinetins (Murashige and Skoog, 1962). Therefore, it is envisaged that GRs play important role in the success of graft union / bud take. The take of bud grafts could be increased by treating the bud wood with GRs. The bud wood is immersed in aqueous solution of GR for specified time and then budded / grafted. Dipping or smearing of GR on to the cut surface of the scion and root stock can also be done, but found to be ineffective (Kulkarni *et al.*, 1988). However, GRs have been seldom tried in grafting of mango, as in case of rooting of layers / cuttings.

A few different methods of application of PGRs are adopted to gain success in graftage. They include pre-soaking of seeds, soaking scion sticks (Kulkarni *et al.*, 1988) smearing of PGR on to the cut surface of scion and root stock (Kulkarni *et al.*, 1989). Application of PGR to graft union at the time of grafting as well as pre-treatment of rootstock and scion sticks (Kannan and Rao, 1964) resulted in better graft success. Immersion of bud wood in aqueous solution of PGR for specified period like 12 hours prior to budding (Gur and Samish, 1962).

Mango is mainly and commercially propagated by different methods of grafting *viz.*, approach grafting, veneer grafting, soft wood grafting and epicotyl grafting. Though different budding methods have given considerable success, they are not followed commercially. But some information is available on the effect of GRs, on both budding and grafting of mango.

2.3.1. Use of auxins in grafting/budding

Application of optimal concentration of IAA paste above the bud shield improved the bud take in mango and increased the callus formation in seedling root stocks (Gur and Samish, 1962). They obtained highest take in mango cv. 'Gondu' by immersing the bud wood in IAA (200 ppm) for 12 hrs and budding them on 3-year-old root stocks during the rest period and found that concentration lower or higher than 200 ppm was partially harmful and reduction of bud take when IAA was applied to the root stock.

The grafting success increased by all concentrations of IAA (200 – 1000 ppm) in case of epicotyl grafting. The number of leaves and total leaf area per plant increased with IAA concentration. IAA was useful for increasing the success of epicotyl grafting upto 750 ppm and thereafter there was decrease in the success (Narwadkar and Anserwadkar, 1985). Mathod (1983) also found that final success of grafting in epicotyl grafting increased upto 750 ppm and at 1000 ppm there was slight decrease in the success and IAA was better than

GA_3. In a study at Parbhani, Narwadkar and Anserwadkar (1985), tested GA_3 and IAA at 200 ppm and found that GA_3 was ineffective in providing success in epicotyl grafting as compared to IAA and control.

Treatment of defoliated scion sticks of 'Mallika' with NAA (750 - - 1000 ppm) improved the graft take in veneer and wedge grafts on 3-year old root stocks (Satyanarayana, 1982). IBA and NAA proved to be ineffective and even detrimental at higher concentration (1000 ppm) in case of veneer grafts at Sanga Reddy (Kulkarni *et al.*, 1988). Contrary to the promising effects of auxins, Ram and Din (1971) and Ramakrishna (1985) observed that auxin treatments were either ineffective or detrimental in graft take in different mango cultivars at Pantnagar and Sanga Reddy.

2.3.2. Use of gibberellins grafting

In certain fruit crops like mango, citrus etc., the seedling root stocks take much time (> 1 year) to reach of graftable size. This time can be reduced with the aid of GRs. GA_3 is effective in this respect, when it is applied to nursery-rootstock plants. Such GA_3 treated seedling plants make very rapid growth and reach graftable size with in 3 or 4 months of treatment. However, the information on the use of GA in grafting of mango is meager.

Kannan and Rao (1964) reported that GA_3 at 1000 ppm applied to graft union of inarch grafts at the time of grafting as well as pre-treatment of rootstock and scion with GA_3 resulted in acceleration of graft take in mango. Satyanarayana (1982) observed that the treatment of scion shoots of mango with GA_3 at 750-1000 ppm did not affect success in veneer and wedge grafts of cv. 'Mallika'. However, GA_3 at 1000 ppm resulted in increase in number of leaves, length and diameter of scion sticks in case of veneer grafts. On the contrary, the length and breadth of scion shoots of wedge grafts treated with 1000 ppm GA_3 were very poor compared to control. Treatment with GA_3 showed negative effect on graft union of veneer as well as wedge grafts (Mathod, 1983; Ramakrishna, 1985). GA_3 proved to be either ineffective or detrimental at different concentrations (100 to 1000 ppm) in veneer grafts at Sanga Reddy (Kulkarni *et al.*, 1988).

2.3.3. Use of cytokinins in grafting

Cytokinins (CKs) play a major role in cell devision and cell enlargement in plants (Letham, 1967; Ram *et al.*, 1983) and also retard the shoot senescence (Richard and Lang, 1957; Nooden and Leopold, 1978). Kinetin and BAP are often employed in grafting of mango. The primary effect of BAP may be attributed to its well known cell division, cell enlargement and formation of callus cells (Letham, 1967; Ram *et al.*, 1983). CKs are also known for their role in delaying senescence (Thomas and Stoddart, 1980). In case of failure of graft take the scion shows wilting symptoms. If this senescence could be avoided or at least delayed, graft take could be improved. BAP acts in this way and thereby improves graft take. Improvement of graft take with BAP was observed in grafts in veneer grafts of mango (Tiwari, 1970; Ram and Din, 1971; Mathod, 1983; Ramakrishna, 1985; Amarnath Reddy, 1987; Kulkarni *et al.*, 1988). Another remarkable effect of BAP was the acceleration of graft take in mango. Over 80 per cent of graft take was observed with in 30 days of grafting in BAP treated grafts against 40-45 days in control (Amarnath Reddy, 1987).

The graft take of 'Banganapalli' cultivar was improved by dipping the scion sticks in BA solution (10 or 100 ppm) before grafting by veneer method (Kulkarni *et al.*, 1988). Ramakrishna (1985) observed that treatment of scion shoots with BAP at 10 ppm in veneer graft of mango both in nursery and containers recorded maximum percentage graft take, earlier sprouting of grafts, greater scion length, more number of nodes and branches whereas NAA alone or in combination with BAP had little effect at Sanga Reddy. Improvement of graft take with BAP has shown that cytokinin may play an important role in the formation of graft union where BAP could have acted in retarding the shoot senescence (Nooden and Leopold, 1978) or by accelerating the wound healing and callusing needs to be established.

Mathod (1983) stated that treatment of 'Alphonso' scion shoots with kinetin at 500 ppm for an hour before grafting resulted in higher percentage of success in veneer grafting whereas it adversely affected the success of grafts in wedge grafting. Treatment with kinetin at 500 ppm to veneer grafts appreciably increased the length and girth of new shoots on scions, besides increasing the number of leaves and leaf area per plant. Wedge grafts treated with kinetin (500 ppm) took least number of days (22 days) for graft union as against 48 days in control. In case of scions treated with GA_3 in combination with kinetin, some unappreciable success in graft take was observed in both methods of grafting. Kinetin, which delayed graft union in veneer grafting, with GA showed earliness in union, but showed a slight delay for union in case of wedge grafts. Similarly GA in combination with kinetin showed an appreciable increase in length of new growth on scions of veneer grafts.

Tiwari (1970) reported that a mixture of kinetin (25 ppm) and IBA (1000 ppm) provided better success of 74 per cent in veneer graft of mango. Ram and Din (1971) also observed maximum success with a combination of BAP (75 ppm) plus IBA (5000 ppm) in veneer grafts of mango Cv. 'Dashehari' at Pantnagar.

3. Role of PGRs on Plant Growth

The vegetative growth of mango is not continuous, but cyclic. It occurs as intermittent short lasting flushes of shoots from apical or lateral resting buds. Mango exhibits strong apical dominance. While in young and pre-bearing trees strong apical dominance and moderate to vigorous vegetative growth are welcome but they are not required in mature and bearing trees, particularly in those trees / varieties which exhibit over vigorousness and periodicity of bearing. Vegetative and reproductive phase are antagonistic to each other. It is postulated that excessive and untimely vegetative growth phase pushes the trees into unproductive stages or into alternate bearing tendency. Hence, it becomes desirable and essential to curb the excessive and vigorous vegetative growth to promote fruit production. However, in some cases like newly planted young plants it is essential to push the growth for development of strong scaffold frame work to sustain flowering and fruiting in the future. But in other cases control of vegetative growth is essential. Control of vegetative growth vis-a-visa to reduce the plant height is desirable in high density plantings as well as in normal density plantings to keep the trees at manageable size and restrict their spread with in the allotted space in the orchard.

The role of endogenous hormones in plant growth is well known and well documented, while phytohormones such as auxins, gibberellins and cytokinins are known to induce and increase the growth, through cell division, cell elongation and cell differentiation, inhibitors like ABA and ethylene are implicated as inhibitory factors of growth.

Davenport and Nunez-Elisea (1997) proposed that there exist two physiological stimuli that regulate growth and development in mango. The first switch regulates initiation of bud-break, but had no regulatory role in determining its vegetative or reproductive fate. It is proposed that the ratio of promotive and inhibitory components determine when shoot initiation should occur. A root generated cytokinin may accumulate in stem tips over time. An inhibitor found in leaves that decrease with age may antagonise it. Decreases in the availability of auxin to stem tips, with leaf age has been reported (Veen and Jacobs, 1969; Davenport *et al.*, 1980). Davenport and Nunez-Elisea (1997) suggested that a rising ratio of cytokinin to auxin acts, to simulate initiation of bud break in the resting stems of mango, once a critical ratio is reached.

3.1. Role of Gibberellins in plant growth

Gibberellins are necessary for normal shoot elongation in woody plants like mango, as they cause cell elongation and help in extension of growth. Endogenously gibberellins are synthesized in developing leaves and rapid shoot growth is associated with the synthesis of gibberellins (Davenport *et al.*, 2000). The most abundant gibberellic acids in apical stem buds were GA_3 and GA_{19} (Davenport *et al.*, 2000). Endogenous GA3 influences the release of the resting of the buds and it decreases in stem tips or leaves with increasing age of stems. However, investigation of Davenport et al (2000) proved otherwise. They observed that endogenous GA3 increased in the buds with increasing age of the stem. The concentrations of other gibberellic acids were greater in leaves of elongating shoots. These results are consistent with the concept that rapid shoot growth is associated with the synthesis of gibberellic acids in the plant.

Exogenously applied GA_3 delayed bud break but does not determine whether the resulting flush of growth is vegetative or reproductive. The role of exogenous GA_3 in delaying bud-break in mango is not known, but it is proposed that GA_3 may enhance or maintain the synthesis or activity of endogenous auxin, thereby maintain a high auxin / cytokinin ratio, similar to responses to GA_3, that maintains apical dominance (Davenport *et al.*, 2000).

Exogenous spray application of GA_3, irrespective of concentration, time of application, variety and location of trial, increases vegetative growth and shoot elongation (Singh and Singh, 1972-74; Kikani, 1986; Singh and Rajput, 1990) and curbs reproduction (Kachru *et al.*, 1971).

3.2. Role of growth inhibitors and retardants in growth

Abscisic acid (ABA) is known to inhibit the growth of shoots in woody plants. ABA may inhibit the shoot growth by decreasing gibberellin levels in shoots. ABA is antagonistic to gibberellins in two ways: i) Prevents gibberellin action at the level of transportation by suppressing gibberellin – induced m-RNA, ii) Promotes the synthesis of protein (building blocks) inhibitors (Jacobson and Chandler, 1982).

Vigour of the trees (here root stocks) appears to be associated with endogenous levels of ABA, cytokinins and auxins. Low levels of ABA and higher levels of cytokinins (t-ZR and DHZR) in the plant sap and leaves and IAA in leaves were evident with vigorous root stock seedlings of 'Muvandan' and Olour and reverse trends were associated with the above hormones in less vigorous rootstock seedlings of 'Vellaikolamban' and 'Kurukkan' at Bangalore (Murti *et al.*, 2003).

Cycocel and Daminozide at different concentrations (1000-4000 ppm) applied to differentially aged mango trees inhibited the vegetative growth of trees in West Bengal

(Maiti *et al.*, 1972a). However, Kikani (1986) did not get alteration in vegetative growth with the application of cycocel in mango.

B_9 (Alar) or MH applied (1000-7000 ppm) delayed the appearance of subsequent flushes from 35 days in the control to 52-114 days (B_9) and 90-114 days (MH), besides increasing the diameter of shoots of 'Banganapalli' mango trees at Bhubaneswar, Odissa (Das and Panda, 1976). On the other hand, MH (1500 ppm) coupled with ringing increased the diameter of the shoots, reduced the shoot length and vegetative growth in 'Langra' mango trees, when applied as foliar sprays in mid-October and first week of November at Bhubaneswar (Rath and Das, 1979). The combined treatment of pruning and BA markedly increased shoot production, when applied as foliar spray (10 ppm) at 15 days interval from mid-July at Bangalore (Murti and Upreti, 1998).

Paclobutrazol (PBZ) is a synthetic growth retardant and is a broad spectrum plant growth retardant. It is a strong inhibitor of gibberellins synthesis in plants, hence often referred to as "anti-gibberelline'. PBZ is well known as suppressor of vegetative growth and promoter of flowering. The major biochemical responses to the application of PBZ (Cultar) are the suppression of gibberellins synthesis by inhibiting the oxidation of Kaurene to Kaurenoic acid in gibberellin pathway (Jones, 1973). The reduction of gibberellin synthesis reduces cell division and cell elongation (expansion) and as a direct morphological consequence, vegetative growth is reduced (Dalzial and Lawrence, 1984).

PBZ treatments alter hormonal balance in treated plants of mango. There has been increased in the levels of ABA and cytokinins (t-ZR and DHZR) in the treated plants (Murti *et al.*, 2001). It is possible that the increase in ABA and cytokinins as a result of PBZ application could be related to increased availability of cytokinins as a result of inhibited gibberellin biosynthesis (Murti *et al.*, 2003).

Pre-flowering (before fruit bud differentiation / 90-120 days before bud break) application of PBZ, particularly soil application, significantly suppressed vegetative growth of woody plants including mango in terms of reduction in tree height, annual growth, length of terminal shoots, length of internodes, number of leaves, leaf area, percentage of lateral shoots, emergence of new shoots, number of shoots per tree and number of dormant shoots. Application of PBZ resulted in high xylem, phloem ratio and decrease in xylem sap flow in treated plants (Murti *et al.*, 2001). The above effects of PBZ were observed particularly more in 'off' year and after post-harvest pruning in mango (Williams, 1984; Kulkarni, 1988; Voon *et al.*, 1991; Srihari, 1995; Murti and Upreti, 1998, Chandrasekhar, 1999; Singh and Singh 2003; Padhiar and Patel, 2003; Rajkumar *et al.*, 2007a; Karuna and Mankar, 2008; Singh, 2008; Singh *et al.*, 2010; Salvi *et al.*, 2011; Sarkar and Rahim, 2012; Bhagwan *et al.*, 2013; Srilatha, 2015).

4. Flowering and Plant Growth Regulators in Mango

4.1. Physiology of flowering in mango

Flowering is an important and essential event in life of the plants, for fruit production. The events / sequence of flowering in mango are as follows: Once mango trees enter into reproductive phase (bearing phase) they flower and fruit more or less every year. During the flowering there will be a simultaneous vegetative flush of shoots to a lesser extent. After the fruit harvest, there will be another vegetative flush of shoots, which is heavy. During autumn – winter period, another vegetative flush appears, which is small and optional. All these growth flushes are potential flowering shoots. After growing to certain length / for some time, these growth flushes enter into a quiescent stage and proceed to mature. After a few months of quiescent period, flower induction and initiation begin in these matured shoots in response to changes in nutrients (particularly carbohydrates), endogenous phytohormones, environmental conditions (particularly temperature) and soil water stress. It is believed that during vegetative flushing gibberellin levels will be high, while during pre-flowering and flowering periods, their levels will be lowered and the levels of flower promoters rise in the would be flowering shoots.

4.2. Role of PGRs on flowering

Singh (1961) postulated that flower induction in mango is governed by the rhythms of 'on' and 'off' year conditions of the tree, rather the age, length and maturity of shoots. However, many mango researchers did not agree with Singh (1961) and believed that flowering in mango is under the control of phytohormones, besides nutrition and environmental conditions, notably temperature and soil water stress. Growth active substances seemed to play a great role (s) in controlling and modifying most of the physiological processes in plants including flowering. All classes of phytohormones have extremely wide action spectrum and interact quantitatively to regulate different processes of plants such as flowering. Several attempts have been made to correlate the qualitative and quantitative changes in the levels of endogenous growth regulators with flower initiation and flowering in mango trees. In mango, vegetative growth and flowering seem to be controlled by the quantitative changes in the levels of endogenous growth promoters and growth inhibitors, besides their balances / ratios. The roles of different classes of phytohormones in mango flowering have been reviewed by many authors (Chacko, 1984; Chacko and Pal, 1985; Pandey, 1989).

The roles of phytohormones were determined based on their relative levels in leaves, shoots and particularly in shoot tips just prior to flower initiation / flower bud differentiation, during panicle emergence and flowering and effect of exogenously applied synthetic growth regulators at the above different phases of flowering.

A brief account of probable role of different plant growth regulators and the effects of exogenous application of growth regulators in mango flowering is presented in the following pages.

4.2.1. Auxins in flowering

Though several workers concluded that auxins play a role in floral induction in woody plant species including mango, there is no fool – proof evidence to support this conclusion. Auxin levels in leaves, shoots and shoot tips, prior to and during flowering as well as during vegetative growth were measured (Chacko *et al.*, 1972c; Chen, 1987; Paulas and Shanmugavelu, 1982) who recorded fluctuations in the levels of auxins at different stages of flowering and vegetative phase. Many recorded higher levels of auxins during pre-flowering and flowering periods. Chacko *et al.* (1972c) observed increase in the levels of auxin-like substances in the stems of two cultivars of mango from September to December. Floral initiation occurs during this period (Northern hemisphere). They concluded that a high level of auxins was necessary for floral initiation in mango. Jagdande *et al.* (2000) and Jagadande and Choudhari (2001b) also concluded that higher level of auxin-like substances increased in the shoot tips at fruit bud initiation, fruit bud differentiation and panicle emergence stages which is responsible for flowering. Thus auxins appear to play an important role in the flower initiation and flowering in mango. This is supported by the increased flowering due to exogenous application of auxins (Ram, 1996). In contrast, Chen (1987) using physico-chemical purification and detection methods reported that higher levels of IAA diffused from developing vegetative shoots and the levels of diffusible IAA were lower at the early fruit bud formation and full bloom in 'Irwin' mango in China. The synthesis and transmit of auxins in leaves gradually decrease with age (Veen and Jacobs, 1969; Davenport *et al.*, 1980). With return of the apical buds to quiescent status and the distribution of leaves along the distal intercalary unit sources of auxins from leaves become diffused and are likely to decrease with age. This may account for lower levels of auxins found at sampling times other than during vegetative shoot development (Chen, 1987).

Several studies have compared auxin levels in shoots of mango in relation to flowering in 'on' versus 'off' year of some of alternate bearing cultivars (Chacko *et al.*, 1972c; Choudhury *et al.*, 1972; Singh and Singh, 1972; Dhatt, 1973; Mishra, 1976; Lal and Ram,

1977; Nabway *et al.*, 1984). Chacko *et al.* (1972c) reported higher levels of extractable auxin-like compounds in shoots of 'on' year trees during autumn. Lal and Ram (1977), however, found no measurable differentiation in free IAA, between shoot tips of 'on' and 'off' years as determined by an Avena Coleoptile bioassays. Studies of Chacko *et al.* (1972c), Choudhury *et al.* (1972) and Singh and Singh (1972) revealed that shoots of mango in 'on' year contained auxins at higher level than those of 'off' year, corresponding with the period of fruit bud differentiation (in North India). Shoot tips of 'Dashehari' mango in 'on' year contained about 35 times more of auxin than those in 'off' year (Lal and Ram, 1977).

The conflicting results mentioned above regarding the role of auxins in flowering of mango likely reflect inconsistencies in purification and analytical methodologies. These range from use of relatively inefficient purification and separation methods, use of bioassays, susceptible to multiple components influence (both negative and positive) and use of field grown trees from varying backgrounds and environments. The lack of consistent correlation between both endogenous levels and responses to exogenous auxin application with mango flowering most likely reflect the lack of direct role for this hormone in mango floral induction (Davenport and Nunez-Elisea, 1997).

Several lines of evidence indirectly implicate auxin involvement in mango flowering process probably through regulation of the timing of shoot and root initiation. It is also possible that auxin is required for ethylene production, which is essential for flower induction. Ethylene production is closely related to internal levels of auxin. Thus, auxin (IAA) at higher levels could induce the formation of ethylene which, in turn, would suppress further auxin synthesis and stimulate IAA oxidase, peroxidase and polyphenol oxidase enzymes and set a limit for the auxin.

Exogenous auxin can both inhibit and promote flower initiation in a number of fruit crops. In several plants, auxins are promoted at low doses and inhibitory at higher doses. Inhibition at higher doses might simply be due to a general growth inhibitory activity (Jacobs, 1985) or an auxin-induced ethylene biosynthesis (Krekule *et al.*, 1985). Manipulation of flowering by sprays of synthetic substances at appropriate times and intervals may be helpful to induce flowering in mango, particularly during 'off' year and 'off' season.

Inadequate supply of auxins to the buds results either in delayed or no flowering. Pre-flowering sprays of NAA on 9-year-old 'Sunderja' cultivar caused earliest panicle emergence and shortened flowering span, which may be due to the attainment of certain balance of auxin in the tree with the additional supply of NAA at Jabalpur (Baghel *et al.*, 1991). Sen and Maiti (1965) obtained increased flowering in mango with IAA, NAA, MH and TIBA sprays. The spraying of IAA was the most effective when applied in May on 'Langra' and in 'November' in 'Himsagar' cultivars of mango in West Bengal. The aqueous sprays of auxins, IAA and 2, 4, 5-T at probable time of flower bud differentiation (FBD) failed to increase panicle formation in the 'off' year phase of mango trees. There was no specific response to MH and TIBA (Singh and Singh, 1963). Likewise, Dhatt (1973) did not obtain any enhancement in the panicle initiation in mango following application of NAA.

4.2.2. Gibberellins in flowering

Gibberellins, endogenous or exogenous are considered to be anti-flowering. It has been suggested that gibberellins enhanced vegetative growth and inhibited flowering in mango (Tomer, 1984; Chacko, 1984; Chen, 1987) and this is supported by earlier studies of Kachru *et al.*, (1971) with exogenous application of GA_3 on mango.

Gibberellin levels fluctuate in plants as per the growth and developmental stages. It was reported that levels of gibberellins are high during vegetative growth and low at flowering. Higher levels of gibberellins are antagonistic to the formation of flower primordia and low levels of endogenous gibberellins favour fruit bud initiation (Kachru *et al.*, 1971, Chen, 1987). Higher levels of gibberellins at the quiescent stage might have been utilized for bud burst and again higher level of gibberellin at flowering and fruiting are required for cell elongation (Paulas and Shanmugavelu, 1989). It was suggested that gibberellin primarily acts on cell elongation indirectly releasing the flowering response in some cases, rather than directly on flowering itself (Stowe and Yamaki, 1957). Chen (1987) reported the highest levels of gibberellin in mango xylem sap during leaf differentiation and lower concentration during rest, panicle emergence and full bloom in 'Irwin' mango in China. Dhatt (1973) recorded lower levels of gibberellin in reproductive shoots of 'Langra' and 'Totapuri Red Small' cultivars as compared to the vegetative shoots during panicle initiation period and identified as a complex of gibberellins, consisting of GA_3, GA_4 and GA_7. Later on Mishra (1976) proposed that higher levels of gibberellin suppressed the differentiation of vegetative shoots into flowering buds.

Tongumpai *et al.*, (1991) also claimed that flowering of 'Khiew Saweier' mangoes is related to the reduced level of gibberellin in the shoot tips. Low content of gibberellins allows accumulation of non-structural carbohydrates particularly starch, in the leaves and buds leading to the formation of flower initials (Protacio *et al.*, 2000b).

Gibberellin – like substances in shoot tips of 'Dashehari' mango were found to be higher in the 'off' year than in the 'on' year (Chacko, 1984). Pal and Ram (1978) earlier also reported that gibberellins were at higher level in the vegetative 'off' year than in the following 'on' year of alternate bearing cultivars. In 'Dashehari' mango, Chacko (1968) reported a lower level of gibberellin – like substances in the 'on' year shoots at fruit bud differentiation time as compared to vegetative shoots.

The role of gibberellins in the flowering of mango shoots appears to be quantitative. Absence or presence of these compounds at low concentration in 'on' year shoot tips from October to February corresponding with the period of FBD indicate that higher levels of gibberellins were not accompanied by a beneficial effect on FBD and flowering of mango shoots. On the other hand, presence of these compounds at higher levels from October to February in 'off' year was accompanied by inhibitory effect on flowering (Pal and Ram, 1978).

The exogenous application of GA_3 at higher concentration has been reported to inhibit flowering and to induce vegetative growth in mango (Kachru *et al.*, 1971, 1972). Foliar spray of gibberellin is effective in suppressing flowering even under induction conditions (Kachru *et al.*, 1971) however, GA_3 did not inhibit flowering when applied after the period of FBD.

The most commonly observed response of mango to GA_3 application prior to bud break in the autumn is delayed flowering. Kachru *et al.* (1971) first reported a delay in flowering which was quantitatively correlated with the concentration of GA_3 applied in lanolin to apical buds. Bud-break of resting buds treated with the highest levels of GA_3 was delayed until March and grew vegetatively whereas resting buds treated with lower concentration initiated growth earlier (during the cool inductive period) and produced reproductive shoots. Similar responses were observed by Tomer (1984). Nunez-Elisea and Davenport (1991) reported quantitatively correlative delays in inhibition of axillary shoots. When GA was applied to de-blossomed stems, higher GA_3 concentrations (upto 250 ppm) caused longer delays in shoot initiation. GA_3 did not inhibit induction of flowering, as evidenced by the production of axillary panicles, so long as cool inductive temperatures were present during axillary shoot initiation. Late inhibiting buds which grew during warm spring temperature formed vegetative shoots (Shawky *et al.*, 1978; Turnbull *et al.*, 1996).

4.2.3. Cytokinins in flowering

The phytohormones such as cytokinins (CKs) play an important role in the physiology of growth and differentiation in woody plants. The role of CKs has been emphasized in flowering of other plants. But only a limited evidence for their involvement in floral initiation in mango is known. Relationships between flowering in mango and the endogenous levels of CKs in leaves (Paulas and Shanmugavelu, 1989; Kurian *et al.*, 1992), stem tips (Agrawal *et al.*, 1980) and xylem sap (Chen, 1987) as well as the effects of CK application exogenously on bud break and shoot development (Singh and Singh, 1974; Chen, 1985; Singh and Rajput, 1990; Oosthuyse, 1991) have been reported.

The levels of CKs were higher in the reproductive shoots of 'Langra' mango compared to vegetative shoots (Dhatt, 1973). CK content was the highest at the quiescent stage and at bud stages. The least content was at the fruiting stage. At flowering stage, it was higher than at the fruiting stage, but lower than at the quiescent and bud burst stages (Paulas and Shanmugavelu, 1989).

It was concluded that higher levels of CK at the time of FBI / FBD favour flowering in mango. The elevated CK levels found prior to and during the flowering response to applied BA led to the conclusion that CKs are involved in flowering of mango (Chen, 1985, 1987). However, such a response can also be explained if cultivars are involved in stimulation of bud break not necessarily in flowering *per se*.

Agrawal *et al.*, (1980) reported relatively higher levels of total measurable CKs in 'on' year shoot tips than in 'off' years ones at the time of FBD in 'Dashehari' cultivar during October which subsequently declined until January and then increased to October concentration in February of the 'on' year of mango. These results suggested that flowering in mango is associated with the levels of endogenous CKs and is associated with high levels of endogenous CKs. The CKs are present in different ratios and their balance is critical for flowering. Chen (1987) found increased levels of CKs and drastically reduced gibberellin levels during flowering in 'Irwin' mangoes in the xylem sap compared to vegetative growth stages.

The primary CKs found in higher plants like mango are zeatin, dihydro zeatin, isopentenyl adenine and their ribosides. They are translocated from roots and accumulate in resting stems (Hendry *et al.*, 1982a, b). The lowest levels of putative trans zeatin and its riboside were translocated from roots during the vegetative growth and resting stages. Whereas, the highest levels were measurable during the early flowering and full bloom stages (Chen, 1987). In contrast, Paulas and Shanmugavelu (1989) observed no significant differences in CK levels of both fourth and fifth leaves of shoots during resting bud and flowering stages.

Exogenous CKs cause promotion and inhibition of floral initiation in a variety of species, although promotion effects are much more frequent than inhibitive ones (Bemier, *et al.*, 1981a,b). Precocious bud break and flowering in mango shoots were reported following exogenous application of synthetic cytokinin (Thidiazuron) (Nunez-Elisea *et al.*, 1990). Kurian *et al.* (1992) reported a substantial reduction in several CKs in mango leaves in the limp red leaf stage from trees treated with paclobutrazol. Concurrent with this response was suppression of bud initiation and reduced internodal length of shoots for a couple of years of application of paclobutrazol.

Despite the above, the role of CKs in flowering is still unsolved. The few available reports also are conflicting to draw a definite conclusion, primarily due to sampling of different organs at non-comparable times and / or conditions.

4.2.4. Growth inhibitors and retardants in flowering

Mango is a vigorous tree exhibiting more vegetative growth than necessary for production. Vegetative and reproductive growth in general are antagonistic to each other. Vegetative growth utilizes most of the photosynthates soon after their synthesis without a chance for accumulation of photosynthates. Accumulation of starch and carbohydrates are required for induction of flowering in mango and other woody plants. It was conceived that by inhibiting vegetative growth, the accumulation of carbohydraes can be achieved, pushing the plant into flowering (Haribabu and Jyothi, 2004). For this purpose growth inhibitors / retardants like Cycocel, MH, PBZ etc are generally employed.

4.2.5. Abscisic acid in flowering

Abscisic acid (ABA) is the endogenous phytohormone, a growth inhibitor in plants, performing several physiological functions. Its relative levels in plants influence different physiological processes like vegetative growth, flowering and fruiting of plants. The levels of ABA were estimated in different parts of the plant like leaves, shoots etc. at different stages.

Chacko (1968) recorded high levels of ABA in the reproductive shoots of 'Dasheheri' mango as compared to the vegetative shoots at the time of FBD at Delhi. Sen and Choudhary (1969) and Choudhary *et al.*, (1972) also reported that during the active stage of vegetative growth, the inhibitor level was low and the level rose sharply during growth cessation tage in mango. The experimental findings of Dhatt (1973) also indicated higher level of ABA at the time of FBD. Mishra (1976) claimed that higher level of ABA in relation to GA favoured flowering in cv. 'Langra', but FBD was drastically retarded with very high level of ABA – like substances. Chen (1987) in China found higher levels of total free ABA during early FBF and at full bloom in 'Irwin' mango. Contrary to the above reports,

Jagdande and Choudhari (2001a) found higher activity of ABA at all the stages of shoot development in all the cultivars of mango. The content of ABA increased as the time progressed. Hence, it was inferred that ABA like activity may check vegetative growth, there by providing conditions suitable for FBI and higher levels of ABA are conducive to the process of flowering in mango (Jagdande and Choudhari, 2001a). According to Ram (1999) higher levels of inhibitors might be antagonising vegetative growth by counter acting gibberellin activity, its synthesis or both.

The levels of ABA seem to differ in 'on' and 'off' years and also between cultivars. Chacko *et al.* (1972b) reported relatively higher level of ABA-like substances in the shoots of 'Dashehari' and 'Totapuri Red Small" during the period of FBI in the 'on' year than in the shoots of 'off' year, indicating that ABA may be involved in the induction of flowering in mango.

The differences in ABA equivalent were very distinct especially during the critical period of transformation of vegetative tissues to reproductive ones. Higher levels of ABA activity was helpful for the phenomenon of floral differentiation in mango. The higher levels of ABA may help in early cessation of shoot growth and for creating a suitable quantitative balance of different hormones for induction of flowering.

It is clear that higher levels of ABA are conducive to the process of flowering in mango. Absence of adequate levels of ABA in mango throws the trees into unfruitfulness partially. Manipulation of ABA levels to attain threshold level, by spraying of synthetic substances at appropriate time and intervals would be helpful to induce flowering, particularly in the 'off' season. However, spraying of the growth substances is not always useful in inducing flowering. Hence, besides, ABA, other factors may be involved in the control of growth, flowering and development of mango trees.

4.2.6. Synthetic growth retardants in flowering

Several attempts have been made to use the plant growth retardants to stimulate early and / or profuse flowering especially in the 'off' years of alternate bearing cultivars of mango (Maiti and Sen, 1968; Maiti *et al.*, 1971; Das and Panda, 1976; Vacillauime, 1991). Basically, growth retardants like Alar, cycocel and paclobutrazol have promoted flowering and fruiting by checking the vegetative growth of treated plants (Maiti and Sen, 1968; Maiti *et al.*, 1971, 1972; Sergent *et al.*, 1997).

4.2.6.1. Alar and Cycocel

Early experiments tested the efficacy of foliar application of the growth retardants: Alar and cycocel for increasing flowering in 'off' year of alternate bearers of mango (Maiti *et al.*, 1972a; Mukhopadyay, 1978; Rath and Das, 1979; Suryanarayana, 1980; Rath *et al.*, 1982; Ou and Yen, 1985). Later studies examined their potential use in stimulating early flowering (Suryanarayana and Rao, 1977; Chen, 1985; Kurian and Iyer, 1993, 1994) and enhanced flowering, but inconsistent flowering was achieved in response to these compounds, especially cycocel.

Two sprays of Alar (500 ppm) at fortnight intervals from mid-October gave highest number of panicles per shoot (13.3) compared with 1.6 in control in 25-year-old 'Alphonso' mango trees at Dharwad (Rao and Ravishankar, 1992). Alar (500 ppm) however, did not induce flowering in 'Dashehari' mango (Pal *et al.*, 1984). Alar (200 ppm) coupled with ringing reduced the length and diameter of shoots and number of leaves per shoot, increased the percentage of flowering shoots and advanced flowering in 'Langra' when sprayed twice in mid-October and first week of November at Bhubaneswar (Rath and Das, 1979).

Cycocel sprayed thrice at fortnight intervals starting from mid-May when the trees were in active vegetative phase, induced flowering in treated trees in West Bengal (Maiti *et al.*, 1971). In another trial 7 monthly sprays of cycocel (500 ppm) from May onwards, not only increased the percentage of flowering shoots, but also induced the axillary buds to flower, on the treated shoots, which flowered along with apical buds and thereby significantly increased the number of panicles per shoot. The effect was considered to be due to potential loss of apical dominance brought out by repeated sprays of cycocel (Maiti *et al.*, 1971).

CCC caused early and complete cessation of shoot growth and significantly promoted flowering of 'Langra'. The results confirmed that elongation of growth of shoot is inversely related to flowering and any manipulation treatment which causes cessation or retardation of shoot growth is likely to promote flowering. As in the case of girdling, the growth retardant effects of the chemical were also associated with accumulation of carbohydrates. The relation between carbohydrate accumulation in shoots and consequently high C/N ratio and flowering, was however, indirect. Accumulation of surplus photosynthates especially reserve polysaccharides was primarily due to their non-utilization in vegetative growth of shoots. In any case, such a nutritional status of tissues may favour biosynthesis of flowering stimulus. Changes in the levels of inhibiter (ABA) are also related to growth and flowering of 'Langra' mangoes. If growth could be retarded more effectively, thereby inducing a prolonged growth cessation phase, a rise in the level of the inhibitor would also be expected. In view of the above it was postulated that in mango cycocel might have promoted flowering through its direct effects : (i) Causing an increase in levels of ABA and ii) reducing the levels of endogenous gibberellins (Chaodhuri and Rudra, 1971).

The application of cycocel increased the formation of inflorescences in 'Totapuri Red Small" and in the first 'off' year of 'Langra' and 'Malda' but no such effects were exhibited during the second year of 'Langra' indicating that persistence of cycocel is very short, confined to one season only (Dhatt, 1973).

A single spray of CCC (2500, 5000 and 7500 ppm) in October on 'Pairi' mango trees advanced flowering date by 2 to 21 days with increased concentration at Akola (Dalal *et al.*, 2005). Reduction of vegetative growth due to CCC (500 and 1000 ppm) on 25-year-old 'Dashehari' trees at FBD i.e. first fortnight of October was attributed to inhibitory effect of CCC on the biosynthesis of gibberellins at subapical meristem at Ludhiana (Singh *et al.*, 2004). However, CCC did not show any effect on flowering. Earlier, based on 5 experiments, Pal *et al.* (1979) also reported that even 5 sprays of cycocel or Alar at 500 ppm at monthly intervals starting from September failed to stimulate flowering in mango, which may be due to immature conditions of the shoots.

Application of CCC (500 ppm) followed by ethephon (200 ppm) application increased over all flowering and also increased flowering in fruited terminals of the previous year as well as on current terminals of October flush, that grew after fruit harvest (Kikani, 1986). This suggests for getting flowering cycocel may be used with other growth regulators like ethephon as sole application of CCC is ineffective.

The effect of CCC on induction of flowering in mango may be attributed to its established inhibitory effect on gibberellin synthesis, since gibberellins are antagonistic to flowering in mango. It is of significance to note that CCC application influenced the activation of α-amylase, peroxidase and IAA oxidase enzymes. Ravishankar (1987) concluded that CCC may be acting via influencing the activities of these enzymes, with a possible effect on ethylene content.

4.2.6.2. Triazoles

Excessive and prolonged vegetative growth results in sparse or no flowering in woody species. Post-harvest pruning, pruning for canopy management, tree size control and rejuvenation of old and senile trees results in vigorous and excessive vegetative growth at the cost of flowering. Some cultivars of mango by nature are over vigorous producing little crops. In the alternate and biennial bearers of mango particularly in 'off' years, trees produce excessive vegetative growth and less flowering. Excessive growth generally occurs due to higher levels of nitrogen and gibberellins in the shoots, leading to poor flowering. Vegetative growth thus is antagonistic to flowering. So, reduction / retarding of vegetative growth may results in flowering. It is established that excessive vegetative growth is mainly due to gibberellins in plants. Hence it is conceived that reduction of levels of gibberellins in the shoots will induce flowering. In other words flowering results when the endogenous gibberellins decline during the flower inductive period. Decline in endogenous gibberellin levels under natural growing conditions is either slow or nil depending upon the prevailing growing conditions and rhythm of bearing. It is possible, in such situations, to reduce the endogenous gibberellin levels through the use of anti-gibberellin hormones such as triazoles, which are reputed as inhibitors of biosynthesis of gibberellins / reduce the levels of gibberellin in the shoots thereby pushing the trees into reproduction.

The flower induction cycle of mango tree is in the autumn season (September to October) in different mango growing regions of India. But at times, the flowering is prevented / delayed by the flower inhibitors (probably gibberellins) until the build up of sufficient flower promoters to counteract the action of flower inhibitors or sustenance of high levels of gibberellins in the shoots. Fruited shoots generally do not produce new shoots after fruit harvest (required for next season flowering) and flowering in the ensuing season especially in alternate bearers of mango and remain in quiescent condition during next one or more seasons, probably due to inadequacy or imbalance of phytohormones in the shoots in spite of favourable environmental conditions.

In the above situations, triazoles appear to be helpful in achieving the flower induction stage much earlier and make the fruited shoots to produce new shoots for ensuing season for flowering. Keeping the above in view, several attempts have been made to study and utilize the triazole compounds (paclobutrazol and uniconazol) to induce flowering in cultivars showing excessive and vigorous vegetative growth and in alternate / biennial bearing cultivars during 'off' year, for early and regular flowering in mango.

In India, first attempt in this direction was made by Kulkarni (1988), followed by others both in India and abroad, with different cultivars, doses and methods of application (Rukayah, 1990; Burondkar and Gunjate, 1991, 1993; Charnivichit *et al.*, 1991). The triazole growth retardants – uniconazol and paclobutrazol, the well known gibberellin inhibitors, reduce the putative vegetative promoters, thought to be gibberellins and thereby suppress vegetative growth and cause early, regular and profuse flowering in mango (Kulkarni, 1988;

Rukayah, 1990; Burondkar and Gunjate, 1991; Charnvichit *et al.*, 1991). Between the two triazole compounds, uniconazole is reported be more effective than paclobutrazole. However, paclobutrazole is more commonly used in mango and is formulated and sold as 'Cultar'. PBZ promotes flowering and regulates the time of flowering (Kulkarni, 1988), induces flowering and fruiting regularly on newly produced shoots after fruit harvesting and pruning (Ram, 1996; Ram *et al.*, 1996, 1997).

PBZ reduced vegetative growth, thereby induced early flowering and contributed to higher production in rejuvenated old and over crowded and over vigorous mango trees (Lal *et al.*, 2003; Mistry and Patil, 2003). PBZ reduced the juvenile period in mango seedlings (Kulkarni, 1988; Chusri *et al.*, 2008). Soil application of PBZ prior to FBD inhibited vegetative growth, thereby controlled tree vigour and contained tree size and enhanced flowering in mango in high density planting systems (Kulkarni, 1988; Burondkar and Gunjate, 1993; Kurian and Iyer, 1993; Murti and Upreti, 2000). Soil application of PBZ on mango trees enhanced flowering in young as well as older mango trees (Haw, 1986; Hongasbhanich *et al.*, 1986; Kulkarni, 1988). PBZ also reduced the alternate bearing habit of mango (Hiller and Rudge, 1991; Burondkar and Gunjate, 1993; Hoda *et al.*, 2001).

Several studies indicated that PBZ induced flowering by inhibiting vegetative growth *vis-a-vis* biosynthesis of gibberellins, the vegetative promoter and flower inhibitor. According to Voon *et al.* (1991) PBZ could intensify flowering by blocking the gibberellin synthesis by interrupting Kaurene biosynthesis path way. However, the PBZ inhibition of tree vigour and promotion of flowering in mango is not only associated with gibberellin inhibiting property of PBZ, but also to its influence on conducting tissues (phloem and xylem) and levels of other endogenous hormones, PBZ treatment caused an enhancement in the levels of ABA and CKs [Zeatin riboside (t-ZR) and dehydrated zeatin riboside (DHZR)] in the xylem sap and leaves and phenols in leaves, concomittant with decreases in leaf IAA content (Murti *et al.*, 2001). Kurian *et al.* (1992) observed reduce CK levels in leaves of PBZ treated mango trees with inhibition of bud initiation in treated plants. Elevated concentration – dependent level of phenolic compounds was also found in resting apical buds of PBZ treated trees (Kurian *et al.*, 1994). These suggest that lower activity and high phenolic levels in buds contributed to the inhibition of vegetative growth in PBZ treated trees. Studies revealed that PBZ also increased the activation of auxin, CKs and ethylene besides its own activity (Ram *et al.*, 1996). PBZ enhanced 1-amino cyclopropane 1-carboxylic acid (ACC) levels which coincided with the development of floral initials in treated trees indicating possible involvement in the enhancement of the ethylene biosynthetic pathway in floral initiation in mango trees (Protacio *et al.*, 2000a).

Starch content increased in stems of PBZ treated trees suggesting that PBZ promotes flowering by increasing the starch accumulation in shoots (Protacio *et al.*, 2000a). According to Upreti *et al.* (2014) PBZ-induced flowering is accompanied by an increase in starch in leaves, concomitant with an increase in soluble sugars (sucrose, glucose and fructose) in apical buds. The inhibition in the activity of amylase in leaves in association with induction in the activities of AI, SS and SPS following PBZ treatment contributed to higher levels of sucrose, glucose and fructose in apical buds and induction in FBF in mango (Upreti *et al.*, 2014). Higher total non-structural carbohydrates (TNC) in the shoots of the PBZ – treated trees two weeks before flowering compared with control has been reported (Yeshitela *et al.*, 2004). They also stated that the increased number of panicles in the PBZ treated plants was due to low utilization of tree reserves in vegetative growth parameters and consequently no assimialte limitations compared with over / excessive vegetative growth on the control trees.

Physiological maturity of the shoots by the time of FBD in autumn is one of the pre-requisite for flowering. Under normal conditions, some times this maturity is either slow or delayed. In such cases PBZ appears to help to cut short the time required for maturity by reducing vegetative growth and manipulating the endogenous factors of maturity. According to Kulkarni (1988) it is possible that PBZ caused on early reduction of endogenous gibberellin levels in the shoots causing them to mature earlier than those of untreated trees, favouring early flowering.

4.2.7. Ethylene in flowering

Mango trees in Philippines are induced to flower by exposing the trees continuously to smoke from burning leaves, the process called '**smudging**'. Possible causal factors included heat (Bueno and Valmayor, 1974) and Carbondioxide [CO_2] (Gonazales, 1933). Rodriguez (1932) investigating smoke induced flowering of pineapple, first proposed that 'ethylene' generated by burning material may stimulate flowering. Dutcher (1972a,b) later confirmed that smoke from smudge fires contained ethylene which stimulated flowering in mango trees. These results coupled with the first successful use of ethephon (an ethylene releasing chemical compound) in 1968 by F. Mannel and others (Bondad, 1972, 1976) to promote mango flowering, led to the hypothesis that endogenous 'ethylene' plays an integrated role in the floral inductive process (Bonded, 1976; Chadha and Pal, 1986). Patil (1984) also reported that flowering in mango was controlled by ethylene. Numerous investigations have since shown 'ethylene' to be an effective floral promoter of some mango cultivars under specific conditions found in the low latitude tropicals (Singh and Singh, 1963; Choudhuri and Rudra, 1971; Chacko and Randhawa, 1971; Chacko *et al.*, 1972a, b, 1974a; Maiti *et al.*, 1972b; Sen *et al.*, 1973; Nunez-Elisea *et al.*, 1979; Rath and Das, 1979; Chen, 1985; Das *et al.*, 1989 etc).

It is clear that endogenous ethylene is involved in flowering in mango. Saidha *et al.* (1983) observed a gradual increase in internal leaf ethylene production as the season of floral initiation approached. Ethylene production by flowering shoots was upto five fold than vegetative shoots, suggesting a higher internal leaf ethylene levels to be an important factor in the flowering of mango. There was a gradual increase in leaf ethylene levels from June to November with a sharp increase in December, coinciding with panicle emergence. Based on the above, a possible role of ethylene in the flowering of mango was suggested (Saidha *et al.*, 1983). Higher ethylene levels found in flowering shoots of different mango cultivars compared to vegetative shoots further support the above contention.

Shoots of normal year had higher leaf ethylene level than shoots produced in the 'lean' year. Correspondingly there was increased flowering in the 'normal' year compared to the 'lean' year. Thinning of shoot clusters appeared to give an internal impetus to the internal leaf ethylene level and also provided better conditions for growth and differentiation of the remaining shoots, there by increase the percentage of flowering shoots (Saidha *et al.*, 1983).

Involvement of endogenous ethylene in the process of flowering was also supported by observations that indirectly link symptoms of ethylene production to the flowering. Extrusion of latex from terminal buds at the time of inflorescence initiation and epinasty of mature leaves located near the apex during expansion of the panicle provide indirect evidence to the role of ethylene in flowering (Davenport and Nunez-Elisea, 1990, 1991). Both are classic symptoms of plants exposed to high levels of either endogenously or exogenously generated ethylene (Abeles, 1975). Indirect – support also comes from reports that potassium nitrate (KNO_3) stimulated flowering in mango is mediated by the increasing levels of endogenous ethylene (Thuckthye, 1978; Lopez, 1984; Lopez *et al.*, 1984). Mosqueda-Vazquez and Avila-Pesendiz, 1985) reported that the efficacy of KNO_3 was negated by calcium chloride ($CaCl_2$) and silver nitrate ($AgNO_3$) compounds which inhibit the synthesis and action of ethylene respectively when sprayed 1-4 hours after the application of KNO_3. Pruning and thinning treatments increased the leaf ethylene level and the perception of flowering. Thus there is sufficient evidence that ethylene in general plays a crucial role (Chacko *et al.*, 1972a) and leaf ethylene levels in particular (Saidha *et al.*, 1983; Ravishankar, 1987) in the regulation of flowering in mango.

In spite of the above, the role of ethylene in flowering is still unsolved. Ethylene may simply stimulate shoot initiation through its activity to inhibit auxin transport in leaves and stems (Riov and Goren, 1979, 1980; Ramina *et al.*, 1986). Davenport and Nunez-Elisea (1991) later reported that the consistent lack of correlation between ethylene production and flowering suggested that floral induction in mango might not be mediated by increased ethylene synthesis in leaves or buds. The requirement for additional factors for induction of flowering, such as inductive conditions, aged leaves and shoots, specific cultivars and / or equatorial location for a flowering response suggest that other factors or signals might be more directly responsible for floral induction in mango (Ona and de-Guzman, 1982; Davenport, 1993). The conflicting reports on the involvement of ethylene in flowering from the studies carried in the sub-tropics and low latitude tropics indicated that more precise experimentation is needed to gain better understanding of its role in flowering of mango.

Following the discovery of 'ethylene' as the active ingredient of smudging (smoking), mango workers in India and Philippines experimented extensively with 'ethephon' to induce flowering in 'normal' years and 'off' years and to alleviate alternate bearing trend in mango, but obtained mixed / inconsistent results. Inconsistent (Pandey *et al.*, 1973a,b; Sen *et al.*, 1973; Winston and Wright, 1984) or non-responsive results with ethephon spray (Pandey and Narwadkar, 1984; Pandey, 1989; Davenport and Nunez-Elisea, 1990, 1991) have been reported.

Exogenous application of ethrel (2-chloroethyl phosphonic acid) inhibited vegetative growth of shoots and promoted flowering in mango (Chacko *et al.*, 1972a, Sen *et al.*, 1973). Sprays (6 times) of ethephon at 500 ppm given between September and October gave high percentage of flowering and greater number of fruits per panicle in 'Langra' and 'Evain cultivars of mango in Egypt (Rawash *et al.*, 1983). On the other hand, monthly sprays of ethephon (100 and 200 ppm) made 5 times from September were ineffective to induce flowering in mango over 5 years of experimentation (Pal *et al.*, 1979). The influence of ethephon appears to be dose dependent. Low rates of ethephon resulted in limited flowering, while higher rates better flowering and advanced flowering by 1-4 weeks earlier than control, but caused excessive defoliation (Winston and Wright, 1984).

Besides enhancing flowering, ethrel (ethephon) caused precociousness in flowering of mango. A single spray of ethrel (200 or 300 ppm) in October on 'Pairi' cultivar advanced flowering date at Akola (Dalal *et al.*, 2005). Ethrel sprayed twice in September and October on 'Amrapali' mango trees also caused earliness in flowering by 13 days (Kundu *et al.*, 2011). Pruning plus CEPA (400 ppm) advanced flowering time by 16 days in 'Langra' mango cultivar at Bhubaneswar (Rath and Das, 1979).

Ethylene (Ethephon / Ethrel) has been employed to induce flowering in 'off' year to alleviate alternate bearing habit as well as 'off' season flowering in different cultivars of mango, but the results obtained were erratic. 'Off' year is ascribed to many causes. One of the causes attributed is the inability of fruited shoots to differentiate fruit buds, during the following year. The fruited shoots tend to throw out untimely vegetative shoots during October, which is detrimental to FBD and thus contributed to the sustenance of alteration in cropping. Further, this vegetative flush (shoots) have inadequate time to attain physiological maturity to become reproductive. Hence, efforts were made to hasten maturity and to inducc flowering during 'off' year by using some growth regulators like ethrel and PBZ. Autumn sprays of ethrel (240 ppm) induced flowering in 'off' year in mango trees at Delhi (Pandey *et al.*, 1973a,b). Dhatt (1973) reported that 200 ppm ethrel sprays could induce flowering in the first 'off' year of 'Langra' but failed to show any encouraging response in flowering in the second 'off' year.

Exogenous application of ethephon stimulated flowing in mango trees in 'off season' at Delhi (Chacko and Randhawa, 1971) and at Coimbatore (Vijayalakshmi and Sreenivasan, 1998) was reported. Chacko *et al.* (1974a) could reduce juvenile duration of mango seedlings by sprays of ethylene at weekly intervals at Delhi.

Ethrel alone, however, did not promote flower induction and higher productivity in mango, compared with $CaNO_3$ or KNO_3 (Halfe *et al.*, 2003) indicating that besides ethylene some other factors are also required for flowering. Ethrel (250 or 500 ppm) plus uniconazole (500 ppm) increased the number of flowering panicles per tree and resulted in higher yield. This might be due to inhibition of vegetative growth by uniconazole followed by promotion of flowering by ethephon in Egypt (Galila and Elmasry, 1991). GA (500 ppm) with urea combination (10,000 ppm) when applied during June after crop harvest produced vegetative flush both 'on' and 'off' year shoots, but when followed with ethrel sprays (200 ppm) during November – December resulted in increased flowering in the 'off' year shoots of cultivar 'Langra' but decreased flowering in 'on' year shoots (Das *et al.*, 1989). Combined application of ethrel (0.25 ml/litre) with $CaNO_3$ (30 g / litre) induced flowering in Tommy Atkins' mangoes in Brazil (Halfe *et al.*, 2003).

5. Role of PGRs in Sex-Expression and Sex Ratio

Mango is an andromonoecious fruit crop i.e. it produces two types of flowers on the same panicle viz., staminate (male) and hermaphrodite (perfect, bisexual, female) flowers. However, the intensity of each type of flowers varies much depending on the variety, area of cultivation (environment of the area, mainly temperature and light), seasons of flowering, 'on' and 'off' year rhythm of the bearing, nutrition of the tree, kind and status of the phytohormones and their balance, tree age etc. The total number of flowers per panicle may vary from 1000 to 6000 depending on the variety. Generally staminate flowers are greater in number than hermaphrodite flowers. The percentage of perfect flowers may vary from 2 to 78 depending on the variety, location, environmental conditions (light, temperature) age of the tree, crop load, nutrition and endogenous growth substances and their balance. Production of higher number of hermaphrodite flowers is desirable as they are the flowers which will set and result in production. Generally speaking higher the number of perfect flowers greater will be the production of fruits.

Sex ratio is the proportion of staminate to perfect flowers. This ratio (sex-ratio) between male and perfect flowers is also highly variable depending on the variety and is influenced by several factors as in the case of sex-expression. Generally the sex-ratio in mango is less than 50.

The number of hermaphrodite flowers per panicle is one of the important factors affecting fruit yield in mango. Variation in the productivity of different varieties in different regions and years has been attributed to the varying percentages of perfect flowers borne per panicle. It is generally observed that higher the number or percentage of bisexual flowers more is the yield. Hence, it is highly desirable to increase the number or percentage of hermaphrodite flower per panicle to achieve enhanced yields in mango. According to Spencer and Kennard (1955) less than 10 per cent female flowers could influence fruit set / production. Therefore, sex expression towards femaleness is one of the important factors influencing the fruit production. Sexual differentiation of flowers in plants is affected by environmental factors and / or plant growth regulators (endo as well as exogenous).

The modification of sex expression in flowering plants including mango by treatments with external agencies (nutrients and plant growth regulators) has been reported. In plants / cultivars that inherently produce low proportion of perfect flowers and consistently set poorly where growing conditions are good, it may be argued that, nutrients and reserves of food material (especially carbohydrates) may not be the cause. Hence, application of plant growth regulators may be needed and beneficial (Oosthuyse, 2015). All the preset classes of growth regulators were tested to tilt the sex-expression towards femaleness with variable results.

Auxins are known to modify the sex ratio in plants. Mallik *et al.* (1959) were probably the first workers in India to report that aqueous sprays of NAA (50 and 100 ppm) increased the percentage of perfect flowers in cv. Kalapady in Bihar from 4.8 (in control) to 6.8 to 9.9, respectively. However, contrary to the above, Singh and Dhillon (1986a) in 'Dashehari' in Punjab and Vijayalakshmi and Srinivasan (1998) in Alphonso at Coimbatore reported significant reduction in the sex-ratio with foliar application of NAA compared to control treatment. Singh (1971) observed that 2,4,5-T (400 ppm) narrowed the sex ratio in Dusehri at Varanasi. Gibberellins are powerful modifies of sex expressions in plants (Kurian and Iyer, 1993). Gibberellins besides being involved in the phenomenon of apical dominance, flowering and fruiting, have a striking ability to alter the sexuality of plants. Three repeated sprays of GA_3 at 15 days interval from mid-March onwards upto mid-April when trees were in active phase of vegetative growth which corresponded to the time of fruit bud differentiation in cv. 'Baramesi' at Calcutta, given with the objective of achieving summer flowering in May, modified the sex expression, resulting in more male flowers than untreated control. GA_3 treated trees produced less hermaphrodite flowers than the untreated control trees. GA_3 enhanced ratio of male to female flowers compared to control. Suppression of hermaphrodite flowers and promotion of maleness over control with GA_3 sprays, resulted in an increase in the ratio of male to female flowers. This was considered to be due to the regulation of endogenous gibberellin : auxin balance (Maiti, 1973). This view is further supported by the fact that cycocel, a growth retardant, which is known to suppress, gibberellin synthesis in the treated trees, as foliar spray application altered sex – expression towards femaleness (Kachru *et al.*, 1972). Jones (1973) opined that by careful manipulation of gibberellin levels in the plants using either GA_3 or its inhibitors of biosynthesis (like paclobutrazol), the sex ratio can be modified in mango. Further, the knowledge of increasing male flowers in mango by supplying GA_3 just prior to fruit bud differentiation may be effectively utilized in reducing too much hermaphrodite flower formation in biennial cultivars of mango and thereby minimizing the changes of excessive setting of fruits and resultant exhaustion of food resources and energies, which are considered to be one of the most important causes of irregular bearing in mango (Maiti, 1973).

In general growth retardants are known to reverse the effect of gibberellins. Cycocel by suppressing gibberellin synthesis in plants alter the sex expression towards femaleness in mango (Kachru *et al.*, 1972; Singh *et al.*, 2004). Application of BA (50 ppm) + Calcium (2 %) in January and again in March in both during 'off' and 'on' years gave the highest number of perfect flowers per panicle (70.5 %) in 'Langra' at Varanasi (Singh and Rajput, 1990). BA (1000 ppm) applied on 'Irwin' mango trees (growing in a warm climate) after the differentiation of floral organs increased the number of perfect flowers, suggesting that cytokinins are involved in the sexual differentiation in mango (Utsunomiya *et al.*, 1995). On the other hand, BA (50 ppm) prevented fruit bud differentiation in 'Irwin' mango, when applied as soil drench at the end of September, which was transferred to a green house at the end of November and grown then onwards in Japan (Yamashita and Ogawa, 1998).

In 'Himsagar' and 'Langra' cultivars, MH (4×10^{-2} mg/l) sprayed four times at monthly intervals from September to December increased the number of hermaphrodite flowers in West Bengal (Sanyal *et al.*, 1996). This effect of MH was considered to be due to its inhibitive effect on gibberellin biosynthesis as high gibberellin content in flowering shoots / panicles tends to rise maleness (Luckwill, 1964). However, Singh *et al.* (1965) and Subhadrabandhu (1986) expressed that MH, either had no effect on sex ratio or lowered the sex ratio in mango.

Paclobutrazol (PBZ) is a well known inhibitor of synthesis of endogenous gibberellin in plants and hence expected to increase femaleness in plants on application, counteracting the effect of gibberellins on sex expression. Irrespective of method of application PBZ promoted femaleness. Soil application of PBZ increased the proportion of hermaphrodite flowers and narrowed the sex ratio in different cultivars of mango in different growing regions (Burondkar and Gunjate, 1991; Desai and Chundawat, 1991; Voon *et al.*, 1991; Kurian and Iyer, 1993; Burondkar *et al.*, 1997; Vijayalakshmi and Srinivasan, 1998; 2002; Singh, 2000; Padhiar and Patil, 2003; Singh *et al.*, 2005; Karuna and Mankar, 2008; Muhammad *et al.*, 2010; Bhagwan *et al.*, 2013, Satyandra Singh *et al.*, 2014). PBZ by soil or foliar spray application has increased the average number of hermaphrodite flowers over the control in cv. 'Tommy Atkins' in Ethiopia (Yeshiltela *et al.*, 2004). Cultar (PBZ) increased sex – ratio through increasing the hermaphrodite flowers and reduced the sex ratio by reducing the female flowers. Sex ratio varied with the concentrations of PBZ applied and this variation in sex ratio with different concentrations of PBZ might be due to changes in endogenous hormone levels in treated trees (Karuna and Mankar, 2008). Though PBZ produced the short and compact panicles (a disadvantage), it increased hermaphrodite flowers in the panicle (Padhiar and Patel, 2003). The increase in hermaphrodite flowers due to PBZ was attributed to inhibition of gibberellin synthesis, thereby checking the vegetative growth, resulting in starch accumulation in the plant, which ultimately led to higher femaleness in treated trees (Padhiar and Patel, 2003; Bhagwan *et al.*, 2013).

6. Role of PGRs on Fruit Set

One of the causes attributed to low production in mango in the country is the poor fruit set. The lower fruit set and greater fruit drop lead to significant loss in fruit yield (Singh, 1978). Fruit production is entirely dependent on good fruit set and successful retention of fruits on trees upto fruit maturity and harvest. Though mango produces innumerable flowers upto 6000 per panicle (both male and female flowers), less than 0.1 per cent flowers set fruit (Naik and Rao, 1943; Singh, 1954). Extent of fruit set is a varietal character and is dependent on certain variables like number of hermaphrodite flowers, their position on the panicle, pollination, fertilization, compatibility with in a variety and between varieties, pollinators, environmental factors, time of flowering, mineral nutrition, phytohormones synthesized in the embryo and their mutual balance etc. Lower fruit set in mango is attributed to the self incompatibility, predominance of male flowers, poor pollination, reduced pollen germination and pollen tube growth as well as unfavourable environmental conditions during flowering and anthesis (Quintana *et al.*, 1984). Initial fruit set depends largely on the number of hermaphrodite flowers in a panicle, as they are the ones that develop into fruits.

Studies on pollination revealed that about 40-60 per cent of the perfect flowers failed to receive pollen grains in mango (Naik and Rao, 1943; Mukherjee, 1951). While investigating the causes of poor fruit set in mango, Wagle (1928) found that 38.4 per cent of the perfect flowers were shed before pollination / fertilization and 56.9 % after fertilization.

Since, fruit set in mango is very less, efforts were made from time to time to improve fruit set through some approaches like: increasing the number of perfect flowers, providing adequate pollination, compatible pollinizer, optimum environmental conditions, adequate mineral nutrition and supplying needed hormones. Supply of hormones was considered more essential to achieve good to high fruit set in view of the observations that there was deficiency of hormones at the time of anthesis and fruit set.

Auxins were considered to be potent in enhancing the fruit set in plants. When applied to developing panicle, IAA has substantially increased fruit set in mango (Singh and Arora, 1965). NAA was the most effective auxin analogue for improving fruit set at different concentrations in different varieties and locations (Singh *et al.*, 1965; Singh, 1977; Aravindakshan *et al.*, 1979; Ram, 1983; Singh and Ram, 1983; Prakash and Ram, 1986; Sharma *et al.*, 1990). 2, 4-D (20 ppm) sprays also substantially increased fruit set in mango at Varanasi (Singh, 1977).

GA_3 (30 ppm) plus urea (2 %) gave highest fruit set (62.87 %) in 'Dashehari' compared to control at Varanasi (Rajput and Singh, 1983). The increase in fruit set and its percentage due to GA_3 sprays might be due to increased availability of nutrients from leaves (Garcia – Martinez and Garcia – Papi, 1979; Saleem *et al.*, 2008).

Polyamine (PA) sprays on fully grown panicles prior to anthesis increased fruit set in different varieties of mango at different locations (Singh and Singh, 1995; Singh and Janes, 2000). Out of the three major PAs, spermine was more effective in increasing fruit set as compared to control. Exogenous application of PAs not only enhanced fruit set but also increased fruit retention upto harvest (Singh and Singh, 1995). PAs are known to inhibit, ethylene biosynthesis by preventing 1-aminocyclopropane Carboxylic acid (ACC), activity (Tiburacio *et al.*, 1993). The increased fruit set in mango with the exogenous application of PAs and reduction of fruit set with the treatment of inhibitors of biosynthesis of PAs when applied to panicles prior to anthesis clearly indicate an important role of PAs in fruit set (Singh and Janes, 2000).

Mango is a vigorous tree by nature (whether it is a seedling or a graft) which produces lot of vegetative growth at the cost of flowering and fruiting. It is a common observation that a vigorously growing mango tree is less productive because all the photosynthates are utilized for vegetative growth immediately after synthesis. Therefore, it is conceived that to increase the fruit set and yield vegetative growth should be reduced / checked to avoid exhaustion of food reserves and to allow accumulation of food material (particularly carbohydrates) for flowering. Growth retardants enter the scene here to check / retard vegetative growth for inducing higher flowering, fruit set and finally greater production. With this in view, plant growth retardants like CCC, MH, PBZ etc. were evaluated for improving fruit set in mango. Plant growth retardants control tree growth and thereby increase flowers (particularly hermaphrodite flowers) and fruit set. This has been achieved not only by reduction of cell division, but also by lowering the rate of cell division and regulation of the plant height physiologically (Rademacher, 2000). Many of these growth retardants inhibit gibberellin biosynthesis and thereby cell division and cell elongation (Mansuroglu *et al.*, 2009). Growth retardants help the tree to reserve food by reducing the annual vegetative growth which could be useful for inducing fruit set. Cycocel (Chloremquat, CCC) sprayed at 1000-4000 ppm on panicles was the best one in enhancing fruit set, followed by MH at 1500 ppm in mango at Navasari, Gujarat (Chaudhuri *et al.*, 2014).

Soil application of PBZ at different doses and locations increased fruit set as well as fruit retention in mango (Sarkar *et al.*, 1998; Desai and Chundawat, 1994; Sharma *et al.*, 2002; Yeshitela *et al.*, 2004; Sarkar and Rehim, 2012; Bhagwam *et al.*, 2013; Satyaendra Singh *et al.*, 2014; Coelha *et al.*, 2014). However, fruit set decreased with increase in the concentration of PBZ applied (Khettab *et al.*, 2006). Higher fruit set and fruit retention due to PBZ application was probably due to restricted vegetative growth and higher accumulation of carbohydrates in their leaves (Sharma *et al.*, 2002). Treated trees had higher reserve food which enhanced fruit set compared to control trees in which fruits lacked adequate food resulting in poor fruit set (Yeshitela *et al.*, 2004). Reduction in tree growth and vigour resulted in less competition between vegetative and reproductive phases resulting in increase in fruit set (Karuna *et al.*, 2007).

Ethylene plays a negative role in the fruit set and fruit retention in mango being a promoter of abscission. Exogenous spray application of ethrel to fully grown panicles before anthesis significantly reduced fruit set upto 76 per cent in Philippines (Andam, 1983) and in Australia (Singh and Agrez, 2002). The role of ethylene in fruit set and fruit retention was confirmed by using exogenous application of ethrel and inhibitors of its synthesis (AVG, AOA, $COSO_4$) and action (STS) (Singh and Agrez, 2002). In view of the above, it was concluded that for improving fruit set, retention and yield in mango, a single exogenous spray application of cobalt sulphate ($COSO_4$) would be most effective (Singh and Agrez, 2002).

7. Role of PGRs on Parthenocarpy

Parthenocarpic fruits are those which have set with or without pollination, but without fertilization of ovule and are seedless in nature. Parthenocarpic fruits are borne naturally in many fruit crops like banana, pineapple, grape etc., while in single seeded fruits such as mango, litchi etc., parthenocarpy is uncommon and is a rare event.

Though the occurrence of parthenocarpy is not much reported, it is quite common in some varieties of mango (Oppenheimer, 1947; Lakshminarayana and Augilar, 1975). Thimmapaiah and Singh (1983) reported that 'Dashehari' cv. of mango produced parthenocarpic fruits to a certain extent occasionally under subtropical climatic conditions of Lucknow (India). These fruits are locally called 'Seep (meaning 'Unio' a mollusc) and were usually flat, mostly undersized and seedless, with rudimentary cotyledons in some cases. The seed cavity of these fruits was empty except for a discoloured cotyledonary wall, which was some times rudimentary showing the signs of degeneration. Among the North Indian cultivars only 'Dashehari' was conspicuous in this phenoemenon. Other varieties like 'Langra' and 'Chausa' did not exhibit parthenocarpy even under the same agro-climatic conditions. Mallika (Neelum x Dashehari) hybrids also produced to certain extent parthenocarpic fruits similar to its male parent 'Dashehari'. Among the South Indian varieties only cv. 'Thambwa' was known to exhibit natural parthenocarpy (Kulkarni and Rameshwar, 1978). Except for these varieties, the incidence of partheneocarpy is unknown in other varieties of mango in India. However, parthenocarpy has been reported in some of the varieties of mango of other countries. In Mexico, under low temperature conditions cv. 'Haden' was found to be highly parthenocarpic (Lakshminarayana and Augilar, 1975). High incidence of parthenocarpy in 'Ataulf' cv. of mango was considered as an important commercial problem in Mexico (Perez – Barraza *et al.*, 2007). In Australia 'Irwin' was observed to be parhenocarpic (Lim *et al.*, 1996). Extremes of temperature at the time of flowering (particularly at anthesis) has been attributed to the phenomenon of parthenocarpy. In Palestine the incidence of parthenocarpy was reported to be due to high temperature (44°C) and low humidity conditions (15 %) during flowering, while in Australia, low temperature is said to be causing parthenocarpy (Lim *et al.*, 1996).

The fact that Gustafson (1936) and several other workers after him, obtained fruit set without pollination by exogenous application of plant growth regulators (auxins, gibberellins or kinins) and reports of their natural occurrence in a number of fruit crops suggest that these growth substances play a crucial role in the fruit setting and development (van Overbeck, 1962; Crane, 1964; Nitsch, 1965). The developing (mango) fruit (particularly embryo) is a rich source of many growth substances, such as auxin – and gibberellin like substances (Chacko, 1968). Pollination and fertilization might give the necessary stimulus for the production of auxins, gibberellins and kinins in the endosperm of the seed, which are essential for the subsequent fruit enlargement. Lack of pollination and / or fertilization resulted in the cessation of active growth of the ovule (by cell division, cell multiplication and cell differentiation) there by eliminating the source of these growth substances for the enlargement of fruits initially and such fruits become partheneocarpic and are usually abscised.

Though, natural parthenocarpy is a rare phenomenon in mango, it can be induced by the application of growth regulators. However, efforts to induce parthenocarpic fruits in mango with growth regulator (auxin) were not successful (Venkataratnam, 1949). Studies of Checko and Singh (1969) demonstrated the possibility of parthenocarpy in mango with the provision of young ovaries with the supply of Kinin in perforated bags and later supplemented with a mixture of auxin and gibberellic acid. They obtained fully developed parthenocarpic fruits, in self-incompatible 'Dashehari' mango by supply of N6-Benzyladenine (250 ppm) at anthesis followed by application of combination of NAA (10 ppm) and GA_3 (250 ppm) at fortnight intervals. Such growth regulator induced parthenocarpic fruits had under developed ovaries of pin-head size, but normal development of stony endocarp. The disintegrated ovule in side the stony endocarp was enclosed in a thin papery white seed coat. The pulp / stone ratio was much higher in the parthenocarpic fruits (15: 25) than in the normal fruits (5: 6). These parthenocarpic fruits at maturity, though smaller, (65g) were superior in quality as compared to normal seeded fruits (165 g). No parthenocarpy was noted in panicles treated with N6-Benzyaladenine, NAA and GA_3 singly. The successful induction of parthenocarpic fruit development achieved for the first time in the studies of Chacko and Singh (1969), confirm the postulated correlation (Chacko, 1968) between fruit growth and the endogenous growth substances produced in mango seeds. However, parthenocarpic fruit development in mango occurs only when the ovaries were supplied with a kinin at the initial stages and later with a combination of auxin and GA_3.

Venkataratnam (1949) reported parthenocarpic fruitset in three varieties of mango under tropical conditions, when the emasculated flowers were sprayed with 10 ppm beta-naphthoxy acetic acid. The resultant developing fruits locked normal, but dropped off after reaching the marble size. The results of Chacko and Singh (1969) also indicated that application of auxin alone failed to induce parthenocarpy in mango. Similarly, auxins and GA_3 applied either separately or in combination were ineffective in inducing parthenocarpic fruits in mango. The inability of exogenous auxins alone to induce parthenocarpy in mango suggests that a complex of growth hormones such as Kinins and gibberellins as well are involved in this phenomenon. Chacko *et al.*, (1970b) found that ovaries can grow without pollination upto pea-size, indicating that they contain certain growth factors other than auxins, which are necessary for the initial swelling of ovaries. It may be a cytokinin, the occurrence of which is well established in most of the fruits (Lethum, 1967).

Application of Kinin in the initial stages and later on auxin also proved effective. Hence, for fruit development in mango an initial supply of a cell division factor like Kinin and later on factors that promote cell enlargement such as auxin and gibberellic acid are necessary (Chacko and Singh, 1969). However, of all the growth regulators *viz.*, GA_3 (100 ppm) and NAA (40-50 ppm) tried on emasculated flowers of 'Thambwa' cv. of mango in February and March under tropical conditions of Sanga Reddy (Andhra Pradesh) GA_3 alone was effective in inducing parthenocarpy (Kulkarni and Rameshwar, 1978).

8. Role of PGR on Fruit Growth and Development

Mango fruit growth and its development, fruit drop, ripening etc. are associated with endogenously synthesized plant growth substances (phytohormones) like auxins, gibberellins, cytokinins, ethylene etc. The relative levels of different classes of growth substances in mango fruits and their sources of biosynthesis such as pericarp and seed have been investigated by a number of workers, who ascertained their roles in fruit set, fruit growth, fruit drop etc.

Fruit growth is a function of endogenously synthesized growth regulating substances as it is correlated with several of them. For instance, the highest rates of fruit growth were correlated with peak levels of growth promoting ones (auxins, gibberellins, cytokinins and polyamines) and low levels of inhibitors (abscisci acid) and ethylene, synthesized in seeds etc. (Chacko, 1968; Chacko *et al.*, 1970a,b,c; Singh and Singh, 1974; Ram and Pal, 1979; Chen, 1981; Ram, 1983; Prakash and Ram, 1984 etc.). Improvement in fruit growth and fruit retention as a result of exogenous application of synthetic growth regulators also suggested the above concept and association of endogenous growth substances in fruit growth and fruit retention in fruit crops including mango.

It is thus, a well established fact that several plant processes including fruit growth and development are mediated through endogenous hormones and are under their control.

Further, endogenous hormones (phytohormones) synthesized mainly in seed and partly in the pericarp of the fruit regulate the fruit growth at different stages by virtue of changes in their levels. Various classes of phytohormones in developing mango fruits have been detected and identified for their role (s) in growth and development of fruits. All the classes of phytohormones *viz.*, auxins, gibberellins, cytokinins, polyamines, inhibitors, ethylene etc. are implicated or associated with fruit growth and development, ripening, abscission and senescence (Ram, 1992; Murti and Upreti, 1995; Malik and Singh, 2004 etc.).

Phytohormones play a role in cell division, cell enlargement and cell differentiation. Their sustainable synthesis, particularly auxins, gibberellins, cytokinins and polyamines is quite necessary for the sustainable growth of fruits. On the other hand, production of inhibitors and ethylene is essential to regulate the growth of the fruits. A good relationship exists between the levels of endogenous growth substances and different phases of fruit growth and development (Krisanapook *et al.*, 2000; Barbosa Martinez *et al.*, 2009).

The possible types of hormone interactions required to bring about fruit growth and development include i) One hormone affecting the biosynthesis of another hormone (polyamines Vs ethylene), ii) Two (or more) hormones classes acting independent of one another with their concerned action being required to bring about normal growth and development for example, each hormonal class affects a different cell type of tissue with in the fruit, but the combined action of all hormones is required for normal fruit growth and development, iii) one class of hormones triggers an event, but another class of hormones is required to bring about completion of the event, iv) a component of the signal transduction pathway of one hormone affecting a component (s) of the signal transduction path way of another hormone.

Auxins, gibberellins and cytokinins are produced less and inhibitors more during the initial 20-25 days after fruit set corresponding to slow growth of fruit growth and the growth promoters rise fast upto 40-45 days after fruit set corresponding with high rate of fruit and seed growth (Ram and Pal, 1979; Ram *et al.*, 1983; Prakash and Ram, 1986; Ram, 1992). Cytokinins would be at higher levels at the time of hardening of endocarp. Towards maturity as well as the initial stages of ripening, ethylene is produced in large quantities. A reverse trend in the synthesis of different classes of phytohormones below or above threshold levels at different stages, either slow down fruit growth or may result in irregular growth and / or fruit fall.

The different classes of growth regulators that are endogenously synthesized in the plants including fruits that play different roles in fruit growth and development are presented class wise below:

8.1. Auxins in fruit growth and development

Auxins are well known for inducing cell division in plants including in fruits and are synthesized by the embryo in the seeds of fruits, Besides embryo, pericarp of the fruit, is also claimed as a source of auxin synthesis. Chacko (1968), Chacko *et al.* (1970a), Singh and Singh (1974), Lal and Ram (1977) and Prakash and Ram, 1984) detected auxins like IAA (indole-3-acetic acid), IAN (indole acetic nitrate) and inhibitors like ABA (Abscisic acid) in developing mango fruits. Prakash and Ram (1984) observed an increase in endogenous auxin level corresponding with rapid growth and high level of inhibitor corresponding with a high rate of fruit drop in 'Dashehari' mango fruits. Thus, auxin levels were found to increase or decrease with the rate of fruit growth. In other words, the rate and progress of fruit growth depended on the increase or decrease in endogenous auxin levels, their deficiency reduced the growth and may cause fruit shedding. The low concentration of these growth promoting substances present in the seed may be responsible for the slow growth rate at the initial stages and initial fruit drop.

8.2. Gibberellins in fruit growth and development

Gibberellins are involved in cell enlargement and are produced mainly in the seeds of fruits, with pericarp being a minor source of synthesis (Ram, 1983). Gibberellin content in the seeds increases rapidly during early seed growth and declined there after resulting in slow growth of the fruit. The peak production of gibberellin - like substances in the fruit occurred at the beginning of endosperm development and corresponded to the time of rapid growth of fruit. The content of gibberellins continued to increase till the fruit attained full size (Chacko *et al.*, 1970c). Nagar (1978) and Ram and Pal (1979) isolated 8 gibberellin - like substances in immature seeds of fruits of 'Dashehari' mango cultivar and claimed their promoting role in fruit growth.

8.3. Cytokinins in fruit growth and development

Cytokinins are an important class of phytohormones whose role (s) in growth and development of fruits has been realized well in various crops including mango. The synthesis of cytokinin - like substances in mango fruits (seeds) has been reported. Ram (1983) noted the presence of cytokinins both in the fruit pericarp and seed. The pattern of synthesis and levels in the activity of cytokinin like substances were similar to that of gibberellins in mango fruits (Krisanapook *et al.*, 2000). The concentration of cytokinins in the panicles and pulp of mango fruit was maximum at 5-10 days after full bloom and decreased rapidly thereafter (Chacko *et al.*, 1972d). The cytokinin levels were high during the early stages of fruit growth, when there was rapid cell ddivision and cell enlargement and the rise in cytokinin content preceeded the faster growth of fruits (Murti and Upreti, 1997). The high activity of cytokinins in the early stages of mango fruit growth was possibly responsible for cell division in the early stages.

Cytokinins exert positive influence on polyamines. The observed high levels of polyamines during initial stages of fruit growth could be from such influence (Suresh *et al.*, 1978; Chen, 1983; Murti and Upreti, 1997).

8.4. Inhibitors in fruit growth and development

Occurrence of growth inhibitors like abscisic acid (ABA) was observed in mango fruit (Chacko *et al.*, 1970b). Their role appears to be to check the growth of fruit at particular stages. The levels of inhibitors fluctuate as per the rate of fruit growth. Ram (1983) estimated the levels of inhibitors at different stages of fruit growth and found that the level of ABA-like inhibitors was high in the initial 24 days of pollination, which corresponded with the period of slow growth and heavy fruit drop and during the period of rapid fruit growth, the level of inhibitors declined and that of promoters raised. However, during the period of maturation and slow fruit growth the levels of growth inhibitors were low.

8.5. Ethylene in fruit growth and development

Besides being involved in fruit ripening, ethylene also influences several other physiological phenomena in plants. However, information on ethylene production and its role in fruit growth and development is scanty. During early stages of fruit growth of mango, large quantities of ethylene are produced (Chacko, 1968). According to Mendoza (1981) the production of ethylene was extremely high in young fruits of mango, which declined rapidly until the fifth week after fruit set. Again the rate of production tended to increase until eleventh week, which was followed by a fall until thirteenth week, when a rise in ethylene production was apparent in fully matured fruit. The high rate of ethylene production observed immediately after fruit set may have a causative role in the high incidence of initial fruit drop, as ethylene is a well known abscission promoting agent. The final rise in ethylene production was associated with fruit ripening.

8.6. Polyamines in fruit growth and development

PAs were observed in various fruits including mango (Malik and Singh, 2005). PAs are synthesized at high rates during the early stages of fruit growth and decrease gradually towards the completion of fruit development. Malik *et al.* (2005) reviewed the role of PAs in mango fruit growth and development.

All the three major polyamines (PUT, SPD and SPM) were detected in the extracts of pericarp and seeds of developing fruits of mango (Murti and Upreti, 2003). PUT is the dominant polyamine in the fruit pericarp during most of the growth phase, while SPM was the dominant in case of seeds. The highest polyamine concentration in fruits of 'Alphonso' was preceded by active growth phase and in case of seed it correlated with active phase of growth. The high concentration of PAs at initial stages of fruit growth and seed was associated with the periods of cell division (Murti and Upreti, 2003). This suggests the critical role of PAs during cell division. Thus, the presence of higher levels of endogenous free PAs, especially SPM and SDM at initial growth and development suggest their possible role in cell division and the concurrent increase in PAs and ethylene during mango fruit development suggests that endogenous PA levels may have increased in response to higher ethylene production (Malik and Singh, 2004). Ethylene and PAs share a common intermediate metabolite *viz.*, S-adenosyl methionine (SAM) in their biosynthesis path way (Evans and Malanberg, 1989). Thus the lower concentration in the pericarp may be a consequence of greater channeling of SAM for ethylene production for fruit maturity and ripening.

9. Role of PGRs on Fruit Drop and Retention

Several investigators have indicated the link between phytohormones and abscission process. Like flowering, fruit, fruit growth, fruit retention and fruit abscission are also under the control of phytohormones. Phytohormones in fruits regulate the formation and activation of abscission zone with in the separation layer (Singh *et al.*, 2005). Deficiencies of the phytohormones (auxins, gibberellins, cytokinins and polyamines) coupled with high levels of growth inhibitors and ethylene cause fruit abscission. Phytohormones play both direct and indirect roles. In many fruit crops a negative relationship between endogenous growth substances or their imbalance and / or their in adequacies were reported to be the causes of fruit drop and failure of fruit retention. Fruit abscission and fruit retention are regulated by complex interaction of different phytohormones which protect the abscission zone or induce the separation process (Singh and Arora, 1965; Prakash and Ram, 1984; Krisanapook *et al.*, 2000).

Developing seeds and pericarp (peel/skin) of the fruit are the synthesizers and rich source of all the known classes of phytohormones. Each class of phytohormones plays a significant and specific role (s) in maintenance of abscission zone / induction and progress of the separation of the plant appendages like fruits.

In mango inadequate levels of endogenous growth promoters (auxins, gibberellins, polyamines and cytokinins) and higher levels of growth inhibitors (ABA) and ethylene were considered responsible for fruit shedding. Hence, enrichment of different growth promoters and reduction of fruit drop promoters (inhibitors and ethylene) in the fruit by exogenous application of deficient growth substances was thought to facilitate retention of fruits in fruit crops. Application of synthetic auxins, gibberellins etc. is recommended to suppress the synthesis of ABA and ethylene and consequently reduce fruit drop and result in higher retention of fruits (Sakhidin *et al.*, 2011). The possibility of preventing fruit drop by means of exogenous supply of synthetic hormones was suggested way back in 1950 by Gokhale and Kanitkar (1951) in India.

When the deficiencies of phytohormones and imbalance among them occur, the application of synthetic plant growth regulators exogenously interfere in the process of their biosynthesis, transport and metabolism or supplement them at sub-optimal concentration (Berghoef and Bruinsma, 1976). The sprays of small quantity of plant growth substances would activate and increase the efficiency of endogenous hormones to balance the metabolic activation of source and sink in the plant.

Exogenously applied growth regulators exhibit 4-fold functions *viz.*, i) provision of growth factor (s) for the development of ovary / fruit ii) mobilization assimilates (influx) into the fruit, iii) inhibition of abscission layer formation in the developing plant appendages (like fruits) and iv) counteracting of adverse effects of endogenous inhibitors and ethylene to prevent fruit shedding.

The fruit drop can be reduced to some extent by exogenous application of different plant growth regulators, notably auxins, gibberellins, cytokinins and polyamines at appropriate concentration at proper stage of fruit growth and development.

9.1. Role of Auxins on fruit drop and retention

Auxins are growth promoters and also inhibitors of abscission of plant appendages (leaves, flowers and fruits). Auxins are synthesized in seeds and their endosperm. Luckwill (1948) pointed out that fruit abscission was under the control of auxins produced in the seed. Among the phytohormones, concentration of auxins is of prime importance in the formation and maintenance of abscission layer. The fruit drop was inversely correlated with the period of low auxin production during the growth and development of fruits (Luckwill, 1957; Leopold, 1958). Lower concentration in endogenous auxins in about-to-abscise fruits and their pedicels as well as reduction in fruit shedding with the exogenous application of auxins support the concept that the auxins play an important role in protection of fruit abscission zone. This observation further gets support from numerous reports of fruit retention with exogenous application of auxins.

Auxins inhibit / prevent abscission (drop) by checking the physiological break down of calcium pectate in the middle lamella of cell walls in the abscission layer (van Overbeek, 1952) and inhibiting the activity of abscission zone by reducing the sensitivity of cells (in the abscission zone) to ethylene (promoter of abscission) (Taylor and Whitelaw, 2001). A direct correlation exists between abscission and endogenous auxin level. Indole acetic acid (IAA) a natural auxin in the plants is postulated to have a direct effect on assimilate partitioning (Patrick, 1979; Agusti *et al.*, 2002) i.e. regulation of the flux of assimilates to the developing fruits, that might be an important factor in determining whether or not a plant appendages like fruit is to be retained intact with plant (Else *et al.*, 2004). High IAA export from fruitlets to abscission zone might control the sensitivity of the abscission zone to ethylene and thus abscission susceptibility. However, fruit abscission / drop is regulated by fluctuating levels of endogenous auxins in the fruit. During the initial stages of fruit growth, the synthesis of auxin is generally low, resulting in high shedding of fruits, while in the middle of fruit growth, preceding the period of rapid growth, synthesis of auxins reaches a peak when the fruit fall is reduced (Murti and Upreti, 1995).

Plant-synthesized auxins are exported / transported to abscission zones for protecting the zones. IAA exports come from the adjacent fruits or shoot tips. Such export of IAA is reduced in about-to-abscise fruits depending on the dominance of adjacent fruits, on differences in time of fruitset, number of fruitlets per panicle as well as vigour of shoots. Fruits that have set first seems to be dominant, having a stronger polar transport through the abscission zone and cause a gradient at the flow junctions. It is hypothesized that unfavourable conditions in combination with plant stresses during fruit set and growth periods result in a decreased export of IAA out of fruits and in enhanced activation of the abscission process which eventually leads to fruit drop (Roemer *et al.*, 2011).

However, Davenport and Nunez-Elisea (1997) felt that despite successful reports of increased fruit retention and prevention of fruit drop of mango using exogenous application of auxins, there had been few studies that examined endogenous auxins in fruits as related to retention of fruits.

9.1.1. Control of fruit drop by auxins

Auxins appear to be important regulator of ethylene sensitivity in plants and abscission of plant appendages. Supplied auxins check the fruit drop by delaying the formation of abscission layer in fruits and other appendages (Brown, 1997), inhibit abscission by preventing physiological breakdown of calcium pectate of middle lamella of cell wall in the abscission layer (Bonner, 1950) and also inhibit fruit drop by strengthening the fruit pedicel. The effect of auxins, however, depends on the kind of auxin used, its concentration employed, the stage of fruit growth at which the auxin is applied, the weather condition at the time of application and thereafter and additives mixed (Meland, 1998; Agusti *et al.*, 2000; Srilatha, 2003).

Auxins and auxin analogues were tried on pre-anthesis panicles or panicles bearing fruitlets / fruits of various sizes / stages to check the fruit drop and improve fruit retention and yield in various fruit crops including mango. Auxin content in mango fruits during the initial 2 to 3 weeks after pollination / fruitset is low and their ability to mobilize food material is, therefore poor, which results in fruit fall. As the fruit develops the amount of auxin(s) rises which is helpful in mobilization of food material from the reserve sources to fruit (Chacko *et al.*, 1972d). At this stage the competition among the developing fruitlets starts and those fruitlets which compete less effectively are forced to drop. Foliar application of growth regulators at this stage proved to be beneficial with regard to fruit retention (Singh, 1977).

9.1.1.1. Role of Naphthelene Acetic Acid (NAA) on controlling fruit drop

Of all the synthetic auxins, tested against fruit shedding in mango, NAA proved and appears to be the most effective auxin (at different concentrations, at different phenological stages, at different locations and on different cultivars) for prevention of fruit drop and enhancement of fruit retention and yield.

Probably, Gokhale and Kanitkar (1951) were the first workers in India to try NAA against fruit drop of 'Alphonso' mango. Fruit Drop in untreated 'Fazli' cultivar mango was recorded as 52 percent. This high fruit drop was reduced to 22 percent by spraying of NAA at 30 and 40 ppm, 6 weeks after petal fall. This shows the effectiveness of NAA in reducing fruit drop (Singh *et al.*, 1959). Srivastava (1962) tried NAA (25-50 ppm) sprays at fruit set and 7 days later and found reduction in the fruit drop. Under Delhi conditions, 40 ppm NAA was found useful for minimizing the fruit drop in 'Dashehari' cultivar (Arora and

Singh, 1964). NAA at 5-10 ppm sprayed on 4 different cultivars of mango was effective in preventing fruitfall with in 12 days of application (Roy *et al.*, 1963). Higher control of fruit drop in 'Kesar' mango was achieved with one ml of 'Planofix' (containing NAA) per gallon (4.5 litres) of water applied 30 days after fruitset in Pakistan (Jagirdar and Chandhury, 1967).

'Langra' mango trees were sprayed with NAA (20 or 40 ppm) from pea-size stage of fruits with or without a second spray one month later. NAA 40 ppm applied once at pea-size stage gave best results in reducing the fruit drop at Varanasi (Maurya *et al.*, 1973). Treating the trees of mango cultivars (Langra and Ewais) with NAA (40 ppm) in April significantly reduced fruit drop, compared with unsprayed control and other growth regulators in Egypt (Rawash *et al.*, 1983). NAA 40 ppm gave best fruit retention when sprayed at pre-bloom stage, pea-size and marble-size stages of fruits with 300, 200 per cent increase respectively in fruit retention, 200 per cent with 60 ppm and 95 per cent with 20 ppm respectively in 'Dashehari' mango at Pantnagar (Singh and Ram, 1983). NAA (5 ppm) applied thrice (at full bloom, pea-size and marble-size stage fruits) gave maximum fruit retention in 'Langra' (Moti *et al.*, 1986). Maximum percentage of fruits was retained till harvest following NAA (150 ppm) sprays during October and November at Jabalpur (Baghel *et al.*, 1987b). Foliar sprays of NAA (10 ppm) at pea-size stage of fruits enhanced fruit retention by 200 to 500 per cent over control in mango in Pakistan (Rizvi, 1989-90). In 'Dashehari' and 'Sindhuri' cultivars of mango fruit retention was enhanced when NAA (10, 15 or 20 ppm) was sprayed at pea-size stage of fruits as well as two weeks later at marble size. Also NAA sprays increased yield by about 95 per cent in 'Sindhuri' and 54 per cent in 'Dashehari' in Pakistan (Naqui *et al.*, 1990). Foliar sprays of NAA (40 ppm) after flowering on previously deblossomed 10-year old ' Tommy Atkins' trees, increased fruit retention and yield in South Africa (Oosthuyse, 1993).

Maximum fruit retention (25 %) in 'Neelum' mango trees with NAA (40 ppm) sprays twice in the fruiting season at weekly intervals increased the number of mature fruits per panicle as well as per tree and final yield in 'Taimour' cultivar of mango in Egypt (Rawash *et al.*, 1998). Eight year-old-trees of 'Bombai' variety of mango showed maximum percentage of fruit set and fruit retention with minimum fruit drop in response to application of NAA (20 ppm) when sprayed at monthly intervals from fourth week of fruit set in Bangladesh. NAA sprays also delayed fruit maturity by 60 days, improved fruit size and yield, delayed ripening with minimum incidence of post-harvest anthracnose disease and spoilage of fruits (Gofur *et al.*, 1998).

Fruit retention was maximum with NAA (40 ppm) in 'Bombai' mango trees in Bihar (Gupta and Brahmachari, 2004). Under Parbhani conditions (Maharashtra) NAA (40, 60 and 80 ppm) sprayed at flowering, pea-size and marble-size stages of fruits, improved fruit retention of which 40 ppm was most effective (Patel *et al.*, 2005). NAA (20 ppm) sprayed at marble size stage raised fruit retention percentage in 'Guarjeet' mango in Gujarat (Chaudhari *et al.*, 2011). At harvest fruit number was higher under NAA (20 ppm) treatment in 'Alphonso' at Bangalore (Upreti *et al.*, 2011). Application of NAA (50 ppm) significantly improved fruit retention, yield and fruit quality of 'Amrapali' mango fruits in West Bengal (Animesh and Ghosh, 2011).

NAA in combination with other growth regulators and urea has also proved beneficial in arresting fruit fall, improving fruit retention and yield in mango. Application of NAA (20 ppm) plus ethrel (ethephon 500 ppm) once in November per season as well as uniconazole (500 ppm) plus ethrel (500 ppm) twice in December per season reduced the percentage of fruit drop in Egypt (Rawash *et al.*, 1998). The response of 'Kesar' mangoes to foliar application of NAA, 2, 4-D and urea in combination was found promising in respect of fruit retention and yield (Patel and Patel, 1993).

9.1.1.2. Role of 2, 4-Dichloro Phenoxy Acetic Acid (2, 4-D) on controlling fruit drop

2, 4-D has been extensively tried to prevent fruit drop in several fruit crops including mango. This auxin proved as a potent preventer of fruit drop in different fruit crops, notably in citrus and mango. However, in mango 2, 4-D comes after NAA in checking the fruit drop. 2, 4-D is highly reputed for its inhibition of fruit shedding by delaying the formation and development of abscission layer through prevention of loss of pectic material in the middle lamella of cell walls of cells of abscission zone (Hendershot and Wilson, 1967). This delaying effect of 2, 4-D on abscission is directly related to a retardation of increase in the activity of endo-and exo-cellulase (Zur and Goren, 1977).

Gokhale and Kanitkar (1951) reported the use of 2, 4-D in the prevention of fruit drop in mango. Aqueous solution of 2, 4-D at 10 ppm applied to 'Fazli' mangoes 6 weeks after full bloom reduced fruit fall from 52 per cent in control to 25 per cent (Singh *et al.*, 1959). Rao (1961) in a trial with NAA, 2, 4-D and GA_3 alone found that 2, 4-D (30 and 40 ppm) was most effective in retarding the fruit drop upto 51 per cent and 53 per cent respectively over the unsprayed control. Gill and Mukherjee (1967) tried several synthetic growth regulators to prevent fruit drop in different mango cultivars and observed that 2, 4-D recorded best results in all the cultivars as compared to NAA and 2, 4-5-T. "Dashehari' mango trees were sprayed fortnightly from pea-size stage until harvest with 2, 4-D (20, 40 and 60 ppm). All the treatments reduced the fruit drop and improved fruit retention at Varanasi (Maurya *et al.*, 1973). Winata (1975) found that spraying of 2, 4-D (20 ppm) before pollination of

'Gadung' and 'Maualegi' cultivars of mango increased the fruit yield by effectively checking the fruit fall. In a 2-year trial, 2, 4-D (20 or 40 ppm) sprayed in November (pre-flowering period), March (flowering time) and May (fruit maturity period) on mango trees enhanced fruit retention and yield (Singh, 1977).

Out of the three levels of 2, 4-D (5, 10 and 20 ppm) applied at pea-size stage of fruits (early April) on 'Hindy Be Sinnara' cultivar in Egypt, 10 ppm reduced the fruit drop from 98 per cent to 92 per cent and increased the final productivity of panicle to 39 per cent (Dahshan and Habib, 1981). Application of 2, 4-D to half-grown fruits of 'Alphonso' mango at Dharwad reduced fruit drop indicating that the time of spraying is not important with 2, 4-D (Kulkarni, 1983). However, best control of fruit drop was obtained with 2, 4-D at 10 ppm sprayed thrice *viz.*, once at pin-head stage (late March) pea-size stage (April) and again at the marble-size stage (end of April) followed by GA_3 (15 ppm) spray on 18-year-old 'SB Chausa' cultivar (Singh *et al.*, 1986).

Spraying of flower panicles with 2, 4-D (40 ppm) at pea-size stage of fruits reduced fruit drop in 'Malda' cultivar by 23 per cent compared with unsprayed control and 2, 4-5-TP in Pakistan (Khan *et al.*, 1993). 2, 4-D (20 ppm) was highly effective in controlling fruit drop when foliar applied in the last week of April or at the stage of pea-size (Singh and Singh, 1996). In Odissa, 2 sprays of 2, 4-D (15 ppm) at panicle emergence and next at post-settting drop stage in 'Neelum' and 'Gulab Khas' were best in preventing fruit drop (Nayak *et al.*, 1997). Application of 2, 4-D (40 ppm) six weeks after fruit set resulted in maximum fruit retention and minimum fruit shedding followed by lower (30 ppm) concentration (Chadha *et al.*, 1979).

Generally, 2, 4-D produced better results at a concentration below 20 ppm. At higher concentration, it is phytotoxic and retards the development of seed as well as the fruit (Arora and Singh, 1964).

9.1.1.3. Role of other Phenoxy Auxins on controlling fruit drop

2, 4-5-trichlorophenoxy acetic acid (2, 4-5-T) and 2, 4, 5-trichlor propionic and (2, 4, 5-TP) were also tested for controlling fruit drop in mango and other fruit crops. However, they were found to be effective in deciduous fruits (apple, pear, peach etc.). Attempts in using these pheroxy auxins for checking fruit fall in mango are relatively few and the results are not encouraging.

2, 4-5-T (20, 40 and 60 ppm) was spray applied on 'Dashehari' mango trees at fortnightly intervals from pea-size stage until fruit harvest. Except the higher rates, all other doses have improved fruit retention (Maurya *et al.*, 1973). The treatment of panicles bearing three week old fruits with 2, 4, 5-T sprays (10, 30, 50 and 60 ppm) did not prove effective in checking the fruit fall of 'Mulgoa' mango (Kalyanasundaram, 1974). However, 2, 4, 5-T at 20 and 40 ppm sprayed on 'Dashehari' mango trees in November to May improved fruit retention at Varanasi (Singh, 1977). Mango trees of 'Langra' when sprayed with 2, 4-5-T (20 and 40 ppm) at pea-size stage of fruit with or without subsequent spray after one month responded with higher fruit retention at Varanasi (Maurya and Singh, 1979). At 20, 40 and 60 ppm, sprayed at pea-size stage of fruits (cv. Dashehari). 2, 4, 5-T reduced fruit drop and retained more number of fruits, but fruit retention decreased with increase in the concentration of the auxin (Maurya *et al.* 1973). At 20 ppm, 2, 4-5T was more effective, when sprayed on inflorescences at pea-size stage to check fruit drop in Pakistan (Arifkhan *et al.*, 1993). Pre-harvest application of 2, 4-5-T at completion of fruit set and a fort night before harvest increased marginally the percentage of fruit retention in 'Amrapali' mango at Sabour in Bihar (Rani and Brahmachari, 2004). However, sprays of 2, 4, 5-TP were ineffective in enhancing fruit retention in mango (Singh, 1954-56). Kennard and Winters (1956) tested 50 to 200 and 800 ppm, 2, 4, 5-TP on 3 and 6 weeks-old fruits without success, instead the auxin increased fruit fall in 'Amini' cultivar along with reduction in fruit size. Singh *et al.* (1961) demonstrated large and significant effect of 2, 4, 5-TP (20-80 ppm) in reducing the post set fruit drop compared with control in 'Dashehari' and 'Langra' cultivars but not in 'Fazli' in which case the auxin was totally ineffective. Thus 2, 4, 5-T and 2, 4, 5-TP were found to be less effective than NAA and 2, 4-D in preventing fruit drop in mango.

9.2. Role of gibberellins in controlling fruit drop

Gibberellins are implicated in growth and development of mango fruits. However, their direct role in fruit abscission has not been elucidated clearly and is ambiguous and the studies involving gibberellins in abscission are inconclusive (Chacko *et al.*, 1970a; Ram, 1983; Davenport and Nunez-Elisea, 1997). Biosynthesis of gibberellins takes place in seeds, young leaves, young shoots and in roots of plants and are probably transported in xylem and phloem vessels to the target organ. In mango major endo-gibberellins have been reported in apical bud leaves and vegetative shoots (Davenport *et al.*, 2000).

Gibberellins intensity caused organ's ability to function as nutrient sinks and increase the biosynthesis of IAA in ovaries, which delays the formation of the separation layer and this enhances fruit retention (Wasty, 1995). Besides, gibberellins are believed to serve as mediators for faster translocation and mobilization of stored photosynthates from source to sink (Looney *et al.*, 1992).

Opinions differed regarding the direct involvement of gibberellins in fruit abscission. Some workers (Chacko *et al.*, 1970a; Ram and Pal, 1979; Chen, 1981; Ram, 1983) did not support the direct involvement of gibberellins in fruit abscission, while others claimed a correlate relationship of depleted endogenous levels of gibberelling with fruit abscission resulting in fruit drop (Bains *et al.*, 1997; Singh *et al.*, 2005).

Attempts were, however, made to elucidate the role of gibberellins in fruit drop through the application of growth retardants, which inhibit the biosynthesis of gibberellins. However, highly variable responses did not allow to draw favourable conclusion with regard to the role of gibberellins in fruit abscission.

9.2.1. Control of fruit drop by gibberellins

Gibberellins are known to be involved in growth and development of fruits, however, but do not appear to be directly linked to the abscission of plant appendages like fruits (Chacko *et al.*, 1970a; Ram, 1983). However, a correlative relationship of depleted endogenous level of gibberellins with mango fruit abscission was claimed (Singh *et al.*, 2005). This strongly implicated and supported the effectiveness of gibberellins in the initial stages of fruit growth, which suggested the possibilities of using gibberellins exogenously to prevent fruit drop. Gibberellins intensity resulted in organ's ability to function as nutritious sink, increase the synthesis of other phytohormones and nullify the action of ABA (Rani and Brahmachari, 2004). However, results of spray applications of GA_3 (20 to 500 ppm) to pre-and post-anthesis panicles to increase fruitset and retention have been inconsistent. Not with standing the above, as in the case of some phenoxy auxins, not much work was done with gibberellins, oriented towards checking fruit drop and improving fruit retention in mango.

Improvement of fruit retention due to GA_3 application at post-bloom stage was reported (Teaotia *et al.*, 1967; Rao *et al.*, 1979; Ram and Sirohi, 1979; Singh and Ram, 1983; Srilatha, 2003). Murthy (1976) was able to achieve reduction in fruit drop in 'Peter', 'Cherukurasam' and Au-Rumani cultivars with GA_3. When GA_3 was applied as a foliar spray at pre-bloom, pea-size and marble-size stages, highest frit retention (160%) was observed at both 20 and 40 ppm and marble-size stages followed by 40 ppm GA_3 spray at pea-size stage (105 %). Singh *et al.* (1986) achieved maximum fruit retention with GA_3 (10 ppm) sprays made at full bloom in 'Dashehari' mango cultivar. Treatment with GA_3 (20 ppm) resulted in better retention of fruits in 'Chausa' cultivar (Prakash and Ram, 1985). GA_3 at 30 ppm recorded higher fruit retention and yield per plant in mango (Sarkar and Ghosh, 2004).

9.3. Role of Cytokinins

Although CKs are not generally associated directly with abscission, Ram (1983) and Ram *et al.*, (1983) concluded that low CK levels during fruit development may contribute to fruit loss. CKs are required for fruit growth and development and their retention, however, their role in abscission is not clear and inconclusive (Davenport and Nunez-Elisea, 1997). CKs are found in both pericarp and seed (Ram *et al.*, 1983; Singh *et al.*, 2005). Application of synthetic cytokinin (CPPU) demonstrated the role of cytokinin, by improving fruit retention by maintaining fruit abscission zone intact (Notodimedjo, 2000).

9.3.1. Control of fruit drop by cytokinins

Cytokinins are important during the initial stages of fruit growth in mango. Although CKs are not generally associated with abscission, Ram (1983) concluded that low cytokinin levels during fruit development may contribute to fruit loss through fruit fall.

Studies in East Java showed that exogenous application of 10 ppm of synthetic cytokinin *viz.*, CPPU [N (2-chloro-4-peridyl)-N-phenyl urea], 14 days after full bloom resulted in an increased number of fruits per panicle and number of fruits retained per tree in cv. 'Aurunmanis' mango (Notodimedjo, 2000). Post bloom application of CPPU and GA_3 markedly increased fruit retention in 'Tommy Atkins' mango in South Africa (Oosthuyse, 1995). A single spray of application of CPPU effectively improved fruit retention besides leading to greater fruitset per inflorescence than irrigation in Vietnam (Roemer, 2011). Among different PGRs, CPPU showed promising in fruit retention under the conditions of shortage of water availability in mango in Vietnam (Hegemann *et al.*, 2011).

9.4. Role of Polyamines

Based on the results of studies in olive (Lui *et al.*, 1999), mango (Singh and Singh, 1995; Murti and Upreti, 1999; Singh and Janes, 2000; Murti *et al.*, 2008) and grape (Aziz *et al.*, 2001), polyamines (PAs) were implicated to play a significant role in the abscission of plant organs.

The role of PAs in fruitlet abscission of mango was investigated by estimating the endogenous PAs and their levels in intact and about-to-abscise fruitlets and their pedicels and by studying the effects of exogenous application of synthetic PAs and inhibitors of PA biosynthesis, by which it was demonstrated that PAs play a key role in mango fruitlet abscission (Malik and Singh, 2003b). The association of initial fruitlet abscission with reduced levels of endogenous PAs and increased ethylene biosynthesis in mango fruitlets and their pedicels (Malik *et al.*, 2003b) were also demonstrated.

Among the three major PAs, obviously the role of spermine (SPM) seems to be more critical in mango fruitlet abscission, as the about-to-abscise fruitlets had very low levels of SPM and fruitlets were smaller and lighter due to cessation of their growth, which is attributed to lower levels of PAs during the phase of cell division, affecting the normal process of growth of embryo and fruit. In general, lower levels of endogenous PAs in the abscised fruits and reduced fruit drop following the exogenous spray application of PAs as well as the acceleration of abscission with the inhibitors of PAs support their role in fruit abscission. Lower levels of endogenous PAs in about-to-abscise fruitlets and their pedicels lead to higher endogenous production of ACC and ethylene resulting in increased fruitlet abscission. On the other hand, higher levels of PAs in intact tissue inhibit production of ACC synthese and corresponding ethylene biosynthesis and thereby reduce the abscission of fruitlets. The above observation supports the key role of endogenous PAs in fruitlet abscission in mango (Malik and Singh, 2003b). PAs show direct binding with cell membrane and maintain membrane integrity through prevention of lipid peroxidation and thereby prevent abscission (Philip and Malamberg, 1989). PAs improve floral organ development, pollination, fertilization and subsequent embryo development resulting in increased fruit retention (Rugini *et al.*, 1986).

9.4.1. Control of fruit drop by polyamines

Rugini *et al.*, (1986) considered PAs as anti-abscission substances, as they prevent the abscission through prevention of ethylene bio-synthesis (Singh and Singh, 1995). Their exogenous application has shown promise in preventing fruit drop in different fruit drops including mango (Singh and Singh, 1995; Sten and Gazit, 2000; Aziz *et al.*, 2001). The increase in fruit retention with exogenous supply of PAs was ascribed to the increase in the levels of endogenous PAs in fruitlets and pedicels. Besides the above growth promoting properties, PAs may have helped improvement of overall reproduction processes resulting in reduced shedding of fruits. The endogenous deficiencies of PAs in fruitlets and their pedicels can be elevated by the exogenous application of PAs and thereby check the abscission of fruitlets. Exogenous application of PAs improved fruit retention by inhibiting the endogenous biosynthesis of ethylene (Brown, 1997). However, the response of mango is significantly influenced by the type of PA, its concentration, time of application and physiological stage of fruits at the time of application (Malik and Singh, 2003a). Exogenous application of PAs reduced fruitlet abscission, when applied 2 weeks after fruitset. Aqueous solution of PAs applied to flower panicles at full bloom stage resulted in higher fruit retention in 'Dashehari' mango in Punjab (Singh and Singh, 1995) than spraying prior to anthesis. Out of the three major PAs, SPM was more effective than other two in enhancing fruit retention, when sprayed on full growth panicles prior to anthesis (Singh and Singh, 1995). Exogenous application of the three PAs (Putrescine, Spermine and Spermidine) at full bloom and frutiset stages reduced fruitlet abscission in "Kensington Pride' mangoes in Australia (Malik, 2003; Malik and Singh, 2003a). Among the above three PAs, SPM was more effective in fruit retention (Upreti *et al.*, 2011).

9.5. Role of Growth inhibitors (Abscisic acid)

Plants produce growth promoters as well as growth inhibitors for promoting growth and for checking / suspending the growth either temporarily or permanently. One of the growth inhibitors endogenously synthesized in plants is 'abscissic acid' (ABA) which is most commonly linked to the abscission of plant organs (Addicott, 1970).

Several investigations / studies indicated the occurrence of higher levels of ABA-like inhibitors during the first few weeks of fruit growth after pollination / fruitset in mango (Ram and Pal, 1979; Ram *et al.*, 1983; Prakash and Ram, 1984; Ram, 1992; Murti and Upreti, 1995, 1999) which coincided with slow growth of fruit, low levels of growth parameters and high fruit drop. This indicates the role of ABA in fruit abscission.

Fruit drop was correlated with levels of endogenous ABA in the fruit (Chacko *et al.*, 1970b, 1972d; Prakash and Ram, 1984). The fruit drop was proportional with the increase in the level of ABA and higher levels of ABA occurred during the early fruit drop and with advancing age of the fruits (Chen, 1981).

Inhibitors like ABA play a dominant role in fruit drop when concentrations of growth promoters (auxins, gibberellins etc.) are low in the fruit (Ram and Pal, 1979). It was suggested that inhibitors (ABA – like) might counteract the activity of auxins, gibberellins and cytokinins in fruit abscission (Singh and Singh, 1974; Ram, 1992).

The endogenous content of ABA at pin-head and pea-size stages of fruit growth was higher in abscissed fruits than intact fruits of 'Alphonso' mango (Murti and Upreti, 1999). This may be due to the fact that dropping fruits were under some stress which causes increase in the ABA level. This increased ABA interacted with IAA and stimulated the drop of fruits. Thus, it is clear that inhibitor (ABA) plays a major role in the fruit shedding in mango. In other words, fruit drop, particularly of fruitlets during the initial two weeks (post-setting and pin-head) of fruit drop is mainly due to the presence of inhibitors in them.

A higher level of ABA in fruits and pedicels stimulate ethylene production, which in turn causes abscission. Endogenously synthesiszed ABA in fruits get translocated into pedicels and abscission zone and accentuates biosynthesis of hydrolytic enzymes that initiate the formation of abscission layer leading to the drop of fruits (Sexton and Roberts, 1982).

9.6. Role of Ethylene

Ethylene is involved in the abscission of plant appendages, besides other physiological processes like flowering, fruit growth, fruit ripening etc. Ethylene is a potent accelerator of abscission of plant organs; hence it is often used as 'abscission promoter' for deblossoming malformed panicles and mechanical harvesting of fruits (Chadha *et al.*, 1979).

The involvement of ethylene in organ abscission was first suggested by Gawadi and Avery (1950) who advanced "Ethylene and hormone balance" theory to explain the involvement of ethylene in organ abscission of fruit crops. This theory is based on the extreme sensitivity of plants to ethylene gas. According to this theory abscission of plant organs is controlled by the balance between the amounts of endogenously produced auxins and ethylene. Ethylene causes / accelerates abscission either by reducing the concentration of auxins or by interfering and / or by restricting the action of auxins in maintaining the abscission zone. When the accelerating effect of ethylene surpasses the effect of auxin, the fall of organ occurs.

In fruit crops production of ethylene by fruits was considered to be a cause of fruit drop, since production of ethylene was higher in abscissed fruits (Van lely veld and Nel, 1982; Nunez-Elisea and Davenport, 1986). Motivating role of ethylene in abscission was demonstrated by dose-responsive pre-cautious abscission of fruitlets exposed to elevated levels of the ethylene gas (Nunez-Elisea and Davenport, 1986). Reduced fruit drop in response to inhibitors of ethylene production and its action support this notion (Singh and Ram, 1983; Naqui *et al.*, 1990). Based on the above, ethylene was recognized as the trigger of abscission of various plant organs across all plant species (Abeles *et al.*, 1992).

Van Lely Veld and Nel (1982) were the first to implicate the involvement of ethylene in mango fruitlet abscission based on detection of higher levels of ethylene in abscissed fruitlets followed by others (Naqui *et al.*, 1990). Biosynthesis of ethylene increases during fruitlet abscission, providing substantial evidence to support that ethylene is the main trigger of the process of abscission (Malik *et al.* 2003a). The role of ethylene in mango fruitlet shedding was demonstrated by determining the endogenous ethylene concentration in intact and about-to-abscise fruitlets and their pedicels and studying the effect of exogenous application of ethylene and its inhibitors (Malik *et al.*, 2003a). Studies of Nunez-Elisea and Davenport (1986) indicated a key role of ethylene in the abscission of mango fruitlets, but the events responsible for induction of ethylene bio-synthesis are unknown. Brown (1997) suggested that one role of ethylene during abscission process is to co-ordinate the process of separation of plant appendages.

Thus, it appears that all known groups of phytohormones are involved in fruit growth, fruit retention and fruit drop. A proper balance of these phytohormones retains and carries the fruit to maturity and aids the growth and development of fruits, while an imbalance causes improper growth and development and finally shedding of fruits. Similarly inadequate synthesis of either of the hormones, particularly auxins, gibberellins, cytokinins and polyamines leads to fruit abscission, while enhanced synthesis of inhibitors (ABA) and ethylene promotes abscission. Synthesis of various phytohormones at appropriate time of fruit growth and development is required for carrying the fruits to maturity and retention. Otherwise, there will be abscission of fruitlets / fruits.

9.6.1. Control of fruit drop by ethylene and ethylene inhibitors

Ethylene has exhibited the greatest and immediate impact on flower and fruitlet abscission. It is more often used as an abscission promoter rather than inhibitor of abscission, as an aid in chemical harvesting of fruits. The involvement of ethylene in the abscission of young mango fruits has been documented (Van Lely veld, 1978; Nunez-Elisea and Davenport, 1986; Naqui *et al.*, 1990). Lab studies indicated that high ACC availability and ethylene synthesis in the fruitlets are important contributors of fruitlet abscission. In view of this, Murti *et al.* (2008) suggested that ethylene regulation during the initial phases of fruit growth is important in working out strategies for the control of fruit drop in mango.

Exogenous application of ethylene or ethephon (ethrel) promotes abscission of a variety of immature and mature fleshy fruits, which is exploited in horticulture to manipulate fruit load and assist harvest (Brady and Spiers, 1991). Since ethylene is an abscission promoter, it is not generally employed to check fruit drop, instead its biosynthesis and action inhibitors can be used for preventing fruit drop.

Ethylene biosynthesis inhibitors include amino-ethoxy vinyl glycine (AVG) and amino oxyacetic acid (AOA) while its action inhibitors include, silver thiosulphate (STS) and silver nitrate ($AgNO_3$) and cobalt ions. Ethyene bio-synthesis inhibitors block the conversion of S-adenosyl methionine (SAM) to 1-aminocyclo propane-1- carboxylic acid (ACC) in the biosynthesis path way of ethylene (Yang, 1980). Inhibitors of ethylene action like silver nitrate etc. counteract high levels of ethylene in plant organs. Substantial reduction in young fruit abscission in mango was achieved with exogenous application of both types of inhibitors (Naqui *et al.*, 1990; Singh and Agrez, 2002; Malik *et al.*, 2003a). Studies in

Australia indicated the key role of ethylene in fruit retention. The results also indicated the possibility of increasing initial or final fruit set with a single spray application of both types of inhibitors before anthesis to improve fruit retention and yield in mango. AVG was most effective in enhancing final fruit retention as compared to AOA, cobalt sulphate and STS prior to anthesis on full grown panicles (Singh and Agrez, 2002). Exogenous application of inhibitors of ethylene biosynthesis and its action (Naqui *et al.*, 1990) and polyamines (Sigh and Singh, 1995; Singh and Janes, 2000) increased fruit retention probably by inhibiting endogenous ethylene synthesis and / or its action.

9.7. Control of fruit drop by growth inhibitors and retardants

A few attempts were made to investigate the possibility of employing growth retardants and inhibitors to check fruit drop and thereby improve fruit retention and yield in mango. Rao *et al.* (1976) reported the efficiency of 'Alar', a growth retardant (100 ppm) in checking fruit drop in 'Dashehari' mango, when applied at pea-size stage of the fruit at Lucknow. In a similar study on 'Dashehari', Alar (100 ppm) at pea-size stage was more effective than that sprayed at full bloom and marble size stage under Tarai conditions of Uttarkhand at Pant Nagar (Singh and Ram, 1983). Alar sprays of 1000 ppm reduced fruit drop in field grown 'Neelum' (Rao and Rao, 1985). Alar (B-9) sprayed at 300 ppm at fruit bud differentiation stage recorded increased fruit retention compared with unsprayed control in 'Chausa' cultivar (Prakash and Ram, 1985).

Application of Chloremquat (Cycocel, CCC) 200 ppm at pea-size stage of fruit gave the best results in preventing the fruit drop in cv. 'Dashehari' at Pantnagar (Singh and Ram, 1983). The effect of both Alar and Cycocel was better when applied at pea-size stage of fruit than at other two stages *viz.*, pre-bloom and marble-size stage, which was attributed to their retardation of competition between developing fruits and older fruits. Another reason could be increase in the activity of CK – like substances in the fruits following the application of Alar and Cycocel (Singh and Ram, 1983).

Paclobutrazol (PBZ) is widely employed to alleviate alternate bearing in several fruit crops by improving flowering in the 'off' years. It was also tried to minimize fruit fall, through indirect checking of the concomitant occurrences of growth flushes, which enhance fruitfall, because of competition for assimilates between flushes and fruits (Kulkarni, 1988; Oosthuyse and Jacobs, 1997). Soil application of PBZ (4 OZ a.i/tree) in water applied in rainy season registered lowest fruit drop (17 %) in 25-year old 'Alphonso' mango trees, among the various growth retardants tested (Bhatt and Kumar, 1997). PBZ at 300 ppm as foliar spray twice during fruiting season retained more mature fruits per panicle and per tree and yield per tree in Egypt (Rawash *et al.*, 1998).

Control of fruit drop and increased fruit retention by PBZ is mediated through increased activities of growth promoters (auxins and CKs) and / or through the increased CK-like activity of the growth retardants (Pal *et al.*, 1979; Singh and Ram, 1983).

10. Role of Plant Growth Regulators on Fruit Maturity

Fruit maturity is the final stage of growth leading to development of physiological maturity and it is followed by ripening and senescence in a sequence. The duration of fruit maturity in mango is a varietal character and depends on several factors like commencement of flowing, prevailing environmental conditions etc. The duration of fruit maturity in mango from fruit set to fruit maturity ranges from 90 to 110 days. Fruit maturity is earlier and shorter under tropical conditions of South India compared to subtropical conditions of North India (Singh, 1978; Tandon and Kalra, 1983).

Regulation of fruit maturity is desirable to avoid some problems and gain certain benefits. Advancing the maturity of fruits is essential to avoid damage by insect pests like fruit borers, fruit flies and nut weevils and diseases like anthracnose and disorders like sprongy tissue etc. and damage due to early rains. Attempts have been made to advance fruit maturity with the aid of plant growth regulators to provide opportunity to commercial advantage of early markets. However, not much work was done in this direction.

Advancing flowering is supposed to advance fruit maturity. Any chemical / growth regulator, which advances flowering also advances fruit maturity. Similarly any chemical / growth regulator which delays flowering causes delay in fruit maturity.

GA_3 is the plant growth regulator that is more frequently used successfully to alter the date and intensity of flowering and thereby regulate the fruit maturity in fruit crops. Pre-harvest application of GA_3 (50 or 75 ppm) brought forward fruit maturity by 8-10 days in mango at Sabour (Kumar and Singh, 1993). Application of GA_3, caused inhibition of flowering in mango and delayed the date of anthesis, there by extended the fruit maturity (Kachru *et al.*, 1972; Nunez-Elisea and Davenport, 1991; Salazar – Garcia *et al.*, 1998). They could delay fruit maturity with the application of GA_3 due to delayed flowering and delayed start of anthesis in mango. Shawky *et al.* (1978) and Turnbull *et al.* (1996) achieved delay in fruit maturity of mango with GA_3 foliar sprays.

Cycocel (3000 ppm) induced early maturity in mango cv. Kesar at Navasari by inducing early flowering and suppressing vegetative growth of trees. The reduced endogenous levels of gibberellins, resulting from the action of cycocel might have favoured the early and profuse flowering in mango, which might have resulted in early maturity of fruits (Kurian and Iyer, 1993).

Paclobutrazol is well known for induction of early flowering. This early flowering has hastened fruit maturity by 3 weeks than control which resulted in early harvest of fruit (Padhiar and Patel, 2003, Tendel and Patel, 2011).

Ethrel when sprayed at 200 ppm at pea stage of fruits was considerably effective in reducing the maturity period by 10 days over control in mango at Basti (Pandey *et al.*, 1980).

11. Role of Plant Growth Regulators on Fruit Quality

Increased fruit production without improvement in fruit quality is meaningless. Plant growth regulators (PGR) play important role in fruit quality in respect of both physico-chemical and biochemical parameters. To improve the flowering, fruit set, fruit growth, to check fruit drop and certain physiological disorders PGR are often applied (either through foliage or soil) at different stages like pre-flowering, during flowering, during fruitset, fruit growth etc. to obtain better flowering, higher fruit set, improve fruit growth, hastening or delaying fruit maturity, to improve fruit retention as well as to prevent fruit drop, which concurrently influence fruit quality. PGRs are also applied to influence the fruit quality. The influence of PGR may be in favour of improvement of fruit quality but some times there might be unfavourable influence. In general, there would be considerable improvement in physical parameters (size, weight, colour etc.) and chemical composition (TSS, total sugars, reducing and non-reducing sugars, ascorbic acid, TSS / acid ratio), but there would be a depletion in acidity.

A perusal of available information on the role / effect of PGR in fruit quality, however, is contradictory and inconclusive, which may be attributed to different PGRs, their roles, their concentrations, time of spraying, varieties and location of the trials. Further, many research workers have not gone into the details of roles of PGR in fruit quality.

Application of PGR showed some benefits. The physical characteristics of fruits such as appearance, colour, flavour, aroma, taste and texture of growth regulators treated mango fruits would be more attractive and acceptable than the untreated control fruits. The nutritional value of mango fruits is also affected remarkably after treatment with PGR. The quality of treated mango fruits in terms of ascorbic acid, total soluble solids, total, reduced and non-reducing sugar contents of the pulp was often superior to those of control fruits (Morshed *et al.*, 2006).

11.1. Role of auxins in fruit quality

The most common auxin, which was tried to improve the fruit quality in mango was NAA, followed by 2, 4-D, 2,4, 5-T etc. Suryanarayana and Das (1971) sprayed NAA (0, 20, 40 ppm) at 4 different stages *viz.*, panicle emergence, before anthesis, at pea-size stage and marble-size stages of fruits in Odissa and obtained improvement in fruit size, TSS, TSS/acid ratio with higher levels of NAA (40 ppm) in 'Banganapalli' cultivar. At Varanasi, 'Dashehari' responded positively in producing large fruits with 20 ppm NAA and improvement in fruit quality with increasing TSS, sugars and ascorbic acid contents and a decrease in acidity. The highest dose (60 ppm) was most effective for increase in TSS and ascorbic acid, while lower dose (20 ppm) was most effective for sugar content (Maurya *et al.*, 1973). The increase in sugar content of fruits receiving NAA was attributed to mobilization of soluble carbohydrates into fruits (Leopold, 1958). The increase in all the parameters of fruits *viz.*, weight, volume of fruits due to combined sprays of urea plus NAA at different doses applied twice in October and November in cv. 'Sundarja' was ascribed to accumulation of more food material and auxin into the treated fruits (Baghel *et al.*, 1987b) at Jabalpur. The combination of urea plus NAA besides increasing the physical parameters also improved the percentage of pulp, TSS, reducing and non-reducing sugars and decreased the acidity (Baghel *et al.*, 1989). In a similar study on 'Langra' mango trees with a combination of urea (4 %) and NAA (40 ppm), Sharma *et al.* (1990) in Madhya Pradesh obtained higher number of fruits per plant with rise in TSS, total sugars, non-reducing sugars and ascorbic acid and lower acidity. NAA sprayed twice on bearing 'Langra' trees at 15 days interval during February – March, once prior to flowering and thereafter, enhanced dry matter, reducing, non-reducing and total sugars, TSS, Vitamin A and C contents in fruit pulp on ripening at Varanasi (Rath and Rajput, 1990b). The stone percentage significantly increased with the application of NAA in cv. Parbhani Bhushan (Sharma *et al.*, 1990) and ascorbic acid in 'Alphonso' mangoes with NAA (20 ppm) at Coimbatore (Singh, 1980).

Suryanarayana and Das (1971) did not find much influence of 2, 4-D on fruit quality promotion of Banganapalli fruits in Odissa. On the other hand, Maurya *et al.* (1973) from Varanasi, reported 2, 4-D at 20 ppm as the best treatment for improving fruit quality in 'Dashehari'.

Singh *et al.*, (1976) observed a depressing effect of 2, 4, 5-T (250 ppm) on fruit acidity. Kumar *et al.* (1992) investigated the influence of 2, 4, 5-T on fruit quality of 'Dashehari' in storage, in which the fruits were dipped in aqueous solution for 10 minutes, packed in wooden boxes with paper cuttings as cushioning material and kept at room temperature (34 ± 4°C and 50-65 % RH). The fruit quality appreciably increased during storage by post-harvest dipping of fruits in 2, 4, 5-T (1000 ppm) solution in terms of increase in TSS, total and reducing sugars and decrease in ascorbic acid and acidity.

11.2. Role of gibberllins in fruit quality

While, auxins decreased acidity of traits, gibberellic acid increased acidity of treated fruits. Suryanarayana and Das (1971) noted increased acidity in 'Banganapalli' mangoes in Odissa due to sprays of GA_3 at four different stages of flowering and fruiting. Fruit diameter and weight of fruits increased significantly when GA_3 (150 ppm) was sprayed on fruits at pea-size stage in Gaurjeet mango (Singh *et al.*, 1976). At Basti (Uttar Pradesh) GA_3 100 ppm recorded heaviest fruits with higher diameter in 'Dashehari' mango (Pandey *et al.*, 1980). Fruit diameter, TSS, total, reducing and non-reducing sugars, vitamin A and C contents in fruit pulp of bearing 'Langra' mango trees increased, while sugar / acid ratio decreased with increasing rate of GA_3 from 50 to 150 ppm, with its application twice at 15 days interval prior to flowering, in the ripened fruits (Rath and Rajput, 1990b). 'Dasheri' fruits that received two sprays of GA_3 (> 200 ppm) at 10 days interval after flowering exhibited less TSS, a lower TSS / acid ratio, loss in total carotenoid, amylase and peroxidise activity at harvest. Total acidity, ascorbic acid and total chlorophyll in the peel were significantly higher in treated fruits of 'Dashehari' at Lucknow (Khader, 1991). At Sabour (Bihar) GA_3 (75 and 150 ppm) improved fruit quality and shelf life of 'Amrapali' mango fruits with sprays on fruits in the first week of June (Singh *et al.*, 1995). GA_3 (200 ppm) applied 30 days before harvest and stored at 12°-25°C for 25 days reduced the content of TSS and percentage of weight loss and resulted in higher colour rating in Venezula (Manzano-Mendoza, 1995). Significantly heaviest fruits and maximum quality were achieved with GA_3 (200 ppm). However, maximum pulp weight / fruit was observed with 100 ppm and maximum edible: non-edible ratio were recorded with 200 ppm sprays. The exogenous application of GA_3 might have stimulated cell division and cell elongation in giving large fruits. The probable reason behind all these effects of increase in fruit weight and volume might be due to rapid cell division and cell enlargement and accumulation of more sugars and water under the influence of exogenous application of GA_3 (Rani and Bramhachari, 2004). Fruits dipped in GA_3 solution or treated with vapor guard significantly retarded ascorbic acid degradation in the pulp and chlorophyll degradation in the peel (for which GA is well known) and reduced a-amylase and peroxidase activity during storage at ambient temperature of 34° – 37°C at

Lucknow (Khader, 1992). Post harvest application of GA_3 to delay ripening of 'Lucknow' mangoes retarded the increase in TSS, total sugars, loss of ascorbic acid content and acidity and reduced spoilage of fruits (Jain and Mukherjee, 2001).

11.3. Role of ethylene in fruit quality

Ethylene, a ripening hormone influences the different ripening changes in fruits. The decrease or increase of different chemical parameters in treated fruits may be considered due to ethylene application. However, Conegtion and Rodriguez (1994) recorded no effect of ethylene on chemical characteristics of 'Keitt' fruits, which were harvested 170 days after fruit initiation and immersed in aqueous solution of 500 or 1000 ppm ethephon for 5 minutes in Brazil.

11.4. Role of 1-Methylcyclopropane in fruit quality

1-Methylcyclopropane (1-MCP) retards ripening of fruits caused by ethylene by binding irreversibly to ethylene – binding sites in the fruits. As a result ethylene is unable to bind and elicit – subsequent signal transduction and translation (Serek *et al.*, 1994). Fruits treated with 1-MCP remain insensitive to ethylene (Sisler and Serek, 1997) and there by influence the fruit quality parameters in treated fruits. 1-MCP positively affected the quality of 'Palmer' fruits and treatment with 1-MCP at 150 ppm contributed to a reduction in fruit softening unlike the 18 days of refrigeration storage at room temperature (Hojo *et al.*, 2006). 1-MCP plus controlled atmosphere reduced the incidence of anthracnose, weight and firmness loss, delayed skin and flesh colour development, prevented the increase in soluble solids concentration / titrable acid ratio, ethanol and acetaldehyde content, maintained the ascorbic acid, carotenoid, total phenol and flavonoid content and antioxidant scavenging activity in hot water treated 'Kent' mangoes (Devender *et al.*, 2012). It was suggested that the combination effect of 1-MCP and controlled atmosphere storage can be recommended as an alternate treatment to replace prochloraz application for hot water treated mangoes and can be adopted commercially for organic export markets of mango (Devender *et al.*, 2012).

11.5. Role of paclobutrazol in fruit quality

Similar to auxins, PBZ improves fruit quality, but unlike auxins which when applied during flowering and fruiting, pre-flowering application of PBZ influences the fruit quality. Further, unlike auxins, gibberellins etc. PBZ is most effective through soil application and exhibits residual activity for a couple or more years on the treated crop and this residual activity is supposed to influence the fruit quality. Several investigators reported the improvement of fruit quality of mango with the application of PBZ (Rai and Bist, 1992; Singh and Dhillon,

1992; Vijayalakshmi and Srinivasan, 2000; Nandkumar and Kurupaiah, 2006; Chusri *et al.*, 2008; Bamini *et al.*, 2009 and Sarkar and Rahim, 2012). The improvement of fruit quality of PBZ treated mango trees has been attributed to diversion of photosynthates towards the fruits (Rai and Bist, 1992).

Irrespective of the mode of application, PBZ treatment decreased the acidity compared with control. Foliar application increased the TSS, while soil application did not show much effect on TSS. The soil application of cultar (20 g / tree) significantly increased the TSS / acid ratio, obviously due to decrease in acid content and increased TSS content of the fruits. However, neither method of application influenced fruit weight and pulp to stone ratio significantly (Singh and Dhillon, 1992).

Application of 10 ml PBZ / tree (by soil) had shown the greatest effect in increasing all the parameters of fruit quality of mango; ascorbic acid, carotene, TSS, total, reducing sugars and sugar / acid ratio, except acidity (Vijayalakshmi and Srinivasan, 2000). Cultar (PBZ) improved fruit quality in terms of TSS, total acidity, total chlorophyll, total carotenoids, a-amylase and peroxidase activity, but TSS value was significantly higher and total acidity value was lower than control at 2 g / tree dose of cultar at Ludhiana (Singh and Saini, 2001). Pruning after harvest followed by application of PBZ improved fruit quality in 'Baneshan' mango at Annamalai in Tamilnadu (Nandkumar and Kurupaiah, 2006). A direct relationship between PBZ concentration and fruit weight, TSS and other parameters of fruit quality was observed in mango in Egypt (Khethab *et al.*, 2006).

Paclobutrazol did not show negative influence on fruit quality of mango in Mexico. Higher doses of PBZ resulted in higher content of TSS, less total acidity, less firmness and greatest weight loss in Mexico (Robellede – Martinez *et al.*, 2008). However, Brix value and titratable acidity of fruits did not differ due to PBZ treatments in 'Irwin' mangoes grown in plastic houses in Taiwan (Chusri *et al.*, 2008). The fruit quality in terms of TSS, total, reducing and non-reducing sugars, ascorbic acid and carotenoid contents was enhanced markedly, when treated with PBZ in both 'light' and severely pruned trees during main as well as 'off' years, but physiological loss in weight was lowest in PBZ treated trees of 'Neelum' at Periyakulam in Tamilnadu (Bamini *et al.*, 2009). Spray treatment of PBZ at 750 ppm during mid-October recorded heaviest fruits, highest edible portion, lowest stone to pulp and peel to pulp ratio in mango, besides overall improvement in fruit quality in Bangladesh (Sarkar and Rahim, 2012).

12. Role of Plant Growth Regulators on Fruit Yield

The ultimate goal of any farmer is the achievement of higher crop yields. For this end he adopts several approaches. The productivity of mango in India is very low compared to its potential as well as to some mango growing countries like Brazil, Israel, Australia etc. The low productivity in mango is attributed to several factors like genetical, excessive vegetative growth, irregularity in bearing, less proportion of hermaphrodite flowers, inadequate fruit set followed by heavy fruit drop, incidence of pests and diseases etc. In commercial orchards of mango, it is desirable to get economically higher yields. Besides nutrients, phytohormones also play role in fruit production of mango.

Auxins like NAA were tested for improvement of yields in mango. Exogenous application of NAA on pea and marble size fruits of cv. 'Amrapali' at Nadia, West Bengal (Vejendla *et al.*, 2008) and in 'Keitt' in Ghana (Nkamsah *et al.*, 2012) improved yields by improving fruit retention. The increase in fruit retention and yield was ascribed to NAA which caused cell elongation by enlargement of vocuoles in the cells and loosening of cell walls (Agrawal and Dikshit, 2008).

Gibberellins are believed to serve as mediators for faster translocation and mobilization of stored metabolites and / or photosynthates from source to sink and also to play significant role in increasing auxin synthesis in ovules. And all these stimulate higher fruit set and fruit retention and finally improvement in yield (Looney *et al.*, 1992). Moore (1979) observed stimulation of both cell division and cell elongation as a result of GA_3 application increased tree yield in 'Keitt' mango in Ghana (Nkansah *et al.*, 2012).

Plant growth retardants are synthetic compounds, used to retard the shoot length of plants in a desired way without changing the developmental processes. Various growth retardants have significantly increased the production of flowers, which ensured an increase in fruit production through higher fruit set and reduction in fruit drop.

An application of Cycocel (CCC) had effectively controlled the emergence of vegetative flush by interfering the synthesis of gibberellin (Sarkar *et al.*, 1998). Foliar sprays of CCC (3500 ppm) given prior to fruit bud differentiation increased the number of fruits and yield per tree in 'Kesar' cultivar at Navasari (Chaudhuri *et al.*, 2014). According to Guha (1993) and Sarkar *et al.* (1998) CCC exhibits growth retarding effect leading to higher fruit retention and higher yield. A single spray of CCC in October increased fruit set and yield many folds in mango at Akola (Dalal *et al.*, 2005).

Similarly, foliar application of MH reduced the vegetative growth in 'Langra' mango, resulting in higher fruit yield in Odissa (Rath and Das, 1979). MH limited the vegetative growth, through altering the sink strength with in the plants which might have helped in enhancing the assimilate partioning in favour of reproduction (Guha, 1993).

Several studies over the world with PBZ on mango resulted in higher yields (Davenport, 1987, 1994; Kulkarni, 1988; Burondkar and Gunjate, 1991; Hiller and Rudge, 1991; Tongumpai *et al.*, 1991; Ram and Tripathi, 1993; Desai and Chundawat, 1994; Oosthuyse, 1995; 2015; Wang *et al.*, 1995; Sergent *et al.*, 1997; Sarkar *et al.*, 1998; Patil and Talathi, 1994; Notodimedjo, 2000; Anbu *et al.*, 2001; Hoda *et al.*, 2001; Singh and Saini, 2001; Yeshitela *et al.*, 2004; Baghel *et al.*, 2004; Karuna *et al.*, 2005; 2007; Yadav *et al.*, 2005; Singh and Singh, 2006; Balasubramanyam *et al.*, 2006; Rajkumar *et al.*, 2007b; Karuna and Mankar, 2008; Vejendla *et al.*, 2008; Reddy and Kurian, 2008; Martinez *et al.*, 2008; Bamini *et al.*, 2009; Singh *et al.*, 2010; Muhammad *et al.*, 2010; Nafeez *et al.*, 2010; Tandel and Patel, 2011; Chaudhuri *et al.*, 2014; Sarkar and Rahim, 2012; Kotur, 2012; Husen *et al.*, 2012; Bhagwan *et al.*, 2013; Upreti *et al.*, 2013; Satyendra Singh *et al.*, 2014; Srilatha *et al.*, 2015). PBZ also increased yield both in 'on' and 'off' years particularly in 'on' year in mango (Karuna *et al.*, 2007; Karuna and Mankar, 2008; Kotur, 2012). Application of PBZ to rejuvenated mango trees enhanced early panicle emergence and also enabled the trees to produce shorter and compact panicles with more number of hermaphrodite flowers, which resulted in more fruit set and more fruit yield per tree in cv. Neelum at Periyakulam, Tamil Nadu (Bamini *et al.*, 2009). Further, PBZ reduced the incidence of malformation, leading to increased yield in Pakistan (Muhammad *et al.*, 2010; Nafeez *et al.*, 2010). Contrary to the favourable effect of PBZ on yield, uniconazole, another triazole, when applied in soil reduced yield severely, which was attributed to lack of subsequent growth and flowering following the treatment (Davenport, 1987, 1994).

Increased fruit yields due to soil application of PBZ has been attributed to inhibition of gibberellin synthesis in treated plants, which changes the sink source relationship in favour of fruit production by reallocating the accumulated carbohydrate sources towards flowering and fruiting (Wang *et al.*, 1995; Karuna *et al.*, 2007; Sarkar and Rahim, 2002). Inhibition of geibberellin synthesis cheks the vegetative growth by which food and energy are saved and food particularly carbohydrates is accumulated in the plant (Baghel *et al.*, 2004; Rajkumar *et al.*, 2007b; Satyendra Singh, *et al.*, 2014). This accumulation results to flowering, There will be an increase in the intensity of flowering, increase in the number of flowering shoots (Burondkar and Gunjate, 1991; Tongumpai *et al.*, 1991; Voon *et al.*, 1991; Karuna *et al.*, 2005) increase in the number of panicles and perfect flowers higher fruit set and fruit retention, ultimately increase in yield (Burondkar and Gunjate, 1991; Sarkar *et al.*, 1998; Singh *et al.*, 2010). Fruit growth is extended as a consequence of increase in cell number and cell elongation (Leopold *et al.*, 1972) because of which size and weight of fruit enhance resulting in higher yield (Karuna *et al.*, 2005).

Ethylene reduces fruit yield, probably because of its promotive action on fruit drop. On the other hand, its biosynthesis and action inhibiters promote fruit set and fruit retention and thereby increase yield, when spray applied on fully grown panicles, prior to anthesis in mango (Singh and Agrez, 2002).

13. Role of Plant Growth Regulators on Fruit Ripening

After the completion of physiological maturity, fruit ripens. Ripening of fruit is the final stage of fruit development followed by senescence. Ripening is an inevitable event in the life of the fruit. Ripening causes several changes in the fruit both quantitative and qualitative. Changes occur in physic-chemical, biochemical and physiological parameters, which make the astringent or acidic (sour) fruit to sweet and edible. Fruit ripening is defined as a sequence of changes in colour, flavour, taste and texture, which are brought out by changes in physic-chemical, physiological and biochemical parameters which lead the fruit to acceptable edible stage. Based on the respiratory behaviour of fruits, mango is categorised as a "Climacteric fruit", which undergoes rapid increase in respiration, followed by marked changes in the above qualitative and quantitative parameters. Ripening of fruits shows progressive decline in respiration once they are harvested and ripening begins.

Under natural ripening, all the above changes are mediated through the action of endogenous phytohormones particularly ethylene. Ripening of fruits can also be achieved by exogenous application of synthetic plant growth regulators especially with ethylene. Plant growth regulators are also utilized for regulation (accelerating or delaying) of ripening of fruits.

One of the greatest problems in post-harvest handling of climacteric fruits such as mango, particularly during transit, is their rapid ripening, which does not allow large quantity of fruits to transport and to reach the distant markets without ripening and spoilage. Therefore, their ripening should be regularized (i.e. delayed). Regulation of ripening of mango is necessary because:

i. Mangoes are highly perishable and can hardly be stored for 7 to 10 days without any treatment at ambient condition after harvest in unripe and sound stages

ii. Hard green, unripe mangoes ripen rapidly during transit and reach in half ripe or ripe condition at the destination depending on the maturity of the fruits, storage conditions (temperature), distance from the production centres to consumer / marketing centres and the time taken to reach the destination, besides, how the commodity is harvested and handled after harvest

iii. Mangoes are generally handled at ambient temperatures, which are high in the season, particularly in tropical and subtropical regions, leading to spoilage due to physiological degradation and diseases resulting in huge economic losses

iv. Mangoes, in general are not amenable for cold storage as in the case of temperate fruits (apple, pear, peach, etc.) since they are highly susceptible to chilling injury.

In view of the above, regulation of ripening (particularly delaying) of mangoes is very much desirable for increasing shelf life and for export and would be worthwhile in regulating the marketability leading to subsequent higher commercial return. Any means, to prevent/ delay the ripening during handling, storage and transit would go a long way in extending the availability of mangoes.

13.1. Role of PGRs in delaying ripening

Delaying ripening of mango fruits can be achieved by different approaches, one of which is 'the use of plant growth regulators'. Based on changes in respiration, pH, TSS and cell wall degrading enzymes (PE), activity, treatment with gibberellic acid, 1-methyl cyclone propane and abscisic acid lead to delaying ripening process while those of ethrel accelerate and enhance the process (Singh *et al.*, 2007).

13.1.1. Role of gibberellins delaying ripening

Role of endogenous gibberellins in fruit ripening has not been reported. On the other hand, the influence of exogenous gibberellins on fruit ripening is well documented. Several studies indicated that exogenous gibberellins, delay ripening of mango fruits though the time of application (pre-or post-harvest), concentrations, cultivars and location varied (Burg and Burg, 1962; Dilley, 1969; Mehta *et al.*, 1986; Majumder *et al.*, 1981; Krishnamurthy and Rao, 1982; Paramar and Chundawat, 1988; Khader *et al.*, 1988; Khader, 1989, 1992; Kumar and Singh, 1993; Jain and Mukherjee, 2001; Sudhavani and Shankar, 2002 etc.).

Mature fruits of "Dashehari" were harvested in June at Lucknow. The harvested fruit were dipped in 200 ppm GA3 for 5 minutes. After treatment, the treated fruit were packed in ventilated wooden cases lined with newspaper and held at ambient temperature of 38-42°C. This treatment delayed ripening up to 9 days, while control fruits ripened within 3 days (Khader, 1989). Dipping of "Langra" mango fruits in GA3 (200 or 300 ppm) delayed ripening of fruits (duration not given) in storage at 27-36 °C (Jain and Mukherjee, 2001). Pre-harvest sprays (20 days before harvest) with GA3(15 ppm) + carbendazim (0.05%) + Teepol (0.1%) delayed ripening of "Baneshan" Mango fruits by 22.33 days after harvest in storage at 10-20 °C (Sudhavani and Sankar, 2002). Post-harvest treatment of 1-MCP delayed ripening of "Zhua" Mango fruits delayed ripening for 8 days in storage at 20 °C (Jiang and Joyce, 2000).

Both pre-harvest (foliar / fruit sprays), post-harvest and pre-storage dip treatments of gibberellin delayed ripening of mango fruits. Various reasons for retarding / delaying the effect of gibberellins were expressed by different research workers, like retarding the activity of enzymes of ripening (Burg and Burg, 1962; Krishnamurthy and Rao, 1982; Khader *et al.*, 1988; Kumar, 1998), reduction of accumulation or inhibition of synthesis of ethylene (Dilley, 1969), retarding biochemical processes involved in ripening process (Burg and Burg, 1962; Dilley, 1969), retarding PLW, loss of chlorophyll and decline in ascorbic acid content, colour development, carotene accumulation (Burg and Burg, 1962, Delley, 1969, Kumar, 1998).

13.1.2. Role of auxins delaying ripening

Similar to gibberellins, auxins (IAA, 2, 4-D; 2, 4, 5-T) also delayed ripening of mango fruits either as pre-harvest sprays or post-harvest dip-treatments (Subramanyam *et al.*, 1962; Pandey *et al.*, 1980; Majumder *et al.*, 1981; Yuvaraj and Ughreja, 1998). Auxin treated fruits showed reduced PLW, with or without improvement in colour of the skin, but maintained marketability and showed extension of storage life (Majumder *et al.*, 1981; Yuvaraj and Ughreja, 1998).

13.1.3. Role of polyamines delaying ripening

Polyamines (PAs) are implicated to play role in fruit ripening process (Nair and Singh, 2004b). PAs were detected in fruits during ripening process (Malik and Singh, 2003b). However, the simultaneous increase of ethylene and PAs during ripening of fruits suggested that their biosynthesis from a common precursor may not be competitive and free PAs may have evolved as a response to increased biosynthesis of ethylen (Malik and Singh, 2004). When fruits are transferred to 25°C for ripening the levels of PAs increased in fruits (Wangchai *et al.*, 2000). Application of PAs to mango fruits by pressure infiltration delayed fruit ripening (Galston and Kaura – Samhney 1995; Purwoka *et al.*, 1998; Malik and Singh,

2003b). There was reduction in the colour development, total and non-reducing sugars and total carotenoids in PA treated fruits compared to the untreated control fruits in Australia (Malik and Singh, 2003b).

13.1.4. Role of growth inhibitors and retardants delaying ripening

Delayed ripening of mango fruits following the post-harvest application of MH was reported (Date and Mathur, 1959; Krishnamurty and Subramanyam, 1970; Musa, 1974). However, MH treated fruits showed undesirable flavours (Date and Mathur, 1959). Post-harvest dip treatments with cycocel delayed ripening in the early stages of ripening only. A reduction in carotenoid content, in the cycocel treated fruits and enhanced carotenoid content in Alar treated mango fruits were observed (Subramanyam and Sebastian, 1970; Krishnamurthy and Rao, 1982). Benzylamino purine (BA) delayed ripening of mango fruits, but altered the composition of the fruits on ripening. However, total and reducing sugar contents were increased and organic acids and free amino acid contents were decreased (Passera and Spattoli, 1981).

13.1.5. Inhibitors of ethylene synthesis and action in fruit ripening

The involvement of ethylene in fruit ripening is well acknowledged and documented. Its activity increases as the time passes after detachment and as the fruit ripens. In other words, fruit ripening progresses at a rapid pace due to increased synthesis and activity of ethylene in the fruit tissues. So it is logically envisaged that by reducing the biosynthesis of ethylene and / or lowering its activity, it is possible to retard / delay ripening. It is expected any chemical which has an inhibitory effect on ethylene production and its action will retard / delay ripening of fruits.

13.1.5.1. Role of 1-Methyl cyclo propane in fruit ripening

1-Methyl cyclo propane (1-MCP) is a chemical which found a place among the growth regulators and is used for regulation (delaying) of fruit ripening. It has been recognized as an ethylene bio-synthesis inhibitor that blocks ethylene production by its irreversible binding to the ethylene receiptor protein (Sisler and Serek, 1997; Sisler *et al.*, 1996) and prevents ethylene responses in plants (Fan *et al.* 1999; Selvaraj *et al.* 2001) and severely reduces ripening related changes in fruits and vegetables (Faubion, 2000). 1-MCP totally blocks ethylene – mediated ripening processes in climacteric fruits (Fan *et al.* 1999; Feng *et al.*, 2000). 1-MCP is a non-toxic colourless gas and effective in very low concentration (ppb) (Sisler and Serek, 1997). 1-MCP mediated inhibition of endogenous ethylene directly causes inhibition of respiration climacteric by two-folds in mango fruits. It effectively delays fruit ripening for 8 days in mango (Mathur and Srivastava, 2005). A pre-harvest spray of 1-MCP (10 ppm) at 100 days after flower initiation delayed ripening of fruits of Cv. Carabao", mango cultivar (Castillo-Israel *et al.*, 2014).

Post-harvest application of 1-MCP delayed colour development and fruit softening of 'Zhua' mangoes stored at 20°C (Jiang and Joyee, 2000). Delayed ripening (as indicated by softening and loss of chlorophyll, can be increased and increases in soluble solids levels and aroma volatiles) of treated fruits of different cultivars of mango in different mango growing countries was reported (Harris *et al.* 2000; Jiang and Joyce, 2000, Hofman *et al.*, 2001; Mir *et al.*, 2001; Jacomino *et al.*, 2002; Kluga and Jacomino, 2002; Lalel *et al.*, 2003; Alves *et al.*, 2004; Cocozza *et al.*, 2004; Osuno – Garcia *et al.*, 2007; Singh and Pathak, 2008). The effects of 1-MCP include inhibition of firmness losses, delay in colour changes and decrease in respiration and ethylene production rates, inhibition of ethylene action, reduced fresh weight and higher TTA (Harris, *et al.*, 2000; Mir *et al.*, 2001; Jacomino *et al.*, 2002; Alves *et al.*, 2004). 1-MCP coupled with MAP or HWT also delayed ripening of mango fruits (Jiang and Joyce, 2000; Osuna-Garcia *et al.*, 2007). However, 1-MCP treated fruits showed two-fold incidence of decays (Stem-end rots) (Hofman *et al.*, 2001). Therefore, additional precaution is necessary to reduce the disease incidence in 1-MCP treated fruits.

14. Accelerating of Fruit Ripening by Plant Growth Regulators

Accelerating of fruit ripening in case of mango becomes necessary some times when the grower intends to catch early market; to avoid erratic ripening in case of some cultivars (like Mallika); to hasten ripening in certain slow ripening cultivators and also in case of certain cultivars. Several changes (physic-chemical, physiological and biochemical) occur in fruits during ripening and these changes are brought by ripening enzymes (Polygalacturonase, Pectin Esterase, Cellulase, α-amylase) etc. which are induced and activated by ethylene. Hence, it is remarked that ethylene controls ripening. Fruit ripening can be accelerated by the use of growth regulators like ethylene, ABA, PBZ etc.

14.1. Accelerating of ripening by ethylene

Ethylene is produced endogenously in the plant. It is a natural hormone and is the principal hormone involved in fruit ripening. Ethylene is synthesized in the fruit tissues during ripening, particularly after detachment from the plant. Large quantities of ethylene are produced during fruit ripening. Because of its greater involvement in ripening, ethylene is often referred to as 'ripening hormone'. However, Tucker (1993) pointed out that only some ripening events such as colour and textural changes are under the control of ethylene, while other changes including flavour development may be independent of ethylene. Not withstanding the above view, it may be stated that ethylene brings out several ripening changes through enzymes.

Endogenous synthesis of ethylene is sometimes inadequate for causing ripening. Further, ripening by endogenous ethylene is slow and is not uniform. Therefore, for harvesting and for uniform ripening exogenous application of ethylene becomes necessary and assumes significance from commercial point of view. Ethylene is applied as pre-harvest or post-harvest treatment; either in the form of gas or in the form of liquid (ethephon), as a spray treatment or dip treatment. An advantage of ethylene treatment is the reduction in spoilage as well as in the number of spoiled of fruits and avoidance of fruit fly incidence (Kumar and Singh, 1993; Singh *et al.*, 1995).

Ethephon (2-chloroethane phosphonic acid) is strongly acidic in water solution. When the solution is above 5.0 pH, the molecules of ethephon spontaneously hydrolyze, liberating ethylene gas. Ethephon is commercially available (ethrel, floral CEPA) and is used for enhancing post-harvest ripening of fruits like banana, mango etc. The practical application of ethylene in the form of ethrel as a post-harvest treatment to promote the artificial ripening of fruits was described (Garison, 1968; Russo *et al.*, 1968; Lakshminarayana *et al.*, 1974). Ethrel, on exposure releases ethylene which triggers ripening of fruits and other host of changes associated with ripening of fruits. Ethrel is also included in other post harvest handling treatments to reap additional benefits. Ethrel in hot and cold water resulted in accelerated and uniform ripening with reduced spoilage. Ethrel is more effective in inducing excellent surface and ground colour of the fruits. However, ethrel in hot water induced discolouration particularly at higher concentration of 1000 ppm (Lakshminarayana *et al.* 1975).

Exogenous application of ethephon not only hastens ripening and uniform ripening, but also improves fruit quality (Chaplin, 1988; Lizada, 1991; Kalra *et al.*, 1995). Quality of the ethrel treated fruits was the best with enhanced TSS, but without a fall in TTA (Chikkasubbanna and Hudder, 1982). Colour development of green mangoes is also faster. Ethrel treatments also reduce spoilage (Kumar and Dhawan, 1995; Joon et al., 2001), because of slow release of ethylene (Chundawat *et al.*, 1973; Kumar and Dhawan, 1995).

The ability of 'ethrel' to simulate and hasten the ripening of fruits was first reported by Russo *et al.* (1968) while working on banana. Enhanced ripening of mango fruits was accomplished both by pre-harvest (sprays application) and post-harvest (dip treatments) treatments, of late ripening of mango fruit is achieved by exposing fruits to ethylene gas in air tight chambers. Irrespective of method of application and concentration ethrel accelerated ripening of fruits from 2 to 10 days in different mango cultivars at different locations (Lakshminarayana *et al.* 1975; Pandey and Singh, 1976; Bhuva *et al.* 1978; Gupta *et al.* 1979; Kalyanasundaram and Parthasarathy 1979; Sampson, 1981; Medlicott *et al.* 1987; Kumar and Singh, 1993; Mohmed and Abu-Goukha, 2003 etc.).

Pre-harvest spray of ethrel (500 ppm) advanced fruit maturity by 8-10 days and ripening by 10-14 days compared to control at Sabour (Kumar and Singh, 1993). Enhancement of ripening, early pigmentation and maturation were probably due to release of ethylene and strong mobilizing forces (Crane *et al.*, 1970; Iwahori and Lyons, 1970).

14.2. Accelerating of ripening by abscisic acid

Post-harvest treatment of mango fruits with abscisic acid (ABA) either by dip treatment or vaccume infiltration, accelerated ripening as judged by higher levels of TSS, total and reducing sugars, carotene content and increased activity of hydrolytic enzymes [amylase, cellulase and invertase] and reduction in total acidity (Majumder *et al.*, 1981; Palijwala *et al.*, 1989). The higher level of sugars and hydrolytic enzymes in ABA-treated fruits might be the result of an enhanced rate of ripening, probably due to the increased production of ethylene (Libermann *et al.*, 1978). Promotion of ripening in ABA-treated fruits was also associated with decreasing levels of acids and elevation of sugars and sugar / acid ratio. The action of ABA is linked with protein synthesis. It was also suggested that an interaction between ABA and other hormones may have a significant role to play in regulating the process of ripening in mangoes. Blocking effect of ABA in hastening of ripening in mango by vaccum infiltration of cycloheximide (a protein synthesis inhibitor) do support the role of ABA in fruit ripening (Palijwala *et al.* 1989). Application of ABA accelerated ethylene biosynthesis through promoting the activities of ethylene biosynthesis enzymes (L-amino cyclopropane-1-carboxylic acid synthase (ACS), accumulation of 1-amino cyclo propane-1-carboxylic acid oxidase (ACO) and accumulation of 1-amino cyclo propane-1-carboxylic acid (ACC) and activity of endo-polygalacturonase and reduced pectin easterase activity in the pulp. The activity of ethylene bio-synthesis and softening enzymes were significantly delayed and / or suppressed in the pulp of nordihydrogualaritic acid (NDGA) (an inhibitor of ABA) treated fruit. The ABA-treated fruits had higher total sugars and sucrose as well as degradation of total organic acids and citric and fumeric acids compared with NGDA treatment. The above results suggest that ABA is involved in regulating mango fruit ripening and its effects are at least in part, mediated by changes in ethylene biosynthesis (Zaharah *et al.* 2003).

14.3. Accelerating of ripening by paclobutrazol

Paclobutrazol (PBZ) was not usually used for accelerating / advancing fruit ripening. However, incidentally, foliar application of PBZ (2000 – 3000 ppm) prior to the commencement of flowering advanced ripening of 'Dashehare' mango fruits. Advancement of ripening of fruits may be due to its inhibiting effect on gibberellins biosynthesis in leaves and fruits, facilitating the rise of ethylene evolution and its concentration to threshold level needed for triggering the ripening process (Khader, 1990). Another probable reason may be; the advancement of flowering might have advanced fruit maturity and subsequent ripening.

14.4. Accelerating of ripening by jasmonates

Jasmonates are a class of oxylipins, regarded as naturally occurring plant growth regulators, that induce a wide variety of higher plant responses (Sembdner and Porthier, 1993). Jasmonic acid and its methylester (methyl jasmonate) were detected in a variety of plant organs, with the largest amount being in fruits (Meyer *et al.*, 1984).

Fan *et al.*, (1998) suggested that jasmonates might be involved together with ethylene in regulating the early stages of ripening of climacteric fruits. Lalel *et al.* (2003) from Australia concluded that methyl jasmonate might be important in promoting the ripening of mangoes and might be related to the increased ethylene biosynthesis. Their results indicated that methyl jasmonate (MJ) treatment increased ethylene production in 'Kensington pride' mango cultivar in the climacteric stage of fruit when MJ was applied as vapour. Exogenous application of MJ (10^{-5} M) significantly promoted the biosynthesis of ethylene, fatty acids, ripening and aroma volatiles during ripening of mangoes. The results suggested that MJ is involved in early stages in modulation in mango fruit ripening (Lalel *et al.*, 2003).

15. Role of PGR on Storage Life of Fruits

Extension of shelf life, in other words, 'Storage' of fruits is achieved by delaying their ripening, checking their biochemical activity and senescence, so as to keep them fresh in condition after they have fully matured, for as long period as possible without impairment in their quality and nutritive value. The principal aim of storage of fruits is to control the rate of respiration, transpiration, ripening, bio-chemical changes and disease infections and to preserve the commodity in its most acceptable and usable form for consumption and utilization. The object of storage is not only to prolong the post-harvest life of the fruit but also to preserve / maintain the characteristics like flavour, taste, freshness and appearance as well as their nutritive value. Several methods of storage of fruits are in vogue and practiced.

One of them is the storage with the help of plant growth regulators particularly where the availability of cold and controlled / modified storage facilities are either lacking / limited. A number of synthetic plant growth regulators have been tested to prolong the storage of mango fruits, as they retard ripening and senescence.

15.1. Role of auxins on storage life of fruits

Generally harvested fruits are dipped in aqueous solutions of auxins for prolonging their storage life. NAA (100, 200 and 300 ppm) as a dip treatment was effective in prolonging the storage life of 'Kesar' mangoes in Gujarat (Wavhal and Athale, 1988). The shelf life of mango fruits could be significantly prolonged by 4-8 days with NAA treatment (200 ppm) packed in polyethylene bags or with potassium permanganate ($KMnO_4$) (Pruthi, 1992). Hormonal treatment with NAA (20 ppm) was more effective in prolonging the shelf life of fruits by 6 days (Yeashmin *et al.*, 2011).

Fruits dipped in 2, 4-D (40 ppm) for 5 minutes enhanced the storage life of 'Dashehari' fruits at room temperature (Garg *et al.*, 1976). Mango fruits of cv. "S.B. Chausa' were successfully stored for a period of 20 days at room temperature, following the treatment with 2, 4-D / 2, 4, 5-T. The greatest delay in fruit storage was observed with 2, 4-D (40 ppm) followed by 2, 4, 5-T (60 ppm) (Deol, 1985). Kumar *et al.* (1992) observed extension of storage life of mango fruits with post-harvest treatment of 2, 4, 5-T besides improving fruit quality in terms of TSS, total sugars and reducing sugars. Storage with 2, 4-D (10 and 30 ppm) was more effective in prolonging the shelf life by 6 days (Yeashmin *et al.*, 2011).

Date and Mathur (1959) observed that 2, 4, 5-T at 1000 and 1500 ppm retarded ripening and showed better prospects of extending storage of mango fruits. Dipping of 'Dashehari' mango fruits in 2, 4, 5-T (1000 ppm) and storing at room temperature proved best to extend shelf-life, in retarding the rate of ripening, loss in weight and spoilage percentage (Garg *et al.*, 1976). Dipping of 'Dashehari' fruits in 2, 4, 5-T (1000 ppm) for 10 minutes reduced the loss in weight and rotting. This positive response was attributed to the increase in affinity for free water, which can retain more water against fruit evaporation resulting in lesser water loss (Garg *et al.*, 1976).

Morshed *et al.* (2006) in Bangladesh studied the effect of different auxins (IAA, NAA and 2, 4-D) at 10, 20 and 30 ppm for the extension of shelf life and quality characteristics of mango cv. Fazli. The shelf life of auxin treated fruits was prolonged by 26.67 per cent. NAA (10 and 30 ppm) and 2, 4-D (20 ppm) were more effective in prolonging the shelf life of fruits by 4 days.

15.2. Role of gibberellins on storage life of fruits

GA_3 is known for its anti-senescent properties which cut down the evapo-transpiration of fruits resulting into physiological loss of weight of fruits and thereby extending the storage life of fruits. Several studies have shown that post-harvest treatment of fruits with GA_3 with or without other storage techniques extended the storage life. Mango fruits treated with GA_3 (50 ppm) and sealed in perforated polyethylene bags were stored longer than those in control (Joshi and Roy, 1985; Singh, 1988; Ahmad and Singh, 2000).

Pre-harvest sprays of GA_3 (100-300 ppm) significantly reduced ripening of mango fruits for upto 6 days of storage at ambient temperature, allowed retention of more ascorbic acid during storage (Khader, 1991). The storage life of 'Baneshan' mango fruits was considerably increased by pre-harvest sprays of GA_3 (15 ppm) coupled with carbendazim (0.5 %) and fruits under this treatment were free from pathogenic infection even at advanced stage of ripening (Sudhavani *et al.*, 2000). Dipping of mature 'Amrapali' fruits in GA_3 (50 ppm) and subsequently storing at ambient temperature (33-37°C) in perforated polyethylene bags provided a shelf life of 11 days against 7 days in control (Ahmed and Singh, 1999).

GA_3 (200 ppm) treated fruits of 'Bangalora' and 'Neelum' stored in ventilated polyethylene bags with ethylene absorbent showed significant increase in the extension of storage life for 18 and 17 days respectively (Srinivasa Reddy and Haripriya, 2002).

15.3. Role of growth inhibitors and retardants on storage life of fruits

In general growth inhibitors and retardants were found to be ineffective in extending the storage life of treated mango fruits. Dipping of 'Dashehari' fruits in Maleic Hydrazide (MH) at 1000 ppm for 5 minutes and kept at room temperature (32°-38°C) and relative humidity of 76-78 per cent proved ineffective in increasing the shelf life of fruits, but higher dose (1500 ppm) retarded ripening and showed better prospects of extending the storage life of mangoes (Garg *et al.*, 1976). Ram *et al.* (1979) also found the ineffectiveness of MH in extending the storage life of mango fruits at Pantnagar. On the other hand, Singh and Gupta (2004) reported promising response of mango fruits to MH. 'Dashehari' mango fruits, treated with MH (1000 ppm) and stored in perforated polyethylene bags remained stored for 12 days and were acceptable at the end of the storage period. Similarly dipping mango fruits in aqueous solution of MH (1000 ppm) and wrapping in news paper recorded a storage life of 11 days (Singh and Gupta, 2004).

Cycocel and Alar also did not affect the storage life of mango fruits (Garg *et al.*, 1976; Ram *et al.*, 1979). On the other hand, application of PBZ @ 2-3 g ai/tree in October after pruning severely recorded higher shelf life of mango fruits (of 11.52 days) over control at Periakulam, Tamil Nadu (Bamini *et al.*, 2009).

15.4. Role of polyamines on storage life of fruits

A short life of fruits is one of the major impediments in the market and to increase both the internal and international trade. The ripening process involves numerous rapid biochemical changes and is triggered by ethylene, whether derived from endogenous or exogenous source. Synthesis of ethylene and polyamines (PAs) are inter-related, since they share a common pre-cursor (SAM). Therefore, it is envisaged that, there is a possibility of PAs to regulate the process of fruit ripening and storage. With this idea PAs are used to both hinder the ripening and extend the storage life. PAs are reported to delay senescence and fruit ripening and extend shelf life and maintain fruit quality.

PA treatment maintained significantly higher fruit firmness in storage than control and SPM and SPD performed better than PUT. But PA treated fruits stored at room temperature showed blemishes and rotting in Indochina (Purwako *et al.*, 1998). Post-harvest application of PAs (PUT, SPM and SDM) with Tween 20 (0.01 %) as surfactant on mango fruits exhibited a positive influence on the shelf life of the fruit without adverse effect on fruit quality (Malik *et al.*, 2003b). In a further trial, they confirmed the above observation that exogenous application of PA improved shelf life of mango fruits without deleterious effect on fruit quality of 'Arumanin' in Australia. PA application retarded fruit softening and colour development in treated fruits during storage and maintained firmness (Malik *et al.*, 2006). Retardation of fruit softening with the application of PAs may be due to inhibition of polygalacturonase activities, presumably through binding to pectic substances (Kramer *et al.*, 1991). PA application seems to have reduced chlorophyll breakdown and carotenoid biosynthesis. PA treatment especially PUT was reported to delay colour development during storage in a number of fruits (Valero *et al.*, 2002). Pre-storage dip treatment of PAs improved shelf life of mango fruit without impairing fruit quality in storage in a cool room ($13 \pm 1^{\circ}C$; 85 + 90% RH) in Australia (Malik and Singh, 2005). Dip treatment of SPM (0.01 %) + modified atmosphere storage extended the shelf life mango fruits upto 28 days by delaying ripening at Lucknow (Pushpa *et al.*, 2011). Similarly SDM (0.01 %) could act as a potent 'ethylene' blocker and extended the shelf life of mango fruits in modified atmosphere storage upto 25 days (Kiladi and Dikshit, 2011).

15.5. Role of 1-Methyl cyclopropane on storage life of fruits

Ethylene enhances and accelerates ripening of climacteric fruits including mango. The ripened fruits do not keep long. Studies have shown that 1-Methylcyclopropane (1 MCP) could inhibit ripening by binding irreversibly to ethylene-binding sites in the fruit. As a result ethylene is enable to bind and elicit subsequent signal transduction and transition (Serek *et al.*, 1994). Fruits treated with 1-MCP remain insensitive to ethylene. 1-MCP treated fruits showed extended storage life without undesirable effects on fruit quality (Wang *et al.*, 2006).

Among various other treatments, 1-MCP was more effective for extending the shelf life of 'Lucknow Safeda' fruits for 3 days higher over non-treated fruits, as it delayed ripening resulting in more firmness, imparted attractive bright yellow colour, palatability and TSS / acid blend coupled with minimum spoilage (Pandey and Singh, 2007). 'Palmer' fruits treated with 1-MCP under refrigeration storage extended the storage life, decreased weight and delayed ripening, but promoted shriveling and loss of quality (Hojo *et al.*, 2009). Exposure of pre-cooled 'Alphonso' fruits to vapour of 1-MCP (1000 ppb) extended the storage life by 10-12 days under cold storage conditions by delaying ripening (13 days) without affecting fruit quality (Balamohan *et al.*, 2011). Fumigation of 1-MCP (250 ppb) was most effective in delaying softening of 'NAM Dok Mai' mangoes in Thailand. Treatment with higher concentration of 1-MCP (500 and 1000 ppb) resulted in lower respiration rate and ethylene production. 1-MCP also prolonged the shelf life of fruits upto 15 days of storage at 20°C and delayed colour changes (Penchaiya *et al.*, 2006).

16. Role of Plant Growth Regulators on Physiological Disorders

16.1. Role of PGRs on spongy tissue

'Spongy tissue' is a physiological ripening disorder of mango particularly in 'Alphonso' cultivar. It is the development of a non-edible sour patch with or without air packets in the mesocarp (pulp) of fruits, which renders the fruit unfit for human consumption as well as processing. It is a serious problem in domestic and over seas trade and processing industry as the affected fruit exhibits no out side symptoms.

Ripening of fruits is due to the action of endogenous ethylene. Sometimes the ripening of mango fruits is inhibited by the inactivation of ripening enzymes such as catalase, peroxidase and amylase (Mattoo *et al.*, 1968). Exogenous application of ethephon to fruits or leaves releases ethylene in side the fruit tissues. It is possible that this ethylene, released endogenously by ethephon might be interacting with the endogenous inhibitors in the fruits, which inactivate the ripening enzymes during the process of ripening (Mattoo *et al.*, 1968) and results in 'spongy tissue'. Therefore, Lal *et al.* (1985) tested the effect of post harvest application of ethylene treatment on the occurrence of 'spongy tissue' in 'Alphonso' mango at Dapoli (Maharashtra, India) by dip-treatment of application. Dipping of fully matured fruits immediately after harvest in 500, 750 or 1000 ppm ethephon solution for 15 minutes and then ripening in wooden crates between layers of paddy straw and news paper as cushioning material, reduced the incidence of 'spongy tissue' besides advancing the ripening of fruits by 2 to 3 days.

Cultar (Paclobutrazol) reduced the incidence of 'spongy tissue' non-significantly and to small extent in 'Alphonso' mango fruits at Dapoli (Shinde *et al.*, 2010).

1-Methyl cyclopropane, which is reputed to be a ripening inhibitor, was found to exhibit significant effect on the occurrence of 'spongy tissue' in 'Alphonso' mangoes. 1-MCP is known for reduction of "Spongy tissue". It significantly reduced the incidence of "Spongy tissue" from 20-25% in control to 9.9 percent in 1-MCP treated fruits (Burondkar *et al.*, 2013).

16.2. Role of PGRs on chilling injury

'Chilling injury' (CI) or low temperature breakdown is one of the major problems of storage of tropical and subtropical fruits and vegetables such as mango, citrus etc. It is a storage disorder, when mangoes are stored for longer periods below critical temperature of 10°C (Hatton *et al.*, 1965). It is responsible for great economic loss during storage and also in prolonged refrigerated shipment, as the chilling damaged fruits are unmarketable due to discolouration of the peel and poor quality of pulp. It is the most important obstacle to the expansion of trade of tropical fruits like mango, banana in the world market.

The symptoms of CI include uneven ripening of fruits, dull and blemished skin and loss of sweetness. The symptoms develop on transfer of fruits from low temperature to high temperature for ripening. The most common visual symptoms of CI in mango fruits are: dark scaled like discolouration in the peel and pittings (Wardlaw and Leonard, 1936).

The suppression of endogenous ethylene production during storage and ripening results in high incidence of CI (Nair, 2002; Nair *et al.*, 2004). In other wards low levels of endogenous ethylene (lower than threshold level for inducing ripening) during cold storage may result in CI. How ripening or ethylene treatment can alter fruit physiology and indirectly change sensitivity of fruits to CI is still unclear. Published information indicates that ethylene treatment can reduce increase in CI or have no effect on CI development or its alleviation (Burg and Burg, 1962; Nair *et al.*, 2004). Chilling induced ethylene production also followed general pathway of ethylene biosynthesis. The amelioration of CI in mango with exogenous application of ethephon (ethrel) prior to low temperature storage clearly indicates that CI is associated with suppression of endogenous biosynthesis of ethylene (Nair *et al.*, 2004).

The involvement of polyamines (PAs) in CI was also suggested (Zhang *et al.*, 2000; Nair and Singh, 2004). Accumulation of putrescine during exposure to chilling stress is attributed for reduction of CI (Mc Donald and Kushad, 1986; Escribono and Meradio, 1994). Indirect evidence also points to the involvement of PAs in alleviating CI. Post harvest treatments that elevate PAs level in fruit tissues also reduce CI (Kramer and Wang, 1989). The reduction of CI with PAs was related to their antioxidant activity and stabilizing effect on cell membrane (Ballas *et al.*, 1983; Drolet *et al.*, 1986). A study on the effect of exogenous putrescine on CI of mango fruits during cold storage indicated that low temperature ($2 \pm 0.5^{\circ}C$) induced an increase in endogenous PAs level and ethylene production by fruit and maintained super oxide dismutase (SOD) activity in the pericarp at a higher level, thus retarding the malondiealdehyde (MDA) content and maintained membrane permeability, thereby postponed the occurrence of CI. Thus, PAs are associated with CI development in mango. Accumulation of PAs was investigated by Wangchai *et al.* (2000) in Mango fruits at three stages of ripening of fruits viz., Stage I (Green fruits); Stage II (Greener than Yellow fruits) and Stage III (Yellower than green fruits) and a relationship between PA accumulation and the degree of CI was noted. It appears that formulation of PAs is affected by CI. That increases in the beginning of storage. Decline of PA levels influences the catabolism during ripening. Depletion of endogenous PAs *viz.*, SDM and SPM, in the peel and pulp of chill-injured mango fruits and a substantial reduction in the CI symptoms with pre-storage application of PAs indicated that CI development in 'Kensington Pride' mango fruits seemed to be associated with biosynthesis of PAs (Nair and Singh, 2004).

16.2.1. Prevention of chilling injury

Prevention of CI is essential to avoid economic losses in storage and also prolonging their storage life. It also assumes significance in view of the external and internal trade of mango. During the past 60 years various methods of preventing or alleviating CI were devised both in fruits and vegetables. These include physical methods like heat treatment, cold adaption, intermittent heating, modified / cold storage and chemical methods like treatment with plant growth regulators (ethylene, PAs etc.,) in mango.

16.2.1.1. Role of ethylene in prevention of Chilling injury

Since ethylene is a ripening hormone, it was envisaged to increase resistance of fruits to CI with pre-storage ethylene treatment. Efforts have been made in this direction in mango by using growth regulators (Mohammad and Brecht, 2000; Nair and Singh, 2003; Nair *et al.*, 2004). Exogenous application of ethylene (100 ppm) to half mature fruits of mango for one day at 20 °C in air (Probably in an open room) prevented CI in fruit during 15 days at 5 °C (Mohammed and Bercht, 2000). A two day ethylene treatment though effective in reducing CI, however, it resulted in fruit decay upto 10 per cent (Mohammad and Brecht, 2000). The amelioration of CI with the exogenous application prior to low temperature storage clearly indicated that CI is associated with suppression of endogenous biosynthesis of ethylene (Nair *et al.*, 2004). Pre-storage treatment of ethrel (500 ml / litre) dip treatment for 5 minutes and storage at 5°C for four weeks and then ripened at 22°C reduced CI and improved fruit quality and taste of 'Kinsington Pride' mango fruits in Australia (Nair and Singh, 2003).

16.2.1.2. Role of polyamines in prevention of Chilling injury

The involvement of PAs in the reduction of CI has been suggested based on the successful results of exogenous application of PAs in preventing CI (Zhang *et al.*, 2000; Nair and Singh, 2004); Post-harvest treatments that elevate PA levels in tissues have been shown to lead to the reduction of CI. Temperature conditioning (a method of preventing CI) inhibits the development of CI with concomitant increase in PAs in different fruits (McDonald, 1989; Rodov *et al.*, 1995; Gonzalez *et al.*, 1997). Post harvest treatments with exogenous PAs before cold storage increased internal PAs levels and reduced CI (Kramer and Wang, 1989). The reduction of CI with PAs may relate to their antioxidant activity and stabilizing effect on cell membranes (Ballas *et al.*, 1983; Drolet *et al.*, 1986). Nair and Singh (2004) achieved substantial reduction of CI symptoms on 'ripe fruits of 'Kensington Pride', in 4 weeks storage at 5°C by pre-storage exogenous application of PAs in Australia.

16.2.1.3. Role of methyl jasmonates in prevention of chilling injury

Methyl januonates (MJs) were tested for reduction of CI in mango (Gonzalez *et al.*, 2000, 2001). Exposure of 'Tommay Atkins' fruits to MJ vapour (10 µm) for 24 hours at 25°C reduced CI during subsequent storage at 7°C for 21 days and afterwards 5 days of storage at 20°C. The CI tolerance induced by MJ treatment was correlated with its reduction in percentage ion leakage of mango tissues. MJ treatment prevented CI symptoms without altering ripening process. The overall quality of MJ treated fruits was also better than that of control fruits (Gonzalez *et al.*, 2000). Treatment of 'Kent' mangoes with MJ vapour for

20 hours at 20°C not only reduced the symptoms of CI, but also enhanced the peel colour. 10^{-4} $_M$ was the most effective concentration for reducing CI and decay in fruits stored at lower temperatures. Based on the above results Gonzalez *et al.* (2001) concluded that MJ could be used with advantage to reduce CI and decay symptoms and also to improve colour development of mango fruits without effect on fruit quality.

16.2.1.4. Role of auxins in prevention of chilling injury

According to Wang Bao Gang *et al.*, (2008), 2, 4-D could be used as an effective growth regulator to prevent CI and preserve the post-harvest life of mango fruits. It was observed that 2, 4-D regulate chilling resistance by altering the balance of endogenous levels of enzymes and phytohormones. The activities of catalase, SOD, ascorbate peroxidase and glutathione were higher in 2, 4-D treated fruits than in control. 2, 4-D also enhanced the levels of ABA and H_2O_2 and GA_3 and reduced IBA and Zeatin riboside in the treated fruits.

17. Role of Plant Growth Regulators on Malformation

Mango malformation is one of the main causes for the low productivity of mangoes in many mango growing countries. It is an economically important disease / disorder that has become a limiting factor in the extension of area under mango especially in subtropical regions. Malformation is widely distributed in many mango growing countries particularly in sub-tropical region. It is rare in tropical regions like South India, which is attributed to higher temperature under which the growth and survival of causal agent is questionable. The etiology of malformation has always been controversial. However, majority of the evidence points out to the involvement of a fungus, *Fusarium* spp probably transmitted through an eriophid mite, acting as a vector. Two types of malformation are observed *viz.*, vegetative and floral malformation.

Vegetative malformation (bunchy top): this is common in nursery seedlings and young plants. in this case, the leaves become compact and a bunch is formed at the apex of shoot or in the axil of the leaf, hence growth of shoots is arrested. the affected seedlings produce small shootlets bearing small scaly leaves with a bunch like appearance at the apical portion and in the leaf axils. affected seedlings die at the young age (nirvan, 1956; singh *et al.*, 1961).

Floral malformation: unlike vegetative malformation, floral malformation affects the trees at the bearing stage and affects the production. it affects mainly inflorescences (panicles). the primary, secondary and tertiary rachis are shortened, thickened and are much enlarged or become hyper tropical. such affected panicles are greener and heavier with increased, crowded branching (hassan, 1944). malformed panicles are generally male and rarely bisexual (singh *et al.*, 1961). the ovary of the malformed bisexual flowers is exceptionally enlarged, but underdeveloped and non-functional (mallik, 1963).

The above symptoms of malformation suggest that hormonal imbalance in the affected tissues is responsible for the disorder. The involvement of imbalance of hormones in malformation of mango was suggested by several investigators (Jagirdar and Jefri, 1966; Majumder *et al.*, 1970; Pandey *et al.*, 1974; Abou-Hussein *et al.*, 1975; El-Ghandour *et al.*, 1976; Misra and Dhillon, 1978; Chadha *et al.*, 1979; Singh, 1986b; Singh and Dhillon, 1986a, b, c, Singh, 1996 etc.). Work in South Africa also provided good evidence that hormonal imbalance in mango is responsible for malformation (Nicholson and Van Staden, 1988; Van Stadam and Nicholson, 1989; Van Staden *et al.*, 1989). Apparently the symptoms of this malady seem to be related to the imbalance of endogenous hormones particularly growth promoters and growth inhibitors (Pandey *et al.*, 1973b). The balance between auxins and antiauxins in the plant at the time of fruit bud differentiation was related to the disorder (Jagirdar and Jafri, 1966). Pal *et al.* (1983) were of the opinion that malformation might be due to depleted levels of auxins as well as hormonal imbalance. Dahshan (1987) from Egypt suggested that accumulation of hormones in malformed panicles was associated with an ineffective translocation of hormones from the panicles to the shoots, since the activity of endogenous auxins and gibberellins was higher in malformed tissue than in normal and healthy panicle, but was lower in shoots carrying the malformed panicles than those carrying the healthy panicles.

17.1. Role of auxins in controlling mango malformation

It was postulated that mango malformation was due to depletion of auxins in the malformed tissues (Pandey *et al.*, 1974). Kumar *et al.* (1980) found decreased amounts of IAA and indole acetic nitrite (IAN) in malformed bunchy top affected tissues than in healthy tissues. Pandey (1988) found auxin like substances at higher levels in healthy panicles as compared to malformed panicles and reverse trend in the levels of inhibitors. Rajan (1986) observed enhanced activity of acidic auxins in malformed buds and reported 50 times more content of acidic auxins and IAN in malformed panicles. The alleviation of floral malformation by exogenous application of auxins further indicated that malformation is some how associated with lower levels of auxins (Singh and Dhillon, 1986a).

The reported lower levels of auxins may be due to higher activity of IAA-oxidase (Pal *et al.*, 1983) and / or increase in the activity of peroxidase and polyphenol oxidase in infected tissues (Chattopadhyay and Nandi (1976). Kumar (1977) and Kumar *et al.* (1980) further confirmed this theory by reporting that the activity of IAA oxidase, peroxidase and polyphenol oxidase being 350, 118 and 32 per cent higher respectively in bunch top affected tissues as compared to healthy tissues. On the other hand, Singh and Dhillon (1993) suggested that the lower levels of IAA in malformed panicles may be due to non-conversion of tryptophan (precursor of IAA) into IAA. However, Pandey and Ram (1995) felt that auxin deficiencies may not be the sole cause of malformation, since auxin applications have failed to revert malformed panicles into healthy ones and auxin deficiencies could be an effect of infection of causal pathogen, rather a cause of malformation. Contrary to the above, higher levels of auxin have also been observed in malformed panicles (Abou-Hussein *et al.*, 1975).

17.2. Role of gibberellins in controlling mango malformation

Many mango workers hypothesized that gibberellins have role in malformation of mango. Higher quantities of gibberellins - like substances were found associated with malformation (Mishra and Dhillon, 1978, 1980, Abou-Hussein *et al.*, 1975; Bist and Ram, 1986a, b; Singh, 1986b; Singh and Dhillon, 1990b). However, others reported lower quantities of gibberellins - like substances in malformed panicles (Kumar and Baniwal, 1979; Singh and Dhillon, 1986a,b; 1992). El-Ghandour *et al.* (1976) recorded very low levels of gibberellins, but higher growth inhibiting activity in malformed tissues. However, Kumar and Baniwal (1979) could not detect gibberellins like substances in the bunchy top affected tissues. Fluctuations in the level of gibberellins - like substances in the affected tissues were also reported. The malformed panicles and shoots bearing malformed panicles at full swollen bud stage had lower levels and higher levels of gibberellins at other three stages of panicle development (Singh, 1986b; Singh and Dhullan, 1986c, 1990b; El-Ghandour *et al.*, 1976). The gibberellins content in malformed panicles increased sharply at full bloom in March and decreased in April and reached a minimum in the first week of May (Abou-Hussein *et al.*, 1975; Mishra and Dhillon, 1980; Bist and Ram, 1986a, b). Rajan (1986) found that the gibberellins activity decreased at rapid growth of panicles. The increase in the incidence of malformation also resulted in increased level of gibberellins - like substances which may account for the production of solely male flowers and continuous growth of malformed panicles (Mishra and Dhillon, 1980).

17.3. Role of cytokinins in controlling mango malformation

Besides auxins and gibberellins, endogenous cytokinins were also implicated in the malformation malady. The quantitative and qualitative changes in the cytokinins of healthy and malformed panicles at different periods of their growth were studied (Nicholson and Van Staden, 1988; Bist and Ram, 1986a,b, Van Staden and Nicholson, 1989). Malformed panicles contained higher levels of endogenous cytokinins than the healthy tissues. According to Van Staden *et al.*, (1989) the higher levels of cytokinin in malformed panicles may be due to i) Increased production of Ribosyl-zeatin and glucosyl-0-Zeatin, ii) Blockage of the production of dihydro-zeatin like compounds by the causal fungus. The higher zeatin level in malformed parts may be responsible for excessive cell division leading to malformed growth. The higher level of zeatin in malformed panicles and shoots bearing malformed panicle than in healthy panicles and shoots bearing healthy panicles specially at first stage of panicle development, might have contributed to the continuous growth of malformed panicles even after the flowering season.

Cytokinin like activity detected in the mycelium of *F. moniliformae* as well as in its culture medium suggests that the fungus is capable of synthesizing cytokinins (Van Staden and Nicholson, 1989). This production of cytokinins is additional to the plant endogenous cytokinins, resulting in higher than normal levels of cytokinins. Further more, the fungus may block production of dehydro-zeation like substances in affected tissues. In total, these events would alter the cytokinin : auxin ratio to the extent that the morphology of infected tissue would be altered.

The concentration of cytokinin and its additional kind other than zeatin and isopentynyl adenosine (IPA) has been reported higher in malformed panicles than in healthy ones (Bist and Ram, 1986a; Nicholson and Van Staden, 1988; Van Staden and Nicholson, 1989; Singh and Dhillon, 1990a). The absence of trans-zeatin, dehydro-zectin and ribosyl-dehydro- zeatin and the presence of IPA (Iso pentenyl adenosine) has also been associated with this melady (Van Staden *et al.*, 1989).

17.4. Role of polyamines in controlling mango malformation

Polyamines are involved in the plant responses to environmental challenges by changing their growth and development pattern. PAs regulate the synthesis of ethylene and vice-versa either directly or by metabolic competition for S-adenosyl methiomine (SAM), a common precursor in their biosynthesis path ways (Evans and Malamberg, 1989). Plants are subjected to many stress conditions such as exposure to pollutant (SO_2) (Priebe *et al.*, 1978) and low temperature stress (Wang, 1987) and the K-deficiency necrosis (Richards and Coloman, 1952). Malformation of tissues causes necrosis. This necrosis was attributed to accumulation of Putrascine. Before the appearances of necrosis some other symptoms appear on malformed tissues. This symptoms were assumed due to higher content of spermidine and spermine in the initial stages of malformation. The involvement of PAs as malformation agents was studied in the floral tissues of healthy and malformed ones of certain mango cultivars,

Cadoverine was detected in the normal, malformed and partially malformed dead samples in traces. Putrascine was found generally in malformed and dead samples (Agrawal, 1997).

17.5. Role of abscisic acid in controlling mango malformation

Mishra and Dhillon (1978) observed lower levels of abscisic acid (ABA) and higher levels of promotic activity in malformed panicles. In contrast, Singh (1986b), Singh and Dhillon (1990a, 1992) reported higher levels of ABA in both vegetative and floral malformed tissues. The malformed panicles and shoots bearing affected panicles were reported to have higher levels of ABA than the healthy panicles and shoots bearing healthy panicles (Singh, 1986b, Singh and Dhillon, 1990a).

17.6. Role of ethylene in controlling mango malformation

Singh and Dhillon (1986a, b; 1989c) proposed the involvement of ethylene in mango malformation syndrome, a suggestion supported by the suppression of apical dominance and the shortening and thickening of malformed panicles, which are the characteristics of ethylene damage (Abeles, 1975). Later, Singh and Dhillon (1989b, c) observed that malformation stimulated ethylene production and confirmed their contention. Higher levels of endogenous ethylene in malformed panicle as well as in the leaves and shoots bearing malformed panicles at all stages of panicle development also prompted them to conclude the role of ethylene in causation of floral malformation in mango (Singh and Dhillon, 1989c). The treatment of seed stones of mango prior to sowing with ethephon and PBZ or their application to bearing trees induced malformation like symptoms. This suggested high levels of ethylene and low levels of gibberellins favour the incidence of malformation in mango (Ram and Yadav, 1999).

Another possibility for increased levels of ethylene in malformed panicles is the possibility of 'malformin' which occurs in malformed panicles particularly inducing ethylene production. 'Malformins' are known to stimulate ethylene production in plants (Curtis, 1969). High levels of 'Malformin-A' like substances along with other compounds possessing similar activity were assayed in developing malformed panicles than in healthy ones, where it is absent (Ram and Bist, 1984; Singh and Dhillon, 1989a). Singh and Dhillon (1990,a, b) concluded that floral malformation was associated with malformin stimulated ethylene production and higher levels of ABA, GAs, Zeatin and lower levels of IAA, alternatively, the higher levels of ethylene in malformed tissues may be due to ABA induced production of ethylene.

Ethylene is known to reduce auxin levels in plant tissues and to increase both ABA and cytokinin levels (Abeles, 1975). According to Abeles (1975) wounding causes the release of ethylene. The analysis of malformed panicles and their shoots revealed low levels of auxins (Pandey *et al.*, 1974; Kumar *et al.*, 1980; Rajan, 1986; Dahshan, 1987; Pandey and Ram, 1995), higher levels of gibberellins (Abou – Hussein *et al.*, 1975; Mishra and Dhillon, 1980; Bist and Ram, 1986a,b, Dahshan, 1987), cytokinin (Bist and Ram, 1986a,b; Rajan, 1986; Dahshan and Abdul Raheem, 1987), ethylene (Singh and Dhillon, 1990, b,c) and ABA (Pandey *et al.*, 1974; Mishra and Dhillon, 1978; Pandey and Ram, 1995) in malformed tissues. However, others reported low levels of gibberellins in malformed tissues (El-Ghandour *et al.*, 1976). Further, specific cytokinins of healthy and malformed panicles differed qualitatively and quantitatively (Nicholson and Van Staden, 1988).

Smudging (as in Philippines) the mango trees for inducing flowering resulted in reduced malformation. Smoke from smudges elevates both temperature and carbondioxide around the trees, which in turn resulted in lessened ethylene production (Abeles, 1975). However, smudging induces vegetative malformation (Mallik, 1963).

Furthermore, malformed panicles treated with silver nitrate (600 ppm) were found to grow into fruit bearing healthy panicles (Bist and Ram, 1985). The effect may be due to the inhibitory action of silver ions in ethylene action (Beyer, 1976). Similarly the reduction in malformation by NAA sprays (100-200 ppm) in the first week of October (Majumder *et al.*, 1970) or by the spray of CEPA (400 ppm) prior to fruit bud differentiation and at bud inception stage, may be partly due to auto-inhibition of ethylene (Riov and Yang, 1982). All these reports directly or indirectly propose a role for ethylene in mango malformation.

In conclusion, the studies of Singh and Dhillon (1993) on metabolic changes associated with floral malformation of mango showed that malformin like substances could be implicated with the floral malformation due to stimulated ethylene production (Singh and Dhillon, 1989b) which is manifested in hormonal imbalance (Singh and Dhillon, 1990a) and in disturbance of metabolic system.

17.6.1. Stress ethylene

The basic phenomenon of increased ethylene production in response to stress is commonly called 'Stress ethylene'. Production of 'stress ethylene' can initiate various physiological responses, which include epinasty, abscission, formation of parenchyma etc. (Abeles, 1975). Several biochemical, physiological and morphological changes in the malformed tissues could be explained in terms of 'stress ethylene' (Abeles *et al.*, 1992). Pant (2000) proposed that malformation of mango might be due to 'stress ethylene'. Production of 'stress ethylene' is one of the general phenomenon observed in plant tissues subjected to various unfavourable stress conditions such as chilling (Cooper *et al.*, 1969), contact with noxious chemicals (Abeles and Abeles, 1972), browsing (Hanson and Kende, 1976), water logging (Kawase, 1976; Jackson *et al.*, 1978), pathogenic microorganisms, viruses, insects and application of auxins and ethylene (Libermann, 1979; Yang and Hoffman, 1984; Iwasaki, 1991). The etiologic agents (virus, fungal, bacteria etc.) were proposed to generate 'stress ethylene' which manifests in the form of various symptoms, one of them being malformation (Bains *et al.*, 2003). Estimation of endogenous levels of ethylene revealed higher level of 'stress ethylene' in malformed tissues compared to corresponding healthy tissue. In the light of these facts, it was suggested that the disorder may be due to the production of 'stress ethylens' by mango plants (Pant, 2000).

It is cleared that hormonal imbalance in the plants, especially in the shoots and panicles during the period of fruit bud differentiation causes / results in malformation of panicles in mango. However, it is impossible to assign unequivocal roles of individual hormones as well as for production of hormones by the host and subsequent development of symptoms. For example, whether the production of hormones by the pathogen directly caused the noted changes or whether hormones produced by the host have somehow altered the metabolism of the plant in the presence of pathogen is not clear.

18. Management of Malformation by Plant Growth Regulators

Hormonal imbalance and excess or deficiencies of phytohormones in the malformed tissues are considered to be responsible for the malformation of vegetative as well as floral parts of mango. Malformation renders the affected trees unproductive. However, it is conceived that it is possible to make the thus unproductive trees productive with the aid of plant growth regulators. Hence, attempts are being made in this direction to correct the imbalance between different hormones, particularly between growth promoters and growth inhibitors, to raise the levels or to reduce the levels of hormones in the plant, through exogenous application of different synthetic growth regulators. PGRs have been used to reduce floral malformation and increase production either by delaying panicle emergence or by maintaining hormonal balance at fruit bud differentiation time.

18.1. Use of auxins in management of malformation

The observation that an imbalance between auxins and antiauxins and the depletion of auxins in the malformed tissue at the time of fruit bud differentiation led to the disorder prompted to correct the imbalance and enhance the level of auxins at fruit bud differentiation as a measure of reducing floral malformation and improving yields. Out of the many auxins, NAA was tested extensively and found to be more effective than others in reducing the malady.

The main damaging effect of malformation is the production of excess number of male flowers. Exogenous spraying of 100 ppm NAA during October was more effective in reducing floral malformation than 200 ppm and hence NAA 100 ppm was suggested instead of 200 ppm. Increase in fruit weight, pulp weight and stone weight and eventually in yield was also obtained with 100 ppm NAA (Singh, 1986a).

Three mango cultivars (Dashehari, Chausa and Bombay Green) carrying 100 per cent incidence of floral malformation in the previous year were sprayed with NAA at 100 and 200 ppm respectively in the first week of October. The results indicated that a single spray of NAA (200 ppm) can control floral malformation to a large extent, counteracted the inhibitory effects and also caused 5-fold increase in the percentage of perfect flowers as compared to untreated control (Singh, 1955).

NAA (200 ppm) at first bud differentiation, followed by hand deblossoming resulted in higher percentage of bisexual flowers in malformed panicles. This is probably due to delay in panicle emergence which synchronized with higher temperatures favourably for induction of hermaphrodite flower formation under sub-tropical climate of Ludhiana (Singh and Dhillon, 1986a). A spray of NAA or planofix (containing NAA) @ 200 ppm applied during October resulted in average increase of normal panicles, perfect flowers, fruitset and yield. This increase was accompanied by increased levels of auxins like –substances and depleted levels of inhibitors at Delhi (Dinkar, 1987).

Spraying 200 ppm NAA in the first week of October followed by spraying of 500 ppm ethrel at bud inception stage during February was highly effective in reducing floral malformation at Ludhiana (Singh and Dhillon, 1986a). NAA at 100 or 200 ppm applied in the second week of January reduced floral malformation considerably in Pakistan (Malik and Raza, 1985). In Egypt NAA at 50 ppm appreciably increased the total number of panicles, but had no marked effect on the incidence of malformation, which was attributed to low dose used (El-Beltagy *et al.*, 1979). On the other hand, when adult trees of cv. Taimour were treated with 250 ppm every month from November to March, November application led to the production of fewest malformed inflorescences (Azzouz *et al.*, 1980).

Singh and Dhillon (1986 b) obtained maximum reduction in malformation by using IBA 200 ppm prior to fruit bud differentiation in India. In Egypt, IBA (500 ppm) sprayed after removal of malformed panicles by pruning at full bloom reduced the floral malformation markedly and increased yield (EL-Beltegy *et al.*, 1979). Complete inhibition of malformation was obtained with 2, 4-D (10 ppm) sprays at Pantnagar in cv. 'Dashehari' (Hansraj and Ram, 1996).

18.2. Use of gibberellins in management of malformation

Accumulation of higher levels of gibberellin - like substances in malformed panicles was reported to deter the normal development of panicles and induce maleness in the affected panicles (Mishra and Dhillon, 1980). This may be the reason for the increase of malformation in mango when trees were sprayed with GA_3 before fruit bud differentiation (Azzouz *et al.*, 1980; 1984, Pandey and Pandey, 2000). However, application of GA_3 allowed the tree to escape the incidence of malformation by delaying flowering from a period of cool temperature to a period of high temperature (Kachru *et al.*, 1972; Shawky *et al.*, 1978). Besides detaining flowering, GA_3 (50 ppm) has increased the number of hermaphrodite flowers, pollen viability, fruit set and finally yield (Kachra *et al.*, 1972; Shawky *et al.*, 1978; Das *et al.*, 1988).

On the other hand, GA_3 at 100 or 200 ppm and above increased the percentage of malformed panicles in Egypt (El-beltagy *et al.*, 1979). According to Pandey and Pandey (2000) GA_3 at lower concentration (< 50 ppm) delays flowering and increases the number of perfect flowers, while higher concentrations (100 to 150 ppm) reduce the number of healthy panicles and increase the number of malformed panicles (El-beltagy *et al.*, 1979).

18.3. Use of growth inhibitors in management of malformation

The treatment of 250 ppm cycocel reduced the incidence of malformation in mango in Egypt (El-beltagy *et al.*, 1979). Application of cycocel at 200 or 500 ppm to the Baytolan (fungicide) treated trees (100 ml / tree) was effective in reducing the percentage of malformed panicles and increasing that of healthy panicles (El-baltagy *et al.*, 1979).

It was postulated that higher levels of endogenous gibbrellin – like substances will lead to malformation, hence it was logically thought that reducing the endogenous gibberelline like substances would allow the development of healthy panicles (Abou – Hussein *et al.*, 1975; Mishra and Dhillon, 1980; Singh, 1986a; Singh and Dhillon, 1990b). Dhillon and Singh (1989) observed that spraying of PBZ (1000 ppm) prior to fruit bud differentiation during the first week of October reduced floral malformation, increased number of healthy panicles and yield in Punjab. The foliar spray and soil drench application of Cultar (PBZ) at 10-100 grams per tree to 'Dashehari' prior to fruit bud differentiation during the first week of October has significantly reduced floral malformation and increased yield (Singh and Dhillon, 1992). Galila and El-Masry (1991) reported that Uniconazole (500 ppm) plus ethephon (250 – 500 ppm) applied twice in the month of December to cv. Ewais in Egypt increased the number of healthy panicles and yield per tree, in addition to reduction of floral malformation.

18.4. Use of ethylene in management of malformation

Reduction of floral malformation in mango with the exogenous application of ethephon is possible and the effect was due to ethylene induced ethylene oxidation that had reduced endogenous ethylene levels in the emerging panicles (Singh and Dhillon, 1989c). Ethrel (200 ppm) applied at fruit bud differentiation reduced floral malformation in mango at Ludhiana, while 300 ppm increased the percentage of hermaphrodite flowers highest (Singh and Dhillon, 1986b). A spray application of ethephon (400 ppm) prior to flower bud differentiation and at bud inception in February significantly reduced floral malformation and increased yield (Singh and Dhillon, 1989a). Spray application of 500 ppm ethrel at bud inception stage during February, preceded by treatment during first week of October substantially reduced floral malformation (Singh and Dhillon, 1986b). The higher activity of growth promoters (auxin, gibberellins) and lower levels of phenolics in ethrel treated shoots are responsible for reduction of floral malformation in the treated mango plants. Further, it was postulated that the reduction in floral malformation by exogenous application of ethephon has possible link with some processes associated with endogenous ethyelene consequently alleviating malformation. The reduction of floral malformation with spray application of silver nitrate supports the above hypothesis (Singh, 1986a).

19. Role of Plant Growth Regulators In Salt (Saline) Stress

Several factors have been attributed to poor / low productivity of mango orchards, one of them being salt (saline) stress in the soil and / or due to saline irrigation water. Salinity has been observed in mango orchards that are being grown in arid and semi-arid regions. Salinity is one of the most serious obstacles in growing mango under the above environments, since mango, particularly at younger stages is sensitive to salts in the soil. Soil salinity, therefore, restricts the growth and extension of area under mango. Saline soils are characterized by high exchangeable sodium, high pH, low infiltration, poor aeration and deficiency of nitrogen, zinc and calcium. Salinity affects the growth and survival of young plants. Salinity in soil adversely affects the morphological characteristics of the plant. This is probably due to more accumulation of sodium in the root zone of the plants, which in turn decreases the uptake of other essential nutrients by mango plants, ultimately affecting the growth of the plants (Gupta and Sen, 2003). Salt stress causes membrane damage due to accumulation of sodium and chloride ions. The predominant effect of salinity is the suppression of plant growth. Salinity results in accumulation of salt ion in toxic amounts in the leaves of mango, which in turn inhibit growth, cause injury (leaf chlorosis), nutrient imbalance (Jindal *et al.*, 1979a, b) and reduces the uptake of major nutrients and water (Ahmad and Ahmad, 1997).

With the increasing population and continuous pressure on agricultural land, it has been necessary to find out practical solution for effectively harnessing of salt affected soil for stepping up the production of crops including mango. Therefore, it is quite desirable to overcome the salinity in mango orchards.

A few approaches have been suggested to avoid the salinity in the soils. The general advice is to avoid saline soils for planting saline sensitive crops including mango. The most common practice to alleviate the salt stress in soil is leaching of the salts away from the root zone of the plants by heavy and liberal irrigation. However, as the supply of irrigation water is limited and competition between agricultural and domestic uses intensifies, leaching became an expensive and impractical option. Breeding saline resistant / tolerant genotypes is claimed as a best long term and permanent solution to the problem of soil salinity, but it is likely to require a long time for implementation. Therefore, short-term and new approaches / solution to salt stress are needed to bind over the gap between current productive levels and productive potential of crop. Breeding and use of saline tolerant (at least) root stock is an approach to overcome the problems of salinity. However, this is also a long term programme. Attempts have been made to overcome the adverse effects of salt stress by using growth regulators which has the potential to mitigate the salinity in the soil. Among the plant growth regulators screened, polyamines (PAs) and paclobutrazol (PBZ) were successfully tried to alleviate salt stress in mango.

19.1. Polyamines in salt (saline) stress

Polyamines were found to mitigate the detrimental effects of salt stress in mango (Kanchanalata *et al.*, 2011). PAs have mitigated the salinity stress and reduced membrane injury of mango seedlings, which were subjected to salinity. PAs play a vital role in altering the sodium chloride stress in mango rootstocks. This positive influence was attributed to increased photosynthetic pigments, relative water content, protein accumulation and enhanced activities of antioxidant enzymes and by reducing membrane injury in mango plants. Possible role of PAs has been suggested in plant defense responses to various types of oxidative stresses, like UV radiation (Kramer *et al.*, 1991), salt stress (Aziz *et al.*, 1997), acid rain (Velikova *et al.*, 2000) and heavy metal stress (Groppa *et al.*, 2001). All of these stresses inhibit normal plant growth through accumulation of reactive oxygen species (ROS) such as sulperoxide radicles and hydrogen peroxide (H_2O_2).

19.2. Paclobutrozal (PBZ)

Paclobrutazol reduced the effects of salt stress in guava (Elaidey *et al.*, 1992), peach (El-

Khashab *et al.*, 1997), pomegranate (Saeed, 2005) and mango (Arunkishore *et al.*, 2009), PBZ inhibits gibberellin synthesis, increases cytokinins content, gives brief rise in ABA levels and a decrease in ethylene content (Fletcher and Hostra, 1988; Mackay *et al.*, 1990). Studies also have shown that triazoles have biochemical effects on plants such as detoxification of reactive oxygen (Upadhyaya *et al.*, 1989) and increasing levels of protein (Mackay *et al.*, 1990), antioxidant (Senratna *et al.*, 1988) and chlorophyll (Fletcher and Hostra, 1990). Results of Arunkishore *et al.* (2000) at New Delhi suggested a role for PBZ in promoting the avoidance of salt stress in mango by increasing the levels of photosynthetic pigments, water content, potassium (K^+) uptake and accumulation by reducing defoliation, the membrane injury index and the uptake and accumulation of harmful sodium (Na^+) and chloride (cl^-) ions. Srivastava *et al.* (2010) found that PBZ treatment to mango plants subjected to sodium chloride stress, mitigated the salinity stress and reduced ion leakage of mango seedlings, increased protein content. PBZ treated seedlings showed high antioxidant enzyme activity under saline conditions. PBZ minimized the negative effect of salinity with evidence of less membrane damage by up regulating the endogenous production of protein and antioxidant enzyme. Therefore, it was suggested that PBZ application is beneficial under salt stress conditions as PBZ alter the equilibrium between free radical production and enzymatic defense reaction by enhancing the proline content and free radical scavenging capacity (Srivastava *et al.*, 2010).

References

- Abeles, A.L. and Abeles, F.B. 1972. Biochemical physiology of stress – induced ethylene. *Plant Physiology*, **50** : 496-498.
- Abeles, F.B. 1975. Ethylene in Plant Biology, Academic Press, Newyork. pp. 103.
- Abeles, F.B., Morgan, P.W. and Saltveit, Jr. M.E. 1992. Ethylene in Plant Biology. Academic Press, Newyork.
- Abou-Hussain, M.R., Fadl., M.S. and Ghondour, M.A. 1975. Some physiological features of floral malformation in mangoes. *Egyptian Journal of Horticulture*, **2** : 199-208.
- Acala, P.E. and Sonpedro, A. 1935. Bud differentiation in smudged mango trees. *Philippine Agriculturist*, **24** : 27-48.
- Addicott, F.T. 1970. Plant hormones in the control of abscission. *Botanical Review*, **45**: 485-524.
- Agrawal, A., Rach, S. and Garg, G.R. 1980. Endogenus cytokinins of mango (*Mangifera*

indica L.) shoot tips and their significance in flowering. *Indian Journal of Experimental Biology*, **18**: 504-509.

- Agrawal, S. and Dikshit, S.N. 2008. Studies on the effects of plant growth regulators on growth and yield of Sapota (*Achrus sapota* L.) cv. Cricket ball. *Indian Journal of Agriculture Research*, **12**: 207-211.
- Agrawal. S. 1997. High performance liquid chromatography determination of free and conjugated polyamines in healthy and malformed floral tissues of mango (*Mangifera indica* L.). M.Sc. Thesis. G.B. Pant & University of Agriculture and Technology, Pantnagar, India.
- Agusti, M., Juan, M., Almela, V. and Gariglio, M. 2000. Loquat fruit size is increased through the thinning effect of naphthalene acedic acid. *Journal of Plant Growth Regulators*, **31**: 167-171.
- Agusti, M., Zaragoza, S., Iglesias, D.J., Almela, V., Drimo-Millo, E. and Talon, M. 2002. The synthetic auxin 3, 5-6 – TPA stimulates carbohydrate accumulation and growth in citrus fruit. *Journal of Plant Growth Regulation,* **36**: 141-147.
- Ahmad, A.M. and Ahmad, F.F. 1997. Effect of saline water irrigation and cycocel on growth and uptake of some elements of Taimour and Alphonso mango seedlings. *Annals of Agriculture Science,* **35** : 901-908.
- Ahmed, M.S. and Singh, S. 1999. Effect of various post-harvest treatments on shelf-life of Amrapali mango. *Orissa Journal of Horticulture,* **27**(1): 22-33.
- Ahmed, M.S. and Singh, S. 2000. Studies on extension of storage life of Amrapali mango. *Orissa Journal of Horticulture*, **28**(2): 73-76.
- Al-Khatib, A.Z. 1986. Effect of season and growth regulators on rooting of air-layers of 'Alphonso' and 'Mallika' cultivars of mango (*Mangifera indica* L.). M.Sc. (Ag.) Thesis, University of Agricultural Sciences, Bangalore, Karnataka, India.
- Alves, R.E., Filgueiras, H.A.C. and Almedia, A.D.S. 2004. Post-harvest ripening of 'Tommy Atkins' mangoes on two maturation stages treated with 1-MCP. *Acta Horticulturae,* **645**: 627-632.
- Aly, M.A., Bacha, M.A.A., Al-Queeb, R.S. and Abdul – Rahman, A.O. 2000. Effect of thiourea, KNO_3 and GA_3 on bud break, yield and berry quality of Thompson seedless grape vines. *Journal of King Saudi University of Agricultural Science,* **12**(2) : 1420.
- Amarnath Reddy, V. 1987. Improvement of veneer grafting in mango (*Mangifera indica*

L.). M.Sc. (Ag.) thesis. Andhra Pradesh Agricultural University, Hyderabad, Andhra Pradesh, India.

- Anbu, S., Parthiban, S., Rajanagam, J. and Thangaraj, T. 2001. Induction of off season flowering in mango (*Mangifera indica* L.) using of paclobutrazol. *South Indian Horticulture*, **49**(special) : 384-385.

- Andam, C.J. 1983. Responses of maturing 'Carabao' mango fruits to pre-harvest ethephon application. NSTA Technology Journal. 8(3) : 4-13.

- Andrews, L. and Le Fook, U. 1985. Effects of growth regulators on flowering pattern, flower suppression and fruitset in mango (*Mangifera indica* L. cv. Julie). Joint Proceedings and 1st Annual meeting of the Caribbean Food Crops Society and 32nd Annual meeting of the American Society for Horticultural Science, Tropical Region, Port of Spain, Trinidad, pp. 61-65.

- Animesh, G. and Ghosh, B. 2011. Foliar application of growth regulators and micronutrients on retention, yield and quality of fruit in mango cv. Amrapali. Global conference on Augumenting production and utilisation of Mango : Biotic and Abiotic Stresses, Lucknow. 21-24th June 2011, p. 77 (abstract).

- Antognom, F.S., Fomale, C., Grimmer, E., Komer, E. and Bagui, N. 1998. Long distance translocation of polyamines in phloem and xylem of *Recinus communis* L. plants. *Planta*, **204**: 520-527.

- Aravindakshan, K.C., Ramachandra and Pynadath, J.S. 1979. Effects of planofix on fruitset in mango (*Mangifera indica*) var. Neelum. *Agricultural Research Journal of Kerala*, **17**(1) : 105-107.

- Arifkhan, Md, Malik, A.B., Mokhadoon, M.I. and Haq, A. 1993. Investigation on the efficacy of exogenous synthetic growth regulators on fruit drop in mango (*Mangifera indica* Linn.). *Egyptian Journal of Horticulture*, **20**(1): 1-14.

- Arora, K.S. and Singh, R. 1964. Effect of plant growth regulators on fruit drop, fruit quality and seed germination. *Indian Journal of Agriculture Science*, **34**: 46-55.

- Arunkishore, Srivastava, M., Dubey, A.K., Singh, A.K., Sairam, R.K., Pandey, R.N., Dahuja, A. and Sharma, R.R. 2009. Paclobutrazol minimises the effects of salt stress in mango (*Mangifera indica* L.). *Journal of Horticultural Science and Biotechnology*, **84**(4): 459-465.

- Aziz, A., Brun, O. and Audran, J.C. 2001. Involvement of polyamines in the control of fruitset, physiological abscission in grape vine (*Vitis vinifera*). *Progressive Horticulture,* **19**(3-4) : 200-202.

- Aziz, A., Martin Ianguy, J. and Lawler, F. 1997. Plasticity of polyamine metabolism associated with high osmotic stress in rape leaf discs and with ethylene treatment. *Plant Growth Regulation,* **21**: 153-163.
- Azzouz, S., Anil, N. and Omar, N.A. 1969. Effect of hormones on air layering of mango. *Agriculture Research Review, Cairo,* **47**(4) : 59-67.
- Azzouz, S., Khelifa, A., Handy, Z.M. and El-Masry, H. 1980. Effect of growth regulators on the malformed inflorescence and stunting growth of mango. *Agricultural Research Review,* **58**(3): 1-14.
- Azzouz, S., Mostafa, H.A., Said, G. and El-Masry, H. 1984. Effects of some growth regulators on mango malformation. *Agricultural Research Review,* **62**: 181.
- Baghel, B.S., Sharma, R.K. and Nair, P.K.K. 1987a. Influence of pre-flowering sprays of urea and NAA on fruit retention of mango. *Progressive Horticulture,* **19**(3-4) : 200-202.
- Baghel, B.S., Sharma, R.K. and Nair, P.K.R. 1987b. Efficacy of pre-flowering spray of urea and NAA on physical standards of mango (*Mangifera indica* L.) fruits. *Progressive Horticulture,* **19**(3-4) : 231-234.
- Baghel, B.S., Sharma, R.K. and Nair, P.K.R. 1989. Chemical constituents of mango fruit (*Mangifera indica* L.) as influenced by foliar feeding of urea and NAA. *Gujarat Agricultural University Research Journal,* **15**(1) : 36-40.
- Baghel, B.S., Sharma, R.K. and Nair, P.K.R. 1991. Commencement of mango (*Mangifera indica* L.) flowering as influenced by foliar feeding of urea and NAA. *Scientific Horticulture,* **2**: 9-14.
- Baghel, B.S., Tiwari, R. and Gupta, N. 2004. Effects of cultar and NAA on flowering and fruiting in mango (*Mangifera indica* L.) cv. Langra. *South Indian Horticulture,* **52**(1): 302-304.
- Bagni, N. 1986. The function and metabolism of polyamines in plants. *Acta Horticultura,* **179** : 95-103.
- Bains, B.S., Bajwa, G.S. and Singh, Z. 1997. Abscission of mango fruitlets I. In relation to endogenous concentrations of IAA, GA and ABA in pedicels and fruitlets. *Fruits,* **52**: 159-165.

- Bains, G., Sand, N.K. and Pant, R.C. 2003. Role of stress ethylene in mango (*Mangifera indica* L.) malformation. *Indian Journal of Plant Physiology*, Special Issue. 480-482.
- Bakshi, J.C. 1963. Propagation studies with monoembryionic varieties of mango. *Punjab Horticultural Journal,* **3**: 185-193.
- Balamohan, T.N., Jayakumar, P., Sivakumar, S.S. and Jagadeswari, V. 2011. Post harvest studies in Alphonso mango fruits using 1-Methyl cylcopropane (1-MCP). Global Conference on Augmeting Production and utilization of Mango: Biotic and abiotic stresses, Lucknow, 21-24 June 2011, 145 (Abstract).
- Balasubramanyan, S., Auxicilla, J. and Anbu, S. 2006. Studies on planting cum high density planting in mango cv. Neelum. National Seminar on ATSH, Annamalai University, Annamalai, March 20-21, 2006, 53(abstract)
- Ballas, S.K., Mohandas, N., Morton, L.S. and Shohat, S.B. 1983. Stabilization of erythrocyte membranes by polyamines. *Proceedings of National Academy of Science (USA).* **80**: 1942-1946.
- Bamini, T., Parthiban, S. and Manivanam, M.I. 2009. Effect of pruning and paclobutrazol on yield and quality of mango (*Mangifera indica* var. Neelum). *Asian Science*, **4**(1&2): 65-68.
- Barba, R.C. 1974. Induction of flowering of the mango by chemical spray. *Proceedings of the Crop Science Society of the Philippines,* **5**: 154-160.
- Barbosa-Martinez, C., Ponade, Leon-Garcia, L. and Pelayo, Zaldivar, C. 2009. Marpho-histology of 'Manila' and 'Haden' fruit development. A Comprehensive study and post-harvest implication. *Acta Horticulturae,* **820**: 281-288.
- Basu, R.N., Chaudhury, N., Bose, T.K. and Sen, P.K. 1966. Rooting of mango cuttings. *Indian Agriculturist*, **10**: 147-151.
- Basu, R.N., Ghosh, B., Dutta, P. and Sen, P.K. 1972. Rooting in cuttings of *Mangifera indica* L. *Acta Horticulturist*, **24**: 61-63.
- Basu, R.R., Roychoudhury, N.K., Ghosh, B., Lahiri, B., Bose, T.K. and Sen, P.K. 1979. Physiological studies on rooting of mango cuttings and air-layers. Research Reports on mango. Mango workers' meeting, Panaji, Goa, 126-128.
- Bausher, M.G. and Yelenoskes, G. 1986. Sensitivity of potted citrus plants to top sprays and soil application of paclobutrazol. *Hort Science,* **21**: 141-143.

- Bausher, M.G. and Yelenoskes, G. 1987. Morphological changes in citrus associated with relatively high concentration of paclobutrazol. *Journal of Plant Growth Regulation,* **5**: 139-147.

- Bemier, G., Kinel, J.M. and Sachs, R.M. 1981a. The physiology of flowering. Vol.I Boca Raton, CRC Press.

- Bemier, G.V., Kinel, J.M. and Sachs, R.M. 1981b. The physiology of flowering. Vol.II. Boca Raton, CRC Press.

- Berghoef, J. and Bruinsma, J. 1976. Flower development of *Begonia trancomia* Labin I. effects of growth regulating substances and environmental conditions on the composition of the inflorescence. Z.P flonzen. *Physiology*, **93**: 303-310.

- Beyer, E.M. 1976. A potent inhibitor of ethylene action in plants. *Plant Physiology*, **58**: 268-271.

- Beyer, E.M.Jr. and Morgan, P.W. 1971. Abscission : the role of ethylene in modification of auxin transport. *Plant Physiology,* **48**: 208-212.

- Bhagwan, A., Vanajalatha, K, Sarkar, S.K., Girwani, A. and Misra, A.K. 2013. Standardization of dose and time of soil application of cultar on flowering and yield in mango cv. Banganapalli. *Journal of Eco-friendly Agriculture,* **8**(1): 39-43.

- Bhati, B.S. and Yadav, P.K. 2003. Effect of foliar application of urea and NAA on quality of ber cv. Gola. *Harayana Journal of Horticulture Science*, **32**(1&2) : 32.

- Bhatt, R.J. and Kumar, S. 1997. Response of plant growth regulators on flowering and fruiting in 'Alphonso' mango trees. *Gujarat Agricultural University Research Journal,* **22**(2) : 85-93.

- Bhuva, H.P., Patel, J.C. and Amin, H.D. 1978. Effect of ethrel on ripening of some varieties of mango (*Mangifera indica* L.) fruits. *Indian Journal of Agricultural Research,* **12**(4): 263-265.

- Bid, N.N. and Mukherjee, S.K. 1969. Varietal reaction to etiolation and growth regulators treatment in air-layering of mango (*Mangifera indica* L.). *Indian Journal of Agricultural Science,* **39**: 1013-1019.

- Bist, L.D. and Ram, S. 1985. Effect of some anti-malformins and growth regulators on the control of floral malformation in mango. *Indian Journal of Horticulture,* **42**: 161-174.

- Bist, L.D. and Ram, S. 1986a. Effect of malformation on changes in endogenous gibberellins and cytokinins during floral development of mango. *Scientia Horticulturae*, **28**: 235-241.
- Bist, L.S. and Ram, S. 1986b. Effect of malformation on changes in gibberellins and cytokinins during floral malformation. *Indian Journal of Horticulture*, **42**: 161-174.
- Bondad, N.D. 1972. Effects 2-chloroethane-phosphonic acid in some Philippine horticultural crops. *Philippine Journal of Geography*, **10**: 31-35.
- Bondad, N.D. 1976. Response of some tropical and subtropical fruits to pre- and post-harvest application of ethephon. *Economic Botany*, **30**: 67-80.
- Bonner, J. 1950. Plant biochemistry. Academic Press, New Delhi.
- Borja, V. and Bautista, B.R. 1932. Mango investigation in Muntinglupa, Rizal. *Philippine Journal of Agriculture*, **3**: 111-145.
- Bouchereau, A.A., Aziz, A., Lather, F. and Martin-Tanguy, J. 1999. Review : Polyamines and environmental changes. Recent Developments. *Plant Science*, **140**: 103-125.
- Brady, C.J. and Spiers, J. 1991. Ethylene in fruit antogeny and abscission. p. 235-258. In : A.K. Mattoo and J.C. Sutter (eds). The Plant Hormone Ethylene. CRC Press. Boca Raton, Florida, USA.
- Brown, K.M. 1997. Ethylene and abscission. *Physiologia Plantarium,* **100**: 567-576.
- Bueno, P.B. and Valmayor, R.V. 1974. Potassium nitrate key to mango flowering. *Agriculture Los Banos*, **13** : 4-16.
- Burg, S.P. and Burg, E.A. 1962. Role of ethylene in fruit ripening. *Plant Physiology,* **37**: 179-189.
- Burondkar, M.M. and Gunjate, R.T. 1991. Regulation of shoot growth and flowering in 'Alphonso' mango with paclobutrazol. *Acta Horticulturae,* **291**: 79-84.
- Burondkar, M.M. and Gunjate, R.T. 1993. Control of vegetative growth and induction of regular cropping in Alphonso mango with paclobutrazol. *Acta Horticulturae*, **341**: 206-215.
- Burondkar, M.M., Gunjate, R.T., Magum, M.B., Govekar, M.A. and Waghmer, G.M. 1997. Increase productivity of mango orchards by pruning and application of paclobutrazol. *Acta Horticulturae*, **455**: 367-374.

- Burondkar, M.M., Mane, A.V., Mote, M.S. and Jadav, B.B. 2013. Ripening behaviour, storage and quality of 'Alphonso' mangoes as influenced by 1-methyl cyclopropane. *Acta Horticulturae,* **992**: 597-606.
- Burrows, G.E., Boag, T.S. and Stewart, W.P. 1992. Changes in leaf, stem and root anatomy of chrysanthemum cv. Lilian Hock following paclobutrazol application. *Journal of Plant Growth Regulation,* **11**: 189-194.
- Castillo-Israel Katherine Ann T., Edralina P. Serrano , Elda B. Esguerra , and Florinia E. Merca. 2014. Dual preharvest application of 1-methylcyclopropene (1-MCP) aqueous spray controls mango (Mangifera indica L. cv. 'Carabao') ripening. *Philippine Science Letters*, 7(2):324-330.
- Chacko, E.K. 1968. Studies on physiology of flowering and fruit growth in mango (*Mangifera indica* L.). Ph.D. thesis. P.G. School of Indian Agricultural Research Institute, Delhi.
- Chacko, E.K. 1984. Physiology of vegetative and reproductive growth in mango (*Mangifera indica* L.) trees. Proceedings of the 1st Australian Mango Research workshop. pp. 54-71. Cairns, Queensland, CSIRO, Melbourne, Australia.
- Chacko, E.K. and Pal, R.N. 1985. *Mangifera indica.* In CRC Handbook of Flowering. vol. V. pp. 211-230 (ed) Halevy, A.H., CRC Press, Florida, U.S.A.
- Chacko, E.K. and Randhawa, G.S. 1971. Towards an understanding of the factors affecting flowering in mango. *Andhra Agriculture Journal,* **18**: 226-236.
- Chacko, E.K. and Singh, R.N. 1969. Induction of parthenocarpy in mango (*Mangifera indica* L.) using plant growth regulators. *Hort Science*, **4**(9) : 121-123.
- Chacko, E.K., Kachru, R.B. and Singh, R.N. 1970a. Changes in the level of acidic and natural growth promoters during fruit development in Dashehari mango (*Mangifera indica* L.). *Journal of Horticultural Science*, **45**: 341-349.
- Chacko, E.K., Kohili, R.R. and Randhawa, G.S. 1972a. Flower induction in mango (*Mangifera indica* L.) by 2-chloroethane phosphonic acid and its possible use in control of Biennial bearing. *Current Science*, **41**(15): 500-501.
- Chacko, E.K., Kohli, R.R. and Randhawa, G.S. 1972b. Studies on the effect of 2-chloroethyl phosphoric acid (Ethrel) on mango I. Flower induction in off year in Langra trees. *Indian Journal of Horticultue*, **29**: 1-4.

- Chacko, E.K., Kohli, R.R. and Randhawa, G.S. 1974b. Investigation on the use of 2-chloroethyl phosphonic acid (Ethephon, CEPA) for the control of biennial bearing in mango. *Scientific Horticulturae*, **2**: 188-190.
- Chacko, E.K., Kohli, R.R., Doreswami, R. and Randhawa, G.S. 1974a. Effect of 2-chloroethyl phosphonic acid in flower induction in juvenile mango (*Mangifera indica* L.) seedlings. *Physiologia Plantarium*, **321:** 188-190.
- Chacko, E.K., Singh, R.N. and Kachru, R.B. 1970b. Physiology of flowering and fruit growth in mango (*Mangifera indica* L.). Partition, bioassay and characterization of naturally occuring auxins and inhibitors in immature fruits. *Indian Journal of Experimental Biology*, **8** : 135-158.
- Chacko, E.K., Singh, R.N. and Kachru, R.B. 1970c. Gibberllin-like substances in developing fruits of the mango (*Mangifera indica* L.). *Journal of Horticultural Science*, **45**: 371-378.
- Chacko, E.K., Singh, R.N. and Kachru, R.B. 1972c. Studies on the physiology of flowering and fruit growth in mango (*Mangifera indica* L.). VII. Naturally occurring auxins and inhibitors in the shoots of flowering (on) and vegetative (off) mango trees. *Indian Journal of Horticulture*, **29**: 115-125.
- Chacko, E.K., Singh, R.N. and Kachru, R.B. 1972d. Studies on the physiology of flowering and fruit growth in *Mangifera indica* L. VI Hormonal control of fruit development and its possible significance to biennial bearing. *Acta Horticulturae*, **24**: 155-163.
- Chadha, K.L. and Pal, R.N. 1986. *Mangifera indica*. In : Halevy, A.C. (ed.) CRC Hand book of Flowering. Vol.5, CRC Press, Boca Raton, Florida. pp. 211-230.
- Chadha, K.L., Pal, R.N. and Sahey, R.K. 1979. Effect of abscission promoting chemicals on deblossoming of mango. *Indian Journal of Horticulture*, **20**(1): 30-33.
- Chadha, K.L., Pal, R.N., Prakash, OM., Tandon, P.L., Singh, H., Singh, N.P., Rao, M.R.K. and Beche, Lal 1979. Studies on mango malformation, its causes and control. *Indian Journal of Horticulture*, **36** : 359-365.
- Chandrasekhar, R. 1982. Studies on certain aspects of propagation in mango and cashew. M.Sc.(Ag.) thesis. Andhra Pradesh Agricultural University, Hyderabad, Andhra Pradesh, India.
- Chandrasekhar, R. 1999. Studies on the performance of component tree crops of mango

(*Mangifera indica* L.) with guava (*Psidium guajava* L.) in two interplanting systems and crop regulation in mango. Ph.D. Thesis. Acharya N.G. Ranga Agricultural University, Hyderabad.

- Chaplin, G.R. 1988. Advances in post harvest physiology of mangoes. *Acta Horticulturae,* **231**: 639-648.
- Charnvichit, S., Tongampai, P., Saguamsupyakom, C., Phavaphutanam, L. and Subhadrabandu, S. 1991. Effect of paclobutrazol on canopy size control and flowering of mango. cv. Nam Dok Mai Twai No. 4 after hand pruning. Acta Horticulturae. 291 : 60-66.
- Chatterjee, B.K. 1982. Effect of different concentrations of growth regulators on rooting and survival percentage of mango air layers. *Punjab Horticultural Journal,* **22** (3/4): 128-130.
- Chattopadhyay, N.C. and Nandi, B. 1976. Peroxidase and polyphenoloxidese activity in malformed inflorescenes caused by *Fusarium moniliformae* var. *subglutans. Biologia Plantarum,* **18**: 321-326.
- Chaudhari, P.M. and Patel, B.N. 2012. Effect of pre-sowing treatments, sowing position and duration of germination of mango stones. *Bioinfolat,* **9**(3): 277-279.
- Chaudhuri, D., Patel, N.L., Patel, N.B., Tandel, Y.N. and Gurjar, T. 2014. Effect of foliar spray of growth retardants on flowering, fruiting and yield of mango. (*Mangifera indica* L.) cv. Kesar. *Trends in Bioscience,* **7**(22): 3765-3767.
- Chaudhuri, N.N., Shah, N.R., Samdarva, N.R. and Patel, R.G. 2011. Influence of hormones on fruit retention, early maturity and quality of mango cv. Kesar. Global conference on augmenting production and utilisation of Mango. Biotic and Abiotic stresses, held at Lucknow. 21-24 June, 2011 (abstract), pp. 3-31.
- Chen, W.S. 1981. Physiological studies of fruiting in mango trees. II. Effect of endogenous growth substances on fruiting. Proceedings of National Science Council. Part B. Life Scienes, Talpd. Republic of China. 5 : 49-55.
- Chen, W.S. 1983. Cytokinins of the developing mango fruit, isolation, identification and changes in levels during maturation. Plant Physiology, **71**: 356-361.
- Chen, W.S. 1985. Flower induction in mango (*Mangifera indica* L.) with plant growth substances. Proceedings of National Science Council Part B. Life Sciences Taipel, Republic of China. 9 : 9-12.

- Chen, W.S. 1987. Endogenous growth substances in relation to shoot growth and flower bud development of mango. *Journal of American Society for Horticultural Science*, **112**: 360-363.

- Chhonkar, V.S. and Singh, N. 1967. Effects of plant regulators on air-layering in cashewnut. *Indian Journal of Horticulture*, **24**(1-2) : 26-29.

- Chikkasubbanna, V. and Huddar, A.G. 1982. Ripening of Alphonso mangoes (*Mangifera indica* L.) for processing. *Indian Food Packer*, **36**: 38-42.

- Choudhary, J.M., Basu, R.N. and Sen, P.K. 1972. Changes in endogenous growth regulating substances in the *Mangifera indica* L. shoots in relation to growth and development. *Acta Horticulturae*, **24**: 196-205.

- Choudhury, J.M. and Rudra, P. 1971. Physiological studies on chemical control of growth and flowring in mango (*Mangifera indica* L.). *Indian Agriculture*, **15**: 127-135.

- Chundawat, B.S., Gupta, O.P. and Singh, S.P. 1973. Effect of chloroethyl phosphonic acid (Ethrel) on ripening of mango (*Mangifera indica* L.). *Haryana Journal of Horticultural Science*, **2**(3): 11-16.

- Chusri, O, Kozai, N., Ogata, T., Higuchi, H. and Yonemoto, Y. 2008. Application of paclobutrazol for flowering and fruit production of 'Irwin' mango (*Mangifera indica* L.) in Okinawa. *Tropical Agriculture and Development*, **52**(3): 69-73.

- Cocozza, F. del M., Jorge, J.T., Alves, R.E., Filgueives, H.A.C., Garruti, D dos and Perira, M.E.C. 2004. Sensory and physical evaluation of cold stored 'Tommy Atkins' mangoes influenced by 1-MCP and modified atmosphere packaging. *Acta Horticulturae*, **645**: 655-661.

- Coelho, E.F., Batista, L.D.S. and Alves, A.A.C. 2014. Flowering and fruitset of mango in different doses of paclobutrazol (PBZ). *Enciclopedia Biosfera*, **10**(19): 1117-1124.

- Conegtion, R.C.C. and Rodriguez, J.D. 1994. Influence of ethylene on chemical characteristics of mango cv. Keitt fruits harvested in the pre-climacteric stage. *Scientia Agricola*, **51**(1): 36-42.

- Cooper, W.C. 1936. Transport of root forming hormones in woody cuttings. *Plant Physiology*, **11**: 779-793.

- Cooper, W.C., Rasmussen, G.K. and Waldon, E.S. 1969. Ethylene evolution stimulated by chilling in citrus and Persea spp. *Plant Physiology*, **44**: 1194-1196.

- Crane, J.C. 1964. Growth substances in fruit setting and development. *Annual Review of Plant Physiology*, **15:** 303-326.

- Crane, J.C., Naor, M. and Nelson, N.M. 1970. Growth and maturation of fig fruits stimulated by 2-chloroethyl phosphonic acid. *Journal of American Society for Horticultural Science*, **95**(3): 367-369.

- Crozier, A., Kamiya, Y., Bishop, G. and Yokota, T. 2000. Biosynthesis of hormones and eliciter molecules. p. 850-929. In. B.B. Buchannar W. Gruissem and R.L. Jones (Eds). Biochemical and Molecular Biology of Plants American Society of Plant Physiology. Rockville, Maryland, USA.

- Curtis, R.W. 1969. Stimulation of ethylene and ethane production by malformation. *Plant and Cell Physiology*, **10**: 909-916.

- Cutting, J.G.M., Bower, J.R., Wolstenhome, B.N. and Hoffman, P.J. 1990. Changes in ABA polyphenol oxidase, phenolic compounds and polyamines and their relationship with mesacarp discolouration in ripening avocado (Persea Americana). *Journal of Horticulural Science*, **65**(4) : 465-471.

- Dahshan, D.I. 1987. Physiological studies on malformation of mango panicles. *Annals of Agricultural Science*, **32**(1): 565-575.

- Dahshan, D.I. and Abdul Raheem, M.A.A. 1987. Studies on some physiological and cytological aspects associated with malformation of mango panicles. *Annals of Agricultural Science*, **32**: 1475-1488.

- Dahshan, D.I. and Habib, S.S. 1981. Effect of 2, 4-D sprays on fruit drop of Hindy Be Sinnara mango. Research Bulletin. Faculty of Agricultural Ainshan University. No. 1515, p. 12.

- Dalal, S.R., Gonge, V.S., Jadhao, B.J. and Jagdande, N.D. 2005. Effect of chemicals on flowering and fruit yield of mango var. Pairy *International Journal of Agricultural Sciences*, **1**(1): 24-25.

- Dalzial, J. and Lawrence, D.K. 1984. Biochemical and biological effects of Kauren oxidase inhibition, such as Paclobutrazol. *British Plant Grown Regulator Group Monograph*. 4: 1-14.

- Das, G.C. and Panda, J. 1976. Study of the effect of B-nine (N-dimethylamino succinamic acid) and maleic hydrazide on vegetative shoots of late occurrence in mango (*Mangifera indica* L.). *Orissa Journal of Horticulture*, **4**: 33-36.

- Das, G.C., Rao, D.P. and Lenka, P.C. 1988. Studies on mango malformation in the clone Chiratapudi. *Acta Horticulturae*, **231:** 866-872.
- Das, G.C., Sahoo, S.C. and Ray, D.P. 1989. Studies on effect of gibberellic acid and urea either alone or in comination on the growth and flowering behaviour of some 'on' and 'off' year shoots in Langra mango. *Acta Horticulturae,* **231**: 495-499.
- Date, W.B. and Mathur, P.B. 1959. Effect of post-harvest treatment with growth regulators on the ripening of mangoes (*Mangifera indica* L.). *Horticulture Advance*, **3**: 103-110.
- Davenport, T.L. 1987. Opportunities for tropical fruit crop growth regulation. Proceedings of the 14th Annual meeting of the Plant Growth Regulation Society of America. pp. 406-409.
- Davenport, T.L. 1993. Floral manipulation in mango. In : Chia, L.E. and Evans, D.O. Proceedings of the Conference on Mango in Hawaic Cooperative Extension Service, University of Hawaii, Honolulu. pp. 54-60.
- Davenport, T.L. 1994. Potential problems with the use of uniconazole on mango (*Mangifera indica* L.). *PGRSA Quarterly*, **22**: 143-153.
- Davenport, T.L. and Nunez-Elisea, R. 1990. Ethylene and other endogenous factors, possibly involved in mango flowering. *Acta Horticulturae,* **275**: 441-448.
- Davenport, T.L. and Nunez-Elisea, R. 1991. Is endogenous ethlene involved in mango floral induction. *Acta Horticulturae*, **291**: 85-94.
- Davenport, T.L. and Nunez-Elisea, R. 1997. Reproductive physiology. In : R.E. Litz (ed.). The Mango Botany, Production and Uses. CAB International, Willingford, Oxon., U.K.
- Davenport, T.L., Morgan, P.W. and Jordan, W.R. 1980. Reduction of auxin transport capacity with age and internal water deficits in cotton petioles. *Plant Physiology*, **65**: 1023-1025.
- Davenport, T.L., Pearce, D.W. and Rood, S.B. 2000. Correlatin of endogenous gibberellic acid with initiation of mango shoot growth. *Journal of Plant Growth Regulation*, **19**: 445-452.
- Davies, P.J. 2004. The plant hormones : their nature, occurrence and functions. Plant Hormones. Springer, Netherlands. pp. 1-15.
- Deol, I.S. 1985. A study on chemical changes during the storage life of Samar Bahist chausa mango (*Mangifera indica* L.). *Punjab Horticultural Journal*, **25**(1-4): 5-11.

- Desai, M.M. and Chundawat, B.S. 1994. Regulation of flowering in mango by paclobutrazol. *Indian Journal of Horticulture*, **51**(1): 9-15.
- Dett Weiler, C. 1942. Effect of hetero-auxin upon the formation, of growth substances in higher plants. *Planta,* **33**: 258-277.
- Devender, D.S.F., van Terry, L.A., Polenta, G.A. and Koresten, L. 2012. Combination of 1-methyl cyclopropane treatment and controlled atmosphere storage retains overall fruit quality and bioactive compounds in mango. *Journal of the Science of Food and Agriculture*, **92**(4) : 821-830.
- Dhatt, A.S. 1973. Studies on the bearing of behaviour of mango (*Mangifea indica* L.). Ph.D. thesis. Punjab Agricultural Universit, Ludhiana.
- Dhillon, B.S. and Singh, Z. 1989. Depletion of indole-3-acetic acid in malformed tissue of mango. *Mangifera indica* L. and its alleviation. *Acta Horticulturae,* **239**: 371-374.
- Dibble, A.R.C., Davies, P.J. and Mustachler, M.A. 1988. Polyamine content of long keeping Alcobaca tomato fruit. *Plant Physiology,* **83**: 338-340.
- Dilley, D.R. 1969. Hormonal control of fruit ripening. *Hort Science,* **4**: 11-14.
- Dinkar, V.D. 1987. Studies on mango (*Mangifera indica* L.) malformation. Ph.D. thesis. Indian Agricultural Research Institute, New Delhi.
- Dohono, W. and Walker, D.R. 1957. Effect of gibberellic acid on breaking of rest period of Filbert Peach. *Science,* **126:** 1178-1179.
- Drolet, G., Dumbroft, E.B., Legge, R.L. and Thompson, J.E. 1986. Radical scavenging properties of polyamines. *Phytochemistry*, **25**: 367-371.
- Dutcher, R.D. 1972a. Factors influencing flower induction in 'Carabao' mango (*Mangifera indica* L.) in the Philippines. Ph.D. Thesis. Cornell University, New York, U.S.A.
- Dutcher, R.D. 1972b. Induction of early flowering in 'Carabao' mango in the Philippines by smudging and ethaphon application. *Hort Science*, **7**: 345.
- Elaiday, A.A.,Sharma, M.I., El-Sammek, A. and Abou El-Khashab, A.M. 1992. Salt tolerance of 'Baladi' guava plants as affected by cultar and ethrel. *Journal of Agricultural Research*, **18**: 373-381.
- El-beltagi, M.S., El-Ghandaur, M.L. and El-Hamawi, H. 1979. Effect of Bay foldan

and some growth regulators on modifying flowering and the incidence of flowering malformation of mango (*Mangifera indica* L.). *Egyptian Journal of Horticulture,* **6**(2): 125-133.

- El-Ghandour, M.A., El-Hanawi, H.A. and Fadh, M.S. 1976. Biological studies for explaining interaction between some native growth regulatiors and abnormal growth of mango seeds. *Egyptian Journal of Horticulture*, **3**: 69-71.
- El-Kheshab, A.M., El-Sammak, A.F., Elaiday, A.A. and Salama, M.J. 1997. Paclobutrazol reduces some negative effects of salt stress in Peach. *Journal of American Society for Horticultural Science.* **122**: 43-46.
- Else, M.A., Stankiewier -Davies, A.P., Crisp, C.M. and Atkinson, C.J. 2004. The role of polar auxin transport through petioles of *Prunas avium* L. in relation to fruit development and retention. *Journal of Experimental Botany*, **55**: 2099-2109.
- Escribano, M.I. and Meradio, C. 1994. The relevance of polyamine levels in Cherimoya (*Annona cherimoya* Mill) fruit ripening. *Journal of Plant Physiology*, **69:** 701-706.
- Evans, P.T. and Malamberg, R.L. 1989. Do polyamines have role in plant development? *Annual Review of Physiology and Plant Molecular Biology,* **40**: 234-269.
- Fan, X., Blankenship, S.M. and Matheis, J.P. 1999. 1-Methyl cyclopropane inhibits apple ripening. *Journal of American Society of Horticultural Science*, **124:** 690-695.
- Fan, X., Mattheis, J.P. and Fellman, J.K. 1998. Response of apples to post-harvest jasmonate treatments. *Journal of American Society of Horticultural Science,* **123:** 421-425.
- Faubion, D. 2000. Could new ethylene inhibitor work on fruit ?. Good Fruit Grower. **50**: 18.
- Feng, X., Blankenship, A.I., Sisler, E.N. and Goren, R. 2000. Control of ethylene responses in avocado fruit with –Methyle cyclopropane. *Post Harvest Biology and Technology*, **20**: 143-150.
- Fletcher, B. and Hofstra, G. 1988. Triazoles as potential plant protectors. In : Sterol synthesis inhibition in plant protection (Berg, D. and Plempel, M. eds.). Ekis Harward Ltd., Cambridge, U.K : p. 321.
- Galan Sauco, V. 2013. World wide Mango production and Market : Current, Situation and Future Prospects. *Acta Horticulturae,* **992:** 37-48.

- Galila, A.S. and El-Masry, H.M. 1991. Effect of ethrel and S-3307 D (a new plant growth retardant) on flowering malformation, fruitset and yield of Ewais mango cultivar. *Anals of Agricultural Science, Cairo,* **36**(2): 645-650.
- Galston, A.N. 1983. Polyamines as modulators of plant development. *Bio Science,* **33:** 382-388.
- Galston, A.W. and Kaur-Sawhney, R. 1995. Polyamines and senescence in plants. pp. 167-181. In : Plant senescence : Its biochemistry and physiology (eds) WW.Thomson, E.A. Nothnagal and R.C. (Haffakin). American Society of Plant Physiology. Rodeville, M.D.
- Galston, A.W., Kaur-shawney, R., Altabella, T. and Tiburio, A.F. 1987. Plant polyamines in reproductive activity and response to abiotic stress. *Botanic Acta,*, **110**: 197-207.
- Garcia – Martinez, J.L. and Garcia, - Papi, M.A. 1979. The influence of gibberellic acid, 2, 4-dichlorophenoxy acetic acid and 6-Benzylaminopurine on fruitset of clumentine mandarin. *Scientia Horticulurae,* **10**(3): 285-293.
- Garg R.C., Ram, H.B., Singh, S.K. and Singh, R.V. 1976. Effect of some growth regulators on the storage behaviour, rate of transpiration and general quality of mango cv. Dashheri. *Progressive Horticulture,* **8**: 51-55.
- Garison, S.A. 1968. Stimulation of tomato ripening by Amchen, 66-329. *Horti. Science,* **3**: 122 (abstract).
- Gawadi, A.G. and Avery, G.S. 1950. Leaf abscission and the socalled 'Abscission layer'. *American Journal of Botany,* **37**: 172-180.
- Gill, A.P.S. and Mukherjee, S.K. 1967. Control of fruit drop in mango. *Proceedings of International Tropical and Subtropical Horticulture,* New Delhi. 454-467.
- Giri, A. and Choudhuri, M.V. 1966. Relation of mango stone weight to its germination and seedling vigour. *Pakistan Journal of Science,* **18**: 148-150.
- Gofur, M.A., Phofique, M.Z., Haleti, M.O.R., Ibrahim, M. and Rehman, M.M. 1998. Effect of application of plant hormones on the control of fruit drop, yield and quality characterstics of mango (*Mangifera indica* L.). *Bangladesh Journal of Scientific and Industries Research,* **35**(4): 493-498.
- Gokhale, V. and Kanitkar, U.K. 1951. Fruit drop in Alphonso mango and its control. *Proceedings of the Indian Science Congress.* p. 151.

- Gonzales, L.G. 1923. The smudging of mango trees and its effects. *Philippine Agriculturist*, **12**: 15-27.
- Gonzales, L.G. 1933. The influence of smuding on the respiration and catalase activity of the mango. *Philippine Agriculturist*, **21**: 535-540.
- Gonzalez A.G.A., Buta, J.G. and Wang, C.Y. 2001. Methyl josmonate reduces chilling injury symptoms and enhances colour development of 'Kent' mangoes. *Journal of the Science of Food and Agriculture*, **81**(13): 1244-1249.
- Gonzalez, A.G.A., Fortiz, J., Cruz, R., Baez, R. and Wang, C.Y. 2000. Methyl jasmonate reduces chilling injury and maintains post-harvest quality of mango fruit. *Journal of Agricultural and Food Chemistry*, **48**(2) : 515-519.
- Gonzalez, A.G.A., Zacarias, L., Mulas, M. and Lafunte, M.T. 1997. Temperature and duration of water dips influence chilling injury decay and polyamine content in 'Fartune' mandarins. *Post-harvest Biology and Technology*, **12**: 61-69.
- Griggs, D.L., Hedden, P., Temple Smith, K.E. and Rademacher, W. 1991. Inhibition of gibberellin of 2B-hydroxylases by acyleyclohexan edione derivatives. *Phytochemistry*, **30**: 2513-2517.
- Groppa, M.D., Tomato, M.L. and Benarides, M.F. 2001. Polyamines as protectors against cadmium or copper induced oxidatie damage in sunflower leaf discs. *Plant Science*, **161** : 481-488.
- Guha, D. 1993. Regulation of tree growth and yield of golden delicious apple with cycocel and ethrel. *South Indian Horticulture*, **41**(6): 333-340.
- Gupta, N.K. and Sen, N.L. 2003. Studies on initial establishment of mango seedlings in saline environment. *South Indian Horticulture*, **51**(1-6): 106-109.
- Gupta, O.P., Dhawan, S.S. and Chauhan, K.S. 1979. Effects of post-harvest application of ethrel (2-chloro-ethyl phosphonic acid) on ripening of mango. *Research Reports on mango. Mango workers meeting, Panaji-Goa, India.* 2-5 May 1979, pp. 304-305.
- Gupta, R.K. and Brahmachari, U.S. 2004. Effect of foliar application of urea, potassium nitrate and NAA on fruit retention, yield and quality of mango cv. Bombai. *Orissa Journal of Horticulture*, **32**(2): 7-9.
- Gur, A. and Samish, R.M. 1962. The role of growth substances in the budding of mango. *Horticulture Advance*, **6**: 1-18.

- Gustaffan, 1936. Inducement of fruit development by growth promoting chemicals. *Proceedings of National Academy of Science,* **22**: 628-636.

- Haissig, B.E. 1972. Meristamatic activity during adventitious entitious root primordia development. Influence of endogenous auxin and applied gibberellic acid. *Plant Physiology,* **49**: 886-892.

- Halfe, O.M., Dolfino, F.I., Mendinca, Y. and Arauja – Neto, S.E., de. 2003. Flowering and production of mango cv. Tommy Atkins using ethrel, potassium nitrate and calcium nitrate. *Revista-de-gencias. Agrarias,* **39**: 145-153.

- Hanson, A.D. and Kende, H. 1976. Biosynthesis of wound ethylene in morning – glory flowertissue. *Plant Physiology,* **57**: 538-541.

- Hansraj, D. and Ram, S. 1996. High pressure chromatography mango malformation. Plant Physiology and Biochemistry. Unpublished. Quoted by Sant Ram 2001. Mango Malformation. *Indian Journal of Horticulture,* **58**(1&2): 78-90.

- Haribabu, R.S. and Jyothi, G. 2004. Mango and paclobutrazol. *Agribios News Letter,* **2**(9): 32-34.

- Hartman, H.S. and Kester, D.E. 1972. Plant Propagation : Principles and Practices. 2nd Edition. Prentice Hall of India Private Limited, New Delhi. 331-451.

- Hasdisene, C. and Tongumpai, P. 1986. Effect of paclobutrazol on vegetative growth, flowering and fruiting of mango 'Nam Doc Mai Twai No. 4'. *Proceedings of the National Conference Kesetsort University,* **24**: 205-302.

- Hassam, A.S. 1944. Notes on *Eriophyes mangifera* S.N. (Acarina). Bulletin Society. *Fouad. Entomol.,* **28**: 179-180.

- Hatton, T.T., Reeder, W.P. and Campbell, C.W. 1965. Ripening and storage of Florida mangoes. USA Marketing Researh Report No. 275, Agriculture Research Service, Washington, D.C.

- Haw, K.C. 1986. The potential use of PP_{333} on fruits in Malaysia with particular reference to mangoes. In: Anon, cultar, tropical crops workshop, Malaysia. Jan. 20-22, 1986. ICI Fernhurst, Hoslemere, ICI, Surrey, U.K.

- Hegemann, M., Roema, M.S., Winterhagen, P., Hegde, M., Wunsche, J. 2011. Application of plant growth substances during pre-mature fruit drop of mango for identifying the key physiological factor. Tropentag. October 5-7, 2011. Bonhi.

- Heldt, H.W, Piechulla, H. and Heldt, F. 2011. Plant biochemistry. Academic Press Burlington. Massachusetts, U.S.A.
- Hendershot, C.H. and Wilson, W.C. 1967. Abscission of tree fruits. *Annual Report of Florida Agricultural Experiment Station*. pp. 142-146.
- Hendry, N.S., Van Staden, Y. and Allan, P. 1982a. Cytokinins in citrus. I. Fluctuations in the leaves during seasonal and developmental changes. *Scientia Horticulturae*, **16:** 9-16.
- Hendry, N.S., Vanstaden, J. and Allan, P. 1982b. Cytokinins in citrus II. Fluctuations during growth in juvenile and adult plants. *Scientia Horticulturae,* **17**: 247-256.
- Herris, D.Q., Leberry, J.A., Wills, R.B.H. and Spohn, L.J. 2000. Effect of fruit maturity on efficacy of 1-Methylcyclopropane to delay the ripening of bananas. *Post-harvest Biology and Technology,* **20**: 303-308.
- Hiller, G.R. 1991. Promotion of regular fruit cropping in mango. *Acta Horticulturae,* **291**: 51-59.
- Hiller, G.R. and Rudge, T.G. 1991. Promotion of regular fruit cropping in mango with cultar. *Acta Horticulturae,* **291**: 51-59.
- Himabindu, A. and Srihari, D. 2013. Influence of pre-sowing treatments on germination, growth and vigour of mango cv. Alphonso. *The Asian Journal of Horticulture*, **8**(1): 122-125.
- Himabindu, A., Srihari, D. and Madhavi, M. 2014. Effect of pre-sowing treatments on germination, growth and vigor of mango (*Mangifera indica* L.) cv. Totapuri. *Environment and Ecology,* **32**(4a): 1588.
- Hiron, W.P. and Wright, S.T.C. 1973. The role of endogenous abscisic acid in the response of plants to stress. *Journal of Experimental Botany*, **24:** 769-781.
- Hoda, M.N., Singh, S. and Singh, J. 2001. Effect of cultar on flowering, fruiting and fruit quality of mango cv. Langra. *Indian Journal of Horticulture*, **58**(3): 224-227.
- Hofman, P.G., Jobin, D.M., Meiburg, G.F., Manish, A.J. and Joyce, D.C. 2001. Ripening and quality responses of avocado, custard apple, mango and papaya fruits to 1-Methyl cyclopropane. *Australian Journal of Experimental Agriculture*, **41**(4): 567-572.
- Hojo, E.T.D., Abren, C.M.P., de Correa, A.D. and Boas, E.V. de B.V. 2006. Performance of 'Palmer mangoes treated with methyl cyclopropane and stored under refrigeration *Revista Brasileiva de Amazen amento,* **31**(2): 181-186.

- Hojo, E.T.O., Abreu, C.M.P., Dofsmer, S.A., Hojo, R.H., Correa, A.D., Boas, E.V. de B.V. 2009. Evaluation of quality of 'Palmer' mangoes fruit treated with 1-Methyl cyclopropane and stored under refrigeration and environmental conditions (Portuges). *Revista Brasileirade Fruiti cultura,* **31**(1) : 28-38.
- Hongasbhanich, N. 1986. Effects of paclobutrazol on vegetative growth, flowering and fruitsetting in mango 'Nam Dok Mai Twai No. 4". In : Anon : Cultar, Tropical crops workshop. Malaysia, January 20-22, 1986, ICI, Fernheirst, Haslemere, ICI, Survey, U.K.
- Husen, S, Kuswanto, Ansari, S. and Basuki, N. 2012. Induction of flowering and yield of mango hybrids using paclobutrazol. *Journal of Agriculture and Food Technology*, **2**(9): 153-158.
- Ioannoidis, N.E. and Kotzalasis, K. 2007. Effects of polyamines on the functionality of photosynthetic membrane *in vivo* and *in vitro*. *Biochim. Biophys. Acta.,* **1767**: 1232-1382.
- Iwahari, S. and Lyans, J.M. 1970. Maturation and quality of tomatoes with pre-harvest treatment of 2-chloro-ethyl phosphonic acid. *Journal of American Society for Horticultural Science*, **95**: 88-91.
- Iwasaki, H. 1991. The biochemistry of ethylene biosynthesis. In : The Plant Hormone Ethylene (Eds : A.K. Mattoo and J.C. Sultie). C.R.C. Press, Boca Raton. Ann. Arbor. Boston, London, pp. 337.
- Jackson, M.B., Galas, K. and Combell, D.J. 1978. Effect of water logged soil conditions on the production of ethylene and on water relationships to tomato plants. *Journal of Experimental Botany*, **29**: 183-193.
- Jacobs, W.P. 1985. The role of auxin in inductive phenomenon. *Biologia Plantarium*, **27**: 300-309.
- Jacobson, J.V. and Chandler, R.M. 1982. Gibberellins and abscisic acid in germinating cereals. In : P.J. Davies (Ed.) Plant hormones and their role in plant growth and development (pp. 164-173), Bostan, Martinus, Nijhoft Publishers.
- Jacomino, A.P., Kluge, R.A., Brackmann, A. and Castro, P.R.C. 2002. Control e amadurecimentoe senescention de mamad. Com-1-Methyl cyclopropane. *Scientia Agriculture.* **59**: 303-308.
- Jagdande, N.D. and Choudhari, K.G. 2001a. Seasonal changes in abscisic acid content and its role in flowering of mango (*Mangifera indica* L.). *The Orissa Journal of Horticulture,* **29**(1): 46-49.

- Jagdande, N.D. and Choudhari, K.G. 2001b. Seasonal changes in auxin content and its role in flowering of mango (*Mangifera indica* L.). *Orissa Journal of Horticulture,* **29**(2): 10-12.
- Jagdande, N.D., Chandhari, K.G. and Dhemre, J.K. 2000. Seasonal changes in hormonal contents and its role in flowering of mango (*Mangifera indica* L.). *Orissa Journal of Horticulture*, **28**(2): 49-52.
- Jagirdar, S.A.P. and Chaudhury, M.R. 1967. Fruit drop control in mango by hormonal spray. *West Indian Journal of Agricultural Research*, **5**: 33-42.
- Jagirdar, S.A.P. and Jafri, N.R. 1966. Malformation of mango inflorescence. *Agriculture Pakistan,* **17**: 351-357.
- Jain, S.K. and Mukherjee, S. 2001. Post-harvest application of GA_3 to delay ripening in mango (*Mangifera indica* L.) cv. Langra. *Journal of Eco-Physiology,* **4**(1/2) : 27-30.
- Jauhari, O.S. and Nigam, V. N. 1962. Air-layering of mango. Fruit Nursery practices in India. Directorate of Extension. Ministry of Food and Agriculture, New Delhi. pp. 77-81.
- Jiang, Y. and Joyce, D.C. 2000. Effect of 1-Methyl cyclopropane alone and in combination with poly-ethylene bags on the post harvest life of mango fruits. *Annals of Applied Biology,* **137**(3): 321-327.
- Jindal, P.C., Singh, J.P. and Gupta, O.P. 1979b. Effect of salinity on the mineral nutrients in mango seedlings. *Indian Journal of Agricultural Sciences,* **49:** 106-109.
- Jindal, P.C., Singh, J.P. and Gurta, O.P. 1979a. Haryana Journal of Horticultural Science, **8**: 39-41.
- Jones, R.L. 1973. Gibberellins: their physiological role. *Annal Review of Plant Physiology,* **24** : 571-598.
- Joon, M.S, Kumar, J., Sharma, R.K. and Singhrot, R.S. 2001. Comparison of calcium carbide and Ethephon on ripening of mango. *Haryana Journal of Horticultural Science,* **30**(3&4): 181-182.
- Joshi, G.D. and Roy, S.K. 1985. Effect of integrated post-harvest handling in biochemical changes in 'Alphonso' mango fruits. *Progressive Horticulture*, **17**: 56-63.
- Kachru, R.B., Singh, R.N. and Chacko, E.K. 1971. Inhibition of flowering in mango (*Mangifera indica* L.) by gibberellic acid. *Horti. Science*, **6**: 140-141.

- Kachru, R.B., Singh, R.N. and Chacko, E.K. 1972. Inhibition of flowering in *Mangifera indica* L. by gibberellic acid. *Acta Horticulturae*, **24**: 206-209.
- Kakkar, R.K. and Rai, V.K. 1983. Plant polyamines in flowering and fruit ripening. *Phytochemistry,* **33**: 1281-1288.
- Kakkar, R.K., Nagar, P.K., Ahuja, P.S. and Rai, V.K. 2000. Polyamines and plant morphogenesis. *Biol. Plant.*, **43**: 1-4.
- Kalra, S.K., Tandon, D.K. and Singh, B.P. 1995. Mnago. In : D.S. Salunkhe and D.K. Kadam (Eds.). Hand Book of Fruit Science and Technology, Production, Composition, Storage and Processing. Market Dekkar, Inc., New York. pp. 123-160.
- Kalyanasundaram, P. 1974. Studies on fruit drop in Mulgoa mango (*Mangifera indica* L.). *Indian Journal of Horticulture*, **31**(1): 47-51.
- Kalyanasundaram, P. and Parthasarathy, V.A. 1979. Effect of thiobendazole and ethrel on ripening and stem-end-rot of mango cv. Padiri. *Annamalai University Agricultural Annual,* **7/8** : 161-164.
- Kanchanalata, Srivastava, M., Dubey, A.K., Singh, A.K. and Sairam, R.K. 2011. Effect of polyamines on seedlings of two mango (*Mangifera indica* L.) rootstocks under salt stress. *Indian Journal of Plant Physiology,* **16**(NS) (3&4) : 258-267.
- Kannan, K. and Rao, V.N.M. 1964. Effect of gibberellic acid on seed germination, seedling growth and graft take in the mango. *Madras Agriculture Journal,* **51**: 77-78.
- Karuna, K. and Mankar, A. 2008. Effect of urea, paclobutrazol and bio-regulators on vegetative growth and productivity of Langra mango. *Orissa Journal of Horticulture,* **36**(2) : 88-92.
- Karuna, K., Manka, A. and Singh, J. 2007. Effect of urea and growth substances on yield and quality of mango cv. Langra. *Orissa Journal of Horticulture,* **35**(1): 67-70.
- Karuna, K., Mankar, A. and Singh, J. 2005. Effect of urea and growth substances on yield, physico-chemical characteristics of mango. *Horticultural Journal*, **18**(3): 131-133.
- Kawase, M. 1976. Ethylene accumulation in flooded plants. *Physiologia Plantarum,* **36:** 236-241.
- Kennard, W.C. and Winters, H.W. 1956. The effect of 2, 4, 5-TP application on the size, maturation and quality of 'Amini' mango (*Mangifera indica* L.). *Proceedings of American Society for Horticultural Science,* **67**: 290-297.

- Khader, S.E.S.A. 1989. Delaying ripening by post-harvest treatment of GA_3 in mango. *Indian Journal of Horticulture*, **46**(4): 444-448.
- Khader, S.E.S.A. 1990. Orchard application of paclobutrazol on ripening, quality and storage of mango fruits. *Scientia Horticulturae*, **41**: 329-335.
- Khader, S.E.S.A. 1991. Effect of pre-harvest application of GA_3 on post-harvest behaviour of mango fruits. *Scientia Horticulturae,* **47**(3-4) : 317-321.
- Khader, S.E.S.A. 1992. Effect of gibberellic acid and vapor gard, on ripening, amylase and peroxidase activities and quality of mango fruits during storage. *Journal of Horticultural Science*, **67**(6) : 855-860.
- Khader, S.E.S.A., Singh, B.P. and Khan, S.A. 1988. Effect of GA_3 as a post-harvest treatment of mango fruit on ripening, amylase and peroxidase activity and quality during storage. *Scientia Horticulturae*, **36** (3-4): 261-266.
- Khan, M.A., Malik, A.B., Makdoom, M.L. and Haq. A. 1993. Investigation on the efficiency of exogenous synthetic growth regulators on fruit drop in mango (*Mangifera indica* L.). *Egyptian Journal of Horticlture,* **20**: 1-14.
- Khettab, M.M., Haseeb, G.M., Shaban, A.E. and Arafa, M.A. 2006. Effect of paclobutrazol and potassium nitrate on flowering and fruiting of Ewais and Sidik mango trees. *Bulletin of Faculty of Agriculture, Cairo University*, **57**(1): 107-123.
- Khobragade, H.M., Patil, B.N., Patil, S.P. and Belorkar, P.V. 1999. Performance of mango root stocks under nursery conditions. *Journal of Soils and Crops*, **9**: 244-246.
- Kikani, K.P. 1986. Integrated approach to the use of plant groth regulators and urea in mango (*Mangifera indica* L.) cv. Langra. Ph.D. Thesis, Gujarat Agricultural University, Ahmedabad, Gujarat (Thesis Abstracts 1989). 13: 136-139.
- Kiladi, B. and Dikshit, A. 2011. Enhancing of shelf life of chausa mango by use of spermidine. Global conference on Augmenting Production and Utilization of Mango : Biotic and Abiotic Stresses, Lucknow. 21-24, June, 2011, 150-151 (Abstracts).
- Kluge, R.A. and Jacomina, A.P. 2002. Shelf life of peaches treated with 1-Methyl cyclopropane. *Scientia Agriculturae*, **59**: 69-72.
- Konigshofer, H. and Lechner, S. 2002. Are polyamines involved in the synthesis of heat shock proteins in cell suspension culture of tobacco and alfadfa in response to high temperature stress ?. *Plant Physiology and Biochemistry*, **40**: 51-59.

- Kotur, S.C. 2012. Effect of paclobutrazol application on nutrient dynamics, vigour and fruit yield in 'Alphonso' mango (*Mangifera indica* L.). *Journal of Horticulture Science,* **7**(2): 134-137.
- Kramer, G.F. and Wang, C.V. 1989. Correlation of induced chilling injury with increased spermine and spermidine levels in Zucchini squash. *Physiologia Plantarium,* **76**: 479-484.
- Kramer, G.F., Norman, H.A., Krizek, D.T. and Mirecki, R.M. 1991. Influence of UV radiation on polyamines, liped peroxidation and membrane lipids in cucumber. *Phytochemistry,* **30:** 2104-2108.
- Kramer, G.F., Wang, C.Y. and Conway, W.S. 1991. Inhibition of softening by polyamines application in 'Golden Delicious' and 'Mcintosh' apples. *Journal of American Society of Horticultural Science,* **116** : 813-817.
- Krekule, J.J., Paviova, L., Souckova, I. and Machackove. I. 1985. Auxin in flowering of short-day and long-day *Chenoing* species. *Biologia Plantarium,* **27**: 310-317.
- Krisanapook, K., Pheraphytanon, L., Kaewladdakom, V., Jutamanee K. and Pichakum, A. 2000. Studies on fruit growth, levels of GA – like substances and cytokinin – like substances in fruit of mango cv. Khew Sawoey. *Acta Horticulturae,* **509**: 697-704.
- Krishnamurthy, S. and Rao, K.P.G. 1982. Regulation of ripening by chemicals in 'Alphonso' mango. *Scientia Horticulturae,* **16**: 179-183.
- Krishnamurthy, S. and Subramanyam, H. 1970. Effects of maleic hydrazide and 2, 4-5-Trichloro propionic acid on ripening and quality of mango fruit. *Pesticide Science,* **1** : 63-65.
- Kulkarni, P.B. 1983. Studies on regulation of vegetative growth, flowering and fruit drop in mango (*Mangifera indica* L.) cv. Alphonso. *Thesis Abstracts. HAU.* **9**(4) : 344-345.
- Kulkarni, V.J. 1988. Chemical control of tree vigour and the promotion of flowering and fruiting in mango using paclobutrazol. *Journal of Horticultural Science,* **63**: 557-566.
- Kulkarni, V.J. and Rameshwar, A. 1979. Studies on mango dwarfing Research Reports on mango. Mango Workers' Meeting, Panaji, Goa 2-5th May. pp. 107-108.
- Kulkarni, V.J. and Rameswar, A. 1978. Natural and gibberellic acid induced parthenocarpy in mango cv. Thambva. *Current Science,* **47**(10): 353-354.
- Kulkarni, V.J., Ratnam, K.K. and Ramakrishna, G. 1988. Propagation studies in mango. *Acta Horticulturae,* **231**: 186-191.

- Kumar, A. and Dhawan, S.S. 1995. Effect of post-harvest treatments on the enhancement of ripening of mango (*Mangifera indica* L.) fruit cv. Dashehari. *Haryana Journal of Horticultural Science*, **24**(22) : 109-115.
- Kumar, A., Taylor, M., Altabeta, T. and Tiburcio, A.F. 1997. Recent advances in polyamine research. *Trends in Plant Science*, **2:** 124-130.
- Kumar, D. 1998. Effect of post-harvest treatment on shelf-life and quality of mango. *Indian Journal of Horticulture,* **55**(2): 134-138.
- Kumar, J. 1977. Status of auxins, gibberellin – like substances and oxidative enzymes in bunchy top stage of mango malformation affected seedlings. M.Sc. Thesis. G.B. Pant University of Agriculture and Technology, Pantnagar, India.
- Kumar, J. and Beniwal, S.P.S. 1979. Note on gibbellin – like substances and inhibitors in mango seedlings affected with bunchy top stage of mango malformation. *Pantnagar Journal of Research*, **41**: 13.
- Kumar, J., Beniwal, S.P.S. and Ram, S. 1980. Depletion of auxins in mango seedlings affected with bunchy top stage of mango malformation. *Indian Journal of Experimental Biology*, **18:** 286-289.
- Kumar, P. and Singh, S. 1993. Effect of GA_3 and ethrel on ripening and quality of mango. Cv. Amrapali. *Horticultural Journal*, **6**(1): 19-23.
- Kumar, R., Kaushik, R.A. and Chharia, A.S. 1992. Effect of post-harvest treatment on the quality of mango during storage. *Haryana Journal of Horticultural Science*, **21**(1-2): 46-55.
- Kumar, S. and Sharma, V.P. 2000. Rooting of Kurukkans mango layers affected by etiolation and juvenility. *Haryana Journal of Horticultural Sciences,* **29**(3/4) : 145-147.
- Kundu, S., Ghosh, B., Mishra, J., Datta, P., Ghosh, P. and Kundu, P. 2011. Induction of flowering in mango cv. Amrapali in West Bengal. Global Confernce on Augmentating Production and Utilization of Mango : Biotic and Abiotic stresses, Lucknow. 21-24 June 2011 (Abstract).
- Kurian, R.M. and Iyer, C.P.A. 1993. Chemical regulation of tree size in mango cv. Alphonso. II Effect of growth retardants on flowering and fruit set. *Journal of Horticultural Science,* **68**: 355-360.
- Kurian, R.M., Iyer, C.P.A. and Murti, G.S.R. 1994. Total phenols of stem apical bud in relation to tree vigour in mango (*Mangifera indica* L.). *Gartenbauwissenschaft,* **59**: 268-270.

- Kurian, R.M., Murti, G.S.R. and Iyer, C.P.A. 1992. Changes in cytokinin levels in mango (*Mangifera indica* L.) leaf extracts following soil drenches with paclobutrazol. *Garlenbauwissenchaft*, **57**: 84-86.

- Kusano, T., Berberich, T., Tateola, C. and Takahashi, Y. 2008. Polyamines essential factor for growth and survival. *Planta,* **228**: 367-384.

- Lakshminarayana, S, Velasco Gardens, T.T. and Sarmientolopez, L. 1974. Mango variedades *Keitt y Kent, treatments expost-Cosecha, Tecno, Ali (Mexico),* **9**: 57-66.

- Lakshminarayana, S., Shetty, M.S. and Krishna Prasad, C.A. 1975. Accelerated ripening of Alphonso mangoes by the application of ethrel. *Tropical Science*, **17**: 95-101.

- Lal, B., Tripathi, S. and Mishra, D. 2003. Effect of paclobutrazol on bearing behaviour of rejuvenated mango trees. National Seminar on Mango. Gujarat Agricultural University, Junagadh, June. 14-15.

- Lal, B.L., Gunjate, R.T. and Salvi, M.J. 1985. Effect of post-harvest ethephon dippng of fruits on occurrence of spongy tissue disorder in Alphonso mango. *Indian Journal of Plant Physiology*, **28**(1): 85-87.

- Lal, K. and Ram, S. 1977. Auxins of mango shoot tip and their significance in flowering. *Pantnagar Journal of Research*, **2**: 31-35.

- Lalel, H.J.D., Singh, Z. and Tan, S.C. 2003. The role of methyl jasmonate in mango ripening and biosynthesis of volatile compounds. *Journal of Horticultural Science and Biotechnology*, **78**(4): 470-474.

- Lanuza, E.A. 1939. Notes on bud differentiation in 'Carabao' mango (*Mangifera indica* L.). *Philippine Journal of Agriculture*, **10:** 131-151.

- Ledin, R.B. and Ruehle, G.D. 1954. Mango selection, propagation and culture. Annual Report Florida Agriculture Experiment Station. 1954. pp. 291-292.

- Leopold, A.C. 1958. Auxin uses in the control of flowering and fruiting. *Annual Review of Plant Physiology*, **9**: 281-301.

- Leopold, A.C. and Kriedemann, E.T. 1983. Plant growth and development. Tata-Macgrow Hill Pub. Co., Ltd., New Delhi.

- Leopold, A.C., Brown, K.M. and Emerson, F.H. 1972. Ethylene in the wood of stressed trees. *Hort. Science*, **7**: 175.

- Letham, D.S. 1967. Chemistry and physiology of Kinetin-like compounds. *Annual Review of Plant Physiology*, **13**: 349-364.
- Libermann, M. 1979. Biosynthesis and action of ethylens. *Annual Review of Plant Physiology*, **30**: 533-591.
- Libermann, M., Baker, J.E. and Sieger, M. 1978. Influence of plant hormones on ethylene production in apple, tomato and avocado during maturation and senescence. *Plant Physiology*, **60**: 214-251.
- Lim, T.K., Luders, L. and Poffley, M. 1996. Studies on fruit deformity and nubbins in Irwin mango. In : Proceedings of International Conferene on Tropical Fruits. Malaysian Agricultural Research and Development Institute (MARDI), Kaulalumpur, pp. 287-293.
- Liu, K., Fu, H., Bei, Q. and Luan, S. 2000. Inward potassium channel in guard cells as a target of polyamine regulation of stomatal movements. *Plant Physiology*, **124:** 1315-1325.
- Lizada, M.C.C. 1991. Post-harvest physiology of mango – A review. *Acta Horticulturae*, **291** : 437-453.
- Looney, N.E., Granger, R.L. and Haris, R.P. 1992. Influence of gibberellin A-4, gibberellin A4 plus gibberllin A7 and gibberellin 4 plus Iso-minus gibberellin A7 on apple fruit quality and tree productivity. Effects of fruit russet and tree yield compounds. *Journal of Horticultural Science*, **67**: 615-618.
- Lopez, M.R. 1984. El nitrato depotassiocomo promoter delasitesis endogena deetilenoy la induction florden mango (*Mangifera indica* L.) cv. Manita. M.Sc. Thesis, Univesidad Autonoma Chapingo, Chapingo, Mexico.
- Lopez, M.R., Mosqueda -Vazquez, R. and Pimienta, B.E. 1984. El nitrato de potassio como promoter deetileno Y la induction floral del mango (*Mangifera indica* L.) cv. Manila. Numero Zode Aguscalientes. Ags. pp. 110.
- Luckwill, L.C. 1948. Hormone content of seed in relation to endosperm development and fruit drop in apples. *Journal of Horticultural Science*, **24**: 32-44.
- Luckwill, L.C. 1964. New way to initiate fruit bud formation. *The Grower*, **62**: 791-793.
- Luckwill, L.C.1957. Hormonal aspects of fruit development in higher plants. *Symposium of Society of Experimental Biology*, **11**: 63-85.
- Lui, J.J., Guo, Y., Hu, F.M. and Zhou, C.J. 1999. Changes in superoxide radical and

polyamine levels during the leaf abscission phase in olive. *Scientia Silvae Simica,* **35**: 113-115.

- Lukshminarayana, S. and Augilar, F.R. 1975. Effect of orchard heating in reducing parthenocarpic fruits in 'Haden' mango. *Proceedings of Florida state Horticulture Society,* **83**: 502-505.
- Mackay, C., Hael, J., Hofstra, G. and Fletcher, R. 1990. Uniconazole induced changes in abscisic acid and total amino acids and proline in *Phaseolus vulgaris. Pesticide Biochemistry and Physiology*, **37**: 74-82.
- Maiti, S.C. 1973. Effect of gibberellic acid (GA_3) sprays on sex expression of mango (*Mangifera indica* L.). *Science and Culture*, **39**: 150-151.
- Maiti, S.C. and Sen, P.K. 1968. Effects of growth retardants on flowering and fruiting of Langra mango (*Mangifera indica* L.). *Current Science*, **37**: 560-567.
- Maiti, S.C., Basu, R.N. and Sen, P.K. 1972a. Chemical control of growth and flowering in *Mangifera indica* L. *Acta Horticulturae*, **24**: 192-195.
- Maiti, S.C., Maiti, S.K. and Sen, P.K. 1972b. Effect of ethylene producing chemicals on growth and flowering of mango (*Mangifera indica* L.). *Indian Agriculturist,* **16**: 201-203.
- Maiti, S.C., Mukhopadyaya, A.K. and Sen, P.K. 1971. Effect of growth retardants on flowering and apical dominance of mango (*Mangifera indica* L.). *Current Science*, **40**: 388.
- Majumder, G., Modi, V.V. and Palejwala, V.A. 1981. Effect of plant growth regulators on mango ripening. *Indian Journal of Experimental Botany,* **191**: 885-886.
- Majumder, P.K., Sinha, G.C. and Singh, R.N. 1970. The effect of exogenous application of Alfa-napththyl acetic acid on mango (*Mangifera indica* L.) malformation. *Indian Journal of Horticulture*, **20**: 130-131.
- Malamberg, R.L., Watson, M.B., Galloway, G.L., Yu, W. and Yu, W. 1998. Molecular genetic analysis of plant hormones. Critical Reviews. *Plant Sciences*, **17**: 199-224.
- Malik, A.U, Agrez, V. and Singh, Z. 2003a. Fruitlet abscission of mango in relation to ethylene. *Journal of Horticultural Science and Biotechnology*, **78**: 458-462.
- Malik, A.U. 2003. Fruitlet abscission and fruit ripening in relation to polyamines. Ph.D. thesis. Curtin University of Technology, Perth, Western Australia.

- Malik, A.U. and Singh, Z. 2003a. Abscission of mango fruitlets as influenced by biosynthesis of polyamines. *Journal of Horticultural Science and Biotechnology*, **78**: 721-727.
- Malik, A.U. and Singh, Z. 2003b. Increased biosynthesis of free polyamines in response to higher ethylene production during ripening of *Mangifera indica* L. Australian Post-harvest Conference, Brisbane, Australia, 1-3 October, 2003, pp. 117-119.
- Malik, A.U. and Singh, Z. 2004. Endogenous free polyamines of mango fruit in relation to development and ripening. *Journal of American Society of Horticultural Science*, **129**: 280-286.
- Malik, A.U. and Singh, Z. 2005. Pre-storage application of polyamines improve shelf-life and fruit quality of mango. *Journal of Horticultural Science and Biotechnology*, **80**(3): 363-369.
- Malik, A.U., Singh, Z. and Dhaliwal, S.S. 2003b. Exogenous application of putrescine affects mango fruit quality and shelflife. *Acta Horticulturae*, **628**: 121-127.
- Malik, A.U., Singh, Z. and Khan, A.S. 2005. Role of polyamines in fruit development, ripening, chilling injury, storage and quality of mango and other fruits : A Review. Proceedings of International Conference on Mango and Date Palm – Culture and Export, 20th – 27th June, 2005.
- Malik, A.U., Singh, Z. and Tan, S.C. 2006. Exogenous application of polyamines improves shelf life and fruit quality of mango. *Acta Horticulture*, **699**: 291-296.
- Malik, M.M. and Raza, M. 1985. Effect of different doses of NPK and NAA and time of deblossoming on the intensity of malformation of mango inflorescence. *Journal of Agricultue Research, Paleistina*, **23**: 97.
- Mallik, P.C. 1963. Mango malformation – Symptoms, causes and cure. *Punjab Horticultural Journal*, **3** : 292-299.
- Mallik, P.C., Sahay, R.K. and Singh, D.L. 1959. Effect of hormones on the sex-ratio in the mango. *Current Science*, **28**: 410.
- Mansuroglu, S., Karaguzel, O., Ortausme, V. and Sayan, M.S. 2009. Effect of paclobutrazol on flowering, leaf and colour of *Consolida orientalis*. *Pakistan Journal of Botany*, **41**: 2323-2332.
- Manzano, Mendoza, J.E. 1995. Response of several mango cultivars to gibberellic acid application. *Harvest and Post-harvest technologies for fresh fruits and vegetables*. **477**: 487.

- Marie, C.P. 2011. Dwarfing potential of indigenous mango varieties. M.Sc. Thesis, Kerala Agricultural University, Thrissur, Kerala, India.
- Martinez, A.R., Perez, A.L.A. and Moreno, J.R. 2008. Effects of paclobutrazol and KNO_3 over flowering and fruit quality in two cultivars of mango. *Intercencia*, **35**(7): 518-522.
- Mathod, J.C. 1983. Effect of growth regulators on success of Veneer and Wedge grafts of Alphonso mango (*Mangifera indica* L.). M.Sc.(Ag.) Thesis. University of Agricultural Science, Dharwad, Karnataka, India.
- Mathur, K. and Srivastava, G.C. 2005. Effect of 1-MCP on malic enzyme activity and ethylene production in mango during ripening. *Indian Journal of Plant Physiology,* **10**(3) (NS) : 273-275.
- Mattoo, A.K., Modi, V.V. and Reddy, V.V.R. 1968. Oxidation and cortenogenesis regulation factors in mango. *Indian Journal of Biochemistry*, **5**: 111-114.
- Maurya, A.N. and Singh, J.N. 1979. Effect of three growth regulators on fruit retention and quality of mango (*Mangifera indica* L.) cv. Langra. *Journal of the National Agricultural Society of Ceylon*, **16**: 53-56.
- Maurya, A.N., Singh, S.M. and Singh, A.R. 1973. Effect of plant regulators on the fruit retention and quality of Dashehari mango. *Punjab Horticultural Journal,* **13**: 117-121.
- Mc Donald, R.E. 1989. Temperature conditioning affects polyamines of lemon fruit stored at chilling temperatures. *Hort Science*, **24**: 475-477.
- Mc Donald, R.E. and Kushad, M.M. 1986. Accumulation of putrescine during chilling injury of fruits. *Plant Physiology*, **82**: 324-326.
- Medlicott, A.P., Sigrist, J.M.M., Reynolds, S.B. and Thompson, A.K. 1987. Effects of ethylene and autylene on mango fruit ripening. *Annals of Applied Biology,* **111**(2): 439-444.
- Mehta, P.M., Shivaraj, S. and Raju, P.S. 1986. Influence of fruit ripening retardants on succinate and dehydrogenases in papaya fruit with emphasis on preservation. *Indian Journal of Horticulture*, **43**: 169-173.
- Meland, M. 1998. Thinning apples and pears in a Nordic climate. III. The effect of NAA, ethephon and lime sulphur on fruit set, yield and return bloom of the apple cultivars. *Acta Hort.*, **463**: 517-525.

- Mendoza, D.B. Jr. 1981. Development physiology of carobao mango (*Mangifera indica* L.). fruit. Ph.D. Thesis, University of Philippines, Langra, Philippines.

- Meyer, A., Miersch, D., Buttner, C., Dutthe, W. and Sembadner, G. 1984. Occurrence of the plant growth regulator. Jasmonic acid in plants. *Journal of Plant Growth Regulation*, **3**: 1-8.

- Mir, N.A., Curell, E., Khan, N., Whitaker, N. and Baudt, R.M. 2001. Harvest maturity, storage temperature and 1-MCP application treatment alter firmness, retention and chlorophyll fluroscence of 'Redy Chief Delicious' apples. *Journal of American Society of Horticultural Science*, **126**: 618-624.

- Mishra, K.A. 1976. Studies on bearing behaviour of *Mangifera indica* L. and its malformation. Ph.D. Thesis. Punjab Agricultural University, Ludhiana.

- Mishra, K.A. and Dhillon, B.S. 1978. Levels of abscisic acid like substances in healthy and malformed panicles of mango (*Mangifera indica* L.). *Science and Culture*, **44**: 419-420.

- Mishra, K.A. and Dhillon, B.S. 1980. Levels of endogenous gibberellins in healthy and malformed panicles of mango. *Indian Journal of Horticulture*, **37**: 33-6.

- Mistry, P.M. and Patil, B.N. 2003. Rejuvenation of old and over crowded Alphonso mango by heading bark plus paclobutrazol under South Gujarat Agro-climatic condition. National Seminar on Mango. Gujarat Agricultural University, Junagadh, June 14-15, 2003.

- Mohamed, H.E. and Abu-Gouka, A.B. 2003. Effect of ethrel in aqueous solution and ethylene released from ethrel on mango fruit ripening. *Journal of Horticultural Science and Biotechnology*, **78**(4): 568-573.

- Mohammad, M. and Brecht, J.K. 2000. Influence of ethylene treatments on ripening and chulling injury alleviation in mango (cv. Palmer) fruit. *Acta Horticulturae*, **509**: 437-446.

- Moore, T.C. 1979. Biochemistry and Physiology of Plant Hormones 2nd edition. Springer Verlage, Newyork. USA.

- Morgan, P.W. and Gausman, H.W. 1966. Effects of ethylene on auxin transport. *Plant Physiology*, **41**: 45-52.

- Morshed, M.H., Ibrahim, M., Islam, M.A., Helati, M.O.H. and Islam, M.S. 2006. The efficacy of plant hormones on the shelf life and quality of mango (*Mangifera indica* L.) cultivar. Fazli. *International Journal of Sustainable Agricultural Technology*, **2**(8): 5-10.

- Mosqueda – Vazquez, R. and Avila – Resendiz, G. 1985. Induction floral del mango con applicaious de KNO_3 Y su inhibition al applicar $AgNO_3$ o $CoCl_2$. *Horticulture. Mexicana,* **1**: 98-101.
- Moti, S., Chaudhury, A.S. and Prasad, M. 1986. A note on the ethylene plant regulators on fruit retention in mango (*Mangifera indica* L.). *Haryana Journal of Horticultural Science,* **15**(3-4) : 221-223.
- Muhammad, N., Muhammad, F., Saeed, A., Mueen, A.K., Moazzam, J. and Muhammad, N.A. 2010. Paclobutrazol soil drenching suppresses vegetative growth reduces malformation and increases production in mango. *International Journal of Fruit Science,* **10**: 131-140.
- Mukherjee, S.K. 1951. Pollen analysis in *Mangifera* in relation to fruit set and taxonomy. *Journal of Indian Botanical Society,* **30**: 49-55.
- Mukherjee, S.K. and Bid, N.N. 1965. Effect of etiolation and growth regulator treatments on the success of air-layering. *Indian Journal of Agricultural Sciences,* **35**: 309-314.
- Mukherjee, S.K. and Majumder, P.K. 1963. Standardization of root stocks of mango I. Studies on the propagation of clonal root stocks by Stooling and air layering. *Indian Journal of Horticulture,* **20**: 204-205.
- Mukhopadyaya, A.K. 1978. A note on the effect of growth retardants and L-methionine on flowering of mango (*Mangifera indica* L.). *Horticulture Abstract,* **48**: 254.
- Mullins, P.D.F. 1985. Delaying of flowering in Haden mango trees. Technical Communications in Horticultural Science. Series No. 200. Department of Agriculture, Republic of South Africa. Citrus and Subtropical Research Institute, Nels Fruit. RST. pp. 6-8.
- Muralidhar, B.M., Reddy, Y.T.N., Akshitha, H. J. and Srilatha, V. 2015. Effect of pre-sowing treatments on germination, growth and vigor of polyembryonic mango seedlings. *Environment and Ecology,* **33**(3): 1014-1018.
- Murashige, T. and Skoog, P. 1962. A revied medium for rapid growth and bioassays with tobacco tissue cultures. *Physiologia Plantarium,* **11**(3): 7-11.
- Murthy, V.K. 1976. Effects of growth regulators on fruitset and fruit drop of mango (*Mangifera indica* L.). M.Sc. (Ag.) Thesis. A.P.A.U., Tirupati, Hyderabad, India.
- Murti, G.S.R,, Upreti, K.K. and Kurian, R.M. 2003. Physiological studies on vigour regulation in mango. National Seminar on 'Physiological Intervention for improved

crop production and quality: Opportunities and Constraints, held at Tirupati. 12-14 December 2003 (Abstract).

- Murti, G.S.R,, Upreti, K.K., Kurian, R.M. and Reddy, Y.T.N. 2001. Paclobutrazol modifies tree vigour and flowering in mango cv. Alphonso. *Indian Journal of Plant Physiology,* **6**(4) (NS) : 355-360.

- Murti, G.S.R. and Upreti, K.K. 1995. Changes in some endogenous growth substances during fruit development in mango. *Plant Physiology and Biochemistry, New Delhi.* **22**(1): 44-47.

- Murti, G.S.R. and Upreti, K.K. 1997. Endogenous cytokinins in relation to fruit development in mango (*Mangifera indica* L.) cv. Alphonso. *Indian Journal of Plant Physiology,* **2**(2) : 98-100.

- Murti, G.S.R. and Upreti, K.K. 1999. Endogenous hormones and polyamines in relation to fruitlet retention in mango cv. Alphonso. *Journal of Plant Physiology,* **26**: 1.

- Murti, G.S.R. and Upreti, K.K. 2000. Physilogy of mango. In : Hemantranjan (ed.) Advances in Plant Physiology. vol. 3, p. 397-419. Scientific Publishers, Jodhpur, India.

- Murti, G.S.R. Upreti, K.K. 1998. Changes in the levels of endogenous hormones in relation to shoot vigour in mango (*Mangifera indica* L..). Plant Physiology and Biochemistry, New Delhi. 25(2) : 167-171.

- Murti, G.S.R. Upreti, K.K. 2003. Changes in the levels of common endogenous polyamines in the pericarp and seeds of mango fruits during development. *Indian Journal of Plant Physiology (NS)* **8**(2): 111-114.

- Murti, G.S.R., Upreti, K.K. and Jayaram, H.L. 2008. Fruit drop in mango cv. Alphonso. Changes in ethylene production and ACC content. *Indian Journal of Horticulture,* **65**(1) : 97-99.

- Musa, S.K. 1974. Peliminary investigations on the storage and ripening of 'Totapuri' mangoes in Sudan. *Tropical Science,* **16**: 65-73.

- Nabway, S.El., Abou Rawash, M., El-Hammadu, A., Khalifa, A. and Masry, H.El. 1984. Indole and phenol contents in 'Langra' and 'Ewais' mango cvs. in relation to flowering. *Egyptian Journal of Horticulture,* **11**(1): 31-40.

- Nafeez, M., Faqueer, M., Ahmed, S., Alamkhan, M., Jamil, M. and Naveed Aslam, M. 2010. Paclobutrazol soil drenching suppresses vegetative growth, reduces malformation and increases production in mango. International Journal of Fruit Science. 10 : 431-440.

- Nagar, P.K. 1978. Gibberellins in mango (*Mangifera indica* L.) fruit. *Plant Science,* **10**: 95-100.
- Naik, K.C. and Rao, M.M. 1943. Studies on blossom biology and pollination in mangoes. *Indian Journal of Horticulture,* **1**(2): 107-109.
- Nair, S. 2002. Biochemical aspects and control of chilling injury in mango fruit. Ph.D. Thesis, Curtain Universit of Technology, Australia.
- Nair, S. and Singh, Z. 2003. Pre-storage ethrel dip reduces chilling injury, enhances respiration rate, ethylene production and improves fruit quality of 'Kensington Pride' mango. *Journal of Food, Agriculture and Environment,* **1**(2) : 93-97.
- Nair, S. and Singh, Z. 2004b. Chilling injury in mango fruit in relation to biosynthesis of free polyamines. *Journal of Horticultural Science and Biotechnology,* **79**(4): 515-522.
- Nair, S., Singh, Z. and Tan, 2004a. Chiling injury in relation to ethylene biosynthesis in 'Kensington Pride' mango fruit. *Journal of Horticultural Science and Biotechnology,* **79**(1): 82-90.
- Nandkumar, S. and Krurupaih, P. 2006. Effect of pruning and paclobutrazol application on yield and quality of mango cv. Baneshan. National Seminar on ATSH, Annamalai University, Annamalai. March 20-21, 2006.
- Naqui, S.S.M., Alam, S.M. and Mumtaz, S. 1990. Effect of cobalt and silver ions and naphthalene acetic acid on fruit retention in mango (*Mangifera indica* L.). *Australian Journal of Experimental Agriculture,* **30**: 433-435.
- Narwadkar, P.R. and Anserwadkar, 1985. Effect of growth regulators on the success of epicotyl grafting in mango. *Acta Horticulturae,* **231**: 175-178.
- National Horticulture Board. 2013. Indian Horticultural Data Base. (ed. R.K. Tiwari) Available online at http://nhb.gov.in/area %20 production.html.
- Nayak, P., Rath, S. and Dora, D.M. 1997. Fruit retention in two mango cultivars as influenced by growth regulators and urea treatments : A note. *Orissa Journal of Horticulture,* **25**(2) : 107-108.
- Nicholson, R.D. and Van Staden, J. 1988. Cytokinin and mango flower malformation. I. Tentative identification of the complement of healthy and molformed inflorescences. *Journal of Plant Physiology,* **132**: 720-724.

- Nickell, L.G. 1983. Plant growth regulating chemicals. CRC Press, Boca Raton.
- Nirvam, R.S. 1956. Bunchy top of young mango seedlings. *Science and Culture*, **18**: 335-336.
- Nitsch, J.P. 1965. Physiology of flower and fruit development. pp. 1537-1627 (ed W. Ruhland). Encyclopedia of Plant Physiology. XVII Springer Vertag. Berlin.
- Nkansah, G.O., Ofosu-Anim, J. and Mawuli, A. 2012. Gibberellic acid and Napththalene acetic acid affect fruit retention, yield and quality of Keitt mangoes in the Coastal Savana Ecological zone of Ghana. *American Journal of Plant Physiology*, **7**(6): 243-251.
- Nooden, L.D. and Leopold, A.C. 1978. Phytohormones and the endogenous regulation of senescence and abscission. In : Phytohormones and related compounds : A comprehensive treatise. Vol.II, pp. 329-369, (Eds). D.S. Letham, P.B. Goodwin and T.J.Y. Higgins, Elsevier, Amsterdam. 648 pp.
- Notodimedjo, S. 2000. Effect of GA_3, NAA and CPPU on fruit retention, yield and quality of mango cv. 'Arumanis' in East Java. *Acta Horticulturae*, **509**: 587-600.
- Nunez-Elisea, R. and Davenport, T.L. 1991. Flowering of Keitt mango in response to deblossoming and gibberellic acid. *Proceedings of Florida State Horticulture Society*, **104:** 41-43.
- Nunez-Elisea, R. and Davenport, T.L. 1992. Requirement for mature leaves during floral induction and floral transmission in developing shoots of *Mangifera indica* L. *Acta Horticulturae*, **296**: 33-37.
- Nunez-Elisea, R. and Davenport, T.Z. 1986. Abscission of mango fruitlets as influenced by enhanced ethylene biosynthesis. *Plant Physiology*, **82**: 991-994.
- Nunez-Elisea, R., Becerul – Roman, A.E. and Diaz-Montenagro, D. 1979. Efectos del ethrel sobrela floracian in mango cv. Haden. *Proceedings of the Tropical Region of the American Society for Horticultural Science*, **23**: 138-140.
- Nunez-Elisea, R., Celdira, M.L. and Davenport, T.L. 1990. Tridiazuron effects on growth initiation and expression in mango (*Mangifera indica* L.). *Hort Science*, **25**: 1167 (abstract).
- Ona, Ma, L.D. and de-Guzman, E.V. 1982. Note : floral gross morphology and histology of 'Carabao' mango (*Mangifera indica* L.) treated with a chemical flowering inducer. *Philippines Agriculturist*, **65**: 201-203.
- Oosthuyse, S.A. 1991. Effect of promalin on flowering and lateral branching of young

mango trees in the nursery. South African Mango Growers' Association Year Book. 11 : 54.

- Oosthuyse, S.A. 1993. Effect of spray application of KNO_3, urea and growth regulators on the yield of 'Tommy Atkins' mango. *South African Mango Growers' Association Year Book*, **13** : 58-62.
- Oosthuyse, S.A. 1995. Effect of post bloom aqueous spray application of GA_3, NAA and CPPU on fruit retention, fruitset and yield in 'Tommy Atkins' and 'Heidi' mango. *South African Mango Growers' Association Year Book*, **15**: 31-33.
- Oosthuyse, S.A. 2015. Spray application of KNO_3, low biuret urea and growth regulators and hormones during and after flowering on fruit retention, fruit size and yield of mango. *Acta Horticulturae*, **1075**: 135-141.
- Oosthuyse, S.A. and Jacobs. G. 1997. Effect of soil applied, paclobutrazol on fruit retention, fruit size, tree yield and tree revenne in 'Sensation' and 'Tommy Atkins' mango. *South African Mango Growers' Association Year Book*, **17** : 57-62.
- Oppenheimer, C. 1947. The acclimolization of new tropical and sub-tropical fruit trees in Palestine. *Bulletin of Agricultural Experiment Station*, **44**: 1.
- Osuma – Garcia, J.A., Caceres – Morales, 1., Monstaloo – Gonzalez, E. Mata – Montes de Oca, M. and Tovar – Gomez, B. 2007. Effect of 1-Methyl cyclopropane (1-MCP) and hot water treatment on the physiology and quality of 'Keitt' mangoes (Spanish). *Revista Chagingo Sene Horticultura*, **13**(2) : 157-163.
- Ou, S.K. and Yen, C.R. 1985. Effect of growth regulators on the flowering of 'Irwin' mango at high temperature. In : Forcing culture of fruit trees. Taichung District Agriculture Improvement Station, pp. 137-148.
- Padhiar, B.V. and Patel, B.M. 2003. Response of certain mango cultivars to Paclobutrazol on fruiting, yield and quality. National Seminar on Mango held at Gujarat Agricultural University, Junagadh, Gujarat, June 14-15-1993 (Abstract) p. 33.
- Padma, M. and Reddy, Y.N. 1998. Effect of pre-sowing treatments of stones and kernels of mango (*Mangifera indica* L.) on germination. *Journal of Research, Acharya N. G. Ranga Agricultural University*, **26**(2) : 17-21.
- Pal, R.N., Chadha, K.L. and Rao, M.R.K. 1984. Effect of different plant growth regulators and other chemicals on flowering behaviour of mango. *Indian Journal of Horticulture*, **41**(1) : 8-15.

- Pal, R.N., Katre, S.K., Tandon, D.K. and Chadha, K.L. 1983. Activity of IAA oxidase, catalase and amylase in morphactin induced malformation of mango inflorescence. *Scientia Horticulturae,* **19**: 271-277.
- Pal, R.N., Rao, M.R.K. and Chadha, K.L. 1979. Effect of growth retardants and other growth regulators for contrl of fruit drop in mango. *Indian Journal of Horticulture,* **36:** 218-219.
- Pal, S. and Ram, S. 1978. Endogenous gibberellins of mango shoot tips and their significance in flowering. *Scientia Horticulturae,* **9**: 369-379.
- Palijwala, V.A., Amin, B., Parikh, H.R. and Modi, V.V. 1989. Role of abscisic acid on the ripening of mango. *Acta Horticulturae,* **231**: 662-667.
- Pandey, D. and Pandey, S.N. 2000. Effect of growth regulators and zinc sulphate on floral malformation in mango. *Annals of Agriculture Research,* **21**: 129-132.
- Pandey, G. 1988. Role of molformins and auxins in the floral malformation of mango (*Mangifera indica* L.). Ph.D. Thesis, G.B. Pant University of Agriculture and Technology, Pantnagar, India.
- Pandey, G. and Ram, S. 1995. Effect of malformation on changes in endogenous levels of auxins during floral development of mango. *Indian Journal of Horticulture,* **52**: 92-96.
- Pandey, G. and Singh, B.P. 2007. Comparative response of 1-Methylcyclo propane and $KMnO_4$ on post-harvest behaviour of mango. *Indian Journal of Horticulture,* **64**(2): 123-126.
- Pandey, I.C., Tripathi, B.M. and Upadhyay, N.P. 1980. Effect of plant regulators on fruit ripening and quality of mango. *Punjab Horticulture Journal,* **20**(1-2): 24-27.
- Pandey, R.M. 1989. Physiology of flowering in mango. *Acta Horticultuae,* **231**: 361-380.
- Pandey, R.M. and Narwadkar, P.R. 1984. Studies on the induction of growth and flowering in Dashehari mango. *Indian Journal of Horticulture,* **41**: 171-176.
- Pandey, R.M. and Singh, R.N. 1976. Chemical regulations of ripening in mango. *Indian Journal of Horticulture,* **33**(3-4) : 209-211.
- Pandey, R.M., Rathore, D.S. and Singh, R.N. 1974. Hormonal research on mango malformation. *Current Science,* **43**: 694-698.
- Pandey, R.M., Singh, R.N. and Sinha, G.C. 1973a. Usefulness of ethrel in regulating flower bearing behaviour in mango. *Science and Culture,* **39**: 148-150.

- Pandey, R.M., Sinha, G.C., Singh, R.N. and Majumder, P.K. 1973b. Some biochemical aspects of vegetative malformation in mango (*Mangifera indica* L..) I. carbohydrate reserves and nitrogenous fractions. *Indian Journal of Horticulture,* **30**(3-4): 475-480.

- Pant, R.C. 2000. In 'Stress ethylene' the cause of mango (*Mangifera indica* L..) malformation. *Physiology and Molecular Biology of Plants*, **6**: 8-14.

- Pantastico, F.B. and Manuel, F.C. 1978. The Philippines recommends for mango. 1979. Report of the Philippines Council for Agriculture, PCARR. pp. 43.

- Paramar, P.B. and Chundwat, B.S. 1988. Effect of various post-harvest treatments on the physiology of Kesar mango. *Acta Horticulturae*, **231**: 679-684.

- Passera, C. and Spattoli, P. 1981. Effect of benzyl amino purine on mango fruit ripening. *Food Chemistry*, **7**: 195-201.

- Patel, A.S., Tidke, S.N., Patil, M.A., Shinde, B.N. and Gore, A.K. 2005. Effect of chemicals and growth regulators on fruitset and fruit retension of Parbhani Bhushan mango. *Soils and Crops*, **15**(1) : 64-66.

- Patel, M.J. and Patel, B.M. 1993. Influence of L-NAA, 2, 4-D and urea sprays on fruit retention, yield and quality of mango cv. Kesar. Paper presented at Golden Jubilee Symposium Horticultural Research – Changing Scenario. May 24-28, 1993, Bangalore, (abstract): 207.

- Patel, R.M. and Singh, M.P. 1982. Effect of growth regulators on rooting in air layering of mango (*Mangifera indica* L..). *JNKVV Research Journal,* **16**(4): 351-352.

- Patil, H.K. and Talathi, J.M. 1994. Economic viability of paclobutrazol application in mango cv. Alphonso. *South Indian Horticulture*, **47**(1-6): 145-146.

- Patil, P.B. 1984. Studies on physiological and bio-chemical factors associated with frui bud differentiation and flowering in mango. Ph.D. Thesis. University of Agricultural Sciences, Bangalore, India.

- Patrick, J.W. 1979. Auxin promoted transport of metabolites in stems of *Phaseolus vulgaris* L. Further studies on effect of remote from the site of hormone application. *Journal of Experimental Botany*, **30**: 1-13.

- Paulas, D. and Shanmugavelu, K.G. 1989. Physiological and biochemical changes in the leaf tisues from quescent to fruiting stages of mango. *Acta Horticulturae,* **231**: 394-398.

- Pawan, S. 1984. Propagation of certain varieties of mango (*Mangifera indica* L.) by air-layering. M.Sc. (Ag.)Thesis, University of Agricultural Science, Bangalore, Karnataka, India.

- Pearse, H.L. 1938. Experiments with growth controlling substances. I. The reaction of leafless cuttings to treatment with root forming substances. *Annals of Botany, London*, **2**: 227-236.

- Penchaiya, P., Jansasithorn, R. and Kaulavariarat, S. 2006. Effect of 1-MCP on physiological changes in mango. 'Nam Dokmai'. *Acta Horticulturae,* **712**(2): 717-721.

- Penz, L.Z., Tang, J. Xo, E., Song, N.J. and Lei, T. 1991. Effect of paclobutrazol on the morphological and anatomical characteristics of the new organs in citrus trees. *China Citrus,* **20** : 7-10.

- Perez – Barraza, M.H., Yazquez-Valdivia, V. and Osuna-Garcia, J.A. 2007. 'Ataulto' mango parthenocarpic fruits incidence in commercial orchards of Nay-Rit. Revista Chapingo *Serie Horticltura,* **13**(2): 149-158.

- Philip, T.E. and Malamberg, R.L. 1989. Do polyamines have a role in plant development. *Annual Review of Plant Physiology and Plant Molecular Biology*, **40**: 235-269.

- Prakash, S. and Ram, S. 1984. Naturally occurring auxins and inhibitors and their role in fruit growth and drop of mango 'Dasehari'. *Scientia Horticulturae*, **22**(3): 241-248.

- Prakash, S. and Ram, S. 1985. Effect of various growth regulators and chemicals on fruit retention in mango cv. Chausa (abstract) In : II International Mango Symposium, 20-24, May, Bangalore, India, p. 54.

- Prakash, S. and Ram, S. 1986. Effect of various concentrations of auxins and their time of application on fruit retention in mango (*Mangifera indica* L.) cv. Chausa. *Progressive Horticulture,* **16** : 167-174.

- Preibe, A.H.K. and Jagar, H.J. 1978. Role of polyamines in SO – polluted pea plants. *Journal of Experimental Botany*. **29**: 1045-1050.

- Protacio, C.M., Bugante, R.D. Jr., Quinto Motin Yawe, G. and Paclmo, G. 2000a. Regulation of flowering in 'Carabao' mango trees by paclobutrazol. *Philippine Journal of Crop Science*, **25**: 27-33.

- Protacio, C.M., Quinto, J.E., Serrano, E.P., Marquez, I.P., Rodriquez, F.M. 2000b. Unvavelling the mechanism of mango flowering. *Acta Horticulture*, **820**: 259-270.

- Pruthi, J.S. 1992. The Mango Book. I. Production, post-harvest technology and export marketing published by APEDA, Ministry of Commerce, Government of India, New Delhi.
- Purwako, B.S., Kesmayanti, N., Susanta, S. and Nasutian, M.Z. 1998. Effect of polyamines on quality changes in papaya and mango fruits. *Acta Horticulturae,* **464**: 510.
- Pushpa, K., Kiladi, B. and Dikshit, A. 2011. Effect of spermine on the shelf life of mango cv. Dashehari under low temperature. Global conference on Augumenting Production and utilisation of mango : Biotic and Abiotic Stresses, Lucknow, 21-24 June 2011 : 153p (Abstract).
- Quintana, E.G., Nanthachai, P., Hiranpradit, D.B., Mendoza, Jr. and Kets, S. 1984. Growth and development of mango. In : Mango fruit development, post harvest physiology and marketing in ASEAN Medoza. D.B. (Ed.) Wills, R.B.H. ASEAN Food Handling Bureau, p. 21-39.
- Rademacher, W. 1991. Biochemical effects of plant growth retardants. In : Gausman, H.W. (ed.). Plant Growth Regulators. Marcel Dekker, Newyork. pp. 169-200.
- Rademacher, W. 2000. Growth retardants. Effects on gibberellin biosynthesis and other metabolic pathways. *Annual Review of Plant Physiology and Molecular Biology,* **51**: 501-531.
- Rai, N., and Bist, L.D. 1992. Effect of soil and foliar applied paclobutrazol on vegetative growth, flowering, fruitset and yield of oriental peach. *Scientia Horticulture,* **50:** 153-158.
- Rajan, S. 1986. Biochemical basis of mango (*Mangifera indica* L.) malformation. Ph.D. Thesis, IARI, New Delhi.
- Rajan, S. and Ram, S. 1983a. Some factors affecting root regeneration in mango cuttings in mist and hot bed. *Progressive Horticulture* **15**(1-2) : 11-16.
- Rajan, S. and Ram, S. 1983b. Some new concepts in the rooting of mango. *Progressive Horticulture,* **15**(1-2) : 155-156.
- Rajan, S. and Ram, S. 1988. Studies on the root regeneration in mango air-layers. *Acta Horticulturae,* **231**: 192-197.
- Rajkumar, M., Reddy, Y.N., Chandra Sekhar, R. and Srihari, D. 2007b. Studies on the effect of foliar application of chemicals and growth regulators on the fruit yield of unpruned mango (*Mangifera indica* L.) trees cv. Baneshan. *Orissa Journal of Horticulture,* **38**(2) : 44-47.

- Rajkumar, M., Reddy, Y.N., Chandrasekhar, R. and Srihari, D. 2007a. Effect of pruning, paclobutrazol and chemicals on production of vegetative flushes and sunsequat flowering in mango (*Mangifera indica* L.) cv. Baneshan in Andhra Pradesh. *Orissa Journal of Horticulture*, **35**(2) : 6-11.
- Rajput, C.B.S. and Singh, J.N. 1983. Effect of urea and GA_3 sprays on the growth, flowering and fruiting characters of mango. *Progressive Horticulture,* **15**(3): 174-177.
- Rajput, C.B.S. and Singh, J.N. 1989. Effect of urea and GA_3 sprays on the growth, flowering and fruiting characters of mango. *Acta Horticlturae*, **231**: 301-305.
- Ram and Din, I.B. 1971. Improvement on the existing propagation techniques in mango. Annual Report 1970-71, Govind Ballabi Pant University of Agriculture and Technology, Pantnagar, India.
- Ram, S. 1983. Hormonal control of fruit growth and fruit drop in mango cv. Dashehari. *Acta Horticlturae,* **134**: 169-178.
- Ram, S. 1992. Naturaly occurring hormones of mango and their role in growth and drop of the fruit. *Acta Horticulturae*, **321**: 343.
- Ram, S. 1996. High density orcharding in mango. Research Bulletin, No. 122. Directorate of Experiment Station. G.B.P.U.A & T, Pantnagar, India.
- Ram, S. 1997. Propagation. In : The mango. Botany, Production and Uses. RE. Litz. (Ed.). CAB International, Willingford, U.K., p. 367.
- Ram, S. and Bist, L.D. 1984. Occurrence of malformin – like substances in malformed panicles and control of floral malformation in mango. *Scientia Horticulturae,* **23:** 331-336.
- Ram, S. and Pal, S. 1979. Studies on the naturally occurring gibberellins in mango (*Mangifera indica* L.) in fruit. *Journal of Horticultural Science,* **54**: 209-215.
- Ram, S. and Sirohi, S.C. 1979. Studies on fruit drop in mango. Research reports on mango. Mango Workers Meeting, Panaji, Goa. May 2-5, 1979, pp. 422-423.
- Ram, S. and Tripathi, P.C. 1993. Effect of cultar on flowering and fruiting in high density – Dashehari mango trees. *Indian Journal of Horticulture,* **50**: 292-295.
- Ram, S. and Yadav, V.K. 1999. Mango malformation – A Review. *Journal of Applied Horticulture, Lucknow*, **1**(1): 70-78.

- Ram, S., 1999. Hormonal physiology of flowering in 'Dashehari' mango. *Journal of Applied Horticulture (Lucknow).* **1**(2): 84-88.
- Ram, S., Pujari, K.H. and Singh, D.K. 1996. Role of paclobutrazol in fruit production. In : *Proceedings of National Seminar on Plant Bioregulators in Horticulture*, February 28 – March 2 at B.C.K.V.V. Kalyani, West Bengal. 139-150.
- Ram, S., Singh, C.P. and Kumar, S. 1997. A success story of high density orcharding in mango. *Acta Horticulturae,* **455**(1): 375-382.
- Ram, S., Sirohi, S.C. and Rathore, V.S. 1983. Naturally occurring cytokinins in mango (*Mangifera indica* L.). *Australian Journal of Plant Physiology*, **10**: 65-73.
- Ramakrishna, C. 1985. Studies on certain aspects of veneer grafting in mango (*Mangifera indica* L.). M.Sc.(Ag.) Thesis, Andhra Pradesh Agricultural University, Hyderabad, Andhra Pradesh, India.
- Ramchand, P. 1984. Studies on embedded pot layering in mango (*Mangifera indica* L.) and Sapota (*Achrus sapota* L.). M.Sc.(Ag.) Thesis, Andhra Pradesh Agricultural University, Hyderabad, Andhra Pradesh, India.
- Rameshwar, A. and Kulkarni, V.J. 1979. Studies on mango dwarfing. Research reports on mango. Mango Workers' Meeting, Goa, 2-5th May, p. 108.
- Ramina, A., Masia, A. and Vizzoto, G. 1986. Ethylene and auxin transport and metabolism in peach fruit abscission. *Journal of the American Society for Horticultural Science,* **111**: 760-764.
- Rani, R. and Brahmachari, V.S. 2004. Effect of growth substances and calcium compounds in fruit retention, growth and yield of Amrapali mango. *Orissa Journal of Horticulture*, **32**(1) : 15-18.
- Rao, A.N. 1983. Effect of growth regulators on rooting of air layers in chicku, seedless lemon and pomegranate. M.Sc.(Ag.) Thesis, Punjabrao Krishi Vidyapeeth, Akola, Maharashtra, India.
- Rao, M.M. and Ravishankar, H. 1992. Chemical induction of flowering directly on the fruited shoots in 'off' phase. Alphonso mango trees. *Karnataka Journal of Agricultural Science,* **5**(2) : 180-182.
- Rao, M.N., Rameshwar, A. and Rao, S.N. 1979. Studies on the effect of different growth regulators, irrigation and urea sprays on fruit drop in mango. Mango research reports. Mango Worker's Meeting, Panaji, Goa, May 2-5, 1979, pp. 421-422.

- Rao, M.R.K., Pal, R.N. and Chadha, K.L. 1976. Control of fruit drop in mango. Paper presented at the Fruit Research Workshop held at Hyderabad, May 24-28, 1976, pp. 218 (Abstract).
- Rao, S.N. 1961. Studies on the effect of GA and other plant growth regulators on fruitset, fruit drop and total yield of citrus, guava, mango and other fruit crops. Paper presented at the Fourth Horticultural Research Worker's Conference, Poona.
- Rao, S.N. and Rao, M.R. 1985. Effect of certain growth regulators and anti-transipirants on flowering, fruitset and fruit drop in some mango cultivars. Proceedings of II International Symposium on mango, Bangalore. May 20-24, 1985 (Abstracts).
- Rao, S.N., Swamy, G.S., Sreeramulu, P. and Reddy, R.N. 1963. Growth regulators in propagation of mango by air-layering. *Punjab Horticultural Journal*, **3:** 175.
- Rath, S. and Das, G.C. 1979. Effect of ringing and growth retardants on growth and flowering of mango. *Scientia Horticulturae*, **10:** 101-104.
- Rath, S. and Rajput, C.B.S. 1990a. Effects of growth retardants sprays on growth and fruiting of mango (*Mangifera indica* L.). *Orissa Journal of Agriultural Research,* **2** : 232-235.
- Rath, S. and Rajput, C.B.S. 1990b. Effect of beta-naphoxy acetic acid and gibberellic acid on chemical composition of mango fruits. *Orissa Journal of Agricultural Research*, **3**(2): 155-156.
- Rath, S., Das, G.C. and Singh, R.L. 1982. Manipulation of flowering in mango by forcing the dormant buds. *Bangladesh Horticulture*, **10**: 39-41.
- Ratnam, K.K. 1985. Studies on propagation of mango (*Mangifera indica* L.) by air-layering. M.Sc. (Ag.) Thesis, Andhra Pradesh Agricultural University, Hyderabad, Andhra Pradesh, India.
- Ravishankar, H. 1987. Studies on physiological and biochemical aspects into the causes and control of alternate bearing of mango (*Mangifera indica* L.). Ph.D. Thesis, University of Agricultural Sciences, Dharwad.
- Rawash, M.A. El-Nasr, N.A.; El-Nasy, H. and El-Bead, S. 1998. Effect of spraying of some chemical substances on flowering, fruitset, fruit drop, yield and fruit quality of 'Taimour' mango trees. *Egyptian Journal of Horticulture,* **25**(1) : 83-99.
- Rawash, M.A., El-Hammdy, A., El-Nabawy, S., Kalifa, A.S. and El – Marry, H. 1983. Regulation of flowering and fruiting in mango tree using some growth regulators. Annals of Agriculture Science, Cairo. 28 : 227-240.

- Reddy, K.M. and Singh, R.N. 1987a. Propagation of mango by cuttings. I. Effect of bottom heat, propagation duration and wounding treatment on rooting of hardwood cuttings of mango. *Journal of Research, Andhra Pradesh Agricultural University*, **15**(1): 24-30.
- Reddy, K.M. and Singh, R.N. 1987b. Propagation of mango by cuttings. II. Influence of month and age of the source plant on rooting of hardwood cuttings of mango under bottom heat. *Journal of Research, Andhra Pradesh Agricultural University*, **15**(2) : 129-135.
- Reddy, Y.N. and Majumder, P.K. 1978. Synergism of phenol and flavonoids with IBA in regeneration of mango (*Mangifera indica* L.) and guava (*Psidium guajava* L.) cuttings. *Vatika*, **1**(1) : 37-44.
- Reddy, Y.T.N. and Kurian, R.M. 2008. Cumulative and residual effects of paclobutrazol on growth, yield and fruit quality of 'Alphonso' mango. *Journal of Horticulture Science*, **3**(2): 119-122.
- Richard, A.E. and Lang, A. 1957. Effect of kinetin on protein content and survival of detached Xanthium leaves. *Science*, **125:** 650-656.
- Richards, F.J. and Colomom, R.G. 1952. Occurrence of putrescine in potassium deficient barley. *Nature*, 460.
- Riov, J. and Goren, R. 1979. Effect of ethylene on auxin transport and metabolism in mid rib sections in relation to leaf abscission of woody plants. *Plant Cell and Environment*, **2**: 83-89.
- Riov, J. and Goren, R. 1980. Effect of ethylene on auxin transport and metabolism in mid rib in relation to leaf abscission of woody plants. *Israel Journal of Botany*, **28**: 60.
- Riov, J. and Yang, Y.K. 1982. Auto-inhibition of ethylene production in citrus peel discs. Suppression of 1-amino cyclopropane-1-carboxlic acid synthesis. *Plant Physiology*, **69**: 682-690.
- Rizvi, A.R. 1989-90. Chemical control of plant growth. Studies on minimizing fruit drop in mango. Annual Report (1988-89) of Atonic energy. Agricultural Research Centre, Thandojam, pp. 189-193.
- Robellede – Martinez, A., Angel-Prez, A.L., del Moreno, J.R. 2008. Effects of paclobutrazol and KNO_3 over flowering and fruit quality in two cultivars of mango Manila. *Inter Ciencis,* **30**(7): 518-522.
- Roberts, D.R., Walker, M.A., Thompson, J.E. and Dumbroft, E.N. 1984. The effects of inhibitors of polyamine and ethylene biosynthesis on senescence ethylene production and polyamine levels in cut Carnation flowers. *Plant Cell Physiology,* **25**: 315-322.

- Rodov, V., Ben-Yehoshura, S., Albagh, Q. and Fang, D.O. 1995. Reducing chilling injury and decay of stored citrus fruits by hot water dips. *Post-harvest Biology and Technology,* 5: 119-127.

- Rodriquez, A.C. 1932. Influence of smoke and ethylene on the fruiting of pineapple (*Ananus sativus* Schulf). *Journal of the Agricultural University of Puerto Rico,* **16**: 5-18.

- Roemer, M.G. 2011. Pre-mature fruit drop in mango (*Mangifera indica* L.) in Northern Vietnam. Ph.D. Thesis, University of Hohenbeim, Vietnam.

- Roemer, M.G., Hegele, M., Wunscho, J.N. and Huang, P.T. 2011. Possible physiological mechanism of pre-mature fruit drop in mango (*Mangifera indica* L.) in Northern Vietnam. *Acta Horticulturae,* **903**: 999-1004.

- Roy, R.S., Chhonkar, V.S. and Prasad, S.N. 1963. Effect of plant growth regulators on fruit drop in mango. *Punjab Horticultural Journal,* **3**(2-4): 209-213.

- Rugini, E., Bongi, G. and Mencuccini, M. 1986. Effects of putresciene, Larginine and cobalt on fruitset, ethylene content and apparent partheneocarpy in Olives (Olea european). *Acta Horticulturae,* **179**: 365-368.

- Rukayah, A. 1990. The effects of uniconazole on flowering and fruiting of mango clone MA 125. Abstracts of contributed papers, XXIII International Horticulture Congress Firenze, Italy, Abstract No. 1255.

- Russo, L. Jr., Soastal, H.C. and Leopold, A.C. 1968. Chemical regulation of fruit ripening. *Bio Science,* **18**: 109.

- Saeed, W.T. 2005. Pomegranate cultivars as affected by paclobutrazol, salt, stress and change in fingerprints. Bulletin of the Faculty of Agriculture. Cairo University, Giza, Egypt. 56 : 586-616.

- Saidha, T., Rao, V.N.M. and Santha Krishnan, P. 1983. Internal leaf ethylene levels in relation to flowering in mango (*Mangifera indica* L.). *Indian Journal of Horticulture,* **40**: 139-145.

- Sakhidin, Rurwoko, B.S., de Silvia, J.A.T., Poerwant, R. and Soesanto, S. 2011. Indole-3-acetic acid, 1-Amino cyclopropone-1-carboxylic acid and carbohydrate in relation to fruit deop in mango tree. *Journal of Fruit and Ornamental Plant Research,* **19**(2): 41-49.

- Salazor – Garcia, S., Cossio – Vargas, L.F., Gonzalez – Durian, 1.J.L. and Lovatt, C.J. 2007. Folion applied GA_3 Advances Fruit maturity and allows off – season harvest of 'Hass' avacade. *Hort. Science,* **42**: 257-261.

- Salazor – Garcia, S., Lord, F.M. and Lovatt, C.J. 1998. Inflorencence and flower development of the 'Hass" avacade (*Persea annericana* Mill) during 'on' and 'off' crop year. *Journal of American Society of Horticultural Science,* **125**: 537-544.
- Saleem, B.A., Malik, A.U., Pervez, M.A. and Khan, A.S. 2008. Growth regulators application affect vegetative and reproductive behaviour of 'Blood Red' sweet orange. Pakistan Journal of Biology, **40**(5) : 2115-2125.
- Salomon, E. and Renveni, O. 1994. Effect of paclobutrazol treatment on the growth and firtst flowering of infact and autografted seedings of mango. *Scientia Horticulture,* **60**: 81-87.
- Salvi, B.R., Chavan, S.A., Dalvi, N.V., Deshpande, R.S. and Munj, A.V. 2011. Physiological stress necessary for flowering of Alphonso mango. Global Conference on Augumenting Production and Utilization of Mango. Biotic and Abiotic Stresses, Lucknow, 21-24, June, 2011. pp. 86 (Abstract).
- Sampson, V.R. 1981. Effect of 2-chloro, ethyl phosphomic acid and mango ripening. Anais da Escola Superior de Agriculture "*Luiz de Queiros*", **38**(1): 93-98.
- Sandhu, M.K. 1979. Effect of pre-treatment of stock plants of mango with cycocel, ethrel and morphactin on the rooting of cuttings and air-layers. *Scientia Horticulture,* **10**(4): 363-368.
- Sandhu, M.K. 1992. Pre-treatments of stock plants with different growth regulating chemicals to induce rooting of mango cuttings. *Acta Horticulturae,* **321:** 356-365.
- Sandhu, M.K. and Bose, S. 1980. Effect of ethylene on rooting of cuttings and air-layers of mango, guava and water apple. *Indian Journal of Horticulture,* **37**(4): 335-337.
- Sanyal, D.I. Biswas, B. and Mitra, S.K. 1996. Studies on irregular bearing of mango (*Mangifera indica* L.). I. Effect of chemicals. South Indian Horticulture, **44** (1-2): 1-4.
- Sarkar, B.C. and Rahim, M.A. 2012. Vegetative growth, harvesting time, yield and quality of mango (*Mangifera indica* L.) as influenced by soil drench application of paclobutrazol. *Bangladesh Journal of Agriculture Research,* **37**(2): 335-348.
- Sarkar, S. and Ghosh, B. 2004. Effect of growth regulators on fruit morphology, retention and yield of mango cv. Amrapali. *Indian Agriculturist,* **48**(3/4): 185-188.
- Sarkar, S.K., Gautham, B., Srihari, D. and Seethambaram, Y. 1998. Regulation of tree vigour in mango. *Indian Journal of Horticulture,* **55**(1): 37-41.

- Satyanarayana, B.N. 1982. Effect of growth regulators on veneer and wedge grafting of mango (*Mangifera indica* L.). M.Sc. (Ag.) Thesis, University of Agricultural Sciences, Bangalore, India.
- Satyendra Singh, N, Singh, C.P., Singh, A.K., Vandana, D. and Sharma, A. 2014. Response of the cultar on growth, flowering and yield behaviour of mango (*Mangifera indica* L.). cv. Dashehari. *Progressive Horticulture*, **46**(1): 61-64.
- Selveraj, S., Bauchot, A.D. and John, P. 2001. Internal browing in cold-stored pineapples is suppressed by a post-harvest application of 1-Methyl Cyclopropane. *Post-harvest Biology and Technology*, **23**: 167-170.
- Sembdner, G. and Porthier, B. 1993. The biochemistry and the physiological and molecular action of Jasmonates. Annual Review of Plant Physiology and Plant Molecular Biology. 44 : 569-589.
- Sen, P.K. and Bose, T.K. 1966. Physiological studies on regeneration of roots. VII. Relation of chemical composition to the regeneration of roots in air layers of mango and litehni. *Indian Agriculturist,* **10**(2): 75-86.
- Sen, P.K. and Choudhary, J.M. 1969. Studies on growth substances in shoots of two varieties of mango, Langva (Biennial) and Baramasi (Perpetual) in relation to growth and fruit bud formation. *Proceedings of Indian Science Congress Association,* Calcutta.
- Sen, P.K. and Maiti, S. C. 1965. Chemical control of growth and flowring. I. Effect of growth regulators on flowering in mango *In*: "Growth and Development in Plants" (Ed) Asana and Nanda, pp. 54-60.
- Sen, P.K., Bandopadhyay, M., Roy, S.S. and Basu, R.N. 1973. Use of ethrel in controlling non-uniform bearing of mango (*Mangifera indica* L.). *Indian Agriculturist*, **17**: 285-288.
- Senratna, T., Mackay, C., McKersie, B. and Fletcher, R. 1988. Uniconazole induced chilling tolerance in tomato and its relationship to antioxidant content. *Journal of Plant Physiology*, **133**: 56-61.
- Serek, W., Sisler, E.C. and Reid, M.S. 1994. Novel gaseous ethylene binding inhibition prevents ethylene effects in potted flowering plants. *Journal of the American Society of Horticultural Science*, **119**: 1230-1233.
- Sergent, E., Ferrari, D. and Leal, D.F. 1997. Effect of potassium nitrate and paclobutrazol on flowering induction and yield of mango (*Mangifera indica* L.). cv. Haden. *Acta Horticulturae,* **455**: 180-184.

- Sexton, R. and Roberts, J.A. 1982. Cell biology of abscission. *Annual Review of Plant Physiology*, **33**: 133-162.
- Shabini, A.E.A. 2010. Improving seed germination and seedling growth of some mango root stocks. *American – Eurasian Journal of Agriculture and Environmental Science,* **7**(5): 535-541.
- Shalini, P., Badge, T.R. and Bharati, B. 1999. Growth of mango (*Mangifera indica* L.) seedlings as influenced by stone treatment. *Journal of Soils and Crops,* **9**(2): 227-230.
- Sharma, M.K., Joolka, N.K. and Sharma, N. 2002. Effect of tricontanol and paclobutrazol on photosynthetic efficiency, carbohydrates, metabolism and leaf nutrient status of non-parell almond. *Progressive Horticulture,* **34**(1): 117-118.
- Sharma, T.R., Nair, P.K.R. and Nema, M.K. 1990. Influence of foliar sprays of urea, KNO_3 and NAA on chemical composition of mango cv. Langra. *Punjab Horticultural Journal,* **30**(1-4): 53-56.
- Shawky, I., Zidan, Z., El-Toni, A. and Dahsan, D.I. 1978. Effect of GA_3 sprays on time of blooming and flowering malformation in Taimour mango. *Egyptian Journal of Horticulture,* **5**: 123-152.
- Shinde, A., Jadhav, B.B., Shinde, K.A. and Shelke, R. 2010. Effect of paclobutrazol (Cultar, Boultar and Austar) on flowering and yield of Alphonso mango. *Annals of Plant Physiology,* **24**(2): 185-187.
- Shirol, A.M., Hanamasetti, S.I., Kanamadi, V.C., Thamaiah and Patil, S. 2005. Studies on pre-soaking method and season of grafting of sepota root stock Kirni. *Karnataka Journal of Agricultural Science,* **18**(1): 96-100.
- Singh, A.K. and Rajput, C.B.S. 1990. Effect of GA, BA and calcium on vegetative growth and flowering in mango (*Mangifera indica* L.). *Research Development Report,* **7**: 1-11.
- Singh, A.R. 1977. Effect of foliar sprays of nitrogen and growth regulators on the flowering and fruiting of mango (*Mangifera indica* L.). *Punjab Horticultural Journal,* **17**: 34-50.
- Singh, A.R. 1980. Effect of folior sprays of nitrogen and growth regulators on the physico-chemical composition of mango (*Mangifera indica* L.). *Plant Science* : 75-81.
- Singh, A.R. and Singh, J.R. 1972-74. Effect of foliar application of nitrogen and plant regulators on the growth characters of mango. *Horticulture Advance,* **9**: 15-20.

- Singh, B.K. 1988. Post-harvest studies of mango (*Mangifera indica* L.). Ph.D. Thesis, Rajendra Agricultural University, Patna, Bihar, India.
- Singh, C.P., Singh, L.S. and Singh, A.K. 1992. Certain physiological aspects of malformation in mango (*Mangifera indica* L.). *Acta Horticulturae*, **296**: 39-49.
- Singh, D.B. 2008. Effect of paclobutrazol on reduction in tree vigour and advancement of floweirng in young mango grafts. *The Horticultural Journal*, **21**(1): 18-20.
- Singh, I.P., Singh, B.P. and Singh, S.P. 1986. Effec of 2, 4-D and GA_3 on fruit drop and fruit quality of mango (*Mangifera indica* L.). c.v. Sumer Bahist Chausa. *Haryana Journal of Horticultural Sciences,* **15**(1): 29-32.
- Singh, K.K., Singh, S. and Chadha, K.L. 1959. Control of fruit drop in mango. *Current Science,* **28**: 346.
- Singh, L.B. 1961. Biennial bearing in mango. Effect of gibberllic acid and maleic hydrazide. *Horticulture Advance*, **5**: 96-106.
- Singh, L.B., Singh, S.M. and Nirvan, R.S. 1961. Studies on mango malformation : Review, symptoms, extent, intensity and cause. *Horticulture Advance*, **5**: 197-207.
- Singh, N. P., Melhi, C.S. and Sharma, R.C. 2005. Effect of plant bio-regulators (PBRs) on flowering, fruit yield and quality in mango cv. Dashehari. *Horticultural Journal*, **18**(1): 10-12.
- Singh, N.D. 1996. Studies on the effect of agrochemicals on vegetative growth, chemical composition and control of malformation in commercial mango variety Dashehari. *Advances in Horticulture and Forestry*, **5**: 1-9.
- Singh, N.P., Malhi, C.S. and Dhillon, W.S. 2004. Effect of plant bioregulators on the promotion of flowering in mango cv. Dusehri. *Journal of Research, Punjab Agricultural University*, **41**(3): 341-344.
- Singh, N.P., Srivastava, R.P., Singh, H. and Rajput, M.S. 1979. Comparison of different methods of propagation of mango all the year round. Research Reports on Mango. Mango Workers' Meeting, Panaji, Goa, 2-5th May. pp. 85-90.
- Singh, P.V. and Gupta, R. 2004. Effect of carbendazim, GA and MH on quality and shelf life of mango cv. Dashehari. Paper presented at 1st Indian Horticulture Congress held at New Delhi. 6-9 November, 2004, pp. 353 (Abstract).

- Singh, R. and Arora, K.S. 1965. Some factors affecting fruit drop in mango (*Mangifera indica* L.). *Indian Journal of Agricultural Science*, **35**: 196-205.
- Singh, R. and Singh, R.N. 1972. Lateral bud growth in *Mangifera indica* L. in relation to auxin and inhibitor content of shoots and fruits. *Acta Horticulturae*, **24**: 175-184.
- Singh, R. and Singh, R.N. 1974. Growth promoting and growth inhibiting substances in developing fruits of biennial and regular bearing varieties of mango (*Mangifera indica* L.). *Indian Journal of Horticulture*, **31**: 16-22.
- Singh, R., Singh, P., Pathak, N., Singh, V.K., and Dwivedi, V.N. 2007. Modulation of mango ripening by chemicals : physical and biochemical aspects. *Plant Growth Regulation*, **53**(2): 137-145.
- Singh, R.N. 1954. The sex-ratio and fruit set in three varieties of mango. *Science*, **119**: 389-390.
- Singh, R.N. 1971. Biennial bearing in fruit-trees. ICAR Technical Bulletin (Agriculture) No. 3. Indian Council of Agricultural Research, New Delhi.
- Singh, R.N. 1978. Mango. Indian Council of Agriculture Research, New Delhi.
- Singh, R.N., Majumder, P.K. and Sharma, D.K. 1965. Studies on the bearing behaviour of some south Indian varieties of mango (*Mangifera indica* L.) under north Indian conditions. *Tropical Agriculture*, **42:** 171-174.
- Singh, R.S. and Ram, S. 1983. Studies on the use of plant growth substances for fruit retention in mango cv. Dashehari. *Indian Journal of Horticulture*, **40**: 188-194.
- Singh, S. and Singh, A.K. 2006. Regulation of shoot growth and flowering in mango cv. Gulabkhas by paclobutrazol. *Annals of Agriculture Research*, **27**(1): 4-8.
- Singh, S. and Singh, B. 1963. Atternate bearing in mango II regulation of growth and bearing with some plant growth regulators. *Punjab Horticultural Journal*, **3**: 137-147.
- Singh, S. and Singh, J. 2003. Regulation of shoot growth, flowering and fruiting with cultar in mango cv. Dashehari. *Orissa Journal of Horticulture*, **31**(1): 40-43.
- Singh, S., Kumar, D., Brahmchari, V.S. and Singh, D.N. 1995. Effect of pre-harvest spray of GA_3 and ethrel on storage life of mango cv. Amrapali. *Orissa Journal of Horticulture*, **23**(1-2): 112-118.
- Singh, S., Singh, K.K. and Gill, A.S. 1961. Further studies on the control of fruit drop

in mango with plant regulators. Proper presented at 4th Horticultural Research Workers' Conference, Poona.

- Singh, S.M. 1955. Malformation diseases of mango (*Mangifera indica* Linn.). *Current Science*, **24**: 168-169.
- Singh, S.N. and Teotia, S.S. 1951. Effect of some hormones on the rooting of mango. *Science and Culture*, **17**: 203-210.
- Singh, U.R. 1954-56. Effect of 2,4,5-T and 2,4,5-TP on the fruit drop in mango (*Mangifera indica* L.) variety. Kerala Bhagalpur. Annual Report, Research Station, Saharanpur, U.P. pp. 65-68.
- Singh, U.R., Pandey, I.C., Upadhyay, N.P. and Tripathi, B.M. 1976. Effect of plant regulators on ripening and quality of mango (*Mangifera indica* L.). cv. Gaurjeet. *Progressive Horticulture,* **8**: 83-88.
- Singh, V.K. and Pathak, N. 2008. Response of 1-methyl cyclopropane on activities of pectin hydrolases and associated post-harvest physiological changes in different mango cultivars. *Indian Journal of Horticulture,* **65**(1): 20-24.
- Singh, V.K. and Saini, J.P. 2001. Effect of paclobutrazol on yield, photosynthetic efficiency and quality mango. In : K.K., Bora Karan Singh and Aravind Kumar (Ed.) Production and Developmental Physiology. Pointor Publications, Jaipur, India, Chapter, 38 : 254-264.
- Singh, V.K., Singh, A.K., Deepak, R. and Shukla, K.C. 2010. Effect of paclobutrazol on growth and yield of mango. *JNKVV Research Journal*, **44**(2): 163-165.
- Singh, V.P. and Singh, D. 1996. Control of fruit drop. *Former and Parliament*, **31**(4): 4.
- Singh, V.P., Malhi, C.S. and Dhillon, W.S. 2004. Effect of plant bioregulators on the promotion of flowering in mango cv. Dushehari. Journal of Research, *Punjab Agriculture University*, **41**(3): 341-344.
- Singh, Z. 1986a. Biochemical aspects and conrol of mango malformation. Ph.D. Thesis, Punjab Agricultural University, Ludhiana, India.
- Singh, Z. 1986b. Hormonal physiology of mango malformation (*Mangifera indica* L.). *Hort. Science,* **21**: 673.
- Singh, Z. 2000. Effect of (2RS, 3RS) paclobutrazol on tree vigour, flowering, fruit set and yield in mango. *Acta Horticulturae*, **509**: 459-462.

- Singh, Z. and Agrez, V. 2002. Fruit set, retention and yield of mango in relation to ethylene. *Acta Horticulturae,* **575**: 805-811.
- Singh, Z. and Dhillion, B.S. 1990a. *In vivo* role of indole-3-acetic acid, gibberellic acid, zeatin, abscisic acid and ethylene in floral malformation of *Mangifera indica* L. *Journal of Phytopathology,* **128**: 235-245.
- Singh, Z. and Dhillion, B.S. 1993. Mango malformation. *In : Advances in Horticulture,* **vol. 4.** Fruit crops. Part-4 (Eds) K.L. Chadha and O.P. Pareek. Malhotra Publishing House, New Delhi. 2047-2064.
- Singh, Z. and Dhillon, B.S. 1986a. Effect of naphthalene acitic acid, ethnel, di kagulac and hand deblossoming on flord malformation, flowering, yield and fruit quality of mango (*Mangifera indica* L.). *Acta Horticulturae,* **175**: 307-313.
- Singh, Z. and Dhillon, B.S. 1986b. Effect of plant regulators on floral malformation flowering, productivity and fruit quality of mango (*Mangifera indica* L.). *Acta Horticulturae,* **175**: 315-319.
- Singh, Z. and Dhillon, B.S. 1986c. Hormonal physiology of malformation of mango (*Mangifera indica* L.). *Hort. Science,* **21**: 673- .
- Singh, Z. and Dhillon, B.S. 1989a. Presence of malformin-like-substances in malformed floral tissue of mango. *Journal of Phytopathology,* **125**: 117-123.
- Singh, Z. and Dhillon, B.S. 1989b. Hormonal changes associated with vegetative malformation of mango (*Mangifera indica* L.). *Journal of Phytopathology,* **125**: 193-197.
- Singh, Z. and Dhillon, B.S. 1989c. Reletionship of endogenous and exogenous ethylene with floral malformation of mango (*Mangifera indica* L.). *Acta Horticulturae,* **231**: 367-370.
- Singh, Z. and Dhillon, B.S. 1990b. Floral malformation, yield and fruit quality of *Mangifera indica* L. in relation to ethylene. *Journal of Horticultural Science,* **65**: 215-220.
- Singh, Z. and Dhillon, B.S. 1992. Effect of paclobutrazol on floral information, yield and quality of mango (*Mangifera indica* L.). *Acta Horticulturae,* **296**: 51-53.
- Singh, Z. and Janes, J. 2000. Regulation of fruit set and retention in mango with exogenous application of polyamines and their biosynthesis inhibitors. *Acta Horticulturae,* **509**: 675-680.
- Singh, Z. and Malik, A.U. and Davenport, T.L. 2005. Fruit drop in mango. *Horticulture Review,* **31:** 111-153.

- Singh, Z. and Singh, L. 1995. Increased fruit set, retention in mango with exogenous application of polyamines. *Journal of Horticultural Science,* **70**(2): 271-277.
- Sinha, R.B., Mallik, P.C. and Chhonkar, V.S. 1960. An observation on metabolic effect of Indole Butyric acid in mango and lilchi cuttings. *Indian Journal of Horticulture,* **17**: 225-229.
- Sisler, E.L. and Serek, M. 1997. Inhibitors of ethylene responses in plants at receptor level : Recent development. *Physiologia Plantarium,* **100:** 577-582.
- Sisler, E.L., Serek, M. and Dupille, E. 1996. Comparision of 1-Methyl cyclopropene and 3, 3-dimethyl cyclo-propane as ethylene antagonists in plants. *Plant Growth Regulation,* **10**: 169-174.
- Sivakumaran, S. and Hall, M.A. 1978. Effects of age and water stress an endogenous levels of plant growth regulators in *Euphorbia lathyrus* L. *Journal of Experimental Biology,* **29**: 195-205.
- Spencer, J.L. and Kennard, W.C. 1955. Studies on mango (*Mangifera indica* L.) fruit set in Puerto rico. *Tropical Agriculture, Trinidad,* **32**: 323-330.
- Srihari, D. 1995. Physiological and Biochemical studies on fruit bud differentiation and management of alternate bearing problem in mango (*Mangifera indica* L.). Ph.D. Thesis, University of Agricultural Sciences, Dharwad, Karnataka, India.
- Srilatha, V. 2003. Studies on fruit retention, quality and nutrient removal in mango (*Mangifera indica* L.). M.Sc. (Ag.) Thesis, A.N.G.R. Agricultural University, Tirupati, Hyderabad, Andhra Pradesh, India.
- Srilatha, V. 2015. Canopy management intervention in mango (*Mangifera indica* L.). Ph.D. Thesis, University of Agricultural Sciences, Bangalore, India.
- Srilatha, V., Reddy, Y.T.N., Upreti, K.K. and Jagannath, S. 2015. Tree vigour and hormonal changes in response to pruning and paclobutrazol application in mango (*Mangifera indica* L.). cv. Dashshari. *Environment and Ecology,* **33**(4): 1413-1417.
- Srinivasa Reddy, N. and Haripriya, K. 2002. Extension of storage life of mango (cvs) Bangalora and Neelem. *South Indian Horticulture,* **50**(1-3): 7-8.
- Srivastava, M., Arun Kishore, Dahuja, A. and Sharma, R.R. 2010. Effect of paclobutrazol and salinity on ion leckage, proline content and activities of antioxidant enzymes in mango (*Mangifera indica* L.). *Scientia Horticulturae,* **125**: 785-788.

- Srivastava, R.P. 1960. Studies in the response of plant growth regulaors in air-layered shoots of mango. *Journal of Science and Research. Banaras Hindu University*, **11**: 1-3.
- Srivastava, R.P. 1962. Studies on the flowering season, sex dishibution, fruit development and the influence of Naphthalene acitic acid spray on the fruit drop of 'Dashehari' mango. Banaras Hindu University *Journal of Science Research*, **12**(1): 104-115.
- Staurt, N.W. and Marth, P.C. 1938. Composition and rooting of Ameican holly cuttings as affected by treatment with indole butyric acid. *Proceedings of American Society of Horticultural Science*, **37**: 1099-1104.
- Sten, R.A. and Gazit, S. 2000. Application of the polyamine putrescine increased yield of 'Mauritius' litchi (*Litchi sinensis* Sonn). *Journal of Horticultural Science and Biotechnology*, **75**: 612-614.
- Stowe, B.B. and Yamaki, T. 1957. The history and physiological action of gibberllin. *Annual Review of Plant Physiology*, **8**: 181-217.
- Subhadrabandu, S. 1986. Studies on plant growth regulator effects on tropical and subtropical tree fruits of Thailand. *Acta Horticulturae*, **175**: 291-297.
- Subramanyam, H. and Sebastian, K. 1970. Effect of succinic acid, 2, 2-dimethyl hydrazide on carotene development in alphonso mango by fumigation. *Tropical Science*, 11: 120-125.
- Subramanyam, H., Moorthy, N.V.N., Dalal, V.B. and Srivastava, H.C. 1962. Effect of a fungicidal wax coating with or without growth regulator on the storage behaviour of mangoes. *Food Science* (Mysore) **11**: 236-239.
- Sudhavani, V. and Shankar, C.R. 2002. Effect of pre-harvest sprays on the shelf life and quality of Baneshan mango fruits under cold storages. *South Indian Horticulture*, **50** (1-3): 173-177.
- Sudhavani, V., Shankar, R.C. and Rao, B.V. 2000. Effect of pre-harvest sprays on the storage behaviour of Baneshan mango. *Andhra Agricultural Journal*, **47** (1&2): 108-112.
- Sudhir Kumar. 1998. Effect of etiolation, variety and growth regulaors on mango air-layers. Paper presented at National symposium on mango production and export. 25-27, June 1998, held at Lucknow, India.
- Sulladmath, U.V. and Kologi, S.D. 1969. Synergistic influence of 3-IBA and NAA on the rooting of air layers of chicke var. Kalipetti. *South Indian Horticulture*, **17**: 9-17.

- Suresh, M.R., Ramakrishna, S. and Adiga, P.R. 1978. Regulation of argining de carboxylase and putracine level in *Cucumil saliva* cotyledons. *Phytochemistry,* **20**: 1477-1488.
- Suryanarayana, V. 1980. Amino acid changes in mango shoots as affected by growth retendants in relation to flowering. *Plant Biochemistry Journal*, **7**: 78-82.
- Suryanarayana, V. and Das, R.C. 1971. Effect of growth regulators on development and quality of fruits in mango (*Mangifera indica* L.). *South Indian Horticulture*, **19:** 29-35.
- Suryanarayana, V. and Rao, V.N.M. 1977. Ascorbic acid changes in shoots of mango cv. Mulgova an affected by growth retardants in relation to flowering. *Indian Journal of Plant Physiology,* **20:** 88-90.
- Suryanarayana, V. and Venkateswara Rao, K. 1984. Effect of growth regulators on air-layering of mango, lemon and ber. *Andhra Agricultural Journal*, **31**(2): 126-131.
- Talakurha, A.G. and Dutt, D.R. 1945. Indole Acetic acid on rooting in gootees of mango. *Current Science*, **9**: 77.
- Tandel, Y.N. and Patel, N.L. 2011. Effect of chemicals on growth, yield and economics of mango (*Mangifera indica* L.). *Karnataka Journal of Agricultural Science*, **24**(3): 362-365.
- Tandon, D.K. and Kalra, S.K. 1983. Studies on maturity of developing mango fruit cv. Dashehari. *Plant Physiology and Biochemistry*, **10**: 113-119.
- Taylor, J.E., and Whitelaw, C.A. 2001. Signals of Abscission. *New Phytologist,* **151**: 323-340.
- Teotia, S.S. 1962. Newer methods of propagation of mango and their economics. Fruit Nursery Practices in India. Directorate of Extension. Ministry of Food and Agriculture, New Delhi. Pp. 67-76.
- Teotia, S.S., Singh, R.N., Upadhyay, S.K. and Srivastava, V.S. 1967. Effect of growth regulators on fruit retention in mango (*Mangifera indica* L.). var. Langra and Dashehari. Paper presented at International Symposium on Sub-tropical and Tropical Horticulture, New Delhi, December, 1967: 225-229.
- Thimmapaiah and Singh, H. 1983. Natural parthenocorpy in Dashehari cv. of mango. *Indian Journal of Horticulture,* **40**(3-4): 195-198.
- Thomas, H. and Stoddart, Jr. 1980. Leaf senescence. *Annual Review of Plant Physiology,* **31**: 83-111.

- Thuck-thye, T. 1978. Ethylene and induction of flowering by potassium nitrate in mango (*Mangifera indica* L.). M.Sc. Thesis, University of Philippines, Los, Banos, The Philippines.
- Tiburcio, A.F., Campos, J.I., Figuoras, X. and Besford, R.T. 1993. Recent advances in the understanding of polyamines functions during plant development. *Plant Growth Regulation,* **12**: 331-340.
- Tiwari, J.P. 1970. Studies on the effect of time and plant regulators on success and anatomy of vener grofling in mango. M.Sc. (Ag.) Thesis, G.B. Pant University of Agriculture and Technology (U.P. Agricultural University), Pantnagar, India.
- Tomer, E. 1984. Inhibition of flowering in mango by gibberellic acid. *Scientia Horticulturae,* **24**: 299-303.
- Tongumpai, D., Hongsbhanich, N. and Voon, C.H. 1989. Cultar for flowering regulation of mango of Thailand. *Acta Horticulturae,* **239**: 375-378.
- Tongumpai, D., Jutamonee, K. and Subhadrabandhu, S. 1991. Effect of paclobutrazol on flowering of mango. cv. Khew Sawoey. *Acta Horticulturae,* **291**: 67-70.
- Tripathi, V.K. and Shukla, P.K. 2006. Effect of plant regulators on growth, yield and quality of strawberry cv. Chander. *Journal of Asian Horticulture,* **2**(4): 280.
- Tucker, G.A. 1993. Introduction. In : G.B. Seymour, J.E., Taylore and G.A. Tucker (Eds) Biochemistry of fruit ripening. Chapman and Hall, London, pp. 1-5.
- Turnbull, C.G.N., Anderson, K.L. and Winston, E.C. 1996. Influnce of gibberllin treatment on flowering and fruiting patterns in mango. *Australian Journal of Experimental Agriculture,* **36**: 603-611.
- Upadhyaya, A., Davis, T., Walser, R., Galbraith, A. and Sankhla, N. 1989. Uniconazole induced alleviation of low temperature damage in relation to antioxidant activity. *Hort. Science,* **24**: 955-957.
- Upreti, K.K., Murti, G.S.R. and Jayaran, H.L. 2011. Chemical control of fruitlet abscission in mango. Paper presented at global conference on angmenting production and utilization of mango : Biotic and Abiotic stresses. Held at Lucknow, June 21-24, 2011 (Atstract) pp. 3-32.
- Upreti, K.K., Prasad, S.R.S., Reddy, Y.T.N. and Rajeswara, A.N. 2014. Paclobutrazol induced changes in carbohydrates and some associated enzymes during floral initiation

in mango (*Mangifera indica* L.) cv. Totapuri. *Indian Journal of Plant Physiology (NS)*, **19**(4): 317-323.

- Upreti, K.K., Reddy, Y.T.N., Shiva Prasad, S.R., Bindu, G.V., Jayaram, R.L., and Rajan, S. 2013. Hormonal concept in response to paclobutrazol induced early flowering in mango. cv. Totapuri. *Scientia Horticulturae*, **150**: 414-418.
- Utsunnomiya, N., Sogawa, K. and Sasak, K. 1995. Sex expression of cv. Irwin mango flowers affected by temperature, soil water stress and plant bioregulators. *Acta Horticulturae*, **394:** 183-187.
- Vacillauime, C. 1991. Towards control of flowering in mango in Cameroon. Use of growth regulator – paclobutrazol. *Fruits*, **46**: 187-198.
- Valero, D., Mawtinez, Romero, D. and Serrano, M. 2002. The role of polyamines in the improvement of shelf life of fruit. *Trends in Food Science and Technology*, **13**: 228-234.
- Valmayor, R.T. 1972. The Philippine mango industry - its problems and progress. *Acta Horticulturae*, **24**: 19-23.
- Van Lely Veld, L.J. 1978. Peroxidase activity and isoenzyme in abscission and normal mango (*Mangifera indica* L.) fruits. *Z. Planzon Physiology*, **89:** 453-456.
- Van Lely Veld, L.J. and Nel, E. 1982. Ethylene connection and polyphenol oxidase activity in mango (*Mangifera indica* L.) fruit abscission. *Z. Planzon Physiology*, **107**: 179-182.
- Van Overbeck, J. 1952. Agricultural application of growth regulators and their physiological basis. *Annual Review of Plant Physiology*, **3**: 87-102.
- Van Overbeek, J. 1962. Endogenous regulators of fruit growth. pp. 37-39. In : S.G. Younkin (ed.). Plant Science Symposium on fruit set. Campbell Soup. Co., Cander, New Jersy, U.S.A.
- Van Staden, J. and Nicholson, R.D. 1989. Cytokonins and mango flower malformation. II. The cytokinin complement produced by *Fusarium moniliformae* and the ability of the fungus to incorporate (8^{-14c}) adenine into cytokinins. Physiological and Moleculr Plant Physiology. 35 : 423-431.
- Van Staden, J. Bailey, A.D. and Macrae, S. 1989. Cytokinin and mango flower malformation III. The metabolism of $(H)^{+}$ iso-petenylademine and (8-14) zeatin by *Fusarium moniliformae*. *Physiology and Molecular Plant Pathology*, **35**: 443-448.

- Varner, J.E. and Ho, D.T.H. 1976. Hormanes, p. 714-765. In: J. Bonner and J. Varner (eds). Plant Bio-chemistry. Elsevier, Ltd. Austardam.
- Veen, H. and Jacobs, W.B. 1969. Transport and metabolism of indole -3-acetic acid and coleus petiol segment of increasing age. *Plant Physiology*, **44**: 1157-1162.
- Vejendla, V., Maity, P.K. and Banik, B.C. 2008. Effect of chemicals and growth regulators on fruit retention, yield and quality of mango cv. Amrapali. *Journal of Crop and Weed*, **4**(2): 45-46.
- Velikova, V. Yordanov, I. and Edireva, A. 2000. Oxidative stress and some antioxidant systems in acid rain treated bean plants. Protective role of exogenous polyanines. *Plant Science*, **151**: 59-66.
- Venkata Rao, 2002. Studies on nursery and production techniques in polyembryonic root stock of mango. M.Sc. (Hort.) Thesis, University of Agricultural Sciences, Bangalore, India.
- Venkata Rao, and Reddy. Y.T.N. 2005a. Effect of asmopriming on germination seeding, growth and vigor of mango (*Mangitera indica* L.) stones. *Karnataka Journal of Horticulture*, **1**: 29-35.
- Venkata Rao, and Reddy. Y.T.N. 2005b. Performance of polyembraynoic mango root stocks under nursery conditions. *Indian Journal of Horticulture*, **62**(5): 298-299.
- Venkata Rao, Reddy, Y.T.N. and Srinivas, N. 2006a. Effect of osmopriming treatments on growth of mango (*Mangitera indica* L.) seedlings under nursery conditions. *Journal of Asian Horticulture*, **3**(1): 17-21.
- Venkata Rao, Reddy, Y.T.N., Srinivas, N. and Shiragur, M. 2006b. Growth of mango (*Mangitera indica* L.) seedlings as influenced by stone treatments. Journal *of Asian Horticulture*, **2**(3): 160-165.
- Venkataratnam, L. 1949. Hormone induced set and pathenocarpy in mango. *Current Science*, **18**: 409.
- Vidya, A., Swamy, G.S.K., Prakash, N.B., Gangadharappa, R.C., Jagadesh, R.C., Jagadesh, S.L. and Mukesh, L.C. 2015. Effect of chemicals, organic and gibberllic acid on germination, growth and graft take in mango (*Mangitera indica* L.). *Mysore Journal of Agricultural Science*, **49**(1): 13-18.
- Vijayalakshmi, D. and Srinivasan, P.S. 1998. Induction of flowering in 'off' year mango cv. Alphonso as influenced by chemicals and growth regulators. *Annals of Plant Physiology*, **12**(2): 93-97.

- Vijayalakshmi, D. and Srinivasan, P.S. 2000. Improving the quality of "off" year Alphonso mango through chemicals and growth regulators. *Orissa Journal of Horticulture*, **28**(1): 31-33.
- Vijayalakshmi, D. and Srinivasan, P.S. 2002. Impact of chemicals and growth regulators on induction of flowering in "off" year mango cv. Alphonso. *Orissa Journal of Horticulture*, **30**(2): 32-34.
- Voon, C.H., Pitakpaivan, C. and Tan, S.J. 1991. Mango cropping manipulation with cultar. *Acta Horticulturae*, **291:** 219-228.
- Vu, J.C.V. and Yelenosky, G. 1992. Growth and photosynthesis of sweet orange plant treted with paclobutrazol. *Journal of Plant Growth Regulation*, **11**: 85-89.
- Wagle, P.V. 1928. Studies in the shedding of mango flowers and fruits. *Memoris of Department of Agriculture. India (Bot)* **15**(8): 219-247.
- Waldon, R.A., Cordeiro, A. and Tiburcio, A.F. 1997. Polyamines, small moleules triggering pathways in plant growth and development. *Plant Physiology*, **113**: 1009-1013.
- Wang Bao Gang, Wang, Jianthul and others 2008. Reduced chilling injury in mango fruit by 2,4-dichlorphenoxy actic acid and the antioxidant response. *Post harvest Biology and Technology*, **48**(2): 172-181.
- Wang, B.C., Jiang, W.B., Lin, H.N., Lin, L. and Wang, J.H. 2006. Enhancing the post harvest qualities of mango fruit by vacuum infiltration treatment with 1-Methyl cyclopropane. *Journal of Horticultural Science and Biotechnology*, **81**(1): 163-167.
- Wang, C.Y. 1987. Use of ethylene biosynthesis inhibitors in Horticulture. In : manipultion of ethylene. Responses in Horticulture (Ed. M.S. Reid). pp. 187-194.
- Wang, L.H. and Lin, C.H. 1992. Effect of paclobutrazol on physiological and biochemical changes in the primary roots of pea. *Journal of Experimental Botany*, **43**: 1367-1372.
- Wang, S.Y., Sun, T., Ji, Z.L. and Faust, M. 1995. Effect of paclobutrazol on water stress induced ethylene biosynthesis and polyamine accumulation in apple seedling leaves. *Phytochemistry*, **24:** 2185-2190.
- Wangchai, Uthaibutra, J. and Iwohari, S. 2000. Endogenous polyamines in 'Nam Dok Mai' mangoes with different ripening stages and its retetion to chilling injury during storage. *Acta Horticulture*, **509**: 429-436.

- Wardlaw, C.W. and Leonard, E.R. 1936. The storage of west Indian mangoes. Imperial College of Tropical Agriculture, Trinidad. *Low Temperature Research Station Memorandum,* **No.3.** 47 pp.
- Wasty, E.H. 1995. Growth regulators and flowering. Academic Bookshop, Egypt. pp. 560-580.
- Wavhal, K.N. and Athale, P.W. 1988. Studies to prolong shelf life of mango fruits. *Acta Horticulturae,* **231**: 771-775.
- Weaver, R.J. 1972. Plant growth substances in Agriculture. S chand and company Ltd. Ramnagar, New Delhi. 110055.
- Werner, H. 1993. Influence of paclobutrazol on growth and leaf nutrient content of mango (cv. Blanco). *Acta Horticulture,* **341**: 225-231.
- Wester, P.J. 1920. The mango, *Philippine Bureau of Agriculture Bulletin,* **18:** 3-70.
- Williams, M.W. 1984. Use of bio-regulators to control vegetative growth of fruit trees and improve fruiting efficiency. *Acta Horticulturae,* **146**: 97-104.
- Winata, L. 1975. The effects of the chemicals, 2, 4-D and wax up on the quality of three varieties of mango in Java. M.Sc. Thesis, Boger Agricultural University, Bogor, Indonesia.
- Winks, C.W. 1988. Maroocby Horticultural Research Station Report. No.5, 1986-87. Department of Primary Industry, Queensland Government, Bisbane.
- Winston, E.C. 1992. Evaluation of puclobutrazol on growth, flowering and yield of mango cv. Kensington Pride. *Australian Journal of Experimental Agriculture,* **32**: 97-104.
- Winston, E.C. and Wright, R.M. 1984. Mango flower induction, ethephon, potassium nitrate and cincturing. Proceedings of the 1st Australian Mango Research Workshop, CSIRO, Mulbourne, pp. 202-210.
- Yadav, R.K., Rai, M., Yadav, D.S. and Asati, B.S. 2005. Use of paclobutrazol in horticultural crops - A review. *Agricultural Review,* **26**(2): 124-132.
- Yamashito, K. and Ogawa, S. 1998. Promotion of flowerbud differentiation and improvement of bisexual flower formation in 'Irwin' mango. *Japanese Journal of Tropical Agriculture,* **42**(3): 153-158.
- Yang, S.F. 1980. Regulation of ethylene biosynthesis. *Hort. Science,* **15**: 238-242.

- Yang, S.F. and Hoffman, N.E. 1984. Ethylene biosynthesis and its regulation in higher plants. *Annual Review of Plant Physiology and Plant Molecular Biology*, **35**: 155-189.
- Yeashmin, F., Ibrahim, M. and Morshed, M.H. 2011. Studies on the reduction of post harest losses and enhance the storage life of mango (*Mangitera indica* L.) cultivar of aswina and its quality characteristics as affected by plant hormone. *International Journal of Sustainable Agricultural Technology,* **7**(12): 13-18.
- Yelenosky, G., Vu., C.V.J. and Bansher, M.G. 1993. Paclobutrazol induced dwarfing of 'Valencia' orange trees. *Proceedings of the Florida State Horticultural Society,* **106**: 329-332.
- Yellesh Kumar, H.S. 2007. Enhancement of germination, growth and graft take in mango. M.Sc. (Hort.) Thesis, University of Agricultural Science, Dharwad, Karnataka, India.
- Yellesh Kumar, H.S., Swamy, G.S.K., Patil, C.P., Kanmadi, Y.C. and Kumar, P. 2007. Effect of pre-soaking treatments on germination, growth, vigour index and vigour of root stocks in mango. *Journal of Asian Horticulturae*, **3**(3): 157-161.
- Yeshitela, T.P., Robbertse, P.J. and Stassen, P.J.C. 2004. Paclobutrazol suppressed vegetative growth and improved yield as well as fruit quality of 'Tommy Atkunis' mango ((*Mangitera indica* L.) in Elihiopia. *New Zealand Journal of Crop and Horticulture Science,* **32**(3): 281-293.
- Yuvaraj, K.M. and Ughreja, P.P. 1998. Effect of post-harvest treatments on ripening changes and storage life of mango (*Mangitera indica* L.) fruits cv. Kesar. Proceedings of National Symposium on Mango, Lucknow. 25-27, June, 19983 P. 80.
- Zaharah, S.S., Zora Singh, Symons, G.M. and Reid, J.B. 2003. Modification of abscisic acid in triggering ethylene biosynthesis and softing during ripening in mango fruit. *Post-harvest Biology and Technology,* **75**: 37-44.
- Zhang, Z.Q., Li, X.P., Hong, H.J. and Ji, Z.L. 2000. Alleviation of chilling injury of mango fruit in cold storage by using exogynous putrescine. *Fujian Journal of Agricultural Science*, **15**(2): 32-36.
- Zimmerman, P.W. and Wilcoxon, F. 1935. Several growth substances, which cause initiation of roots and other responses in plants. *Contib. Bayee Thompsan Insti.*, **7**: 209-229.
- Zur, A. and Goren, R. 1977. Reducing pre-harvest drop of 'Temple orange' fruits by 2,4-D. Role of cellulase in the calyx abscission. *Scientia Horticulturae*, **7**(3): 237-248.

19

Plant Growth Regulators in Mulberry (*Morus alba* L.)

T.R. Ahlawat, Binayak Chakraborty, Suchismita Jena and Dharmishtha Patel

1. Introduction

Morus belongs to the family Moraceae and comprises of 10–16 species commonly referred to as mulberries. They are grown wild as well as cultivated commercially in many temperate and subtropical parts of the world. Due to increasing demand of silk and the limited arable land in the country, stress has been laid on the higher production and improvement of foliage quality to meet the growing demand of sericulture industry of the country. Mulberry (*Morus alba* L.) leaf is the main basic food plant of the silkworm (*Bombyx mori* L.), which converts leaf and protein into silk. The production of quality foliage can be increased by the increasing assimilation rate of the plant and directing its movements to the foliage through the application of the plant growth regulators (Geetha and Murugan, 2017). It has been estimated that nearly 60% of the production cost of silkworm cocoon is incurred by mulberry leaf production (Das and Krishnaswami, 1965). Thus, mulberry varieties with higher leaf yield are important to sustain profitability in sericulture (Vijayan, 2009).

Mulberry (*Morus alba* L.) fruit has a long history of use as an edible fruit and traditional medicine in many countries, especially in Asia. A great diversity in nutritive compounds such as fatty acids, amino acids, vitamins, minerals, and bioactive compounds, including anthocyanins, rutin, quercetin, chlorogenic acid and polysaccharides have been identified in mulberry depending on the cultivar and maturity stage. Further, the extracts and active components of mulberry fruit have demonstrated numerous biological activities, including antioxidant, neuroprotective, antiatherosclerosis, immunomodulative, antitumor, antihyperglycemic and hypolipidemic activities in *in vitro* and *in vivo* studies. They have attracted the attention of pharmaceutical companies and generated interest in researchers (Yuan and Zhao, 2017).

The usage of plant growth regulators in mulberry is somewhat different. Under favourable conditions, plant growth promoters increase the seed germination, sprouting of plants and enhance the leaf yield. While under unfavourable conditions, plant growth retardants are used to reduce the yield losses by reducing the evaporation and transpiration.

2. Role of PGRs in Seed Propagation

Mulberries can be propagated sexually through seeds. Saito and Morhashi (1981) showed that irradiation by red light enhances seed germination in mulberry by increased GA_3 synthesis. Whereas, Petkov (1994) reported that GA_3 at 100-500 ppm with IBA at 100 ppm + urea 1000 ppm gave best germination percentage (96%) on 21st day after treatment as compared to untreated and water soaked controls.

Koyuncu (2005) conducted an experiment to determine the effects of cold stratification (0, 20, 40, 60, 80 or 100 days), application of GA_3 (0, 250, 500, 1000 or 2000 mg L^{-1}) and the combination (GA_3 + stratification) on seed germination of black mulberry. Pre-treatment with GA_3 significantly enhanced the germination of non-stratified black mulberry seeds. Application of 1000 mg L^{-1} GA_3 proved more effective than any of the other concentrations of GA_3 applied. Seeds stratified for 100 days showed 88% germination. The combined treatment of 250 mg L^{-1} GA_3 and 100 days of stratification yielded 96% germination of seeds.

Ponnaiyan and Vezhavendan (2005) observed that when seeds of Indian mulberry (*Morinda citrifolia* L.) were nicked and then treated with GA_3 at 1000 ppm for a period of 24 hr, it resulted in the highest germination of 27% and 81% at the end of 3rd and 9th week after sowing respectively. Further, Singh and Rai (2005) showed that GA_3 at 800 ppm followed by 1000 and 1200 ppm induced significantly higher percentage of germination (91.06, 87.9 and 80.4% respectively) against water treatment. Soaking for 24 h was best in this respect. Every increase in concentration of GA_3 significantly decreased the time taken for germination (58.3 to 9.0 days), increased the height of seedlings (16.05 to 19.23 cm) and number of leaves per plant (9.24 to 13.10).

Song *et al.*, (2016) studied the effects of plant growth regulators on seed germination and seedling growth of mountain mulberry seeds at DMZ Vicinity, Korea. Different concentrations of PGRs including GA_3, kinetin, ABA and NAA were used for experimentation. Highest germination rate was observed with the application of kinetin. The entire growth characters were found better in the treatments of 2.15 mg kinetin and 10 mg GA_3. NAA treatment only showed better growth of seedling width compared to the control. Kinetin was the most effective PGRs for the germination and growth of mountain mulberry near the DMZ.

3. Role of PGRs in Cuttings

Propagation of mulberry through cuttings is still one of the most cost-effective methods for clonal regeneration. Hardwood cuttings of mulberry planted in August under mist chamber condition was found effective in increasing the success rate of the cuttings (Singh *et al.*, 2015). Auxin application has been found to enhance the histological features like formation of callus and tissue and differentiation of vascular tissue (Mitra and Bose, 1954). Adding a small percentage of certain phenoxy compounds to either IBA or NAA increased rooting and produced root systems better than those obtained when phenoxy compounds are used alone (Davis and Haissig, 1990).

IAA at 150 ppm recorded the highest rooting (100%), length of the longest root per plant (19.20 cm), average number of roots per plant (10.60), survival (74%), number of leaves per plant (14.30), fresh weight of shoot (10.02 g/plant) and weight of root (1.34 g/plant) in S36 cultivar. IAA at 200 ppm resulted in higher rooting (100%), length of the longest root per plant (21.16 cm), average number of roots (10.20), survival (86%), number of leaves per plant (13.50), fresh weight of shoot (9.94 g/plant) and fresh weight of root (1.66 g/plant) (Anon., 1977-78).

Suzuki (1987) studied the effects of ABA and ethephon on rooting of nodes of mulberry hardwood stem sections at Kyoto University of Industrial Arts and Textile Fibers, Japan. Abscisic acid and ethephon at optimum concentrations (0.1, 1 ppm) promoted rooting of preformed primordia of *Morus alba* L. cv. Minamisakari hardwood stem sections (cuttings). At these same optimum concentrations, ethephon stimulated the onset of new shoot elongation, rooting and defoliation, whereas ABA delayed it. At 10 ppm, ABA depressed new shoot elongation and defoliation greatly and at 50 ppm, inhibited bud break completely. In contrast to this, 10 ppm ethephon only slightly depressed new shoot elongation but markedly stimulated defoliation.

Black mulberry (*Morus nigra*) is difficult to propagate. Koyuncu and Senel (2003) studied rooting of black mulberry hardwood cuttings at University of Suleyman Demirel, Isparta, Turkey. The dormant cuttings were prepared from one-year-old shoots during January, February and March. Hardwood cuttings treated with 5 g L^{-1} IBA were rooted in perlite using bunches (BP) and rows (RP) as well as in hydroponics (HP). The best rooting percentage was obtained from bunch planting from January cuttings.

IBA was evaluated for rooting of four local cultivars of mulberry *Morus spp.* cvs. Beyrudi, Hatuni, Sami and Yabani in greenhouse experiments at Mustafa Kemal University, Antakya, Turkey. Cuttings of 20-25 cm in length were taken during the dormant period (January 15 in the first year and February 20 in the second year) from one-year-old shoots. Basal sections (1-2 cm) of the cutting were treated with various concentrations of IBA solution for 10 sec before planting. After 93-118 days of cutting, rooting percentage, root number per cutting, average root length were determined. The best rooting results were obtained from cuttings treated with 5000 ppm IBA, while, highest rooting was obtained with 5000 mg. L^{-1} IBA in cv. 'Beyrudi' (Polat, 2008).

Thakur *et al.* (2009) studied the effect of IBA in combination with co-factor, sucrose and fungicide on the rooting of branch cuttings of mulberry plants in early spring season. The IBA concentration ranged from 0.01% to 0.10% plus each containing 0.2% p-HBA, 2% sucrose, 5% captan. Treatment of cuttings with chemical formulation of 0.02% IBA + 0.2% p-HBA + 2% sucrose + 5% captan maximized rooting (96.05%), number of leaves per cutting (7.46), primary root number (12.56), primary root length (7.91 cm), fresh root weight (8.62 g) and dry root weight (5.64 g).

Singh *et al.* (2014) studied the effect of various concentrations of IBA and NAA on the rooting of stem cuttings of mulberry under mist house condition in Garhwal Hill region at Horticultural Research Centre, HNB Garhwal University, Garhwal, Uttrakhand. Stem cuttings of *Morus alba* were treated with 1000, 1500 and 2000 mg L^{-1} IBA and NAA solutions by quick dip method. These cuttings were rooted in a rooting media of 1:1 mixture of sandy soil and farm yard manure in plastic root trainers inside a mist house. Among all the treatments, number of sprouted cuttings, length of the roots per cutting, percentage of rooted cutting, lengths of longest sprouts of root were higher in IBA 2000 mg L^{-1}.

Husen *et al.* (2015) studied the effect of IBA on clonal propagation of mulberry stem cuttings. Soft stem cuttings were treated with various concentrations of IBA (1000, 2000 and 3000 mg L^{-1}) by quick dip method and cultured in mist chamber. 3000 mg L^{-1} IBA was the most effective IBA dose for evoking the highest rooting response. IBA increased rooting phenomenon and activated carbohydrate metabolism. Further, exogenous application of IBA induces changes in enzyme activities and in their cofactor contents, allowing for the establishment of a favourable hormone/energy balance required for root-primordium initiation and development.

Edizer *et al.* (2016) conducted an experiment on effects of growth regulators on propagation through hardwood cuttings of black mulberry at Gaziosmanpasa University, Turkey. Hardwood cuttings (15-20 cm length) were treated with 0, 4000, 6000, and 8000 ppm IBA and 3000 ppm IBA + 150 ppm NAA solutions by quick dip method. Cuttings were kept for about 3 months in perlite in greenhouse fogging with bottom heating. Rooting was not observed in control plants. The highest rooting ratio (66.7%), average number of roots per cutting (3 root/cutting), average root length (43.4 mm) and ratio of survived cutting (43.33%) were observed in 8000 ppm IBA application and they also found that the rooting percentage increased with increase in IBA doses.

Devana Pallavi *et al.* (2018) studied the effect of IBA and NAA on rooting and growth of mulberry cuttings at Indira Gandhi Krishi Vishwavidyalaya, Raipur, Chattisgarh. Different concentrations of IBA (1000, 1500, 2000 ppm) and NAA (500, 1000, 1500 ppm) were used for experimentation. Maximum number of roots and shoots per cutting, number of leaves per shoot, length of roots were recorded under IBA 2000 ppm followed by IBA 1500 ppm.

4. Role of PGRs in Budding

Kako (2012) investigated the effect of IBA and kinetin on budding success in mulberry at University of Duhok, Duhok, Iraq. Cuttings were made in early February (2 February) and treated with different IBA doses (0, 1000, 2000, 3000 and 4000 mg L^{-1}). The maximum rooting percentage (80.72 %) was observed in hardwood cuttings of 12-14 mm diameter subjected to 4000 ppm IBA. Whereas, the maximum length (180.50 cm) and diameter of transplants (14.28 mm) were observed in cuttings of 10-11 mm diameter treated with 2000 ppm IBA.

5. Role of PGRs in Micropropagation

Tissue culture techniques such as micropropagation provide a fast and dependable method for production of a large number of uniform plantlets in a short time and offer potential means not only for rapid mass multiplication of existing stocks but also for the conservation of important, elite and rare plants. Considerable work has been done in the last few years on *in vitro* propagation of *M. nigra* (Yadav *et al.*, 1990). The successful regeneration of plants *in vitro* has been achieved in several mulberry species by axillary shoot proliferation.

The use of ABA in tissue culture is limited because ABA suppresses the metabolism of callus (Ohnishi, 1986b). But it is useful for preservation of callus up to 20 weeks at 28°C. Successful root formation was found from calli in the presence of 1 μM each of IAA, IBA and IPA supplemented individually. Enomoto (1987) developed an efficient *in vitro* culturing method for germplasm preservation on a modified MS medium supplemented with 4.4 μM BA. This new culture method was most successful when the media were plant hormone free. Rooting of explants was also observed on the medium without supplemental regulators. Sharma and Thrope (1990) showed that MS medium supplemented with 2.5 μM BA was optimum for *in vitro* raised seedlings of *Morus alba*. Lakshmi Sita and Ravindran (1991) were able to produce gynogenic plants from ovary culture of *M. indica*. They were successfully raised on MS media supplemented with a combination of 4.4 μM BAP and 4.6 μM Kinetin. Four plants were raised from a single ovary within three weeks of culture. Islam *et al.* (1993) studied *in vitro* propagation as an aid for cloning *Morus laevigata* Wall. Single node explants were excised and cultured on MS medium containing cytokinins. The number of shoots per explant increased with an increase of cytokinins up to 10 μM and declined in 15 μM. Maximum number of shoots was achieved in 10 μM BA followed by 10 μM kinetin. Maximum shoot length was also recorded in 5 μM BA.

Machii and Yamanouchi (1993) observed female flower formation from immature leaves which were cultured on MS medium supplemented with 4.4 μM BA, 20 g L^{-1} fructose and 4.0 g/L^{-1} gelrite. The highest percentage of flower-bud formation was observed in genotypes Kanadasanso (6.1%) and Shin-ichinoso (2.4%). The ovaries of these flowers were swollen 70 days after culture initiation. The combination of GA_3 and BAP sometimes increased the frequency of bud break in mulberry. Pattnaik and Chand (1997) reported rapid clonal propagation through *in vitro* culture of apical shoot buds and nodal explants from mature leaves in MS medium when supplemented with 2.2 – 4.4 μM BAP, in an attempt to induce bud break and multiple shoot formation. Adding GA_3 along with BAP enhanced bud-break frequency.

Bhau and Wakhlu (2001) studied the effect of genotype, explant type and growth regulators on organogenesis in *Morus alba* at Jammu University, J & K, India. Plantlets of mulberry (*Morus alba* L. vars. Chinese White, and Kokuso-27) were produced from callus cultures. For callus induction, leaf, internodal segments, and petiole explants of Chinese White, Kokuso-27 and Ichinose varieties were grown on MS basal medium fortified with 2,4-D and 6-benzylaminopurine (BA). Best callogenesis was obtained on MS medium containing a combination of 1 mg/l 2, 4-D and 0.5 mg/l BA (95-100%). Calluses formed shoots on MS medium supplemented with 1 mg/l BA. Supplementation with 0.1 mg/l 2,3,5-tri iodobenzoic acid (TIBA) in this medium enhanced shooting response. Presence of TIBA in the medium also improved the long-term organogenic potential of the callus. Regenerated shoots produced roots on Murashige & Skoog (MS) medium containing either 0.5 mg/l IBA or NAA.

Anis *et al.* (2003) studied micropropagation of mulberry through *in vitro* of shoot tip and nodal explants at Aligarh Muslim University, Aligarh, India. A high frequency of sprouting (80%) from nodal and (70%) from shoot tip explants and shoot differentiation was observed in primary cultures of *Morus alba* L. on MS medium supplemented with BAP and Kn. *In vitro* proliferated shoots were multiplied rapidly by culture of shoot tips and nodal explants on MS with BAP (2 mg/l) and NAA (0.2 mg/l) as supplements. This combination proved best for multiple shoot formation. Multiplication was also achieved by culture of both the kinds of explants on MS fortified with BAP (2 mg/l) + NAA (0.2 mg/l) + aspargine (25 mg/l) + glutamine (1 mg/l). This medium facilitated the elongation of shoots and sprouting of axillary buds of *in vitro* grown shoots. About 80% rooting was obtained from shoots cultured on the MS supplemented with NAA (1.0 mg/l). Plants with well developed roots were transferred to soil with a survival frequency of 70%.

Bhau and Wakhlu (2003) studied rapid micropropagation of mulberry at University of Jammu, J & K, India. Multiple shoots were initiated from nodal and shoot tip explants collected from mature trees of *Morus alba* L. cultivars Chinese White, Kokuso-27, Ichinose and *M. multicaulis* Perr. cultivars Goshoerami and Rokokuyaso after 2 weeks of culture. Nodal explants were more responsive than shoot tip explants. They reported an increase in shoot multiplication with MS medium containing BAP at 1.5 mg dm^3 with sucrose. BAP was the most effective cytokinin for shoot induction and sucrose was the ideal carbon source examined for shoot multiplication. Nodal explants rooted on an auxin-supplemented medium and the acclimatized plants were successfully transplanted in the field.

Mohammad *et al.* (2003) conducted a trail wherein primary cultures of *Morus alba* L. on MS medium supplemented with BAP (2 mg/l) and NAA (0.2 mg/l) proved best for multiple shoot formation. Multiplication was also achieved by MS medium fortified with BAP (2 mg/l), NAA (0.2 mg/l), aspergine (25 mg/l) and glutamine (1.0 mg/l). This medium facilitated the elongation of shoots and sprouting of axillary buds *in vitro* grown shoots. About 80% rooting was obtained from shoot cultured on MS medium supplemented with NAA (1.0 mg/l). Plants with well-developed roots were transferred to soil with survival frequency of 70%.

A PGR combination of zeatin (2.3 μM) and 2,4-D (2.3 μM) resulted in the highest number (29%) of cell divisions. Whole plants were obtained after culture of microcalli on MS medium containing TDZ at 4.5 μM and IAA at 17.1 μM. The regenerated shoots were rooted on MS medium supplemented with 4.9 μM IBA. With low revival rate during acclimation regenerated plants were established in the green house. Mesophyll tissue of mulberry is a good source material for protoplast isolation and culture (Pavan *et al.*, 2005). Apical buds of mulberry (*Morus indica* L.) were cultured on LSBM fortified with 8.88 μM BAP in combination with 2 μM TIBA which was the most suitable medium for initiation (94% response) and multiplication of shoots (10.6) in variety S54 (Kavyashree, 2007).

Kamareddi (2008) studied the development of synthetic seed in mulberry cv. M-5 and evaluation under controlled conditions at College of Agriculture, Dharwad. In one of the experiments he studied the effect of BAP and 2, 4-D on sprouting of axillary shoot bud explants *in-vitro*. One cytokinin *i.e.*, BAP (0.5, 1.0, 2.0 mg/l) and one auxin *i.e.*, 2, 4-D (0.3, 0.5, 1.0 mg/l) were used as supplements to MS media. 2, 4-D @ 0.3 mg/l showed a high frequency of sprouting (82.5%), highest length of shoot (5.3 cm) and higher number of leaves (4.92). He also studied the effect of cytokinin (BAP) and gibberelic acid on induction of multiple shoots *in vitro*. Different concentrations of BAP (0.5, 1.0, 2.0, 4.0 mg/l) and one combined application of BAP with GA_3 (BAP 4.0 mg/l + GA_3 0.05 mg/l) were used as supplement to MS media. BAP @ 0.5 mg showed a high frequency of multiple shoot induction (78.75%) and higher number of multiple shoots induced per explants (3.30) and highest length of shoots (4.12 cm). Further, he studied the influence of different concentrations of auxins on *in vitro* rooting. Four auxins namely IAA (1.0 mg/l), NAA (1.0 mg/l), IBA (1.0 mg/l) and 2, 4-D (1.0 mg/l) were supplemented to the culture medium. MS medium supplemented with IBA (1.0 mg/l) showed highest per cent of rooting (87.50), higher number of roots (8.70) and highest length of primary root (3.67 cm).

Umate (2010) reported that a combination of zeatin and 2, 4-D induced highest percentage of cell divisions (29%) followed by zeatin and NAA (10%). Zaki *et al.* (2011) studied regeneration and production of multiple shoots from nodal segments with axillary buds of a field grown mature tree of *Morus nigra* L. at University of Kashmir Hazratbal Srinagar, Jammu and Kashmir. To induce axillary shoot proliferation, explants were inoculated on MS medium supplemented with various concentrations of cytokinins like BAP (0.2 and 0.5 mg/l) and Kn (5.0, 5.1 and 5.2 mg/l). It was observed that BAP (5.2 mg/l) was effective in inducing shoot regeneration and BAP (1.5 mg/l) for multiple shoots generation on MS medium. *In vitro* raised shoots were transferred on MS basal medium adjuvanted with various concentrations of auxins including IAA, IBA and NAA. As far as root induction is concerned

NAA (0.5mg/l) was found to be more effective than IAA and IBA. In order to develop a single step protocol for production of clonal plants, the nodal segments were subcultured on MS medium supplemented with various auxin-cytokinin combinations. Complete plantlet development was achieved on MS medium supplemented with BAP (1.0mg/l) + NAA (2.0mg/l) and the regenerated micro plantlets were successfully established in the soil under field conditions after a few days of indoor acclimatization.

Adding auxins such as IAA, 2, 4-D and NAA enhanced the development of callus and adventitious roots in mulberry. They also increased the protein and rutin content. Adding cytokinin such as BA and KN retarded callus and adventitious root development as well as the protein and rutin contents. The highest level of rooting was produced when adventitious roots were grown in a 34/66 ammonium/nitrate full-strength standard MS medium containing 5mg/l IAA. Roots of *M. alba* had the highest levels (242.2 µg/g fresh tissue) of rutin (Yew *et al.*, 2011).

6. Role of PGRs in Vegetative Growth

Human interest in mulberry cultivation originated with the growth of civilization and their fascination with quality fabric which ultimately led to silk. Mulberry leaves being the sole food for silkworm (*Bombyx mori*) bear great importance in sericulture as well as in forestry (Zaki *et al.*, 2011). Mulberry leaves are simple, alternate, stipulate, petiolate, entire or lobed. Numbers of lobes vary from 1 to 5.

IAA (Indole 3-Acetic Acid) at 1000 ppm, improved leaf lobation and induced femaleness (Minimizawa, 1974). Foliar application of 10 ppm GA_3 increased leaf yield and leaf attributes in different genotypes S54 (797.4 g/plant), S799 (537.4 g/plant), S36 (611.8 g/plant), K2 (625.9 g/plant) and local (458.9 g/plant) (Anon., 1977-78). Foliar spray of 100 ppm GA_3 in S-146 variety of *M. alba* recorded an increase of 29.75% in leaf area, 5.86% increase in leaf moisture content and 11.92% increase in foliage moisture retention capability over control under Doon valley conditions (Anon., 1977-78).

Ohnishi (1986a) reported that foliar sprays of gibberellins on mulberry significantly increased leaf emergence and shoot growth. Ethylene spray to axillary bud stimulates the onset of new shoot elongation at low concentration but continuous application of ethylene in solution form was found to depress shoot elongation (Suzuki and Kitano, 1989).

Suzuki (1986) studied the effect of ethrel, auxin and abscisic acid on leaf (petiole) abscission of new shoots of mulberry at Kyoto Kogei-Sen-i University, Kyoto, Japan. Application of 1000 ppm ethrel to debladed petioles of mulberry (*Morus alba* L. cv. Shin-ichinose) 3 days after deblading stimulated the abscission of the blades more significantly than when the application was made concurrently with deblading. Abscission of the debladed petioles of mulberry (*M. alba* L. cv. Ichinose) was depressed when a part of the petiole, 1cm in length, was allowed to remain and the cut surfaces were treated with a lanolin paste containing 500 ppm auxin (IAA). Incubation of the detached shoot tips of mulberry (*M. alba* L. cv. Ichinose) in 500 ppm ethrel accelerated the defoliation and abscission of the debladed petioles compared with the incubation in 500 ppm abscisic acid (ABA). Based on these results, they suggested that leaf senescence and abscission in mulberry are regulated by the basic auxin-ethylene control system, as demonstrated in other plants.

Suzuki *et al.* (1988) investigated lateral bud outgrowth on decapitated shoots of low-pruned mulberry at Kyoto Kogei-Sen-i University, Matsugasaki, Kyoto, Japan. NAA applied as a spray reduced the number of lateral buds sprouting from both defoliated, decapitated erect shoots and intact horizontally trained shoots in 10 yr old plants. The application of NAA at 500 ppm reduced the lateral growth of plants.

Fotadar *et al.* (1990) conducted an experiment at Regional Sericultural Research Station, Pampoor (Kashmir) during 1984-86 on mulberry varieties with the application of indole butyric acid at 100 ppm. Results indicated an increase in plant height over control. Maximum height of 144.86 cm was recorded by the application of 100 ppm indole butyric acid.

An experiment was conducted at Sericulture Research Station, Berhampore, West Bengal using three concentrations of five commercial growth regulators miraculan, paras, planofix, vipul and biozymes during 1991. Results indicated that plant height and leaf weight was significantly increased at 500 and 250 ppm concentration of planofix over control (Anon., 1991a) and the average number of branches were significantly more in biozyme while, number of leaves were more in planofix as compared to miraculan, paras and vipul treatments (Anon., 1991b).

Studies were conducted on the effect of four commercial growth regulators *viz.*, planofix, biozyme crop, vipul and miraculan along with control during 1992 and results revealed that leaf yield was increased by 29 per cent in planofix treated leaves over control. They further reported that hormone treated leaves had lower transpiration (Anon., 1992).

Das and Setua (1993) studied the effects of commercial growth regulators *viz*, planofix

(NAA based), biozyme crop (Kinetin+BA+Auxin precursors), miraculan and vipul (tricontanol) at Saktipur, West Bengal. Growth regulators were sprayed separately on mulberry variety Sjat Saktipur. The results revealed that there was significant increase in plant height in planofix treated leaves, as well as, the average leaf yield was maximum in planofix sprayed plants (8733 kg/ha) per season, whereas, more number of leaves per plant (351) were observed in biozyme crop spray followed by planofix (333).

The ratio of auxins and cytokinins influences the outgrowth of plants. High auxin:cytokinin ratio promotes activation of shoot branching whereas, low auxin:cytokinin ratio promotes activation of lateral roots. Auxin and cytokinin compound, a biologically derived plant growth regulator yielded 6904 kg/ha/yr over control resulting in a net profit of Rs.5425. The auxin and cytokinin formulation increased growth and yield of mulberry by 17%. They have also shown that leaf yield can be increased by 29% on spraying 250 ppm of NAA in cultivar S1 and also in case of local variety, which yielded a net profit of Rs.1870/- per hectare per crop (Das *et al.*, 1995).

Ingle (1995) studied the effect of foliar application of plant hormones on mulberry at Marathwada Agricultural University, Parbhani, India. Different concentrations of Gibberellic acid (25, 50 ppm), IBA (50, 100 ppm), Planofix (50, 100 ppm), IAA (50, 100 ppm) and Kinetin (25, 50 ppm) were used for experimentation. Foliar application of gibberellic acid (50 ppm) was found better in increasing the number of leaves per plant. Use of gibberellic acid acid 25 ppm, planofix 50 ppm and IAA 100 ppm was also effective in increasing the leaf yield of mulberry.

Das *et al.* (2002) found that, foliar spray of kinetin based plant growth regulator (Biozyme 0.05%) in mulberry plants before the onset of water logging, partially compensated the water logging effect and increased leaf yield by 30%. It also improved the chlorophyll content, sugar content and photosynthetic rate significantly.

Foliar spray of kinetin based PGRs (Biozyme) on mulberry plants before onset of water logging, showed that biozyme partially compensated the water logging effect and increased the leaf yield by 30% and improved the chlorophyll, sugar content and photosynthetic rate significantly (Dhiraj and Kumar, 2012).

Sashi Kanta (2016) studied the effect of foliar spray of B-995 on growth modification of mulberry leaf at Punjab Institute of Technology, Dinanagar, Punjab. Plants were sprayed with B-995 (N- dimethyl amino succinamic acid) at 500, 1000 and 1500 ppm. They reported that B-9 produced thicker and darker leaves with high nutritive value. The numbers of lateral branches were highest in 1500 ppm followed by 1000 and 500 ppm.

7. Role of PGRs in Flowering

Mulberry plants can be monoecious or dioecious. Inflorescence is catkin with pendent or drooping peduncle bearing unisexual flowers and is always auxiliary. Male catkins are usually longer than female catkins. Male flowers are loosely arranged and after shedding pollens, the inflorescence dries and falls off. Female inflorescence is usually short and the flowers are very compactly arranged. Mulberry is primarily a wind pollinated crop.

Jaiswal and Kumar (1980) found that 100 or 300 ppm of GA_3 can induce male inflorescences, whereas, Ethrel at 400 ppm can induce femaleness in mulberry according to Ogure *et al.* (1980). Kumar *et al.* (1990) have shown that GA_3 at 2.5 ppm in pollen germination medium containing 15% sucrose and 50 ppm boron gave highest pollen germination and pollen tube growth.

Dennis (2004) conducted an experiment on in vitro modification of sex expression in mulberry using ethrel and silver nitrate at St. Thomas College, Palai, Arunapuram, Kottayam, Kerala, India. Medium was supplemented with 5μM 6-benzyl amino purine (BA). Ethrel applied at different concentrations (500, 1000, 1500, 2000, 2500, 3000 μg/l) produced female, male and mixed inflorescences in male mulberry nodal cuttings. They reported that 2-chloroethylphosphonic acid (ethrel) at 2000 μg/l in MS medium supplemented with 5 μM BAP maximized the female inflorescence (13.6%). However, silver nitrate at 2500 μg/l in MS medium supplemented with 5 μM BAP maximizing the male inflorescence (22.4%). Bisexual flowers were also observed along with male and female flowers in silver nitrate treated plants.

Foliar spray of Indole Acetic Acid (IAA) and Gibberellic acid as plant growth regulators significantly improved leaf lobation and sex expression of Kajli and Mysore Local cultivars of mulberry (Govinda and Basavaiah, 2006).

The height of plants increased and the number of days increased with increasing concentration of GA_3. When applied at a concentration of 100, 200 and 400 ppm GA_3 decreased the production of female and increased the male and mixed flowers, respectively. Pollen viability was moderate and seed set percentage decreased with increasing concentrations of GA_3 (Geetha and Murugan, 2017).

References

- Anis, M.; Faisal, M. and Singh, S. K. 2003. Micropropagation of mulberry (*Morus alba* L.) through *in vitro* culture of shoot tip and nodal explants. *Pl. Tissue Cult.*, **13**(1) : 47-51.
- Anonymous. 1977-78. Annual Administrative Report. CSR&TI, Berhampore, West Bengal, India.
- Anonymous. 1991a. Effect of commercial growth regulators on plant height of mulberry. Annual Report, CBR & TI, Berhampor, pp. 68-69.
- Anonymous. 1991b. Effect of selected commercial growth regulators on morphological characters and leaf yield of mulberry. Annual Report, CSR & TI, Berhampor, pp. 69-70.
- Anonymous. 1992. Effect of growth regulators on leaf yield of mulberry. Annual Report, CSR & TI, Berhampor, pp. 62-63.
- Bhau, B. S. and Wakhlu, A. K. 2001. Effect of genotype, explant type and growth regulators on organogenesis in *Morus alba. Pl. Cell, Tissue Organ Cult.*, **66**(1): 25-29.
- Bhau, B. S. and Wakhlu, A. K. 2003. Rapid micropropagation of five cultivars of mulberry. *Biologia Plantarum*, **46**(3): 349-355.
- Das, B. C. and Krishnaswami, S. 1965. Some observations on Interspecific hybridization in mulberry. *Indian J. Sericulture*, **4**: 1-8.
- Das, C. and Setua, M. 1993. Effect of growth regulators on the morphological parameters of mulberry variety - S. Annual Report, CSR & TI, Berhampore, pp. 27-30.
- Das, C.; Sengupta, T.; Chatterjee, S.; Misra, A.K.; Sen, S.K. and Saratchandra, B. 2002. Reversal of water logging effect through foliar spray of commercial plant growth regulator in mulberry. *Sericologia,* **42**: 433-440.
- Das, C.; Sengupta, T.; Sen, S.K.; Quadri, S.M. and Sharatchandra, B. 1995. Growth regulators for mulberry. *Indian Silk* (November): 23- 24.
- Davis, T.D. and Haissing, B.E. 1990. Chemical control of adventitious root formation in cuttings. *Pl. Growth Regul. Soc. Am. Quart.,* **18**: 1-18
- Dennis, T.T. 2004. *In vitro* modification of sex expression in mulberry (*Morus alba*) by ethrel and silver nitrate. *Pl. Cell, Tissue Organ Cult.*, **77**: 277-281.
- Devana, Pallavi; Sharma, G.L. and Naik, E.K. 2018. Effect of IBA and NAA on rooting

and growth of mulberry cuttings. *Int. J. Current Microbiol. Applied Sci.*, **7**(11): 305-308.

- Dhiraj, K. and Kumar, R. V. 2012. Application of foliar nutrients to increase productivity in sericulture. *J. Ento.*, 9 (1): 1-12.

- Edizer, Y.; Gokcek, O. and Saracoglu, O. 2016. Effects of growth regulators application on propagation with hardwood cuttings of the black mulberry. *J. Agric. Faculty Gaziosmanpasa Uni.*, **33**(3): 92-96.

- Enomoto, S. 1987. Preservation of genetic resources of mulberry by means of tissue culture. JARQ, **21**: 205-210.

- Fotadar, R. K.; Ahsan, M. M.; Dhar, K. L. and Dhar, A. 1990. Effect of indole butyric acid on height of mulberry. *Sericologia,* **30**(3): 347-361.

- Geetha, T. and Murugan, N. 2017. Plant Growth Regulators in Mulberry. *Annual Res. Review in Biol.,* **13**(3): 1-11.

- Govinda, M. V. R. and Basavaiah 2006. Effect of growth regulators on leaf lobation and sex expression of Kajli and Mysore local cultivars of mulberry. **In**: *Proceedings of National Conference on New Strategies in Research and Development of Sericulture*, pp: 27.

- Husen, A.; Iqbal, M.; Siddiqui, S.N.; Sohrab, S.S. and Masresha, G. 2015. Effect of Indole-3-Butyric Acid on clonal propagation of mulberry (*Morus alba* L.) stem cuttings: Rooting and associated biochemical changes. *Proc. Natl. Acad. Sci.*, India, Sect. B *Biol. Sci.,* DOI 10.1007/s40011-015-0597-7.

- Ingle, S.P. 1995. *Studies on the effect of foliar application of plant hormones on mulberry and economic traits of Bombyx mori* L. M.Sc. thesis submitted to Marathwada Agricultural University, Parbhani, India.

- Islam, R.; Zaman, A.; Joarder, O. I. and Barman, A. C. 1993. *In vitro* propagation as an aid for cloning of *Morus laevigata* Wall. *Pl. Cell Tissue Organ Cult.,* **33**: 339-341.

- Jaiswal, V. S. and Kumar, A. 1980. Induction of male inflorescence on the female plants of *Morus nigra* L. by GA_3. *Indian J. Exptl. Biol.*, **18**: 911-913.

- Kako, S. M. 2012. The effect of auxin IBA and Kinetin in budding success percentage of mulberry (*Morus spp.*). *Int. J. Pure Appl. Sci. Technol.,* **13**(1): 46-52.

- Kamareddi, S. 2008. Development of synthetic seed in mulberry (*Morus indica* L.) cv.

M-5 and evaluation under controlled conditions. M.Sc. thesis submitted to College of Agriculture, Dharwad.

- Kavyashree, R. 2007. A repeatable protocol for *in vitro* micro-propagation of mulberry variety S54. *Indian Journal of Biotechnology*, **6**: 385-388.
- Koyuncu, F. 2005. Breaking seed dormancy in black mulberry (*Morus nigra* L.) by cold stratification and exogenous application of gibberellic acid. *Acta Biologica Cracoviensia Series Botanica,* **47**(2): 23-26.
- Koyuncu, F. and Senel, E. 2003. Rooting of black mulberry (*Morus nigra* L.) hardwood cuttings. *J. Fruit Ornamental Pl. Res.*, **11**: 53-57.
- Kumar, R.; Dandin, S.B.; Ravindran, S.; Jain, A.K. and Jolly, M.S. 1990. Studies on the reproductive biology of mulberry. 1. Sucrose, boron and growth hormone requirements for pollen germination and pollen tube growth. *Sericologia*, **30**: 479-484.
- Lakshmi Sita, G. and Ravindran, S. 1991. Gynigenic plants from ovary cultures of mulberry (*Morus indica*), **In:** *Horticulture New Techniques and Applications,* (Eds.) Prakash, J. and Pierik, K.L.M., Kluwer Academic Publishers, London, pp. 225-229.
- Machii, H. and Yamanouchi, H. 1993. In vitro flower bud formation from immature leaf cultures in mulberry. *J. Seric. Sci. Japan.,* **62**: 85-87.
- Minimizawa, K. 1974. Studies on the mechanism of localised sex differentiation in mulberry. VI. NAA treatment. *J. Seric. Sci. Japan*, **42**:1 46-149.
- Mitra, G. C. and Bose, N. 1954. Rooting and histological responses of detached leaves to B-Indolebutyric acid with special reference to *Boerhavia diffusa* Linn. *Phytomorphology*, **7**: 370.
- Mohammad, A., Mohammad, F. and Singh, S.K. 2003. Micropropagation of mulberry (*Morus alba* L.) through *in vitro* culture of shoot tip and nodal explants. *Pl. Tissue Cult.*, **13**(1): 47-51.
- Ogure, H.; Narasimhan, N. K. and Matsushima, M. 1980. Effects of ethrel and gibberellins on sex expression in mulberry. *J. Seric. Sci. Japan*, **49**: 355-361.
- Ohnishi, T. 1986a. Effect of gibberellic acid treatment on the growth of mulberry. *J. Seric. Sci. Japan,*, **55** (4): 293-295.
- Ohnishi, T. 1986b. Influence of gibberellins treatment on the growth of mulberry leaves under different conditions of soil water content. *J. Seric. Sci. Japan.*, **55**:289- 292.

- Pattnaik, S. K. and Chand, P. K. 1997. Rapid clonal propagation of three mulberries, *Morus cathayana* Hemsl, *M. lhou* Koiz and *M. serrata* Roxb. Through *in vitro* culture of apical shoot buds and nodal explants from mature trees. *Pl. Cell Rep.*, **16**: 503-508.
- Pavan, U.; Venugopal, R.K.; Kiranmayee, K.; Jaya Sree, T. and Sadanandam, A. 2005. Plant regeneration of mulberry (*Morus indica*) from mesophyll-derived protoplasts. *Pl. Cell, Tissue Organ Cult.*, **82**: 289-293.
- Petkov, Z. 1994. Effect of some growth regulators on seed germination in mulberry. *Sericologia.*, **34**: 477-482.
- Polat, A. A. 2008. Effect of Indole butyric Acid on rooting of mulberry cuttings. *Acta Horticulturae*, **774**: 351- 354.
- Ponnaiyan, P. and Vezhavendan, S. 2005. The effect of growth regulators and its combination with nicking on the germination of Indian Mulberry (*Morinda citrifolia* L.). *Int. J. Noni Res.*, **1**(1): 23-26.
- Saito, H. and Morohashi, Y. 1981. Actions of phytochrome acid and low temperature on mulberry seeds. *J. Seric. Sci. Japan*, **50**: 415-421.
- Sashi Kanta 2016. Effect of foliar spray B-995 on mulberry on growth. *Int. J. Innovative Res. Review*, **4**(2): 26-28.
- Sharma, K.K. and Thrope, T.A. 1990. *In vitro* propagation of mulberry (*Morus alba* L.) through nodal segments. *Scientia Hort.*, **42**: 307-320.
- Singh, D. R. and Rai, R. B. 2005. Influence of gibberllic acid on seed germination in Noni (*Morinda citrifolia* Linn.) of Andamans. *Int. J. Noni Res.*, **1**(1): 31-35.
- Singh, K. K.; Choudhary, T. and Kumar, A. 2014. Effect of various concentrations of IBA and NAA on the rooting of stem cuttings of mulberry (*Morus alba* L.) under mist house condition in Garhwal Hill region. *Indian J. Hill farming*, **27**(1): 74-77.
- Singh, K.K.; Jat, D.K. and Mehta, S.K. 2015. Rootability of hardwood cuttings of mulberry (*Morus alba* L.) influenced by planting time and growing conditions under valley condition of Garhwal Himalayas. *Pl. Archives*, **15**(2): 1031-1036.
- Song, M. J.; Kim, K. H. and Hur, Y. J. 2016. Effects of plant growth regulators on seed germination and seedling growth of mountain mulberry seeds (*Morus bombycis* Koidz). *J. Korea Soc. Environ. Restoration Technol.*, **19**(1): 101-109.

- Suzuki, T. 1986. Effects of ethrel, auxin and ABA on leaf (petiole) abscission of new shoots of mulberry. *J. Seric. Sci. Japan*, **55**: 309-313.
- Suzuki, T. 1987. Effects of ABA and ethephon on rooting of nodes of mulberry hard wood stem sections. *Nippon Nogeikagaku Kaishi*, **61**: 37-40.
- Suzuki, T. and Kitano, M. 1989. Lateral bud development and shoot growth in low pruned *Morus alba* as affected by stem orientation. *J. Seric. Sci. Japan*, **58**: 1-4.
- Suzuki, T.; Kitano, M. and Kohno, K. 1988. Lateral bud outgrowth on decapitated shoots of lowpruned mulberry (*Morus alba* L.). *Tree Physiol.*, **4**(1): 53-60.
- Thakur, I.K.; Chauhan, K.C. and Singh, C. 2009. Effect of different growth regulators formulations on the rooting potential of mulberry (*Morus alba* Linn.). *Indian Forester*. **135**(12): 1701-1706.
- Umate, P. 2010. Mulberry improvements via plastid transformation and tissue culture engineering. Department of Botany, Kakatiya University, Warangal, India. *P Umate. Plant Signal Behav.*, **5**(7): 785-787.
- Vijayan, K. 2009. Approaches for enhancing salt tolerance in mulberry (*Morus* L.) A review. *Pl. Omics J.*, **2**(1): 41-59.
- Yadav, U.; Malan, L. and Jaiswal, V. S. 1990. Micropropagation of *Morus nigra* from shoot tip and nodal explants of mature trees. *Scientia Hort.*, **44**: 61-67.
- Yew, L.; Dong, E.L.; Hak-Soo, L.; Seong, K.K.; Woo, S. L.; Soo, H.K. and Myoung, W.K. 2011. Influence of auxins, cytokinins and nitrogen on production of rutin from callus and adventitious roots of the white mulberry tree (*Morus alba* L.). *Pl. Cell Tissue Organ Cult.*, **105**: 9-19.
- Yuan, Q. and Zhao, L. 2017. The mulberry (*Morus alba* L.) fruit- A review of characteristic components and health benefits. *J. Agric. Food Chem.*, **65**(48): 10383-10394.
- Zaki, M.; Kaloo, Z. A. and Mohammad, S. S. 2011. Micropropagation of *Morus nigra* L. from nodal segments with axillary buds. *World J. Agril. Sci.*, **7**(4): 496-503.

20

Plant Growth Regulators in Papaya

R.K.Tarai, S.Pradhan, B.K.Sethy and A.K.Sahoo

1. Introduction

Papaya (*Carica papaya* L.) is known as a wonder fruit of the tropics. It is a native of Tropical America and belongs to family *Caricacea*. It gives one of the highest productions of fruits per hectare and an income next to banana as well as it is cheaper source of vitamins (vitamin A) and minerals. Papaya is the second rich in vitamin A after mango. Both ripe and raw fruits of papaya are used in the preparation of various products and preserves like tuti-fruity, petha, jam, jelly, marmalade, syrups, wines, nectar, pickles, toffee, dehydrated flakes and baby foods several types of medicines, dental paste and face cream are prepared from papain which are now available in the market. Papaya fruits are used for the treatment of piles, dyspepsia of spleen and liver, digestive disorders diphtheria and skin blemishes. One of the major limiting factors is dioecious nature of the plant, varying degrees of sexes resulting into a large number of unproductive male plants in the progeny, which leads to continuous slackening of the improvement programme. Various nutritional aspects and growth regulators have been recommended for growing papaya crop which differs with the environment, genotype and other factors. The use of plant growth regulators has assumed an integral part of modern fruit production to improve the quality and production of fruits, and it has resulted in outstanding achievements in a number of fruit crops with regard to improvements in yield and quality (Jain and Dashora, 2011). Because of its diverse effects, it is possible to use certain growth regulating chemicals at particular stages of fruit growth and development to exhibit maximum effects. Occasionally, they are needed to be supplemented exogenously for additional stimulus for plants such as papaya, which require quick responses for increased growth, fruit set and yield (Singh and Singh 2009).

2. Effect of PGRs in Propagation

Although several vegetative methods are possible for propagation, yet papaya is commercially grown by seed in all over India as well as well as abroad. Vegetative propagation offers a means of avoiding the difficulty arising from dioecious nature and standardizing the quality. To perpetuate the papaya plant, true to the type like other fruit crops, utilization of the vegetative multiplication is an alternative method. Budding and micro propagation are recent techniques for propagating papaya plants, which has a potential scope in this direction but papaya is still commercially grown by seeds. A wide variety of pre sowing seed treatment, which includes the plant growth regulators and growing media are used to enhance the seed germination and seedling growth in several fruit crops.

The viability of papaya seeds is poor. Germination percentage of seeds declined rapidly with aging. Unlike other horticultural crops, vegetative propagation of papaya on commercial scale is having an inherent limitation due to difficulty in obtaining sufficient number of lateral shoots. This is primarily due to the strong apical dominance that occurs as a result of the presence of auxin in the meristem zone, thereby limiting the development of lateral meristem (Taiz and Zeiger, 2004). To overcome apical dominance and to induce lateral shoot production, use of growth regulators like GA3 and BA was employed in crop like passion fruit (Couto, 1983), papaya (Norton and Norton, 1986) and cherry (Lang, 1996). Cytokinin is strongly linked with the onset of growth of lateral buds and its application stimulates cell division and growth in many species. After decapitation, auxin levels in lateral buds increase, level of ABA decreases and transport of nutrients and cytokinin from roots increases the lateral bud production (Taiz and Zeiger, 2004).

3. Effect of PGRs in Seed Germination

Proper seed germination and seedling growth are most important considerations in successful seedling production under nursery technique of papaya cultivation. The germination rate and seedling growth can be modified by pre-sowing seed treatment of papaya with growth regulators and micronutrients. *Carica* and *Vasconcellea* seeds are very similar in structure (Badillo, 2000) and both groups require treatment to promote germination. In *Carica papaya* the sarcotesta and inhibitors present in the fruit generally prevent germination, but the effect can be reduced by the application of gibberellic acid and/or potassium nitrate (Pérez *et al.*, 1980 and Yahiro &Oryoji, 1980). Palanisamy and Ramamoorthy (1987) extracted the fully ripened fruits of the papaya cultivar CO-5 and dried it in the shade for 3 days after removing the sarcotesta. To determine dormancy the seeds were tested for germination at 5 day intervals up to 35 days after extraction. In a further trial the seeds were dried for 3 days and treated in different solutions (100 and 500 ppm GA, 0.2% KNO_3, 200 ppm thiourea or 0.5% citric acid) to break dormancy. In the first trial, germination percentage increased with time from 10% on day 0 to 96% on day 35. In the second trial, the highest germination percentage, 98%, was obtained after soaking the seeds in 100 ppm GA for 12 hrs at room temperature. Germination in the non-treated control was 25%.

Sen *et al.*, (1990) treated the papaya seeds with 200 ppm GA_3 at 10^{-3} M sodium phosphate (dibasic), sodium chloride, ferulic acid, thiourea, GA_3 tannic acid, or water for 8 hr or 16 hr. Further treatments consisted of coating the seeds in one of two gum arabic pastes. Paste-I comprised of potassium dihydrogen phosphate, urea and ammonium nitrate each at 1 g/100 g seeds. Paste-II was similar to paste-I with the addition of 100 ppm GA_3 and 100 ppm Agromin (a micronutrient formulation). The seeds were dried and stored in polyethylene bags. Seeds were sown in seed pans in the nursery at 9 or 140 days after treatment. Untreated seeds were sown as a control. All treatments, except the paste treatments in the 140 day sowing, increased percentage of germination as compared with the control. In the 9-day sowing, percentage germination was highest with ferulic acid for 16 hr (95%) followed by chilling for 8 hr (93.2%) and paste-II (92%). In the 140 day sowing, germination percentage was highest with tannic acid for 8 hr (66.7%) followed by 200 ppm GA_3 for 8 hr (58.5%); no germination occurred with the paste treatments. For sowing, seedling height and leaf number were greatest with the 8 hr treatments with 200 ppm GA_3 or sodium phosphate. Zhao *et al.* (2004) revealed that the seed germination of papaya was promoted most with 50 mg GA_3 per litre, resulting in 83.7% germination. Anburani and Shakila (2010) found that the papaya seeds soaked in GA_3 200 ppm for 12 hr were reduced the time taken for 50% germination, increased root and shoot length and vigour index followed by seed treatment with thiourea 2000 ppm for 12 hr and the least values were recorded in the control. Ramteke *et al.* (2015) revealed that GA_3 at 200 ppm is most effective for better germination of papaya seeds as well as growth of papaya seedlings (height of the seedling, average leaf area, stem diameter, length of tap root, fresh and dry weight of seedlings as well as survival per cent of seedlings). The promising effect of GA_3 as pre-sowing treatment to the seeds replaced the dormancy mechanism of the seeds resulting in early germination (Khan, 1981). Gibberellic acid acts on the embryo and causes synthesis of hydrolyzing enzymes particularly amylase and protease and this hydrolyzed food is utilized for growth of embryo and thereby enhanced the germination.

4. Effect of PGRs in Vegetative Growth and Development

Guha and Chaturvedi (1972) observed that plant height was enhanced by GA_3 treatments; this effect was associated with greater internode lengths and became more marked as GA_3 concentration increased in papaya. The increase in plant height of papaya might be due to the fact that GA_3 promotes vegetative growth by active cell division and elongation and therefore the height must have increased. Another probable reason for the increase in plant height might be due to the osmotic uptake of water and nutrients under the influence of GA_3, which maintains a swelling force against the softening of cell wall or due to the stimulus caused by the GA_3 and therefore the plant height, have increased.

In an experiment Ram (2017) observed maximum plant height of 110.00, 158.22, 188.00, 217.00, 244.11 & 295.00 cm at 60, 90, 120, 150, 180 DAT & at maturity respectively with application of RDF + Gibberellic acid 150 ppm. However, maximum plant girth, plant spread (E-W & N-S) and petiole length (59.00 cm, 258.21 & 260.00 cm and 89.44 cm, respectively) were observed under the treatment of RDF + 2, 3, 5-Triiodobenzoic acid at 150 ppm, while the minimum (43.53 cm, 227.03 & 226.00 cm and 72.30 cm) was recorded under the RDF + Control (T_0).

Ghanta and Mitra (1998) studied the effect of different plant growth regulators i.e. GA_3 (25 ppm and 50 ppm), NAA (50, and 100 ppm), TIBA (25 ppm and 50 ppm), Ethrel (300 ppm and 500 ppm) and MH (200 ppm and 500 ppm) sprayed at 45 days after transplanting and observed that plant height and internodal length of papaya were increased by NAA, GA_3 and ethrel, while decreased by MH and TIBA. All chemicals increased the girth and number of leaves per plant. Flowering was hastened by 10 to 25 days by the chemicals. Morales *et al.* (1999) reported that GA_3 at 30 ppm concentration showed positive increase in height of plant, number of leaves, shoot girth and length of the internodes in papaya as compared to 10 and 20 ppm GA_3. They further advocated that growth of papaya was increased in the tune of shoot height, leaf area and shoot dry weight with the use of 30 ppm of GA_3.

Hazarika *et al.*, (2016) conducted a trial on effect of plant growth regulators on growth characters of papaya and concluded that the treatment of GA_3 at 200 ppm showed maximum value with respect to different growth parameters *viz.*, plant height, plant girth and plant spread. They recorded maximum plant height (123.16 cm), girth (28.73 cm), E-W and N-S spread (215.63 cm and 223.29 cm, respectively) with the application of 200 ppm GA_3, whereas the lowest was recorded in control. This might be due to GA_3 application which increased cell division and cell elongation, thereby increasing the vegetative growth by overcoming genetic dwarfism (Singh and Singh 2009). Deb *et al.* (2010) soaked the papaya seeds with GA_3 (at 100, 150 and 200 ppm), sodium thiosulphate (at 20, 25 and 30 ppm) and micronutrient mixtures (4 different combinations of borax and zinc sulphate). Among the 11 treatments, GA_3 at 150 ppm was found best in respect to seed germination (72.2%) followed by sodium thiosulphate at 20 ppm (68.1%). But maximum seedling growth (52.32 cm seedling height and 5.18 cm seedling girth) was observed under GA_3 at 200 ppm. Borax at 0.50% along with ZnSO4 at 0.25% also markedly improved the seedling growth.

Anjanawe (2011) conducted an experiment to study the effect of plant growth regulators and growth media on seed germination, growth and vigour of papaya (*Carica papaya* L.) seedling cv. Barwani Red. Maximum seed germination (65.4%) was found under the application of 100 ppm GA_3 along with FYM :soil: sand (1:1:1),while rest of parameters i.e. completion of germination (23.75), first appearance of plumule(12.69), appearance of first true leaves(18.25), plant height(17.41cm), stem diameter (0.441cm), number of leaves per plant(10.42), average leaf area (41.28 sq.cm), fresh weight of stem and leaves (13.45g), dry weight of stem and leaves (1.036g), number of primary roots per plant (8.83), number of lateral roots per primary root (76.32), length of the longest root (25.36cm), diameter of the thickest root (0.343cm),were found superior with application of 200 ppm GA_3. Sen and Ghunti (1976) observed in papaya cv "Ranchi" that the largest number of leaves (7.66) on seedlings from chilled dry seeds followed closely by seedlings from non-chilled seeds treated with NAA at 200 ppm (7.5). Babu *at al.* (2010) concluded that the papaya varieties viz., Coorg Honey Dew, Pant Papaya, Co-2 and Co-7 possess better seed germination, seedling growth and vigour and were highly suitable for commercial cultivation in north eastern region of India. Their germination and seedling health can be improved by treating the seeds with GA_3 100 ppm. Maximum plant height in GA might have occurred due to cell division and cell elongation, which in turn would have increased the internodal length.

Hazarika *et al.*, (2016) recorded maximum leaves per plant with the application of 200 ppm BA, which was significantly higher than other treatments. The maximum number of leaves with BA may be due to an increase in cell division caused by cytokinins and also due to the higher supply of assimilates mediated by BA application (Dwivedi *et al.*, 1999). Senthilkumar *et al.* (2012) conducted an experiment on the "effect of plant growth regulators in induction of lateral shoots for grafting in papaya" with eight months old field grown papaya trees of 9-1(D) and Co.2. Among the treatments, BA 250 ppm produced a higher number of 25.00 buds in 9-1(D) and 42.00 buds in Co.2 per plant at 20 days after spraying. In respect of number of graftable lateral production, plants sprayed with GA 125 ppm + BA 125 ppm registered the maximum number of laterals viz., 10.83 laterals in 9-1(D) and

21.50 laterals in CO-2 at 20 days and 7.50 laterals in 9-1(D) and 17.50 laterals in Co.2 at 35 days after spraying.

Ramteke *et al.*, (2015) obtained maximum number of leaves per plant with GA_3 at 100 ppm (4.22 and 8.20) followed by GA_3 at 200 ppm (4.14 and 7.80) whereas, control recorded the minimum (3.68 and 7.65) at 25 and 50 days of sowing respectively. Increase in number of leaves might be due to the reason that GA_3 helps in invigoration of physiological process of plant and stimulatory effect of chemicals to form new leaves at a faster rate. Similarly, maximum stem girth was recorded at 25 days after sowing when treated with GA_3 200 ppm (2.31 mm) while maximum stem girth at 50 days after sowing were also associated in GA_3 200 ppm (4.28 mm). The increase in diameter as a result of GA_3 application might be due to the fact that GA_3 increase somatic uptake of nutrients causing cell elongation and thus increasing height of the plant (Feucht and Watson, 1958).

Das *et al.*, (2018) stated that the effect of plant growth regulators, decapitation and their combination was found to be significant on some of the shoot parameters. More number of shoots after 15 and 45 days (660.00 and 80.16 respectively) were obtained from Detopping + 500 ppm BAP + 100 ppm GA spray and plants took minimum number of days (5.66) for 1st shoot initiation. In respect to height of 1st shoot initiation from ground level, injection of 500 ppm BAP + 500 ppm GA (T3) in to the base of stem followed by detopping registered maximum height (153.83 cm). Injection of 500 ppm BAP + 100 ppm GA induced lateral shoots with maximum length (18.49 cm).

5. Effect of PGRs on Flowering and Fruiting

The fruiting is regulated at different times of flower emergence and at different intensities to maintain the productivity of the papaya plants. Subhadrabandhu *et al.* (1997) found that GA_3 treated seedlings grew more vigorously than control plants. Over the 90 days of growing period, maximum growth rate was recorded in all treatments. Plants treated with NAA flowered about 6 days earlier than those in the control treatment, while plants treated with GA_3 flowered about 20 days earlier than those in the control in papaya. Kumar and Prasad (1997) reported that 50 ppm GA_3 was the most effective in increasing the number of female flowers (94.24) and percentage fruit set (53.06%) in papaya. Pusdekar and Pusdekar (2009) reported that Ethrel, Cycocel and MH induced early flowering in Papaya cv. CO-2. Treatment with all the plant growth regulators significantly increased female flowers, fruits per tree and fruit set percentage. NAA was found to be most effective amongst the plant growth regulator. All the treatments bored fruits at lesser height from ground level and required minimum days for fruit maturity from flowering as compared to control. MH was found to be the most effective plant growth regulator regarding the weight of the fruit, volume of fruit, cavity width and TSS of fruits, whereas CCC(1000 ppm) and CCC(1500 ppm) recorded pronounced effect regarding fruit thickness. CCC(500 ppm) recorded maximum ascorbic acid content in fruit. Acidity content in fruit was significantly lower in ethrel (0.12-0.15%) treatment. Highest yield of fruits was recorded in all the concentrations of MH treatments (600, 400 and 200 ppm) and lowest in control.

Syamal *et al.*, (2010) reported that gibberellic acid- GA_3 (100 and 150 ppm) increased the vegetative growth of papaya plant in terms of height and physical characters of fruit in terms of length, diameter and weight while 2, 3, 5 - triodo benzoic acid i.e. TIBA (100 and 150 ppm) had a greater effect on increasing the girth of stem, spread of plant, number of fruits per plant and yield per plant. Among the PGRs, TIBA (100 and 150 ppm) and ethrel (200 and 300 ppm) proved to be best particularly in the reduction of number of days taken for sex differentiation and promotion of femaleness, respectively while both TIBA and ethrel were found to be best for improving the quality of fruit. Therefore, it is worthwhile to manipulate these plant growth substances for getting desirable characters like vegetative growth, appropriate sex form and better quality fruits in papaya.

The effectiveness of plant growth regulators on tissue cultured papaya cv. Red Lady was observed by Hazarika *et al* (2016). Different concentration of growth regulators i.e BA (100, 150 and 200 ppm), GA_3 (100, 150 and 200 ppm), NAA(100,150 and 200 ppm), CCC (500, 750 and 1000 ppm) and etherel (200, 300 and 400 ppm) were sprayed at 30, 45 and 60 days after planting. They found that the growth regulators significantly increased the fruit set of papaya. Application of GA_3 (200 ppm) resulted in highest fruit set (76.46 %) and lowest fruit set was recorded in control (47.72 %).

Bhogave and Raut (2014) reported that the vegetative growth of papaya plant in terms of plant height, number of leaves, flowering characters like total number of female and hermaphrodite flowers per plant increased by foliar application of GA3 at 150 and 200 ppm. The foliar application of NAA found to be effective in respect of increasing plant height, total number of female and hermaphrodite flowers per plant next to GA_3 treatments. Reduced plant height, increasing the girth of stem and spread of plant had greatly influenced by 2, 3, 5 - Triiodo benzoic acid - TIB A (150 and 200 ppm). Among the Plant growth regulators used, TIB A (150 and 200 ppm) proved to be best particularly in the reduction of number of days require to first flowering.

As far as the days to first flowering and days to first fruiting is concerned the treatment RDF + 2, 3, 5-Triiodobenzoic acid 150 ppm treated plants came early flowering (64.08) and fruiting (69.08) as compared to other treatments, while the maximum days to first flowering (78.08) and fruiting (84.05) were observed under the treatment RDF + Control (Ram, 2017).

6. Effect of PGRs in Sex-Expression

Papaya being polygamous grow in three sexes. The female plants produce pistillate flowers that eventually develop into fruits while male plants grow staminate flowers that only develop pollens. On the other hand, hermaphrodite plants produce bisexual flowers that may develop into either fruits or just pollens. This polygamous characteristic has been considered the stumbling block of papaya production. Sex expression in plants has been reported to be influenced by auxins. Exogenous application of plant growth regulators like auxin can alter plant sex ratio of flowers and play a significant role in sex expression and the eventual yield of plants (Chaudhari, 1994). Papaya showed no change in sex expression with sprays of GA_3 or ethephon, applied 21 days after germination (Subhadrabandhu, 1986).

According to Han and Murray (2014), the vegetative forms of male, female and hermaphrodite papaya plants are phenotypically identical. However, the flower and inflorescence morphology of each sex type is unique. Gynodioecious varieties like Sun Up, Sun Up Diminutive mutant, and dioecious AU9 were used to test the response of papaya to gibberellic acid (GA_3). Exogenous applications of GA_3 on female and hermaphrodite flowers of papaya did not yield any sex reversal phenotype but caused a significant increase in peduncle elongation and inflorescence branch number in all treated plants. An increase in flower number was seen in females but not hermaphrodites or males. There was an increase in plant height for all treated plants except Sun Up Diminutive mutant, suggesting that the mechanism causing the dwarf phenotype is independent of gibberellins. Gibberellin metabolism genes were identified in the papaya genome, none of which mapped to the sex-determining region of either the male or hermaphrodite specific region of papaya chromosome. They concluded that a transacting regulatory element that enhances gibberellin biosynthesis plays a role in the extreme length of the male papaya peduncle.

7. Effect of PGRs in Parthenocarpy

To improve the productivity of vegetable papaya in subtropical regions, fruit setting rate (parthenocarpic ability) and fruit productivity between sex types (females and hermaphrodites) and among cultivars and effect of gibberellins (GAs) on fruit swelling, was studied by Rimberia *et al.* (2006). In both sex types, the number of fruits per tree correlated more closely with fruit yield than with individual fruit weight. Females produced higher number of fruits per tree, thus attaining a higher fruit yield than hermaphrodites. A variation in parthenocarpic ability was observed among cultivars, and this ability was higher in female plants than in hermaphrodites. These results suggest that it is possible to grow female cultivars with high parthenocarpic ability. However, parthenocarpic fruits were significantly smaller than those produced by pollination. GA treatment was found to be effective for promoting fruit swelling under greenhouse conditions. In the greenhouse production of papaya, GA treatment was more efficient than hand pollination. Based on these results, they suggested that in subtropical regions, efficient production of papaya fruit for use as a vegetable may be realized by selection and cultivation of female cultivars with high parthenocarpic ability and promotion of fruit swelling by GA treatment.

8. Effect of PGRs on Yield

Shanmugavelu *et al.*, (1973) advocated that 200 ppm of GA_3 and 50 ppm of TIBA increased fruit size and enhanced the fruit yield in papaya. Further, they reported that GA_3 at 50 ppm recorded maximum weight, size and volume of papaya fruit. Hore *et al.* (1989) recorded that the largest size fruits with MH at 250 ppm, while the maximum fruit weight (1.69 kg) and yield (68.71 kg/plant) was registered under application of GA_3 at 100 ppm in papaya.

Bhogave *et al.*, (2014) studied on foliar application of gibberellic acid (150 and 200 ppm) and 2, 3, 5-Triiodobenzoic acid (TIBA) (150 and 200 ppm) on papaya and resulted maximum fruit length, number of fruits per plant, flesh thickness associated with low fruit cavity, less number of seeds per fruit and yield under GA_3 at 150 ppm concentration. Fruit weight was significantly influenced by the spraying of 150 and 200 ppm of 2, 3, 5-Triiodobenzoic acid (TIBA). Hazarika *et al.* (2016) recorded highest vegetative growth, yield attributing characters and yield with GA_3 at 200 ppm. It also recorded maximum value with respect to different growth parameters, viz. plant height, girth and E-W and N-S spread. Treatment with 200 ppm GA_3 also showed superiority in different yield attributing characteristics, such as fruit set percentage, number of fruits per plant, fruit length, fruit diameter, fruit circumference, fruit weight, and fruit volume. The increase in yield with gibberellic acid might be due to an increase in flower number, better fruit setting percentage and the production of a higher number of fruits with maximum fruit weight in addition to better vegetative growth. In addition, GA may have affected the auxin metabolism, which may have indirectly aided in fruit enlargement and thus the production of fruits in higher number. Quality parameters, such as TSS, acidity, total, reducing and non-reducing sugars of fruits were recorded maximum with 400 ppm of Ethel.

Kumar and Prasad (1997) studied the effect of different plant growth regulators and concluded that among various treatments, 50 ppm GA_3 was the most effective in increasing the fruit size (20.65 cm × 17.57 cm), fruit weight (1.725 kg), volume (1850 cc), flesh thickness (3.70 cm) and fruit yield (540.25 q/ha) in papaya.

Ghanta and Mitra (1998) studied the effect of different plant growth regulators *viz.*, GA_3 (25 ppm and 50 ppm), NAA (50, and 100 ppm), TIBA (25 ppm and 50 ppm), Ethrel (300 ppm and 500 ppm) and MH (200 ppm and 500 ppm) sprayed at 45 days after transplanting and observed that all the plant growth regulators at different concentrations produced marked increase in number of fruit, fruit yield, fruit weight, fruit size, pulp and seed ratio. The highest fruit yield of 45.63 kg/plant and 114.07 t/ha were recorded under the treatment 25 ppm GA_3 as compared to control (28.2 kg/plant and 76.6 t/ha, respectively) in papaya.

Vishwakarma *et al.*, (2000) noticed that GA_3 at 50 ppm concentration showed significant effect on the fruit length, fruits size, fruit diameter, pulp: peel ratio, fruit weight and yield of papaya fruit as compared to 2,4-D (10, 25, 50 ppm) and control. Singh *et al.* (2011) conducted a field trial on effect of different plant growth regulators on papaya. Foliar spray of NAA at (100 ppm & 150 ppm) and GA_3 at 200 ppm were provided to the papaya plant and they reported that yield and yield attributing characters were found maximum under NAA at 100 ppm over control.

9. Effect of PGRs on Fruit Quality

Shanmugavelu *et al.*, (1973) concluded that total sugar, reducing sugar and non- reducing sugar, TSS and carotene content of fruits were influenced significantly at 50 ppm concentration of GA_3 as compared to control. Baskaran and Sathiamurty (2008) conducted a field experiment to investigate the influence of four growth retardants *viz.*,Alar, CCC, ethrel and paclobutrazol as foliar sprays on different morphological characters such as flowering, fruit set, yield and fruit quality on papaya cv. CO-2. Among the four growth retardants, Alar had profound influence on quality of fruits. The effects varied depending upon the concentrations of growth retardants. Alar improved fruit weight and volume, while paclobutrazol reduced it. Pedicel length is an important trait, was increased by paclobutrazol. The edible quality of fruit was increased by paclobutrazol except TSS and TSS : acid ratio.

Bhattacharyya and Rao (1981) concluded that application of CCC (chlormequat) at 1000-5000 ppm, ethrel (ethephon) at 100-300 ppm and TIBA at 50-200 ppm applied at flowering stage of papaya, significantly increased total sugar content of papaya fruit, however the highest total sugar (9.23) was recorded under the plants treated with TIBA at 100 ppm concentration.

Ram (2017) obtained increased physical parameters of papaya with respect to fruit length, fruit diameter, fruit volume, fruit weight and pulp weight as well as pulp: peel ratio were increased considerably with the application of treatment RDF + GA3 at 150 ppm, whereas a reduction in peel weight was observed under the treatment RDF + TIBA at 150 ppm as compared to water spray (control). The quality parameters of fruit in terms of total soluble solids, total sugar, reducing sugar and ascorbic acid contents were proved to be best showing a higher status with the treatment RDF + Gibberellic acid at 150 ppm, while acidity of fruit decreased by the application of same treatment.

Hazarika *et al.*, (2016) found maximum fruit length, (20.43 cm), diameter (14.43 cm), circumference (33.32 cm), fruit weight (1183.34 g), fruit volume (1 423.34 cc), pulp weight (995.31 g) and peel weight (93.56 g) when treated with GA_3 at 200 ppm. The exogenous application of GA_3 increased the cell size of the fruit by the proliferation of the sink demand, resulting in enhanced phloem unloading and carbon assimilate metabolism in the fruit and a greater supply of assimilates and photosynthates to the fruits (Ahmed *et al.*, 2012).

Hazarika *et al.*, (2016) also observed that application of 400 ppm ethrel resulted in the highest TSS (16.91 %), lowest titrable acidity (0.128 %) and highest total and reducing sugars (10.17 and 7.58%). The increase in TSS may be the result of a higher accumulation of metabolites and a quick conversion of starch into soluble sugars during the fruit development in response to growth regulators . Similarly, the reduction in titratable acidity with ethrel may be due to its action on the fast conversion of organic acids and starch into reducing and non-reducing sugars and their derivatives through higher respiration and carbon assimilation activity during rapid ripening process. Highest ascorbic acid content was found under different concentrations of gibberellic acid, and the lowest was in the control. The perceptive increase in ascorbic acid with gibberellic acid may be due to catalytic influence of gibberellic acid on its biosynthesis from its precursor glucose-6-phosphate or the inhibition of its conversion to dehydroascorbic acid by ascorbic acid oxidase or both. The softening of fruit with ethrel may be explained through its action on cell wall hydrolysis and changes in complex substances to simpler ones, which occurs during ripening under the control of ethylene (Jain and Dashora 2011).

10. Effect of PGRs on Post-Harvest life

$CaCl_2$ @ 4% showed its beneficial impact on shelf-life of papaya fruits upto 10.67 days without any loss in either physical or physico-chemical properties. Similarly, application of the growth regulator GA_3 @ 100 mg/l also increased shelf-life in papaya fruits upto 12 days, with the best physical and physico-chemical attributes (Ramesh *et al.* (2014). Postharvest dipping of papaya fruits either in GA_3 at 100ppm or $CaCl_2$ at 2% treatment preserves the fruits with extended shelf life upto 9 days without any decline in fruit quality (Rajkumar and Manivannan, 2007).

11. Effect of PGRs on Micropropagation

It is well known that conventional asexual propagation techniques, such as grafting and rooting of cutting (Soomark and Tai, 1975) have not resulted in efficient method for large scale production of papaya plants. Thus, the only alternative left to overcome the above limitations is vegetative propagation of papaya through tissue culture. A clonal propagation technique using mature female plants is highly desirable for commercial practice, especially in sub-tropics where some dioecious lines perform better than hermaphroditic ones. Singh *et al.* (2010) carried out an experiment on use of plant growth regulator on Selection-1 cultivar of mature female papaya (*Carica papaya* L.) by using shoot tip as an explant and media supplementation with the main aim to assess the effect of growth regulators (auxins, cytokinins and silver nitrate) on *in vitro* regeneration of female papaya. Twenty six media were used for shoot regeneration while for root regeneration four media were tested supplemented with different concentration of IBA. Based on the results of this study, for shoot proliferation, MS basal medium supplemented with BAP (2.0 mg l-1) + NAA (0.1 mgl-1) was the best followed by MS medium fortified with BAP (2.0 mgl-1) + NAA (0.5 mgl-1) while MS medium supplemented with IBA (2.0 mgl-1) gave best rooting response.

Anandan *et al.*, (2011) developed an efficient protocol for micro-propagation of papaya var. Co-7. Shoot bud induction occurred on culturing epicotyl segments in MS (Murashige and Skoog) medium, followed by transfer onto shoot multiplication, elongation and rooting media. Adventitious buds were induced on MS medium supplemented with varying concentration of thidiazuron (TDZ: 0.5-10 μM). The highest shoot regeneration frequency and maximum number of shoots per explant was occured on MS medium amended with 2.5 μM TDZ. The explants with initiated shoot buds produced multiple shoots in MS medium supplemented with 5.0 μM 6-benzylaminopurine (BAP) and 0.05 μM naphthaleneacetic acid (NAA). Maximum shoot elongation was observed at 1.5 μM gibberlic acid (GA3). Upto 60.25% of the regenerated (elongated) shoots were rooted on MS medium containing

2.5 μM indole-3-butyric acid (IBA) within 3 weeks after subculturing. The in vitro-raised plantlets were successfully hardened first under culture room conditions, then in greenhouse with 72% survival rate. The identified regeneration system could be efficiently used in various in vitro manipulation studies in papaya.

A study was conducted by Radzuan *et al.* (2018) to test the effect of 16 different concentrations and combinations of Benzylaminopurine (BAP) and Napthalene-Acetic Acid (NAA) on 3 different types of *Carica papaya* explants, namely leaf, stem and root. The 16 different treatments were from combinations of BAP and NAA with concentrations of 0.0 mg/L, 0.5 mg/L, 1.0 mg/L and 2.0 mg/L respectively. Experiment involves seed sterilization and sub culturing of explants in different medium concentrations and combinations of BAP and NAA. The experiment had been conducted for 8 weeks, and the observation was recorded once in 4 weeks. Throughout the experiment, the development of callus, regeneration of shoot and formation of root were observed. The optimum medium concentration for shoot regeneration was MS medium + BAP 2.0 mg/L from stem explant. For callus induction, stem explant was observed to be the most responsive explant when cultured on MS medium + BAP 0.5 mg/L and NAA is 0.5 mg/L. The function of auxin could stimulate the formation of adventitious roots. Da Silva (2013) claimed that the rhizogenesis was possible from young papaya leaves in response to several auxins, but not to phloroglucinol.

12. Effect of PGRs on Stress Management

The ability of papaya plants to withstand drought conditions is not related to the regulation of leaf area through leaf abscission, the arrest of growth or the decrease of net CO_2 assimilation but to an ability to increase ion (Na +,K + and Cl^-) content, which might contribute to osmotic adjustment (Mahouachi *et al.* 2006). The involvement of ABA in the regulation of the responses of some woody plants to water stress is now-a-days well established (Sauter *et al.* 2001). Under water stress conditions, ABA promotes the closure of stomata to minimize water loss through transpiration and mitigates stress damage through the induction of many stress-responsive genes and enzymes involved in the biosynthetic pathways of compatible osmolytes. These responses collectively contribute to increase plant tolerance to stress (Xiong and Zhu 2003). The involvement of Jasmonic acid in plant responses to water deficit has been suggested because this stress induces the expression of several genes that also respond to Jasmonic acid (Mason and Mullet, 1990).

ABA accumulates progressively in leaf and root tissues in a parallel manner to the increase of drought intensity, showing a maximum accumulation at the end of stress period.

Nevertheless, rehydration reverted ABA concentration to control levels in both leaf and root tissue as fast as in 24 hour. involvement of ABA as an accumulative hormonal signal that could be involved in the induction of several physiological responses in papaya under progressive water stress such as the reduction in gas exchange parameters and leaf abscission. The involvement of IAA in the regulation of water stress-induced responses of papaya seedlings has to be neglected, although its steady pattern can be associated to a balanced IAA metabolism to keep control levels even under stress conditions.(Mahouachi *et al.*, 2007).

Papaya is very much sensitive to salinity. Potassium nitrate (KNO3) and nitric oxide are substances that can induce saline stress tolerance in papaya. The effects of sildenafil citrate (SC) and KNO3 on the papaya seeds germination of under saline stress was studied by Zanotti *et al.*(2013). Papaya seeds of hybrids Calimosa and JS12 x Waimanalo were immersed in distilled water or KNO3 solutions and sildenafil citrate and followed germination in distilled water or NaCl solution at –0.3 MPa. Seeds of the hybrid JS12 x Waimanalo had higher percentages of germination and normal seedlings. The control of the two hybrids did not show saline stress tolerance during germination. Seeds of the hybrid JS12 x Waimanalo, when treated with KNO_3 or SC showed stress tolerance. The growth regulators tested were important in the induction of saline tolerance in papaya seeds. The highest tolerance of papaya seeds under saline stress may be associated to the increased activity of the antioxidant enzymes or increased seeds respiratory rate (Zheng *et al.*, 2009).

References

- Ahmed W, Tahir F M, Rajwana I A, Raza S A and Asad H U. 2012. Comparative evaluation of plant growth regulators for preventing premature fruit drop and improving fruit quality parameters in 'Dusehri' mango. *International Journal of Fruit Science* **12**: 372–89.

- Anandan,R.,Thirugnanakumar,S.,Sudhakar,D. and Balasubramanian, P. 2011. *In vitro* organogenesis and plantlet regeneration of (*Carica papaya* L.). *J. Agricultural Technology*,**7**(5): 1339-1348.

- Anburani, A. and Shakila, A. (2010).Influence of seed treatment on the enhancement of germination and seedling vigour of papaya. *Acta Horticulture,* **851:** 295-298.

- Anjanawe,S.(2011).Effect of Plant Growth Regulators and Growing Media on Seed Germination and Growth Vigour of Papaya (*Carica papaya* L.) Seedling cv. Barwani Red. *M.Sc (Ag.) Thesis submitted at the K.N.K. College of Horticulture, Department of Fruit Science, Rajmata Vijayaraje Scindia Krishi Vishwa Vidyalaya*, Gwalior, Madhya Pradesh.

- Babu, K. D.; Patel, R. K.; Singh, A.; Yadav, D. S.; Singh, A.; Yadav, D. S.; De, L.C. and Deka, B. C. 2010. Seed germination, seedling growth and vigour of papaya under North East Indian condition. *Acta Horticulture*, **851**: 299-306.
- Badillo, V. 2000. *Carica* L. vs. *Vasconcellea* St. Hill (Caricaceae) con la rehabilitación de este último. *Ernstia*, **10**(1): 74-79.
- Baskaran,A and S. Sathiamurthy,S. 2008. Effect of growth retardants on quality of papaya (*Carica papaya* Linn) cv. CO 2. *The Asian J. Hort. Journal.*, **3**(2): 313-314.
- Bhattacharya, R.K. and Rao, V.N.M. 1981. Fruit drop of CO-2 papaya as affected by growth regulators. *Current Res.*,**10** (3):49-54.
- Bhogave, A.F. and Raut, U.A. 2014. Studies on effect of plant growth regulators on vegetative growth and flowering of Papaya. *Ecology, Environment and Conservation* **20**:S387-S390.
- Bhogave, A.F., Raut, U.A. and Kharde, R.P. 2014. Effect of plant growth regulators on yield of papaya. *Bioinfolet.* **11** (1a): 119.121.
- Chaudhary, A. S., Singh, M. and Singh, G. N. 1994. Effect of plant growth regulators on the quality of loquat fruit cv. Thames Pride. *Recent Horti.*, **1** (1):44-48.
- Choudhury, B. And Patil, A.V. 1962. Effect of Plant Regulator Sprays on Sex, Fruit Set and Fruit Development in Cucumber (*Cucumis sativus* L.). *Proc. Acad. Bihar Agricultural Science* 9 and 10:28- 34.
- Couto F. 1983. Production of papaya and passion fruit. In: Report of Agriculture, Petrolina. **9**(102):15-18.
- Da Silva, J.A.T.2013. In vitro rhizogenesis in papaya (*Carica papaya* L.). *J. Plant Develop*, **20**: 51-55.
- Das,S., Wangchu, L., Raghavan,M. and Langstieh,L.B. 2018. Effect of plant growth regulators, detopping and their combination on lateral shoots initiation in papaya (*Carica papaya*) var. vinayak. *International Journal of Chemical Studies*, **6**(3):3085-3088.
- Dwivedi M P, Negi K S, Jindal K K and Rana H S. 1999. Influence of photoperiod and bio regulators on vegetative growth of strawberry under controlled conditions. *Advances in Horticulture and Forestry*, **7**: 29–34.

- Ghanta, P. K. and Mitra, S. K. 1998. Effect of some chemicals on the performance of papaya (*Carica papaya* L.). *Progressive Hort.*, **30** (1/2): 8-13.
- Guha, D. B. and Chaturvedi, K. N. 1972. Preliminary studies on the effect of gibberellic acid and CCC (2-chloroethyl trimethyl ammonium chloride) on the vegetative growth, flowering and fruit-set in transplanted seedlings of papaya. *Punjab Horti. J.*, **12** (2/3):111-113.
- Han, J. and Murray, J.E. 2014.The Effects of Gibberellic Acid on Sex Expression and Secondary Sexual Characteristics in Papaya. *Hort. Science,* **49**(3):378–383. 2014.
- Hazarika, T.K., Sangma, B.D., Mandal,D., Nautiyal, B.P. and Shukla, A.C. 2016. Effect of plant growth regulators on growth, yield and quality of tissue cultured papaya *(Carica papaya)* cv. Red Lady. *Indian J. Agricultural Sci.*, **86** (3): 404–408.
- Hore, J. K., Bandyopadhyay, A. and Sen, S. K. 1989. Effect of some chemicals on growth and yield of papaya. *Environment and Ecology*, **7** (3): 690- 693.
- Jain,M.C. and Dashora,L. K. 2011. Effect of plant growth regulators on physic-chemical characters and yield of guava cv. Sardar under high density planting system. *Indian Journal of Horticulture,* **68**: 259–61.
- Khan, A. A. 1981. Hormonal regulation of primary and secondary seed dormancy. *Israel J. Bot.*, **24**: 207-224.
- Kumar, D. and Prasad, J. 1997. Fruit maturity and quality of papaya as influenced by growth regulators and chemicals. *Environment and Ecology*, **15** (4):932-934.
- Lang GA. 1996. Early training of cherry trees- bud management. *Good Fruit grower*, 47(4):49.
- Mahouachi,J., Arbona, V. and Aurelio Gomez-Cadenas,A. 2007. Hormonal changes in papaya seedlings subjected to progressive water stress and re-watering. *Plant Growth Regul.*, **53**:43–51.
- Mahouachi,J., Socorro, A.R. and Talon, M. 2006 .Responses of papaya seedlings (*Carica papaya* L.) to water stress and rehydration: growth, photosynthesis and mineral nutrient imbalance. *Plant Soil.*, **281**:137–146.
- Mason,H.S. and Mullet, J.E. 1990. Expression of two soybean vegetative storage protein genes during development and in response to water deficit, wounding and jasmonate acid. *Plant Cell*, **12**:569–579.

- Morales, P. J. P., Santos, B. M. and Colon, W. 1999. Effect of ethanol and gibberellic acid combinations on papaya (*Carica papaya* L.) seedling growth. *Proceeding of the 35th Annual Meeting, Caribbean Food Crops Society,* Castries St. Lucia, 183-189.

- Norton ME, Norton CR. 1986. An alternative to *in vitro* axillary shoot propagation enhancement on whole plants. *J. Hortic. Sci.*, 1986; 61:423-427.

- Palanisamy, V.; Ramamoorthy, K. 1987. Seed germination studies in papaya. *Progressive Horticulture*, **19**(3-4): 253-255.

- Pérez, A., Reyes, M. & Cuevas, J. 1980. Germination of two papaya varieties: effect of seed aeration, K-treatment, removing of the sarcotesta, high temperature, soaking in distilled water and age of seeds. *Journal of Agriculture of the University of Puerto Rico,* **64** (1), 164-172.

- Pusdekar, G. G.; Pusdekar, M. G. 2009. Effect of plant growth regulators on flowering and fruit quality in papaya (*Carica papaya* L.) cv. CO-2. *New Agriculturist*, **20** (1/2): 107-110.

- Radzuan,L.,Hasbullah N.S., Patah,N.A., Idris, F.K.A. and Lassim,M.M. 2018. Effects of Plant Growth Regulators on Shoot Regeneration and Callus Induction of *Carica papaya. 12th International Conference on Advances in Agricultural, Chemical, Biological & Medical Sciences* (AACBMS-18) August 6-8, 2018 Pattaya (Thailand).

- Rajkumar,M. and Manivannan,K. 2007. Effect of chemicals and growth regulator on storage behavior of papaya (*Carica papaya* cv. CO-2). *Acta Horticulturae*,**740** :41.

- Ram, H. 2017. Effect of plant growth regulators on growth, yield and quality of papaya (*Carica papaya* L.) cv. Red Lady under Net House. *M.Sc. (Hort.) Thesis submitted to the Department of Fruit Science, College of Agriculture, Indira Gandhi Krishi Vishwavidyalaya,* Raipur, Chhattisgarh.

- Ramesh,D., Prasanna Kumar,B.,Rajasekhar,M. and Salomi Suneetha, D.R. 2014. Effect of chemicals and growth regulators on post-harvest shelf-life and quality in papaya (*Carica papaya* L.) cv. Red Lady, *J. Hort. Sci.*,**9**(1):66-73.

- Ramteke,V., Paithankar,D.H., Ningot, E.and Kurrey, V.K. 2015. Effect of GA3 and propagation media on germination, growth and vigour of papaya cv. Coorg Honey Dew. *The Bioscan*, **10** (3):1011-1016.

- Rimberia,F.K., Adaniya, S., Kawajiri.M., Urasaki, N., Kawano, S., Etoh,T. and Ishimine,Y. 2006. Effect of gibberellin treatment on parthenocarpic ability and promotion of fruit swelling in papaya. *Journal of Applied Horticulture,* **8**(1): 58-61.

- Sauter A, Davies,W.J.,and Hartung,W. 2001. The long distance abscisic acid signal in the droughted plant: the fate of the hormone on its way from root to shoot. *J. Exp. Bot.*, **52**:1991--1997.
- Scheldeman,X. 2002. Distribution and potential of cherimoya (*Annona cherimola* Mill.) and highland papayas (*Vasconcellea* spp.) in Ecuador. PhD Thesis. Faculty of Agricultural and applied Biological Sciences, University Ghent. Belgium. p 165.
- Sen, S. K.; Hore, J. K.and Bandyopadhyay, A. 1990. Pre-sowing seed treatment and its role on germination, seedling growth and longevity of papaya. *Orissa Journal of Agricultural Research,* **2** (3-4): 160-164.
- Sen, S.K. and Ghunti, P. 1976. Effect of pre-sowing seed treatments on the germination and seedling growth in papaya. *Orrisa J. Hort.*, **4**(1/2): 38-43.
- Senthilkumar, S., Kumar, N., Soorianthasundaram, K. and Jey Kumar, P. 2012. Effect of plant growth regulators in induction of lateral shoots in papaya. *Paper presented in the International Conference on Agricultural & Horticultural Sciences held on 14-15th September, 2012* at Hyderabad International Convention Centre, India.
- Shanmugavelu, K. G., Rao, V. N. M. and Srinivasan, C. 1973. Studies on the effect of certain plant regulators and boron on papaya (*Carica papaya* L.). *South Indian Hort.*, **21**(1): 19-26.
- Singh A and Singh J N. 2006. Studies on the influence of biofertilizers and bioregulators on flowering, yield and fruitquality of strawberry Cv. Sweet Charlie. *Annals of Agricultural Research,* **27**: 261–64.
- Singh A and Singh J N. 2009. Effect of biofertilizers and bioregulators on growth, yield and nutrient status of strawberry cv. Sweet Charlie. *Indian Journal of Horticulture,* **66**: 220–224.
- Singh, R., Yadav, R.C., Yadav, N.R., Yadav, A. and Sharma, P. 2010. Effect of plant growth regulators and silver nitrate on female papaya using shoot tip as explants. *Indian J. Hort.*, **67**(Special Issue), November 2010: 51-54.
- Singh, V. K., Singh, P. K. and Singh, A. K. 2011. Effect of different plant growth regulators on fruit yield of papaya (*Carica papaya* L.) cv. Pusa Dwarf. *New Agriculturist,* **22** (1): 31-33.
- Soomark, S. and Tai E. 1975. Plant regeneration from shoot tips and callus of papaya. *In Vitro,* **13**: 564-68.

- Subhadrabandhu, S. 1986. Studies of plant growth regulator effects on tropical and subtropical tree fruits of Thailand. *Acta Hortic.*, **175**: 291-298

- Subhadrabandhu, S., Thongplew, U. and Wasee, S. 1997. Effect of 1-naphthyl acetic acid (NAA) and gibberellic acid (GA3) on sex expression and growth of papaya (*Carica papaya* L.). *Kasetsart J., Natural Sci.*, **31**(1):72- 80.

- Syamal, M.M., Bedanga, B. and Pakkiyanathan, K. 2010. Influence of plant growth substances on vegetative growth, flowering, fruiting and fruit quality of papaya. *Indian J.Hort.*,**67** (2):173-176.

- Taiz L, and Zeiger E.2004. Plant physiology. Sinaur Associates, Inc., Publishers, Sunderland, pp. 517-539.

- Vishwakarma, S. A., Sachan, B. P. and Kumar, A. 2000. Effect of 2, 4-D and GA3 on sex expression in papaya (*Carica papaya* L.). *New Botanist*, **27** (1/4):151-158.

- Xiong, L. and Zhu,J.K.(2003).Regulation of abscisic acid biosynthesis. *Plant Physiol.*, **133**:29–36.

- Yahiro, masaki and Oryosi, yuichi 1980. Effect of gibberellins and cytokinine treatment on the promotion of germination in papaya *(carica papaya* L*)* seeds. *Memoris of the Faculty of Agriculture. Kagoshima univ.*, **16**, 45-51.

- Zanotti, R.F., Lopes, J.C., Motta, L.B., Freitas, A.R. and Mengarda, L.H.G. 2013. Tolerance induction to saline stress in papaya seeds treated with potassium nitrate and sildenafil citrate. *Ciencias Agarias*, **34** (6): 3669-3674.

- Zhao ChunXiang; Huang XiuQing; Chen YingYi and Yang Yong 2004. Effects of plant growth regulators on the vitality of papaya seed germination and the growth vigour of seedlings. *South China Fruits*, **33** (1): 36-37.

- Zheng, C. F.; Jiang, D., Liu, F.L., Dai,T.B., Liu,W., Jing,Q., Cao, W. X.2009. Exogenous nitric oxide improves seed germination in wheat against mitochondrial oxidative damage induced by high salinity. *Environmental and Experimental Botany, Elmsford*, **67** (1): 222-227.

21

Plant Growth Regulators in Passion Fruit

S. Pradhan, R.K. Tarai and I. Panigrahi

1. Introduction

Passiflora edulis, the passion fruit, is a tropical fruit but widely distributed throughout the tropics and subtropics of world, has several common names (such as sour passion fruit, yellow passion fruit, black passion fruit, and purple passion fruit), and presents a wide variability with the different rind colours of its fruits (Purseglove, 1968; Bernacci *et al.*, 2008). Currently, nearly 400 species of the genus *Passiflora* are known, most of which its origin has in the Tropical America, with 120 of them being native from Brazil (Bernacci *et al.*, 2003). Leaves of this species contain alkaloids with potential use for the pharmaceutical industry (Lotschert & Beese, 1989; Biasi *et al.*, 2000). The economic importance of passion fruit has increased significantly over the past three decades, reflecting worldwide production of approximately 805,000 tons (Anonymous, 2010). The increased economic importance of passion fruit relates to the numerous food products, recognized pharmacological properties (Tommonaro *et al.*, 2007), and remarkable ornamental value (Abreu *et al.*, 2009; Viana *et al.*, 2010).

The role of plant growth regulators in various physiological processes such as seed germination, flowering, fruiting, seed development, fruit ripening and yield *etc.* in different crop plants is well established (Kiran *et al.*, 2011). A comprehensive use of the growth regulators on passion fruit in different cultivation process is represented in this chapter.

2. Role of Plant Growth Regulators on Propagation

Passion fruit is a perennial plant that can be propagated by seeds, cuttings, air layering or grafting. Commercial passion fruit producers mostly use seedlings to establish plantations (Isutsa, 2004), because they do not spread the woodiness virus (Nakasone and Paull, 1998). However, seed propagation results in undesirable variability, inadequate and seasonal supply (Isutsa, 2004). Vegetative propagation through cuttings and grafting is occasionally practiced, but the risk of spreading the woodiness virus exists (Nakasone and Paull, 1998). In general, auxins are reported to favour rooting of stem cuttings of plants (Thimman and Behnke, 1950). It is well established that IBA plays an important role in rooting of cuttings. But the concentration of IBA needs to be standardised according to the fruit crop and maturity of cuttings.

Koch *et al.*, (2001) studied the effect of different concentrations of IBA (0, 250, 500 and 1000 mg L^{-1}) and ethanol on vegetative propagation of *Passiflora actinia*. The rooting (%) was high, an average of 90% in the experiment with IBA and 70 per cent in the experiment with ethanol. However, Meletti *et al.* (2007) studied the influence of season, the presence of leaves and IBA on the rooting of the sweet passion flower (*Passiflora alata* Curtis) cuttings. The treatment 3000 ppm IBA showed better results, independently of the type of cuttings. Cuttings without leaves showed higher percentage of rooting (80.90%), and the half leaved cuttings showed good results for root, dry weight and longest root length. Roncatto *et al.* (2008) reported that IBA (500 mg L^{-1}) was found better for rooting of cuttings of *Passiflora alata* and *P. nitida* under different seasons. But the rooting (76.7%) was better for *Passiflora edulis* f. flavicarpa in spring (October) without using IBA.

Herbaceous cuttings of plants of *Passiflora setacea*, *P. coccinea*, *P. amethystina*, *P. edulis*, *P. edulis* x *P. setacea* and *P. coccinea* x *P. setacea* were collected and treated with IBA (0, 250, 500, 750 and 1000 mg L^{-1}), and then planted in trays with moistened substrate and held under intermittent mist. The *P. amethystina* species presented better development than other species, showing rooting potential, with 88.67 per cent of cuttings with roots and sprouts. The increasing doses of IBA, increased the efficiency of rooting of cuttings of various species studied (Vaz *et al.*, 2009). Kumar *et al.* (2008) studied propagation of passion fruit *cv.* Kaveri by cuttings under Coorg conditions. Two separate methods for enhancing the rooting in passion fruit were tried. In the quick dip method (higher concentrations of IBA and NAA for 3-5 seconds) highest rooting and root characters were obtained in cuttings treated in October with NAA at 200 ppm while in 12 hours dip, highest rooting response in respect of rooting success and root characters were recorded with 80 ppm NAA treated in September.

Under Imphal (India) condition, Bemkaireima *et al.* (2012) reported that amongst the different IBA treatments (250, 500,750 and 1000 ppm), 750 ppm was found to be the best concentration for rooting of passion fruit var. Purple cuttings, under semi-controlled greenhouse conditions. Many parameters *viz.*, number of leaves per cutting (5.50), number of roots per cutting (30.36), root diameter (4.93 mm), leaf NPK content (4.13, 0.26 and 2.19% respectively) and leaf dry matter (1.31%) was found significantly superior in IBA (750 ppm) treated cuttings. Contrastingly, under Brazil condition, Carvalho Pires *et al.* (2010) noted IBA (500 mg/L) as the most effective for rooting and vegetative growth of cuttings of the species (*P. coccinea*, *P. quadrangularis*, *P. serrato digitata*, *P. edulis* f. *flavicarpa*), as well as the hybrid of sour passion fruit and the Australian purple passion fruit. This result was similar with the report of Roncatto *et al.*, (2008).

3. Role of Plant Growth Regulators on Seed Germination

The passion fruit can be propagated through both sexual and asexual methods but commercially, is propagated through seeds. Reports showed the lower germination of the seeds, which necessitates the pre-germination treatments to increase seed germination. Seeds are viable for short period under ambient condition of storage temperature and micro organisms lead to loss of viability in recalcitrant seeds. Coneglian *et al.*, (2000) observed that seed subjected to gibberellic acid (GA_3) treatment (seeds layered between paper soaked in 0, 150 or 300 mg GA_3/litre for 5 days or immersed in a solution of 0 or 300 mg GA_3/litre for 24 h) showed higher percentage and emergence rate.

The germination of *Passiflora alata* was improved when extracting the aril and administrating GA_3 infusion in gerbox or roll of paper before germination (Ferreira *et al.*, 2005). Delanoya *et al.* (2006) recommended removing the basal point of seeds or removing the basal point in combination with pre-soaking seeds for 48 h in 50 ppm GA_3 for enhancing the germination in *P. mollissima*. While, pre-soaking seeds for 24 h in 400 ppm GA_3 and removal of the basal point in combination with pre-soaking seeds for 48 h in 50 ppm GA_3 are suggested methods to improve germination of *P. nov* sp. Removing the apical point of *P. tricuspis* seeds resulted in maximal germination. They concluded that no unique treatment gave satisfactory results for the three species tested which is due to species specific response of *Passiflora*.

Under Arabhavi (Belgaum, Karnataka) condition, Gurung *et al.* (2014) reported the highest germination (84%) and germination index (2.02) in passion fruit seeds treated with thiourea (1%). But GA_3 (500 ppm) treatment resulted in maximum seedling height and number of leaves in the seedlings at 30, 60 and 90 days respectively. Seedling vigour index, shoot fresh and dry weight were also recorded maximum in same treatment. Pre-sowing seed treatment with GA_3 (100 ppm, 30 min.) was found better for seed germination and higher seedling vigour in passion fruit (Mehta *et al.*, 2016). The use of growth regulators for seed treatment is therefore recommended as an approach to increase the germination and seedling growth and vigour of passion fruit.

4. Role of Plant Growth Regulators on Micropropagation

Micropropagation confers many advantages as it is rapid, produce uniform, true-to-the-type, disease-free plants (Dornelas and Vieira, 1994; Kawata *et al.*, 1995; Faria and Segura, 1997). Tissue culture studies in *Passiflora* initiated as early as 1966 (Da Silva *et al.*, 2011), and since then an increasing number of reports on tissue culture based techniques applied to the genus have been published (as reviewed by Drew, 1997; Vieira and Carneiro 2004; Vieira *et al.*, 2005; Zerbini *et al.*, 2008; Alexandre *et al.*, 2009; Ozarowski and Thiem, 2013).

Although *Passiflora edulis* f. *flavicarpa* is resistant to *Fusarium* wilt and nematodes but other fungi, viruses and bacteria, cause significant losses in yellow passion fruit production (Gardner, 1989; Drew, 1997). The use of meristem culture for the production of virus-free plants has been reported with partial success (Chiari *et al.*, 1994). Several studies also reported that the plant regeneration of passion fruit could be obtained *via* organogenesis from a wide range of species and types of explants such as leaf, hypocotyl, root and cotyledon using different media composition with various growth regulator concentration (Moran Robles, 1978, 1979; Drew, 1991; Vaz *et al.*, 1993; Carvalho & Segura, 1994; Dornelas & Vieira, 1994; Fernando *et al.*, 2007; Lombardi *et al.*, 2007; Nhut *et al.*, 2007; Braglia *et al.*, 2009; Da Silva *et al.*, 2011; Pinto *et al.*, 2010; Huh *et al.*, 2017). *In vitro* culture can also be useful for clonal propagation of superior genotypes and disease resistant rootstocks, and as a tool for genetic transformation.

In most cases plant regeneration by organogenesis was induced using achieved BA with a wide range of concentrations. This cytokinin is essential for *in vitro* regeneration of *Passiflora* species, regardless of the type of explants, and the response to it varies with species and genotype (Ozarowski and Thiem, 2013). Cytokinins are very effective in initiation of direct or indirect shoot formation. Moreover, inducing the growth of adventitious shoots usually depends on the interaction between auxins and cytokinins. However, it is known that high concentrations of cytokinins are not always preferred in plant *in vitro* culture (George *et al.*, 2008). In the studies of micropropagation of *Passiflora* taxa, cytokinins were used in a wide range from 2.2 to 44.4 μM of BA, with or without generally low concentrations of auxins

(Ozarowski and Thiem, 2013). Cytokinin at low concentration is suitable for the proliferation of shoots and development of leaves. However, cytokinin and auxin are not preferred to the differentiation and growth of the roots. Yellow passion fruit might be efficiently multiplied by using different media for propagation of shoots and differentiation of roots, respectively (Costa *et al.*, 2018).

Biasi *et al.*, (2000) induced organogenesis from internodal segments of yellow passion fruit (*Passiflora edulis* f. *flavicarpa*) in a medium supplemented with 1.0 - 4.0 mg L^{-1} benzylaminopurine (BAP). BAP concentrations varying from 1.0 or 2.0 mg L^{-1} was found better for organogenesis. However, higher concentrations were observed more efficient for the adventitious buds induction but detrimental for subsequent shoot development. Da Silva *et al.* (2011) standardized a modified MS medium supplemented with BA (4.44 μM) to induce shoot organogenesis from the incubated root explants of a wild passion fruit species (*Passiflora cincinnata*) and three *P. edulis* commercial populations (FB 100, FB 200 and FB 300). Root explants of *P. cincinnata* responded earlier than those of *P. edulis*. Shoots elongation was done in liquid medium containing GA_3 (2.89 μM). Recently, Huh *et al.* (2017) reported that young leaf explants of purple passion fruit resulted in the highest embryogenic callus formation in MS medium supplemented with BAP (1 mg L^{-1}) and 2,4-D (2 mg L^{-1}). The highest shoot regeneration and elongation was found in the medium supplemented with BAP (1 mg L^{-1}) and GA_3 (1 mg L^{-1}) while, half strength medium with IAA (1 mg L^{-1}) resulted in the highest rooting.

Drew (1991) standardised the *in vitro* propagation of E23 (F1 hybrid of *P. edulis* and *P. edulis f. flavicarpa*) using various media composition. Adult tissue explants did not respond to media having low (<10 μM) concentrations of growth regulators. However, the adult buds on intact stem sections responded well to MS basal medium with 2iP (150 μM), adenine sulphate (200 μM) and IAA (17.1 μM), and further developed into shoots on MS medium having 4.9 μM 2iP (1 mg 1^{-1}) and 5.7 μM IAA (1 mg 1^{-1}). But the juvenile shoots of E23 and *Passiflora* species: *edulis f. flavicarpa, edulis, alata, caerulea, mollissima, coccinea, herbertiana* and *suberosa* grew rapidly on MS medium having kinetin (10 μM) and IAA (5 μM). Rapid multiplication was observed on MS medium containing BA (20 μM), kinetin (10 μM), IAA (5 μM), and roots initiated on MS with IAA (5 μM).

Garcia *et al.*, (2011) observed development of pre-existing meristems from nodal segments of *Passiflora suberosa* in response to a-naphthaleneacetic acid (NAA), picloram (PIC), and 2,4-dichlorophenoxyacetic acid (2,4-D) in combination with MS and MSM (modification in the mineral formulation of MS medium (Monteiro *et al.*, 2000)) basal media. Using BA, direct organogenesis was obtained from internodal segments. Regardless of the BA concentration, MSM medium gave the highest regeneration rates. Indirect organogenesis was achieved from all explant types on media supplemented with BA, used alone or in combination with NAA. However, the highest regeneration efficiency was obtained from internodal segments cultured on MSM medium with BA (44.4 μM). Thidiazuron, PIC, 2,4-D, and NAA lead to induction of compact, friable, or mucilaginous non-morphogenic calli.

Dornelas and Vieira (1994) observed that NAA (1.0 mg l^{-1}) inhibited shoot regeneration even in media containing 4.0 mg 1^{-1} BA in *Passiflora edulis* var. flavicarpa. MS basal media containing BA (2.0 mg 1^{-l}) was observed as optimum for shoot regeneration from cotyledonary and hypocotyledonary explants, while minimum of 1.0 mg 1^{-1} BA was essential to promote organogenesis from hypocotyl derived tissues. Isutsa (2004) obtained proliferation of yellow passion fruit shoot tops on MS medium containing 22.2 μM BAP, while *in vitro* rooting (62%) with MS containing IBA (24.5 μM). But purple passion fruit proliferated only on MS medium containing both BAP (22.2 μM) and GA_3 (11.6 μM) and rooting in NAA (21.5 μM). Kawata *et al.* (1995) reported formation of multiple shoot primordial from excised leaf tissues in MS media supplemented with 3% (w/v) sucrose, BA (1 μM) and IBA (1 μM). Similarly, Faria and Segura (1997) reported multiple shoot formation from the explants (shoot apices) of yellow passion fruit on solidified MS medium containing BA (0, 2, 5 and 20 μM). But the presence of IAA (2 μM) in the induction medium did not affect the sprouting per cent.

Shoot formation occurred through distinct pathways in response to 6-benzyladenine in cell suspense cultures of *Passiflora alata* Curtis (sweet passion fruit) (Pacheco *et al.*, 2012). Da Silva *et al.* (2009) induced somatic embryogenesis from zygotic embryos of *Passiflora cincinnata*. Zygotic embryos formed calli on media with different concentrations of 2,4-D and BA (4.5 μM). A concentration of 18.1 μM 2,4-D resulted in the largest number of somatic embryos. Similarly, Pinto *et al.* (2011) reported the highest frequencies of embryogenic calli from mature zygotic embryo explants of genotype FB-200 (*Passiflora edulis* Sims.) on MS medium supplemented with 2,4-D (72.4 μM) and BA (4.4 μM). Differentiation of somatic embryos was promoted on MS medium supplemented with activated charcoal and indole-3-acetyl-L-aspartic acid (IAA-Asp) either with or without 2,4-D. Prammanee *et al.* (2011) developed a method to propagate virus-free plants using a tissue culture technique involving the apical meristem of purple passion fruit. Shoot tips (2 mm) and the shoots were regenerated by culturing on MS medium supplemented with different concentrations of BA (0.5, 1.0, 1.5, 2.0, 2.5 and 3.0 mg/l) and NAA (0.2, 0.5, 1.0, 1.5 and 2.0 mg/l), while rooting was promoted using different concentrations of IBA (0.0, 0.2, 0.4, 0.6, 0.8 and 1.2 mg/l).

However, thidiazuron (TDZ) has been used alternatively to induce organogenic responses in passion fruit cultured explants. Trevisan and Mendes (2005) observed that culture media supplemented with BAP (0, 2.2, 4.4, 6.6 μmol L^{-1}) and TDZ (0, 1.1, 2.2, 3.4 μmol L^{-1}) were effective for *in vitro* organogenesis of passion fruit by the induction of adventitious buds from leaf discs. Vieira *et al.* (2014) reported that the presence of BA was essential for shoot formation from leaf explants and improved the response of hypocotyl segments as compared to the cytokinin-free control. However, after transfer to shoot elongation medium, the highest number of elongated shoots was obtained from hypocotyl segments that had been induced on BA + TDZ medium and the same was also observed for root explants.

Ethylene also plays a major role in the *in vitro* organogenesis of the explants. Ethylene inhibitors (silver thiosulfate, STS and aminoethoxyvinylglycine, AVG) have beneficial effects on *in vitro* development of axillary buds of yellow passion fruit and these can be used as an alternative approach to evaluate sensitivity of *Passiflora* species to ethylene. On the counterpart, ethylene precursor (1-aminocyclopropane-1-carboxylic acid, ACC) supplemented medium significantly inhibited the organogenic responses of the *in vitro* cultured axillary buds (Reis *et al.*, 2003).

5. Role of Plant Growth Regulators on Growth and Development

The main limiting factor for passion fruit rootstock seedlings production for grafting is the long time duration for seedling development. Several researchers have tried to reduce the time duration for the seedling to reach the grafting point. Oliveira *et al.* (2005) studied the effect of different combinations of growth regulators *viz.* BA, GA_{4+7} + Phenylmethyl-purineamine (GA_{4+7}+CK), GA_3 and Chlormequat chloride (CCC) in concentrations of 0 mg L^{-1}; 25 mg L^{-1}; 50 mg L^{-1}; 75 mg L^{-1} and 100 mg L^{-1} on *Passiflora alata* seedlings development. Different growth parameters were (stem length and the diameter, leaf number, length *etc.* were evaluated. Amongst the different treatments, GA_{4+7} + Phenylmethyl-purineamine treatment increased stem length and diameter and leaf number. But a non-significant difference amongst the treatments was observed for root, leaves, stem and total dry mass. Thus, Oliveira *et al.* (2005) advocated that, GA_{4+7} + Phenylmethyl-purineamine can be effectively used to reduce the time to obtain the grafting point, from 3-8 months to 2 months. Morales-Payan and Stall (2004) studied the effect of spraying (15 days after emergence) of GA, acetylthioproline (AP), BA and glycine-rich complex of amino acids and short-chain peptides (APC) on growth of purple passion fruit seedlings. Regardless of rate, BA did not affect passion fruit seedling growth. But increasing APC, AP and GA rates reduced the time from seedling emergence to the adequate transplanting stage. Spraying of APC (300-400 mg L^{-1}) gave the best result in term of shortening the time which was adequate for transplant production by 26% as compared to control plants.

6. Role of Plant Growth Regulators on fruit Maturity and Ripening, Yield and Quality

Various plant growth regulators and chemicals have been employed to delay the ripening process, reduce post-harvest losses to improve and maintain colour and quality by slowing down the metabolic activities of fruit which ultimately leads to an increased shelf life and maintain the quality marketability of the fruit for a longer period.

Brassinosteroids (BRs) are growth-promoting natural products found at low levels in pollen, seeds, and young vegetative tissues throughout the plant kingdom. BRs play diverse roles in developmental programs, including cell expansion, vascular differentiation, etiolation, and reproductive development. Detailed studies of BR biosynthesis and metabolism, coupled with the recent identification of BR-insensitive and BR-deficient mutants, have greatly expanded our view of steroids playing critical role in plant growth and development (Clouse and Sasse, 1998). Gomes *et al.* (2006) studied the effects of successive brassinosteroid analogue (BR) applications (0.1 mg l^{-1} of brassinosteroid analogue BB-16) on yellow passion fruit orchards in the first year of production under Campos dos Goytacazes (RJ, Brazil) condition. Various treatments were applied including control, BR-1 (one BR application shortly after the first flowers appeared), BR-2, BR-3, BR-4 and BR-5 (BR application in two, three, four and five consecutive weeks after the appearance of the first flowers, respectively) sprayed on the leaves and branches of each plant. BR-3 treatment produced the highest number of fruits per plant compared to the control and the soluble solids content was 1° Brix greater than the control (Table 1). The same treatment resulted in 65% increase in the estimated yield of the passion fruit plants, corresponding to 20 t ha^{-1} compared to the control yield of 12 t ha^{-1}. The results also concluded that spraying of BB-16 during reproductive development have potential to increase number of fruits per plant.

Table 1. Effect of successive foliar applications of brassinosteroid analogue (BB-16) on number of fruits per plant and juice soluble solids content in Yellow Passion fruit.

Treatment	Number of fruits per plant[a]	Soluble Solids Content[a] (°Brix)
Control	53.5 ± 4.83[b]	13.65 ± 0.45[b]
BR-1	69.6 ± 4.69[ab]	13.40 ± 0.42[b]
BR-2	60.0 ± 4.73[b]	14.55 ± 0.46[a]
BR-3	81.5 ± 8.35[a]	14.70 ± 0.48[a]
BR-4	59.9 ± 4.02[b]	13.95 ± 0.54[ab]
BR-5	67.0 ± 2.56[ab]	14.30 ± 0.51[ab]

Source: Gomes et al., 2006

Mean of 105 fruits for each treatment (for soluble solid contents). [a]Means ± S.E. (means followed by the same letter are not significantly different, Tukey test, 5%). (Original table produced from).

References

- Abreu, P. P., Souza, M. M., Santos, E. A., Pires, M. V., Pires, M. M., and de Almeida, A. A. F. 2009. Passion flower hybrids and their use in the ornamental plant market: perspectives for sustainable development with emphasis on Brazil. *Euphytica*, **166**(3): 307-315.

- Alexandre, R. S., Otoni, W. C., Dias, J. M. M., Bruckner, C. H. and Lopes, J. C. 2009. *In vitro* propagation of passion fruit. ***In***: Alexandre, R. S., Bruckner, C. H., Lopes, J. C. (Eds.). Propagation of Passion fruit: Morphological, physiological and genetic aspects. EDUFES, Pp. 117-184.

- Anonymous, Tropicals, I. T. I. 2010. Available *via* www. passionfritjuice.com.

- Bemkaireima, K., Angami, T. and Singh, M. S. 2012. Response of different size and growth regulator on cuttings of passion fruit *var.* Purple (*Passiflora edulis* var. edulis Sims). *Asian Journal of Horticulture*, **7**(2): 515-520.

- Bernacci, L. C., Soares-Scott, M. D., Junqueira, N. T. V., Passos, I. R. D. S. and Meletti, L. M. M. 2008. *Passiflora edulis* Sims: the correct taxonomic way to cite the yellow passion fruit (and of others colors). *Revista Brasileira de Fruticultura*, **30**(2): 566-576.

- Bernacci, L. C., Vitta, F. A. and Bakker, V. V. 2003. Passifloraceae. ***In***: Wanderley, M. G. L., Shepherd, G. J., Giulietti, A. M., Melhem, T.S. (Ed.). *Flora Fanerogâmica do Estado de São Paulo*. Sao Paulo: RiMa/ FAPESP. **3**: 247-274.

- Biasi, L. A., Falco, M. C., Rodriguez, A. P. and Mendes, B. M. J. 2000. Organogenesis from internodal segments of yellow passion fruit. *Scientia Agricola*, **57**(4): 661-665.

- Braglia, L., De Benedetti, L., Giovannini, A., Nicoletti, F., Bianchini, C., Pipino, L. and Mercuri, A. 2009. *In vitro* plant regeneration as a tool to improve ornamental characters in *Passiflora* species. *Acta Hort.*, **855**: 47-52.

- Carvalho Pires, M., Peixoto, J. R. and Yamanishi, O. K. 2010. Rooting of passion fruit species with indole-3-butyric acid under intermittent misting conditions. ***In***: *I International Symposium on Tropical Horticulture*, **894**:177-183.

- Carvalho, J.L. and Segura, J. 1994. Micropropagation of *Passiflora edulis*. ***In***: International Congress of Plant Tissue and Cell Culture, Firenze. **Annals**. Firenze: IAPTC, 1994. p.63.

- Chiari, T., Chiari, A. and Biricolti, S. 1994. Preliminary results in meristem culture of *Passiflora edulis*. ***In***: International Congress of Plant Tissue and Cell Culture, 8., Firenze, 1994. Annals. Firenze: IAPTC, p.208.

- Clouse, S. D. and Sasse, J. M. 1998. Brassinosteroids: essential regulators of plant growth and development. *Annual Review of Plant Biology*, **49**(1): 427-451.

- Coneglian, R. C. C., Rossetto, C. A. V., Shimizu, M. K. and Vasconcellos, M. D. S. 2000. Effect of extraction methods and gibberellic acid in the germination of seeds of sweet passion (*Passiflora alata* Dryand). *Revista Brasileira de Fruticultura*, **22**(3): 463-467.

- Costa, S. I., Mello-Farias, P. C., Copatti, A. S., de Albuquerque, I. C. and Dias, C. S. 2018. Cytokinin and auxin on *in vitro* multiplication of yellow passion fruit. *Journal of Experimental Agriculture International*, **19**(6): 1-6.

- Da Silva, C. V., de Oliveira, L. S., Loriato, V. A. P., da Silva, L. C., de Campos, J. M. S., Viccini, L. F. and Otoni, W. C. 2011. Organogenesis from root explants of commercial populations of *Passiflora edulis* Sims. and a wild passion fruit species, *P. cincinnata* Masters. *Plant Cell, Tissue and Organ Culture*, **107**(3): 407-416.

- Da Silva, M. L., Pinto, D. L. P., Guerra, M. P., Floh, E. I. S., Bruckner, C. H. and Otoni, W. C. 2009. A novel regeneration system for a wild passion fruit species (*Passiflora cincinnata* Mast.) based on somatic embryogenesis from mature zygotic embryos. *Plant Cell, Tissue and Organ Culture*, **99**(1): 47-54.

- Delanoy, M., Van Damme, P., Scheldeman, X. and Beltran, J. 2006. Germination of *Passiflora mollissima* (Kunth) L. H. Bailey, *Passiflora tricuspis* Mast. and *Passiflora* nov sp. seeds. *Scientia Horticulturae*, **110**(2): 198-203.

- Dornelas, M. C. and Carneiro Vieira, M. L. 1994. Tissue culture studies on species of *Passiflora*. *Plant Cell, Tissue and Organ Culture*, **36**(2): 211-217.

- Drew, R. A. 1991. *In vitro* culture of adult and juvenile bud explants of *Passiflora* species. *Plant Cell, Tissue and Organ Culture*, **26**(1): 23-27.

- Drew, R. A. 1997. Micropropagation of *Passiflora* species (Passion fruit). ***In***: *High-tech and micropropagation V.* Springer, Berlin, Heidelberg. Pp. 135-149.

- Faria, J. L. and Segura, J. 1997. Micropropagation of yellow passion fruit by axillary bud proliferation. *Hort. Science*, **32**(7): 1276-1277.

- Fernando, J. A., Vieira, M. L. C., Machado, S. R. and Appezzato-da-Gloria, B. 2007. New insights into the *in vitro* organogenesis process: the case of *Passiflora*. *Plant Cell, Tissue and Organ Culture*, **91**(1): 37-44.

- Ferreira, G., Oliveira, A. D., Rodrigues, J. D., Dias, G. B., Detoni, A. M., Tesser, S. M., and Antunes, A. M. 2005. Effect of aril in *Passiflora alata* seed germination in differents substrates and submited to previous germinative treatments with gibberellin. *Revista Brasileira de Fruticultura*, **27**(2): 277-280.

- Garcia, R., Pacheco, G., Falcao, E., Borges, G. and Mansur, E. 2011. Influence of type of explant, plant growth regulators, salt composition of basal medium, and light on callogenesis and regeneration in *Passiflora suberosa* L. (Passifloraceae). *Plant Cell, Tissue and Organ Culture*, **106**(1): 47-54.

- Gardner, D. E. 1989. Pathogenicity of *Fusarium oxysporum* f. sp. passiflorae to banana poka and other *Passiflora* spp. in Hawaii. *Plant Disease*, **73**(6): 476-478.

- George, E. F., Hall, M. A. and De Klerk, G. J. 2008. Micropropagation: uses and methods. ***In***: *Plant propagation by tissue culture.* Springer, Dordrecht. Pp. 29-64.

- Gomes, M. D. M. A., Campostrini, E., Leal, N. R., Viana, A. P., Ferraz, T. M., do Nascimento Siqueira, L. and Zullo, M. A. T. 2006. Brassinosteroid analogue effects on the yield of yellow passion fruit plants (*Passiflora edulis* f. flavicarpa). *Scientia Horticulturae*, **110**(3): 235-240.

- Gurung, N., Swamy, G. S. K., Sarkar, S. K. and Ubale, N. B. 2014. Effect of chemicals and growth regulators on germination, vigour and growth of passion fruit (*Passiflora edulis* Sims.). *The Bioscan*, **9**(1), 155-157.
- Huh, Y. S., Lee, J. K. and Nam, S. Y. 2017. Effect of plant growth regulators and antioxidants on in vitro plant regeneration and callus induction from leaf explants of purple passion fruit (*Passiflora edulis* Sims). *Journal of Plant Biotechnology*, **44**(3): 335-342.
- Isutsa, D. K. 2004. Rapid micropropagation of passion fruit (*Passiflora edulis* Sims.) varieties. *Scientia Horticulturae*, **99**: 395-400.
- Kawata, K., Ushida, C., Kawai, F., Kanamori, M. and Kuriyama, A. 1995. Micropropagation of passion fruit from sub cultured multiple shoot primordia. *Journal of Plant Physiology*, **147**(2): 281-284.
- Kiran, A., Sanjay, S. and Chavda, J. C. 2011. Effect of post-harvest treatments on quality of Jamun (*Syzygium cuminii* Skeels) fruits during storage. *Asian Journal of Horticulture*, **6**(2): 297-299.
- Koch, R. C., Biasi, L. A., Zanette, F. and Possamai, J. C. 2001. Vegetative propagation of *Passiflora actinia* by semihardwood cuttings. *Semina: Ciências Agrárias*, **22**(2): 165-167.
- Kumar, S., Chithiraichelvan, R., Karunakaran, G. and Sakthivel, T. 2008. Studies on propagation of passion fruit cv. Kaveri by cuttings under Coorg conditions. *Indian Journal of Horticulture*, **65**(1): 106-109.
- Lombardi, S. P., Passos, I. R. D. S., Nogueira, M. C. S. and Appezzato-da-Gloria, B. 2007. *In vitro* shoot regeneration from roots and leaf discs of *Passiflora cincinnata* mast. *Brazilian Archives of Biology and Technology*, **50**(2): 239-247.
- Lotschert, W. and G. Beese 1989. Collins Guide to Tropical Plants. *William Collins Sons and Co. Ltd. London*, p. 256.
- Mehta, S., Prasad, B. and Das, A. K. 2016. Pre-sowing seed treatments with different chemicals and growth regulator on seed germination and growth of passion fruit. *Advance Research Journal of Crop Improvement*, **7**(2): 201-206.
- Meletti, L. M. M., Barbosa, W., Pio, R., Tucci, M. L. S., Costa, A. B. and Feldberg, N. P. 2007. Influence of season, leaf presence and indolebutyric acid on the rooting potential of sweet passion-fruit (*Passiflora alata* Curtis) cuttings. *Revista UDO Agricola*, **7**: 68-73.
- Monteiro, A. D. A., Higashi, E. N., Gonçalves, A. N. and Rodriguez, A. P. M. 2000. A novel

approach for the definition of the inorganic medium components for micropropagation of yellow passion fruit (*Passiflora edulis* sims. f. flavicarpa Deg.). ***In:*** *Vitro Cellular and Developmental Biology-Plant*, **36**(6): 527-531.

- Morales-Payan, J. P. and Stall, W. M. 2004. Passion fruit (*Passiflora edulis*) transplant production is affected by selected biostimulants. ***In:*** *Proc. Fla. State Hort. Soc.*, **117**: 224-227.
- Moran Robles, M. 1979. Potencial morfogenetico de los entrenudos de *Passiflora edulis* var. flavicarpa y de P. mollissima en cultivo *in vitro*. **29**(3): 224-228.
- Moran Robles, M. J. 1978. Multiplication vegetative, *in vitro*, des bourgeons axilaires de *Passiflora edulis* var. flavicarpa Deg. et de P. mollissima Bailey. *Fruits,* **33**: 693-699.
- Nakasone, H. Y., and Paull, R. E. 1998. Tropical Fruits. Cab International, Wallingford, UK.
- Nhut, D. T., Khiet, B. L. T., Thi, N. N., Thuy, D. T. T., Duy, N., Hai, N. T. and Huyen, P. X. 2007. High frequency shoot formation of yellow passion fruit (*Passiflora edulis* f. flavicarpa) *via* thin cell layer (TCL) technology. ***In***: *Protocols for Micropropagation of Woody Trees and Fruits*. Springer, Dordrecht. Pp. 417-426.
- Oliveira, A. D., Ferreira, G., Rodrigues, J. D., Ferrari, T. B., Kunz, V. L., Primo, M. A. and Poletti, L. D. 2005. Effect of plant regulators on *Passiflora alata* seedlings development. *Revista Brasileira de Fruticultura*, **27**(1): 9-13.
- Ozarowski, M. and Thiem, B. 2013. Progress in micropropagation of *Passiflora* spp. to produce medicinal plants: a mini-review. *Revista Brasileira de Farmacognosia*, **23**(6): 937-947.
- Pacheco, G., Garcia, R., Lugato, D., Vianna, M. and Mansur, E. 2012. Plant regeneration, callus induction and establishment of cell suspension cultures of *Passiflora alata* Curtis. *Scientia Horticulturae*, **144**: 42-47.
- Pinto, A. P. C., Monteiro-Hara, A. C. B., Stipp, L. C. L. and Mendes, B. M. J. 2010. *In vitro* organogenesis of *Passiflora alata*. *In Vitro Cellular and Developmental Biology-Plant*, **46**(1): 28-33.
- Pinto, D. L. P., de Almeida, A. M. R., Rego, M. M., da Silva, M. L., de Oliveira, E. J. and Otoni, W. C. 2011. Somatic embryogenesis from mature zygotic embryos of commercial passion fruit (*Passiflora edulis* Sims) genotypes. *Plant Cell, Tissue and Organ Culture*, **107**(3): 521-530.

- Prammanee, S., Thumjamras, S., Chiemsombat, P. and Pipattanawong, N. 2011. Efficient shoot regeneration from direct apical meristem tissue to produce virus-free purple passion fruit plants. *Crop Protection*, **30**(11): 1425-1429.
- Purseglove, J.W. 1968. Tropical Crops. Dicotyledons: Passifloraceae, *Longman Scientific and Technical, London*, Pp. 420-429.
- Reis, L. B., Neto, V. P., Picoli, E. T., Costa, M. G. C., Rego, M. M., Carvalho, C. R. and Otoni, W. C. 2003.. Axillary bud development of passion fruit as affected by ethylene precursor and inhibitors. *In Vitro Cellular and Developmental Biology-Plant*, **39**(6): 618-622.
- Roncatto, G., Nogueira Filho, G. C., Ruggiero, C., Oliveira, J. C. D. and Martins, A. B. G. 2008. Rooting of herbaceous cutting of different passion fruit plant species. *Revista Brasileira de Fruticultura*, **30**(4): 1094-1099.
- Thimann, K. V. and Behnke-Rogars, J. 1950. The use of auxins in the rooting of woody cuttings. Maria Moors Cabot Foundation Publ. No. 1. *Harvard Forest, Petersham, MA, USA.*
- Tommonaro, G., Segura Rodriguez, C. S., Santillana, M., Immirzi, B., De Prisco, R., Nicolaus, B., and Poli, A. 2007. Chemical composition and biotechnological properties of a polysaccharide from the peels and antioxidative content from the pulp of *Passiflora liguralis* fruits. *Journal of Agricultural and Food Chemistry*, **55**(18): 7427–7433.
- Trevisan, F. and Mendes, B. M. J. 2005. Optimization of *in vitro* organogenesis in passion fruit (*Passiflora edulis* f. flavicarpa). *Scientia Agricola*, **62**(4): 346-350.
- Vaz, C. D. F., Peixoto, J. R., Junqueira, N. T. V., Braga, M. F., Santos, E. C. D., Fonseca, K. G. D. and Junqueira, K. P. 2009. Rooting of wild species of passion fruit using five doses of indolebutyric acid. *Revista Brasileira de Fruticultura*, **31**(3): 816-822.
- Vaz, F. D. U., Dos Santos, A. V. P., Manders, G., Cocking, E. C., Davey, M. R. and Power, J. B. 1993. Plant regeneration from leaf mesophyll protoplasts of the tropical woody plant, passion fruit (*Passiflora edulis* fv flavicarpa Degener.): the importance of the antibiotic cefotaxime in the culture medium. *Plant Cell Reports*, **12**(4): 220-225.
- Viana, A. J. C., Souza, M. M., Araujo, I. S., Correa, R. X. and Ahnert, D. 2010. Genetic diversity in *Passiflora* species determined by morphological and molecular characteristics. *Biologia Plantarum*, **54**(3): 535–538.
- Vieira, L. M., Rocha, D. I., Taquetti, M. F., da Silva, L. C., de Campos, J. M. S.,

Viccini, L. F. and Otoni, W. C. 2014. *In vitro* plant regeneration of *Passiflora setacea* DC (Passifloraceae): the influence of explant type, growth regulators, and incubation conditions. *In Vitro Cellular and Developmental Biology-Plant*, **50**(6): 738-745.

- Vieira, M. L. C., Carneiro, M. S. 2004. *Passifloru* spp., Passion fruit. ***In***: Litz, R. E. (ed) Biotechnology of fruit and nut crops. *CABI Publishing*, Oxford, Pp: 435–453.
- Vieira, M. L. C., Oliveira, E. J., Matta, F. P., Padua, J. G., & Monteiro, M. 2005. Biotechnological methods applied to passion fruit breeding. *Passion fruit: germplasm and breeding. Embrapa Cerrados, Planaltina*, 411-453.
- Zerbini, F. M., Otoni, W. C. and Vieira, M. L. C. 2008. Passion fruit. ***In***: Kole, C. and Hall, T. C. (eds). A compendium of transgenic crop plants, V.5, Tropical and Subtropical Fruit and Nuts, 1st Edn. John Wiley and Sons, Wallingford, Pp: 213–234.

22

Plant Growth Regulators in Phalsa (*Grewia subinaequalis* D.C.)

D.S. Mishra, Sanjay Singh and A.K. Singh

1. Introduction

Since the discovery of auxin, there has been spectacular progress in the area of chemical regulation of plant growth and development (Misra and Mishra, 2003). At present there are five groups of PGRs viz., auxins, gibberellins, cytokinins, ethylene and abscisic acid. However, some others like brassinosteroids, jasmonic acid, oligosaccharides and fusicoccin are found useful in one or other ways for growth and development of horticultural crops. These plant growth regulators are organic compounds other than nutrients, which in small amounts promote, inhibit or otherwise modify any physiological process in plants. These growth regulators can be natural or synthetic.

Phalsa (*Grewia subinaequalis* D.C.) a member of family Tiliaceae, is one of the oldest fruits known to Indian. Phalsa has been mentioned in Vedic literature as having certain medicinal properties. It is capable of growing under neglected and water scarcity conditions where only a few other crops could survive. Besides, it is an important catch crop in commercial orcharding (Mishra *et al.,* 2016). The mildly acidic fruits are rich in Vitamin A, C and minerals. Flowering in phalsa starts from February-March and continues till May. After 40-45 days of flowering, the fruits start ripening. Fruits are small-sized and ripen over a period of about a month. Gradual but steady ripening of few fruits in a cluster during summer necessitates frequent harvesting (Salunkhe and Desai, 1984). Therefore a number of pickings are required at 2–3 days intervals which are very expensive. Phalsa fruits have short shelf life; therefore, its fruits are suitable for local market or need to be processed immediately after harvesting. Problems in phalsa whether related to the propagation or production may be tackled with the use PGRs are presented here.

2. Use of PGRs in seed Germination

In commercial crop production, where high yield and uniform produce are two main considerations, slow and variable germination can be a major problem. Rapid, uniform and complete germination are prerequisites for achieving good plant populations from the phalsa seeds. In general, storage of seeds results in partial loss of viability due to disturbances in metabolism and deficiencies of certain PGRs (Misra and Mishra, 2003). GA, Kinetins, auxins, ABA and other regulatory substances are vitally important and their applications to seeds intensify the metabolism and improve germination and plant productivity. However, reports with regard to use PGRs for enhanced germination and seedling growth is lacking in phalsa. So, there is need to study the effects of GA, Kinetins, auxins (IAA) for improving seed germination, seedling vigour and plant productivity.

3. Use of PGRs in Rooting of Cutting and Layering

Stem cuttings and air layers which are means of multiplication of phalsa did not root easily. However, treatments with auxins like IBA and NAA may improve rooting of difficult-to-root hardwood cuttings of phalsa. Effect of auxins has been found to be positive for induction of rooting and for longer roots in phalsa stem cuttings. As phalsa stem cuttings are difficult to root, a treatment with auxins like IAA, IBA and NAA improve rooting of difficult-to-root hardwood cuttings of phalsa (Yadava and Rajput, 1969), ground layering and air-layering (Mohammed and Chauhan, 1970). Only 20 % of semi-hard wood cuttings taken from spring flush were rooted and grew normally when treated with 1,000 ppm NAA. Cuttings from 6 months old shoots during July-August gave profuse rooting when treated with 1000 ppm IBA through quick dip method. Rooting in cuttings (72.5 %) were obtained when shoots were ringed 15 days before planting and treated with 2500 ppm IBA (Singh and Kumar, 1967). Srivastava *et al.*, (1994) reported that 2000 ppm IBA gave best results for growth and survival of *G. subinaequalis* cutting. Effect of auxins (IBA and NAA @ 100-300 ppm) on rooting of semi-hardwood and hardwood cuttings of *G. asiatica* was investigated. The best rooting (82.5%) was obtained following treatment of hardwood cuttings with IBA at 200 ppm (Kathrotia and Singh, 1995). For stooling, phalsa plants are headed back during February-March and newly emerged shoots are ringed except one unringed nurse shoot followed by application of 10000 ppm IBA paste on the upper end of the ring. These treated rings are covered with moist soil by mounding around the bush. The rooted shoots are separated after 2 months and planted in the field after proper hardening in the nursery beds (Singh and Kumar, 1967).

4. Use of PGRs in *in vitro* Regeneration

In vitro micro-propagation has been found an useful tool for large scale production of true-to-type and disease free planting material, unhindered by seasonal constraints. Phalsa is conventionally propagated by hard wood cuttings, which is hard to root. Therefore, pose problem in faster multiplication of elite planting material. Thus, for boosting faster multiplication, phalsa may be propagated by *in vitro* techniques of micropropagation. In this regard Biswas *et al.* (2005) attempted to standardize the *in vitro* micro-propagation protocols for phalsa plant. Phalsa explants prepared using different explants (leaf, nodes, internode, and zygotic embryos), cultured on MS medium supplemented with various cytokinins alone or along with auxins and incubated under a 10-hour photoperiod at ambient temperature. *In vitro* propagation of phalsa tissues through both organogenesis and somatic embryogenesis was also achieved. Of these, single shoots were developed from nodal explants as a result of bud break on MS medium supplemented with BAP, kinetin, and zeatin separately. Somatic embryos were developed from the zygotic embryos when cultured on MS medium with 0.023 μM BA + 0.022 μM zeatin, for 2 weeks following a pulse treatment on NN medium supplemented with 5% sucrose, 0.11 μM BAP, 0.22 μM 2,4-D, and 29.20 μM L-glutamine. Somatic embryogenesis was also observed on modified basal medium supplemented with 13% sucrose, 58.40 μM L-glutamine, and 1.75 μM IAA. Profuse callusing was a major problem for in vitro studies with this species, irrespective of media composition. However, Wani *et al.* (2016) achieved multiple shoot induction in phalsa from nodal explants with axillary buds, on culturing in Woody Plant Medium (WPM) fortified with 3% (w/v) sucrose, 2×10^{-5} M Kinetin (Kn) and 1×10^{-5} M indole-3-butyric acid (IBA) giving rise to an average of 4.25 ± 0.71 micro-shoots per explant. More than 90% of the explants formed micro-shoots with mean shoot length of 10.5 ± 1.96 cm leading to whole plant regeneration. Healthy regenerated shoots showed prolific rooting of more than 95% on WPM supplemented with 4.8×10^{-6} M indole-3-butyric acid (IBA). Following simple hardening procedures, rooted plantlets, were transferred to soil-sand (1:1; v/v) with about 92% success.

5. Use of PGRs in plant Growth, Flowering, Fruit Set, Ripening, Yield and Quality

Sprays of PGRs have proved effective in achieving the enhanced fruit set, yield and uniform ripening in phalsa. Chundawat and Singh (1980) investigated the effect of PGRs on growth and fruiting of phalsa has brought out that GA treatments except in combination with 2,4-D and PCPA increased the vegetative growth significantly and 2,4-D, PCPA and OCLPA (80 ppm) reduced it considerably. The treatments which reduced vegetative growth also showed epinastic effect on leaves and had considerable effect on yield. In general all thc treatments reduced the number of flowers per branch and the percentage of fruit set. However, the effect on fruit drop was most marked and maximum reduction was obtained in GA 60 ppm followed by 2,4,5-T (5 ppm) and GA (60 ppm) alongwith urea (1%). GA treatments were effective in giving early maturity. GA (60) ppm plus urea one per cent, NAA (50 and 75 ppm) and OCLPA (80 ppm) reduced the duration of harvesting while GA (60 ppm) + 2,4-D (2.5 ppm) and 2,4- D, at 2.5 ppm increased it and PCPA delayed the start of harvesting. Application of plant regulators increased the yield of phalsa greatly and the most effective treatment was GA (60 ppm) + 2,4,5-T (5 ppm) followed by 2,4-D (2.5 ppm). Other favourable treatments were OCLPA 40 ppm, GA 60 ppm plus 2,4-D 2.5 ppm, NAA 50 ppm GA 60 ppm followed by 2,4,5-TP 5 ppm and PCPA 40 and GA 60 ppm plus urea one per cent. PCPA 40 ppm only considerably reduced the yield. The effect on yield is mainly due to their effect on fruit size and fruit drop. Accordingly maximum weight and volume of fruit was obtained with GA (60 ppm) + 2,4,5-T (5 ppm) followed by other GA combinations.

The results of experiment conducted by Kacha *et al.* (2014) revealed that an application of NAA 150 ppm significantly increased number of flowers per shoot (151.21), number of fruits per shoot (60.74), 100 fruits weight (49.80 g), juice percentage (57.78 per cent) and the maximum yield of fruits (1.71kg/plant and 5800 kg/ha) followed by NAA 200 ppm. The quality of fruits in terms of TSS (25.23 per cent), reducing sugar (2.01 per cent) and total sugar (5.74 per cent) were significantly higher in treatment with Ethrel 1000 ppm followed by Ethrel 750 ppm. A spray application of gibberellin A_4/A_7 (10–20 ppm and 10 to 40 ppm) at fully swollen flower bud stage during the last week of April increased per cent fruit set and retention. Fruit size, weight and pulp weight were highest with the application of gibberellinA_4/A_7 (20 ppm). However, exogenous application of gibberellinA_4/A_7 (5 and 10 ppm) increased TSS, TSS/acid ratio and total sugar content of the fruit. The treatments of 50 per cent flower thinning followed by spray of gibberellinA_4/A_7 (20 ppm) significantly increased fruit weight and size, TSS/acid ratio and total fruit sugars. Considering all aspects of fruit quality, the treatment of 50 per cent flower thinning followed by 20 ppm GA4/A7 proved the best (Singh *et al.*, 2000). Gradual but steady ripening of few fruits in a cluster during summer necessitates frequent harvesting (Salunkhe and Desai, 1984). Therefore a number of pickings are required at 2–3 days intervals which are very expensive. Pre-harvest application of ethephon or etherel (1000 ppm) when few fruits start to change their colour reduces the number of pickings from 7 to 3 and improve fruit quality in terms of total soluble solids, vitamin C content and colour development due to increase in anthocyanin pigment (Kacha *et al.*, 2012). Similar result was obtained by Singh *et al.* (1979) through application of 1000 ppm of ethrel which resulted in maximum (71.1%) ripening in 5 days after application. Further studies made by Kacha *et al.* (2014) revealed that ethrel 1000 ppm also significantly reduced the span of harvesting (9.76 days) and number of pickings (3.57) followed by ethrel 750 ppm. An application of GA3 150 ppm significantly reduced acidity (2.55 per cent) and increased ascorbic acid content (39.50 per cent).

Improvements in yield and fruit quality were observed by the application of 1000 ppm SADH at early stage of plant growth. SADH (1000 ppm) also reduced the plant height (Chundawat and Gupta, 1974). Increase in fruit size and improvement in fruit quality were also noted by spraying plants with 60 ppm $GA_{3,}$ or 5 ppm 2,4,5-T, or 2.5 ppm 2,4-D (Singh and Singh, 1981). Treatment with GA_3 at 10 ppm or 2,4,5-T at 5 ppm increased the fruit set and yield (Randhawa *et al.*, 1959). Spraying of GA_3 at 60 ppm once at the beginning of flowering and another after 15 days and ethrel at 1000 ppm at the start of berry ripening increased the fruit retention and yield. The treatment also reduced the harvesting span and increased the total soluble solids of the fruit (Singh *et al.*, 1986). Fruit weight and volume were significantly increased by NAA (150 ppm) and NAA (150 ppm) + ethrel (480 ppm). While ethrel application resulted in increased TSS content in fruit (Rema and Sharma, 1993).

References

- Biswas, B., Joshee, N., Yadav, A. and Yadav, A.K. 2005. In Vitro Propagation and Somatic embryogenesis in Phalsa. *Hort. Science*, **40**(4): 1104.

- Chundawat, B.S. and Gupta, O.P. 1974. Effect of growth retardants (B-nine and cycocel) on vegetative growth and yield of phalsa (*Grewia asiatica* L.). *Haryana Journal of Horticulture Sciences*, **3** (3-4): 113-115.

- Chundawat, B.S. and R. Singh. 1980. Effect of growth regulators on phalsa (*Grewia asiatica* L.). Growth and fruiting. *Indian J. Hort.*, **37**:124–131.

- Kacha, H. L., Viradia, R. R., Leua, H. N., Giriraj Jat and Tank, A. K. 2012. Effect of NAA, GA_3 and ethrel on yield and quality of phalsa (*Grewia asiatica* L.) under South-Saurashtra condition. *Asian Journal of Horticulture*, **7** (2): 242-245.

- Kacha, H.L., Jat, G. and Patel, S.K. 2014. Performance of various plant growth regulators on yield and quality of phalsa (*Grewia asiatica* L.) *HortFlora Res. Spectrum*, **3**(3): 292-294.

- Kathrotia, R. K. and Singh, S. P. 1995. Regeneration of roots in phalsa (*Grewia asiatica* Mast) stem cuttings as influenced by maturity of wood and root promoters. *Advances in Horticulture and Forestry*, **4**:35-41.

- Mishra, D.S., Saxena, D. and Chand, S. 2016. Phalsa (*Grewia subinaequalis* D.C.). In: Underutilized Fruit Crops: Importance and Cultivation Part II (Eds.) Ghosh, S.N., Singh, A. and Thakur, A., Jaya Publishing House, Delhi. pp. 1029-1043.

- Misra, K.K. and Mishra, D.S. 2003. Role of plant growth substances in forest nursery stock production. In: *Forest Tree Seeds and Nursery Management: Current Trends* (Eds.) Singh, V. and Lavania, S.K. Bishen SMPS Publ&Dis of Scientific Books, Dehra Dun. pp. 193-209.

- Mohammed Shafat and Chauhan, K.S. 1970. Vegetative propagation of phalsa (*Grewia asiatica* L.). *Indian Journal of Agriculture Sciences,* **40**: 581-586.

- Randhawa, G.S., Malik, R.S. and Singh, J.P. 1959. Note on the clonal propagation of phalsa (*Grewia asiatica* L.). *Indian Journal of Horticulture,* **16**: 199.

- Rema, J. and Sharma, V.P. 1993. Effect of plant growth regulator on yield and quality of phalsa (*Grewia subinaequalis* L.). *South Indian Horticulture,* **39** (6): 327-331.

- Salunkhe, D.K. and Desai, B.B. 1984. Post harvest Biotechnology of Fruits, Vol. II, CRC Press, Boca Raton, Fl, p.129.

- Singh, H. and Singh, R. 1981. Effect of plant growth regulators and manuring on fruit quality phalsa (*Grewia subinaequalis*). *Indian Journal of Horticulture,* **38** (3-4): 157-161.

- Singh, I.S., Singh, H.K. and Chauhan, K.S. 1979. Effect of pre-harvest application of ethaphone on ripening and quality of phalsa (*Grewia asiatica*). *Haryana Journal of Horticulture Sciences,* **8**(3-4): 196-197.

- Singh, M., Gautam, R.K.S., Prasad, Y. and Singh, A.K. 1986. Studies on the effect of plant regulators and their different concentrations on yield and quality of phalsa (*Grewia subinaequalis* D.C.). *Haryana Journal of Horticulture Sciences,* **15**(3-4): 196-199.

- Singh, S.M. and Kumar, H. 1967. Influence of ringing source shoot and concentrations of indole-3-butyric acid on the performance of phalsa (*Grewia asiatica* L.) stem cutting. *Indian Journal of Agriculture Sciences,* **37**: 151-154.

- Singh, Z., Grewal, G.P.S. and Singh, L. 2000. Effects of gibberellinA_4/A_7, and blossom thinning on fruit set, retention, quality, shoot growth and return bloom of phalsa (*grewia asiatica* L.). *Acta Hortic.* **525**: 463-466.

- Srivastava, S.R., Mahajan, S., Nema, B.K. and Chandra, A. 1994. Effect of IBA, sugar and captam on shoot growth of phalsa (*Grewia subinaequalis* D.C.). *JNKVU Research Journal,* **28-29** (1-2): 80-82.

- Yadava, U.L. and C.B.S. Rajput. 1969. Anatomical studies of rooting in stem cuttings of phalsa (*Grewia asiatica* L.). *Hort. Sci.,* **1**:19–22.

- Wani, T. A., Rana, S., Bhat, W.W., Pandith, S.A., Dhar, N., Razdan, S., Suresh Chandra, Sharma, N. and Lattoo, S.K. 2016. Efficient *in vitro* regeneration, analysis of molecular fidelity and *Agrobacterium tumifaciens* - mediated genetic transformation of *Grewia asiatica* L. *J. Plant Biochem. Physiol.,* **4**(2):167. doi:10.4172/2329-9029.1000167.

23

Effect of Plant Growth Regulators on Propagation and Fruit Production in Phalsa (*Grewia subinaequalis* DC)

A.K. Pandey, T. R. Ahlawat, Dharmistha Patel and Suchismita Jena

1. Introduction

Phalsa (*Grewia subinaequalis* DC.; syns. *G. asiatica* L.) is a subtropical fruit belonging to family Tiliaceae and is a native of South Asia. In India, it is commercially cultivated in Punjab, Haryana, Rajasthan, Uttar Pradesh and Madhya Pradesh. It is a quick growing very hardy shrub which thrives well in arid and semi-arid region as well as in salt affected wastelands. It is considered as a nutritive fruit rich in antioxidants due to the presence of vitamin C, phenolics, flavonoids, tannins and anthocyanins. The flavour is pleasantly acidic and somewhat grapelike. Fruits are very delicious, tasty and used for table purpose. They are excellent for making juice and squash, ready to serve, nectar, syrup.

2. Effect of Plant Growth Regulators on Propagation

2.1. Effect of plant growth regulators in rooting of cuttings

Phalsa cuttings are difficult to root and only 20% of semi-hardwood cuttings from spring flush, treated with 1000 ppm NAA and planted in July rooted and grew normally. Treatment with auxins (IAA, IBA, NAA) improved rooting in difficult-to-root hardwood cuttings of phalsa (Yadava and Rajput, 1969), ground layers and air layers (Mohammed and Chauhan, 1970). Singh *et al.* (1961) recorded 68.8 per cent success from hard wood cuttings treated with 100 ppm IBA. Singh and Kumar (1967) obtained 72.5 per cent rooting in cuttings when shoots were ringed 15 days before planting and treated with 2500 ppm IBA. A report from Rajasthan Agricultural University, Udaipur claimed 76.2 per cent success with hard wood cutting when treated with 250 ppm ethrel for five minutes followed by 3000 ppm IBA for two minutes (Anon., 1981). The best rooting was obtained in hard wood cuttings treated with IBA at 200 ppm however; semi-hard wood cuttings did not root as well as hard wood cuttings (Kathrotia and Singh, 1995). Treatment of hard wood cutting with IBA (2000 and 3000 ppm) significantly increased the root length, number of roots/cutting and percentage survival (Shrivastava, 1996). Shrivastava *et al.* (1998) observed the best results in terms of growth and survival of phalsa cuttings with 2000 ppm IBA.

2.2. Effect of plant growth regulators on rooting in layering

Ground layering and air-layering have been tried in phalsa by some workers. Randhawa *et al.*, (1959) obtained 80 per cent success through ground layering using seradix-L-30. Success up to 80-90 per cent was recorded by certain workers from air layers treated with IBA or NAA at 5000 ppm and 10000 ppm. Mohammed and Chauhan (1970) recorded 85 per cent success, when air-layers were treated with a mixture of IBA and NAA at the rate of 1500 ppm each, but when the same mixture was used for ground layering success was 50 per cent only.

2.3. Effect of plant growth regulators on rooting in stooling

Sharma *et al.*, (1987) obtained fairly high percentage of success in stooling of phalsa. The plants were headed back 15 cm above the ground level in January. New shoots emerged out with the onset of spring. A ring of bark (2.0 cm) was removed from the base of these shoots in August and the shoots were then treated with IBA and NAA in lanolin paste. The shoots were earthed up after 7 days. Rooting occurred in about 45 days. Maximum percentage (79 %) of rooting and the longest roots were obtained with IBA 7500 ppm followed IBA 5000 ppm (72 % rooting), while 62 per cent was noted with NAA 10000 ppm.

2.4. Grafting

Inarching, shield budding and ring budding were tried by Randhawa *et al.* (1959a) at different time intervals. Cent per cent success was reported through Inarching done in March, 90 per cent success through shield budding in June and 82 per cent success through ring budding in June.

3. Crop Regulation and Use of Growth Regulators

Application of growth substances has proved effective in increasing the fruit set and yield in phalsa. Treatment with GA_3 at 10 ppm and 2,4,5-T at 5 ppm increased fruit setting and also the yield (Randhawa *et al.*, 1959b). Studies at Allahabad Agricultural Institute showed that the use of cycocel at 250 ppm sprayed twice at an interval of 7 days after 50 per cent fruit set increased the fruit size. It also delayed fruit maturity by 3-5 days. Application of 1000 ppm ethrel resulted in maximum ripening (71.1%) in 5 days after application (Singh *et al.*, 1979). Chundawat and Gupta (1974) recorded improvement in yield and fruit quality with the use of 1000 ppm SADH at early stage of plant growth and it also reduced the plant height. Singh and Singh (1981) also observed an increase in fruit size and improvement in fruit quality by spraying GA_3 (60 ppm), 2,4,5-T (5 ppm) and 2,4-D (2.5 ppm).

Spraying of GA_3 at 60 ppm once at the beginning of flowering and again after 15 days and ethrel at 1000 ppm when ripening of berries had just started, increased fruit retention percentage and yield. These treatments also reduced the harvesting span and increased the total soluble solids content of fruit (Singh *et al.*, 1986). Application of NAA and ethephon showed marked effect on fruiting, yield and quality of phalsa fruits. The results of the experiment conducted by Bhuva *et al.* (1995) with cycocel 2000 ppm, GA_3 60 ppm and ethephon 150 ppm revealed that treatment with GA_3 increased the fruit size and improved the fruit quality. Spraying with NAA 10, 20 and 30 ppm, in general, increased the shoot length and diameter. The lowest concentration of NAA (10 ppm) led to a maximum increase in number of nodes and leaf area (Parihar *et al.*, 1999). Sprays of 10 ppm gibberellic acid have increased fruit-set. At 40 ppm, there is increased fruit size but decreased fruit-set. A spray application of GA_{4+7} (10-20 ppm) at the fully swollen flower bud stage increased the percentage of fruit set and rates of 10-40 ppm increased fruit retention. Fruit size, weight and pulp weight were highest with the application of GA_{4+7} at 20 ppm. Considering all aspects of fruit quality 50 % flower thinning followed by 20 ppm GA_{4+7} proved best (Singh *et al.*, 2000). Rema and Sharma (1991) pruned 80 cm high healthy bushes during the third week of January and maintained 7 shoots on each bush. Fruit weight and volume were significantly increased by treatment with 150 ppm NAA or 480 ppm ethrel + 150 ppm NAA during full bloom. A significant increase in total soluble solids was recorded after treatment with 480 ppm ethrel during full bloom, as well as after applying 1920 ppm ethrel or 960 ppm ethrel + 2.5 ppm 2,4-D one week before harvest. Lakra *et al.* (2018) reported that pruning phalsa at 90 cm height and spraying GA_3@200 ppm + NAA@200 ppm resulted in maximum number of canes per bush (13.67), number of sprouted shoots per cane (30.30), length of shoots per cane (84.64 cm), number of fruiting nodes per shoots (27.72), number of leaves per shoots (23.46), leaf area (240.78cm^2) and leaf area index (7.65cm^2).

4. Effect of Plant Growth Regulators on Flowering and Fruit Development

Phalsa plant starts fruiting after 15 to 18 months of planting but good yield is obtained only after three years of planting. Flowering commences in February-March and continues for almost one month. Although phalsa flowers are hermaphrodite but cross pollination is essential for better fruit set. Gill *et al.* (2001) found that fruit set (61.60%) in open pollination was significantly higher than self-pollination (23.00%). Irrespective of modes of pollination, there were non-significant differences on fruit set in tall (45.00%) and dwarf (38.01%) types of phalsa. The main insect pollinators for phalsa are *Apis florea, A. mellifera, A. dorsata, Megachile bicolor* and *Chalicodoma cephalotes.* They were observed foraging both nectar and pollen, while others foraged for nectar only. It has been found that use of plant growth regulators increased the size of fruits. Prasad and Bajpai (1963) found that NAA (25-75 ppm) at full bloom stage and again 10 days later in phalsa var. Sharbati increased fruit size. Further, maximum diameter was noticed with 50 ppm NAA. Two sprays of 2.5 ppm 2, 4-D at flowering and 8 days after flowering increased yield by 74%.The spray of 60 ppm GA and 5 ppm, 2,4,5-T increased yield by 64 %. These sprays also reduced the time of harvesting (Tripathi, 2009). Whereas, Kacha *et al.* (2014) reported that application of NAA 150 ppm significantly increased number of flowers per shoot (151.21), number of fruits per shoot (60.74), 100 fruits weight (49.80 g), juice percentage (57.78 per cent) and the maximum yield of fruits (1.71 kg/plant and 5800 kg/ha) followed by NAA 200 ppm. Gibberellic acid has been reported to improve fruit set and increase fruit size (Randhawa *et al.,* 1959b). Exogenous application of gibberellins A_4/A_7 (5 and 10 ppm) increased TSS, TSS/ acid ratio and total sugar content of fruits. Fifty per cent flower or cyme thinning followed by foliar spray of gibberellins GA_4/GA_7 (20 ppm) significantly increased fruit weight and size (Singh *et al.,* 2000). Alam and Kumar (2017) revealed that maximum number of shoots/ plant (86.37), number of fruits/ node (17.33), number of fruiting nodes/ shoot (16.33) and fruit yield (65.93 q/ha) were recorded by the application of GA_3 at 150 ppm. These findings are in accordance with those of Singh *et al.* (2017) wherein they recorded maximum shoot length, number of shoots per plant, number of leaves per shoot, inter-nodal length, number of fruits per node, number of fruiting nodes per shoot, fruit yield per plant with the foliar application of GA_3 @150 ppm.

5. Effect of Plant Growth Regulators on Harvesting, Yield and Quality

Phalsa fruits ripe 40 – 55 days after flowering. Fruits are harvested in April and May when they turn to crimson red from green. For distant markets, mature crimson red fruit are harvested while for local market, dark purple colour ripe fruits are harvested. Fruits for storage and transport should be harvested at colour turning stage. Harvesting is done manually as fruits are very small and all fruits do not mature at a time. Harvesting continue for almost one month and is done several times at alternate days due to gradual and steady ripening of fruits on clusters. Therefore a number of pickings are required at 2–3 days intervals which prove to be expensive. Preharvest application of ethephon or ethrel (500ppm) at colour break stage reduces the number of pickings from 7 to 3 and improves fruit quality in terms of total soluble solids, vitamin C content and colour development due to an increase in anthocyanin pigments. The average fruit yield of phalsa is 5 to 10 kg per plant (Salunkhe and Desai, 1984). In phalsa, GA_3 (40 ppm) was found to increase the fruit size and total yield (Randhawa *et al.,* 1959b). Chundawat and Singh (1980) stated that NAA at 50 ppm was most effective for increasing the yield. The increase in fruit yield might be due higher rate of cell division, enlargement, photosynthesis and thereby increase in enzymatic activities as well as involvement in biosynthesis of auxins at critical stage which induced higher fruit set. This in turn helped in increasing the number of fruits per tree resulting into higher fruit yield. As soon as phalsa fruits are harvested, they should be sent to market immediately for sale due to its perishable nature.

Sometimes growers do not get satisfactory market price for the production. So it would be desirable to delay or to hasten the maturity of fruits. To serve the above purpose growth promoters or growth retardants are used, which promote or delay the maturity of fruits. Ethrel (2-chloro ethyl phosphonic acid) hastens the maturity and significantly reduces the number of pickings in phalsa (Singh *et al.*, 1979). The span of harvesting and number of pickings were significantly influenced by the application of growth regulators. Ethrel 1000 ppm also significantly reduced the span of harvesting (9.76 days) and number of pickings (3.57) followed by Ethrel 750 ppm (Kacha *et al.*, 2014). As per Kacha *et al.* (2014), the quality of fruits in terms of TSS (25.23%), reducing sugar (2.01%) and total sugar (5.74%) was significantly better with Ethrel 1000 ppm followed by Ethrel 750 ppm. As per Kaur *et al.* (2018) ethrel 1000 ppm improved fruit quality as well as yield parameters and also fetched premium price in the market due to their attractive colour appearance. Application of GA_3 150 ppm significantly reduced acidity (2.55%) and increased ascorbic acid content (39.50%) in phalsa fruits. Chundawat and Singh (1980) concluded that NAA 50 ppm significantly increased vitamin C and reducing sugar contents of phalsa fruit. The increase in vitamin C content was due to catalytic influence of NAA on bio synthesis of vitamin C from sugar or inhibition in activity of oxidative enzymes or both (Vatsal, 2014).

Debnath *et al.*, (2011) found that GA_3 100 ppm was the most effective in improving yield per plant (3.05 kg), yield per hectare (7.63 t ha^{-1}) and hundred fruit weight (61.48g). Ethrel 500 ppm application recorded maximum total soluble solids content (25.72 %). While, maximum reducing sugar (18.91%), TSS to acid ratio (10.98), pulp weight (51.45g), pulp to stone ratio (5.85g), minimum titratable acidity (2.26 %) and stone weight (8.83g) was recorded with GA_3 100 ppm. Kinetin 30 ppm resulted in maximum shelf life (51.46 hr) of the fruits.

Kamboj *et al.*, (2017) on spraying phalsa plants with GA_3 (60 ppm) and NAA (50 ppm) before flowering and after fifteen days of flowering reported an improvement in physical characters and yield attributes of phalsa with the application of GA_3 60 ppm. Kaur *et al.*, (2018) revealed that GA_3 150 ppm treated plants yielded fruits with maximum fruit length (1.26 cm), fruit breadth (1.43 cm), fruit weight (1.20 g), reducing sugars (7.49 %), ascorbic acid (31.10 mg/100 g pulp) and yield (3.48 kg/plant). Whilst, minimum stone length (0.50 cm), stone breadth (0.33 cm) and stone weight (0.045 g) were recorded in fruits obtained from plants treated with NAA 25 ppm.

References

- Alam, T. and Kumar, A. 2017. Effect of GA_3 on growth, yield and quality of phalsa (*Grewia subinequalis* D. C.). *Plant Archives,* **17**(1): 608-610.
- Anonymous 1981. Annual Report. All India Coordinated Fruit Improvement Project, pp. 3031.
- Bhuva, H.P., Jadav, R.G., Katrodia, J.S., Patel, R.L. and Raja, K.R.V. 1995. Studies on production potential through single and double cropping of phalsa (*Grewia subinaequalis* DC). *Indian J. Hort.*, **52** (1): 19-23.
- Chundawat, B.S. and Gupta, O.P. 1974. Effect of growth retardants (B-9 and Cycocel) on vegetative growth and yield of phalsa (*Grewia subinaequalis* DC). *Haryana J. Hort. Sci.*, **3**: 113-115
- Chundawat, B.S. and Singh, R. 1980. Effect of growth regulators on Phalsa (*Grewia asiatica* L.) I. Growth and fruiting. *Indian J. Hort.*, **37**(2): 124-131.
- Debnath, A., Vanajalatha, K., Momin, U. and Reddy, M. 2011. Effect of NAA, GA_3, kinetin and ethrel on yield and quality in phalsa (*Grewia subinaequalis* D.C.). *The Asian J. Hort.*, **6**(2): 474-477.
- Gill, S.S., Kaushik, H.D. and Sharma, S.K. 2001. Effect of modes of pollination on fruit set and insect pollinators of phalsa (*Grewia subinaequalis* D. C.). *Res. on Crops*, **2**(2): 193-196.

- Kacha, H.L., Jat, G. and Patel, S. K. 2014. Performance of various plant growth regulators on yield and quality of phalsa. *HortFlora Res. Spectrum*, **3**(3): 292-294.
- Kamboj, N., Gandhi, N., Singh, S. and Singh, V. 2017. Effect of Gibberellic acid and Naphthalene acetic acid on the physical parameters and yield of Phalsa. *Int. J. Adv. Res. Sci. Engineer.*, **6**(1): 785-790.
- Kathrotia, R.K. and Singh, S.P. 1995. Regeneration of roots in phalsa (*Grewia asiatica* Mast) stem cuttings as influenced by maturity of wood and root promoters. *Adv. Hort. For.*, **4**: 35-41.
- Kaur, G., Singh, M. and Singh, M. 2018. Influence of bio-regulators on the fruit quality and yield attributes of Phalsa (*Grewia subinaequalis* DC) *J. Exp. Agric. Int.*, **21**(3): 1-6.
- Lakra, S., Kerketta, A., Ekka, R.A. and Saravanan, S. 2018. Effect of pruning and plant growth regulators on vegetative growth of phalsa (*Grewia asiatica* L) *Int. J. Curr. Microbiol. App. Sci.*, **7**: 2882-2885.
- Mohammed S. and Chauhan, K.S. 1970 .Vegetative propagation of phalsa (*Grewia asiatica* Mast). *Indian J. Agric. Sci.*, **40**: 581-586.
- Parihar, M. C., Singh, R. and Gupta, S.C. 1999. Effect of naphthalene acetic acid on vegetative growth of phalsa (*Grewia asiatica* L.). *Haryana J. Hort. Soc.*, **28**: 200.
- Prasad, A. and Bajpai, P.N. 1963. Effect of Alpha NAA on fruit set, size responding and yield of phalsa var. Sharbati. *Kanpur Agri. College J.*, **22**(1): 59-61.
- Randhawa, G.S., Malik, R.S. and Singh, J.P. 1959a. Note on clonal propagation in the phalsa (*Grewia asiatica* L). *Indian J. Hort.*, **16**(2): 119-120.
- Randhawa, G.S., Singh, J.P. and Khanna, S.S. 1959b. Effect of gibberellic acid and some other plant regulators on fruit size, total yield and quality in phalsa (*Grewia asiatica* L.). *Indian J. Hort.*, **16**(3): 202-205.
- Rema, J. and Sharma, V.P. 1991. Effect of plant growth regulators on yield and quality of phalsa (*Grewia subinaequalis* L.). *South Indian Hort.*, **39**(6): 327-331.
- Salunkhe, D.K. and Desai, B.B. 1984. *Postharverst Biotechnology of Fruits*, Vol. II, CRC Press, Boca Raton, Florida, Miami, USA, p. 129.
- Sharma, B.B., Saxena, S.K. and Sharma, V.P. 1987. *Proceedings of Symposium on Himalayan Horticulture in the Context of Defence Supplies.* Organized during October 28-30, 1987 by DRDO at Tezpur, Assam, Abstr. No. C4, p. 49.

- Shrivastava, S.R. 1996. Effect of IBA, sugar and captan on rooting of phalsa (*Grewia subinaequalis* DC.). *Recent Horti.*, **3**(1): 52-54.
- Shrivastava, S.R., Sunil Mahajan, Nema, B.K., Amar Chandra, Mahajan, S. and Chandra, A. 1998. Effect of IBA, sugar and captan on shoot growth of phalsa (*Grewia subinaequalis* Dc.). *JNKVV Res. J.*, **28-29** (1-2): 80-82.
- Singh, B., Yadav, A.L. and Meena, A.K. 2017. The effect of foliar feeding of plant growth regulators on growth and fruiting behaviour of phalsa. *Int. J. Agric. Innov. and Res.*, **5**(5): 806- 808.
- Singh, H. and Singh, R. 1981. Effects of plant growth regulators and manuring on the fruit quality of phalsa. *Indian J. Hort.*, **38**: 157-161.
- Singh, I.S., Singh, H.K. and Chauhan, K.S. 1979. A note on the effect of pre-harvest application of ethephon on ripening and quality of phalsa (*Grewia asiatica* L.). *Haryana J. Hort. Sci.*, **8**: 196-196.
- Singh, J.P., Godara, P.S. and Singh, R.P. 1961. Effect of type of wood and planting dates on the rooting of phalsa. *Indian J. Hort.*, **18** (1): 46-50.
- Singh, M., Gautam, R.K.S., Prasad, Y. and Singh, A.K. 1986. Studies on the effect of plant growth regulators and their different concentrations on yield and quality of phalsa (*Grewia subinaequalis* DC). *Haryana J. Hort. Sci.*, **15**(3-4): 196-199.
- Singh, S.M. and Kumar, H. 1967. Influence of ringing of source shoots and concentration of indole-3-butyric acid on phalsa stem cuttings. *Indian J. Agric. Sci.*, **37**: 151-154.
- Singh, Z., Grewal, G.P.S. and Singh, L. 2000. Effects of gibberellin A4/A7, and blossom thinning on fruit set, retention, quality, shoot growth and return bloom of Phalsa (*Grewia asiatica* L.). *Acta Hort.*, **525**: 463-466.
- Tripathi, P. 2009. Phalsa (*Grewia subinaequalis*) cultivation. https://www.researchgate.net/publication/304765037 (accessed on 26.11.18).
- Vatsal, R.H. 2014. Effect of plant growth regulators on flowering, fruit set, fruit yield and quality of red guava (*Psidium guajava* L.). M. Sc. thesis submitted to Anand Agricultural University, Anand, Gujarat, India.
- Yadava, U.L. and Rajput, C.B.S. 1969. Anatomical studies of rooting in stem cuttings of phalsa (*Grewia asiatica* L.). *Hort. Sci.*, **1**: 19–22.

24

Plant Growth Regulators in Pomegranate

R.K. Tarai, U.S.Nayak, A.K.Panda, V.K. Sharma, V.S. Supe, M.B. Shete and C.Y. Dhumal

1. Introduction

Pomegranate (*Punica granatum* L.) is a fruit of sub-tropical region to semi-arid region. Pomegranate (*Punica granatum* L.) is a favourite table fruit, valued very much for its refreshing juice consisting of nutritional and medicinal properties. The fruit is rich in anthocyanins. Maharashtra state is the largest producer of this fruit crop in India. The cultivation of any fruit crop has certain loopholes in its production technology which are affected by several edaphic, genetic, environmental problems. To overcome those problems in fruit crops, growth regulators have been advantageously used in the recent times to increase the fruit production and to improve the quality of several fruit crops.

Plant growth regulators have remained an important component in horticulture from time immemorial because they were effective means of quantitative as well as qualititative improvement in growth and development of crops. Plant growth and development as well as the responses to environmental factors, are highly regulated by complex and coordinated action of the endogenous hormones. In addition to this other plant growth regulators are also reported to be very helpful in this direction by altering the growth and development of plant. The use of plant growth regulators has become an important horticultural tool for growth, flowering and fruiting in many fruit trees. Plant biostimulants are known to improve fruit size, appearance and aril quality by direct effects on fruit growth and development or indirectly by regulating crop load, tree vigour and canopy architecture. Now a day, commercial formulations of the hormones are available in the market and their use is relatively cheap and economical to the farmers. Plant growth regulators are reported to play a significant role in pomegranate (Chaudhari and Desai, 1993). A brief description about effect of plant growth regulators on plant growth and development, fruit yield and quality, propagation, storage of pomegranate have been discussed below:

2. Effect of Plant Growth Regulators in Propagation

Propagation of pomegranate is done by seeds, cuttings, air layering and grafting. In cultivated pomegranate, seed germination is not a problem. In general, 60-75% seed germination has been observed in most of the cultivated varieties. The seedling population shows high degree of variability, consequently a number of seedling origin varieties were selected in India. But for commercial cultivation, vegetatively propagated materials are preferred and seedling plants are avoided. In pomegranate, mainly air layered saplings are used for planting. The art of propagation by cuttings is the most preferable technique popular in the field of horticulture and also in pomegranate. In cuttings, growth substances applied exogenously are found to enhance early and good root formation. Various classes of growth regulators such as auxins, cytokinins, gibberellins and ethylene influence root initiation in cuttings. Of these, auxins have greater effect on root formation in cuttings. In horticulture, auxins, especially NAA and IBA, are commonly applied to stimulate root initiation in cuttings of plants. Recently, in some parts of India, rooted hard wood, semi-hard wood and soft wood stem cuttings are also utilized in pomegranate. The most convenient and cheapest method of obtaining fully developed stronger trees in considerably lesser time is by cuttings. In order to reduce the high mortality of rooted cuttings under field conditions it is highly desirable to build a healthy and well developed root system by treating with plant growth regulators (Sharma *et al.*, 2009). Generally, Quick dip and prolonged dip method methods (Panwar *et al.*, 2001) using plant growth regulators (Reddy and Reddy, 1990) are practiced to induce rooting in stem cuttings of Pomegranate.

The best and most commonly used chemical for rooting in cuttings is indole butyric acid (IBA), which is decomposed relatively slowly by the auxin-destroying enzyme system in plants. Because this compound also moves very slowly in the plant, much of it is retained near the site of application. Another highly active auxin frequently used for root promotion is naphthalene acetic acid (NAA). As NAA is more toxic than IBA, there is a greater danger of injury to treated plants. The amides of both compounds are also effective rooting agents.

Gurjar and Patel (2007) noticed that the number of sprouts per cutting was highest when hardwood cuttings of pomegranate cv. Ganesh were treated with 4000 ppm IBA and planted in a mixture of soil, sand and leaf mould. In the quick dip method 10 sec to 1 min treatment was beneficial for inducing root in pomegranate stem cutting (Saroj *et al.*, 2008; Singh, 2014). Ghosh *et al.* (1988) reported that in pomegranate, IBA was better than NAA in inducing rooting in cuttings. Application of IBA at 5000 ppm resulted in maximum rooting (83.3%). Improved percentage of rooting and also root and shoot growth was noticed in hardwood cuttings treated with IBA 750 ppm in pomegranate (Panwar *et al.*, 2001). Upadhyay and Badyal (2007) observed that pomegranate hardwood cuttings treated with IBA at 2000 ppm recorded maximum survival percentage (81.33%) under Palampur condition. Sharma *et al.* (2009) obtained maximum rooting, root number and root length is observed with IBA 500 ppm + Borax 1% both in semi hard and hard wood cuttings. Hard wood cuttings respond better to the hormonal treatment as compared to semi-hard wood cuttings.

In an experiment conducted at University of Agricultural Sciences, Bangalore, Rajanna (1981) found that easy-to-root 'Kabul Red' pomegranate cuttings rooted better than difficult-to-root 'Basein Seedless' because of higher per cent of total sugars and total phenols. The phenolic compounds which are supposed to synergise auxin action were higher in easy to root pomegranate cultivar 'Kabul Red' than in difficult-to-root 'Basein Seedless' cuttings.

Kumari *et al.*, (2013) carried a study at Gonawila, Sri Lanka to investigate the effect of plant growth regulators on the development of roots and shoots in hard wood cuttings of pomegranate. The growth regulators used were Indole-3 Butyric Acid and Naphthalene Acetic Acid. Results showed that Indole -3 Butyric Acid at 500 ppm + 1% Borax for 15 mins. recorded the highest survival percentage, rooting percentage, mean root length, number of roots per cutting, shoot length and number of shoots. The possible explanation for high survival percentage in field may be due to, better development of root system with good quality root and shoots parameters enabling the rooted cuttings to make better growth under field conditions after plantation and thereby accounted the highest field survivability.

Damar *et al.*, (2014) conducted an experiment in Madhya Pradesh to study the role of plant growth regulators and biofertilizers on rooting of pomegranate cuttings. Data indicated that among growth regulators, the earliest sprouting of cutting (8.33 days) as well as significantly highest percentage of success (79.00 %), number of leaves per shoot (13.43), length of root (11.33 cm), diameter of root (1.23 cm), fresh weight of roots (2.30 g), dry matter percentage of roots (44.2 %) and number of roots per cutting (10.08) were observed in IBA 2000 ppm. This may be due to increased level of auxins resulted in earlier completion of physiological processes in rooting and sprouting of cuttings. In case of different combinations of biofertilizers and growth regulators, percentage of success of cutting, number of shoots per cutting, number of leaves per shoot, total number of leaves per cutting, diameter of roots, dry matter percentage of roots, number of roots per cutting, length of root and fresh weight of roots were found significantly superior under treatment combination of PSB + 2000 ppm IBA. This can be attributed due to increased level of growth promoting substance, available P_2O_5 and other nutrients with the application of PSB due to synergistic effect of both biofertilizer and growth regulators in various ways.

Singh (2014) conducted an experiment in Uttarakhand, India to study the effect of different IBA concentrations on the rooting of pomegranate (*Punica granatum* L.) Cv. Ganesh hardwood cuttings under mist house condition. The hardwood cuttings of *Punica granatum* L. cultivar Ganesh were collected from healthy vigorous shoots of 4-6 year old plants. Rooting media consisted of one part of sandy soil and one part of FYM. The stem cuttings of *Punica granatum* L. treated with IBA solutions of different concentrations i.e. 100 ppm, 200 ppm, 300 ppm, 400 ppm, 500 ppm and control by quick dip method. Among all the treatments, maximum number of sprouted cuttings (7.33), average length of sprout (20.53 cm), average number of leaves (25.33), percentage of rooted cutting (73.33), number of primary roots (29.26), and average length of roots (24.88 cm) was noticed in 500 ppm concentration of IBA. The enhanced hydrolytic activity in presence of applied IBA coupled with appropriate planting time might be responsible for the increased percentage of rooted cuttings. The increase in length and diameter of cuttings may be due to auxin application. Auxin has been found to enhance the histological features like formation of callus and tissue and differentiation of vascular tissue.

Simranjeet *et al.*, (2016) studied the effect of IBA, PHB and time of planting on rooting of pomegranate cuttings of Cv. Ganesh at Amritsar, Punjab, India. Uniform sized cuttings of pencil thickness, 20 cm long were taken from healthy pomegranate in August and January. The growth regulators (IBA and PHB) was applied by adopting quick dip method (for 1 min.) technique. Results of the study showed that the maximum percentage of sprouted cuttings (85.45%), survival percentage (85.88%), maximum number of roots (18.58), maximum root length (11.13cm) and maximum root weight (1.80 g) were recorded in treatment combination IBA 1000 ppm + PHB 750 ppm. The month of August proved to be the best time of planting for success of cuttings.

Mehta *et al.*, (2018) studied the effect of IBA Concentration and time of planting on rooting in pomegranate cuttings during the Winter season 2016-17 at Kota, Rajasthan,

India. The dates of planting were 25th December, 10th January and 25th January and the three concentrations of IBA were 500ppm, 1000ppm and control. Results of the investigation showed that highest number of sprouted cuttings (7.11), number of sprouts per cutting (4.00), number of leaves on new shoots (10.89), percentage of rooted cuttings (71.11%) was recorded when planting was done on January 25th planting. In case of IBA concentration, highest number of sprouted cuttings (7.33), number of sprouts per cutting (4.66), length of longest sprout (7.28 cm), number of leaves on new shoots (10.66) and percentage of rooted cuttings (73.33%) was observed under 500ppm IBA concentration.

According to Kaur and Kaur (2016), IBA 1000 ppm + PHB 750 ppm proved to be the best in terms of maximum percentage of sprouted cuttings (86.40%), survival percentage (86.73%), number of shoots (7.71), average shoot diameter (1.29 cm), number of leaves (126.00), total leaf area (259.48 cm^2), maximum number of roots (24.65), maximum root length (15.40 cm) and maximum root weight (2.40 g) under Amritsar condition.

An experiment was conducted at Regional Horticultural Research and Extension Center (RHREC), UHS, GKVK, Bengaluru, in the summer of 2014. Ten treatments of growth regulator formulations were imposed, including the Control, under Completely Randomized Design (CRD). Semi-hardwood cuttings of pomegranate were prepared from healthy twigs 20 to 30cm long, bearing 4 to 5 nodes and a thickness of 0.75 to 1.0cm. The basal 6 - 8cm portion of the cuttings was dipped with a combination of different concentrations (500, 1000 or 1500 ppm) of indole butyric acid (IBA), and 500, 1000 or 1500 ppm of naphthalene acetic acid (NAA) for 8 hours. The cuttings were planted in a plastic bag (9 x 5 inches) containing a mixture of sand, soil and FYM in equal proportions. Then, the plastic bags were placed in polyhouse during March. Results showed that highest sprouting percentage (68.00%), highest rooting percentage (60.40%), maximum number of sprouts (2.09), greatest diameter of sprout (0.59cm), longest roots (18.9cm) and widest diameter of root (1.62mm) were recorded in cuttings treated with IBA 1500ppm + NAA 1500ppm, whereas, longest sprout (34.50cm), highest number of primary roots (10.30) and secondary roots (62.70), highest fresh weight of sprout (9.03g), highest dry weight of shoot (4.66g), highest fresh weight of root (2.70g) and the highest dry weight of root (1.63g) per cutting was recorded with IBA 1500ppm + NAA 1000ppm (Ahmad Seiar, 2017).

Stover (2007) has reported that air layering can be advantageously useful in pomegranate propagation to minimize the time for fruiting earlier than cutting planting. To increase the success percentage in air layering, growth regulators are employed, which help in callus formation, root initiation, root development and ultimately survival of air layers. The exogeneous application of IBA induce rooting in stem cuttings and in air layers due to their ability to achieve to active cambium regeneration, cell division and cell multiplication (Rymbai and Reddy 2010). Tomar and Tomar (2012) conducted field trial during 2009-10 and 2010-11 at research orchard of Horticulture, College of Agriculture, Gwalior in sandy clay soil with different concentrations of IBA and NAA on rooting and survival percentage of pomegranate *(Punica granatum L)* air layers revealed that growth regulator IBA has better effect on rooting and survival percentage of pomegranate as compared to NAA. Also, application of IBA with 2000 ppm concentration is best for successful propagation of pomegranate air layers and among different concentration of NAA, 2000 ppm is the best. Evidence suggests that auxin might have increased rooting and ensured length of roots (Hartmann *et al.*, 2002).

3. Role of Plant Growth Regulators on Micro-Propagation

Conventional breeding techniques of woody fruit trees is often difficult and slow because of high levels of heterozygosity and the long generation time between successive crosses as reviewed from time to time (Naik and Chand, 2003; Singh *et al.*, 2007). These difficulties necessitate the development of rapid and efficient regeneration protocol for *in vitro* propagation of elite genotypes. Pomegranate is propagated vegetatively by the rooting of hard wood cuttings, although the establishment of new plants requires one year. Micropropagation in fruit tree would help in overcoming difficulties of vegetative propagation, producing true to-type plants and rapid and mass production of planting materials (Samir *et al.*, 2009). Hence, several studies have been conducted on micropropagation of pomegranate trees over the past several years. Protocols have been developed for regeneration of *P. granatum* L. plantlets *in vitro* through either organogenesis from callus derived from leaf segments, cotyledons (Murkute *et al.*, 2002; Raj and Kanwar, 2008; Kanwar *et al.*, 2010), anthers (Naik *et al.*,1999) or through embryogenesis from various seedling explants, petals and immature zygotic embryos (Kanwar *et al.*, 2010).

Patil *et al.*, (2011) conducted an experiment on propagation of pomegranate cv. Bhagwa through nodal segments cultured on two different media at full strength Murashige and Skoog (MS) and Woody Plant Medium (WPM). The media was prepared as a basal medium supplemented with 0.2 to 2 mg/L 6-benzylaminopurine (BAP), 0.1 to 1 mg/L 1-naphthalacetic acid (NAA), 0.5 to 2.5 mg/L silver nitrate (AgNO3) and 10 to 50 mg/L adenine sulphate for establishment stage. For proliferation stage, 0.1 to 0.5 mg/L BAP and NAA was tested. For rooting stage 0.0, 0.25 and 0.5 mg/L 3-Indolebutyric acid (IBA) and NAA on MS and WPM medium were tested. The nodal explants grown on MS medium containing 1.8 mg/L BAP, 0.9 mg/L NAA, 1 mg/L silver nitrate and 30 mg/L adenine sulphate had the highest proliferation rate (10 to 15 shoots/explants) in establishment stage. The same trend was found concerning the maximum leaves numbers (15 to 20 leaves/explants) on proliferation medium containing 0.4 mg/L BAP and 0.3 mg/L NAA. The plantlets grown on MS medium were found to have better survival compared to WPM medium. 0.5 mg/L NAA and 0.5 mg/L IBA showed equal rooting response in both the medium, whereas thick root formation was observed in the medium containing IBA.

4. Role of Plant Growth Regulators on Improvement of Vegetative Growth

Pomegranate is a shrub that naturally tends to develop multiple trunks and has a bushy appearance. When domesticated, it is grown as a small tree that grows up to 5m. Under natural conditions, it can sometimes grow up to more than 7m. Most of the pomegranate varieties are deciduous trees. However, there are several evergreen pomegranates in India. Rahemi and Atahosseini (2004) studied the effect of NAA and NAD on vegetative growth and recorded significant increase in leaf area index of pomegranate cv. Shisheh Cup with the application of NAA at 50mg/l and NAD at 100 mg/l. Sharma and Ananda (2004) reported that application of GA_3 either alone at 10-40 ppm or at 10 ppm in combination with BA at 5 ppm resulted in significantly higher vegetative growth over the control.

Phawa *et al.*, (2017) conducted a field experiment to study the effect of plant growth regulators on growth of pomegranate cv. Kandhari in Allahabad (Uttar Pradesh). The results revealed that application of GA_3 at 75 ppm showed maximum plant height (194.90 cm), spread of plant (194.20 cm), canopy volume (3.81 m^3), internode length (8.07 cm), shoot length (15.57 cm), plant girth (3.24 cm), leaf area (7.37 cm^2). The maximum increase in plant spread and canopy volume with the spray of GA_3 might be due to beneficial effect of GA_3 in cell elongation and enlargement. The increased uptake of water and nutrients due to persuasive swelling forces leading the softening of cell wall and thereby favoured better development of plants resulting in greater height and number of branches per plant and ultimately the greater plant spread and canopy volume. The maximum increase in plant girth due to GA_3 application might be due to fact that gibberellins regulate the growth of plants by causing cell division and cell elongation, thereby increasing the vegetative growth by overcoming genetic dwarfism. The increase in leaf area was probably due to the fact that gibberellin stimulates the activity of auxin resulting in production of more number of leaves thereby increasing leaf area.

A two years trial was conducted on 13 years old Manfalouty pomegranate trees grown in sandy soil under drip irrigation system from well at El Maghara Station, Desert Research Center, North Sinai Governorate, Egypt. Three concentrations of each Naphthaleneacetic acid (NAA) i.e. 5000, 10000 and 15000 ppm, Paclobutrazol (PP333) i.e. 500, 1000 and 1500 ppm and Maleic hydrazide (MH) i.e. 500, 1000 and 1500 ppm and hand removal (control) treatments were trunk applied on 1st February after suckers removal from the trunk. The results indicated that all NAA, PP333 and MH treatments reduced number of suckers, sucker length and suckers weight of Manfalouty pomegranate trees. Moreover, 1500 ppm PP333 treatment proved to be the most effective application treatment for controlling the growth of suckers with some advantage over the traditional hand removal (control treatment) of growth of suckers but gave an intermediate positive effect on yield and fruit quality. Briefly, the results showed that the reduction of concentration level of PP333 and MH led to an enhancement of yield and fruit quality. In addition, NAA gave an intermediate positive effect for controlling the growth of suckers than PP333 and MH respectively. Moreover, NAA treatments gave a high positive effect on yield and fruit quality as compared with hand removal treatment (control). Generally, 15000 ppm NAA treatment proved to be the most efficient treatment on enhancing yield and fruit quality excepted 10000 ppm NAA treatment induced high positive effect on fruit total sugar , T.S.S. and ascorbic acid content. Moreover, economic study indicated that some growth regulators treatments attained the highest values of both total revenue per feddan (L.E/ feddan), net revenue per feddan (L.E/ feddan), technological efficiency, economic efficiency, and high sealing price at harvest time, because of increasing the fruit quality which leads to the best revenue value but the best treatment is Paclobutrazol PP333 at 500ppm . This treatment attained the highest values of both technological efficiency and economic efficiency. It leads to improve the "Manfalouty" pomegranate fruit quality and raising the net income farmer which reflect positively on pomegranate in El maghara in North Sinai governorate (Salama *et al.*, 2016).

For several decades, auxin and cytokinin were the only two hormones known to be involved in the control of shoot branching through apical dominance, a process where the shoot apex producing auxin inhibits the outgrowth of axillary buds located below. Various growth retardants have been used to restrict the vegetative growth of plant such as AMO 1618, ancymidol, paclobutrazol, B-Nine, chlormequat etc. The reduction in vegetative growth by growth retardants is due to the systemic inhibition of GA3 biosynthesis pathway at the sub-apical meristem, which ultimately reduced cell elongation and rate of cell division and decreased the shoot growth. These growth retardants also inhibit the gibberellin precursor by blocking the oxidation of ent-kaurene to ent-kaurenoic acid.

Digrase *et al.*, (2016) conducted a research to study the effect of plant growth regulators and chemicals on plant growth of cv. Bhagwa under Parbhani conditions. The results indicated that, application of GA_3 75 ppm showed maximum height (170.75 cm), spread of plant (152.50 cm). The effectiveness of GA_3 might be due to its role in promoting growth and stimulated the rapid cell elongation in meristematic zone of vegetative plant organs. Due to cell elongation and division, the length and spread of branches might have increased, hence GA_3 played an important role in enhancing the height and spread of the plant. GA_3 treated plants might be attributed to the initial rapid mobilization of the food reserves from the plants during the rapid extension of vegetative growth since the plants which received GA_3 treatments made extensive vegetative growth than other plants.

5. Role of Plant Growth Regulators on Flowering/Crop Regulation

Use of plant bio-regulators for induction or delaying in flowering in different fruit crops differ as per the purpose of users. In many woody plants including fruits GA inhibits flower formation. In these cases growth retardants *viz.*, paclobutrazol, SADH which inhibits GA biosynthesis are used to promote flowering such as in pears and mango. Ethephon proved to be the most effective promoter of flower bud formation. However, its use on bearing trees was limited because ethephon also caused thinning (Byers, 2003). Pomegranate flowers develop into one of two types of flowers normally produced by pomegranates: hermaphrodite flowers ("vase shape") and male flowers ("bell shape"). Both types have several hundred stamens. The bell-shape flower has a poorly developed or no pistil and atrophied ovaries containing few ovules and is infertile. Therefore, the bell shape flower is referred as a male flower and will drop without fruit set. The vase-shape flower is fertile with a normal ovary capable of developing fruit. The stigma of the hermaphrodite is at the anthers height or emerging above them. This position allows for self-pollination as well as pollination by insects.

Some of the fruits like guava, pomegranate, lemon, etc., if left without any treatment, give several light harvests of the variable quantities and qualities from the various flowering flushes throughout the year. In Pomegranate, there are three distinct flowering season i.e. February-March (Ambe Bahar), June-July (Mrig Bahar) and October-November (Hasth Bahar) with the corresponding harvest period during rainy, winter and spring season, respectively. A good quality production can be ensured by regulating the crop in such a way that they could produce only one crop instead of two or three in a year. Plants are forced to produce only one crop instead of two or three crops with good quality production. The main objective of crop regulation is to force the tree for rest and produce profuse blossom and fruits during any one of the two or three flushes. It can be achieved through with holing irrigation water, root exposure, root and shoot pruning, deblossming, spray of chemical and other plant growth regulators. Most of the chemical treatments were found to alter the sex expression. GA induced significantly more male flowers while Ethrel application appeared to induce less male flowers. GA also reduced the percentage of hermaphrodite flowers, while Ethrel, Carbaryl and MH induced more hermaphrodite flowers. In many crops, GA_3 has been found to favour maleness (Krishnamoorthy 1981).

The study carried out by Ahire *et al.* (1993) at Mahatma Phule Agricultural University, Rahuri on regulation of flowering in pomegranate indicated that hand removal of flowers induces more flowers. MH (1000 ppm). Carbaryl (7000 ppm). GA (20 ppm) and NAA (250 to 500 ppm) and Ethrel (250 to 500 ppm) produced less flowers than control. Ethrel at 500 ppm was the most effective in thinning of flowers. GA induced more male flowers and less hermaphrodite flowers. Ethrel, MH and Carbaryl induced more hermaphrodite flowers.

Desai *et al.*, (1993) carried experiment to study crop regulation on pomegranate cv. Ganesh at Mahatma Phule Krishi Vidyapeeth, Rahuri. It involved Ethrel, GA, NAA, MH, Carbaryl and also mechanical removal of flowers. The treatments were found to influence fruit drop, total fruit number, its quality, yield and number of fruits in different grades. On the basis of net increase in monetary returns over control application of 250 ppm NAA followed by 0.7% carbaryl on the same day after 75 days of commencement of cropping was found to be the best. It also resulted in higher grade fruits of better quality with no phyto-toxic effects.

Anawal *et al.*, (2015) studied the effects of plant growth regulators on induction of flowering on pomegranate Cv. Bhagwa. The growth regulators were sprayed at the time of initiation of new sprouts. Results revealed that application of ethrel at 250 ppm gave significantly maximum number of hermaphrodite flowers (25.22). This might be due to greater accumulation of dry matter which is certainly suggestive to better photosynthetic activity, other metabolic activities and uptake of nutrients from soil. Similarly, due to ethrel spray 50 per cent flowering was observed in 12.67 days and 100 per cent flowering in 30.44 days. They also observed that NAA at 100 ppm resulted in maximum length and diameter of the flowers. The significant increase in size of flowers might be due to drawing of photosynthates to the flower as a consequence of intensification of the sink.

Phawa *et al.*, (2017) conducted a field experiment to study the effect of plant growth regulators on flowering of pomegranate Cv. Kandhari in Allahabad Uttar Pradesh. The results revealed that application of GA3 at 75 ppm showed maximum number of flowers per plant (29.01). The minimum days to flowering (23.67) were recorded under 40 ppm NAA. The increase in number of flowers following GA_3 application might be due to fact that the plants sprayed with GA_3 remained physiologically more active to build up sufficient food reserve for developing flowers. The increase in flowering may also be due to enhanced photosynthesis which increased the potential of trees to develop more flower buds.

6. Role of Plant Growth Regulators on Fruiting Behaviour

The fruit develops from the ovary and is a fleshy berry. The nearly round fruit is crowned by the prominent calyx. It is well known that competition between fruits is one of the principal factors affecting fruit growth, ripening and fruit quality. Fruit competes with each others for assimilates according to their relative sink strength and the availability of assimilates. When crop density is higher as under high fruit set conditions, competition for assimilates is intense, which result not only in decreased fruit size and quality but also promotes alternate bearing. Crop density, a measure of fruit crop, influences fruit growth and ultimately fruit size at harvest. It is largely dependent upon flowering and fruit set.

7. Role of Plant Growth Regulators on Flower and Fruit Thinning / Sex Expression

Normally, in pomegranate, the flower drop is mainly constituted by male (60.7%) followed by intermediate (12.1%) and hermaphrodite (9.5%) flowers. Here, the total number of flowers dropped may not be an indication of efficacy of the chemical to thin the flowers. Because, the production of flowers differed with the chemicals and the male flowers also dropped naturally. Therefore, the percentage drop of hermaphrodite flowers which mostly set the fruits, and the drop of intermediate flowers are to be considered.

Krishnamoorthy (1981) reported that Maleic hydrazide (MH) acts as an anti-auxin and it affects the reduction in flower production in many fruit crops. Among various chemicals used for thinning, ethephon is considered to be an effective agent and has been extensively used for fruit thinning in a number of fruit crops, and therefore, it is known as a flower thinner (Roberts and Hooley, 1988). Its application increases the activity of cellulase at abscission zone (at the pedicel end) and weakens the cell wall, which leads to fruit abscission. Many investigators concluded that Ethrel had a pronounced effect on sex expression in pomegranate. Chaudhari and Desai (1993) observed that the application of Ethrel at 250 ppm concentration significantly reduced the male flowers.

Ahire (1991) reported that GA_3 @ 20 ppm on 75th day after first irrigation of *Mrig* bahar increased the percentage of male flowers and reduced the hermaphrodite flowers.

While, Ahire *et al.*, (1993) found that Ethrel 500 ppm was more effective than 250 ppm in reducing the number of male flowers. The study carried out by Ahire *et al.*(1993) at Mahatma Phule Agricultural University, Rahuri on regulation of flowering in pomergranate indicated that hand removal of flowers induces more flowers. MH (1000 ppm). Carbaryl (7000 ppm). GA (20 ppm) and NAA (250 to 500 ppm) and Ethrel (250 to 500 ppm) produced less flowers than control. Ethrel at 500 ppm was the most effective in thinning of flowers. GA induced more male flowers and less hermaphrodite flowers. Ethrel, MH and Carbaryl induced more hermaphrodite flowers. However, the hermaphrodite flowers were significantly increased with the spray of higher concentrations of Ethrel as was observed by Choudhari and Desai (1992) and Ahire *et al.* (1993) in cv. Ganesh.

Choudhari and Desai (1992) investigated the effect of plant growth regulators for flower thinnmg with Ethrel (250, 500 ppm), NAA (250, 500 ppm), MH (1000 ppm) and Carbaryl (0.7%) applied on trees after removal of male and intermediate flowers and retained only 50 hermaphrodite flowers. They observed that NAA 500 ppm and 250 ppm affect flower thinning to the extent of 15 per cent and 12 per cent respectively as compared to control (5%). Choudhari and Desai (1992) also reported that Maleic hydrazide at 1000 ppm induced flower drop to the extent of 4 per cent when sprayed on 45th day after the commencement of the cropping.

Hussein *et al.* (1994a) stated that NAA can be employed for thinning fruit-lets in pomegranate cv. Manfalouty, when applied at 10 or 20 ppm in mid May or June and found a positive effect on fruiit thinning as compared to hand thinning treatment.

Choudhury and Desai (1993) found that the application of GA_3 at 20 ppm significantly increased the number of male flowers and decreased the hermaphrodite flowers in pomegranate. However, Desai *et al.* (1993) found that 20 ppm GA_3 applied in combination with 250, 500 ppm Ethrel or following hand removal of flowers significantly checked the induction of flowers after fruit set in cv. Ganesh.

Dong *et al.* (1999), while studying the effect of cultural practices on early and high production of pomegranate in China speculated that spray of 80 ppm GA_3 resulted in longer flowering time. However, they did not make any observation on the effect of this chemical on flower thinning.

Chaudhari and Desai (1993) found the possibility of exploitation of Maleic hydrazide at 1000 ppm in commercial cultivation of pomegranate, and observed a marginal influence of 1000 ppm Maleic hydrazide in reducing male flowers with a significant effect on inducing more hermaphrodite flowers in cv. Ganesh. Growth retardant cycocel also called as chlormequat or CCC had popularly been used as shoot growth control chemical in fruit crops, and the literature on the effect of CCC on fruit thinning in pomegranate is not much available. However, Venkatesan and Mohideen (1994) studied the effect of CCC at 250, 500 and 1000 ppm applied thrice at flowering, 40 and 90 days after flowering, and observed no positive effect of CCC on fruit thinning.

8. Role of Plant Growth Regulators on Fruit Drop

Abscission of mature fruit before harvest is a problem in many fruit crops producing countries. Increasing production and the desire for orderly fresh fruit marketing have made ontree storage of fruit an economic necessity. The synthetic auxin 2, 4-D is often incorporated into citrus production practices to reduce preharvest drop of mature fruits. AVG is a naturally occurring compound which blocks ethylene synthesis and limits pre-harvest drop. Shaybany and Sharif(1973) reported that ethephon (ethrel) sprayed 18 days before harvest resulted in increasing fruit drop in Rabbab pomegranate with increasing concentrations from 0 to 2000 ppm. Whereas, Desai *et al.* (1993) observed that in pomegranate cv. Ganesh, the spray of 250 ppm Ethrel resulted in lower fruit drop as compared to 500 ppm Ethrel applied at 75 days after commencement of cropping. Hussein *et al.* (1994a) while studying the crop regulation in pomegranate cv. Manfalouty with NAA at 10 or 20 ppm, Sevin 500 or 1000 ppm and hand thinning, but they did not find the effect of NAA on finit drop at different concentrations. Desai *et al.* (1993) carried out studies on crop regulation on pomegranate cv. Ganesh using Ethrel, GA3, NAA, MH, Carbaryl and hand removal of flowers. However, they obtained an at par effect on the control of fruit drop with 20 ppm GA3 in combination with 250 ppm Ethrel, 500 ppm Ethrel or following hand thinning treatment. Whereas, Dong *et al.* (1999) did not observe any the effect of 80 ppm GA3 on fruit drop in pomegranate under China conditions. Maleic hydrazide 1000 ppm alone or in combination with 500 ppm NAA had a significant influence on the fruit drop in pomegranate cv. Ganesh when sprayed after 75 days of the commencement of cropping (Desai *et al.* (1993). Venkatesan and Mohideen (1994) carried out a study on pomegranate cv. Ganesh with CCC at 250, 500 and 1000 ppm concentrations, applied thrice at flowering, 40 and 90 days after flowering. But, they did not observe any effect of CCC on fruit drop with any concentration used in the experiment.

Devkare (2003) reported that application of 6-BA @ 10 ppm resulted in maximum fruit drop (23.01%) when applied at 60th day after flower initiation, however early application resulted in better retention of fruits/tree (43.90 %) with increased fruit weight.

9. Role of Plant Growth Regulators on Fruit Set

Venkatesan and Mohideen (1994) investigated the effect of plant growth regulators on yield of pomegranate cv. Ganesh and revealed that NAA 25 ppm resulted in the highest number of fruits per tree (64 fruits) followed by 2, 4-D 20 ppm (60 fruits) as compared to control (43.3 fruits). But in terms of fruit weight per tree, 2,4-D 20 ppm ranked first followed by NAA 25 ppm treatment which recorded a yield of 22.52 and 21.02 kg per tree respectively as compared to only 9.21 kg under control.

Anawal *et al.*, (2016) carried an experiment in Karnataka to study the response of various plant growth regulators on fruit set cv. Bhagwa. There were six treatment combinations viz., NAA at 40, 50, 60 ppm and GA3 at 40, 50, 60 ppm. The results indicated that NAA 40 ppm was found effective in increasing number of fruits per tree (62.44).

Ghosh *et al.*, (2009) conducted an investigation to study the effect of plant growth regulators on pomegranate Cv. Ruby. The growth regulators were sprayed three times starting at 50% flowering stage and, subsequently, at 21 days' interval. Results revealed that application of NAA at 25 ppm gave significantly high fruit set (44.3%) and fruit retention (44.1%) which resulted in highest fruit yield of 7.8 kg/plant at the age of 7 years. Beneficial role of NAA in reducing fruit drop may be explained from the fact that it maintains on-going physiological and biochemical process of inhibition of abscission.

10. Role of plant Growth Regulators on Fruit Maturity

Patel (1993) observed that maturity period of pomegranate was reduced by spraying of CCC at 500 ppm. Pawar (2001) studied effect of Cycocel growth regulator on pomegranate cv.Mridula at 20th , 40th and 60th days after fruit set and reported that CCC 700 ppm concentration was helpful for earlier fruit maturity with increased fruit weight. The increase in fruit weight with the application of CCC may be attributed to cessation or suppression of vegetative growth that lead to diversion of metabolites to the fruit.

11. Role of Plant Growth Regulators on Fruit Characters

Desai *et al.*, (1993) observed that application of ethrel @ 250 and 500 ppm alone and in combination with NAA 100 ppm and GA 20 ppm influenced average weight of pomegranate fruit.

Devkare (2003) observed non-prominent effect of brassinosteroids @ 2ml. /l. on all the quantitative as well as qualitative characters under study. On the contrary he further reported reduced fruit length, diameter and mean fruit weight when brassinosteroids @ 2ml. /l. was applied at 60th and 90th day after flower initiation. Application of ethrel @ 500 ppm at 60th and 90th day after flower initiation resulted in fruit drop to the extent of 4 %. Further he also reported that application of ethrel @ 500 ppm at 90th day after flower initiation resulted in reduced rind thickness (0.64 mm), rind percentage (35.13) and acidity (0.40 %) as compared to rest of the treatments.

Manasa (2017) reported that 2,4-D at 40 ppm significantly increases fruit size characteristics viz., length (8.56 cm), diameter (8.62 cm), weight of fruit (223.24 g) and volume of fruit (244.80 ml) and minimum was recorded under control treatment.

12. Role of Plant Growth Regulators on Fruit Yield and Quality

Application of plant bio-regulators resulting in better quality and yield in fruit crops because it had improved the internal physiology of developing fruit in terms of better supply of water, nutrients and other compounds vital for their proper growth and development which resulted in improved size, quality and ultimately greater yield.

Desai *et al.,* (1993) observed a significant reduction in crop load and overall fruit size with 1000 ppm maleic hydrazide treatment alone or in combination form of 500 ppm NAA + 1000 ppm Maleic hydrazide in cv. Ganesh. However, highest unmarketable fruit numbers were recorded in control followed by 1000 ppm Maleic hydrazide alone. The spray of Melic hydrazide was found to have no phytotoxic effects and the fruit quality in term of TSS and acidity was not impaired. Whereas, 1000 ppm MH in combination with 500 ppm NAA gave highest TSS of 16.1% as compared to 14.5 per cent in control.

Patel (1993) observed that the quality of pomegranate cv. G-137 was improved by spraying of CCC at 500 ppm. CCC at 50 ppm accelerated fruit maturity from 140 to 137.3 days.

Pawar (2001) studied effect of Cycocel at 300, 500 and 700 ppm concentration with other growth regulators in pomegranate cv. Mridula at 20th, 40th and 60th days after fruit set and reported that CCC 700 ppm concentration increased TSS and sugars.

Devkare (2003) conducted experiment to study the effect of growth regulators on fruit size and quality of pomegranate. The growth regulators were sprayed 30, 60 and 90 days after flower initiation. He observed that maximum number of fruits/tree (58.58) and highest yield

(13.39 kg) was obtained when sprayed with CCC @ 250 ppm. There was increase in reducing and total sugars with decrease in acidity. The quality parameters *viz.*, TSS, reducing sugars and total sugars were mostly influenced by the application of CPPU @ 2 ml/l. The highest TSS was recorded by all the treatments of CPPU @ 2ml/l at all the intervals studied. The CPPU applied at 60th and 90th days after flower initiation recorded maximum TSS (20.70%). The CPPU applied at 90th days after flower initiation recorded highest reducing sugar (13.92 %) and total sugars (15.65 %).

Pawar *et al.*, (2005) recorded highest yield in cv. Mridula of pomegranate in Maharashtra with 75 ppm GA_3, while, Mohamed (2004) from Assiut (Egypt) recommended 150 ppm of GA_3 for getting heaviest fruit with lowest fruit-splitting in cv. Manfalouty of pomegranate. Beneficial effect of plant growth regulators in improving TSS content in pomegranate was also observed by Mohamed (2004). Improvement in TSS of fruits due to auxins (NAA, 2,4-D) and GA spray may be explained from the fact that application of these growth regulators after fruit set probably improved the physiology of leaves, thereby causing better translocation of vital components in the fruit and utilization of photosynthates by the developing fruit (Pandey, 1999).

Ghosh *et al.*, (2009) investigated to harness beneficial effects of plant growth regulators on yield and fruit quality in pomegranate cv. Ruby. Seven treatments with three growth regulators, viz., NAA at 25, 50 ppm; GA_3 at 10, 20 ppm; 2,4-D at 5, 10 ppm and control (water spray) were sprayed three times starting at 50% flowering stage and, subsequently, at 21 days' interval. Results revealed that application of NAA at 25 ppm gave significantly high fruit set (44.3%) and fruit retention (44.1%) which resulted in highest fruit yield of 7.8 kg/plant at the age of 7 years, as against 1.7 kg in the control. Fruit weight and quality improved significantly due to growth regulator sprays.

Khalil and Aly (2013) carried an experiment in Egypt to study the effect of foliar application of growth regulators on fruit quality parameters of pomegranate cultivar 'Manfalouty'. The sprays were taken on 2nd and 8th weeks from full bloom. Application of 300 ppm paclobutrazol significantly reduced the extent of fruit cracking (4.50 % to 15.20 % 1st and 2nd seasons respectively). Yield (33.20 kg fruits/tree) and fruit weight (340 g) were increased significantly with application of paclobutrazol. Spraying with paclobutrazol also significantly increased the grain per cent (57.50 % and 56.60 % 1st and 2nd seasons respectively), fruit juice per cent (30.14% and 32.11% 1st and 2nd seasons respectively), TSS per cent (15.28 % and 17.11 % 1st and 2nd seasons respectively) and acidity per cent (1.45 % and 1.38 % 1st and 2nd seasons respectively). The impact of higher rates of paclobutrazol in enormously suppressing vegetative growth, especially during peak fruit development stage, contributed to the superior yield. Foliar spray of paclobutrazol treatment could control excessive vegetative growth, indicating positive effects on fruit growth, yield and fruit quality, which is possibly related to the allocation of more carbohydrates to the fruits at the expense of reduction in shoot growth. In both seasons, the highest fruit length and diameter (9.23 and 9.90cm, respectively) were obtained by spraying 40 ppm NAA followed by spraying 80 ppm GA3 (8.71 and 9.35cm). This can be attributed to nature of auxins (NAA) to stimulate cell division, cell elongation and membrane permeability to water uptake. Gibberellins are involved in cell division and cell elongation. They are known to influence fruit size. Gibberellic acid is also reported to promote growth by increasing plasticity of the cell wall followed by the hydrolysis of starch into sugars which reduces the cell water potential, resulting in the entry of water into the cell and causing elongation. Application of 80 ppm GA3 significantly increased total anthocyanin per cent (0.49 % to 0.59 %).

A study was made to know the response of various plant growth regulators at different concentrations namely, NAA (40, 50, 60 ppm) and GA3 (40, 50, 60 ppm) on fruit set and yield of pomegranate cv. Bhagwa. The results indicated that NAA 40 ppm was found effective in increasing number of fruits per tree (62.44), fruit length (8.66 cm), fruit diameter (8.71 cm), fruit weight (262.23 g), fruit volume (255.44 ml), TSS (16.76°B), total sugars (15.58 %), reducing sugars (13.83 %), non-reducing sugars(1.75 %) against control (Anawal *et al.*, 2016).

A field experiment was conducted by Kumar *et al.* (2016) to evaluate the effect of micronutrients and plant growth regulators on fruit quality in pomegranate cv. Jodhpur Red. The treatments comprising of two micronutrients (H3BO3 & ZnSO4 at 0.20, 0.40, 0.60 %), two growth regulators (2,4-D & NAA at 10, 20, 40 ppm) and water were applied fifteen and thirty days after fruit set. Application of 2,4-D at 40 ppm resulted in more total soluble solids. H3BO3 resulted in more ascorbic acid content when applied at fifteen days after fruit set. 2,4-D when applied at fifteen days after fruit set, significantly increased reducing sugar content. Application of micronutrients and growth regulators applied fifteen days after fruit set was more effective in improving fruit quality as compared to thirty days after fruit set.

Digrase *et al.*, (2016) conducted a research to study the effect of plant growth regulators and chemicals on fruit quality of cv. Bhagwa under Parbhani conditions. The results indicated that, application of GA_3 75 ppm showed maximum length of fruit (8.67 cm), maximum diameter of fruit (9.20 cm) and the highest mean fruit volume (308.00 ml). Treatment combination consisting of GA_3 75 ppm + boron 0.3 per cent produced higher number of fruits (120 / plant), fruit weight (320 g) and yield (13.96 Mt/ha.) of pomegranate fruits on

6 year old tree. Increase in yield parameters of interaction effect might be due to cumulative effect of increased performance of yield characters contributed by application of GA_3 and boron foliar application of essential growth regulators and chemicals might provide required nutrition for resulting in rapid fruit development and increasing number of fruits ultimately increasing weight, yield per plant and yield per hectare.

Rathod *et al.*, (2017) studied the effect of ethrel and banana pseudostem sap on fruit yield and yield attributes of pomegranate cv. Bhagwa at Junagadh, Gujarat, India. The experimental results showed that ethrel @ 250 ppm showed significantly minimum fruit drop (20.39 %), maximum number of fruits per plant (86.57), fruit set (37.13 %), fruit retention (79.71 %), minimum days taken to first harvesting (177.58 days). The experimental results also showed that ethrel @ 250 ppm showed significantly maximum weight of 100 arils (40.62 g), fruit weight (277.75 g), fruit volume (230.42 ml), fruit length (7.17 cm), fruit circumference (26.01 cm) and fruit yield (23.90 kg/plant and 26.55 t/ha).

Manasa (2017) studied the effect of plant growth regulators on yield and quality of pomegranate cv. Bhagwa, at Latur, Mahrashtra, India during *Mrig Bahar* of 2016- 2017. Three plant growth regulators at different concentrations were sprayed at four different times i.e. at fruit set, 20th, 40th and 60th days after fruit set. The yield and quality of pomegranate fruits were significantly influenced by various plant growth regulators at different time of application. Among the various plant growth regulators used, maximum number of days required for maturity was recorded in the control treatment i.e.192 days and minimum number of days for maturity were recorded under NAA 10 ppm treatment i.e., 160 days. The maximum days to maturity after fruit set (176.10) was recorded by spraying at 60th and 40th days after fruit set. The minimum (174.20) was recorded by 20th days after fruit set. In general maximum yield /ha was recorded by application of 2, 4-D 40 ppm and GA3 75 ppm 143.53 q/ha and 137.02 q/ha respectively

13. Role of Plant Growth Regulators on Physico-Chemical Fruit Characteristics

Shaybany and Sharifi (1973) found that with increasing concentrations of ethephon from 0 to 2000 ppm there is decrease in the percent of TSS, pH, vitamin C content and TSS: acid ratio of the juice of Rabbab pomegranate when applied at 18 days before harvest.

Desai *et al.*, (1993) opined that Ethrel in combination form (500 ppm Ethrel + 20 ppm GA_3) gave higher average fruit weight (210.5 g/fruit) than when applied alone at 250 ppm (194.2 g/ fruit) and 500 ppm (207.3 g/fruit) respectively. They have also found higher fruit yield (18.7 kg/plant) with the treatment (NAA 250 ppm + 0.7% Carbaryl) in comparison to control (16.9 kg/plant).

According to Venkatesan and Mohideen (1994), GA3 predominantly produced deleterious effects on fruit characters of pomegranate, especially the higher concentrations at 100 and 200 ppm gave fruits of lower length (6.7 and 6.3 cm), diameter (10.2 and 10.0 cm), volume (130.0 and 93.5 ml/fruit) and fruit weight (120.6 and 84.1 g/fruit) in comparison to lower concentration of 50 ppm and control. Marked increase in rind thickness (5.11 cm) and consequently highest rind content (65.6%) were recorded in 200 ppm GA3 treated fruits, while the rag percentage was significantly reduced over control. However, intense pink colour of aril was found with GA3 50 ppm application.

Venkatesan and Mohideen (1994) also observed that the application of Cycocel had less marked effect on yield but have negative effect on fruit quality in Ganesh cultivar of pomegranate. They also found that CCC at higher concentration at 500 and 1000 ppm significantly reduced the aril weight (99.1 and 85.8g), aril percent (44.7 and 39.3%o), rag per cent (15.0 and 15.6%) and aril: rind ratio (1.10 and 0.87 w/w) as compared to lower concentration (250 ppm) and control.

To evaluate the effects of plant growth regulators on quality and quantity of pomegranate fruits, 2,4-dichlorophenoxyacetic acid (2,4-D), naphthalene acetic acid (NAA) and naphthalene acetamide (NAD) were applied to 'Shisheh Cup' pomegranate trees. In the first experiment 2,4-D (15, 30, 45 mg/L), NAA (25, 50, 100 mg/L) and NAD (50, 100, 200 mg/L) were applied to the uniform branch units of pomegranate trees at full bloom, 45 and 90 days after full bloom. In the second experiment 2,4-D (15, 30, 45 mg/L) were applied at full bloom, 30, 60 and 90 days after full bloom to the uniform branch units of pomegranate trees. In both experiments treatments were applied with a hand sprayer until the drip point and each treatment was applied to 3 trees and 4 uniform branch units per tree. In both experiments plant growth regulators significantly increased fruit L/D ratio, fruit weight, fruit volume, (expectedly NAD) rind weight, aril weight, and leaf area index. For the best effects of plant growth regulators on qualitative and quantitative characters in pomegranate fruits, it was recommended to apply 2,4-D (30 mg/L), NAA (50 mg/L) and NAD (100 mg/L) (Rahemi and Atahosseini, 2003).

Reddy and Prasad (2012) studied the effect of plant growth regulators on pomegranate cv. Ganesh. Nine treatments with three growth regulators, viz., NAA at 20, 30 and 40 ppm; 2,4-D at 20, 30 and 40 ppm, GA_3 at 25, 50 and 75 ppm and control (water spray) were sprayed three times starting at full bloom and subsequently at 45 and 90 days after fruit set. Results revealed that application of 2,4-D at 40 ppm gave significantly high fruit size in length (8.85 cm), breadth (8.91 cm.) and volume (291.50 ml) and higher fruit weight (262.23 g). This can be attributed to nature of auxins to increase the osmotic pressure of cell sap which is responsible for uptake of water and consequently results in increased growth. Higher aril percent (72.03 %), maximum number of fruits (64.00) which resulted in highest fruit yield of 16.78 kg/plant.

Studies on the effect of plant growth substances on quality components were carried out on pomegranate by Goswami *et al.*, (2013). Total 15 treatments were used, out of which only ethrel 200 ppm treatment was found effective on all the five quality traits. Application of ethrel 200 ppm has improved the TSS (17.18 %), reducing sugar (10.83 %), non-reducing sugar (1.68 %), total sugar (12.50 %), and ascorbic acid content (14.85%).

Kishor *et al.,* (2017) conducted an experimental programme to study the effect of some plant bio-regulators and chemicals on fruit physico-chemical traits of 2 years old young pomegranate cv. Bhagwa under Lucknow subtropical conditions. The observation revealed that the application of 2,4-D @ 15 ppm was better for improvement of fruit weight (101.67 g), fruit length (6.00 cm), fruit volume (92.00), specific gravity (1.10), number of aril/fruit (363.33), number of segment/fruit (5.00), acidity (0.45%), vitamin C (10.20 mg/100g), total sugar (6%) and non-reducing sugar (1.93%). While the maximum T.S.S: Acid ratio and reducing sugar was recorded when treated with NAA @ 50 ppm. The biggest fruit (4.73 cm diameter) increase by the application of NAA @ 75 ppm.. Maximum total soluble solids (12.57 ^{0}Brix) were recorded under treatment GA_3 @ 75ppm. The maximum pericarp thickness (4.73 mm) was observed with spray of GA_3 @ 75 ppm. Spray of GA3 increased the thickness of the fruit cuticle and the epidermal layers.

Kumar *et al.,* (2017) studied the effect of micronutrients and plant growth regulators on fruit yield in pomegranate Cv. Jodhpur Red at Haryana. Highest fruit yield per acre (50.8q) was obtained with NAA @ 20 ppm when applied fifteen days after fruit set; while 2, 4-D @ 10 ppm recorded maximum fruit yield when applied thirty days after fruit set.

14. Role of Plant Growth Regulators on Controlling Fruit Cracking

An experiment was carried out by Sharma and Belsare (2011) to study the effect of foliar applied bio-regulators like NAA, GA3, CPPU and nutrients like boron and calcium, in mid-May and mid-June on yield, fruit cracking and quality in pomegranate cultivar 'G-137'.

The extent of fruit cracking was reduced significantly with the application of boron at 0.2% and CPPU at 10 ppm. Increased fruit size, fruit weight and juice contents were noticed with the treatment of CPPU at 5 ppm. However, total soluble solids, sugar contents, titratable acidity, TSS:acid ratio were significantly increased following the applications of GA_3 at 50 ppm and NAA at 20 ppm. CPPU applied at 5 ppm and GA_3 at 50 ppm caused a significant reduction in russet formation on fruits.

Kumar *et al.,* (2016) conducted a field experiment at Haryana to evaluate the effect of micronutrients and plant growth regulators on fruit cracking Cv. Jodhpur Red. The sprays were applied at fifteen and thirty days after fruit set. 2, 4-D @ 10, 20, 40 ppm and NAA @ 10 ppm recorded significantly lower incidence of cracking than all other treatments.

15. Role of Plant Growth Regulators on Post Harvest Management

The fresh fruit and juice market has grown steadily worldwide and has especially boomed in the US in recent years due to its increased consumer awareness of the potential health benefits (Seeram *et al.,* 2006). Generally, pomegranate fruits are ready for harvest in 125-180 days after anthesis and possess typical characteristics of non-climacteric fruits. In the recent past, attempts have been made to prolong the shelf life of pomegranate fruits and isolated arils following controlled atmosphere (CA), cold storage and other storage methods (Defilippi *et al.* 2006; Porat *et al.,* 2007). Anwal *et al.* (2016) studied to extent the shelf life of pomegranate using plant growth regulators and found maximum shelf life of fruits with NAA 40 ppm followed by NAA 50 ppm and minimum was obtained in control treatment (Anwal *et al.* 2016).

With the aim of reduce chilling injury (CI) in pomegranate fruit (*Punica granatum* L.) stored at low temperature (2^0C), putrescine (Put) or spermdine (Spd) at 1 mM under pressure-infiltration or immersion were applied to pomegranates prior to cold storage. Non-treated fruit developed rapidly CI, with main symptoms being skin browning, increased electrolyte leakage and weight loss. During storage losses of firmness and colour and increases in soluble solids concentrations/acidity ratio and respiration rate were observed. All these changes were significantly delayed by polyamine treatments, with both pressure-infiltration and immersion being similarly effective. The reduction in CI severity was correlated with increased levels of free endogenous Put and Spd in the skin, which might induce acclimation of pomegranate to cold temperature and a mechanism of protection against CI, with a net increase in shelf-life (Mirdehghan *et al.,* 2007).

Pomegranates were treated after harvest with methyl jasmonate (MeJa) or methyl salicylate (MeSa) at two concentrations (0.01 and 0.1 mM), and then stored under chilling temperature for 84 days. Control fruits exhibited chilling injury (CI) symptoms manifested by pitting and browning, the severity being enhanced as storage time advanced, and accompanied by softening and electrolyte leakage (EL). The CI symptoms were significantly reduced by MeJa or MeSa treatments, without significant differences among treatments or applied dose. In addition, both treatments significantly increased total phenolics and anthocyanins with respect to controls. Hydrophilic (H-TAA) and lipophilic (L-TAA) total antioxidant activity decreased in control arils, but in both MeJa and MeSa treated fruits H-TAA increased while no significant changes occurred for L-TAA. Results would suggest that both MeJa and MeSa have potential postharvest applications in reducing CI, maintaining quality and improving the health benefits of pomegranate fruit consumption by increasing the antioxidant capacity (Mohammad Sayyari *et al.*, 2011).

16. Role of Plant Growth Regulators on Stress Management

Freezing injury is one of the most important restrictions in commercial pomegranate production. Pomegranate is mostly grown in the margins of deserts in Central and Northeastern Iran where the night temperature during winter may drop to –20°C or lower. Plant growth regulators offer an approach to increase or prolong cold hardiness in fruit trees in hazardous areas. Foliar spray of PBZ moderately reduced freezing injury in pomegranate plants with evidence of less electrolyte leakage through upregulating soluble carbohydrates and proline contents. However, there were no remarkable differences in cold hardiness of PBZ-treated and untreated plants. Further studies on optimizing the time of application and dosage of PBZ may be needed. Although pomegranate has a thick and inedible skin, it is worth measuring the residue levels of PBZ in fruits following foliar application. Physiological effects of PBZ on plants including changes in growth and morphology, antioxidant activity, increasing levels of proline and carbohydrates play key roles in withstanding the low temperature stress (Soluklu *et al.*, 2014).

References

- Abubakar, A.R., Ashraf, N. and Ashraf, M. 2013. Effect of plant bio-stimulants on growth, chlorophyll content, flower drop and fruit set of pomegranate cv. Kandhari Kabuli, *International Journal of Agriculture, Environment & Biotechnology*, **6**(2): 305-309.
- Ahire, G. Z., Choudhari, S. M. and Masalkar, S. D. 1993. Crop regulation in pomegranate;

1. Effect of growth regulators on flower induction, sex expression and flower drop. *Annals of Arid Zone,* **33**(2): 97-98.

- Ahire, G.Z. 1991. Studies to control late season flower production in pomegranate (*Punica granatum* L.). M.Sc. (Agri.) Thesis submitted to Mahatma Phule Krishi Vidyapeeth, Rahuri.
- Ahmad Seiar, Y. 2017. Effect of growth regulators on rooting of cuttings in pomegranate (*Punica granatum* L.) cv. 'Bhagwa'. *J. Hortl. Sci.,* **11**(2): 156-160.
- Anawal, V., Narayanaswamy P. and Yallesh Kumar H. S. 2015. Effects of plant growth regulators on induction of flowering in pomegranate (*Punica granatum* L.) cv. Bhagwa. *International Journal of Scientific Research*, **4**(12): 7-9.
- Anawal, V.V., Narayanswamy, P. and Ekabote, D. 2016. Effects of plant growth regulators on fruit set and yield of Pomegranate cv. Bhagwa. *International Journal of Current Research,* **8**(4):29008-29010.
- Byers, R.E. 2003. Flower, fruit thinning and vegetative fruit balance. p. 409-436. In: D.C. Ferree and I.J. Warrington (eds.), *Apples botany production and uses*. CABI, Publishing, Wallingford, UK.
- Chaudhari, S. M. and Desai, U. T. 1992. Effect of plant growth regulators on flower sex in pomegranate {*Punica granatum* L.). *Indian Journal of Agricultural Sciences,* **63**(1): 34-35.
- Choudhari, S. M. and Desai, U. T. 1993. Efficacy of plant grow1;h regulators for flower thinning in pomegranate. *Annals of Arid Zone,* **31**(3): 221.
- Damar, D., Barholia, A. K., Lekhi, R. and Haldar, A. 2014. Effect of growth regulators and biofertilizers on survival of pomegranate (*Punica granatum* L.) stem cuttings, *Plant Archives,* **14**(1) : 347-350.
- Defilippi BG, Whitaker BD, Hess-Pierce BM, Kader AA 2006. Development and control of scald on 'Wonderful' pomegranates during long-term storage. *Postharvest Biology and Technology*, **41**: 234-243
- Desai, U.T., Ahire, G. Z., Masalkar, S. D. and Choudhari, S. M. 1993. Crop regulation in pomegranate: II. Effects of growth regulators on fruit set, yield and fruit quality. *Annals of Arid Zone*, **32** (3): 161-164.
- Devkare, A.V. 2003. Effect of Plant Growth Substances on Fruit Size and Quality of Pomegranate (*Punica granatum* L.) cv. Mridula. M.Sc. (Agri.) Thesis submitted to

Mahatma Phule Krishi Vidyapeeth , Rahuri.

- Digrase, S.S., Tambe, T.B., Kadam, A.S. and Kalalbandi, B.M. 2016. Effect of different plant growth regulators and chemicals on growth and yield of pomegranate (*Punica granatum* L.) cv. Bhagwa. *Advance Research Journal OF Crop Improvement*, **7(1)** : 96-99.
- Dong, R. B., Li, Y. L., Zhao, X. C, Wu, K. X. and Sun, Z. C.1999. Cultural experiment for early high production of pomegranate. *China Fruits*, **1**: 43.
- Ghosh, D.A., Bamdyopadhyay, A. and Sen, S.K. 1988. Effect of NAA and IBA on adventitious root formation in stem cuttings of pomegranate (*Punica granatum* L.) under intermittent mist. *Indian Agriculturist*, **32**: 239-243.
- Ghosh, S.N., Bera, B. and Kundu, A. 2009. Effect of plant growth regulators in yield and fruit quality in pomegranate cv. Ruby. *Journal of Hort. Sci.*, **4**(2):158-160.
- Goswami, J.D.; Patel, N.M.; Bhadauria, H.S. and Wankhade, V.R. 2013. Effect of plant growth regulators on quality traits of pomegranate cv. Sindur. *The Asian Journal of Horticulture*, **8**(1): 361-363.
- Gurjar, P.K.S. and Patel R.M. 2007. Effect of rooting media, type of stem cutting and growth regulator on rooting and growth of pomegranate cv. Ganesh. *Bharitiya Krishi Anusandhan Patrika*, **22**(1): 62-66.
- Hartmann H T, Kester D E, Davis F T and Genev R L 2002. Plant Propagation: Principles and Practices. Prentice Hall, Englewood Cliffs 880.
- Hussein, M. A., El-Sese, A. M., El-Mahdy, T. K. and Abd- El-Sabour, B. 1994 (a). Physiological studies on thinning effects on yield and iruit quality of Manfalouty pomegranate. A: Flowering behaviour and Sevin, NAA and hand thinning influences on fruit thinning, yield and certain fruit characteristics. *Assiut Journal of Agricultural Sciences*, **25**(3): 27-40.
- Kanwar K, Joseph J, Raj D. 2010. Comparison of *in vitro* regeneration pathways in *Punica granatum* L. *Plant Cell, Tissue Organ Culture*, **100**(2): 199-207.
- Kaur,S. and Kaur, A. 2016. Effect of IBA and PHB on rooting of pomegranate(*Punica granatum* L.) cuttings cv. Ganesh.*Biological Forum – An International Journal*, **8**(2): 203-206.
- Khalil, H. A. and Aly, H.S.H. 2013. Cracking and fruit quality of pomegranate (*Punica granatum* L.) as affected by pre-harvest sprays of some growth regulators and mineral nutrients. *Journal of Horticultural Science and Ornamental Plants*, **5**(2): 71-76.

- Kishor, S., Maji, S., Meena, M.L., Dwivedi, D. H., Kishor, S. and Kumar, S. 2017. Effect of plant bio-regulators and chemicals on fruit physico-chemical traits of pomegranate (*Punica granatum* L.) cv. Bhagwa. *Journal of Pharmacognosy and Phytochemistry*,**6**(4): 1573-1575.

- Krishnamoorthy, H. N. 1981. Plant growth substances including applications in agriculture. Tata Hill, Publisher Company. New Delhi.

- Kumar, K., Joon, M. S., Yadav, R. and Daulta, B.S. 2016. Effect of growth regulators and micronutrients on fruit quality in pomegranate. *International Journal of Environment, Agriculture and Biotechnology*, **1**(4): 627-630.

- Kumar, K., Joon, M.S., Yadav, R. And Daulta, B.S. 2016. Effect of Growth Regulators and Micronutrients on Fruit Quality in Pomegranate. *International Journal of Environment, Agriculture and Biotechnology*,**1** (4): 627-630.

- Kumar, K., Pinder, R., Jitender, S., Dabas, S., Yadav, B. and Rana, S. 2017. Effect of growth regulators and micronutrients on fruit cracking and fruit yield in pomegranate. *Indian Journal of Agricultural Research*, **51**(3): 272-276.

- Kumari, G.G.S.,Kumari, S.M.., Vithana, M.D.K., , and Mannanayake, M.A.D.K. 2013. Effect of Plant Growth Regulators on Hard Wood Cuttings of Pomegranate (*Punica granatum* L.). Proceedings of 12th Agricultural Research Symposium organised by *Department of Horticulture and Landscape Gardening, Faculty of Agriculture and Plantation Management, Wayamba University of Sri Lanka, Makandura, Gonawila. (NWP)*PP. 127-131.

- Manasa, P. 2017. Effect of Plant Growth Regulators on Yield and Quality of Pomegranate (*Punica granatum* L.) cv. Bhagwa. M.Sc. (Agri.) Thesis submitted Department of Horticulture College Of Agriculture, Latur Vasantrao Naik Marathwada Krishi Vidyapeeth, Parbhani, (M.S.) India.

- Mehta, S.K., Singh, K.K., and Harsana, A.S. 2018. Effect of IBA concentration and time of planting on rooting in pomegranate (*Punica granatum* L.) cuttings. *Journal of Medicinal Plants Studies*; **6**(1): 250-253

- Mirdehghan, S.H., Rahemi M., Rahemi D. S., Castillo A, D., Mart´ınez-Romero A, Mar´ıa Serrano B, D. Valero A. 2007. Pre-storage application of polyamines by pressure or immersion improves shelf-life of pomegranate stored at chilling temperature by increasing endogenous polyamine levels. *Postharvest Biology and Technology*, **44** (2007) 26–33.

- Mohamed, A.K.A. 2004. Effect of gibberellic acid (GA3) and benzyladimine (BA) on splitting and quality of Manfalouty pomegranate fruits. *Assiut J. Agril. Sci.*, **35**:11-21.

- Mohammad Sayyari, Mesbah Babalar, Siamak Kalantari, Domingo Martínez-Romero, Fabián Guillén, María Serrano and Daniel Valero.2011. Vapour treatments with methyl salicylate or methyl jasmonate alleviated chilling injury and enhanced antioxidant potential during postharvest storage of pomegranates. *Food Chemistry,*124 (3): 964-970.
- Murkute AA, Patil S, Patil BN, Kumari M. 2002. Micropropagation in pomegranate, callus induction and differentiation. *South Indian Hort.*, **50**(1, 3): 49-55.
- Naik SK, Pattnaik S, Chand PK.1999 . *In vitro* propagation of pomegranate (*Punica granatum* L. cv. Ganesh) through axillary shoots proliferation from nodal segments of mature tree. *Scientia Horticulturae,* **79:** 175-183.
- Naik,S.K. and Chand, R. 2003. Silver nitrate and amino ethoxy-vinyl glycine promote in vitro adventitious shoot regeneration of Pomegranate (*Punica granatum* L.). *Journal of Plant Physiology*, **160**: 428-430.
- Pandey, V. 1999. Effect of NAA and GA3 spray on fruit retention, growth, yield and quality of ber cv. Banarasi Karka. *Orissa J. Hort.*, **27**:69-73
- Panwar, R.D., Kaushik, R.A., Survinder, S. and Gupta, R.B. 2001. Effect of IBA on rooting of hardwood cuttings in pomegranate. *Haryana Journal of Horticultural Sciences,* **30**(1&2): 72.
- Patel, D.S. 1993. Use of growth regulators for improvement of quality in pomegranate (*Punica granatum* L.) cv. G-137. M.Sc. (Agri.) Thesis submitted to Mahatma Phule Krishi Vidyapeeth , Rahuri.
- Patil, V.M., Dhande, G.A., Thigale, D.M. and Rajput, J.C. 2011. Micropropagation of pomegranate (Punica granatum L.) 'Bhagava' cultivar from nodal explant. *African Journal of Biotechnology*, **10**(79):18130-18136.
- Pawar, P.S. 2001. Effect of plant growth regulators on maturity, yield and quality of pomegranate (*Punica granatum* L.) cv. Mridula. M.Sc. (Agri.) Thesis submitted to Mahatma Phule Krishi Vidyapeeth , Rahuri.
- Pawar, P.S., Jagtap, D.D., Garad, B.V. and Shirsath, H.K. 2005. Effect of plant growth regulators on maturity, yield and fruit weight of pomegranate cv. Mridula. *Adv. Pl. Sci.*, **18**:167-170.
- Phawa, T., Prasad, V.M. and Rajwade, V.B. 2017. Effect of Plant Growth Regulators on

Growth and Flowering of Pomegranate (*Punica granatum* L.) Cv. Kandhari in Allahabad Agro-Climatic Conditions. *International Journal of Current Microbiology and Applied Science,* **6** (8) : 116-121.

- Porat R, Weiss B, Fuchs Y, Zandman A, Kosto I. 2007. Storage of pomegranate fruit in large modified atmosphere bags in harvest bins . *Alon Hanotea,* **61**:714-717.
- Rahemi and Atanosseini, A. 2004. Effect of plant growth regulators on fruit characteristics and leaf area of pomegranate cv. Shisheh cup. *Acta Hort.,* **662**: 313-317.
- Raj, D. and Kanwar, K. 2008. Efficient in vitro shoot multiplication and root induction enhanced by rejuvenation of microshoots in *Punica granatum* cv. Kandhari Kabul. *Paper presented in the National seminar on physiological and biotechnological approaches to improve plant productivity* held at CCSHAU, Hisar, India, 24.
- Rajanna, N. 1981. Studies on the propagation of Basein seedless pomegranate (*Punica granatum* L.) by cuttings. *M.Sc. (Agri.) Thesis,* University of Agricultural Sciences. Bangalore.
- Rathod, M.J, Ramdevputra, M.V, Nurbhanej, K.H and Patel, M.S. 2017.Effect of Ethrel and banana pseudostem sap on fruit yield and yield attributes of pomegranate (*Punica granatum* L.) cv. Bhagwa. *International Journal of Chemical Studies;* **5**(5): 392-396.
- Reddy, P.A. and Prasad, D.M. 2012. Effect of plant growth regulators on fruit characters and yield of pomegranate (*Punica granatum* L.) cv. Ganesh. *International Journal of Plant, Animal and Environmental Sciences,* **2** (2): 91-93.
- Reddy, Y.T.N. and Reddy. Reddy, Y.T.N. 1990. Effect of basal wounding, growth regulators and polythene covering on rooting of pomegranate cuttings. *Journal of Maharashtra Agricultural University,* **15**: 153-155.
- Reddy,P.A. and Prasad,M.D. 2012. Effect of plant growth regulators on fruit characters and yield of pomegranate (Punica granatum l.) cv.Ganesh. *International Journal of Plant, Animal and Environmental Sciences,* **2** (4-6):91-93.
- Roberts JA & Hooley R 1988 Plant Growth Regulators. Blackie, Chapman and Hall, New York. 151-169.
- Rymbai H and Reddy G S 2010. Effect of IBA, time of air layering and rooting media on air-layers and plantlets survival under different growing nursery conditions in guava. *Indian J Hort.,* **67**: 99-104.
- Samir Z, El-Agamy, Rafat AA, Mostafa, Mokhtar M, Shaaban, Marwa T, El-Mahdy

2009. *In vitro* Propagation of Manfalouty and Nab Elgamal Pomegranate Cultivars. *Research. J. Agric. Biol. Sci.,* **5**(6): 1169-1175.

- Saroj, P.L., Awasthi, O.P., Bhargava, R. and Singh, U.V. 2008. Standardization of pomegranate propagation by cutting under mist system in hot arid region. *Indian J. Horticulture,* **65**: 25-30.
- Seeram NP, Schullman RN, Heber D. 2006a. Commercialization of pomegranates: Fresh fruit, beverages and botanical extracts. In: Seeram NP, Schull- man RN, Heber D (Eds) Pomegranates Ancient Roots to Modern Medicine, CRC Press Taylor and Francis Group, Boca Raton, Florida, pp 187-196
- Sharma, G. and Ananda, S.A. 2004. Effect of pre-bloom foliar application of plant bioregulators on growth, fruiting and quality of apple under warmer agroclimatic conditions. *Acta Hort.,* **662**: 353-357.
- Sharma, N. and Belsare, C. 2011. Effect of plant bio-regulators and nutrients on fruit cracking and quality in pomegranate (*Punica granatum* L.) 'G-137'. *Acta Hort.,* **890** (1): 345-352.
- Sharma, N., Anand, R. and Kumar, D. 2009. Standardization of pomegranate (*Punica granatum.L)* propagation trough cuttings. *Biological Forum,* **1**(1): 75-80.
- Shaybany, B. and Sharifi, H. 1973. Effect of pre-harvest applications of ethephon on leaf abscission, fruit drop and constituents of fruit juice in pomegranates. *Journal of Horticultural Sciences.* **48**(3): 293-296.
- Simranjit Kaur, Amarjeet Kaur and Gurpinder Kaur. 2016. Effect of IBA, PHB and Time of Planting on Rooting of Pomegranate (*Punica Granatum* L.) cuttings cv. Ganesh. *Asian Journal of Science and Technology,* **7**(11) : 3757-3761.
- Singh N.V., Singh S.K., Patel V.B. 2007. *In vitro* axillary shoot proliferation and clonal propagation of 'G-137' pomegranate (*Punica granatum*). *Indian Journal of Agric. Sci.,* **77**(8): 505-508.
- Singh, K.K. 2014. Effect of IBA Concentrations on the Rooting of Pomegranate (*Punica Granatum* L.) cv. Ganesh Hardwood Cuttings under Mist House Condition. *Plant Archives,* **14**(2): 1111-1114.
- Soluklui, A.A.G., Ershadi, A., Tabatabaee, Z.E. and Fallah, E. 2014. Paclobutrazol-Induced Biochemical Changes in Pomegranate (Punica granatum L.) cv. 'Malas Saveh' under

Freezing Stress. *International Journal of Horticultural Science and Technology*, **1**(2): 181-190.

- Tomar,K.S and Tomar, S.S. 2012. Effect of different dosage of growth regulators on rooting & survival percentage of pomegranate *(Punica granatum L.)* air layers. *Green Farming*, **3** (6): 701-703.

- Upadhyay SK and Badyal J. 2007. Effect of growth regulators on rooting of pomegranate (*Punica granatum* L.) cutting. *Haryana Journal of Horticultural Sciences*, **36** (1/2) : 58-59.

- Venkatesan, K. and Mohideen, M. Kader. 1994. Effect of growth regulators on fruit characters and yield of pomegranate *(Punica granatum* L.) cv. Ganesh. *South Indian Journal Horticulture*, **42**(4): 239-244.

25

Plant Growth Regulators in Sapota

T. R. Ahlawat, Kapil Mohan Sharma,
Dharmishtha Patel and Suchismita Jena

1. Introduction

Sapota [*Manilkara achras* (Mill.) Fosberg] a tropical fruit popularly known in India as chiku, belongs to family Sapotaceae and is native to Tropical America. Sapota cultivation in India can be traced back to 1898 to a village called Gholwad in Thane district of Maharashtra. Sapota is mainly cultivated for its sweet and delicious fruits in India. In Central and South American countries, it is grown for the milky latex produced from bark and fruits, which yields a commercial product called *Gutta parcha*. It is the base for preparing chewing gum (Chickle). Sapota fruits are a good source of digestible sugar which ranges from 12 to 18 per cent and an appreciable amount of protein, fibre, minerals, calcium and iron.

Different groups of plant growth regulators like auxins, gibberellins and growth retardants at various concentrations have been reported to influence flowering, fruit set, retention development and quality characters in sapota as discussed below:

2. Role of Plant Growth Regulators on Seed Propagation

In India, sapota is commercially propagated by grafting or layering and *khirnee* is used as rootstock. Sapota seedlings grow at a very slow rate and therefore are not preferred as rootstock. It is however, imperative to raise plants of F_1 generation through seeds in sapota hybridization programmes. There is a distinct possibility of improving germination and hastening growth by the use of plant growth regulators. Farooqui *et al.* (1971) studied the effect of growth regulators on germination of sapota seeds and recorded 70 per cent germination in seeds treated with 50 ppm GA_3. Tendulkar (1978) recorded 38 per cent germination in *khirnee* seeds when soaked in 250 ppm gibberellic acid for 24 hours prior to sowing as against 18 per cent in control.

Ethylene was long ago shown to be effective in breaking seed dormancy (Vacha and Harvey, 1927). Nikam *et al.* (1986) reported that *khirnee* seeds soaked in 500 ppm ethephon solution for 24 hours had better germination (54.66 per cent) as compared to control (18.17 per cent). In a similar experiment, Pampanna *et al.* (1995) discovered that pre-soaking *khirnee* seeds in 400 ppm ethrel resulted in the highest germination.

Pampanna and Sulikeri (2001) studied the effect of gibberellic acid (GA) and ethrel on seed germination and seedling growth of sapota cv. 'Kalipatti'. Cracked seeds were soaked for 24 hours in 100 ml solution of GA, ethrel and GA + ethrel (200, 300 and 400 ppm). Pre-soaking of seeds in 400 ppm GA + 400 ppm ethrel resulted in high (90%) and early (12.15 days) germination and the highest emergence rate. Seedlings raised from seeds pre-soaked in 400 ppm GA produced the highest shoot and root length. The number of leaves per seedling and seedling vigour were highest in seedlings raised from seeds pre-soaked in GA + ethrel. It might be due to the stimulating effect of imbibitions on subsequent seed germination caused by action of GA_3.

Srinivas (2007) conducted an experiment at Agricultural College, Bapatla to improve seed germination and subsequent growth in rayan seedlings. Pre-soaking seeds in GA 400 ppm gave the highest germination (44%), maximum height of seedlings (4.46 cm) and maximum number of leaves (3.62).

Higher germination by GA pre-soaked seeds may be due to the fact that, exogenous application of GA antagonises the effect of inhibitors (Brian and Hemming, 1958; Wareing *et al.*, 1968) and increases endogenous gibberellin like substances (Mathur *et al.*,1971). Gibberellin induces *de novo* synthesis of proteolytic enzymes like α-amylase and ribonuclease. Amylases in turn hydrolase starch in the endosperm providing essential sugars for the initiation of growth processes (Copeland and Mc Donald, 1995). Other enzymes, activated by GA weaken the seed coat and allow the root tip to burst through. Gibberellin also enhances cell elongation, which enables the radicle to push through the endosperm and seed coat (Hartmann and Kester, 1972). GA_3 is also known to overrule photodormancy, thermo-dormancy, dormancy imposed by incomplete embryo development, mechanical barriers and presence of germination inhibitors (Diaz and Martin, 1971).

3. Role of Plant Growth Regulators in Asexual Propagation

Air layering and inarch grafting are the preferred techniques for multiplying sapota in India. Plant growth regulators are used on a commercial scale in the propagation of several fruit crops. In an early study, air layering of sapota without the application of IBA or NAA at 500 ppm was unsuccessful. Gunny cloth and straw proved to be better wrapping materials than plastic film (Alam and Faruque, 1974).

Ramsundar and Abdul Khader (1986) conducted an study at Coimbatore, Tamil Nadu on rooting of sapota cuttings, where softwood shoots of sapota were etiolated by wrapping with black polyethylene over 10 cm below the apical bud. After 5 days the shoots were girdled below the etiolated section of the shoot. Non-etiolated shoots were also girdled. Some 45 days later, cuttings taken from all shoots were treated with IBA (2000 ppm) and rooted under mist. The rooting results were assessed 160 days later. Survival was assessed at 1 month after potting of the rooted cuttings. Only etiolated and ringed shoots treated with IBA rooted, the rooting percentages being 13.3, 29.7 and 39.4 with 500, 1000 and 2000 ppm IBA, respectively and survival 40, 60 and 80%, respectively.

Seedling derived 15 year old trees of sapota were invigorated by pruning branches 9-11 cm in diameter, at a height of 1.5 m above the ground in the month of February. In the following year, new shoots developing from the stumps were etiolated using black moist cloth wrapped with alkathene. Shoots were layered in August using IBA and NAA, alone or in combination. Of the growth regulator treatments, invigorated shoots after etiolation treatment and application of 10,000 ppm IBA gave 91.7% rooting success and 90.9% survival of air-layers (Chatterjee *et al.*, 1989).

Rayan seedlings attained graftable size of seedlings within 16 months on application of 11.5 g N/m^2 (Urea 25 g/m^2) at monthly interval from 3rd month to 11th month after sowing and GA_3 200 mg/l at 3, 6 and 9 months of age of (Anon., 2004).

IBA is the most common exogenously applied growth plant growth regulator for rooting in horticultural plants and has the ability to promote adventitious root formation (De Klerk *et al.,* 1999). It is relatively stable and insensitive to the auxin degrading enzyme systems (Epstein and Ludwig-Muller, 1993). IBA thus provides auxins for a longer period of time which stimulate root development by inducing root initials that differentiate from cells of the young secondary phloem, cambium and pith tissue.

Tissue culture technique offers an efficient *in vitro* regeneration method for fast multiplication of important trees and an attractive alternative for species in which conventional propagation is impractical. The ratio of cytokinin to auxin determines fate of callus. If cytokinin: auxin ratio is high it promotes shoot proliferation whereas, low cytokinin: auxin ratio enhances root formation (Jain, 2013).

Hosomi *et al.*, (1984) studied the effect of plant growth regulators on growth and triterpene ester production in tissue culture cells of sapota and they found that the best growth and accumulation of triterpene esters was obtained on a medium which contained 0.05-0.5 mg/l of 2,4dichlorophenoxyacetic acid and 0.5-1.0 mg/l of kinetin.

4. Role of Plant Growth Regulators in Plant Growth and Development

Sapota is a slow growing, long lived evergreen tree. Very few attempts have been made to study the effect of plant growth regulators on the vegetative growth of sapota. Agrawal and Dikshit (2008) studied the effect of plant growth regulators Cycocel (0, 200 and 400 ppm), GA_3 (50 ppm) and NAA (100 ppm) on growth and yield of sapota fruits cv. 'Cricket Ball' at Raipur, Chhatishgarh. Shoot length was significantly reduced by the application of Cycocel @ 400 ppm at FBD stage. Nevertheless, it increased the number of leaves and leaf area. Foliar spray of 100 ppm NAA at flowering stage increased the shoot length, number of leaves and leaf area. Enhanced vegetative growth of sapota was found under GA_3 @ 150 ppm with increased shoot length, maximum number of leaves per shoot and earliness in sprouting of new shoots. This can be attributed an increase in the number of cells produced and the elongation of individual cells under GA_3 treatments (Sachs *et al.*, 1960). However, the maximum shoot girth was recorded with treatment Cycocel @ 450 ppm (Bhujbal *et al.*, 2012).

5. Role of Plant Growth Regulators in Flowering and Fruit Set

Under tropical conditions, flowers are seen almost throughout the year. However, there are two main seasons of flowering and hence two harvesting seasons. Bhuva and Chundawat (1978) reported that the flowering and fruiting in sapota in South Gujarat is a continuous process with two peak seasons of crop in September– November and February–March. Although, sapota trees bear many flowers, the number of those reaching maturity is small. Most of the flowers either fail to set fruit or there is shedding of fruits. Flower shedding and low fertility in sapota was earlier reported by Cheema *et al.* (1954) and Hayes (1957).

Ramamurthy and Ramalingam (1954) also reported the failure of fruit set when flowers were bagged with or without emasculation.

Lenka *et al.*, (1996) observed two distinct peak flowering periods on 5 sapota cultivars (February-April and October-November) during their experimental studies at Bhubaneswar in 1993-94. The crop requires 55.3 to 60 days from flower bud initiation to anthesis. The percentage of flower buds which developed into flowers ranged from 50.00 in 'Cricket Ball' to 72.72 in 'Kalipatti'. The gap between flowering and fruit set was 50.0 to 54.5 days. The peak period of anthesis was at 04.00 h and continued up to 08.00 h. Kadam *et al.* (2005) studied the influence of growth regulators on flowering behaviour of sapota under semi-arid zone of Maharashtra at Rahuri. Twelve year old sapota trees were sprayed with Cycocel (200 and 400 ppm) at FBD stage and GA (50 ppm) and NAA (100 ppm) at flowering stage. They reported the minimum flower drop (58.46% and 58.34%) and the maximum fruit retention (41.54 and 41.66%) under 100 ppm NAA during 1997-98 and 1998-99. Nambisan *et al.* (2007) at Rahuri, Maharashtra reported the highest number of flowers per shoot under 200 ppm CCC and 30 ppm 2,4-D (13.8 and 13.9) in sapota cv. 'Kalipatti'. Whereas, GA_3 at 30 ppm resulted in the highest fruit set percentage (13.9%). These growth regulators were sprayed at the flowering stage and three months thereafter.

Agrawal (2004) assessed the effect of plant growth regulators on growth, flowering, yield and quality of sapota cv. 'Cricket Ball' under Chhattisgarh plains. Cycocel (0, 200 and 400 ppm) was applied at FBD stage followed by other growth regulators (GA/NAA-0, GA-50 and NAA-100 ppm) at flowering stage. Further GA 50 ppm and NAA 100 ppm were sprayed either at pea stage or lag phase of fruit development. The maximum number of flowers and fruit set was observed by the application of CCC 400 ppm at FBD stage. Further, the application of 100 ppm NAA at flowering as well as pea stage of fruit development decreased flower as well as fruit drop and increased fruit set and retention as compared to GA 50 ppm. Moreover, spraying NAA at pea stage was found to be more effective than lag phase of fruit development.

Nagargoje *et al.* (2007) studied the effect of NAA (50 and 100 ppm) on flower drop and fruit set in sapota cv. 'Kalipatti' at Parbhani, Maharashtra. Trees were sprayed at 50% flowering stage and at pea stage of fruit development. Foliar spray of NAA at 100 recorded the lowest fruit drop and the highest fruit set at both the stages. It reduced flower drop to 52.42% and increased fruit set to 47.58% as compared to NAA at 50 ppm and water sprayed trees. Ingale *et al.* (2008) observed maximum fruit set in sapota cv. 'Kalipatti' at Navsari, Gujarat when trees were subjected to foliar sprays of NAA at 100 ppm + GA_3 at 100 ppm.

Foliar sprays of NAA at 100 ppm on sapota cultivar 'Kalipatti' at Parbhani, Maharashtra resulted in the maximum number of flowers (54.0) and highest fruit set (43.13%). Sprays were made at the time of flowering, pea stage of fruit development and lag phase of fruit growth (Chavan *et al.,* 2009). At Parbhani, Maharshtra, foliar application of CCC at 450 ppm gave the maximum number of flowers per tree in sapota cv. 'Kalipatti'. This can be attributed to the suppressive effect of CCC on vegetative growth which created a favourable C/N ratio in terminals. Whereas. NAA at 200 ppm proved most effective for increasing fruit set (Bhujbal *et al.,* 2013).

Plant growth regulators may have induced a higher rate of cell division, cell elongation, cell enlargement, photosynthesis and thereby an increase in enzymatic activities. Further, their involvement in biosynthesis of auxins at critical stages may have increased fruit set. Higher fruit set under NAA and GA treatments might be due to their involvement in translocation and mobilization of stored metabolites or photosynthates from source to sink (Kaur, 2017).

6. Role of Plant Growth Regulators in Controlling Fruit Drop

Among the various causes of fruit drop, the simplest one is a decline in the level of endogenous auxins (Addicot and Lynch, 1995). Among the synthetic anxins tested in this regards, NAA was found to be the most effective in sapota (Ravoof, 1963 and Das and Mahapatra, 1976).

Agrawal (2004) reported that CCC 400 ppm at FBD stage and NAA 100 ppm at flowering as well as pea stage of fruit development reduced fruit drop in sapota cv. Cricket Ball under Chhattisgarh plains. Nagargoje *et al.* (2007) conducted a trial to study the effect of NAA on fruit drop in sapota cv.'Kalipatti' at Marathwada Agricultural University, Parbhani. Foliar sprays of NAA (50 and 100 ppm) were done at 50% flowering stage and at pea stage of fruit development. They reported the lowest flower and fruit drop in sapota cv. 'Kalipatti' under NAA at 100 ppm.

Panigrahi *et al.*, (2011) conducted an experiment to study the effect of NAA, GA_3 and Ethrel on fruit retention in ten years old trees of sapota cv.'Cricket Ball'. Trees were subjected to foliar sprays of plant growth regulators at 50 per cent flowering and pea stage of fruit development. Of the different plant growth regulators, NAA @ 200 ppm followed by GA_3 @ 100 ppm and NAA @ 400 ppm proved most effective in reducing fruit drop. NAA at 200 ppm substantially reduced flower as well as fruit drop in sapota variety 'Kalipatti' ultimately increasing fruit set and fruit retention, respectively (Bhujbal *et al.*, 2013). Probably, NAA reduced fruit drop by interfering in the enzymatic breakdown of cell walls in the abscission zone (Modise *et al.*, 2009).

7. Role of Plant Growth Regulators in Fruit Retention

Fruit retention is the single most important factor governing fruit yield. In coastal climate, sapota flowers almost throughout the year. Flowers and fruits tend to drop at different stages of development from setting to maturity which can have an adverse impact on fruit yield and profitability of growers (Patil and Narwadkar, 1974; Farooqui and Rao, 1976; Kawadiwale, 1988).

Das and Mahapatra (1975) studied fruit retention in 12 year old trees of sapota cv. 'Cricket Ball' at Bhubaneswar, Odisha when treated with GA, Ethrel, CCC, B995 and NAA at different concentrations. Trees were sprayed before flowering and again at the pea stage. The maximum fruit retention was observed in sapota trees sprayed with NAA at 300 ppm. Rathod and Amin (1981) have also reported higher fruit retention with NAA as compared to GA in sapota. Ray *et al.* (1992) investigated the effect of SADH, NAA and GA_3 when applied as sprays just before anthesis and one month later in sapota cv. 'Cricket Ball' at Bhubaneswar, Odisha. They recorded the maximum fruit retention with 100 ppm SADH.

At Rahuri, Maharashtra the lowest fruit drop (85.66% and 85.62%) and the highest fruit retention (14.34% and 14.38%) was observed when sapota trees were treated with 100 ppm NAA at pea stage during 1997-98 and 1998-99, respectively (Kadam *et al.*, 2005). Nagargoje *et al.* (2007) reported the lowest percentage of fruit drop (74) and the highest percentage of fruit retention (26) when sapota trees of cv. 'Kalipatti' were sprayed with 100 ppm NAA at 50% flowering stage and at pea stage of fruit development in Parbhani, Maharashtra.

Trees of sapota cv. 'Kalipatti' at Parbhani, Maharashtra were subjected to foliar sprays of GA_3 (50, 100 and 150 ppm) and NAA (50, 100 and 150 ppm). These sprays were made at the time of flowering, pea stage and at the lag phase of fruit development. Foliar application of NAA at 150 ppm resulted in the highest number of flowers (54.0), highest fruit set (43.13%), highest retention (over fruit set) at pea stage (18.77%) and at lag phase (17.21%). It also yielded the maximum number of fruits (2633.00) and fruit yield (115.02 kg/tree) followed by NAA 100 ppm (Chavan *et al.*, 2009).

In the above studies, NAA was more effective in improving fruit retention as compared to GA. Higher fruit retention under NAA might be due to proper supplementation of nutrients and preventing the formation of an abscission layer by inhibiting enzymatic activities. On the other hand, GA had a positive effect on fruit retention by reducing immature fruit drop.

8. Role of PGRs Fruit Growth and Maturity

Studies on periodical changes in fresh weight of developing fruits showed that it followed a double sigmoid pattern of growth. There were 2 periods of active growth, separated by a period of retarded growth. Stage I (1st to 12th fortnight) of fruit growth accounted for 23.6% of the total weight of fruit attained at maturity, while stage II accounted for only 14.4%. Stage III (17th to 21st fortnight) accounted for the maximum amount of growth (62%) as measured in terms of fresh weight increase (Sulladmath *et al.*,1979). Statistical and graphical analyses of the data confirmed the double sigmoid nature of the growth curve. The fruit took about 10.5 months from fruit set to maturity. This can be attributed to a very slow growth rate in the initial stages of fruit development and a long period of the lag phase (2 months).

Experiments were conducted during 1997-1998 at Dapoli, Maharashtra by Raut *et al.* (2004) to evaluate maturity indices for sapota cv. 'Kalipatti'. Fruits attained full maturity after 8 months from fruit set. The fruit weight, volume, length, breadth and the length:breadth ratio increased appreciably with maturity after fruit set. An increase in total soluble solids and sugars with a corresponding decrease in acidity and ascorbic acid were observed during the course of fruit development in sapota.

9. Role of Plant Growth Regulators on Early and Uniform Ripening

There are reports which indicate the potential use of ethrel for induction of early and uniform ripening in sapota. Research done in this regard is reviewed as under.

Sapota fruits were dipped for 1 min in ethrel [ethephon] at 250- 7500 ppm, 2,4D at 5- 15 ppm or wax emulsion at 39% or were kept in dry paddy straw. Ripening was greatly accelerated by ethrel and all fruits ripened within 3 days of treatment compared with 26% under untreated controls. TSS and total sugar contents, assessed 4 days after treatment, were also greatly enhanced by ethrel (Das and Mahapatra, 1976).

Sapota fruits, cv. 'Oval', were treated (dipping and exposure to gas) with 1000, 2500 and 5000 ppm of ethrel [ethephon] after harvest and stored for 10 days in air-tight boxes. Fruits attained edible ripeness on the eighth day after harvest in control and on the third day in the 5000 ppm treatment and on the fourth day in the 1000 and 2500 pppm treatments. Ethrel caused significant increases in reducing sugars, total sugars, total soluble solids (TSS), ascorbic acid and total titratable acidity, while a reduction in non-reducing sugars, sugar/acid ratio and TSS/acid ratio was noticed. The effect of ethrel increased with increase in concentration from 1000 to 5000 ppm. Thus, ethrel (5000 ppm) can be used for quick ripening and for improving the quality of sapota fruits (Suryanarayana and Goud, 1984).

Madhavi *et al.*, (2005) carried out an experiment, where mature fruits of sapota cv. Pala were dipped in water (control) or aqueous solutions of ethrel [ethephon] at 500, 1000, 2000 or 3000 ppm for 5 minutes. The fruits were then dried in the shade and packed in polyethylene bags. Ethrel treated fruits showed more uniform ripening with higher total soluble solids and total sugars than untreated fruits. Ripening and spoilage increased with an increase in ethrel concentration. Ethrel at 1000 ppm resulted in higher percentage of marketable fruits. Total soluble solids, total sugars and the TSS:acid ratio increased, whereas the ascorbic acid content and acidity decreased with an increase in storage duration. The best results were obtained with ethrel at 1000 ppm.

Bons *et al.*, (2016) studied the effect of various levels of ethephon on ripening of sapota fruits under North Indian conditions on 'Kalipatti' and 'Cricket Ball' cultivars at PAU, Ludhiana. Fruits were dipped in solutions of ethephon @ 500 ppm, 750 ppm and 1000 ppm for two minutes to induce ripening of fruits. Dipping fruits in ethephon 1000 ppm resulted in better ripening of fruits after three days with pleasant flavour, high TSS, lower acidity and acceptable sensory quality.

The increase in soluble solids and sugars upon ripening could be due to hydrolysis of starch and organic compounds (Lelievre *et al.*, 1997). The decline in acidity may be due to utilization of acids in the process of ripening in the presence of reduced supply of sugar as a substrate of respiration which might be due to lower rate of starch degradation during the ripening.

10. Role of Plant Growth Regulators on Fruit Yield and Quality

Although, sapota starts bearing from the third year of planting, economical yields can be obtained from the seventh year onward. Fruit yield in sapota is determined by a multitude of factors mainly age of the tree, variety, prevailing agroclimatic conditions, management practices and plant protection measures.

Ray *et al.*, (1992) at Bhubaneswar, Odisha observed maximum fruit weight (108.7 g) in sapota cv. 'Cricket Ball' when sprayed with 100 ppm GA_3 at anthesis and one month later. Reddy and Khan (2001) conducted a research experiment to determine the effect of Paclobutrazol (1.25, 2.50 and 5.00 g), Cycocel [chlormequat-250, 500 and 750 ppm] and Chamatkar [Mepiquat chloride-1000, 2000 and 3000 ppm] on fruit yield of sapota cv. 'Kalipatti' at Bangalore, Karnataka. All growth regulator treatments were at par and significantly increased the fruit yield compared to the control. Growth regulator treatments had a more pronounced effect on the second year fruit yield. The highest fruit yield (56.3 kg/plant) in the second year was obtained with 5 g paclobutrazol.

Ray *et al.*, (1992) also observed highest fruit TSS, total sugar content, ascorbic acid content and decreased acidity under 200 ppm NAA in sapota cv. 'Cricket Ball'. The possible reason for increased total soluble solids with the application of NAA might be due to accumulation of metabolites and quick conversion of starch into sugars during the fruit development and an increased in vitamin C content was due to catalytic influence of NAA on bio synthesis of vitamin C from sugar or inhibition or activity of oxidative enzymes or both. The reason for decreased titrable acidity with the application of NAA might be due to early maturity and rapid utilization of organic acids during respiration after harvesting, because sapota is a climacteric fruit.

Agrawal (2004) reported that CCC 400 ppm when applied at FBD stage and two sprays of NAA 100 ppm, once at flowering and again at pea stage of fruit development resulted in significantly more number of fruits as well as yield per tree.

Kadam *et al.,* (2005) recorded the maximum number of fruits per tree (631.9) and fruit yield (44 kg/tree) in sapota when sprayed with NAA at 100 ppm at pea stage of fruit development in Rahuri, Maharashtra.

Nambisan *et al.*, (2007) studied the effects of CCC [chlormequat] (100, 150 or 200 ppm), NAA (10, 20 or 30 ppm), 2,4D (10, 20 or 30 ppm) and GA_3 [gibberellic acid] (10, 20 or 30 ppm) on the yield of sapota in Rahuri, Maharashtra. Growth regulators were sprayed at the flowering stage and 3 months thereafter. Foliar spray of 2,4D at 30 ppm resulted in the highest number of fruits per 100 shoots (59.96). The total fruit weight per 100 shoots was highest under 30 ppm NAA (4.1), 20 ppm 2,4D (4.1) and 30 ppm 2,4D (4.3).

Cycocel (0,200, 400 ppm) was sprayed at FBD stage, superimposed with GA (50 ppm) and NAA (100 ppm) sprays at different stages of fruit development to promote flowering in sapota cv. 'Cricket Ball' at Raipur, Chhattisgarh. Application of Cycocel (400 ppm) at FBD stage produced significantly more number of fruits as well as fruit yield per tree. Physical characters of fruits *viz*, length, diameter and weight of fruits were also increased considerably under the same treatment (Agrawal and Dikshit, 2008).

Ingale *et al.*, (2008) at Navsari, Gujarat reported maximum number of fruits per tree under NAA at 100 ppm + GA_3 at 100 ppm in cv. 'Kalipatti' during the spring season. However, fruit yield was highest under NAA at 100 ppm + GA_3 at 75 ppm. Application of NAA at 100 ppm + GA_3 at 100 resulted in the maximum fruit weight. The total yield in terms of weight was increased by During the postmonsoon, spraying NAA at 100 ppm + GA_3 at 25 ppm was more effective in increasing fruit weight and pulp weight. NAA may have increased fruit weight and yield by causing cell elongation, by enlargement of vacuoles and loosening of cell wall after increasing cell wall plasticity (Agrawal and Dikshit, 2008).

Stern *et al.*,(2007) reported that NAA treatments encourage cell expansion in the fruit mesocarp, which in turn, causes an increase in fruit volume and yield. Pulp weight may have increased due to an expansion in cell size and intracellular spaces coupled with accumulation of sugars, water and other soluble solids in greater amounts as a result of translocation of metabolites towards fruit development (Kaur, 2017).

Chavan *et al.*, (2009) recorded maximum number of fruits (2633) and fruit yield (115.02 kg/tree) in sapota cv. 'Kalipatti' at Parbhani, Maharashtra when sprayed with 100 ppm NAA at flowering, pea stage and lag phase of fruit development.

The number of fruits and yield per tree increased with increase in concentration of NAA (Patil *et al.*, 2011). Bhujbal *et al.* (2012) reported maximum number of fruits per tree under Cycocel at 450 ppm and the maximum fruit weight under GA_3 at 150 ppm in sapota cv. 'Kalipatti' at Parbhani, Maharashtra. Navsari Agricultural University has recommended sapota growers in Souith Gujarat to spray 1 g Gibberellic acid in first week of November, December and January for higher production and better economic returns. The resulting winter season fruits are superior in size (Anon., 2014). Yuvraj (2015) conducted an experiment at Navsari, Gujarat to assess the effect of biofertilisers, GA_3 and micronutrients on growth, yield and quality of sapota cv. 'Kalipatti'. They recorded significantly higher yield under the treatment comprising of 100% RDF+ Biofertilisers (*Azospirillum*+PSB) @ 40 ml/ tree + GA_3@50 ppm+ micronutrient mixture grade 4 @ 0.5%. Soil application of RDF was made in 2 equal splits *i.e.* in June and October. Liquid formulations of *Azospirillum* and PSB@ 40 ml/ tree were applied to the soil in the month of July. Foliar spray of GA_3@50 ppm and micronutrient mixture grade 4 @ 0.5% was applied in the month of November and December receptively.

Gondaliya (2016) conducted an investigation to examine the influence of pre-harvest sprays ($CaCl_2$, GA_3, CPPU) on fruit growth of sapota cultivar 'Kalipatti' at Navsari, Gujarat. Twenty year old sapota trees were sprayed 15 days before probable optimum harvesting stage during the winter season. They observed that pre-harvest spray of GA_3 50 ppm or CPPU 10 ppm can be used for increasing fruit growth characters like length, girth, weight and volume of sapota fruits.

Higher fruit retention and increased fruit weight, fruit diameter, fruit volume and number of fruits per tree, low percentage of fruit drop taken together contributed to an increase in fruit yield. It may also be due to inhibition of vegetative growth, resulting in better flowering, fruit set and ultimately higher fruit retention as well as translocation of extra metabolites towards the reproductive growth or sink *i.e.* fruits.

11. Role of Plant Growth Regulators on Extending Shelf Life

Sapota is a highly perishable fruit and cannot be stored for long under ambient conditions. The shelf life of sapota is very short due to its climacteric nature and many other reasons such as quick ripening, rapid loss of moisture, microbial spoilage and fruit sensitivity to cold storage (Siddiqui *et al.*, 2014). The rapid softening of fruits is primarily due to high activity of many oxidative enzymes and liberation of ethylene (Bhutia *et al.*, 2011). Plant growth regulators can be effectively employed for prolonging the shelf life of sapota by retarding the process of ripening.

Kumbhar and Desai (1986) at Rahuri, Maharashtra treated sapota fruits with GA or IBA, packed them in polyethylene bags and stored them at room temperature for up to 11 days. The best shelf life was observed when fruits were dipped for 3 minutes in a solution of 75 ppm GA and packed in polyethylene bags with 1.2% vent holes. Weight loss after 11 days under this treatment was 13.89% as compared to 90% in control (no chemicals and no bags).

Sapota trees of cv. 'Oval' at Bhubaneswar, Odisha were subjected to preharvest sprays of 2 or 4% $CaCl_2$, 1 or 2% $Ca(NO_3)_2$, 100 or 200 ppm GA_3 or water, 1 month before harvesting fruits. After harvest, some of the fruits were dipped in the same solutions or in Bavistin [carbendazim] at 1000 ppm for 10 min, before storage in polyethylene bags at room temperature. A combined preharvest spray application and postharvest dip in GA_3 at 100 ppm extended shelf life to 36 days (compared with 8 days in the water sprayed control), and reduced physiological weight loss and percentage rotting. Postharvest application of Bavistin at 1000 ppm (in addition to the GA_3 treatments) also extended the shelf life (to 32 days) and further reduced fruit spoilage, but increased fruit acidity, compared with control treatments (Bhanja and Lenka, 1994).

Patel and Katrodia (1996) conducted an experiment, wherein they treated sapota fruits with 150 or 200 ppm gibberellic acid (GA), 2% Waxol, 100 ppm kinetin or 2000 ppm maleic hydrazide in combination with 500 ppm Bavistin [carbendazim] to delay ripening and extend shelf life. Dipping in 150 ppm GA + fungicide solutions for 8-10 minutes, followed by drying and packing in wooden boxes using paddy straw as packing material, gave the best results (maximum delay in ripening, and lowest weight loss, spoilage, and total soluble solids contents). These treatments also helped in safe transportation of fruits to distant markets without any adverse effect on fruit quality.

A study was conducted by Madhavi and Srihari (2002) at Hyderabad, Andhra Pradesh to determine the effect of postharvest application of growth regulators in extending the shelf life of sapota fruits cv. 'Pala' during storage at low temperature (12°C). Fruits were dipped for 5 minutes in GA_3 (100 and 200 ppm), kinetin (100 and 200 ppm), 2,4D (2 and 4 ppm), or distilled water (control). Fruits treated with 4 ppm 2,4D and 200 ppm GA_3 exhibited the longest shelf life of 32 days and 31.33 days, respectively, as compared to control (20 days).

An investigation was carried out at Bhubaneswar, Odisha by Choudhury *et al.* (2003) to study the effect of Topsin M [thiophanatemethyl - 0.1 and 0.2%], Bavistin [carbendazim - (0.1 and 0.2%], $CaCl_2$ (1 and 2%), GA_3 (100 and 200 ppm) and water on sapota cv. 'Pala' fruits. All the treatments extended shelf life, reduced physiological loss in weight and rotting as well as augmented biochemical attributes such as total soluble solids, acidity and sugar. GA_3 (200 ppm) proved most effective when applied as preharvest spray and postharvest dip.

Sahoo and Munsi (2004) assessed the effect of GA_3 and $AgNO_3$ with two wrapping materials *viz.*, perforated polyethylene bags and ordinary newspaper on storage life and quality of sapota fruits cv. 'Cricket Ball' at Kolkata, West Bengal. Postharvest dip in GA_3 and $AgNO_3$ solutions were found to extend the storage life of fruits, but extension periods differed depending upon the treatments. Sapota fruits treated with 50 ppm and 100 ppm of $AgNO_3$ and GA_3 respectively and then kept in polyethylene bags of 200 gauge with 20% vent emerged as the best treatment for extending storage life with less degradation.

An experiment was conducted by Sudha *et al.* (2007a) in Coimbatore, Tamil Nadu, India, to evaluate the influence of pre and postharvest chemical treatments on the physical characteristics of sapota fruits cv. 'PKM 1'. Fruits were treated with GA_3 and chemicals like $CaCl_2$, Ca $(NO_3)_2$ and KCl as preharvest spray and postharvest dip. Preharvest spray of 50 ppm GA_3 with postharvest dipping of GA_3 50 ppm along with 0.2% bavistin [carbendazim] recorded the lowest physiological loss of weight and shrinkage percentage, longest shelf life and lowest spoilage percentage. Preharvest spray of 50 ppm GA_3 with postharvest dipping of $CaCl_2$ 1% along with 0.2% bavistin recorded highest firmness.

Sudha *et al.*, (2007b) also investigated the effect of packaging and hormonal treatments on increasing the shelf life and quality of fruits of sapota cultivars 'PKM 1' and 'CO_2'. The treatments comprised vacuum packaging (T_1), packaging with 1% (T_2), 2% (T_3) and without (T_4) ventilation in polyethylene bags, preharvest spray of gibberellic acid (GA_3) at 50 ppm+postharvest dipping in GA_3+0.2% Bavistin [carbendazim]+T_1 (T_5), preharvest spray of GA_3+postharvest dipping in GA_3+T_2 (T_6), preharvest spray of GA_3+postharvest dipping in GA_3+T_3 (T_7), preharvest spray of GA_3+postharvest dipping in GA_3+T_4 (T_8), and control (T_9). A shelf life of 24.5 days was observed in T_5, followed by 14.5 days in T_6. Treatment T_5 recorded the highest ascorbic acid (19.36 mg/100 g), total soluble solids (23.40 °Brix) and total sugar contents (17.28%) and the least sugar:acid ratio (64.72).

Patel (2009) conducted an experiment at Navsari, Gujarat to assess the effect of post harvest dipping of $GA_3$150 ppm on extending the shelf life of different sapota cvs 'Kalipatti, PKM-1, CO-2, Singapore and Kirtibharti'. They found that the postharvest dipping in an aqueous solution of $GA_3$150 ppm before cold storage resulted in regulation of ripening and also maintained desirable qualities like sugars, TSS, less ripening percentage and less physiological loss in weight at 12°C. Dipping sapota fruits in 1% calcium hydroxide for 5 minutes and wet rubbed after drying improved the appearance of fruits. Dipping them again in 4 ppm 2, 4-D or 300 ppm GA_3, or 100 ppm kinetin extended the shelf life without any adverse effect on fruit quality as compared to farmers practice *i.e.* wet or dry rubbing only (Anon., 2011). Bhujbal *et al.* (2013) recorded the maximum total sugars in sapota cv. 'Kalipatti' when treated with GA_3 at 100 ppm.

The increased in total soluble solids and sugars might be due to hormonal synthesis, which increased the metabolites available for total soluble solids formation. Increase in TSS during ripening may have resulted from an increase in concentration of organic solutes as a consequence of water loss (Ryall and Pentzer, 1982). It also might be due to the alteration or transformation in cell wall structure and break down of complex carbohydrates, pectic substances, hemicellulose or other polysaccharide into simple sugars (Kittur *et al.*, 2001).

The possible reason for increased ascorbic acid content with the application of GA_3 might be due to catalytic influence of growth regulators on its bio-synthesis from its precursor glucose-6- phosphate or inhibition of its conversion to dehydro ascorbic acid by the enzyme ascorbic oxidase (Kaur, 2017). It is also attributed to early maturity and rapid utilization of organic acids during respiration after harvesting as sapota is a climacteric fruit. The conversion of organic acids into soluble sugars and long chain polysaccharides may also have contributed to a decrease in acids (Lelievre *et al.*, 1997). Plant growth regulators may have increased the activity of enzymes like amylases, which hydrolyses complex polysaccharides into simple sugars and thereby increased total sugar, reducing sugar and non reducing sugar content in fruits. The decrease in firmness, during ripening may be due to breakdown of insoluble protopectin into soluble pectin or by cellular disintegration leading to membrane permeability (Brinston *et al.*, 1988). Softening of flesh could also be due to the degradation of soluble pectin by the activity of endo-polyglacturonase in fruits (Martin-Cabrejas *et al.*, 1994).

References

- Addicot, F.T. and Lynch, R.S. 1955. Physiology. *Ann. Rev. Pl. Physiol.*, **6**: 211-238.
- Agrawal, S. 2004. Studies on the effect of plant growth regulators on growth, flowering, yield and quality of sapota (*Achras sapota* L.) cv. Cricket Ball under Chhattisgarh plains. Ph.D. thesis submitted to IGAU, Raipur, Chhattisgarh, India.
- Agrawal, S. and Dikshit S.N. 2008. Studies on the effect of plant growth regulators on growth and yield of sapota (*Achras sapota* L.) cv. Cricket Ball. *Indian J. Agric. Res.*, **42**(3): 207-211.
- Alam, T. and Faruque, A.H.M. 1974. Success of air layerage of sapota (*Achras sapota*) with the use of IBA and NAA. *Bangladesh Horticulture*, **2**(2): 45-48.
- Anonymous 2004. *Activities and Achievements*. Department of Fruit Science, ASPEE College of Horticulture and Forestry, Navsari Agricultural University, Navsari.
- Anonymous 2011. *Salient Research Achievements of Navsari Agricultural University* (2004-2014), Navsari Agricultural University, Navsari, pp. 69.
- Anonymous 2014. *Salient Research Achievements of Navsari Agricultural University* (2004-2014), Navsari Agricultural University, Navsari, pp. 70.
- Bhanja, P.K. and Lenka, P.C. 1994. Effect of pre and postharvest treatments on storage life of sapota fruits cv. Oval. *Orissa J. Hort.*, **22** (1/2): 54-57.
- Bhujbal, D.S., Naik, D.M. and Kale, S.A. 2012. Effect of growth regulators on growth and yield of sapota (*Manilkara achras* (Mill.) Fosberg). *The Asian J. Hort.*, **7**(2): 351-353.
- Bhujbal, D.S., Naik, D.M. and Kale, S.A. 2013. Studies on effect of growth regulators

on flowering, fruiting and quality of sapota. *Int. J. Agric. Sci.*, **9**(1): 289-292.

- Bhutia, W., Pal, R.K., Sen, S. and Jha, S.K. 2011. Response of different maturity stages of sapota (*Manilkara achras* Mill.) cv. Kalipatti to in-package ethylene absorbent. *J. Food Sci. Technol.*, **48**(6): 763-768.
- Bhuva, H.P. and Chundawat, B.S. 1978. Flowering and fruit growth in sapota. (*Achras sapota* L.). *Udyanika,* **2**: 54-56.
- Bons, H.K.; Boora, R.S. and Rattanpal, H.S. 2016. Studies on effect of various levels of ethephon on ripening of sapodilla fruits (*Manilkara achras* (Mill) Fosberg). *Int. J. Agric. Envi. and Biotech.*, **9**(4): 563-566.
- Brian, P. W. and Hemming, H. G. 1958. Complementary action of gibberellic acid auxins in pea internode extension. *Ann. Bot.*, **2**: 1-17.
- Brinston, K.; Dey, P.M.; John, M.A.; Pridhan, J.B. 1988. Post-harvest changes in *Mangifera indica* mesocarp walls and cytoplasmic polysaccharides. *Phytochem.*, **27**: 719-723.
- Chatterjee, B.K., Ray, D.P., Chakladar, B.P. and Mukherjee, S.K. 1989. Effect of different concentrations of growth regulators, etiolation and invigoration on the rooting of air-layers of sapota (*Achras zapota* Linn. [*Manilkara zapota*]). *Indian J. Hort.*, **46**(4): 426-429.
- Chavan, S.R., Patil, M.B., Phad G.N. and Suryawanshi A.B. 2009. Effect of growth regulators on flowering and yield of sapota [*Manilkara achras* (Mill.) Fosberg]. *The Asian J. Hort.*, **4**(1): 119-120.
- Cheema, G.S., Bhat, S.S. and Naik, K.C. 1954. *Commercial Fruits of India with Special Reference to Western India*. Macmillan and Company, Calcutta, pp. 359–378.
- Choudhury, S., Ray, D.P., Das, B.K. and Sahu, G.S. 2003. Effect of pre and postharvest chemical treatments on ripening, quality and storage life of sapota [*Manilkara achras* (Mill.) Fosberg] cv 'Pala'. *Orissa J. Hort.*, **31**(2): 54-57.
- Copeland, L.O. and Mcdonald, M.B. 1995. *Principles of Seed Science and Technology.* III Edition-Chapman and Hall Publications, New York, pp. 127-146.
- Das, R.C. and Mahapatra, S.K. 1975. Effects of growth substances on retention and growth of sapota fruits (*Achras sapota* L.). *Pl. Sci.*, **7**: 93-94.
- Das, R.C. and Mahapatra, S.K. 1976. Effect of growth substances and fungicidal wax emulsion on ripening of sapota (*Achras zapota*). *Prog. Hort.*, **8**(3): 75-80.

- De Klerk, G.J., W. Van der Krieken and J.C. De Jong 1999. The formation of adventitious roots: new concepts, new possibilities. *In Vitro* Cell. *Dev Biol Plant*, **35**: 189-199.
- Diaz, D. H. and Martin, G. C. 1971. Peach seed dormancy in relation to inhibitors and applied growth substance. *J. Amer. Soc. Hort. Sci.*, **97**(5): 651-654.
- Epstein, E. and Ludwig-Muller 1993. Indole-3-butyric acid in plants: occurrence, synthesis, metabolism and transport. *Physiol. Plant*, **88**: 382-389.
- Farooqui, A. A. and Rao, M. M. 1976. Studies on fruit set in some sapota varieties in relation to intra and inter-varietal pollination. *Mysore J. Agric. Sci.*, **10**:28-34.
- Farooqui, A. A.; Nalawadi, U. G. and Sulladmath, U. V. 1971. Effect of growth regulators on the germination of sapota (*Achras sapota*) seeds. *Mysore J. Agril. Sci.*, **5**(3): 341-343.
- Gondaliya, P. J. 2016. Influence of preharvest spray and postharvest dip treatments on physic-chemical properties and shelf life of sapota cv. Kalipatti fruits. M.Sc. thesis submitted to Navsari Agricultural University, Navsari, Gujarat, India.
- Hajam, M.A., Hassan, G.I., Bhat, T.A., Bhat, I.A., Rather, A.M., Parray, E.A., Wani M.A. and Khan, F.A. 2017. Understanding plant growth regulators, their interplay: For nursery establishment in fruits. *Int. J. Chem. Stud.*, **5**(5): 905-910.
- Hartmann, H. T. and Kester, D. E. 1972. Plant propagation principles and practices, 2nd edition, Prentice Hall of India Pvt. Ltd., New Delhi, pp. 117-141.
- Hayes, W.B. 1957. *Fruit Growing in India*. Kitabistan, Allahabad.
- Hosomi, K., Yoshimoto, T., Fukuzumi, T. and Itoh, Y. 1984. Studies on tissue culture of *Achras sapota [Manilkara zapota]*. II. Effect of plant growth regulators on growth and triterpeneester production of tissue culture cells. *J. Japan Wood Res. Soc.*, **30**(5): 404-408.
- Ingale, K.D. Dixit, C.K. Patil, R.V. and Wakchaure, N.Y. 2008. Effect of plant growth regulators on fruit set and yield of sapota cv. Kalipatti. *J. Maharashtra Agric. Uni.*, **33**(1): 79.
- Jain V. K. 2013. *Fundamentals of Plant Physiology*, S. Chand and Company Pvt. Ltd., New Delhi, pp. 301-358.
- Kadam, D.D.; Garad, B.V.; Jadhav, Y.R.; Mangave, K.K. and Patgaonkar, D.R. 2005. Influence of growth regulators on flowering behaviour of sapota under semi-arid zone of Maharashtra. *J. Soils and Crops*, **15**(1): 111-114.

- Kaur, N. 2017.Comparative efficiency of plant growth regulators on fruit set, yield and quality of sapota cv. Kalipatti. M.Sc. thesis submitted to Punjab Agricultural University, Ludhiana, India.
- Kawadiwale, K.R. 1988. Studies on flowering and fruiting behaviour. chiku (*Achras sapota* L.) under Akola condition. M.Sc. (Ag.) thesis submitted to Dr. Panjabrao Deshmukh Krishi Vidyapeeth, Akola, Maharashtra, India.
- Kittur, F.S.; saroja, N. and Tharanathan, R.N. 2001. Polysaccharide-based composite coating formulations for shelf-life extension of fresh banana and mango. *European Food Res. and Tech.*, **213** (4-5): 306-311.
- Kumbhar, S.S. and Desai, U.T. 1986. Studies on the shelf life of sapota fruits.*J. Maharashtra Agric. Uni.*, **11**(2): 184-187.
- Lelievre, J. M.; Tichit, L.; Dao, P.; Fillion, L.; Nam, Y.W.; Pech, J.L. and Latche, A. 1997. Effects of chilling on the expression of ethylene biosynthesis gene in 'Passe-crassane'pears (*Pyrus communis*) fruits. *Pl. Molecular Bio.*, **33**(5): 847-855.
- Lenka, P.C. Das, D.K. and Samal, B. 1996. Studies on floral biology and physical characteristics of sapota cultivars. *Orissa J. Hort.*, **24**(1/2): 42-46.
- Madhavi, M. and Srihari, D. 2002. Effect of postharvest application of growth regulators on low temperature storage life of sapota [*Manilkara achras* (Mill.) Fosberg] fruits cv. Pala. *South Indian Hort.*, 50(1/3): 185-189.
- Madhavi, M. Srihari, D. and Babu, J.D. 2005. Effect of postharvest ethrel treatment on ripening and quality of sapota cv. Pala fruits. *Indian J. Hort.*, **62**(2): 187-189.
- Mahaveer, S., Pency, D.S., Deva, R.M. and Sahu, O.P. 2017. Effect of plant growth regulators on fruit crops. *Journal of Pharmacognosy and Phytochemistry,* **6**(2): 331-337.
- Martin-Cabrejas, M.A.; Waldran, K.W.; Selvendaran, R.R.; Parket, M.L. and Maotes, G.K. 1994. Ripening related changes in cell wall of 'Spanish' pear. *Physiologia Plantarum*, **91**(4): 671-679.
- Mathur, B.D.; Couvillon, G.A.; Vines, H.M. and Hendershoot, C.H. 1971. Stratification effects on endogenous gibberellic acid in peach seed. *Hort. Sci.*, **6**: 538-539.
- Modise, D. M., Likuku, A. S., Thuma, M. and Phuti, R. 2009. The influence of exogenously applied 2, 4-dichlorophenoxyacetic acid on fruit drop and quality of naval oranges (*Citrus sinensis* L.). *African J Biotechnol.*, **8**(10): 2131-37.

- Nagargoje, B.S.; Kachave, D.B.; Shinde, S.B. and Jadhav, K.M. 2007. Effect of NAA on flower drop, fruit set, fruit drop and fruit retention in sapota variety Kalipatti. *J. Soils and Crops*, **17**(1): 173-175.

- Nambisan, J.V. Desai, U.T. Kshirsagar, D.B. and Kamble, A.B. 2007. Effect of plant growth regulators on yield of sapota cv. Kalipatti. *J. Maharashtra Agric. Uni.*, **32**(2): 280-281.

- Nikam, B. C.; Gunjate, R. T.; Lad, B. L. and Salvi, M J. 1986. Studies on improving germination of seeds of rayan. *Maharashtra Agric. Univ.*, **11**(3): 299-300.

- Pampanna, Y. and Sulikeri, G.S. 2001. Effect of growth regulators on seed germination and seedling growth of sapota. *Karnataka J. Agri. Sci.*, **14**(4): 1030-1036.

- Pampanna, Y.; Sulikeri, G. S. and Hulmani, N. C. 1995. Effect of growth regulators on seed germination and growth of seedling of sapota [*Manilkara achras* (Mill.) Fosberg.]. *Karnataka J. Agri. Sci.*, **8**(1): 60-64.

- Panigrahi, H.K. Singh, P. and Dikshit, S.N. 2011. Studies on the effect of plant growth regulators on fruit retention and fruit drop of sapota (*Manilkara achras* Mill) cv. Cricket Ball. *Journal of Interacademicia*, **15**(1):10-16.

- Patel, A.B. and Katrodia, J.S. 1996. Response of GA to extend the shelf life of sapota fruits after transportation. *GAU Res. J.*, **21**(2): 60-64.

- Patel, R.G. 2009. Extending the shelf life of sapota varieties in cold storage with application of GA_3. M.Sc. thesis submitted to Navsari Agricultural University, Navsari, India.

- Patil, M.B., Munde, G.R., Nainwad, R.V. and Mane, S.S. 2011. Studies on effect of plant growth regulator on physical characters of sapota. *The Asian J. Hort.*, **6**(1): 98-100.

- Patil, V.K. and Narwadkar, P.R. 1974. Studies on flowering, pollination, fruit set and fruit drop in chiku. *Punjab Hort. J.*, **14**(12): 39-42.

- Ramamurthy, K. and Ramalingam, B. 1954. A note on hybridization sapota (*Achras zapota* L.). *Indian J. Hort.*, **8**:57-60.

- Ramsundar, V. and Abdul Khader, J.B.M.M. 1986. Studies on rooting of sapota cuttings. *South Indian Hort.,* **34**(1): 40-43.

- Rathod, R.P. and Amin, H.D. 1981. Effect of NAA and GA on chiku (*Achras sapota* L.) Proceedings of National Symposium on Tropical and Sub-Tropical Fruit Crops organized during January 21-24, 1981 at IIHR, Bangalore.

- Raut, V.U., Joshi, G.D. and Khandekar, R.G. 2004. Studies on maturity indices of sapota [*Manilkara achras* (Mill.) Fosberg] fruits. *J. Soils and Crops*, **14** (2): 284-288.
- Ravoof, A.A. 1963. Preliminary studies on the effect of boric acid, gibberellic acid and growth regulators on fruit set, maturity and composition in the sapota. *Madras Agric. J.*, **51**(2): 88-89.
- Ray, D.P.; Samant, P.K.S.; Dora, D.K.; Sahu, P. and Das, B.K. 1992. Effect of plant growth regulators on fruit set, retention, development and quality of sapota (*Achras sapota* L.) cv. Cricket Ball. *Indian Agriculturist*, **36**(1): 913
- Reddy, Y. T. N. and Khan, M. M. 2001: Effect of growth regulators on water relations and fruit yield of rainfed sapota (*Achras sapota*). *J. Applied Hort.*, **3**(1): 5657.
- Ryall, A. L. and Pentzer, W. T. 1982. *Handling, Transportation and Storage of Fruits and Vegetables*. Vol. 2, AVI Pub. Co., Westport, Connecticut, pp. 1-40.
- Sachs, R.M., Lang, J., Bertz, C.F. and Roach, J. 1960. Shoot histogenesis: Sub-apical merestematic activity in a caulecent plant action of gibberellic acid and AMO 1618. *Amer. J. Bot.*, **47**: 260-266.
- Sahoo, S. and Munsi, P.S. 2004. Effect of GA_3 and $AgNO_3$ with wrapping materials on storage life of cricket ball sapota (*Acharas sapota* L.). *Indian Agriculturist*, **48**(1/2): 103-112.
- Siddiqui, M.W., Moasosang Longkumer, Ahmad, M., Kalyan Barman, Thakur, P.K. and Jahangir Kabir 2014. Postharvest biology and technology of sapota: A concise review. *Acta Physiologia Plantarum*, **36**(12): 3115-3122
- Srinivas, G. 2007. Standardization of soft wood grafting in Sapota [*Manilkara achras* (Mill.) Fosberg]. M.Sc. thesis submitted to ANGRAU, Bapatla, India.
- Stern, R.A., Flaishman, M. and Ben-Arie, R. 2007 Effect of synthetic auxins on fruit size of five cultivars of Japanese plum (*Prunus saliciana* Lindl.), *Scientia Horticulturae*, **112:** 304-309.
- Sudha, R.; Amutha, R.; Muthulaksmi, S.; Rani, W. B.; Indira, K. and Mareeswari, P. 2007a. Influence of pre and post harvest chemical treatments on physical characteristics of sapota (*Achras sapota* L.) var. PKM 1. *Res. J. Agric. Biol. Sci.*, **3**(5): 450-452.
- Sudha, R.; Ponnuswami, V. and Kavino, M. 2007b. Influence of packaging treatments on the physical and biochemical characteristics of sapota fruits. *Indian J. Hort.*, **64**(2): 222- 225.

- Sulladmath, V.V., Rao, M.M. and Advani, M. R. 1979. Studies on the pattern of growth of developing 'Kalipatti' sapota (*Manilkara achras* (Mill) Fosberg) fruit. *J. Maharashtra Agric. Uni.*, **4**(1): 55-60.

- Suryanarayana, V. and Goud, P. V. 1984. Effect of post-harvest Ethrel treatment on ripening of sapota fruits. *Andhra Agril. J.*, **31**(4): 308-311

- Tendulkar, S. 1978. Studies on growth of rootstocks and propagation of sapota (*Manilkara achras*). M.Sc. (Agri) thesis submitted to University of Agricultural Science, Bangalore, India.

- Tsomu, T.; Patel, H. C.; Thakkar, R. M.; Ajang, M. and Vasara, R. P. 2015. Response of post-harvest treatments of chemical and plant growth regulators on biochemical characteristics of sapota fruit cv. Kalipatti. *The Bioscan*, **10**(1): 33-36.

- Vacha, G. A. and Harvey, R. B. 1927. The use of ethylene, propylene and similar compounds in breaking the rest period of tubers, bulbs, cuttings and seeds. *Pl. Physiol.*, **2**: 187-194.

- Wareing, P. F.; Good, J. And Manuel, J. 1968. Some possible physiological roles of Abscisic acid. **In**: *Biochemistry and Physiology of Plant Growth Substances*, (Ed.) Wightman, F. and Setterfield, G., Runge Press, Ottawa, Canada, pp. 1561-1579.

- Yuvraj, K. R. 2015. Effect of biofertilizers, GA_3 and micronutrients on growth, yield and quality of sapota cv. Kalipatti. Ph.D. Thesis submitted to Navsari Agricultural University, Navsari, Gujarat.

26

Plant Growth Regulators in Strawberry

A.K. Godara, Vikas Kumar Sharma, A.K. Panda and R.K. Tarai

1. Introduction

The cultivated strawberry (*Fragaria × ananassa* Duch.) is a widely adapted small fruit, grown from the low-latitude tropics and subtropics (Maxico, Colombia, Costa Rica, Guatemala) to high-latitude, cold, continental areas (Poland, Russia, Germany, Belgium, Scandinavia). Strawberry is basically a fruit plant of temperate climate, but during the recent years, there has been phenomenal increase in its area, production and cultivation in the non-traditional region of India (Sharma and Sharma, 2004). In India, higher returns per unit area and short duration (six month) of crop have attracted large number of Indian farmer of Maharashtra, Punjab, Haryana, Himachal Pradesh, West Bengal, Delhi, Uttar Pradesh, Uttrakhand and Jammu and Kashmir states (Bhat *et al.*, 2005; Anon., 2015). Strawberries have traditionally been a popular delicious fruit for its flavour, taste, fresh use, freezing and processing. Now, strawberries are available as fresh fruit throughout the year rather than being a traditionally seasonal crop. Maximization of fruit production with improvement in the fruit quality has been continued to be the primary aim of the strawberry growers. To achieve this goal, it needed to adopt the latest developed orchard practices. Now-a-days, use of plant growth regulator in strawberry for increasing production and improvement in the quality of the fruit has been recognized as one of the essential production practices.

Plant growth regulators (PGR's) are manmade or naturally derived plant hormone enhancers or disruptors. Plant growth regulators function as chemical messengers for intercellular communication. They are extremely important agent in the integration of developmental activities and work together co-coordinating the growth and development of cells. However, many of these functions are still unknown processes. Environmental factors often exert inductive effects by evoking changes in hormones in metabolism and distribution within the plant. Apart from it, they also regulate expression of intrinsic genetic potential of plants (Suman *et al.*, 2017). Different strawberry cultivars responded differently to the applied PBRs (Zurawicz *et al.*, 2004; Tripathi and Shukla, 2010; Suvalaxmi *et al.*, 2016; Vishal *et al.*, 2016) at different concentration (Kumar *et al.*, 2012; Camacaro *et al.*, 2013; Saravanan *et al.*, 2013; Jasrotia *et al.*, 2014 Thakur *et al.*, 2015) and further regulated by environmental factors (Hancock *et al.*, 2013).

The phytohormone auxin, plays a critical role in regulating many aspects of plant growth and developmental processes. At the cellular level, auxin regulates the cell division, extension and differentiation and on a whole plant level, auxin plays an essential role in processes such as apical dominance, lateral root formation, tropic responses, fruit growth and development, vascular differentiation and embryogenesis. The effects of other hormones have been less studied. However, according to a few reports, cytokinin may affect axillary bud differentiation (Waithaka *et al.*, 1980; Braun and Kender, 1985; Hou and Huang, 2005). Plant growth regulators were found to be very effective in increasing the vegetative growth, flowering and yield of berry fruits in temperate, tropical as well as subtropical regions. In most of the studies, high concentration of gibbereliic acid increased vegetative growth and runner production in strawberry whereas cycocel, a growth retardant was very effective in improving fruit quality (Kumra *et al.*, 2018). Mechanism of gibberellins action in the apex of responsive plant resulted in increased protein synthesis, cell division and auxin production. Cycocel is an onium compound, which blocks GA biosynthesis at the step between geranylgeranyl pyrophosphate and copalyl pyrophosphate. Plants treated with this compound, have shorted internodes and enhanced photosynthesis. Plant growth retardants generally have great effects on expending of elongation cells, where inhibition of GA synthesis rapidly causes reduction in stem elongation and leaf expansion (Tanimoto, 1983).

2. Role of Plant Growth Regulators in Propagation

Axillary scapes (a *scape* is a long internode forming the basal part or the whole of a peduncle) of strawberry carry white flowers and joined together in cymes. After fertilization, the receptacle of the flower becomes fleshy and forms the edible part. The surface of the receptacle is covered with dry achenes that are, botanically, fruits (Hamdouni *et al.*, 2001). The seed propagation is mainly used in breeding programme to develop new varieties. The *in vitro* germination of the strawberry plant achenes constitutes a necessary stage in the production of sterile seedlings essential to any process in biotechnology as genetic transformation. However, the germination of achenes is often poor (Hamdouni *et al.*, 2001). The achenes of strawberry germinate better when taken from mature fruits (Hanke, 1993), in particular of the basal and median part of the fruit (Chellappa, 1959; Gupta, 1962), and if they have been stored a minimum of time (Adam and Wilson, 1967). The strawberry hard pericarp simply acts like a physical barrier to expansion and germination of the embryo. Strawberry achenes also contain one or more endogenous germination inhibitors. Various substances have been tested, which makes the tissues surrounding the embryo permeable to water and gases necessary to this physiological process. The hormones like gibberellic acid (Chellappa, 1959; Gupta, 1962; Nakamura, 1972) and ethephon (Hanke, 1993) improve and speed the rate of germination and percentage in strawberry.

Negi and Singh (1972) observed 40-44 per cent germination of strawberry seeds with gibberellic acid at 100 ppm. After a short period (18 days) of chilling treatment, gibberellic acid at 100 ppm promoted significant increase in germination of Freya and Vigour strawberry (Thompson, 1969). According to Lata *et al.* (2018) recorded highest germination (86%) with ethrel 5000 ppm after 30 days of experiment followed by 200 ppm gibberellic acid (84.33%) and 1000 ppm ethrel (80.67%).

3. Role of Plant Growth Regulators in Micro-Propagation

Strawberry is commercially propagated by runners (Biswas *et al.*, 2008); therefore the health of daughter plant depends on their mother plants. So to maintain the production potential, biotechnological intervention is required. The tissue culture plants are used by the nurserymen for multiplication by runner production under protected cultivation, which are used by farmers in the field for fruit production. Strawberry is perhaps the first fruit crop in which micro-propagation technique has been standardized (Sharma and Singh, 1999). The method for avoiding virus infection, meristem culture of shoot tip of health mother stock is to be used, as propagation material. The micro-propagated strawberry plants were comparatively better in different characters (crown size, number of runners, flowering time and yield of berries) than conventionally propagated runner plants (Karhu and Hakala, 2002).

In vitro production of plants involves the application of plant growth regulator such as auxin for process initiation. Nevertheless, these auxins are known to be associated with genetic instability in plants, a phenomenon called somaclonal variation (Karp, 1989; Phillips *et al.,* 1994). The hormone concentration that stimulates the rapid differentiation could affect the magnitude of abnormality such as hyper-flowering, fruit malformation, small plant, lower yields, runner-less female fertile plants, as mentioned by Swartz *et al.* (1981) and Anderson *et al.* (1982). Different explants like meristem-tip, anthers, immature embryos and first axillary buds of stolon have been used to produce millions of plants in a year (Boxus, 1989). Plant growth regulators *viz.,* BAP, KIN, 2,4-D and GA_3 in different combinations produced different desired responses in *in vitro* (Ara *et al.,* 2012a; Bhat *et al.,* 2012; Sonmez and Kafkas, 2012; Ara *et al.,* 2012b).

Reproducible protocol for regeneration of complete plantlets from 'Bounty' strawberry (*Fragaria ananassa* Duch.), using a combination of gelled medium and bioreactor system, has been standardized. Sepals, leaf discs, and petiole halves produced multiple buds and shoots when cultured on semi solid-gelled medium containing 4 μM thidiazuron (TDZ) for 4 wk followed by transferring in liquid medium containing 2 μM TDZ in a bioreactor system and cultured for another 4 wk. TDZ induced shoot proliferation at 0.1 μM in the bioreactor system but inhibited shoot elongation. TDZ-induced shoots were elongated and rooted *in vitro* on gelled medium containing 2 μM zeatin. Such bioreactor-derived tissue culture (BC) plantlets obtained from sepal explants were grown *ex vitro* and compared with those propagated by tissue culture on gelled medium (GC) and by conventional runner cuttings (RC), for growth, morphology, anthocyanin content, and antioxidant activity after three growth seasons. The BC and GC plants produced more crowns, runners, leaves, and berries than the RC plants although berry weight per plant did not differ significantly. BC and GC plants produced berries with more anthocyanin contents and antioxidant activities than those produced by the RC plants. *In vitro* culture on TDZ and zeatin containing nutrient media apparently induced the juvenile branching characteristics that enhanced vegetative growth with more crown, runners, leaf and berry production (Debnath, 2009).

Influence of gibberellic acid (GA_3) on strawberry in vitro shoot culture growth and development was investigated. 'Senga Sengana' and 'Elsanta' clones were grown on the medium recommended by Boxus (1999), supplemented with BA (0.5 mg dm^{-3}), IBA (0.1 mg dm^{-3}), glucose (40.0 g dm^{-3}) and GA_3 (0.1 - control, 1.0, 2.0, 5.0, 10.0 mg dm^{-3}). In general, gibberellic acid improved axillary shoot elongation and reduced the growth of callus as well as the formation of roots and the development of adventitious shoots. GA_3 applied at a concentration of 1.0-2.0 mg dm^{-3} significantly increased the number of axillary shoots (mainly crown shoots), whereas under higher (5.0-10.0 mg dm^{-3}) doses it stimulated the development of runners. It seems that a new method of strawberry micropropagation based on the multiplication of axillary crowns and runners, which could also reduce the risk of domination of in vitro cultures with adventitious shoots (Litwinczuk *et al.*, 2009).

In studies on micro propagation of strawberry (Fragaria X annanassa Duch.) var. chandler eight different treatments were considered (T_1 to T_8) to which MS media is supplemented with different combinations of Benzyl amino purine (BAP), Kinetin (Kn) and Gibberellic acid (GA_3). After one week the shoots were observed to be emerging from the nodal explants. After 10 to 20 days of culture T_4 [BA (1.5mg/l) + Kn (0.5mg/l)] showed good result in terms of shoot length. However, the T_3 [BA (1.5mg/l) + Kn (0.1mg/l)] showed best results throughout the culture. It is found from the present study that even though highly disease borne strawberry, can be made disease free and established under aseptic condition. Once disease free, they can be manipulated for quality planting material production. Plant growth regulators viz., BAP, Kn and GA3 in different combinations produced different desired responses in vitro. After 30 days of culture, the treatment T3 [BAP (1.5mg/l) + Kn (0.1 mg/l)] gave the highest number of shoots (3) with 5.5cm shoot length, highest number of leaves (7) and maximum survival rate of explants (70%) and thus it was found to be the best among the various treatment combination; as well as more economically feasible over the rest of the treatments (Bhat *et al.*, 2012).

Effect of different growth regulators on in vitro multiplication of strawberry (Fragaria × ananassa Duch.) cv. Camarosa through runner tip was studied by Sharma *et al.* (2016) with various combinations and concentrations IAA, BAP & Kinetin and IBA & NAA hormones using Murashige and Skoog media (MS). MS media with 0.5 μM IAA + 2.5 μM kinetin took minimum days (11.67) for shoot proliferation and produced maximum number of shoots/ culture (12.67) followed by 0.1 μM IAA + 2.0 μM kinetin. Application of IBA (2 mg/l) in MS media initiated maximum roots in minimal time (7.33). Maximum mean number of roots/ explant (9.33) and length of roots (6.17 cm) was also recorded in MS + IBA (2 mg/ l) followed by MS + IBA (1 mg/ l) media. The rooted plants transferred in cocopeat + perlite + vermicompost (3:1:1) showed maximum transplanting success (94.33%) while minimum from S 6 (63.00%).

4. Effect of Plant Growth Regulators on Vegetative Growth

Formation of new leaves, crown and shoots is indispensable for maintenance and to produce optimum yield in strawberry. Thompson and Guttride (1959) found an increase in length and upright growth of petioles and concluded that GA_3 can substitute for growth promoting substances that are produced normally under long days. The gibberellins regulate the growth of strawberry plants by causing cell division, cell elongation and a corresponding increase in epidermal and parenchyma's cell length in mature petiole and sub-apical meristem of strawberry (Pathak and Singh, 1976; Arteca, 1995). Improvement in the vegetative growth of strawberry plants was reported with the spray of GA_3 75 ppm by Leshem and Koller (1964). Singh and Kaul (1967) studied the effect of GA_3 in November and January on the growth of strawberry plants and reported that height, spread, leaf area and petiole length were mainly increased by increased length of parenchyma cells. GA_3 at 20 ppm in October did not affect vegetative growth but when applied in January, it stimulated the growth (Celestre, 1970; Celestre and Pierandrei, 1971). Number of leaves, internodal length, plant height, plant spread, petiole length, leaf area, total biomass, photosynthetic rate and stomatal conductance increased with 25-450 ppm GA_3 in different cultivars of strawberry (Paroussi *et al.*, 2002a; Sharma and Singh, 2009; Kumar *et al.*, 2011; Jamal *et al.*, 2012; Qureshi *et al.*, 2013). Miranda *et al.* (1990) recorded that there is no significant difference in plant morphology characteristics as a result of different growth regulator treatments (GA_3, NAA, Ergostin or Atenix) except in the case of GA_3, where more vertical growth was noted in cv. Sequeia. This might be attributed to the fact that GA_3 supply induces cell elongation and/or cell division (Rademabher, 2000; Blatt and Crouse, 1970). Throughout the period of plants treated with 20 mg/l^{-1} of gibberellic acid increased number of leaves and crown (Perez-de-Camacaro *et al.*, 2009).

To control vegetative growth, Ramina *et al.* (1985) applied Paclobutrazol as a foliar spray on strawberry plants cv Belrubi, grown under tunnels, before flower emergence at rates equivalent to 6.75, 3.37,1.68, 0.84, 0.42, 0.21 and 0.10 kg ha^{-1}, with and without a surfactant. The chemical did not affect yield, but plant growth was greatly reduced at all application rates. Petioles were shortened, the leaves appeared tightly ad pressed to the crown, leaf surface was greatly reduced and runner formation inhibited. The chemical also affected the dry weight partitioning within the plant, and the leaf photosynthetic efficiency. The presence of the sufactant appeared to be an important factor in attaining a stronger and more uniform control of vegetative growth.

To investigate the effects of paclobutrazol (PBZ) on vegetative and reproductive growth of Strawberry cv. Selva Shakeri *et al.* (2010) conducted experiment in a greenhouse. PBZ was applied in two methods, drench and foliar spray, at 0, 0.5, 1, 1.5 and 2 mg active ingredient per plant. The results indicated that applications of PBZ as foliar spray was more effective than soil drench on vegetative and reproductive growth. With increasing the amount of PBZ leaf area, shoot fresh weight, whole plant fresh weight and petiole length of treated strawberries were decreased, however fresh weight, number and diameter of roots were not affected by

PBZ. Also PBZ had no significant effect on fruit firmness, total soluble solids and vitamin C. The interaction of PBZ and its method of application on characteristics including leaf area, petiole length and diameter, total fresh weight, shoot fresh weight, leaf chlorophyll content, yield, fruit diameter and harvest index were significant. Foliar application of PBZ at the rate of 2 mg active ingredient per plant lead to maximum value of petiole diameter, leaf chlorophyll content and harvest index. Yield (fruit/plant) was also increased (27.16%) when the plants received 1 mg active ingredient per plant PBZ through soil drench method.

5. Effect of Plant Growth Regulators on Flowering and Fruit Set

It is imperative to assess the effect of plant bio-regulators which modifies various physiological processes with the advantage to increase strawberry production. Obviously, the regulation of flowering in strawberry is a complex process that is intertwined with the regulation of vegetative growth including axillary bud differentiation, as well as other developmental processes occurring during the annual life cycle. Many of these factors are related to the genetic properties of strawberry cultivars and correlated with the changes in the content and activity of endogenous plant hormones (Thompson and Guttridge, 1959). Furthermore, some of these factors are influenced by exogenous application of plant growth regulators at the proper stage of growth and development. According to a few reports, cytokinin may affect axillary bud differentiation to promote flowering (Braun and Kender 1985, Hou and Huang, 2005) and exogenous gibberellin application advanced flowering and fruiting by anticipating the endogenous gibberellin which is normally formed at a somewhat later stage of plant development (Leshem and Koller, 1964; Pathak and Singh, 1971; Celestre and Pierandrei, 1971; Soczek, 1966; Moroto *et al.*, 1983; Lopez *et al.*, 1989), whereas, Singh and Kaul (1967) reported delayed flowering with the application of GA_3.

Ozguven and Kaska (1990) applied GA3 at varied rates (12.5, 25 and 50 ppm) at three dates starting from November to April on strawberry. They observed that GA3 at 50 ppm concentration when applied in November advanced the flowering compared to other dates.

The effect of gibberellic acid A_3 (GA_3) on the vegetative growth, flowering characteristics and yield of three strawberry cultivars, the June bearing 'Camarosa' and 'Laguna' and the day neutral 'Seascape', was investigated by Paroussi *et al.*, (2002b) under heated and unheated greenhouse conditions and under short (10 h) and long (16 h) photoperiod. Young plants, potted in mid-November, were kept outdoors for 1 month, and then sprayed once with GA_3 (0, 50, 200 mg l^{-1}) and placed under the different environmental conditions. GA_3 application increased petiole length and leaf area of the strawberry plants in most treatments. It reduced the time needed for inflorescence emergence, accelerated flowering and increased the number of flower buds and open flowers in most growing conditions, the effect being greater on 'Seascape' compared with the other cultivars. GA_3 at the lower concentration (50 mg l^{-1}) did not affect total marketable yield, whereas at 200 mg l^{-1}, combined with long photoperiod, it increased the percentage of aborted flowers plus malformed fruits, resulting in a significant decrease in total marketable yield.

Effects of different plant growth regulators on photosynthesis, metabolism of reactive oxygen species and plant quality of Fragaria×ananassa Duch cv. French 3 were studied by Shan *et al.* (2007). In the experiment, 50 mg/L IAA, 50 mg/L 6-BA and 25 mg/L GA3 were sprayed in late autumn. Results showed that the plant growth regulators applied in late autumn increased significantly net photosynthetic rate, the content of in chlorophyll a and b in leaves, and SOD and CAT activities in leaves, but decreased reactive oxygen and MDA contents, compared with the control without foliar spray of the plant growth regulators. In addition, foliar spray of all three plant growth regulators increased significantly dry mass of shoot and root, as well as the root to shoot ratio. Foliar spray of IAA and 6-BA both increased the flower numbers in the next year as well. It is concluded that foliar spray of 50 mg/L IAA and 50 mg/L 6-BA in late autumn could improve the strawberry plant quality in autumn.

Spraying GA_3 at 75 ppm both during mid-November and mid-February was much more effective in achieving the desirable results than single application of GA_3 either during mid-November or mid-February (Sharma and Singh, 2009).

The influence the exogenous hormone on flowering and fruiting of strawberry was investigated by Isam *et al.*, (2012a). Two strawberry cultivars, Camarosa and Camaroga *cvs.*, were grown under green house and treated with 0 and 50 ppm of auxin (IBA), gibberellins acid (GA3) or cytokinin (6-BA) either singly or in combination by foliage application. The results showed that, there are significant different between the mean of flowering parameter, percentage of fruit set and average of fruit weight among the two cultivars. The result also showed that the number of the flowers per plant was greatly increased by about 138% comparing with the control plants when treated with 50 ppm GA3. 6-BA and IBA shows an antagonized effect on the role of GA3 in producing the flowers. The percentage of fruits set had increased by about 97.24 % and 81.5 % after treated with IBA and GA3, respectively. Combined application of GA3+6BA had increased the fruit weight by 33.85% compared with the control plants. The result also indicated that these two cultivars of strawberry response differently to the exogenous hormones in the producing flowers and fruits. The first bud of the plant treated with IBA+6-BA formed to flower as compared with the plants treated with GA3.

Duration of flowering largely depends on the time of application of plant bio-regulators which hastened the flowering period when applied about a month before the appearance of flower buds and showed an increase in sccondary flowering because of temperature rise. GA cause the production of large number of flowers with rapid elongation of peduncle, leading to full development of flower buds having all reproductive parts functional which increases the fruit set and number of berries per plant. It could be due to the fact that GA application accelerated the development of differentiated inflorescence (Saima *et al.*, 2014).

The results of effect of plant bio-regulators on yield and yield attributing characters of strawberry CV. Chandler showed that at 75 DAT (days after transplanting) the plant height attained maximum of 20.50 cm, spread (25.53 cm) leaf number per plant (23.00), petiole length (0.20 cm), leaf area (136.30 cm^2) and yield (356.56 g/plant) respectively, of strawberry was obtained highest with the application of GA 75 mg/L while, cycocel 750 mg/ lresulted in minimum days taken to first flowering (61.66 DAT) and fruit formation (66.66 DAT). The maximum number of flowers per plant were recorded as 30.22 and number of berries (24.80) with application of GA 75 mg/L (Saima *et al.*, 2014). However, the effect of GA_3 was depended on the growth stage; GA_3 applied about a month before the natural appearance of the flower buds hastened flowering (Honda, 1972), whereas, GA_3 applied at flower opening hastened fruit maturation.

Singh (2016) concluded that for getting substantial higher yield of quality berries with more propagating materials, the plants of strawberry should be treated with GA3 75 ppm in the plains of Uttar Pradesh, India.

6. Role of Plant Growth Regulators on Length and Dry Matter Content of Fruit Peduncle

The strawberry inflorescence peduncle is a modified stem terminated by the primary flower (Taylor, 2002). Length and dry matter content of peduncle (stalk of inflorescence) are recognized as one of important factors in determining the fruit quality and efficient labour cost during harvesting in strawberry production. Under the normal practice, harvesting of strawberry fruit is done manually by hand picking. Recently, the machine was introduced to harvest the strawberry fruits and this requires not only the cultivars with high fruit firmness but also the plant with a long peduncle or inflorescence stalk. The use of harvesting machine requires an appropriate detection of peduncle of the targeted fruit, and in this case a long peduncle fruit could contribute to the success of the process (Hayashi *et al.,* 2010). Exposing a long size peduncle to the sunlight possibly raised the quality of fruits (Sharma and Singh, 2009). At the same time the long peduncle contributes to the greater number of epidermal cells indicating an increasing of photosynthesis in the peduncle (Nishizawa, 1994). The peduncle length, however, greatly increased by about 351% comparing with those in the control plants when applied with 50 ppm GA_3 in singly without combined with the other hormones. Combined application of GA_3 + IBA had increased the dry mater content of peduncle by 15.4% comparing with those in the control plants (Isam *et al.,* 2012b).

Although GA_3 sprayed on the plant has forced the cultivars to open the flower earlier, hastened fruit maturation and increased peduncle length (Honda, 1972; Paroussi *et al.*, 2002a; 2002b), but high concentration of GA_3 (200 mg L^{-1}) caused marked inflorescence abnormalities (Paroussi *et al.*, 2002b). The combinations of GA_3 with IBA which produced the greatest dry matter content in strawberry could be due to the stimulation of IBA on the enhancement of cell division and chlorophyll accumulation, and that led to the higher photosynthetic activity and accumulation of dry matter content in the plant, but oppositely the application of 6-BA could reduce the dry matter content of the peduncle. Spray of GA_3 as single treatment led to extend the peduncle length (Singh and Pathak, 1972), while cytokinin may inhibit the role of gibberellin.

7. Role of Plant Growth Regulators (PGRs) on Fruit Size

Consumers prefer large fruits; thus, there is a significant price difference between large and small fruits. In terms of determining profitability, fruit size in strawberry has become as important as the yield. Fruit size is affected by many factors including irrigation management, soil type and ecological variables, which cannot be controlled by the producer. Moreover, fruit size is inversely proportional to yield and fruit number. The availability of carbohydrates (increases fruit size), flower intensity (decreases fruit size), competition between flowers/fruitlets (decreases fruit size) and abiotic stresses (decrease fruit size) are the most important factors affecting the final fruit size of strawberry. Alternatively, synthetic PGRs can be applied to increase fruit size. Applying synthetic auxins during the cell division stage of fruit development significantly reduces the number of fruits and increases the final fruit size by reducing the competition for carbohydrates between fruitlets. Techawongstein (1989) sprayed the plants of strawberry cv. Tioga with varied rates of NAA (0, 50, 100, 200 and 400 ppm) and reported that NAA (400 ppm) resulted in the largest fruits in terms of diameter, weight and volume. Miranda *et al.*(1990) studied the effect of growth regulators on morphological characteristics and productivity of strawberry cv. Sequcia and found that no appreciable effects were noted on fruit length, diameter and yield of strawberry cultivar Sequcia following the application of GA3, NAA, Ergostin (Folcysteire 5% + Folic acid 0.01%) or Atenik (Nitrophenolate Sodium + Nitroguaicot Sodium) @ 10 ppm each. Banday *et al.* (2005) reported GA_3 (25 ppm) treatment resulted in increased length, diameter and volume of fruits. Singh and Tripathi (2010) observed that spray of aqueous solutions of GA_3 (100 ppm) before bud initiation (65 days after transplanting) recorded increasing in berry length, width, weight and volume at each picking. Khunte *et al.* (2014) revealed that treatment application of 2.50 tones ha^{-1} poultry manure + 100 ppm triacontanol) showed the maximum fruit weight (16.84g), fruit length diameter ratio (1.58) and specific gravity (1.34).

8. Role of Plant Growth Regulators (PGRs) on Fruit Yield

In intensive fruit production, in order to obtain high yields, it is necessary to use high fertilization rates as well as plant protection products. An alternative to that kind of production is presented by natural stimulators of plant growth and development called biofertilizers, biopreparations, biostimulators or phytostimulators. They are preparations of natural (plant or animal) origin, harmless to humans and the environment. Biostimulators contain biologically active substances, *i.e.* plant hormones, enzymes, macro and microelements, and other compounds that stimulate the growth and development of plants (Rajmund *et al.*, 2010).

The application of exogenous hormones can increase the productivity of strawberry but only when grown under long day photoperiod (Isam *et al.*, 2012a). Triacontanol treated plants increased number of root which causes plants to take up more nutrients from soil and increase production per plants (Blarke and Lenz, 1983).

Application of exogenous gibberellic acid (GA_3) is a common agricultural practice in the coastal region of Turkey to break flower bud dormancy early. This treatment extends the strawberry growing season, increasing fruit yield. The effectiveness of GA_3 at 0, 5, 10 and 20 ppm, applied exogenously during the 2nd and 4th week of January was assessed on cv. Camarosa plants. The test-plants were field cultivated in a randomized block design at a Mediterranean coastal side in Turkey. Tests confirmed the efficiency of the experimental treatments to cause early flowering and fruit quality (weight, soluble solid, acidity) was not diminished. From this study, a treatment of 10 ppm GA_3 applied to fruit frost protected plants during the 2nd week of January is recommended for cv. Camarosa (Camacaro *et al.*, 2008).

Strawberry plants (Fragaria × ananassa Duch.) are very susceptible to salinity. In order to investigate the interaction effects of paclobutrazol (PP333) and salinity on flowering, yield and fruit quality of strawberry plants, three levels of NaCl [0, 5, 10 mM] were incorporated into the nutrient solution and four levels of PP333 [0, 10, 20, 30 mg L^{-1}] were sprayed. PP333 application caused reduction in inflorescence length, but increased inflorescence and flower number. Salinity reduced number of achenes per fruit. However 10 mg L^{-1} of PP333 ameliorated the diverse effect of salinity at 10 mM NaCl salinity and increased significantly the number of achenes per fruit. Interaction of 20 and 30 mg L^{-1} PP333 with 5 mM NaCl, showed increase in yield in comparison with 0 and 10 mg L L^{-1} PP333 at the same salinity level. Total soluble solids (TSS) were increased at 30 mg L L^{-1} PP333 in combination with 5 and 10 mM NaCl also increased by this level of PP333. The highest level of PP333 (30 mg L^{-1}) reduced titratable acidity (TA) on salt treated (5, 10 mM) plants. The ratio of TSS/TA tended to rise by application of PP333. Salinity reduced vitamin C, but 20 mg L^{-1} of PP333 significantly increased vitamin C in both levels of salinity (5, 10 mM). From the results of the study, PP333 application is recommended to increase reproductive growth, yield and improve fruit quality in saline condition (Jamalian *et al.,* 2008).

To investigate the ability of ethephon and "organic" Croplife-B to extend strawberry shelf life, Camarosa and Honor cultivars were sprayed pre-harvest with either 1000 or 1500 ppm ethephon for one time and/or Croplife-B weekly at a rate of 300 mL/hectare. The total yield was increased. Quality was maintained, as indicated by unchanged firmness, Total Soluble Solids and Titratable Acidity. During continuous cold storage (2-6 ^{0}C) ethephon and Croplife delayed the appearance and development of gray mould, *Botrytis cinerea,* on fruits. Results suggest ethephon and Croplife-B extend shelf life of strawberries while maintaining yield (Jamal and Wardi, 2011).

Application of chemicals for strawberry forcing or production in subtropical regions in autumn or winter months, inside or outside of greenhouse has been reported as a successful method. Present study was carried out to evaluate the effects of Volk oil (2.5, 5%), Dormex (0.5, 1%), GA3 (50, 100 mg. L-1) and KNO3 on vegetative and reproductive growth of strawberry cv. 'Merak' in subtropic conditions of Darab city in Fars province of Iran. Dormant young rooted daughter plants were potted in 3L plastic pots and after 2 weeks of establishment period the treatments were applied and then plants were grown for 3.5 months in outside conditions. Results indicated that KNO3 caused augmentation of leaf surface area, roots length, number of flowers and inflorescences. Also Dormex at 1% concentration increased flowering period and decreased number of runners. Volk oil at both concentrations improved roots length and number of achenes in primary and secondary fruits (Eshghi *et al.*, 2012).

Paclobutrazol (PBZ) and sodium chloride (NaCl) were sprayed on strawberry (Fragaria × Ananassa Duch. cv. Selva) grown in greenhouse conditions. Plants were treated with two levels of NaCl (0 and 20 mg L-1) and four levels of PBZ (0, 30, 60 and 90 mg L-1). Results showed that PBZ is an important factor for increasing the generative traits and fruit quality, but not on vegetative traits. The highest leaf number, leaf area, petiole length, and total soluble solid (T.S.S.) percent werc observed in control plants, while highest fruit number, fruit weight, fruit set, flower number and yield of strawberry were obtained in plants treated with 90 mg L-1 PBZ. Foliar application of PBZ prior to flowering is recommended to increase the yield of strawberry (Lolaei *et al.,* 2012).

With the purpose of implementing a practice that helps to improve strawberry production in low-yield areas, the use of growth regulators, which may influence the postharvest behavior of fruits, was evaluated by Viasus-Quintero *et al.,* 2013). Strawberry plants (cv. Albion) received applications of the following regulators: (a) GA_3, (b) Mixture of GA_4+GA_7+6BAP, and (c) 6BAP. The products were sprayed at doses of 300, 600 and 900 mg.L^{-1}. Only GA_4+GA_7+6 BAP affected the red color with respect to the control, and there was no other definite effect on fruit quality, due to the fact that the acidity was not affected, while TSS and firmness had fluctuating responses. It is concluded that regulators, in the average of their doses, were able to increase the fruit size and production, but no effect on the organoleptic quality was detected.

Kazemi (2013) investigated the effect of salicylic acid and calcium foliar application on growth, yield and yield components of strawberry plants grown in hydroponic greenhouse condition. Results showed that salicylic acid (0.25 mM) and calcium chloride (2.5 mM) spray either alone or in combination (0.25 mM SA+2.5 mM Ca) affected on vegetative and reproductive growth, significantly. Mean comparisons indicated yield, and quality of strawberry plants was improved in low salicylic acid and calcium chloride concentration.

Hazarika *et al.,* (2015) carried out an experiment to find out the effect of integration of bio-fertilizers with organic manures and inorganic fertilizers on growth, yield and quality of strawberry (Fragaria × ananassa Duch.) cv. Festival. The growth, yield and quality of strawberry fruits were significantly influenced by integration of bio-fertilizers, organic manure and inorganic fertilizers. The maximum growth in terms of plant height, spread and number of crowns/plant were observed in the treatment consisting of 75 % RDF + vermicompost + Azospirillum + PSB + 50 ppm GA3+ 50 ppm BA. This treatment also exhibit superiority in yield attributing characters like number of runners/plant, berry set percentage, berry weight, number of berries/plant and yield/ha. The quality parameters of the fruit like TSS, titrable acidity, ascorbic acid, total sugars and reducing sugars were also influenced by combined use of organic, inorganic and biological sources of nutrients.

A field trial was conducted by Thakur *et al.* (2017) to study the effect of growth regulators on growth, flowering, yield and fruit quality of strawberry cv. Chandler. The results revealed that the plants sprayed with 15 ppm GA4+7 two weeks before flowering significantly reported highest plant height (33.43 cm), leaf area (239.70 cm2), number of flowers (31.94), fruit set (87.45 %), number of fruits per plant (29.02), yield (540.01 g/plant), fruit length (53.63 mm) and fruit diameter (37.19 mm) and fruit weight (23.70 g) as compared to control. This treatment resulted in 196.36 % increase in yield and 56.22 %in fruit weight over control. Plants sprayed with 6 ppm promalin also showed significant improvement in vegetative growth, fruit size, yield and fruit quality, which resulted in 137.92 % increase in yield and 51.81 % increase in fruit weight over control. Hence, it was concluded that foliar spray of 15 ppm GA4+7 applied 2 weeks before flowering is beneficial in improving growth, yield and fruit quality of strawberry plants.

Studies on the effect of plant growth regulators on plant growth, fruit set, yield and quality of strawberry (*Fragaria ananassa* Duch) cv. Camarosa in coastal Andhra Pradesh was made by Venkata Chandana (2017). Different growth regulators were sprayed three times at 30, 60 and 90 days after transplanting. The results indicated that GA3 @ 50 ppm was found to be best treatment for flowering and fruit set and yield in strawberry under open conditions and under shade net in coastal Andhra Pradesh. Triacontanol @ 1 ppm also showed on par results with GA3 @ 50 ppm for vegetative growth and fruit quality parameters in strawberry under open conditions and under shade net in coastal Andhra Pradesh

9. Effect of Plant Growth Regulators on Fruit Ripening

Notwithstanding the economic importance of non-climacteric fruits like grape and strawberry, little is known about the mechanisms that regulate their ripening. Up to now no growth regulator has emerged with a primary role similar to that played by ethylene in the ripening of the climacteric fruits. Strawberries can produce ethylene, although in limited amounts. Two cDNAs coding for enzymes of the ethylene biosynthetic pathway (i.e. *FaACO1* and *FaACO2*), and three cDNAs encoding different ethylene receptors have been isolated. Two receptors (i.e. *FaEtr1* and *FaErs1*) belong to the type-I while the third (i.e. *FaEtr2*) belongs to the type-II group. The expression of both the ACO and the receptor-encoding genes has been studied in fruits at different stages of development and in fruits treated with hormones (i.e. ethylene and the auxin analogue NAA). All the data thus obtained have been correlated to the known data about ethylene production by strawberry fruits. Interestingly, a good correlation has resulted between the expression of the genes described in this work and the data of ethylene production. In particular, similarly to what occurs during climacteric fruit ripening, there is an increased synthesis of receptors concomitant with the increased synthesis of ethylene in strawberries as well. Moreover, the receptors mostly expressed in ripening strawberries are the type-II ones, that is those with a degenerate histidine–kinase domain. Since the latter domain is thought to establish a weaker link to the CTR1 proteins, even the little ethylene produced by ripening strawberries might be sufficient to trigger ripening-related physiological responses (Trainotti *et al.*, 2005). In the non-climacteric strawberry, auxin and abscisic acid (ABA) are thought to be the important. Various studies probably indicate that strawberry produces low levels of ethylene that are sufficient to regulate some ripening aspects (Natalia *et al.*, 2010).

In the non-climacteric strawberry (*Fragaria anannassa*), auxin and abscisic acid (ABA) are thought to be important, but the roles of other hormones suggested to be involved in fruit development and ripening are not clear. Here changes in the levels of indole-3-acetic acid (IAA), ABA, GA_1, and castasterone from anthesis to fully ripened fruit are reported. The levels of IAA and GA_1 rise early in fruit development before dropping to low levels prior to colour accumulation. Castasterone levels are highest at anthesis and drop to very low levels well before ripening commences, suggesting that brassinosteroids do not play an important role in ripening in strawberry. ABA levels are low at anthesis and gradually rise through development and ripening. The synthetic auxin, 1-naphthaleneacetic acid (NAA), can delay ripening, but the application of GA_3, the gibberellin biosythesis inhibitor paclobutrazol, and ABA had no significant effect. IAA and ABA levels are higher in the developing achenes than in the receptacle tissue and may be important for receptacle enlargement and ripening, and seed maturation, respectively. Contrary to a recent report, the biologically active GA_4 was not detected. The pattern of changes in the levels of the hormones are different from those reported in another well studied non-climateric fruit, grape, suggesting that a single consistent pattern of hormone changes does not occur in this group of fruit during ripening (Symons *et al.*, 2012).

It is thought that as the fruit develops the level of auxin in the achenes, and consequently in the receptacle, falls below a critical level, thus permitting ripening (Manning, 1993). Therefore, removing the achenes promotes ripening, while treating strawberries with synthetic auxins delays ripening (Given *et al.*, 1988; Manning, 1994).

Although the plant hormone abscisic acid (ABA) has been suggested to play a role in the ripening of non-climatic fruit, direct genetic/molecular evidence is lacking. A strawberry gene homologous to the *Arabidopsis*ABA receptor gene *PYR1*, named *FaPYR1*, was isolated by Chai *et al.* (2011) and characterized. The 627 bp cDNA includes an intact open reading frame that encodes a deduced protein of 208 amino acids, in which putative conserved domains were detected by homology analysis. Using tobacco rattle virus-induced gene silencing (VIGS), the *FaPYR1* gene was silenced in strawberry fruit. Down-regulation of the *FaPYR1* gene not only significantly delayed fruit ripening, but also markedly altered ABA content, ABA sensitivity, and a set of ABA-responsive gene transcripts, including *ABI1* and *SnRK2*. Furthermore, the loss of red colouring in *FaPYR1* RNAi (RNA interference) fruits could not be rescued by exogenously applied ABA, which could promote the ripening of wild-type fruits. Collectively, these results demonstrate that the putative ABA receptor FaPYR1 acts as a positive regulator in strawberry fruit ripening.

It has been suggested that endogenous ABA could play a role in strawberry color development through up-regulation of ethylene synthesis and PAL activity (Jiang and Joyce, 2003). The plant hormone abscisic acid (ABA) has been suggested to play a role in fruit development, but supporting genetic evidence has been lacking. Jia *et al.* (2011) reported that ABA promotes strawberry (*Fragaria ananassa*) fruit ripening. Using a newly established *Tobacco rattle virus*-induced gene silencing technique in strawberry fruit, the expression of a 9-cis-epoxycarotenoid dioxygenase gene (*FaNCED1*), which is key to ABA biosynthesis, was down-regulated, resulting in a significant decrease in ABA levels and uncolored fruits. Interestingly, a similar uncolored phenotype was observed in the transgenic RNA interference (RNAi) fruits, in which the expression of a putative ABA receptor gene encoding the magnesium chelatase H subunit (*FaCHLH/ABAR*) was down-regulated by virus-induced gene silencing. More importantly, the uncolored phenotype of the *FaNCED1*-down-regulated RNAi fruits could be rescued by exogenous ABA, but the ABA treatment could not reverse the uncolored phenotype of the *FaCHLH/ABAR*-down-regulated RNAi fruits. We observed that down-regulation of the *FaCHLH/ABAR* gene in the RNAi fruit altered both ABA levels and sugar content as well as a set of ABA- and/or sugar-responsive genes. Additionally, we showed that exogenous sugars, particularly sucrose, can significantly promote ripening while stimulating ABA accumulation. These data provide evidence that ABA is a signal molecule that promotes strawberry ripening and that the putative ABA receptor, FaCHLH/ABAR, is a positive regulator of ripening in response to ABA.

The enlargement of receptacle cells during strawberry (Fragaria × ananassa) fruit development is a critical factor determining fruit size, with the increase in cell expansion being one of the most important physiological processes regulated by the phytohormone gibberellin (GA). Csukasi *et al.* (2011) studied the role of GA during strawberry fruit development by analyzing the endogenous content of bioactive GAs and the expression of key components of GA signalling and metabolism. Bioactive GA(1) , GA(3) and GA(4) were monitored during fruit development, with the content of GA(4) being extremely high in the receptacle, peaking at the white stage of development. •Genes with high homology to genes encoding GA pathway components, including receptors (FaGID1(GIBBERELLIN-INSENSITIVE DWARF1)b and FaGID1c), DELLA (FaRGA(REPRESSOR OF GA) and FaGAI(GA-INSENSITIVE)), and enzymes involved in GA biosynthesis (FaGA3ox) and catabolism (FaGA2ox), were identified, and their expression in different tissues and developmental stages of strawberry fruit was studied in detail. The expression of all of these genes showed a stage-specific pattern during fruit development and was highest in the receptacle. FaGID1c bound GA in vitro, interacted with FaRGA in vitro and in vivo, and increased GA responses when ectopically expressed in Arabidopsis. This study thus reveals key elements of GA responses in strawberry and points to a critical role for GA in the development of the receptacle.

To investigate the role of jasmonates (JAs) in the ripening of *Fragaria chiloensis* fruit, two concentrations of methyl jasmonate (MeJA, 10 and 100 μM) were evaluated at 2, 5 and 9 d using an in vitro ripening system. MeJA accelerated fruit ripening by means of a transitory increase in the soluble solid content/titratable acidity ratio, anthocyanin accumulation and an increase in softening at day 5. The expression of several phenylpropanoid-related genes, primarily those associated with anthocyanin biosynthesis, was increased under MeJA treatment, which correlated with an increased accumulation of anthocyanin. MeJA also altered the expression profiles of some cell wall-modifying genes, namely, EG1 and XTH1, and these changes correlated with a transient reduction in the firmness of MeJA-treated fruits. MeJA-responsive elements were observed in the promoter region of the EG1 gene. MeJA also increased the expression of LOX, AOS and OPR3, genes involved in the biosynthesis of JAs, and these changes correlated with the transient activation of fruit ripening observed. Conversely, the expression of ethylene and lignin biosynthesis genes (ACS, ACO, CAD and POD27) increased in MeJA-treated fruits at day 9. The findings suggested that JAs promote the ripening of non-climacteric fruits through their involvement in anthocyanin accumulation, cell wall modification and the biosynthesis of ethylene and JAs (Cristobal *et al.*, 2013).

It has long been known that achene derived auxin is the key phytohormone controlling the growth and ripening of strawberry receptacles. Auxin stimulates receptacle expansion during fruit development and later delay or inhibits fruit ripening (Given *et al.*, 1988; Manning, 1993).

Although brassinosteroid (BR) has been suggested to play a role in strawberry fruit ripening, the defined function of this hormone remains unclear in the fruit. BR content and BR receptor gene *FaBRI1* expression were analysed by Chai *et al.* (2013) during 'Akihime' strawberry fruit development. They found that BR levels increased during the later developmental stages, and the mRNA expression levels of *FaBRI1* increased rapidly from white to initial red stages, suggesting that BR is associated with fruit ripening. This was further confirmed by exogenous application of BR and its inhibitor brassinazole (BZ) to big-green fruit, which significantly promoted and inhibited strawberry fruit ripening, respectively. More importantly, down-regulation of *FaBRI1*expression in de-greening fruit markedly retarded strawberry red-colouring. It was concluded that BR plays a role in strawberry fruit ripening. In addition, both BR content and *FaBRI1* expression reached their peak levels in small-green fruit, suggesting that BR might also be involved in early strawberry fruit development.

Strawberry fruit undergoes extensive softening during ripening due to cell wall disassembly, caused by a coordinated action of cell wall proteins and enzymes including polygalacturonases (PGs). However, the application of 1-MCP reduced PG activity, suggesting that ethylene could play a role in the regulation of polygalacturonase expression during strawberry fruit ripening (Natalia *et al.*, 2009). In the case of strawberry, no single growth regulator appears to play a positive role analogous to that played by ethylene in climacteric fruits. However, it has been suggested that the small amount of ethylene produced by strawberry fruit could be enough to trigger ripening-related physiological responses (Tian *et al.*, 1997; Trainotti *et al.*, 2005).

Strawberry fruit shows a marked softening during ripening and the process is associated with an increment of pectin solubility and a reduction of the molecular mass of hemicelluloses. In this work, we report the activity of beta-xylosidase and the expression of a beta-xylosidase gene in strawberry fruit. Martinez *et al.* (2004) cloned a cDNA fragment encoding a putative beta-xylosidase (FaXyl1) from a cDNA library obtained from ripe strawberry fruit. The analysis of the deduced amino acid sequence revealed that FaXyl1 is closely related to other beta-xylosidases from higher plants. The expression of FaXyl1 was strongly associated to the receptacle tissue although a low expression level was detected in achenes and ovaries. The accumulation of FaXyl1 mRNA is ripening-related, starting in white fruit, reaching the maximum at 25-50% red fruit and decreasing thereafter. The total beta-xylosidase enzyme activity was detected in all ripening stages with the maximum in 25-50% red fruit. The low activity level detected in immature stages, where no expression of FaXyl1 was found, suggests the presence of other beta-xylosidases-like genes. Both the expression of FcaXyl1 and the total beta-xylosidase activity were down regulated by auxins, as occurs for most of the ripening-related processes in strawberry fruit.

10. Effect of Plant Growth Regulators on Fruit Quality

The growth regulators are considered imminent as they influence the ripening, maturity indices and quality of the fruits. The quality of fruits constituted the fruit TSS, acidity, TSS /acid ratio, total sugars, reducing sugars, non-reducing sugars and ascorbic acid during the research investigations. The application of GA_3 75 ppm alone or in combination with urea did not show any effect on the total soluble solids however, improvement in fruit acidity and vitamin C was noted, but decrease in fruit volume, weight and size at peak harvesting periods was recorded (Singh and Kaul, 1967), whereas, Mikhteleva and Petrovska (1975) reported that strawberry plants sprayed with GA_3 50 ppm in spring, had higher levels of reducing sugars, amino acids and ascorbic acid and low contents of starch and fats. Varied concentrations of GA_3 and NAA at 35-51 days after planting significantly increased the TSS level and ascorbic acid of the fruits (Bhautkar, 2001; Saravanan *et al.*, 2013).Gibberellic acid increase the fruit quality by producing firmer fruit with high ascorbic acid and total soluble sugars, whereas, inducing no significant effect on titratable acidity (Ozguven and Yizmaz, 2002; Usenik *et al.*, 2005; Sharma and Singh, 2009; Ouzounidou *et al.*, 2010; Qureshi*et al.*, 2013; Ziba *et al.*, 2013). However, Kaur *et al.* (2009) observed GA3 100 ppm reduced acidity level of fruits. Montero *et al.* (2001) reported the effect of foliar application of gibberellic acid (GA_3 @ 30 or 60 micro g/litre), on the development of fruits of young strawberry cv. Chandler plants and found that the content of phenolic compounds (total polyphenols and anthocyanins), the activities of phenylalanine ammonialyase (PAL) and tyrosine ammonialyase (TAL) improved along with weight, size and colour of fruits when treated with GA_3 60 ppm. Strawberries treated with MJ 1 Mm had higher T.S.S content (Sembner and Parthier, 1993). Firmness of the fruits increased as MJ concentration increased (Lolaei *et al.*, 2013). Foliar application of PBZ (90 mg L^{-1}) is an important factor for increasing the generative traits and fruit quality, but not on vegetative traits (Lolaei *et al.*, 2012), it also increased achenes per fruit in strawberry (McArthur and Eaton, 2003). Cycocel has a potential for effective control and successfully reduced the vegetative growth in strawberry (Barritt, 1974; Brown *et al.*, 1997) but improved fruit quality (Agafanov and Solovei, 1975; Kumar *et al.*, 2011). NAA, in comparison with GA_3 proved to be superior in improving yield and quality of strawberry, except vitamin C and acidity (Mir *et al.*, 2004).However, Roussos *et al.* (2009) observed growth stimulators had no significant effect on fruit organic acid and carbohydrate concentration and on fruit colour, although they enhanced total anthocyanin concentration.

11. Effect of Plant Growth Regulators on Root and Shoot Dry Matter Content

Dry matter content is always used to measure productive in plant. Its dissemination in plant is partly determined by the balanced distribution of assimilates from the source organ to the other various organs. Under its original climatic condition, strawberry plant accumulates dry matter (starch) in root and crown and becomes dormant during short day in autumn and winter seasons (Durner *et al.*, 1984; Maas, 1986). The effect of growth regulators on the dry matter of strawberry are varies depending on the type and concentration of the hormones, time season of application and cultivars (Fouad *et al.*, 1991; Shou-Ming *et al.*, 2007).

Ariel dry matter of strawberry are most significantly increased when GA_3 are used (Engels and Ebert, 1988). It was reported that the application of GA_3 during the late autumn caused the aerial dry matter of strawberry either reduces or increases depending on the GA_3 concentration and cultivars used (Weidman and Stang, 1983). Cytokine (BA) had also significantly affected the root and aerial dry matter of strawberry (Archbold and Strang, 1986). Meanwhile, Shou-Ming *et al.* (2007) and Madhagi *et al.* (2011) indicated that the combined application of GA_3, IAA and 6-BA with various proportions can significantly affected the dry matter of shoots and roots of strawberry.

The effect of interaction among exogenous hormones auxin (IBA), Gibberellins Acid (GA3) and cytokinin (6-BA), photoperiod and cultivars on the total dry matter of strawberry plant was investigated by Madhagi *et al.* (2011) in Cameron Highland, Pahang, Malaysia. The results showed that only GA3 gave the significant effect on the dry matter of strawberry by its own action where, 17.06% dry matter content was produced under application of 50 ppm GA3. The result demonstrated that the photoperiod, variety, IBA and 6-BA factors do not act individually on the production of strawberry dry matter but rather by the mutual interaction among them. It was found that the highest, 21.2%, dry matter content of the strawberry was produced by Camaroga cultivar grown under 15 h photoperiod and treated with 50 ppm IBA and GA3 and thus such combined treatment is recommended as the best condition for increasing productivity of strawberry in Cameron Highland.

Archbold and Strange (1986) studied the effect of BA, at 50, 125 and 250 ppm, on "Redchief" cultivar and showed that increasing BA concentration decreased crown dry weight at a significant level.

12. Effect of Plant Growth Regulators on Runner Production

Strawberry is propagated commercially mostly by runners. Although runner formation is cultivar related, vegetative characteristics, such as leaf number, crown weight, crown diameter, leaf area, number of runner and number of plantlets (daughter plants), is affected by growth regulators (Kender *et al.*, 1971; Franciosi *et al.*, 1980; Ali *et al.*, 2011). Gibberellin sprayed on leaves can be absorbed by plants and transferred to auxiliary buds. This transfer is more prominent in long photoperiods than short days. GA_3 foliar application at different concentration (20-100 ppm) in strawberry plants, increased number of runner and crown/plant, which help us to produce strawberry plant (Chroma and Himelrick, 1983; Duarte and Hermosa, 2007; Kumar *et al.*, 2008; Ziba *et al.*, 2013).

Some cultivars are more sensitive to the application of growth regulators (GA) than others. Exogenous application of gibberellins has been reported to promote runner production in different strawberry cultivars at different concentration (Moore and Scott, 1965; Blatt and Crouse, 1970; Kender *et al.*, 1971; Hossan *et al.*, 2012).

Singh and Phogat (1983) recorded higher number of runners and green leaves with GA_3 25-50 ppm, whereas, GA_3 75 ppm had no effect on runner production in strawberry cv. Majestic. In the contrasting results GA_3 100 ppm produced significantly more number of runners in strawberry cv. Cavalier, while GA_3 50 ppm produced the lesser number in cv. Jeolikota (Singh, 1979). Pipattanawong *et al.* (1996) reported that application of GA_3 50 ppm resulted in a 2 to 3 fold increase in the number of runners in Miyoshi, whereas, GA_3 or GA_3 + BA led to an increase up to 4 to 8 times in Enrai and Summer Berry day neutral strawberries.

Turemis and Kaska (2004) reported the effect of GA_3 on strawberry cultivars Cruz, Vista, Tufts, Pocahontas and Aliso at 30 or 60 days after planting at 50 or 100 ppm. Number of runners and stem diameter increased significantly with GA_3 application, but root length was not affected.

Dale *et al.*, (2006) found that foliar spray of GA3 had increased runner production in day neutral strawberries. Runner production increased linearly with increasing GA concentration to 100 ppm. Nasiri and Ebadi (2006) reported that, GA application at 300 ppm and BA at 1200 ppm in July on "Selva" cultivar had no effect on runner formation, may be due to the inappropriate time of hormone application.

In greenhouse and field studies, benzyladenine (BA) and gibberellic acid (GA3) applied together as a foliar spray increased runner production in day-neutral strawberries (Fragaria x ananassa Duch.) but not when applied separately. Runner production increased linearly with increased BA concentration to 1800 mg-L-1. At high dosages, GA3-treated plants produced elongated internodes that, in the field, led to fewer daughter plants. In Florida, daughter plants derived from plants sprayed with the growth regulators increased yield by up to 10% in fruiting experiments. To induce runnering in the field and greenhouse, a treatment with BA at 1200 mg·L-1 and GA3 at 300 mg·L-1 is recommended (Dale *et al.*, 2006).

Experiments were conducted by Ali *et al.* (2011) to study the effect of Banzyladenin and Gibberellin on increasing runner production and some vegetative traits of three strawberry cultivars in greenhouse. Three strawberry cultivars (Pajaro, Queen Eliza and Paros) were sprayed in May and July. Highest number of plantlets with a standard diameter of 6.15 mm was produced by applying the second treatment (mixture of BA 1200 ppm and GA 300 ppm sprayed in May) and it was chosen as the best treatment. However, highest number of runners in each mother plant and highest diameter of plantlets were observed in the control (0 ppm). Cultivars responded differently to hormones.

A certain level of GA_1 is needed for runner initiation in strawberry and the reduction of GA_1 concentration below this level leads to branch crown formation (Schwechheimer, 2008). The phytohormone gibberellic acid (GA) controls important aspects of plant growth such as seed germination, elongation growth, and flowering. The key components of the GA signaling pathway have been identified over the past 10 years. The current view is that GA binds to a soluble GID1 receptor, which interacts with the DELLA repressor proteins in a GA-dependent manner and thereby induces DELLA protein degradation via the E3 ubiquitin ligase SCF(GID2/SLY1). GA-dependent growth responses can generally be correlated with and be explained by changes in DELLA repressor abundance, where the DELLA repressor exerts a growth restraint that is relieved upon its degradation. However, it is obvious that other mechanisms must exist that control the activity of this pathway (Schwechheimer, 2008).

Ziba *et al.*, (2013) noted that number flower on inflorescence and Runner significantly increased with foliar application of GA_3 at 50 mg/l in Gaviota strawberry plants. GA was sprayed on plants after removing the last flower, and repeated application once more after a week.

Gibberellin enhanced the number of runners in all strawberry varieties by specifically stimulating the stolon forming systems during long days. It is also responsible for increasing the number of runners per crown at higher rates of application (Guttridge and Thompson 1963b;Tafazoli and Vince-Prue 1978; Chroma and Himelrick 1984; Hytonen *et al.*, 2009).

Application of Cytokinin, at 50 to 600 ppm, increased runner formation when it was used in long days and moderate temperatures between 15°C at night and 30°C at day (Waithaka and Dana, 1978; Braun and Kender, 1985; Pritts *et al.*, 1986; Hasse *et al.*, 1989).

Zatyko *et al.*, (1980) reported that strawberry cvs. Gorella and Hakras Romata produced more runners with $AgNo_3$ +GA_3 treatment, GA_3 alone stimulated only shoot elongation.

Maximum crown production was obtained with 1000 ppm Cycocel, whereas greatest number and length of runners were obtained with 50 ppm GA3 (Dwivedi *et al.*,2001).

A study on the effect of plant growth regulators on strawberry (*Fragaria ananassa*) cv. Chandler was conducted by Ram *et al.* (2003). The application of gibberellic acid (GA3; 25, 50 and 75 ppm), 2,4-D (10, 20 and 40 ppm), TIBA (25, 50 and 75 ppm) and NAA (25, 50 and 75 ppm) caused highly significant reduction in plant (runners) survival during summer (May and June) compared to untreated plant.

13. Role of Plant Growth Regulators on Root Growth

Endogenous hormones play an important role in the growth and development of strawberry roots. Wang *et al.* (2009) studied the effect of four types of N fertilizers on the root growth of strawberry (*Fragaria ananassa* Duchesne) and the endogenous enzymes of indole-3-acetic acid (IAA), abscisic acid (ABA), and isopentenyl adenosine (iPA) in its roots and leaves using enzyme-linked immunosorbent assay. Application of all types of N fertilizers significantly depressed ($P \leq 0.05$) root growth at 20 d after transplanting. Application of organic-inorganic fertilizer (OIF) as basal fertilizer had a significant negative effect ($P \leq 0.05$) on root growth. The application of OIF and urea lowered the lateral root frequency in strawberry plants at 60 d ($P \leq 0.05$) compared with the application of two organic fertilizers (OFA and OFB) and the control (CK). With the fertilizer treatments, there were the same concentrations of IAA and ABA in both roots and leaves at the initial growth stage (20 d), lower levels of IAA and ABA at the later stage (60 d), and higher iPA levels at all seedling stages as compared to those of CK. Thus, changes in the concentrations of endogenous phytohormones in strawberry plants could be responsible for the morphological changes of roots due to fertilization.

Marschner (1995) reported that when N is deficient, an increase in abscisic acid (ABA) and a decrease in cytokinins occur and this leads to an increase in root to shoot ratios. Triacontanol treated plants increased number of root which causes plants to take up more nutrients from soil and increase production per plants (Blarke and Lenz, 1983). There is evidence indicating that in root elongation, IAA and ABA are considered a growth inhibitor.

The ability of plants to respond appropriately to nutrient availability is of fundamental importance for their adaptation to the environment. Nutrients such as nitrate, phosphate, sulfate and iron act as signals that can be perceived. These signals trigger molecular mechanisms that modify cell division and cell differentiation processes within the root and have a profound impact on root system architecture. Important developmental processes, such as root-hair formation, primary root growth and lateral root formation, are particularly sensitive to changes in the internal and external concentration of nutrients. The responses of root architecture to nutrients can be modified by plant growth regulators, such as auxins, cytokinins and ethylene, suggesting that the nutritional control of root development may be mediated by changes in hormone synthesis, transport or sensitivity. Recent information points to the existence of nutrient-specific signal transduction pathways that interpret the external and internal concentrations of nutrients to modify root development. Progress in this field has led to the cloning of regulatory genes that play pivotal roles in nutrient-induced changes to root development (Lopez-Bucio *et al.*, 2003).

14. Role of Plant Growth Regulators on Controlling Physiological Disorder and Disease Control

Strawberry as a non-climacteric and perishable fruit that is susceptible to fungal infections especially Botrytis cinerea. Therefore, its storage life is too short. Alternatives to the use of conventional fungicides are needed because of concerns about human health risks. Postharvest application of conventional fungicides to fruits is prohibited. Methyl jasmonate (MJ), a methyl ester of jasmonic acid (JA), is a phytohormone with ubiquitous distribution among plants and generally considered to modulate many physiological events in higher plants, such as defence responses to fungal pathogen, flowering and senescence (Cheong and Choi, 2003; Lolaei *et al.*, 2013). MJ either as a vapor or as an emulsion has been shown to inhibited grey mold infection in strawberries (Moline *et al.*, 1997). Sharma *et al.*, (2009) reported GA_3 (75 ppm) spray either during November or February or at both times reduced the production of malformed and button berries.

15. Role of PGRs on Abiotic Stress

Abiotic stress is becoming more prevalent as the intensity of agriculture and the demand for farmable land are ever increasing. Besides drought and salinity stress, chilling or freezing stress is one of the most important limiting factors of crop production all around the world Huseyin *et al.*, 2009).

Strawberry is considered as a salinity sensitive species and is adversely affected in response to the salt stress in terms of growth and yield (Huseyin *et al.*, 2009b) and because of increasing trend salinization of water source and inevitable using non-traditional water SA mitigate the detrimental effect of saline conditions on growth of strawberry plants, especially at 200 mg/L concentration (Eshghi *et al.*, 2017). Salicylic acid induced the activity of antioxidant enzymes, alleviated membrane lipid peroxidation as well as membrane system damage, thus improving drought resistance of strawberry plants (Sun *et al.*, 2013).

Salicylic acid (SA) is a common plant-produced signal molecule that is responsible for inducing resistance to a number of biotic and abiotic stresses. Karlidag *et al.* (2009) conducted a study to determine the effect of foliar SA applications on fruit-quality characteristics and yield of strawberry under anti-frost heated greenhouse conditions. Spraying of 1 mM SA (1 mM) was done once (SA1), twice (SA2), three times (SA3), or four times (SA4) during the vegetation period with 7 d intervals. Foliar applications of SA positively affected TSS and ascorbic acid (AA) content of strawberry fruits. Salicylic acid treatments had no effect on TA of strawberry fruits. SA4 treatment gave the greatest (colour) a^* in fruit and chlorophyll-concentration values in the leaves. It was shown that SA treatments increased the content of all nutrients in the leaves of strawberry plants, and greater values often were obtained from SA3 and SA4 treatments. The early yield and total yield of strawberry were significantly affected by SA applications, among which SA3 and SA4 resulted in the highest early and total yields. The present study suggests that SA3 and SA4 treatments can ameliorate the deleterious effects of low temperatures on strawberry plants and that SA application may offer an economical and simple method for low-temperature protection (Karlidag *et al.*, 2009).

Pot experiments were conducted to determine the effect of exogenous salicylic acid (SA) application on physiology, growth, chlorophyll and mineral content of strawberry grown under salt stress and greenhouse conditions. Strawberry plants were treated with SA at different concentrations (0.0, 0.25, 0.50 and 1.00 mM). Salinity treatments were established by adding 0 and 35 mM of NaCl to a base complete nutrient solution. Salt stress negatively affected the growth, chlorophyll content and mineral uptake of strawberry plants. However, plants treated with SA often had greater shoot fresh weight, shoot dry weight, root fresh weight and root dry weight as well as higher chlorophyll content under salt stress. The greatest values were obtained with 1.00 mM SA treatment in both saline and non-saline conditions. Leaf water relative content (LWRC) was reduced in response to salt stress while electrolyte leakage was raised. SA treatments induced increases in LWRC and decreases in electrolyte leakage compared to the control under salt stress. With respect to the nutrient content, SA treatments increased almost contents of all nutrients in leaves and roots of strawberry plants under salt stress. The greatest values were often obtained by the 1.00 mM SA treatment. These findings suggest that the SA treatments can ameliorate the negative effect of salinity on the growth of strawberries (Huseyin *et al.*, 2009a).

Most abiotic stress conditions result in increased production of the plant hormone ethylene. High levels of ethylene inhibit growth, cause premature ripening and induce the onset of senescence, potentially reducing the productivity of hydroponically-grown crops. Ethylene levels from leaves of hydroponically-grown strawberry plants can be used as an early indicator of stress conditions (Justin *et al.*, 2006).

PBZ has been used to provide plant protection against abiotic stresses such as water stress (Zhu *et al.*, 2004), and salinity (Kishor *et al.*, 2009), drought, low and high temperature (Fletcher *et al.*, 2000). PBZ has biochemical effects on plants, such as detoxification of active oxygen and increased levels of proline, antioxidants (Senaratna *et al.*, 1988) and chlorophyll contents (Fletcher and Hofstra, 1988).

16. Role of Plant Growth Regulators in Extending Shelf of Strawberry

Strawberry is a non-climacteric fruit, but it's very high rate respiration can attributed to the rapid postharvest deterioration. Hence, the scientific methodologies reducing the rate of respiration will be highly beneficial to extend the up keep the inherent quality of the stored strawberry fruits. SA and Ca treatments have the potential for commercial control of quality properties and increase shelf life of harvested fruit. The effect of salicylic acid treatments on maintains of total phenolics content plausibly may be attributed to delay in senescence process due to the salicylic acid significantly reduced the membrane permeability and MDA content (Kazemi *et al.*, 2011).

Asrey *et al.*, (2004) reported result of an experiment to extend the shelf life of 'Chandler' strawberry (*Fragaria × ananassa*) fruits with 4 plant growth regulators (NAA, TIBA, gibberellic acid and 2,4-D) and 2 calcium salts (calcium nitrate and $CaCl_2$). The pre-harvest foliar spray of crop with GA3, NAA, 2,4-D and calcium nitrate extended the shelf life and helped in developing quality ripe fruits in terms of retaining high reducing sugars, minimum cumulative physiological loss in weight and higher vitamin C retention. Preharvest (14 days prior harvesting) spray with 25 ppm GA_3 was the best treatment, followed by 10 ppm 2,4- D and 1% calcium nitrate, which not only extended the shelf life (up to 9 days) but also reduced the postharvest decay losses (1.55%) and minimized the cumulative physiological loss in weight (7.26%) due to moisture loss and degradative metabolism during ripening without adversely affecting the quality. The preharvest treatment with NAA 25 ppm favoured the higher vitamin C (49.30 mg/100 g pulp) content during storage.

Strawberry is very susceptible to postharvest losses and more than 50% of harvested fruit is lost in Iran. For many years synthetic chemicals have been used to increase fruit storage life but, because of food safety issues, the use of chemicals in postharvest technology is highly restricted. Extending the storage life of harvested products by enhancing natural resistance and antioxidative systems of the crop is an alternative strategy to chemicals in postharvest technology. Effect of postharvest treatment with salicylic acid (at 0, 1 and 2mmol/L) and methyl jasmonate (at 0, 8 and 16μmol/L) on antioxidant content and some defense enzymes activity of Sabrosa strawberry fruit during storage at 1±0.5°C with 90–95% RH. for 14 days followed by 24h at 20°C was studied. Both phytochemicals, separately or in combination, enhanced fruit total antioxidant content, catalase and peroxidase enzymes activity. Catalase enzyme activity was decreased after treatment with methyl jasmonate but the effect of methyl jasmonate in decreasing catalase activity was prevented by salicylic acid. It seems that in order to stablish a strong resistance network and extend fruit postharvest life a positive crosstalk between the two phytochemicals is necessary in "Sabrosa" strawberry fruit (Mohammadreza and Ali, 2015).

References

- Adam, J. and Wilson, D., 1967. Factors affecting the germination of strawberry seed. *Annu. Rep. Long Ashton Res. Stn.*, 90-95.
- Ali, M., Toktam, S.T. and Shokofeh, M., 2011. Effects of banzyladenine and gibberellin on runner production and some vegetative traits of three strawberry cultivars. *African Journal of Agricultural Research*, **6**(18): 4357-4361.
- Anderson, H.M. Abbott, A.J. and Wildshire, S., 1982. Micro-propagation of strawberry plants *in vitro*effect of growth regulators on incidence of multi-apex abnormality. *Scientia Horticulturae*, **16**(4): 331-341.

- Anonymous, 2015. Strawberry farmers seek better price for their produce. The Tribune,Gurgaon.http://www.tribuneindia.com/news/haryana/community/strawberry-farmers-seek-better-price for-their-produce/54069.html.
- Ara,T., Karim, R., Karim, M.R., Ahmad, S., Islam, R. and Hossain, M., 2012(b). Effects of different hormones on *in vitro* regeneration of strawberry (*Fragaria* × *ananassa* Duch.). *International Journal of Biosciences*, **10**(1): 86-92.
- Ara,T., Rezaul, K.M., Rezaul K., Shahrear, A., Rafiul, I. and Monzur, H., 2012 (a). callus induction and shoot regeneration in strawberry (*Fragaria* × *ananassa* Duch.). *International Journal of Biosciences*, **10**(1): 93-100.
- Archbold, D.D. and Strang, J.G., 1986. Effect of BA on growth and yield of Redchief strawberry. *Hort. Science*, 21: 1377-1379.
- Arteca, R.N., 1995. Plant growth substance, principles and applications. Chapman and Hall, New York, USA.
- Asrey, R., Jain, R.K. and Singh, R. 2004. Effect of pre-harvest chemical treatments on shelf life of 'Chandler' strawberry (Fragaria × ananassa). *Indian Journal of Agricultural Sciences*, **74**(9): 485-487.
- Banday, F.A., Sofi, S.A. and Hafiza, A., 2005. Effect of growth regulators on physicochemical characters and yield attributes of strawberry. *Applied Biological Research*, **7**(1/2): 27-30.
- Bhat, A., Sharma, R.M., Singh, A.S. and Massodi, F.A., 2005. Performance of some strawberry (*Fragaria* × *ananassa* Duch.) cultivars under Jammu conditions. *Progressive Horticulture*, **37**(1): 163-165.
- Bhat, R.P., Devi, K.M., Jayalaxmi, H., Sophia, I.and Prajna, P.S., 2012. Effect of plant growth regulators on establishment and growth of strawberry (*Fragaria* × *ananassa* Duch.) var. Chandler *in vitro*. *Agricultural Science Research Journals*, **2**(12): 623-632.
- Biswas, M.K., Islam, R. and Hossain, M., 2008. Micro propagation and field evaluation of strawberry in Bangladesh. *J. Agric. Technol.*, **4**(1): 167-182.
- Blarke, Z. and Lenz, A., 1983. Effect of triacontanol on dry matter accumulation of quality and production of strawberry, *Eruuebsobstbau*, **25**: 360-361
- Blatt, C.R. and Crouse D.N.A. 1970. Effect of gibberellic acid and nitrogen on the strawberry cv. Red Coat. *Hort. Science*, **5**: 437-438.

- Boxus, Ph., 1989. Review on strawberry mass propagation. *Acta Horticulturae*, **265**: 309-320.
- Braun, J.W. and Kender, W., 1985. Correlative bud inhibition and growth habit of the strawberry as influenced by application of gibberellic acid, cytokinin, and chilling during short day length. *J. Am. Soc. Hort. Sci*,, **110**: 28-34.
- Camacaro, M.P., Mogollon, N., Ojeda, M., Gimenez, A. and Colmenares, C., 2008. The effect of gibberellic acid on the growth and flowering of strawberry (*Fragaria × ananassa* Duch.) 'Chandler' vitroplants. *Acta Horticulturae*, **842**: 178-184.
- Celestre, M.R. and Peirandrei, F., 1971. The Effect of gibberallic acid on productivity and ripening date in three strawberry varieties. *Annali dell 'Istituto sperimentale per is la Frutticoltura*, **2**(1): 61-76.
- Celestre, M.R., 1970. The effect of gibberellic acid on the vegetative and reproductive activity of strawberries. *Annali dell 'Istituto spermentale per la Frutticoltura*, **1**(2): 89-96.
- Chai, Y.M., Jia, H.F., Li, C.L., Dong, Q.H. and Shen, Y.Y., 2011. FaPYR1 is involved in strawberry fruit ripening. *Journal of Experimental Botany*, **62**: 5079-5089.
- Chai, Y.M., Zhang, Q., Tian, L., Li, C., Xing, Y., Qin, L. and Shen, Y.Y., 2013. Brassinosteroid is involved in strawberry fruit ripening. *Plant Growth Regul.*, **69**: 63-69.
- Chellappa, T., 1959. The physiology of strawberry seed germination. *Diss. Abstr.*, **20**: 859-860.
- Cheong. J.J. and Choi, Y.D., 2003. Methyl jasmonate as a vital substance in plants. *Trends in Genetics*, **19**: 409-413.
- Chroma M.E. and Himelrick, D.G., 1983. Responses of day-neutral, June-bearing and ever bearing strawberry cultivars to gibberellic acid and phthalimide treatments. *Hort Science*, **22**: 257-264.
- Chroma, M.E. and Himelrick, D.G., 1984. Responses of day-neutral, June-bearing and ever-bearing strawberry cultivars to gibberellic acid and naphthalimide treatments. *Sci. Hort.*, **22**: 257-264.
- Cristobal, M., Concha, A., Nicolas, E., Figueroa, A., Leticia, A., Poblete, A., Felipe, A., Onate, A., Wilfried, S.B., Carlos, R. and Figueroa, N., 2013. Methyl jasmonate treatment induces changes in fruit ripening by modifying the expression of several ripening genes in *Fragaria chiloensis* fruit. *Plant Physiology and Biochemistry*, **70**: 433-444.

- Csukasi, F., Osorio, S. and Gutierrez, J.R., 2011. Gibberellin biosynthesis and signalling during development of the strawberry receptacle. *New Phytologist*, 191: 376–390.
- Dale, A., Elfving, D.C. and Chandler, C.K., 2006. Benzyladenine and gibberellic acid increase runner production in strawberries. *Horticulture Science*, **31**(7): 1190-1194.
- Debnath, S.C., 2009. Characteristics of strawberry plants propagated by *in vitro* bioreactor culture and *ex vitro* propagation method. *Engineering in Life Sciences*, **9**: 239-246.
- Duarte, O. and Hermosa, R., 2007. Effect of gibberellic acid and bud removal on runner production of strawberry (*Fragaria × ananassa*) cv. Chandler. *Proceedings of the Interamerican Society for Tropical Horticulture*, **41**: 110-112.
- Durner, E.F. Barden, J.A. Himelrick, D.G. and Poling, E.B., 1984. Photoperiod and temperature effects on flower and runner development in day-neutral, June bearing and ever bearing strawberries, *J. Am. Soc. Hortic. Sci.*, **109**: 396-400.
- Dwivedi, M.P., Negi, K.S., Jindal, K.K. and Rana, H.S., 2001. Effect of bio-regulators on vegetative growth of strawberry. *Scientific Horticulture*, **6**: 79-84.
- Engels, B. and Ebert, G., 1988. Effect of combined application of GA_{4+7} and BA on photosynthesis, transpiration and leaf growth in strawberry cv. Korona (*Fragaria × ananassa* Duch.). *Erwerbsobstbau*, **30**: 183-185.
- Eshghi, S., Safizadeh, M.R., Jamali, B. and Sarseifi, M., 2012. Influence of foliar application of volk oil, dormex, gibberellic acid and potassium nitrate on vegetative growth and reproductive characteristics of strawberry cv 'Merak'. *J. Biol. Environ. Sci.*, 6(16): 35-38.
- Fletcher, R.A. and Hofstra, G., 1988. In: Berg, K., Plempel, M. (Eds.), Triazoles as potential plant Protectants. Sterol biosynthesis inhibitors: pharmaceutical and agrochemical aspects. Ellis Harwood Ltd., Cambridge, pp. 321–331.
- Fletcher, R.A., Gilley, A., Davis, T.D. andSankhla, N., 2000. Triazoles as plant growth regulators and stress protectants. *Hort. Rev.*, **24**: 55-138.
- Fouad, M., Swartz, H.J. and Buta, J.G., 1991. The role of abscisic acid and plant growth regulators in tissue-culture induced rejuvenation of strawberry *ex vitro*. *Plant Cell Tissue Organ Cult.*, 25: 75-84.
- Franciosi, R., Salas, P., Yamashiro, E. and Duarte, O., 1980. Efecto del acido giberelico en el estoloneo de diferentes cultivares de fresa. *Proc. Amer. Soc. Hort. Sci. Trop. Reg.*, **24**: 127-130.

- Given, N.K., Venis, M.A. and Grierson, D., 1988. Hormonal regulation of ripening in strawberry, a non-climacteric fruit. *Planta*, **174**: 402-406.
- Gupta, P.K., 1962. The occurrence of indigenous auxins and inhibitors and their effect on the germination of strawberry seed. *Diss. Abstr.*,**22**: 41-48.
- Guttridge, C.G. and Thompson, P.A., 1963. The effect of day length and gibberellic acid on cell length and cell number in strawberry petioles. *Plant Physiology*, 16: 604-614.
- Guttridge, C.G. and Thompson, P.A., 1963b. The effects of gibberellins on growth and flowering of *Fragaria* and *Duchesnea*. *J. Exp. Bot.*, **15**: 631-646.
- Hamdouni, A., Lamarti, A. and Badoc, 2001. *In vitro* germination of the achenes of strawberry (*F. ananassa* Duch.) cvs. 'Chandler' and 'Tudla' (*Fragaria* × *ananassa* Duch). *Bull. Soc. Pharm. Bordeaux*, **140**: 31-42
- Hancock, J.F. Sjulin, T.M. and Lobos, G.A., 2013. Strawberries. In J.F. Hancock (Ed.), Temperate Fruit Crop Breeding. Springer, New York.
- Hanke, V., 1993. Untersuchungen zur Keimung von Achänen bei Erdbeere.- *Erwerbsobstbau*,**35**(4): 105-109.
- Hasse, L., Pritts, M. and Eames-Sheavly, M., 1989. Growth regulators affect vegetative and reproductive growth in a day-neutral and June-bearing strawberry cultivar. *Adv. Strawberry Prod.*, **8**: 45-50.
- Hayashi, S., Shigematsu, K., Yamamoto, S., Kobayashi, K., Kohno, Y., Kamata, J. and Kurita, M., 2010. Evaluation of a strawberry-harvesting robot in a field test. *Biosystems Engineering*, **105**(2): 160-171.
- Hazarika, T.K., Ralte, Z., Nautiyal, B.P. and Shukla, A.C., 2015. Influence of biofertilizers and bio-regulators on growth, yield and quality of strawberry (*Fragaria ananassa*). *Indian Journal of Agricultural Sciences*,**85**(9): 1201-1205.
- Honda, F., 1972. The effect of gibberellic acid on growth and flowering in strawberries. *Bulletin of the Horticultural Research Station, D (Kurume)*, **7**: 45-57.
- Hou, Z. and Huang, W., 2005. Immunohistochemical localization of IAA and ABP_1 in strawberry shoot apexes during floral induction. *Planta*, **222**: 678-687.
- Huseyin, K., Ertan, Y. and Metin, T., 2009. Exogenous applications of salicylic acid affect quality and yield of strawberry grown under antifrost heated greenhouse conditions.

Journal of Plant Nutrition and Soil Science, **172**(2): 270-276.

- Huseyin, K., Ertan, Y. and Metin, T., 2009b. Salicylic acid ameliorates the adverse effect of salt stress on strawberry. *Sci. agric. (Piracicaba, Braz.)*,**66**(2): 180-187.
- Hytonen, T., Elomaa, P., Moritz, T. and Junttila, O., 2009. Gibberellin mediates day length-controlled differentiation of vegetative meristems in strawberry (*Fragaria* × *ananassa* Duch). *BMC Plant Biol.*, **9**: 18.
- Isam, A.H., Al-madhagi, Sayed, M.Z.H., Aziz, A., Abdullah, M.Z. and Wan, A.Y., 2012b. The possible role of exogenous hormones in influencing the length and dry matter of strawberry (*Fragaria* × *ananassa* Duch) peduncle. *Advances in Environmental Biology*, **6**(3): 1307-1311.
- Isam, A.H., Sayed, M.Z.H., Aziz, A., Abdullah, M.Z. and Wan, A.Y., 2012a. The influence of exogenous hormone on the flowering and fruiting of strawberry (*Fragaria* × *ananassa* Duch). *Journal of Biology, Agriculture and Healthcare*, **2**(4): 46-52.
- Jamal, A.A.F. and Wardi, A.S.S., 2011. Ethephon and Croplife-B extend strawberry fruit shelf life and increase yield. *Jordan Journal of Agricultural Sciences*,**7**: 436-447.
- Jamal, U.A.F.M., Hossan, M.J., Islam, M.S., Ahsan, M.K. and Mehraj, H., 2012. Strawberry growth and yield responses to gibberellic acid concentrations. *Journal of Experimental Bioscience*,**3**(2): 51-56.
- Jamalian, S., Tehranifar, A., Tafazoli, E., Eshgi, S. and Davarynejad, G.H., 2008. Paclobutrazol application ameliorates the negative effect of salt stress on reproductive growth, yield, and fruit quality of strawberry plants. *Hort. Environ. Biotechnol.*,**49**: 1-6.
- Jasrotia, R.K., Parshant, S.S., Rafiq, B. and Raina, S.V., 2014. Influence of gibberellic acid and blossom removal on flowering and yield of strawberry (*Fragaria* × *ananassa* Duch.) cv. Belrubi. *International Journal of Agricultural Sciences*, **10**(1): 272-275.
- Jia, H., Chai, Y., Li, C., Lu, D., Luo, J., Qin, L., Shen, Y., 2011. Abscisic acid plays an important role in the regulation of strawberry fruit ripening. *Plant Physiol.*, **157**: 188-199.
- Jiang, Y. and Joyce, D.C., 2003. ABA effect on ethylene production, PAL activity, anthocyanin and phenolic contents of strawberry fruit. *Plant Growth Regul.*,**39**:171-174.
- Justin, D., Hogan, E.E., Murray and Marcia, A.H., 2006. Ethylene production as an indicator of stress conditions in hydroponically-grown strawberries. *Scientia Horticulturae*, **110**: 311-318.

- Karhu, S. and Hakala, K., 2002. Micropropagated strawberries on the field. *Acta Horticulturae*,**2**: 182.
- Karlidag, H., Yildirim E. and Turan M., 2009. Exogenous application of salicylic acid affects quality and yield of strawberry grown under anti-frost heated greenhouse condition. *J. Plant Nutr. Soil Sci.*, **172**: 270-276.
- Karp, A., 1989. Can genetic instability be controlled in plant tissue cultures. Newslett. Intl. Assoc. *Plant Tissue Culture*, **58**: 2-11.
- Kazemi, M., 2013. Foliar application of salicylic acid and calcium on yield, yield component and chemical properties of strawberry. *Bull. Env. Pharmacol. Life Sci.*, **2**(11): 19-23.
- Kazemi, M., Zamani, S. and Aran, M., 2011. Effect of salicylic acid treatments on quality characteristics of apple fruits during storage. *Am. J. Plant Physiol*,**6**: 113-119.
- Kender, W.J., Carpenter, S. and Braun, J.W., 1971. Runner formation in ever bearing strawberry as influenced by growth-promoting and inhibiting substances. *Ann. Bot.*,**35**: 1045-1052.
- Khunte, S.D., Kumar, A., Kumar, V., Singh, S. and Saravanan, S., 2014. Effect of plant growth regulators and organic manure on physicochemical properties of strawberry (*Fragaria* × *ananassa* Duch.) cv. Chandler. *International Journal of Scientific Research and Education*,**2**(7): 1424-1435.
- Kishor, A., Srivastav, M., Dubey, A.K., Singh, A.K., Sairam, R.K., Pandey, R.N., Dahuja, A., Sharma, R.R., 2009. Paclobutrazol minimizes the effect of salt stress in mango (*Mangifera indica* L.). *J. Hortic. Sci. Biotechnol.*,**84**: 459-465.
- Kumar, R., Tiku, A.K., Singh, D. and Mir, M.M., 2008. Effect of GA_3, NAA and CCC on growth, yield and quality of strawberry (*Fragaria ananassa* Duch.) cultivar Sweet Charlie. *Environment and Ecology*,**26**(4): 1703-1705.
- Kumra, R., Reena, Saravanan, S., Bakshi, P., Kumar, A., Singh, M. and Kumar, V., 2018. Influence of plant growth regulators on strawberry- A review. *International Journal of Chemical Studies*,**6**(1): 1236-1239.
- Lata, S., Sharma, G., Garg, S. and Joshi, M., 2018. Effect of different chemical treatments on germination of strawberry seeds. *Int.J. Curr.Microbiol.App.Sci.*,**7**(3): 1270-1274.
- Leshem, Y. and Koller, D., 1964. The control of flowering in the strawberry *Fragaria* × *ananassa* Duch. *Ann. Bot.*, **28**: 569-578.

- Litwinczuk, W., Okolotkiewicz, E. and Matyaszek, I., 2009. Development of *in vitro* shoot cultures of strawberry (*Fragaria × ananassa* Duch.) Senga Sengana and Elsanta under the influence of high doses of gibberellic acid. *Folia Horticulturae Ann.,* **21**(2): 43-52.

- Lolaei, A., Behzad K., Mojtaba K.R., Mohammad A.R. and Matin M., 2012. Effect of Paclobutrazol and salinity on vegetative and sexual growth and fruit quality of strawberry (*Fragaria × ananassa* Duch. cv. Selva). *Annals of Biological Research,* **3** (10): 4663-4667.

- Lolaei, A., Teymouri, N., Bemana, R., Kazempour, A. and Aminian, S., 2013. Effect of gibberellin on vegetative and sexual growth and fruit quality of strawberry (*Fragaria × Ananassa* Duch. cv. Selva and Qeen elisa). *International Journal of Agriculture and Crop Sciences,* **5**(14): 1508-1513.

- Lopez, G.S., Pascual, B., Alagarda, J. and Maroto, J.V., 1989. The influence of winter gibberellic acid application on earliness, productivity and other parameter of quality in strawberry cultivation of the Spanish Mediterranean Coast. *Acta Horticulture,* **265**: 217-222.

- Lopez-Bucio, J., Cruz-Ramyrez, A. and Herrera-Estrella, L., 2003. The role of nutrient availability in regulating root architecture. *Curr. Opin. Plant Biol.,* **6**(3): 280-287.

- Maas, J.L., 1986. Photoperiod and temperature effects on starch accumulation in strawberry roots. *Adv. Strawberry Prod.,* **5**: 22-24.

- Madhagi, A.I., Haidar, A., Sayed, H.M., Zain, Ahmad, A.B. and Yusoff, W.A.B., 2011. The interaction effect of photoperiod and exogenous hormone on the dry matter of strawberry (*Fragaria ananassa*). *Agriculture Journal,* **6**(6): 340-346.

- Manning, K., 1993. Soft fruit, in biochemistry of fruit ripening. Chapman &Hall, London.

- Manning, K., 1994. Changes in gene expression during strawberry fruit ripening and their regulation by auxin, *Planta,* **194**: 62–68.

- Marschner, H., 1995. Mineral nutrition of higher plants. 2ndEdition, Academic Press, London, pp 889.

- Martinez, G.A., Chaves, A.R., and Civello, P.M., 2004. b-xylosidase activity and expression of a b-xylosidase gene during strawberry fruit ripening. *Plant Physiol. Biochem.,* **42**: 89-96.

- Miranda, S.H.G., Gloria, B.A.D.A and Castro, P.G.C., 1990. Effect of growth regulators on morphological characteristics an productivity of strawberry Sequoia. *Anais da Escola superior de Agri.,* **47**: 317-334.

- Mohammadreza, A. and Ali, R.H., 2015. Interaction effects of salicylic acid and methyl jasmonate on total antioxidant content, catalase and peroxidase enzymes activity in "Sabrosa" strawberry fruit during storage. *Scientia Horticulturae*, **197**: 490-495.
- Moline, H.E., Buta, J.G. and Saftner, R.A., 1997. Comparison of three volatile natural products for the reduction of postharvest decay in strawberries. *Adv. Strawberry Res.*, **16**:43-48.
- Moore, J.N. and Scott, D.H., 1965. Effect of gibberellic acid and blossom removal on runner production of strawberry varieties. *Proc. Amer. Soc. Hort. Sci.*, **87**: 240-244.
- Nakamura, S., 1972. Germination of strawberry seeds. *J. Jpn. Soc. Hortic. Sci.*, **41**(4): 367-375.
- Natalia, M.V., Claudia, A.B., Pedro, M.C. and Gustavo, A.M., 2010. Effect of ethylene and 1-MCP treatments on strawberry fruit ripening. *Journal of the Science of Food and Agriculture*, **90**(4): 683-689.
- Natalia, M.V., Gustavo, A.M. and Pedro, M.C., 2009. Influence of plant growth regulators on poly-galacturonase expression in strawberry fruit. *Plant Science*, **176**: 749–757.
- Negi, S.P. and Singh, R., 1972. Effect of different chemicals on germination of strawberry seeds. *Indian J. Hort.*, **29**: 265-268.
- Nishizawa, T., 1994. Effects of photoperiods on the length and number of epidermal cells in runners of strawberry plants. *Journal of the Japanese Society for Horticultural Science*, **63**(2): 347-352.
- Paroussi, G., Voyiatzis, D.G., Paroussis, E. and Drogoudi, P.D., 2002b. Growth, flowering and yield responses to GA_3 of strawberry grown under different environmental conditions. *Scientia Horticulturae*, **96**(1/4): 103-113.
- Paroussi, G., Voyiatzis, D.G., Paroussis, E. and Drogoudi, P.D., 2002a. Effect of GA_3 and photoperiod regime on growth and flowering in strawberry. *Acta Horticulture*, **567**: 273-276.
- Pathak, R.K. and Singh, R., 1976. Effect of GA photoperiod on yield and quality of strawberry. *Progressive Horticulture*, **8**: 31-39.
- Perez, D.C.M., Mogollon, N., Ojeda, M., Gimenez, A. and Colmenares, C., 2009. The effect of GA_3 on the growth and flowering of strawberry cv. Chandler *in vitro* plants. *Acta Horticulturae*, **842**: 793-96.

- Phillips, R.L., Kaeppler, S.M. and Olhoft, P., 1994. Genetic instability of plant tissue cultures: breakdown of normal controls. *Proceedings of the National Academy of Science USA*, **91**: 5222-5226.
- Pipattanawong, N., Fujishige, N., Yamane, K., Ijiro, Y. and Ogata, R., 1996. Effects of growth regulators and fertilizer on runner production, flowering, and growth in day-neutral strawberries. *Jpn. J. Trop. Agric.*,**40**(3): 101-105.
- Pritts, M.P., Posner, G.S. and Worden, K.A., 1986. Effects of 6-BA application on growth as development of 'Tristar', a strong day-neutral strawberry. *HortScience*,**21**: 1421-1423.
- Qureshi, K.M., Saman, C., Usman, S.Q. and Nadeem, A.A., 2013. Impact of exogenous application of salt and growth regulators on growth and yield of strawberry. *Pak. J. Bot.*,**45**(4): 1179-1185.
- Rademabher, W., 2000. Growth retardants: Effect on gibberellins biosynthesis and other metabolic pathway. *Ann. Rev. Pl. Physiol. Pl. Mol. Biol.*,**51**: 501-531.
- Rajmund G., Lidia S.P. and Ewa J.T., 2010. The effect of plant stimulant/fertilizer "resistim" on growth and development of strawberry plants. *Journal of Fruit and Ornamental Plant Research*,**18** (1): 111-124.
- Ram, A., Jain, R.K. and Singh, R., 2003. Effect of plant growth regulators on growth and survival of strawberry runners under semi arid region of Punjab. *Indian Journal of Plant Physiology*, **8**(2): 196-198.
- Ramina, A., Tonutti, P. and Tosi, T., 1985. The effect of paclobutrazol on strawberry growth and fruiting. *J. Hort. Sci.*,**60**: 501-506.
- Saima, Z., Sharma, A., Umar, I. and Wali, V.K., 2014. Effect of plant bio-regulators on vegetative growth, yield and quality of strawberry cv. Chandler. *African Journal of Agricultural Research*, **9**(22): 1694-1699.
- Schwechheimer, C., 2008. Understanding gibberellic acid signalling are we there yet. *Curr Opin Plant Biol.*, **11**: 9-15.
- Senaratna, T., Mackay, C., McKersie, B. and Fletcher, R., 1988. Uniconazoleinduced chilling tolerance in tomato and its relationship to antioxidant content. *J. Plant Physiol.*, 133: 56–61.
- Shakeri, F., Ghobadi, S., Baninasab, B. and Mobli, M., 2010. Effects of paclobutrazol concentration and application methods on vegetative and reproductive growth of

strawberry (*Fragaria* × *ananassa* Duch. cv. Selva). *Iranian J. Sci. Technol. Hortic.*, **10**: 85-96.

- Shan, S., Liu, G., Li, S. and Miao, P., 2007. Effects of IAA, GA_3 and 6-BA applied in autumn on plant quality of strawberry. *Journal of Fruit Science*, **24**(4): 545-548.
- Sharma, R.R. and Singh, R., 2009. Gibberellic acid influences the production of malformed and button berries, and fruit yield and quality in strawberry (*Fragaria* × *ananassa* Duch.). *Scientia Horticulturae*, **119**(4): 430-433.
- Sharma, R.R. and Singh, S.K., 1999. Strawberry cultivation- a highly remunerative farming enterprise. *Agro India*, **3**(2): 29-31.
- Sharma, V.K., Godara, A.K. and Goyal, R.K., 2016. Effect of growth regulator in MS media on in vitro multiplication of strawberry (*Fragaria* × *ananassa* Duch.) using runner tip explants. *Current Trends in Biotechnology and Pharmacy*, **10**(1): 36-44.
- Sharma, V.P. and Sharma, R.R., 2004. The Strawberry, ICAR, New Delhi, India.
- Shou-Ming, S., Guo-Jie, L., Shao-Hua, L. and Peng-Fei, M., 2007. Effects of IAA, GA_3 and 6-BA applied in autumn on plant quality of strawberry. *J. Fruit Sci.*, **24**: 545-548.
- Singh, B.K. 2016. Effect of GA3 and NAA on growth, flowering, fruiting, yield and quality of strawberry (Fragaria X ananassa Duch.) cv. Chandler: http://krishikosh.egranth.ac.in/handle/1/76503
- Singh, K., 1979. Studies on runner formation, flowering and fruiting in strawberry. MSc Thesis. Punjab Agriculture University, Ludhiana.
- Singh, O.P. and Phogat, K.P.S., 1983. Effect of plant growth regulators on vegetative growth, yield and quality of strawberry. *Prog.Hort.*, **15**(1-2): 64-68.
- Singh, R. and Kaul, G.L., 1967. Studies on the effect of gibberellic acid on growth and fruiting of strawberry. *Int. Symp. on Sub-trop. and Trop.Hort.* pp 44-48.
- Singh, R. and Pathak, R.K., 1972. Effect of some external factors on the growth and fruiting of strawberry. *Proceedings of the third international symposium on sub tropical and tropical horticulture*. Horticulture Society of India. Banglore pp 57-64.
- Singh, V.K. and Tripathi, V.K., 2010. Efficacy of GA_3 on growth, flowering, yield and quality of strawberry cv. Chandler. *Progressive Agriculture*, **10**(2): 345-348.
- Suman, M., Sangma, P.D., Meghawal, D.R. and Sahu, O.P., 2017. Effect of plant growth regulators on fruit crops. *Journal of Pharmacognosy and Phytochemistry*, **6**(2): 331-337.

- Sun, C.H., Wang, D., Hu, Y.L., Li, X.H., Zhang, W.D., Sun J. and Gao, X.L., 2013. Effects of salicylic acid on physiological characteristics of strawberry leaves under drought stress. *Europ.J.Hort.Sci.*, 78(3): 106–111.
- Suvalaxmi, P., Das, A.K., Sahoo, A.K., Dilip, K.D. and Saudamini, S., 2016. Influence of plant growth regulators on strawberry (*Fragaria ananassa*) Chandler under Odisha condition. *International Journal of Recent Scientific Research*,**7**(4): 945-948.
- Swartz, H.J., Galletta, G.J. and Zimmermann, R.H., 1981. Field performance and phenotypic stability of tissue culture propagated strawberries.*J. Am. Soc. Hortic. Sci.*,**106**: 667.
- Symons, G.M., Chua, Y.J., Ross, J.J., Quittenden, L.J., Davies, N.W. and Reid, J.B., 2012. Hormonal changes during non-climatric ripening in strawberry.*Journal of Experimental Botany*, **63**(13): 4741-4750.
- Tafazoli, E. and Vince-Prue, D., 1978. A comparison of the effects of long days and exogenous growth regulators on growth and flowering in strawberry (*Fragaria × ananassa* Duch).*Journal of Horticultural Science*,53(4): 255-259.
- Tanimoto, E., 1983. Gibberellin-dependent not elongation in Lactuca sativa: recovery from growth retardantsuppressed elongation with thickening by low concentration of GA_3. *Plant Cell Physiol.*, **28**: 963-973.
- Taylor, D.R., 2002. The physiology of flowering in strawberry.*Acta Hort.*,**567**: 245-251.
- Techawongstein, S., 1989. The effect of NAA on fruit quality of strawberry (Fragaria X ananassa Duch.) cv. Tioga. *Kaen Kaset Khon Kaen Agriculture Journal*,**17**(1):30-35.
- Thakur, S., Mehta, K. and Sekhar, R.S., 2015. Effect of GA_3 and Plant Growth Promoting Rhizobacteria on growth, yield and fruit quality of strawberry (*Fragaria × ananassa* Duch.) cv. Chandler.*International Journal of Advanced Research*,**3**(11): 312-317.
- Thakur, Y., Chandel, J.S. and Verma, Pramod. 2017. Effect of plant growth regulators on growth, yield and fruit quality of strawberry (Fragaria x ananassa Duch.) under protected conditions.*Journal of Applied and Natural Sciences*, Vol 9 No.3. (DOI https://doi.org/10.31018/jans.v9i3.1420)
- Thompson, P.A. and Guttride, C.G., 1959. Effect of gibberellic acid on the initiation of flowers and runners in the strawberry. *Nature*,**184**: 72-73.
- Thompson, P.A., 1969. The use of chilling and chemical treatments to promote rapid germination of strawberry achenes.*J. Hortic. Sci.*,**44**(2): 201-210.

- Tian, M.S., Gong, Y. and Bauchot, A.D., 1997. Ethylene biosynthesis and respiration in strawberry fruit treated with diazocyclopentadiene and IAA.*Plant Growth Regul.*, **23**: 195–200.

- Trainotti, L.A., Pavanello, A. and Casadoro, G., 2005. Different ethylene receptors show an increased expression during the ripening of strawberries: does such an increment imply a role for ethylene in the ripening of these non-climacteric fruits. *J. Exp. Bot.*, **56**: 2037-2046.

- Tripathi, V.K. and Shukla, P.K., 2010. Influence of plant bio-regulators on yield and fruit characters of Strawberry cv. Chandler. *Progressive Horticulture*,**42**(2): 186-188.

- Turemis, N. and Kaska, N., 2004. Effect of gibberellic acid (GA_3) on the production and quality of strawberry runners. *Turkish Journal of Agriculture & Forestry*,**21**(1): 41-47.

- Venkata Chandana, A. N. 2017. Studies on the effect of plant growth regulators on plant growth, fruit set, yield and quality of strawberry (*Fragaria ananassa* Duch) cv. Camarosa in coastal Andhra Pradesh. M.Sc. thesis submitted to the Department of Fruit Science, College of Horticulture, Dr. Y.S.R. Horticultural University Venkataramannagudem, West Godavari - 534 101, Andhra Pradesh, 2017.

- Viasus-Quintero,G., Alvarez -Herrera, J. and Alvarado-Sanabria, O., 2013. Effect of applications of gibberellins and 6-benzylaminopurine on the production and quality of strawberry. *Bio. Agro.*,**5**(3):195-200.

- Vishal, V.C., Thippesha, D., Chethana, K., Maheshgowda, B.M., Veeresha, B.G. and Basavraj, A.K., 2016. Effect of various growth regulators on vegetative parameters of strawberry (*Fragaria ananassa* Duch.) cv. Sujatha. *Research Journal of Chemistry and Environmental Sciences*,**4**(4): 68-71.

- Waithaka, K. and Dana, M.N., 1978. Effects of growth substances on strawberry growth. *J. Am. Soc. Hort. Sci.*,**103**: 627-628.

- Wang, B., Lai, T., Huang, Q.W., Yang, X.M. and Shen, Q.R., 2009. Effect of N fertilizers on root growth and endogenous hormones in strawberry. *Pedosphere*,**19**(1): 86-95.

- Weidman, R.W. and Stang, E.J., 1983. Effects of gibberllins (GA_{4+7}), 6-benzyladenine (6-BA) and promaline (GA_{4+7} + 6-BA) plant growth regulators on plant growth, branch crown and flower development Scott and Raritan strawberries. *Adv. Straw. Prod.*,**2**: 15-17.

- Zatyko, J.M., Kin, G., Radics, Z. and Simon, J., 1980. Initiation of strawberry runner formation *in vitro. Fruit Sci. Reports.*, **15**(2): 59-61.

- Zhu, L.H., Van-De-Peppel, A., Li, X.Y. and Welander, M., 2004. Changes of leaf water potential and endogenous cytokinins in young apple trees treated with or without paclobutrazol under drought conditions. *Sci. Hortic.*,**99**: 133-141.

- Ziba, A., Mehrdad, J., Ahmad, R.G. and Abdolrahman, M., 2013. Effect of GA_3 application on fruit yield, flowering and vegetative characteristics on early yield of strawberry cv. Gaviota. *Intl J Agri Crop Sci.*,**5**(15): 1716-1718.

Index

Chapter 1 Plant Growth Regulators

Chapter 2 Brassinosteroids and their Use in Fruit Crops

Chapter 3 Role of Paclobutrazol in Fruit Crops

Chapter 7 Plant Growth Regulators in Ber

Chapter 8 Plant Growth Regulator in Citrus

Chapter 12 Plant Growth Regulators in Grape-Theory and Practical

Chapter 13 Role of Plant Growth Regulators in the Cultivation of Grapes

Chapter 14 Plant Growth Regulators in Guava

Chapter 15 Plant Growth Regulators in Jamun

Chapter 16 Plant Growth Regulators in Kokam

Chapter 17 Plant Growth Regulators in Litchi

Chapter 18 Plant Growth Regulators in Mango

Chapter 19 Plant Growth Regulators in Mulberry

Chapter 20 Plant Growth Regulators in Papaya

Chapter 21 Plant Growth Regulators in Passion Fruit

Chapter 22 and 23 Plant Growth Regulators in Phalsa

Chapter 24 Plant Growth Regulators in Pomegranate

Chapter 25 Plant Growth Regulators in Sapota

Chapter 26 Plant Growth Regulators in Strawberry

For Product Safety Concerns and Information please contact our EU
representative GPSR@taylorandfrancis.com
Taylor & Francis Verlag GmbH, Kaufingerstraße 24, 80331 München, Germany

www.ingramcontent.com/pod-product-compliance
Ingram Content Group UK Ltd.
Pitfield, Milton Keynes, MK11 3LW, UK
UKHW052224190926
475702UK00007B/1184

* 9 7 8 1 0 3 2 2 9 1 7 2 7 *